W0106183

Tissue Engineering

Series Editor—
Anthony Atala, MD
Children's Hospital, Boston

Editorial Advisory Board—
Jeffrey Hubbell, PhD
California Institute of Technology

Robert S. Langer, ScD
Massachusetts Institute of Technology

Antonios G. Mikos, PhD
Rice University

Joseph P. Vacanti, MD
Children's Hospital, Boston

Cell Encapsulation Technology and Therapeutics

Willem M. Kühtreiber, PhD
Robert P. Lanza, MD
William L. Chick, MD
Editors

Foreword by Thomas M.S. Chang

Springer Science+Business Media, LLC

Willem M. Kühtreiber, PhD
BioHybrid Technologies, Inc.
Shrewsbury, MA 05145

Robert P. Lanza, MD
Tissue Engineering &
    Transplant Medicine
Advanced Cell Technology
Worcester, MA 01605

William L. Chick, MD (deceased)
BioHybrid Technologies, Inc.
Shrewsbury, MA 05145

Library of Congress Cataloging-in-Publication Data
Cell encapsulation technology and therapeutics / edited by Willem M.
    Kühtreiber, Robert P. Lanza, William L. Chick.
        p.   cm.
    Includes bibliographical references and index.
    ISBN 978-1-4612-7205-2   ISBN 978-1-4612-1586-8 (eBook)
    DOI 10.1007/978-1-4612-1586-8
    1. Animal cell biotechnology.   2. Microencapsulation.
I. Kühtreiber, Willem M.   II. Lanza, R.P. (Robert Paul)
III. Chick, William L. (William Louis), 1938–1998
TP248.27.A53C455   1998
615′.19—dc21   98-29997
                                                            CIP

Printed on acid-free paper.
© 1999 Springer Science+Business Media New York
Originally published by Birkhäuser Boston in 1999
Softcover reprint of the hardcover 1st edition 1999

All rights reserved. This work may not be translated or copied in whole or in part without the written permission of the publisher
Springer Science+Business Media, LLC, except for brief excerpts
in connection with reviews or scholarly analysis. Use in connection with any form of information storage and retrieval, electronic
adaptation, computer software, or by similar or dissimilar methodology now known or hereafter developed is forbidden.
The use of general descriptive names, trade names, trademarks, etc., in this publication, even if the former are not especially identified, is not to be taken as a sign that such names, as understood by the Trade Marks and Merchandise Marks Act, may accordingly
be used freely by anyone.

COVER PHOTOS—**Left:** Alginate microcapsules made by gelling microdroplets of an alginate solution in a buffered calcium chloride
solution. Courtesy of Kühtreiber et al (Chapter 18). **Center:** scanning electron micrograph of a cross-section of a PAN-PVC hollow
finer as used for cell encapsulation. Courtesy of Jacqueline Sagen et al (Chapter 28.) **Right:** 2A-50 fibroblasts encapsulated in alginate-
PLL-alginate. Courtesy of Colin Ross and Patricia Chang (Chapter 26).

ISBN 978-1-4612-7205-2

Typeset by Impressions Book and Journal Services Inc., Madison, WI.

9 8 7 6 5 4 3 2 1

In memory of William L. Chick, M.D. (1938–1998), cofounder and President of BioHybrid Technologies. Dr. Chick died from complications of diabetes shortly before the completion of the manuscript for this book. He was a pioneer in the field of bioartificial organs and worked on the development of the biohybrid artificial pancreas for over 25 years.

Contents

PART I   FUNDAMENTALS OF CELL ENCAPSULATION

PART II   ENCAPSULATION SYSTEMS

SECTION 1   MICROENCAPSULATION

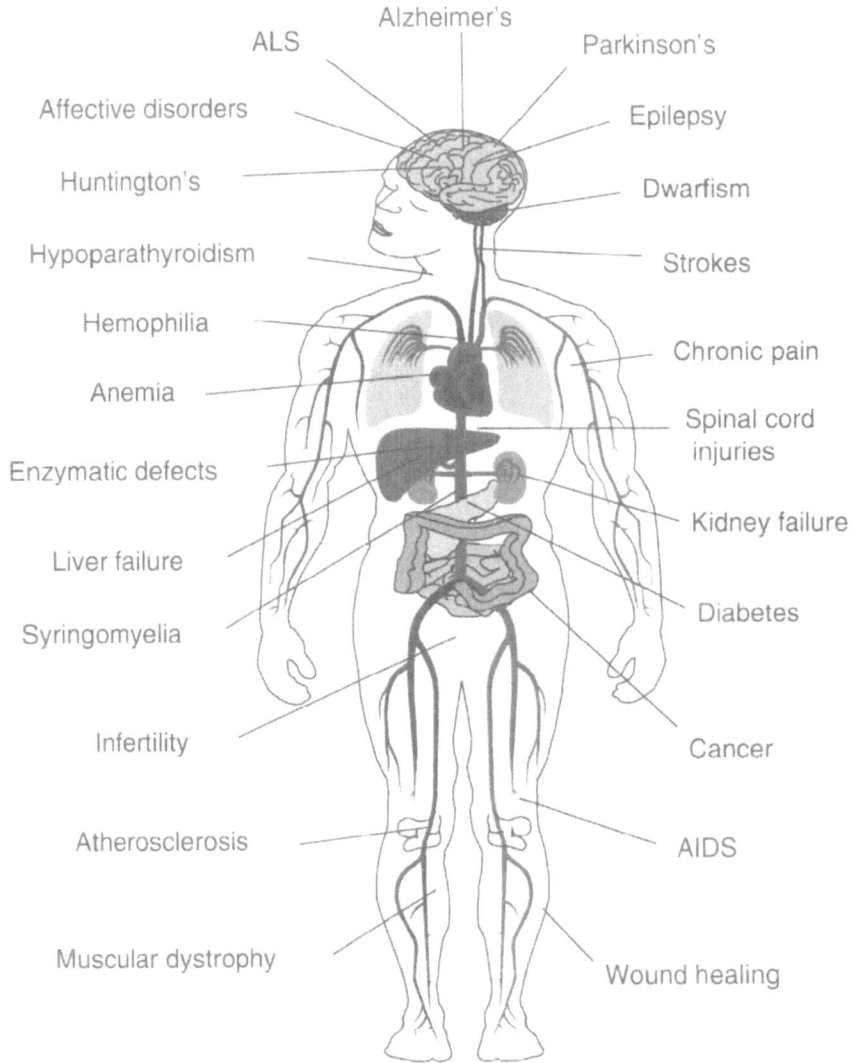

ALS
Affective disorders
Huntington's
Hypoparathyroidism
Hemophilia
Anemia
Enzymatic defects
Liver failure
Syringomyelia
Infertility
Atherosclerosis
Muscular dystrophy

Alzheimer's
Parkinson's
Epilepsy
Dwarfism
Strokes
Chronic pain
Spinal cord injuries
Kidney failure
Diabetes
Cancer
AIDS
Wound healing

Many diseases are caused by the inability of the body to produce the necessary amount of a needed molecule, such as a hormone, factor, or enzyme. Cell therapy offers enormous potential for the treatment of such diseases. Encapsulated cell systems consist of living cells that are immobilized and protected inside micro- or macrocapsules. Such capsules are implanted into a patient, where the cells produce the therapeutic substances that the body cannot produce itself. The figure depicts the most important diseases that lend themselves to such treatment with encapsulated cells.

Foreword

The concept of using encapsulation for the immunoprotection of transplanted cells was introduced for the first time in the 1960s. "[Microencapsulated cells] might be protected from destruction and from participation in immunological processes, while the enclosing membrane would be permeable to small molecules of specific cellular product which could then enter the general extracellular compartment of the recipient. For instance, encapsulated endocrine cells might survive and maintain an effective supply of hormone." (Chang, Ph.D. Thesis, McGill University, 1965; Chang et al., Can J Physiol Pharmacol 44:115-128, 1966). We asked Connaught Laboratories, Ltd., in Toronto to put this concept into practice. In 1980, Lim and Sun from Connaught Laboratories reported on the successful implantation of poly-l-lysine-alginate encapsulated rat islets into a foreign host. [Lim and Sun, Science 210:908-909, 1980]. Now many groups around the world are making tremendous progress in the encapsulation of a multitude of cell types. Kühtreiber, Lanza, and Chick have invited many cell encapsulation groups from around the world to contribute to this book. The result is a very useful reference book in this rapidly growing area. With so many excellent authors describing in detail the different areas of cell encapsulation, my role here will be to briefly discuss a few points.

Mass transfer

Basic research on semipermeable microcapsules containing biologically active materials shows that there is an extremely high surface-to-volume relationship that allows for rapid equilibration of permeant molecules. At the same time, immunocytes and macromolecules such as antibodies are excluded by the microcapsules. Having said this, I should immediately point out that "encapsulation" does not automatically imply good mass transfer. Colton, Goosen, Wang, and others have carried out excellent analyses on the principles of mass transfer. They have shown clearly that, depending on the encapsulation method, there are marked variations in membrane porosity and membrane thickness. Variation in configurations is another important factor. Increasing the diameter of microcapsules or the use of capillary encapsulation can markedly decrease the surface-to-volume ratio. As will be discussed later, in vitro mass transfer does not necessarily transfer to in vivo situations. One of the most important problems is biocompatibility. If the encapsulation material causes fibrosis, the mass transfer into and out of the microcapsules can be greatly reduced.

Can encapsulated cells carry out their function after implantation?

Since Lim and Sun's report, many groups have maintained normal blood glucose levels in diabetic animals after implantation of encapsulated islets. Soon-Shiong's group has maintained a diabetic patient normoglycemic for a brief period of time after implantation. We have shown that encapsulated rat hepatocytes increase the survival of galactosamine-induced acute liver failure rats and can lower elevated bilirubin levels. There are other examples using other types of cells. Thus, on a short-term basis, encapsulated cells can carry out their function after implantation. A more complicated question is how to prepare encapsulated cells that carry out their function for many years after implantation. At a minimum, this requires long-term biocompatibility and, in xenografts, long-term immunoisolation.

What are the factors that determine the long-term function and immunoisolation of encapsulated cells?

Biocompatibility and mass transfer are two of the important factors. Foreign body reaction with resulting fibroencapsulation will quickly decrease mass transfer of oxygen, nutrients, and metabolites, and will result in death of the encapsulated cells. As shown in this book, many groups, such as those of Calafiore, Goosen, Iwata, Neufeld, Poncelet, Rha, Sawhney, Sefton, Zekorn, Zimmermann, and others, are making progress in this very important area, using different materials.

A number of groups have reported animal studies showing that encapsulated islets and other cells are protected from immunorejection. On the other hand, other groups have not obtained similar results. Some emphasize that it is not enough just to exclude immunocytes and antibodies. Complement, cytokines, and other factors may also play an important role. Variation in response between animal species is another factor.

We found that some microcapsules containing hepatocyte xenografts remain free-floating after implantation, whereas others aggregate into clumps. Those that clump do not function for long. Those that remain free are not only protected from immunorejection but show improvements in viability over time. This may indicate recovery of the hepatocyte membranes that were slightly damaged as a result of the isolation procedure. Recovery is possibly due to the entrapment of secreted hepatostimulating factors (MW 140 kD) inside the beads. In addition, the peritoneal cavity is an excellent culture medium, and free microcapsules have large surface-to-volume ratios that allow for optimal exchange of nutrients, oxygen, and waste metabolites.

There is no doubt that encapsulated cells can function well after implantation on a short-term basis. Furthermore, studies by many groups have shown that encapsulated xenograft cells will remain functional after immunoisolation. With research and development by the many excellent groups described in this book, there will be further improvements in biocompatibility and immunoisolation, allowing this approach to become a routine clinical application.

New approaches in configurations and sites of actions

Long-term implantation using microencapsulated cells is an ideal situation. Like all ideal situations, it will take longer to perfect. Meanwhile, several groups are looking into other configurations and other sites of action for more immediately achievable clinical applications.

For example, CytoTherapeutics's ingenious use of capillary fibers to encapsulate cells has allowed the implantation of cells on a short-term basis, while enabling the removal or renewal of the fibers. They are now well into clinical trials. However, this technique can only be utilized in specific applications that do not require large amounts of encapsulated cells. For instance, fibers cannot be used for islets or hepatocytes when the volume of cells required would be too large for this approach.

There are other important approaches. For instance, some years ago Calafiore's group loaded microencapsulated islets into the wall of tubing, allowing them to use an extracorporeal approach. Maki's group is using vascular devices for extracorporeal blood perfusion. Solomon's group is "encapsulating" cells, especially hepatocytes, in an extracorporeal capillary fiber device. Burczak and Ikada's approach is based on the use of ultrafilters. Lanza reports on the use of diffusion chambers for cell encapsulation. All these approaches in configurations and in sites of action will allow for the more rapid development of clinical applications of cell encapsulation.

Should we also look into unexpected approaches in cell encapsulation?

Genetic engineering has resulted in many microorganisms with excellent specialized functions. However, it will be folly to implant these into the body, even in the encapsulated form! On the other hand, one challenge in research is to look into unexpected ways to arrive at new approaches.

For example, we encapsulate genetically engineered E. coli DH5 cells and administer them orally to kidney failure rats (Prakash and Chang, Nature Medicine 2(8):893-887, 1996). During the passage through the intestine, the bacteria remove urea and ammonia and use them as their nitrogen source. This results in the lowering of the elevated urea level to normal. The bacteria remain inside the microcapsules at all times and are excreted with the stool. No bacteria are released into the intestine or the body. This may solve the problem of the lack of a removal system for urea, which is the single major obstacle in the quest to replace dialysis with oral therapy using absorbents and osmotic agents. This research also has implications in other areas, such as the removal of ammonia or unwanted amino acids.

I have just touched on a few points of cell encapsulation to show the complexity and importance of this subject. I look forward, like you, to reading the many chapters in this book, written by excellent research groups from around the world.

McGill University,   *Thomas M.S. Chang,*
Montreal, Quebec   OC, MD, CM, PhD,
Canada   FRCP(C)

Contributors

Glen K. Adaniya, PhD, Laboratory Director, Midwest Reproductive Medicine and Indianapolis Andrology and Laboratory Services, Inc., Indianapolis, IN, USA

Ayman Al-Hendy, MD, PhD, Resident, Department of Obstetrics and Gynecology, Royal University Hospital, University of Saskatchewan, Saskatoon, Saskatchewan, Canada

Julia E. Babensee, PhD, Department of Chemical Engineering and Applied Chemistry, University of Toronto, Toronto, Ontario, Canada

Estela A. Balmaceda, PhD, Biomaterials Science and Engineering Laboratory, Massachusetts Institute of Technology, Cambridge, MA, USA

Giuseppe Basta, MD, DIMISEM, University of Perugia School of Medicine, Perugia, Italy

Reinhard G. Bretzel, Prof. Dr. med., Professor and Chair, Justus-Leibig-Universitaet of Giessen, Medizinische Klinik III & Poliklinik, Giessen, Germany

Suzanne L. Bruhn, PhD, Scientist, CytoTherapeutics, Inc., Lincoln, RI, USA

Branko Bugarski, PhD, Assistant Professor, Department of Chemical Engineering, University of Belgrade, Belgrade, Yugoslavia

Krystyna Burczak, PhD, Associate Professor, Institute for Applied Radiation Chemistry, Technical University of Lodz, Lodz, Poland

Riccardo Calafiore, MD, Director of Islet Transplantation, DIMISEM, University of Perugia School of Medicine, Perugia, Italy

Melissa K. Carpenter, PhD, Scientist, CytoTherapeutics, Inc., Lincoln, RI, USA

Elliot L. Chaikof, MD, PhD, Assistant Professor of Surgery, Emory University School of Medicine, Department of Surgery, Atlanta, GA, USA

Patricia L. Chang, PhD, Professor, Department of Pediatrics, Health Sciences Center, McMaster University, Hamilton, Ontario, Canada

Thomas M.S. Chang, OC, MD, CM, PhD, FRCP(C), Director, Artificial Cells and Organs Research Centre; and Professor of Physiology, Medicine, and Biomedical Engineering, Faculty of Medicine, McGill University, Montreal, Quebec, Canada

William L. Chick, MD, (deceased), formerly President and Scientific Director, BioHybrid Technologies, Inc., Shrewsbury, MA, USA

JoonHo Choi, PhD, Biomaterials Science and Engineering Laboratory, Massachusetts Institute of Technology, Cambridge, MA, USA

John T. Chryssochoos, MD, Department of Surgery, Emory University School of Medicine, Atlanta, GA, USA

Paul de Vos, PhD, Department of Surgery, University Hospital Groningen, Groningen, The Netherlands

Claire Dulieu, PhD, ENSAIA-INPL, Vandoeuvre-les-Nancy, France

Jeffrey Fair, MD, Director of Transplant Surgery, Division of Liver Transplantation, Department of Surgery, University of North Carolina School of Medicine, Chapel Hill, NC, USA

Sudarshan Gautam, MBBS, PhD (BME), Research Fellow, Department of Internal Medicine, University of Michigan, Ann Arbor, MI, USA

Mattheus F.A. Goosen, PhD, Professor and Dean, College of Engineering, Sultan Qaboos University, Sultanate of Oman

John P. Griffin, PhD, Department of Physiology, University of North Carolina School of Medicine, Chapel Hill, NC, USA

Linda Griffith, PhD, Associate Professor of Chemical Engineering, Department of Chemical Engineering, Massachusetts Institute of Technology, Cambridge, MA, USA

Mary K. Hagler, MS, Emory University School of Medicine, Department of Surgery, Atlanta, GA, USA

Christian Hasse, MD, PhD, Department of General Surgery, Philipps-University of Marburg, Marburg, Germany

Gonzalo Hortelano, PhD, Assistant Professor, Department of Pathology, McMaster University, Hamilton, Ontario, Canada

H. David Humes, MD, Chair, Department of Internal Medicine, University of Michigan Medical Center, Ann Arbor, MI, USA

Yoshito Ikada, PhD, Professor, Institute for Frontier Medical Sciences, Kyoto University, Kyoto, Japan

Nahed Ismail, MD, MSc, Department of Microbiology and Immunology, University of Saskatchewan, Saskatoon, Saskatchewan, Canada

Hiroo Iwata, PhD, Associate Professor, Institute for Frontier Medical Sciences, Kyoto University, Kyoto, Japan

Judith A. Kapp, PhD, Professor of Pathology and Ophthalmopathy, Emory University School of Medicine, Atlanta, GA, USA

Sungkoo Kim, PhD, Assistant Professor of Biotechnology, Department of Biotechnology and Bioengineering, Pukyong National University, Pusan, Korea

Hiroshi Kubota, PhD, DVM, Department of Physiology, University of North Carolina School of Medicine, Chapel Hill, NC, USA

Willem M. Kühtreiber, PhD, Director of Islet Physiology, BioHybrid Technologies, Inc., Shrewsbury, MA, USA

Robert P. Lanza, MD, Senior Director, Tissue Engineering & Transplant Medicine, Advanced Cell Technology, Worcester, MA 01605

Rebecca H. Li, PhD, Scientist, CytoTherapeutics, Inc., Lincoln, RI; Adjunct Associate Professor, Brown University, Providence, RI, USA

Jeffery M. Macdonald, PhD, Department of Cell and Molecular Physiology, University of North Carolina, Chapel Hill, NC; Laboratory of Structural Biology, National Institute of Environmental Health Sciences, Research Triangle Park, NC, USA

Takashi Maki, MD, PhD, Associate Professor of Surgery, Beth Israel Deaconess Medical Center and Harvard Medical School, Boston, MA, USA

Anthony P. Monaco, MD, Peter Medawar Professor of Transplantation Surgery, Beth Israel Deaconess Medical Center and Harvard Medical School, Boston, MA, USA

Claudy J.P. Mullon, PhD, Vice President, Research & Development, Circe Biomedical Inc., Lexington, MA, USA

Ronald J. Neufeld, PhD, Professor, Chemical Engineering Department, Queen's University, Kingston, Ontario, Canada

Augustine O. Okhamafe, PhD, Professor of Pharmaceutics and Pharmaceutical Technology, Faculty of Pharmacy, University of Benin, Benin City, Nigeria

Denis Poncelet, PhD, Professor, Bioencapsulation Research Group, ENSAIA-INPL, Vandoeuvre-les-Nancy, France

Satya Prakash, PhD, Research Associate, Artificial Cells and Organs Research Centre, Faculty of Medicine, McGill University, Montreal, Quebec, Canada

Richard G. Rawlins, PhD, HCLD, Professor and Director of Laboratories, Dept. OB/GYN and Rush Center for Advanced Reproductive Care, Rush-Presbyterian-St. Luke's Medical Center, Chicago, IL, USA

Lola M. Reid, PhD, Professor, Department of Physiology, University of North Carolina School of Medicine, Chapel Hill, NC, USA

David H. Rein, PhD, Managing Scientist, CytoTherapeutics, Inc., Lincoln, RI, USA

ChoKyun Rha, ScD, Professor of Biomaterials Science and Engineering, Biomaterials Science and Engineering Laboratory, Massachusetts Institute of Technology, Cambridge, MA, USA

Colin J.D. Ross, MSc, Departments of Biology and Pediatrics, McMaster University, Hamilton, Ontario, Canada

Mathias Rothmund, MD, FACS, Professor, Chairman, Department of General Surgery, Philipps-University of Marburg, Marburg, Germany

Susan Safley, PhD, Department of Pathology, Emory University School of Medicine, Atlanta, GA, USA

Jacqueline Sagen, PhD, The Miami Project to Cure Paralysis, University of Miami School of Medicine, Miami, FL, USA

Amarpreet S. Sawhney, PhD, President, Confluent Surgical, Inc., Boston, MA, USA

Michael V. Sefton, ScD, Professor, Department of Chemical Engineering and Applied Chemistry; and Institute for Biomaterials and Biomedical Engineering, University of Toronto, Toronto, Ontario, Canada

Barry A. Solomon, PhD, Executive Vice President, Circe Biomedical Inc., Lexington, MA, USA

Patrick Soon-Shiong, MD, Chief Scientific Officer, VivoRX, Inc., Santa Monica, CA, USA

Tracy Stockley, PhD, Department of Pediatrics, McMaster University, Hamilton, Ontario, Canada

Wei Wen Su, PhD, Department of Biological and Agricultural Engineering, University of Missouri-Columbia, Columbia, MO, USA

Reinout van Schilfgaarde, MD, Professor and Chairman, Department of Surgery, University Hospital Groningen, Groningen, The Netherlands

Gordana Vunjak-Novakovic, PhD, Research Scientist, Division of Health Sciences and Technology, Massachusetts Institute of Technology, Cambridge, MA; Professor of Chemical Engineering, Tufts University, Boston, MA, USA

G. Taylor Wang, PhD, Centennial Professor and Director, Center for Microgravity Research and Applications, School of Engineering, Vanderbilt University, Nashville, TN, USA

Collin J. Weber, MD, DMSci, Professor of Surgery, Emory University School of Medicine, Department of Surgery, Atlanta, GA, USA

James R. Wright Jr., MD, PhD, Islet Transplantation Laboratory, IWK Grace Health Centre, Professor of Pathology, Associate Professor of Surgery, Dalhousie University, Halifax, Nova Scotia, Canada

Hua Yang, MD, MSc, Islet Transplantation Laboratory, IWK Grace Health Centre; Assistant Professor of Pathology and Surgery, Dalhousie University, Faculty of Medicine, Halifax, Nova Scotia, Canada

Delano V. Young, PhD, Section Head, Cell Culture, Department of Biology, NitroMed, Inc., Bedford, MA, USA

Tobias D.C. Zekorn, Priv. Doz. Dr. med., Abteilung für Innere Medizin, St. Josefshospital, Krefeld, Germany

Andreas Zielke, MD, DMSci, Endocrine Research Laboratory, Department of Surgery, Philipps-University of Marburg, Marburg, Germany

Ulrich Zimmermann, PhD, Chairman, Department of Biotechnology, University of Würzburg, Würzburg, Germany

Part I
Fundamentals of Cell Encapsulation

1
Encapsulation and Immobilization Techniques

Claire Dulieu, Denis Poncelet, and Ronald J. Neufeld

Introduction

Cells are generally found in an immobilized state in natural environments. In biotechnology, immobilization provides protection to the cell from unfavorable conditions, washout, shear, and immunological rejection. Moreover, immobilization provides for an organization of the cells, enabling synergistic interactions between adjacent cells. Multicellular organisms result from higher levels of cell organization and structure that result from a form of cell immobilization. It is therefore not surprising that scientists and engineers seriously consider cell immobilization as a means of solving technological problems involving the handling and processing of cellular materials.

Cell immobilization methods may be classified into three categories, as illustrated in Figure 1.1. This classification is mainly based on the methods used for immobilization. In aggregation, adhesion, and porous carrier adsorption, immobilization results from attachment between cells and/or between cells and the support. In the case of porous carriers, cells first migrate into the porous matrix before being fixed to the carrier. Attachment may be due either to the production of adhesive polymers by the cells or to external ionic or covalent cross-linkers. In some cases, cell selection or genetic modifications are performed to promote cell aggregation.

Cells may also be maintained in a defined volume by use of preformed devices. The barrier may be formed by a membrane dividing the reactor into two parts (membrane reactor) or through the use of hollow fibers. The volume restriction may also be obtained by recycling cells in the reactors after centrifugation, settling, or ultrafiltration. The immobilization device is built in the absence of cells, and immobilization is initiated by addition of the cell suspension.

Bioencapsulation, as the third alternative, involves immobilization of the biologically active component in hydrogel beads or microcapsules. Cells are mixed into the hydrogel or a membrane-forming material, then dispersed dropwise. Gelation or membrane formation is obtained through physical (temperature) or chemical (cross-linker) modifications.

Each method of immobilization has advantages and drawbacks. Selection is a function of the application, the cell line, and other scientific, technical, and economic criteria. For example, the cost of immobilization is more important in food production, compared to some medical or pharmaceutical applications. The food industry may be motivated to reduce investment cost while maintaining strong attachments to traditional technologies. Moreover, food and drug regulations, and the effects of toxic reagents on the cells, may limit the methods and the materials to be applied for immobilization. It is therefore difficult to outline clear guidelines for selecting an immobilization method. In many cases, more than one method, or a combination of methods, may serve the specific requirements.

Table 1.1 summarizes the major reviews on cell immobilization. The review by Willaert and Baron (1996) is particularly exhaustive, with over 1000 references. This chapter will deal more specifically with bioencapsulation. Additional information may also be found through the Bioencapsulation

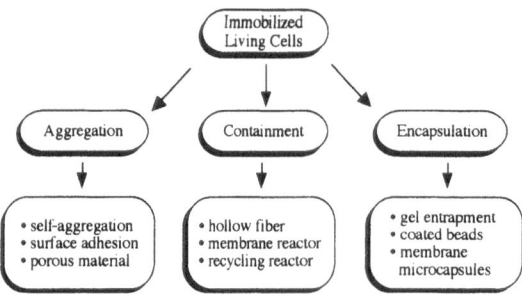

FIGURE 1.1. Classification of the cell immobilization systems.

Research Group Web site located at *http://ensaia .u-nancy.fr/BRG/BRG.html.*

Bioencapsulation Methods

Bioencapsulation methods involve two main steps. First, the internal phase containing cells is dispersed into small droplets, which are then solidified by gelation of the droplets or membrane formation at the droplet surface.

Droplet dispersions are formed by dropwise extrusion, or emulsification of the internal phase into an immiscible external phase. Many methods may be used to solidify the capsules, including gelation by ionic binding, temperature change, or polymerization, or by cross-linking of prepolymers. A membrane coat may then be applied to the gel bead by ionic polymer coating, polymer transacylation, or spray coating, or via direct membrane formation by coextrusion and external layer precipitation or interfacial polymerization or coacervation. These

methods will be described more fully in subsequent chapters of this review.

Diameter and Size Distribution of Microcapsules

Diameter is one of the more important properties of gel beads and capsules. Most formulation processes include the diameter as an input parameter. It is therefore necessary to clearly define the optimum size as a function of the application and to choose a droplet formation process that ensures both correct diameter and minimum size distribution.

The capsule itself must be sufficiently large to contain the cell, cell aggregate, or product in the case of growing cells. There are also practical advantages to larger capsules or beads, as they are easier to handle during washing and settling operations and in reactor operations such as fluidization. In many cases, the cells must be homogeneously distributed within the internal capsular matrix. The probability of finding cells in the microcapsule is therefore related to Poisson's law (Nir et al 1990). To ensure at least one cell per capsule, the mean cell number per capsule must be higher than four.

On the other extreme, excessively large capsules create an internal dead volume. Implantable microcapsules must be smaller than half the internal diameter of the injection needle. In the case of a bioreactor, the shear or abrasion effects on capsules increase dramatically with diameter (Dos Santos et al 1997, Poncelet and Neufeld 1989), and large capsules may lead to mass transfer limitations.

TABLE 1.1. Reference books and reviews on cell immobilization.

Immobilized cells and organelles	Mattiasson, 1983
Immobilized cells and enzymes	Woodward, 1985
Immobilizierte biokatalysatoren	Hartmeier, 1986
Process engineering aspects of immobilized cell systems	Webb et al, 1986
Immobilized cells: principles and applications	Tampion and Tampion, 1987
Bioreactor immobilized enzymes and cells	Moo-Young, 1988
Fundamentals of animal cell encapsulation and immobilization	Goosen, 1993
Wastewater treatment with microbial films	Iwai and Kitao, 1994
Immobilized biosystems: theory and practical applications	Veliky and McLean, 1994
Special issue on immobilized cell technology in food processing	Champagne, 1994
Immobilized living cell systems: modeling and experimental methods	Willaert et al, 1995
Gel entrapment and micro-encapsulation: methods, applications, and engineering principles	Willaert and Baron, 1996

The optimum size is often a compromise. In fermentation, the proposed size is generally around 2 mm to facilitate handling, or 800 μm to reduce the mass transfer limitation. In transplantation, the microcapsule diameter is generally in the range of 300 to 800 μm, and many authors limit the size to less than 500 μm.

Size dispersion may also play an important role in microcapsule behavior. Although uniform diameter, or monodispersed, capsules provide zero-order kinetics for product release, polydispersed capsule preparations result in first or apparent second-order kinetics (Poncelet et al 1988). Also, in a bioreactor, microcapsule mechanical resistance decreases with increasing diameter, and level of mixing or shear, resulting in the mass transfer efficiency. If the capsules are size-dispersed, the level of shear must be limited by the largest-diameter capsules.

It is therefore important to limit the size dispersion as much as possible. Although production of large monodispersed capsules is now feasible, it is still a tedious problem to produce diameters less than 800 μm, especially on a large scale. The size dispersion, expressed as standard deviation may range from 5% of the mean for 3 mm capsules to more than 50% for 300 μm capsules (Poncelet et al 1993).

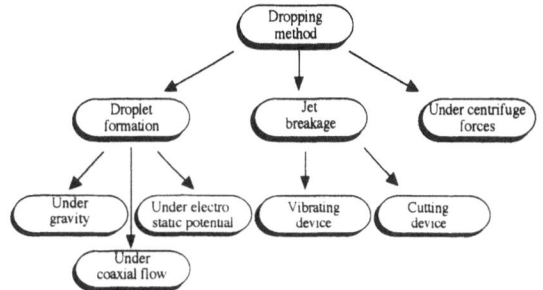

FIGURE 1.2. Dropping methods.

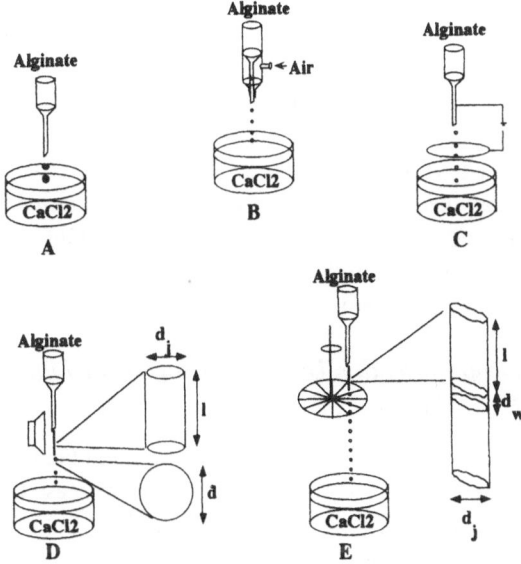

FIGURE 1.3. Dropping devices.

Droplet Formation and Bioencapsulation

Capsules are formulated by droplet extrusion or emulsification. In selecting an appropriate method, the following parameters must be taken into account: desired mean size, acceptable size dispersion, scale of the production, and the maximum level of shear that the cells may tolerate. Although important, tolerable levels of shear for cells is difficult to define, particularly when they are suspended within rheologically complex internal phase media. The following sections outline the various techniques used to form droplets.

Droplet Extrusion

When a liquid is forced through a nozzle or needle, it is extruded initially as individual droplets. With increasing flow rate, the pendent droplet tends to stretch before detachment, becoming more pronounced until the extruded liquid forms a jet or continuous stream from the needle tip. The jet

stream will then break naturally into small droplets. Extrusion methods may be divided in two classes: drop formation/extrusion and jet breakage. The limit between the two techniques or regimes is determined by the minimum jet velocity, $u_{j,min}$:

$$u_{j,min} = 2(\sigma / \rho dj)^{0.5} \qquad (1)$$

where σ is the surface tension, ρ the density of the flowing liquid, and d_j the jet diameter. Figure 1.2 summarizes the different extrusion methods proposed in the literature, and Figure 1.3 illustrates the devices used for producing droplets.

Drop Formation Under Simple Gravity (Figure 1.3A)

The simplest method to form individual droplets is to let a liquid droplet fall from the tip of a needle. The mass of the droplet, m, will then be determined by the equilibrium between the gravity force m g and the forces acting to maintain the droplet attached to the tip (product of the surface tension, γ, and the perimeter of the tip, $2 \pi d_e$, Tate's law):

$$mg = 2\pi d_e \gamma \qquad (2)$$

where d_e is the external tip diameter. The droplet diameter will be given by

$$d = (6m / \pi\rho)^{1/3}. \qquad (3)$$

The real diameter of the capsule needs to be corrected by a swelling or shrinkage factor due to the entrapment or encapsulation process. For example, alginate bead volume is reduced by half during gelation. In contrast, nylon microcapsules swell by a factor of 1.3 during washing (Poncelet et al 1988).

The droplet diameter obtained by extrusion under gravity is typically larger than 2 mm, even for very small needle diameters. Also, the flow remains limited by jet formation at an order of magnitude of mL/h. Interest in this simple system may therefore be limited to laboratory-scale research.

Drop Formation Under Coaxial Air or Liquid Flow (Figure 1.3B)

The application of a coaxial air jet around the needle has been proposed (Lane 1947) to increase the force acting on nascent drops. The air jet may be replaced by a liquid jet (Charwat 1977), permitting a better control of the viscosity, surface tension, and density of the entraining phase, through selection of an appropriate liquid.

Many laboratory studies on cell encapsulation are largely based on air-jet systems, while liquid jets have received limited interest (Dupuy et al 1988). Both methods produce beads or microcapsules ranging from a few micrometers to one millimeter. However, the flow rate remains very limited, to less than 30 mL/h, to avoid formation of a liquid jet. The size dispersion increases drastically when the droplet diameter is decreased (Poncelet et al 1993). For these reasons, the coaxial fluid jet systems have not been considered for scale-up. Even on the laboratory scale, this method is being replaced by the technologies described below.

Drop Formation Under Electrostatic Potential (Figure 1.3C)

Drop formation is greatly improved by replacing the drag force with a high electrostatic potential between the capillary and the collecting solution

(Burgarski et al 1994a, 1994b; Poncelet et al 1994). Alternately, electric potential may be applied between the capillary and a stainless steel ring placed below the capillary. Increasing the electrostatic potential, U, to a critical value, U_{cr}, leads to a decrease in the droplet size, d. For higher values, the liquid exits the tip as a jet that breaks itself into small droplets.

The mass of the droplet detaching from the tip is given by equating the sum of gravity forces, $m g$, plus electric forces, Fe, to the surface tension forces, $2 p d_s \gamma$ (Poncelet et al 1998):

$$mg + Fe = 2\pi d_s \gamma \qquad (4)$$

with

$$Fe = \pi\epsilon_0 (d / 2h)^2 U^2 \qquad (5)$$

where ϵ_o is the electric permittivity of air, h the distance between the pendent droplet and the collecting solution, and d_s the diameter of the droplet detachment section.

In fact, the electric force is relatively small, and thus plays a secondary role in reducing droplet diameter. The primary reason is that charged molecules moving to the droplet surface create a repulsion between molecules at the air-liquid interface, counteracting the surface tension. The resulting decrease of surface tension force is then

$$\gamma = \gamma_0(1 - \epsilon_0 U^2 / d_c) \qquad (6)$$

(where γ_o is the surface tension at $U = 0$), resulting in droplet size reduction with increasing electrostatic potential (Poncelet et al 1998).

When the surface tension approaches zero, the liquid tends to form a jet rather than drops. Equation 6 represents an equilibrium state. In some cases, migration of the molecules to the surface is slower than the rate of drop surface formation. The real surface tension, γ, is then an intermediate value between the value given by Equation 6 and γ_o. Larger drops are therefore obtained when the flow rate is increased. Most encapsulation processes involve an interaction between a polymer and a counter-ion. The counter-ion migrates faster than the higher molecular weight polymer. Smaller drops are indeed obtained if the charge of the droplet has same sign as the counter-ion.

Bead size distribution obtained with an electrostatic generator is generally better than that obtained with coaxial air flow (standard deviation = 15%). However, satellite peaks may be observed in the size distribution profile. Satellites are formed by breakage of the fine filament between the droplet and the needle tip just before separation, resulting in secondary peaks. The flow rate is still limited by the formation of the jet. The electrostatic potential droplet generator is a promising technique to obtain small microdroplets (down to 200 μm), at least at laboratory scale.

Vibrating Capillary Jet Breakage (Figure 1.3D)

If liquid in the capillary exceeds a certain velocity (Equation 1), it exits from the tube as a jet. Capillary jets are unstable and fracture easily, forming small droplets. Experiments have shown that the jet breaks with a specific natural frequency (Savart 1833) equal to

$$f = u_j / \alpha d_j \qquad (7)$$

where u_j and d_j are the linear velocity and the diameter of the jet, and α is a factor equal to

$$\alpha = 4.44 \, (1 + 3\mu / \rho\sigma dj)^{0.5}. \qquad (8)$$

Rayleigh (1878) showed that if an external wave of the natural frequency, f, is applied to the jet, the jet breaks into monodispersed droplets (standard deviation equal to 5%). The jet breaks into cylinders with radius d_j and length, $\lambda = u/f$. From geometric considerations, the final spherical droplet diameter is given by

$$d = 1.15\alpha d_j. \qquad (9)$$

The system is, therefore, simply driven by Equations 7 and 9. One could expect to produce droplets from a few microns to 3 mm at relatively high flow rate (24 L/h for 3 mm diameter carrageenan beads (Hunik et al 1993). However, with increasing viscosity, the pressure required to ensure a jet increases proportionally to the viscosity and to the fourth power of the internal nozzle diameter. The flow rate decreases drastically while producing small droplets with the highly viscous fluids generally encountered in encapsulation processes. The jet diameter, d_j, is 0.8 times the internal nozzle diameter (d_i) for water, but 1.3 times d_i for alginate solution.

Vibrating jet breakage is one of the most efficient techniques to produce large capsules (1 to 3 mm) with a narrow size distribution. A multinozzle system would enable production in the order of hundreds of liters per hour. However, it appears more difficult to use this process for microcapsules less than 800 μm in diameter.

Rotating Systems for Capillary Jet Breakage (Figure 1.3E)

To overcome the limitations of the vibrating system, Prusse et al (1996) proposed a rotating device to cut the jet into small droplets. The diameter of the cutting wires, d_w, is the main parameter determining the effectiveness of this method. To reduce loss due to the cutter, the linear velocities of the jet, u_j, and of the cutting wire, u_w, must be equal. Then, the lost fraction is approximately equal to

$$\text{Lost fraction} \approx 2 \, d_w / \lambda \qquad (10a)$$

where λ is the length of cut jet section. To limit the loss to 5%, λ must be higher than 40 times the cutting wire diameter, d_w. Through geometric considerations, the droplet size, assuming a negligible loss, is equal to

$$d \approx (3d_j^2 \lambda / 2)^{1/3}. \qquad (10b)$$

With a capillary of 400 μm internal diameter and 25 μm wires, the droplet diameter would be equal to 600 μm. The flow rate will only be limited by the pressure applied. Rotating jet breakage appears to be an easy, efficient, and scalable device for producing large quantities of relatively small microcapsules with narrow size distribution.

Rotating Capillary Jet Breakage

Replacing gravity by centrifugal force has been applied in the microencapsulation of food, chemical, and pharmaceutical ingredients. However, very few studies have concerned cell encapsulation. In general, the size of the collecting reservoir is very large with diameters up to 10 meters, complicating recovery of the formed microcapsules and maintaining sterility. Moreover, such systems involve high levels of shear, potentially damaging fragile encapsulants. It would be necessary to address these points before applying the technology to bioencapsulation.

Rotating extrusion devices operate by extruding liquids through nozzles at the periphery of a cylinder, or by dispersing liquids onto a spinning disk. In the first system, the liquid flows through nozzles mainly as jets. Schlameus (1995) reports productivity levels up to 60 kg/h. The droplet diameter ranges from 500 μm to 2 mm with a standard deviation of around 15%. The limitations of jet formation involving pressure and jet diameter have to be considered.

With the spinning disk device, liquid flows on a rotating disk. Liquid exits the disk as droplets, filaments, or films, as a function of the working conditions (Chicheportiche 1993). The filament regime is of particular interest. Filaments break into small droplets in a manner similar to that of jets. Droplets as small as 50 μm may be obtained (unpublished data). No pressure is required to ensure flow, and very limited shear is applied to the cells.

In both devices described, vibration may be applied to facilitate droplet formation. In the case of spinning disks, Chicheportiche (1993) observed a very narrow size distribution (standard deviation of 5% without satellite peak formation).

Emulsification Methods

Although they are quite promising for large-scale production of small capsules, little data exist regarding emulsification methods for cell encapsulation (Audet et al 1989, Poncelet et al 1993). It is necessary to refer to the general theory of emulsification and to a limited number of papers on the encapsulation of chemicals or biochemicals by emulsification methods (Poncelet et al 1989, Ogawa et al 1972).

Emulsification methods provide capsules from a few micrometres to a millimeter in diameter. In all cases, the size dispersion is higher than with extrusion devices, ranging from 30 to 50% of the mean diameter. However, the potential for scale-up is the main advantage. Emulsification is generally performed in a reactor by means of a turbine. However, a more promising technology involves passing the two immiscible phases through a tube containing deflectors or stationary baffles, known as static mixers (Poncelet et al 1993), as illustrated in Figure 1.4. Such a system improves the size distribution and reduces shear. As concerns industrial-scale applications, it permits continuous processing, and the enclosed plumbing enables the maintenance of aseptic conditions, while not being limited by scale.

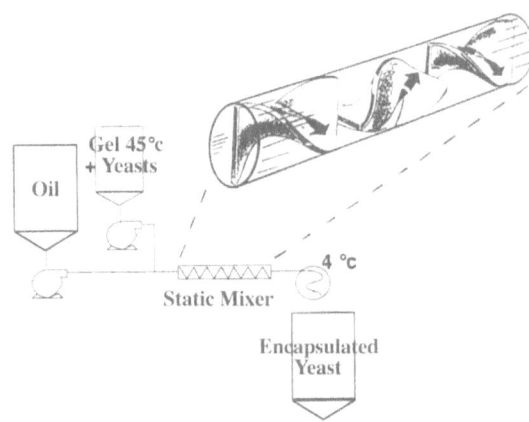

FIGURE 1.4. Static mixer.

Gelation and Membrane-Formation Methods

The most common method for producing capsules is to form gels from liquid droplets. The resulting hydrogel beads are very porous. Polymer coats are often applied to ensure better isolation and retention of the encapsulated material. The gel core of coated beads may also be liquefied, resulting in a liquid droplet retained within a membrane coat. To simplify the encapsulation process, direct membrane formation around liquid droplets has been proposed. Figure 1.5 summarizes the various bioencapsulation procedures as developed in the following paragraphs.

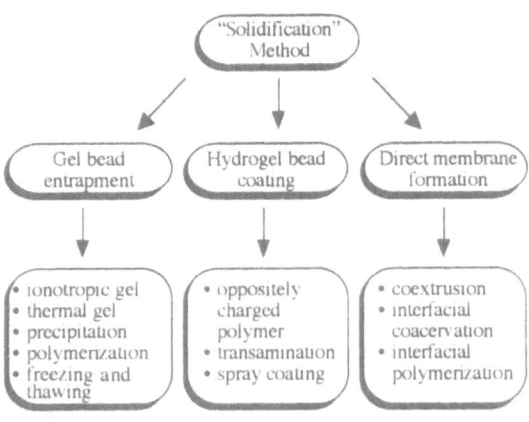

FIGURE 1.5. Bioencapsulation "solidification" methods.

A unique feature of encapsulation, in regards to other cell immobilization technologies, is that a matrix is built around the cells. Thus, the concept of biocompatibility concerns not only the encapsulating material but also the encapsulation process. Materials that are considered to be biocompatible as implantable devices, for example, are not suitable for cell encapsulation, because they may require formulation conditions that are unfavorable for cell survival, as is the case with the polymers used for hollow fiber production. Many chemical encapsulation methods would also be rejected because of the use of organic solvents or cross-linkers. The range of bioencapsulation methods is reduced for these reasons. Moreover, any process has to be carefully optimized to reduce toxic effects and to carefully and reproducibly regulate capsule structures. Cells are very sensitive to their environment; thus, the presence of chemical traces and microstructures within the gel may affect their behavior (Barbotin et al 1990).

Cell growth is possible and common in capsules. Cultivating encapsulated cells in appropriate media may make up any loss resulting from the encapsulation process, or from undesirable release through the porous matrix structure (Groboillot et al 1993). The challenge is to ensure and maintain appropriate cell behavior after encapsulation. Optimizing a bioencapsulation method is in fact a delicate process even if the technology appears simple. As an example, the development of the Moet & Chandon process to produce champagne with encapsulated yeast is the result of a 14-year research program. Also, the artificial pancreas based on islet encapsulation was initially proposed in 1979, but is still in developmental stages.

Hydrogel Bead Entrapment

Entrapment in hydrogel beads is the most commonly used cell immobilization technology, because of its simplicity and gentle formulation conditions. Numerous studies have been conducted to select encapsulation materials, to optimize processing conditions, and to characterize hydrogel beads. A large number of studies have also been devoted to the physiological behavior of hydrogel-entrapped cells.

Gel-forming materials may be classified as natural or synthetic polymers. Natural polymers are composed of polysaccharides or proteins, and synthetic polymers may be preformed, polymerized *in situ* from monomers, or cross-linked from prepolymers.

Ionotropic Gelation

Several charged polymers form gels when introduced dropwise into an oppositely charged multivalent counter-ion solution. The success of ionotropic gelation is mainly due to the mild formulation conditions, involving no pH or temperature changes, reagent toxicity, or residues. It is a simple, fast, and low-cost technique.

Alginate is the most widely used and investigated polymer for cell entrapment, by a large measure. Alginate constitutes a family of unbranched polysaccharides, mainly extracted from algae. It is composed of 1,4-linked β-D-mannuronic (M) and α-D-guluronic acid (G) residues. The monomers are sequenced in homopolymeric blocks (M-M or G-G blocks) with alternating structures (M-G blocks) (Smidsrod et al 1972, 1974). Divalent and trivalent cations, generally calcium, induce gelation by binding mainly to the guluronic blocks. Divalent cations bind within two guluronic blocks, forming a series of electronegative cavities. As it is a cooperative process, stronger gels and improved gelation are obtained with high guluronic alginate (up to 70%) containing long guluronic blocks (up to 15 residues; Martinsen et al 1989). The molecular weight of the chains has a limited impact on the gel properties.

Alginate beads are produced by dropping cell-loaded alginate into calcium chloride solution, in a procedure known as external gelation. Calcium ions diffuse into the alginate drops, forming a three-dimensional lattice of ionically cross-linked alginate. Most studies have been conducted with large beads (2 to 3 mm), produced by droplet extrusion from a syringe. The resulting beads are not homogeneous, because the alginate concentration increases from the center of the beads (2%) to the surface (up to 10%; Skjak-Braek et al 1989). Moreover, Nava Saucedo et al (1996) showed that microchannels are formed between the surface and the bead core.

To obtain higher gel homogeneity with smaller beads on a large scale, Poncelet et al (1992) formed alginate microspheres by internal gelation. Alginate solution containing dispersed insoluble calcium carbonate microcrystals was emulsified in oil, then acidified gently by addition of oil-soluble acetic acid to liberate calcium. Internal gelation may also be obtained by dropping the alginate/calcium carbonate mixture into an acidic solution. The resulting beads are more homogeneous in composition, and the slight pH reduction during formulation (7.5 to 6.5) is unlikely to damage cells.

The major disadvantage of alginate beads is sensitivity to calcium chelators such as phosphate, lactate, or EDTA, and to cations such as sodium or magnesium, which are able to displace calcium. Because these compounds are often required in fermentation media, use of stronger gelling agents such as barium or aluminum ions and higher guluronic content alginates will improve stability of the beads (Martinsen et al 1989). Alternately, chelators are often used to liquefy the alginate when forming liquid-core microcapsules, or when recovery of the encapsulant is desired.

Chitosan is a polyglucosamine polysaccharide, obtained by deacetylation of chitin (Muzzarelli 1977). It forms gels in the presence of polycations such as phosphate (Moore and Roberts 1980). Beads are produced by dropping chitosan solution into phosphate solution (Vorlop and Klein 1981, 1987). Hydrophobic gels may also be obtained by gelation in more hydrophobic anions such as octyl or lauryl sulphate (Vorlop and Klein 1987). Mechanical stability of chitosan beads is comparable to that of alginates (Klein and Kressdorf 1989), and chitosan beads are stable in phosphate buffer. However, chitosan is water soluble only for pH levels lower than 6.5, and chitosan may interact with cell membranes, leading to loss of cell viability or activity. Use in cell encapsulation has been limited thus far.

Pectins are acidic polysaccharides extracted from plant cell walls. The extracted forms are predominantly linear polymers of 1,4 α-D-galacturonate backbone. Acidic groups are partially methoxylated. Pectic acids have less than 5% methoxyl groups. Other pectinic acids are also divided into high-methoxyl (substitution higher than 50%) and low-methoxyl pectins (LM pectinate).

Pectins (pectate and LM pectinate) gel by strong binding of calcium or aluminum ions, as is the case with alginate (Thibault and Rinaudo 1985). Pectate and pectinate gel beads are produced by dropping pectin into calcium (Gemeiner et al 1989, Toth et al 1989) or aluminum ionic solutions (Berger and

Ruhlemann 1988, Navarro et al 1983). Beads are inhomogeneous, like alginates, with decreasing concentration from the surface to the center of the beads (Skjak-Braek et al 1989). Calcium pectate beads are much less sensitive to calcium chelators or competitors than alginate beads (Berger and Ruhlemann 1988). Pectate-alginate mixtures have been proposed to provide highly stable beads (Toth et al 1989).

Polyphosphazene has been proposed as a synthetic polymer to replace alginate (Bano et al 1991). Synthetic materials will reduce problems associated with the purification of natural polymers such as alginate. However, development of this system has been limited thus far.

Thermal Gelation

Thermal gelation is achieved by cooling a warm aqueous polymer solution. Beads are obtained by dropping the polymer solution into cold water, but most often the polymer solutions are emulsified in warm oil and cooled by addition of cold oil or water (Audet et al 1989) or by using heat exchangers (Neufeld et al 1991). Thermally gelled beads are generally less sensitive to destabilizing ions than ionotropic gels. However, the need to heat the polymer solution and encapsulant before gelation can limit its use for fragile cell encapsulation. Some gel manufacturers are responding by developing low-temperature gelling materials. Also, thermally gelled beads may not be suitable for mesotrophic fermentation or implantation, because they would dissolve.

Agar is obtained from red algae and consists of alternating 1,3–linked β-D-galactopyranose and 1,4–linked 3,6–anhydro-α-D-galactopyranose monomers. Substitutions (sulphate, methylether) on the β-D-galactopyranose give different gelation behavior to agar (Clark and Ross-Murphy 1987). Methylether content leads to higher gelling temperature (from 30 to 40°C). Neutral agar is the main gelling fraction, and purification of this fraction provides agarose. The gel structure is maintained mainly by hydrogen bonding.

Carrageenan has often been proposed as an alternative to alginate in fermentation processes. Carrageenans have a structure similar to agar. The 1,4–linked β-D-galactose is only partially in anhydrous form, and the percentage of sulphate substitution is higher. Carrageenans are extracted from red seaweed and are divided in three types (λ, κ, and τ) (Landau 1992, Thomas 1992). κ-carrageenan forms strong gels upon cooling in the presence of potassium ions (Chibata et al 1987). Gelation may also be induced by contact with different ions such as calcium or copper, but thermal gelation is the most common method.

κ-carrageenan beads may be produced by the droplet formation technique (Buitelaar et al 1990, Woodward et al 1988), or, for large-scale production, by emulsification (Audet et al 1990). Higher gel mechanical resistance may be obtained by mixing κ-carrageenan with other gel-forming polymers such as galactomannans (Guiseley et al 1989), locust bean gum (Audet 1989), or taragum (Cairns et al 1986).

Gellan gum is a gel-forming polysaccharide produced by bacterial fermentation (Gibson 1994). It is composed of linear tetrasaccharide units that include two glucoses, one rhamnose, and one guluronic acid (O'Neil et al 1983). Guluronic residues confer anionic charges to the gellan gum. Different substituants may be attached to the chains (Kuo et al 1986) resulting in soft, elastic, and cohesive gels, while de-esterified polymers form strong, hard, and brittle gels (Gibson 1994, Rinaudo 1988). Unsubstituted gels have been reported as stronger than most other gels (alginate, agar, κ-carrageenan; Sanderson et al 1989).

Gellan gelation is achieved by cooling in the presence of stabilizing divalent cations (Norton and Lacroix 1990). Gelation temperature is a function of the gum concentration, ionic strength, and type of counter-ions, and may vary from 35 to 55°C. Stronger gels are obtained with gum mixtures of high gelation temperature. Gellan gum was first proposed for encapsulating thermophilic bacteria (Grasdalen and Smidsrod 1987, Norton and Lacroix 1990); however, by using citrate, phosphate, or EDTA sequesterants, the gelation temperature may be decreased and mesophilic bacteria may be entrapped (Camelin et al 1993) Gellan gum beads may be obtained by droplet extrusion or emulsification methods (Grasdalen and Smidsrod 1987, Norton and Lacroix 1990).

Some polymers can also gel upon slight heating (Perols et al 1997; Markvicheva et al 1991). As an example, polyvinyl capralactam is used to encapsulate enzymes and hybridoma cells (Donova et al 1993, Markvicheva et al 1991). This synthetic polymer is water-soluble, nontoxic, and inexpensive. An increase in temperature from 10 to 40°C permits gelation. Gel characteristics and gel reversibility may be modulated using various stabilizers. For example, monoclonal antibodies produced by hybridoma cells may be concentrated and recovered through bead liquefaction initiated by a temperature drop (Markvicheva et al 1991).

Reticulated Protein Gels

Applications of proteins for cell encapsulation are relatively limited. Collagens, a family of animal fibrous proteins, are rich in glycine and proline, favoring stable triple helix formation (Bornstein and Sage 1980). In the presence of water, collagens swell and gel through ionic, hydrogen-bonding, and other interactions. Collagens dissolve at low pH; thus, encapsulation requires mixing cells with collagen at low temperature, ionic strength, and pH. Gelation is initiated by raising these three parameters. In most cases, reticulation with glutaraldehyde is necessary to obtain strong gels.

Gelatin is a hydrolyzed derivative of collagen. Reversible gelation is obtained by cooling the solution below 30–35°C (De Alteriis et al 1988). Beads are formed by dropping gelatin solution into a cool hydrophobic fluid. Addition of a cross-linker such as glutaraldehyde, formaldehyde, or chromium salt is necessary to stabilize the gel (Sungur and Akbulut 1994).

Synthetic Polymers Formed In Situ

Synthetic polymers are more flexible and have more reproducible characteristics than natural polymers, for the purpose of developing specific properties. Porosity, hydrophobicity, mechanical strength, and stability of the gel may be more precisely modified. In most cases, *in situ* polymerization leads to loss of cell viability as a result of monomer toxicity and the reactive environment. Careful control of the gel formation process is therefore important to maintain high cell viability.

Polyacrylamide gels obtained by free radical linear polymerization of acrylamide (single unsaturation) in the presence of bisacrylamide (double unsaturation) form a three-dimensional network. The ratio between monomer and bisacrylamide determines gel porosity and strength. Free radical reaction is initiated chemically (i.e., by using persulphate) or photochemically (i.e., with riboflavin as photoinitiator). Acrylamide monomers are toxic (Lusta et al 1990), but well-controlled conditions during polymerization (low temperature, minimum polymerization time) enable high cell viability (Skryabin and Koshcheenko 1987). Spherical beads are obtained by dispersing polymerization medium that contains cells into a hydrophobic oil before polymerization (Mosbach 1984, Nilsson et al 1987).

As an alternative to acrylamide, methacrylate monomers can be used and polymerized using a cross-linking agent (Cantarella et al 1983). Again, cell toxicity results in low cell viability. Polymerization using γ-radiation at freezing temperature has been described. This process, while successful in a few cases (Carenza and Veronese 1994) has limited application for cell encapsulation.

Prepolymer resins as gel-forming materials appear to have important advantages (Fukui and Tanaka 1984), because the entrapment procedure is simple and performed under mild conditions, since the use of monomers is avoided. The gel structure may be controlled by varying polymer chain length or hydrophobic/hydrophilic balance. As an example, prepolymerized acrylamide chains, partially substituted with acylhydrazide groups (Freeman 1987), are cross-linked by dialdehydes such as glyoxal and glutaraldehyde. Prepolymer and crosslinker concentrations determine the porosity and strength of the gel. Glyoxal appears to be the best cross-linker both for gel structure and cell viability (Freeman 1987), and several other prepolymers have been proposed. An interesting development has been to clone *E. coli* to increase resistance to acrylamide monomers (Lusta et al 1990).

Freeze-Thawing Polyvinyl Alcohol Gels

Polyvinyl alcohol (PVA) is a low-cost, nontoxic polymer that becomes gelatinous upon freezing. In fact, repeated freeze/thaw cycles strengthen the gel (Nambu 1983). The gel forms by PVA exclusion from water crystals during freezing, resulting in a concentration of the polymer. The high-PVA phase forms a continuous three-dimensional network. Cells are protected during freezing with cryo

protectants such as glycerol. Temperature profiles during both freezing and thawing determine both gel structure and cell viability. PVA gels are stable to 65°C and can be used for thermophilic microorganisms (Varfolomeyev et al 1990).

PVA gels are also formed by cross-linking with boric acid (Ochiai et al 1981). The highly acidic conditions (pH 4) limit the scope of this method. Mixing PVA and alginate has been proposed, mainly to reduce the sticking tendency of PVA gel beads (Wu and Wisecarver, 1992). PVA beads can also be treated with phosphate solution to esterify PVA, strengthening the beads.

Hydrogel Bead Coating

The gel bead structure is porous, allowing for the diffusion of lower molecular weight molecules in and out of the beads. Cell immobilization often requires a reduced permeability, to further isolate cells from the external medium, as in cell transplantation. Beads can also be sensitive to their environment, as are alginate beads, which dissolve readily in citrate or phosphate. The external surface of the beads may also be modified to improve biocompatibility or bead strength. These factors have led researchers to consider applying membrane coatings to beads.

Most types of beads described above may be treated with cross-linkers such as glutaraldehyde to strengthen the external layer. Process conditions are very important to enhancing bead mechanical properties without reducing cell viability. Moreover, the use of dialdehydes can be a problem for food and medical applications, although glutaraldehyde has FDA approval. Diamines or low molecular weight polyamines could also be used to cross-link the external layer of the beads.

High molecular weight polyamine coatings are applied under gentle conditions. Alginate beads are first coated by suspension in poly-L-lysine (PLL) solution (Lim and Sun 1980), and a second alginate coat is then applied by suspending the coated beads in alginate. The PLL/alginate double coat provides a stronger membrane and enhances bead biocompatibility for transplantation (O'Shea et al 1984). The alginate core can be liquefied in citrate if a liquid core is desired. All of these steps are performed at neutral pH and room temperature, and in physiological solution, ensuring mild encapsulation conditions.

Limited alternatives have been proposed to PLL coating. Polyethyleneimine (PEI) has been tested but rejected because of biocompatibility problems following transplantation (Sun and O'Shea 1985). Poly-L-ornithine, poly-L-glutamate (Burgarski et al 1993, Young et al 1993), chitosan (McKnight et al 1988), and modified chitosan (Pandya and Knorr 1991) have all been evaluated as alternatives.

Most studies related to the control of the alginate bead coating involve PLL membranes and were performed by Goosen and coworkers in Canada (Goosen et al 1985, King et al 1987, Okhamafe and Goosen 1993). The main concerns were the molecular cut-off and mechanical strength controlled by selecting PLL molecular weight, concentration, and coating time. Similar studies have also been conducted by Vandenbossche et al (1993).

An innovative process has been proposed by Lévy et al (1996), in which alginate beads containing a protein and PEG-esterified alginate were suspended in an alkaline solution where transamination takes place between the protein and the esterified alginate. Strong covalently bonded membranes result without loss of cell viability, even for fragile cells.

Hydrophobic polymers can also be used to coat beads. The simplest method is to coextrude a hydrogel with a polymer solution (Dupuy et al 1988). Alternately, the beads may be coated by spray application of a polymer solution or a latex (fine polymer dispersion in water) in a fluidized bed reactor (Sun et al 1997).

Direct Encapsulation

A popular encapsulation procedure involves forming alginate beads, coating them with PLL and alginate, and then performing core liquefaction. This procedure is simple and involves mild process conditions, yet requires several steps, which increases cost from a process point of view and increases risk of contamination because of the increased number of operations. It would be desirable to formulate membrane-bound microcapsules in a single step. Most methods require the use of solvents or other toxic reagents or conditions. The following review is limited to techniques that have been successfully applied for cell encapsulation.

The simplest method, at least at laboratory scale, is to coextrude droplets of an internal phase that contains cells in an external polymer solution into a gelification bath (Udulag 1994). The receiving bath may be a gel initiator, or a solvent extraction medium in the case of a hydrophobic external phase. This technology has been developed by Sefton et al, using polyacrylates (Babensee et al 1992, Douglas and Sefton 1990, Sefton et al 1987), for the encapsulation of fragile cells. The main drawback of this method is the control of the process from the hydraulic point of view. Stevenson and Sefton 1993 described the different technological difficulties that must be overcome to obtain spherical microcapsules with a uniform membrane and appropriate diameter. Although scale-up may be difficult, the process is usable with a large variety of membrane materials.

Interfacial polymerization was proposed in 1964 as a process for producing artificial cells (Chang 1964). A diamine solution is emulsified within an organic phase, and an acid dichloride is added to the emulsion to initiate membrane formation. Diamine and dichloride polymerize at the droplet interface, providing microcapsules with thin but strong membrane coats. Initially, the process involved high pH levels (>10), high concentrations of diamine (0.4M), and the use of polar solvents (chloroform). Use of an internal phase containing polyamines such as gelatin, polyethyleneimine, or chitosan improves the biocompatibility of the process (pH 8 and vegetable oil as dispersing phase; Groboillot et al 1993, Larish et al 1994, Poncelet et al 1990). Higher viability was obtained with gelatin (Hyndman et al 1993). Improvements are still needed to ensure high levels of cell viability.

Microcapsules may also be obtained by dropping a charged polymer in a solution of an oppositely charged polymer, in a process known as interfacial coacervation. Polymer coacervate forms at the droplet interface, forming a continuous membrane. Many polymer combinations may be used to create such capsules. Alginate with chitosan was first proposed by Rha et al (1984), and alginate may be replaced by κ-carrageenan (Pandya and Knorr 1991). Dautzenberg (1985) used cellulose sulphate and poly(dimethyldiallylammonium chlorides) to study parameters such as the impact of the polymer molecular weight, degree of substitution, and polymer concentration on the membrane mechanical resistance or molecular cut-off (Dautzenberg 1996). Interfacial coacervation provides molecular weight cut-off values to as low as 3000 daltons. Hunkeler et al (1996) tested 1300 combinations of polymers and defined guidelines for selecting the best combinations. However, multicomponent systems, including for example capsule fillers, improve the microcapsule properties (Hunkeler 1997). Interfacial coacervation is one of the more biocompatible processes for cell encapsulation and may be the most promising alternative to the commonly used alginate-PLL system.

Applications

When considering the possibility of immobilizing cells for a particular application, the cost and technological complexity must be carefully taken into account. Mass transfer limitations, changes to cell behavior, and effects on cell viability may all affect and generally reduce the performance of the immobilized cell system. However, in many cases, immobilization is required to reach specific objectives.

In fermentation, cell immobilization (1) enables continuous operation without relying on cell growth to maintain cell density, (2) simplifies the downstream processing as it facilitates cell separation, and (3) enables the use of mixed and spatially localized microbial cultures to obtain higher yields, especially for secondary metabolites. Immobilization also protects cells from the surrounding environment, stabilizes the cells, reduces inhibition from substrates, and protects implanted cells from immunorejection.

In the next section, the main types of immobilized cell applications are summarized in relation to therapeutics. The review by Willaert and Baron (1996) may be consulted for an in-depth review.

Many antibiotics such as candicidin (Constandinides and Mehta 1991), cyclosporin C (Foster et al 1983) and oxytetracycline (Farid et al 1994) are produced by microbial processes. Immobilization of cells is mainly considered for continuous production of the antibiotics. The main objective is to obtain stable antibiotic production with limited cell growth (Furusaki and Seki 1992). Polyacrylamide, alginate, and carrageenan beads are the most usual immobilization matrices.

Steroids serve as the basis for producing many hormones (Larsson et al 1976). The biotransformation is complex, involving oxygen activation and continuous supply of reductive power. The enzymes involved in these transformations, which include hydroxylases and deshydrogenases, are often unstable. As these biotransformations are generally based on the activity of one or two enzymes, cell viability may be less important. However, maintaining cell structure offers protection for the enzyme. Many transformations of steroids use photo-cross-linked resin beads (Sonomoto et al 1981, Tanaka et al 1984), polyacrylamide (Vlahov et al 1990), and alginate or carrgaeenan beads (Hocknull and Lilly 1990).

Animal and plant cells may be immobilized to produce therapeutics, vaccines, and monoclonal antibodies. Immobilization is required not only to provide protection of cells but also to mimic the natural cell environment. For both types of cells, alginate and agarose beads are the most usual systems of immobilization. However, microcapsules produced by alginate bead coating (Koo and Chang 1993), coextrusion (Uludag et al 1994), or interfacial coacervation (Mansfeld et al 1995) are also commonly used.

Animal cells may also be immobilized as artificial tissues and used for testing different drugs, which makes systematic studies of drugs easier to perform. However, the major application of animal cell encapsulation is the development of artificial organs. Treatment of diabetes with encapsulated pancreatic islets is a major subject, but if successful, it will have important applications in the treatment of many other diseases (Alzheimer's, Parkinson's, hemophilia). PLL-coated alginate beads are used in over 90% of the studies on the subject. However, coextrusion (Sefton et al 1992), interfacial coacervation (Hunkeler et al 1996), and transacylation (Lévy and Edward-Lévy 1996) are very promising technologies. Although their basic principle is relatively simple, development is required to ensure biocompatibility and scalable formulation. This subject will be developed in Part Three.

Conclusions

Cell immobilization and encapsulation has a broad range of applications. Although simple and biocompatible conditions are required for cell encapsulation, technological development thus far is time-consuming, requiring collaboration between scientists and engineers from many disciplines.

A discussion about the application of encapsulated cells is a discussion about the future. It represents an important objective for scientists and industry for the benefit of all.

References

Audet, P and Lacroix, C. 1989. Proc Biochem. 24: 217.

Audet P, Paquin C, Lacroix C. 1990. Appl. Microbiol. Biotechnol. 32: 662.

Babensee JE, DeBoni U, Sefton MV. 1992. J. Biomed. Mater. Res. 26: 1401.

Bano, MC, Cohen, S, Visscher, KB, Allcock, HR, Langer, R. 1991. Biotechnol. 9: 468.

Barbotin JN, Nava Saucedo JE, Thimasset B. 1990. In: Physiology of immobilized cells, deBont JAM, Visser J, Mattiasson B, Tramper J (eds.), Elsevier, Amsterdam, p. 487.

Berger R, Ruhlemann I. 1988. Acta Biotechnol. 8: 401.

Bornstein P, Sage H. 1980. Ann. Rev. Biochem. 49: 957.

Bugarski B, Jovanovic G, Vunjak-Novakovic G. 1993. In: Fundamentals of animal cell encapsulation and immobilization, Goosen MFA (ed.), Boca Raton: CRC Press, p. 267.

Bugarski B, Li Q, Goosen MFA, Poncelet D, Neufeld RJ, Vunjak G. 1994a. AIChE J., 40(6):1026.

Bugarski B, Smith J, Wu J, Goosen MFA. 1994b. Biotechnol. Techn. 7: 677.

Buitelaar RM, Hulst AC, Tramper J. 1990. In: Physiology of immobilized cells, deBont JAM, Visser J, Mattiasson B, Tramper J (eds.), Elsevier, Amsterdam, p. 205.

Cairns P, Morris VJ, Miles MJ, Brownsey GJ. 1986. Food Hydrocolloids 1: 89.

Camelin I, Lacroix C, Paquin C, Prévost H, Cachon R, Divies C. 1993. Biotechnol. Prog. 9: 291.

Cantarella M, Scardi V, Alfani F. 1983. In: Proceedings Biotech. 83, London, p. 1051.

Carenza M, Veronese FM. 1994. J. Control. Rel. 29: 187.

Champagne C. (ed) 1994. Immobilization of cells for application in the food industry. In: Critical Reviews in Biotechnology 14: 2.

Chang TMS. 1964. Semipermeable microcapsules. Science 146: 524.

Charwat, AF. 1977. Rev. Sci. Instrum. 48: 1034.

Chibata I, Tosa T, Sato T, Takata I. 1987. Methods Enzymol. 135: 189.

Chicheportiche J-M. 1993. PhD thesis, Paris VI, France.

Clark AH, Ross-Murphy SB. 1987. In: Advances of polymer sciences, Vol. 83, Springer-Verlag, Berlin, p. 57.

Constandinides A, Mehta N. 1991. Biotechnol. Bioeng. 37: 1010.

Dautzenberg H, Loth F, Fechner K, Mehlis B, Pommerening K. 1985. Makromol Chem-Suppl. 9: 203.

Dautzenberg H, Stange J, Mitzner S, Lukanoff B. 1996. In: Immobilized Cells: Basics and Applications. Wijffels RH, Buitelaar RM, Bucke C, Tramper J (Eds) Elsevier Science B. V. p 181.

De Alteriis E, Parascandola P, Pecorrella MA, Scardi V. 1988. Biotechnol. Techn 2: 205.

Donova MV, Kuz'kina IF, Arinbasarova AY, Pashkin II, Markvicheva EA, Baklashova TG, Sukhodolskaya GV, Fokina VV, Kirsh YE, Koshcheyenko KA, Zubov VP. 1993. Biotechnol. Technique. 76: 415.

Dos Santos APM, Leenen EJTM, Ripoll MM, van der Sluis C, van Vliet T, Tramper J, Wijffels RH. 1997. Biotechnol. Bioeng. 56 (5): 517.

Douglas JA, Sefton MV. 1990. Biotechnol. Bioeng. 36: 653.

Dupuy B, Gin H, Ducassou D. 1988. J. Biomed. Mater. Res. 22: 1061.

Farid MA, Eldiwany AL, Elenshashy HA. 1994. Acta Biotechnol. 14: 303.

Foster BC, Coutts RT, Pasutto FM, Dossetor JB. 1983. Biotechnol. Lett. 5: 693.

Freeman A. 1987. Methods Enzymol. 135: 216.

Fukui S, Tanaka A. 1984. Adv. Biochem. Eng./Biotechnol. 29: 1.

Furasaki S, Seki M. 1992. Adv. Biochem. Eng./Biotechnol. 46: 162.

Gemeiner P, Kurillova L, Malovikova A, Toth D, Tomasovicova D. 1989. Folia Microbiol. 34: 214.

Gibson W. 1994. In: Thickening and gelling agents for food, Imeson A. (ed.). Blackie Academic & Professional, London, p. 227.

Goosen MFA, O'Shea GM, Gharapetian HM, Chou S, Sun AM. 1985. Biotechnol. Bioeng. 27: 146.

Goosen MFA. (ed.) 1993. Fundamentals of animal cell encapsulation and immobilization, CRC Press, Boca Raton.

Grasdalen H, Smidsrod, O. 1987. Carbohydr. Polym. 7: 371.

Groboillot AF, Champagne CP, Darling GD, Poncelet D, Neufeld RJ. 1993. Biotechnol. Bioeng. 42: 1157.

Guiseley KB. 1989. Enzyme Microb. Technol. 11: 706.

Hartmeier, W. 1986. Immobilizierte biokatalysatoren, Springer-Verlag, Berlin.

Hocknull MD, Lilly MD. 1990. Appl. Microbiol. Biotechnol. 33: 148.

Hunik JH, Tramper J. 1993. Biotechnol. Prog. 1993 (9) 186.

Hunik JH. 1993. Biotechnol. Prog. 9: 186.

Hunkeler D, Prokop A, Dimar S, Haralson M, Wang TG. 1996. Water soluble polymers for immunoisolation: Complex coacervation and cytotoxicity. Proceedings of the BRG Internat. Workshop: Bioencapsulation V, Potsdam, Germany. Sept. 22–25, 1996.

Hunkeler D. 1997. Polymers for bioartificial organs, Proceedings of International Workshop on Bioencapsulation VI, Barcelona, Spain, Aug. 30–Sept. 1, 1997.

Hyndman CL, Groboillot AF, Poncelet D, Champagne CP, Neufeld RJ. 1993. J. Chem. Technol. Biotechnol. 56: 259.

Iwai, S, Kitao, T. 1994. Wastewater treatment with microbial films. Lancaster: Technomic.

King GA, Daugulis AJ, Faulkner P, Goosen MFA. 1987. Biotechnol. Prog. 3: 231.

Klein J, Kressdorf B. 1989. Angew. Makromol. Chem. 166/167: 293.

Koo J, Chang TMS. 1993. Int. J. Artif. Organs 16: 557.

Kuo M-S, Mort AJ, Dell, A. 1986. Carbohydr. Res. 156: 173.

Landau M. 1992. Introduction to aquaculture, New York: John Wiley.

Lane, WR. 1947. Rev. Sci. Instrum. 24: 98.

Larisch BC, Poncelet D, Champagne CP, Neufeld RJ. 1994. J. Microencapsul. 11: 189.

Larsson PO, Ohlson S, Mosbach K. 1976. Nature 263: 796.

Lévy M-C, Edwards-Lévy F. 1996. J. Microencapsul. 13: 169.

Lim F, Sun AM. 1980. Science 210: 908.

Lusta KA, Starostina NG, Fikhte BA. 1990. In: Physiology of immobilized cells, deBont JAM, Visser J, Mattiasson B, Tramper J (eds.), Amsterdam; Elsevier, p. 557.

Mansfeld J, Forster M, Hoffman T, Schellenberger A. 1995. Enzyme Microb. Technol. 17: 11.

Markvicheva EA, Kuz'kina IF, Pashkin II, Plechko TN, Kirsh YE, Zubov VP. 1991. Biotechnol. Technique. 5: 223.

Martinsen A, Skjak-Braek G, Smidsrod O. 1989. Biotechnol. Bioeng. 33: 79.

Mattiasson B. 1983. Immobilized cells and organelles, Vol. I and II, Boca Raton: CRC Press.

McKnight CA, Goosen MFA, Penney C, Sun D. 1988. J. Bioact. Compat. Polym. 3: 334.

Moo-Young M. (ed.) 1988. Bioreactor immobilized enzymes and cells: Fundamentals and applications. London: Elsevier Applied Science.

Moore GK, Roberts GAF. 1980. Int. J. Biol. Macromol. 2: 7377.

Mosbach K. 1984. Ann. N.Y. Acad. Sci. 434: 239.

Muzzarelli RAA. 1977. Oxford: Chitin, Pergamon Press.

Nambu M. 1983. Koubunshi-Kakou 32: 523.

Nava Saucedo JE, Roisin C, Bienaim C, Ribeiro T, Barbotin J-N. 1996. Int. Workshop on Bioencapsulation V, Potsdam, Germany, Sept. 23–25.

Navarro AR, Rubio MC, Callieri DAS. 1983. Eur. J. Appl. Microbiol. Biotechnol. 17: 148.

Neufeld RJ, Peleg JS, Rokem JS, Pines O, Goldberg I. 1991. Enz. Microb. Technol. 13: 991–996.

Nilsson K, Brodelius P, Mosbach K. 1987. Methods Enzymol. 135: 222.

Nir R, Lamed R, Gueta L, Sahar E. 1990. Appl. Environ. Microbiol. 56: 2870.

Norton S, Lacroix C. 1990. Biotechnol. Techn. 21: 351.

Ochiai H, Shimizu S, Tadokoro Y, Murakami I. 1981. Polymer. 22: 1456.

Ogawa T, Takamura K, Koishi M, Kondo T. 1972. Bull. Chem. Soc. Jap. 45: 2329.

Okhamafe AO, Goosen MFA. 1993. In: Fundamentals of animal cell encapsulation and immobilization, Goosen MFA (ed.), Boca Raton: CRC Press, p. 55.

O'Neil MA, Selverdran RR, Morris VJ. 1983. Carbohydr. Res. 124: 123.

O'Shea GM, Goosen MFA, Sun AM. 1984. Biochim. Biophys. Acta 804: 133.

Pandya Y, Knorr D. 1991. Process Biochem. 26: 75.

Perols C, Piffaut B, Scher J, Ramet JP, Poncelet D. 1997. Enzyme Microbial Technology. 20: 57.

Poncelet D, Poncelet DeSmet B, Neufeld RJ. 1988. 38th Can. Chem. Eng. Conf. Edmonton, Oct 2–5.

Poncelet D, Neufeld RJ. 1989. Biotechnol. Bioeng. 33: 95.

Poncelet D, Poncelet De Smet B, Neufeld RJ. 1990. J. Membr. Sci. 50: 249.

Poncelet D, Lencki R, Beaulieu C, Halle JP, Neufeld RJ, Fournier A. 1992. Appl. Microbiol. Biotechnol. 38: 39.

Poncelet D, Poncelet DeSmet B, Beaulieu C, Neufeld RJ. 1993. In: Fundamentals of animal cell encapsulation and immobilization, Goosen MFA. (ed.), Boca Raton: CRC Press, p. 113.

Poncelet D, Bugarski B, Amsden BG, Zhu J, Neufeld R, Goosen MFA. 1994. Appl. Microbiol. Biotechnol. 42: 251.

Poncelet D, Babak VG, Neufeld RJ, Goosen MF, Burgarski B. 1998. Advanced Colloid Science.

Prusse U, Bruske F, Breford J, Vorlop KD. 1996. Int. Workshop on Bioencapsulation V, Potsdam, Germany, Sept. 23–25.

Rayleigh JWS. 1878. Proc. London Math. Soc. 10, 4.

Rha C, Rodriguez-Sanchez D, Kienzle-Sterzer C. 1984. In: Biotechnology in the marine polysaccharides. Colwell RR, Pariser ER, Sinkey AJ (eds) Hemisphere Publishing Corp, Washington, p. 283.

Rinaudo M. 1988. In: Gums and stabilizers for the food industry. Phillips GO, Williams PA, Wedlock DJ (eds.), IRL Press, Washington, DC, p. 301.

Sanderson GR, Bell VL, Ortega DA. 1989. Cereal Foods World 34: 991.

Savart F. 1833. Annales de Chimie. 53: 337.

Schlameus W. 1995. Encapsulation and Controlled Release of Food Ingredients. 590: 96.

Sefton MV, Broughton RL, Sugamori ME, Mallabone CL. 1987. J. Control. Rel. 6:177.

Sefton MV, Kharlip L, Horvarth V, Roberts T. 1992. J. Control. Rel. 19: 289.

Skjak-Braek G, Grasdalen H, Smidsrod O. 1989. Carbohydr. Polym. 10: 31.

Skryabin GK, Koshcheenko KA. 1987. Methods Enzymol. 135: 198.

Smidsrod O, Haug A, Lian B. 1972. Acta Chem. Scand. 26: 71.

Smidsrod O. 1974. Faraday Discuss. Chem. Soc. 57: 263.

Sonomoto K, Hoq MM, Tanaka A, Fukui S. 1981. J. Ferment. Technol. 59: 465.

Stevenson WTK, Sefton MV. 1988. J. Appl. Polym. Sci. 32: 1541.

Stevenson WTK, Sefton MV. 1993. In: Fundamentals of animal cell encapsulation and immobilization, Goosen MFA (ed.), Boca Raton: CRC Press, p. 143.

Sun AM and O'Shea GM. 1985. Methods Enzymol. 137: 575.

Sun YM, Chang CC, Huang WF, Liang HC. 1997. J. Control. Release 47: 247.

Tampion J, Tampion MD. 1987. Immobilized cells: principles and applications, Cambridge: Cambridge University Press.

Tanaka A, Sonomoto K, Fukui S. 1984. Ann. N.Y. Acad. Sci. 434: 479.

Thibault J-F, Rinaudo M. 1985. Br. Polym. J. 17: 181.

Thomas WR. 1992. In: Thickening and gelling agents for food, Imeson A (ed.), London: Blackie Academic & Professional, p. 25.

Toth D, Tomasovicova D, Gemeiner P, Kurillova L. 1989. Folia Microbiol. 34: 515.

Uludag H, Horvath V, Black JP, Sefton MV. 1994. Biotechnol. Bioeng. 44: 1199.

Vandenbossche GMR, Bracke ME, Cuvelier CA, Bortier HE, Mareel MM, Remon J-P. 1993. J. Pharm. Pharmacol. 45: 115.

Varfolomeyev SD, Rainina EI, Lozinsky VI, Kalyuzhny SV, Sinitsyn AP, Makhlis TA, Bachurina GP, Bokova IG, Sklyankina OA, Agafonov EB. 1990. In: Physiology of immobilized cells, de Bont JAM, Visser J, Mattiasson B, Tramper J (eds.), Amsterdam: Elsevier. p. 325.

Veliky IA, McLean RCJ. 1994. Immobilized biosystems: Theory and practical applications, Blackie Academic London.

Vlahov R, Pramatarova V, Spassov G, Sucholdolskaya GV, Koshcheenko KA. 1990. Appl. Microbiol. Biotechnol. 33: 172–175.

Vorlop KD, Klein J. 1981. Biotechnol. Lett. 2: 9.

Vorlop KD, Klein J. 1987. Methods Enzymol. 135: 259.

Webb C, Black GM, Atkinson B. 1986. Process engineering aspects of immobilized cell systems, Inst. Chem. Eng., Pergamon, Rugby.

Willaert RG, Baron GV. 1996. Reviews in chemical engineering 12(1–2): 1.

Willaert RG, Baron GV, De Backer L. (eds.) 1995. Immobilized living cell systems: Modelling and experimental techniques, Chichester: John Wiley & Sons.

Woodward J. 1985. Immobilized cells and enzymes, Oxford: IRL Press.

Woodward J. 1988. J Microbiol. Methods 8: 91.

Wu K-YA, Wisecarver KD. 1992. Biotechnol. Bioeng. 39: 447.

Young DV. 1993. In: Fundamentals of animal cell encapsulation and immobilization, Goosen MFA. (ed.), Boca Raton: CRC Press, p. 243.

Zhong YP, Dong LC, Hoffman AS. 1988. In: Proceedings of Third World Biomaterials Congress, April 21–25, Kyoto, Japan

2
Mass Transfer in Immobilized Cell Systems

Mattheus F. A. Goosen

Introduction

Immobilized cell/bioactive agent systems have found applications in a variety of areas, including encapsulated cell therapy (Colton 1996, DeVos et al 1996, Lanza et al 1996, Stegemann and Sefton 1996), immobilized biocatalysts (Svec and Gemeiner 1995, Takizawa et al 1996), and polymeric drug-delivery systems (Gan et al 1996, Yao et al 1994). All, however, suffer from specific mass transfer problems. In the case of immobilized cells, oxygen must be able to reach the viable cells at a sufficient rate to keep the cells alive, while the desired product, such as insulin in the case of diabetes treatment, must be able to diffuse out of the capsule, along with low molecular weight waste products. With biocatalysts, whether they be enzymes or cells, the substrate must be able to reach the bead/capsule interior to allow the biochemical reaction to occur, and the desired products must be able to diffuse out of the bead. Similarly with drug-delivery systems, the release of the bioactive agent from the polymer matrix or capsule must be controlled so as to give a constant steady release rate.

In order to be able to fabricate effective encapsulated cell systems, it is essential that a good understanding be obtained of the oxygen/nutrient/product mass transfer process. As applied scientists, engineers are ideally suited for this task, since they can combine experimental and theoretical (i.e., mathematical modeling) studies to give a clearer insight into potential mass transfer bottlenecks.

This chapter will focus on specific mass transfer topics including oxygen transfer to immobilized animal cells, a theoretical study of oxygen transfer limi-

tations for microencapsulated animal cells, modeling of encapsulated animal cell growth, and encapsulation of somatic plant tissue using electrostatics. Experimental and theoretical/modeling studies will be combined in an attempt to give the reader a better insight into common mass transfer problems.

Oxygen Transfer in Bioreactors for Immobilized Animal Cell Culture

Perhaps the most important function of an animal cell bioreactor is to provide adequate oxygen to the cells without damaging them. It is important to note that in the operation of a bioreactor, the gas to bulk liquid oxygen transfer is the only resistance that can be controlled. Varying the aeration rate will affect the gas to liquid oxygen transfer, which in turn will affect the transfer of oxygen from the bulk liquid to the cell. Researchers have, therefore, focused on measuring the mass transfer coefficient $k_l a$, under various operating conditions, for the purpose of developing useful correlations that may be employed as scale-up criteria for animal cell bioreactors. A common technique used in dynamic $k_l a$ measurements involves deoxygenating the reactor contents (Linek et al 1987). Subsequently, gas of a different oxygen concentration is admitted, and the oxygen profile is monitored.

For immobilized cell systems, it is not adequate simply to transfer sufficient oxygen to the bulk liquid (i.e., culture medium). Oxygen must also be transferred from the liquid to the cells. Consider for instance the transfer of oxygen from a gas bubble,

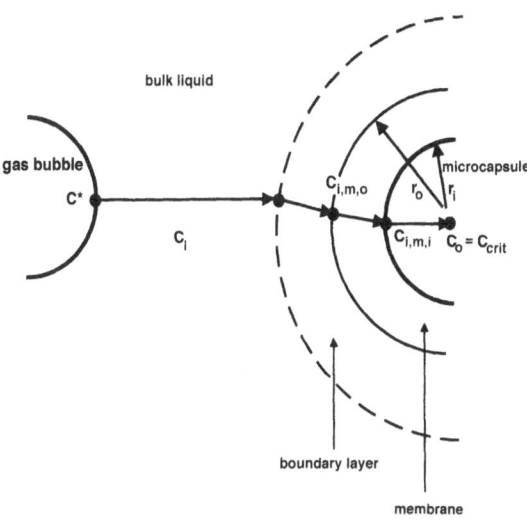

bulk liquid

gas bubble

C^*

C_i

$C_{i,m,o}$

r_o r_i

microcapsule

$C_{i,m,i}$ $C_o = C_{crit}$

boundary layer

membrane

FIGURE 2.1. Schematic diagram of oxygen profile from gas bubble to center of microcapsule (Sharp et al 1998).

through the culture medium, to a microcapsule containing animal cells. The resistances to oxygen transfer from the gas phase to the inside edge of the microcapsule (i.e., gas to liquid, liquid to microcapsule, and transmembrane resistance) (Figure 2.1) can be added together, resulting in the following expression for the resistance to oxygen transfer, R (Sharp et al 1998):

$$R = \frac{1}{V_L k_l a} + \frac{1}{V_L k_s a} + \frac{\left(\frac{1}{r_o} - \frac{1}{r_i}\right)}{4\pi n D_{o_2,m}} \quad (1)$$

where V_L is the volume of the liquid phase, n is the number of microcapsules, r_o and r_i are the outside and inside radius of the microcapsule, respectively, $k_s a$ is the volumetric mass transfer coefficient from a liquid to a solid, and $D_{o_2,m}$ is the diffusivity of oxygen in the membrane. The oxygen transfer rate, OTR, is usually expressed in the form:

$$OTR = \frac{1}{(resistance)}(driving\ force) \quad (2)$$

Employing Fick's Law of Diffusion (McCabe and Smith 1985), it can be shown that the oxygen transfer rate from the gas phase to the inside edge of a microcapsule, OTR_G, is given by

$$OTR_G = \left(\frac{1}{V_L k_l a} + \frac{1}{V_L k_s a} + \frac{\frac{1}{r_o} - \frac{1}{r_i}}{4\pi n D_{o_2,m}}\right)^{-1}$$
$$\times (C^* - C_{i,m,i}) \geq Q = Q_{o_2} x \quad (3)$$

where C^* is the oxygen concentration in equilibrium with the oxygen partial pressure in the gas phase, $C_{i,m,i}$ is the oxygen concentration on the inside of the membrane, Q is the oxygen consumption rate of cells inside a microcapsule (mgO$_2$/Lhr), Q_{o_2} is the oxygen consumption rate per cell, and x is the cell density. To enable the cells in the microcapsule to survive, the oxygen transfer rate, OTR_G, must be greater than (or at least equal to) the oxygen consumption rate of cells inside the capsules, Q. As shown by the work of Heath and Belfort (1987), the oxygen concentration profile within a microcapsule can be represented by

$$C_{i,m,i} - C_o = \frac{Q}{D_{o_2,a}} \frac{1}{6} r^2 \quad (4)$$

where r is the radius of the alginate core. If the concentration of oxygen in the center of the microcapsules, C_o, is equal to a critical oxygen concentration, C_{crit}, the concentration of oxygen at the inner surface of the membrane, $C_{i,m,i}$, can be determined from Equation (4).

The performance of a 1–liter external-loop airlift bioreactor was investigated by studying the gas-liquid oxygen transfer at various aeration rates (0.1 vvm to 1.06 vvm). The influence of suspended alginate beads on the hydrodynamics and mass transfer of the system was examined over a range of microbead loadings (0 to 25% by volume). The intent was to investigate the effect of using various concentrations of cell immobilization matrices on the physical properties of the system. A mathematical correlation was developed for expressing the dependence of $k_l a$ on aeration rate and microbead loading. A mathematical study of the mass transfer resistances from the gas phase to the interior of a microcapsule was also performed, using experimentally determined $k_l a$ values, to enable the determination of the maximum bioreactor microcapsule loading.

According to Van't Riet (1979), the response lag of an oxygen probe can be neglected if the oxygen probe time constant, τ_p, is less than $1/(5\ k_l a)$. The oxygen probe time constant for our studies was 27 $\pm$ 7 sec, and the maximum $k_l a$ value found was 0.00861 sec^{-1}. Since this is approximately equal to $1/(5\ k_l a)$, the oxygen probe response time was therefore not accounted for in the determination of the $k_l a$ values in our investigation.

In order to ensure that the $k_l a$ studies would be measuring the gas-to-liquid mass transfer coefficient and not the mass transfer from the gas to the

Mattheus F. A. Goosen

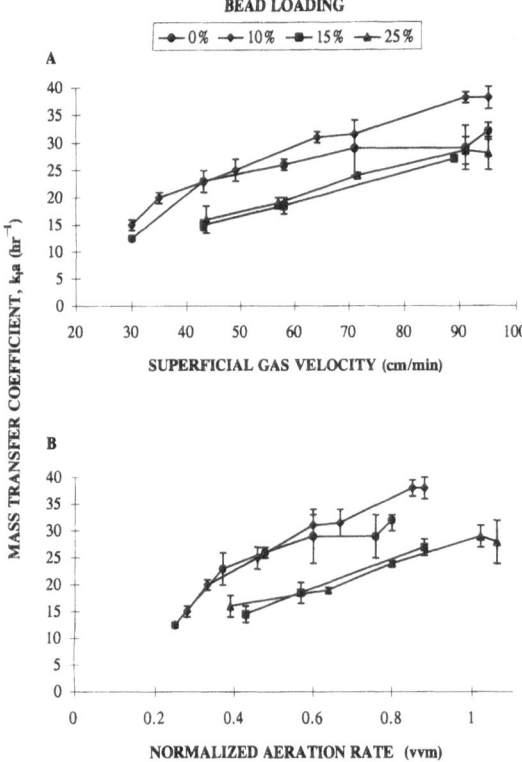

FIGURE 2.2. Mass transfer coefficient, *ka,* versus superficial gas velocity (A) and normalized aeration rate (B). (Sharp et al 1998).

interior of the bead, the $k_l a$ in the presence of alginate beads was compared to that in the presence of ion-exchange resin beads. The $k_l a$ for a 10% loading of alginate beads was determined to be 31.7 ± 0.7 hr^{-1}, and for a 10% loading of ion-exchange resin beads it was 31.4 ± 1.5 hr^{-1}, at an aeration rate of 0.67 vvm. The $k_l a$ values are essentially equal. This suggested that the alginate beads were not acting as a "sink" for oxygen and that the desired quantity, the gas-to-liquid mass transfer coefficient, was being measured.

In two-phase (gas/liquid) systems such as bubble columns and airlift bioreactors, the $k_l a$ is usually found to be an exponential function of the superficial gas velocity, as follows (Deckwer et al 1974):

$$k_l a = \alpha \, v_s^\beta. \tag{5}$$

Figure 2.2A gives the values of $k_l a$ as a function of superficial gas velocity for the various alginate bead loadings studied. All curves showed the same gen-

eral trend; as the superficial gas velocity increased, the $k_l a$ increased. It was apparent, though, that a nonmonotonic relationship between bead loading and $k_l a$ existed. Relative to the value of the $k_l a$ determined in water with no beads present, the $k_l a$ increased slightly (up to 20%) when a small volume percentage (i.e., 10%) of beads was added. As more beads were added (i.e., 25% by volume), the value of the $k_l a$ decreased. Therefore, for our data, other possible correlations for expressing the dependence of $k_l a$ on aeration rates and alginate bead loading were examined.

As a first approach to correlating the $k_l a$ values with an aeration term, the $k_l a$ values were plotted versus the normalized aeration rate, vvm (Figure 2.2B). In the presence of alginate beads, this resulted in a monotonic relationship, whereby increasing the percentage of beads in the reactor decreased the $k_l a$. It is important to note, however, that this was not valid in the absence of beads (i.e., 0% loading). For the range of aeration rates studied, at 0% bead loading, the bioreactor was predominantly operated under turbulent conditions (i.e., Re ≥ 4000). On the other hand, in the presence of beads, 10% to 25%, the bioreactor was mostly being operated in the transition zone. We can speculate that this could affect the $k_l a$ values.

Using vvm instead of the superficial gas velocity accounted for the presence of "internals" (i.e., alginate beads), since vvm normalizes the gas flow rate with respect to the liquid volume. Correlating the $k_l a$ with vvm by performing a linear regression on the data and setting the constant equal to zero, using the statistical package Minitab (Version 5.1.3, Copyright Minitab, Inc., 1985), gave correlation coefficients (R^2 values) ranging from 64.6% for 0% bead loading to 84% for 15% bead loading. This was not considered to be a very good fit of the data because of the low R^2 values. Also, this did not give a single correlation that would account for the presence of the beads, but rather a correlation for each bead loading was required.

It was then proposed that perhaps a term was necessary in the correlation to account for the presence of alginate beads. Thus, a concentration effect, C_E, term was introduced to the correlation. To define the concentration effect, Einstein's equation for the suspension of rigid spheres was used:

$$C_E = 1 + 2.5\Phi \tag{6}$$

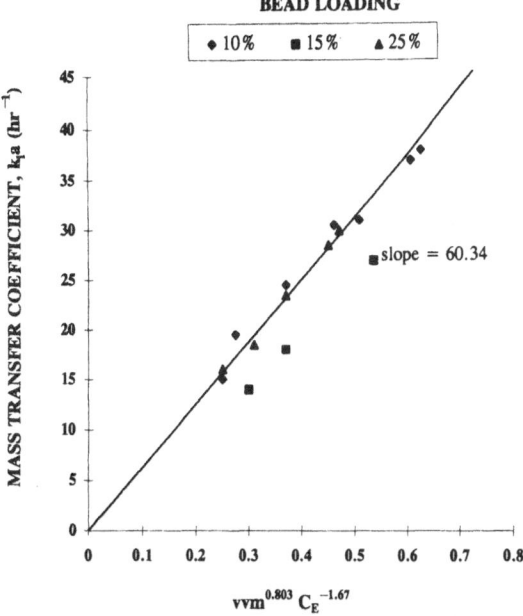

FIGURE 2.3. Correlation between mass transfer coefficient, normalized aeration rate, and bead loading. (Sharp et al 1998.)

where Φ is the volume fraction occupied by the spheres (i.e., alginate beads). A multiple linear regression using Minitab was used to correlate $k_l a$ with vvm and C_E. This multiple linear regression resulted in the following expression:

$$1n\ k_l a = 4.10 + 0.803\ ln\ vvm - 1.67\ 1nC_E \quad (7)$$

where $k_l a$ has units of hr^{-1}.

Rearranging Equation (7) into a more convenient form gave

$$k_l a = 60.34\ vvm^{0.803}\ C_E^{-1.67} \quad (8)$$

This correlation had a coefficient of 0.815, which was indicative of a good fit. There were no trends in the residual plots. Also, the T-ratios for the constant and coefficients in the correlation were very large, which indicated that these parameters were significant to the model. It was, therefore, concluded that this regression line was adequate to explain the data. Equation (8) was used to predict the $k_l a$ value for a given aeration rate (vvm) and alginate bead loading, and gave a good fit (Figure 2.3).

A variety of other correlations was attempted to fit the data. Correlations that did not contain a term

accounting for the presence of beads were rejected because for each bead loading a distinct correlation was required. When attempting to develop a correlation that can be used as a design parameter for bioreactor scale-up, a single correlation that can be applied to all systems is desirable. Correlations involving liquid velocity were not considered to be appropriate because of the dependence of the liquid velocity on the aeration rate. Although used successfully for describing two-phase systems (Moo-Young and Blanch 1983, Siegel and Merchuk 1988), correlations involving the gassed power per unit liquid or reactor volume (P_g/V_L and P_g/V, respectively) did not account for the change in bulk fluid properties because of the presence of alginate beads. It is important to note that the correlation proposed (Equation 8) does not fit the $k_l a$ data when there are no beads present because of the turbulent flow regime at 0% bead loading.

Theoretical Study of Oxygen Transfer to Microencapsulated Insect Cells

For immobilized cell systems, such as microcapsules, ensuring adequate oxygen transfer from the gas phase to the liquid medium does not necessarily ensure that adequate oxygen will reach the immobilized cells. It is possible that the $k_l a$ may not be adequate for a certain microcapsule loading. A theoretical study of oxygen transfer to cells immobilized in microcapsules for various microcapsule loadings in the bioreactor was performed by employing Equations (3) and (4) (Sharp et al 1998). A schematic diagram of the oxygen concentration profile from a gas bubble to the center of a microcapsule is shown in Figure 2.1. Upon arbitrarily specifying the critical oxygen concentration in the center of the microcapsule, the rate of oxygen transfer from the gas phase to the inner surface of the microcapsule membrane (OTR_G, mg/hr) for a certain microcapsule loading was compared to the oxygen demand of the cells ($Q_{o_2} x$) for the same microcapsule loading. The study was made for *Spodoptera frugiperda* cells cultivated in poly-1–lysine/alginate microcapsules at a maximum cell density of 8×10^7 cells/mL capsules (King et al 1987). An oxygen demand of 1.4×10^{10} mmole

O$_2$/cell hr was assumed. These insect cells are usually cultivated at 27°C and 33°C; therefore our study was performed at both temperatures.

It was necessary to estimate several of the parameters used to evaluate the oxygen transfer rate. The diffusivity of oxygen in sodium alginate (the immobilization agent inside the microcapsule) was estimated to be 86% of the diffusivity of oxygen in water (i.e., approximately the same as the diffusivity of oxygen in calcium alginate). According to King et al (1987), the membrane is 5 μm thick and is composed of approximately 90% water. The diffusivity of oxygen through the membrane was therefore assumed to be equal to that of oxygen in water. A critical oxygen concentration in the center of the microcapsule was assumed to be 40% of air saturation.

Figure 2.4 shows the results for the oxygen transfer rate, OTR, attainable for microcapsule loadings of 10%, 15%, and 25% at 33°C and 27°C as a function of the aeration rate. The terminal settling velocity was used to calculate the Reynold's number. Comparing Figures 2.4A and 2.4B—the latter uses the difference between the bead and liquid velocities (determined experimental) to calculate the Reynold's number—indicates that there is not much difference (at most 8%) between the two methods. This suggests that the terminal velocity may be used as a good approximation of the relative velocity between the bead and the liquid, if it is not feasible to determine the liquid and bead velocities experimentally.

At 33°C, for 10% bead loading, the oxygen demand of the cells was achieved at 0.29 vvm, which is the minimum vvm for suspension of the beads (Figure 2.4A). On the other hand, for 25% bead loadings, a vvm of 1.06 is required to meet the oxygen demand of the cells. This is quite a high aeration rate; thus, it may not be feasible to operate at 25% bead loading. Decreasing the temperature to 27°C (Figures 2.4C and 2.4D) increased the oxygen transfer rate, but only slightly (by approximately 8%). This was expected, since a decrease in temperature increases the solubility of oxygen in the bulk liquid, which increases the driving force for oxygen transfer. This, in turn, increases the oxygen transfer rate to the cells. The temperature did not, however, have a very significant effect on the oxygen transfer rate. These results suggest that for this bioreactor the cells will not be oxygen limited at microcapsule loadings of 10% and 15% (by volume); however, there is the potential for oxygen limitation at 25%

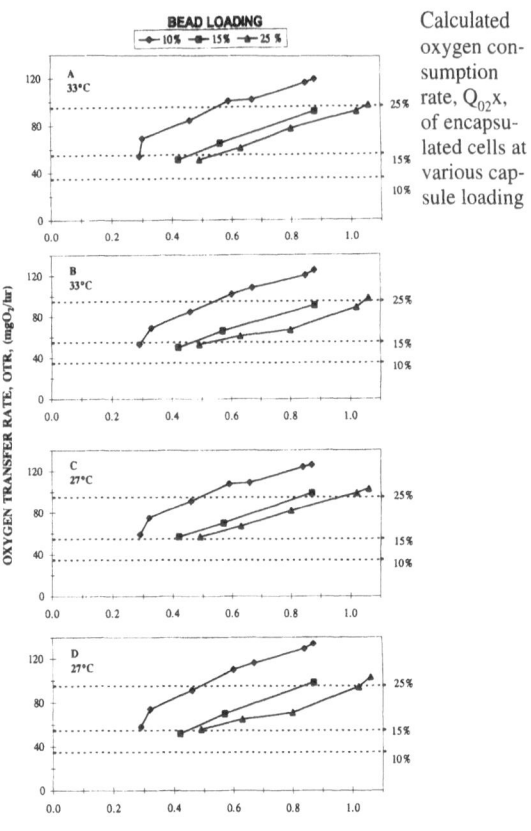

Calculated oxygen consumption rate, $Q_{02}x$, of encapsulated cells at various capsule loading

FIGURE 2.4. Theoretical oxygen transfer rate, OTR, as a function of aeration rate for encapsulated insect cells in bioreactor at 33°C, using the terminal settling velocity to calculate the Reynold's number (A); OTR at 33°C, using the difference between the bead velocity and the liquid velocity to calculate the Reynold's number (B); OTR at 27°C, using the terminal settling velocity to calculate the Reynold's number (C); OTR at 27°C, using the difference between the bead velocity and the liquid velocity to calculate the Reynold's number (D). (Sharp et al 1998).

microcapsule loadings if the reactor is not operated at a minimum aeration rate of 1.06 vvm.

Mathematical Modeling of Encapsulated Animal Cell Growth

In recent studies a mathematical model for animal cell growth in microcapsules was developed and compared to experimental data (Yuet et al 1993, 1995). There have only been a few attempts to address mathematical modeling of microcapsules. Mogensen and Vieth (1973) studied the mass transfer properties of semipermeable microcapsules.

Their model did not account for any cell growth, since they were dealing with enzyme. Heath and Belfort (1987) also provided a mathematical treatment on microcapsules, entrapping biocatalysts, but again no cell growth was taken into account.

The derivation of our model was based on the case in which microcapsules with fluid intracapsular liquid were used in stationary culture. The characteristics of this scenario of cell growth in microcapsules are that the cells initially settle to the bottom of the capsules, and the cell population expands from the bottom up during the culture period. This cell population expansion in the capsules is accounted for in the model by a computation mechanism that initializes the neighboring control volumes with cells when a control volume becomes overcrowded.

The model comprises a set of differential balance equations of the following form:

$$\frac{\partial}{\partial t}(C_i \epsilon) + div(J_i) = S_i \qquad (9)$$

where C_i is the concentration of a nutrient such as glucose and glutamine in a microcapsule, J_i is the diffusive flux of the nutrient, S_i is the rate of consumption of the nutrient, and ϵ is the void fraction inside a microcapsule. Similar balance equations are written for the capsule membrane and the culture medium. The rate of cell growth is expressed as

$$\frac{dX}{dt} = u(t - t_{lag})[\mu X + \lambda \int_o^t X(\eta)\,d\eta] \qquad (10)$$

where

$$u(t - t_{lag}) = \begin{cases} 1 \ if\ t < t_{lag} \\ 0 \ if\ t > t_{lag} \end{cases} \qquad (11)$$

X is the cell density, μ is the specific growth rate, t_{lag} is the lag time, and λ is the cell death constant. Equation (10) describes the change in cell density from the beginning of the exponential growth phase to the end of the culture (Bailey and Ollis 1986). The lag phase can be included mathematically by introducing a step function. The lag time will have to be determined experimentally. Since in an unstructured model a cell is treated as a single-component body, Equation (10) is sufficient to characterize the growth of a cell population. The dependence of the specific growth rate (μ) on nutrient concentrations is expressed as

$$\mu = \mu_{max}\prod_{i=1}^{n}\left(\frac{C_i}{K_{C_i} + C_i}\right) \qquad (12)$$

where C_i is the concentration of rate-limiting nutrient i, K_{c_i} is the saturation constant for nutrient i, n is the total number of rate-limiting nutrients, and μ_{max} is the maximum specific growth rate. Frame and Hu (1988) developed a contact-inhibition model to describe the dependence of specific growth rate on cell density:

$$\mu = \mu_{max}\left[1 - \exp\left(-B\frac{X_{max} - X}{X}\right)\right] \qquad (13)$$

where X_{max} is the maximum cell density, and B is an adjustable parameter. The degree of influence of cell density on μ depends on the value of B; if a relatively large value of B is used, then the specific growth rate will remain more or less unaffected by the cell density until it is very close to X_{max}. Thus, the incorporation of Equation (12) into the kinetic expression with a large value of B should allow us to control the cell density as it approaches the maximum cell density, while leaving the specific growth rate unaffected during most of the culture. A complete expression for μ is therefore proposed as follows:

$$\mu = \mu_{max}\prod_{i=1}^{n}\left(\frac{C_i}{K_{C_i} + C_i}\right)$$
$$\times \left[1 - \exp\left(-B\frac{X_{max} - X}{X}\right)\right]. \qquad (14)$$

The mechanism of cell population expansion may be treated in several ways. The mechanism that was adopted here conforms to the control-volume (CV) formulation. Cell growth in a neighboring CV may begin before the maximum cell density is reached. Equations (10) and (13) constitute the basic model equations for cell growth in this study. The rate of consumption of nutrient i by the cells (i.e., S_i) can be related to the rate of cell growth by the following equation:

$$S_i = \frac{1}{Y_{X/C_i}} \mu X u\,(t - t_{lag}) \qquad (15)$$

where Y_{X/C_i} is the yield factor for nutrient i to cells. Note that the term S_i as expressed in Equation (15) is the rate of consumption of nutrient i based on a unit volume in the interior phase. Equation (15) also assumes a constant yield factor. For details on estimation of model parameters, solution of model equations, derivation of discretization equations, cell population expansion,

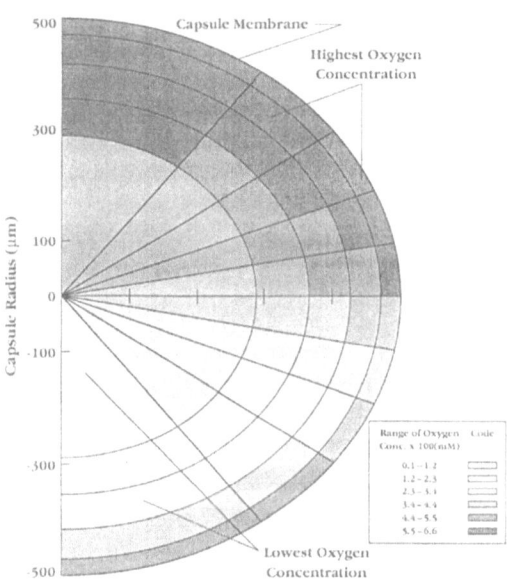

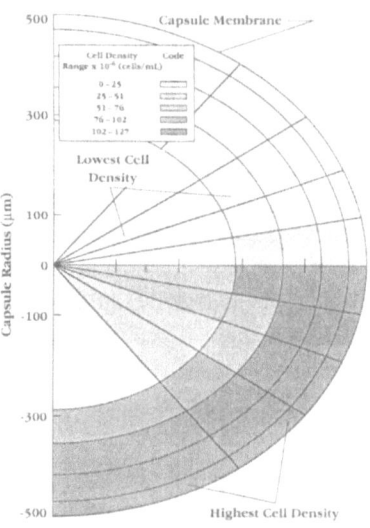

FIGURE 2.6. Simulated cell distribution in a microcapsule after 14 days of stationary cell culture. Capsule diameter 1000 μm. (From Yuet et al 1995. Reproduced by permission of Marcel Dekker, Inc.)

FIGURE 2.5. Simulated oxygen distribution in a microcapsule after 7 days of stationary cell culture. Capsule diameter 1000 μm. (From Yuet et al 1995. Reproduced by permission of Marcel Dekker, Inc.)

and the numerical solution algorithm, see the paper by Yuet et al (1995).

According to this model, the highest rate of growth ($\mu \leq 0.026$ h^{-1}) was found in the boundary region at the top of the cell mass. This is the region that receives the most abundant supply of nutrients and oxygen, both from across the membrane and from the upper half of the capsule. The cells at the bottom of the capsule close to the membrane have virtually stopped growing since they have reached the maximum cell density and there is no more space available for further cell division. The cells at the central region of the population are probably suffering from the lack of nutrients and oxygen and therefore have only a very low specific growth rate ($0.0104 < \mu \leq 0.0156$ h^{-1}).

The distribution of oxygen in the microcapsule is depicted in Figure 2.5. The maximum concentration, which is found in the upper half of the capsule where no cells are present, is only about 0.066 mM, or 26% of the saturation value (0.25 mM). This value is very close to that in the external culture medium (0.068 mM). The low oxygen concentration should not be too surprising, considering the low value of k_la. The oxygen concentration within the cell mass itself is even lower, decreasing to-

wards the center of the cell mass. The distribution of oxygen within the cell population is mainly a result of the consumption of oxygen by the cells located in the outer layer of the capsule, as well as the reduced effective diffusivity through the cell matrix. The transport of oxygen into the microcapsules, however, does not seem to be limited, as the concentration of oxygen in the upper half of the capsule is almost the same as that in the external medium. The lack of oxygen within the cell mass is therefore more likely caused by the high oxygen demand, which is in turn caused by the extraordinarily high intracapsular cell density, particularly in the outer layer of the cell mass. Oxygen deficiency caused by high oxygen demand, but not the diffusion limitation, was also noted by other investigators (Chang and Moo-Young 1988, Gosmann and Rehn 1988). The distribution of cell density in the microcapsule at the end of the culture period (Figure 2.6) shows that the highest density occurs close to the inner surface of the capsule membrane, which confirms earlier experimental work (King et al 1987).

Besides using microcapsules with fluid (low-viscosity) intracapsular liquid in stationary culture, there are two other scenarios for employing microcapsules in cell culture: (1) microcapsules with viscous intracapsular liquid or regular gel beads (without a membrane) used in stationary or suspension culture, and (2) microcapsules with low-viscosity intracapsular liquid in suspension culture. In suspension culture, although the microcapsules are moving around in the bioreactor as the result of the flow of medium, the cells entrapped in the microcapsules or gel beads have only little or even negligible movement with respect to the capsules. The term "stationary" refers to no mixing, while "suspension" refers to mixing.

The scenarios of calcium alginate gel beads and microcapsules with fluid intracapsular liquid in suspension culture were simulated using the same model. In the case of calcium alginate gel beads, the diffusivities of nutrients in calcium alginate were taken as 75% of those in water, while in the two cases of fluid intracapsular liquid microcapsules in suspension, $k_l a$ was taken as 0.6 h^{-1} and 4.0 h^{-1} separately. A $k_l a$ value of 4.0 h^{-1} is more realistic since the culture medium has to be agitated (in a shake flask, for example) in order to keep the microcapsules in suspension; the value of 0.6 h^{-1} was included for comparison. With $k_l a$ equal to 4.0^{-1}, the cell density, after 14 days of culture, is approximately 1.4×10^8 cells ml^{-1}, while with $k_l a$ equal to 0.6 h^{-1}, the cell density reaches 7.1×10^7 cells ml^{-1}. The intracapsular cell density alone, however, does not provide a complete picture of what is actually happening in the microcapsule or the culture medium. A further examination of the oxygen and nutrient concentration in the system reveals a rather interesting situation regarding the effect of nutrients on the specific growth rate.

Let us take a brief look at the case where $k_l a$ is 0.6^{-1} (Figure 2.7), keeping in mind that the medium is changed every 24 h. During the first 3 days of culture, μ stays at about 0.032 h^{-1} because of the abundant supply of oxygen and nutrients. In the meantime, however, the oxygen concentrations in the intracapsular region as well as in the medium are decreasing rapidly. At the beginning of day 4, there is a boost of oxygen concentration caused by the fresh supply of medium, which is saturated with oxygen. The oxygen concentration drops sharply during day 4, and by the end of the day, the intracapsular oxygen is reduced to 0.005 mM, while μ drops to about 0.023 h^{-1}. Beginning from day 5, the specific growth rate is boosted at the beginning of each day because of the addition of fresh medium. However, the growth rate drops abruptly during the first 4 or 5 h because of the rapid deple-

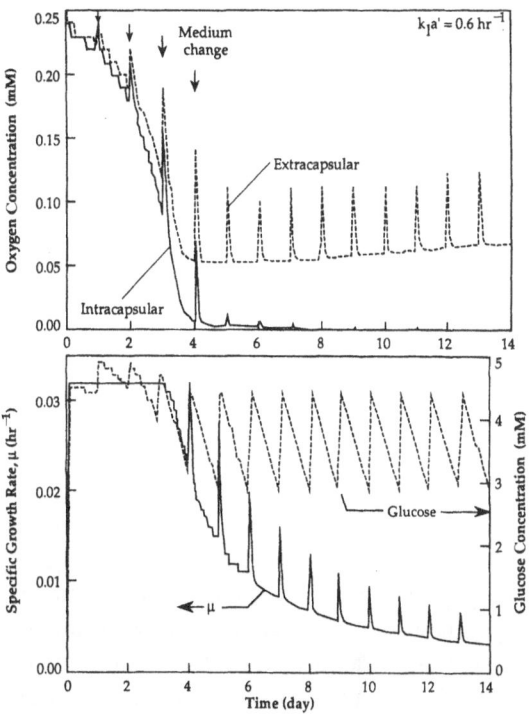

FIGURE 2.7. Simulated time-variation of oxygen concentration in the medium and the interior phase, and the specific growth rate and intracapsular glucose concentration for $K_l a = 0.6$ h^{-1}. Microcapsules with fluid intracapsular liquid in suspension. The medium is changed every 24 h. (From Yuet et al 1995. Reproduced by permission of Marcel Dekker, Inc.)

tion of intracapsular oxygen. Neglecting the pulses, the specific growth rate decreases slowly from about 0.025 h^{-1} to about 0.004^{-1}h at the end of the culture. Note that the intracapsular glucose concentration never drops below 2.8 mM. This result suggests that from the second half of day 4 to the end of the culture, the cell growth in the microcapsule is limited by the lack of oxygen. We now conclude this chapter by looking at the encapsulation of plant cells using electrostatics.

Encapsulation of Somatic Plant Tissue Using Electrostatics

A major concern in cell and bioactive agent immobilization has been the production of very small microbeads so as to minimize the mass-transfer resistance problem associated with large-diameter beads (>1000 μm). Klein et al (1993) reported production of alginate beads with diameters from 100 to 400 μm using compressed air to quickly pass the cell/gel solution through a nozzle. Few attempts have been made in the application of electric fields to the production of μm-size polymer beads for cell immobilization (Goosen et al 1986). In recent articles (Bugarski et al 1994, 1993; Goosen et al 1996, 1997), the mechanism of droplet formation using an electrostatic droplet generator was investigated with a variable-voltage power supply. Animal cell suspensions were successfully extruded using the electrostatic droplet generator. The application of this technology to plant cell immobilization has only recently been reported (Goosen et al 1997).

Somatic embryogenesis is a new plant tissue culture technology in which somatic cells (i.e., any cell except a germ or seed cell) are used to produce an embryo (i.e., plant in early state of development) (Teng et al 1994). The technique of somatic embryogenesis in liquid culture, which is believed to be an economical method for future production of artificial seeds, may benefit from cell immobilization technology. Encapsulation may aid in the germination of somatic embryos by allowing for higher cell densities, protecting cells from shear damage in suspension culture, allowing for surface attachment in the case of anchorage-dependent cells, and being very suitable for scale-up in bioreactors.

The long-term objective of the project reported in this section is the development of an economical method for the mass production of artificial seeds using somatic embryogenesis and cell immobilization technology. The specific aim of the present study was to investigate the production of small alginate microbeads using an electrostatic droplet generator. In a preliminary study, callus tissue from carnation leaves was also immobilized and cultured.

Extrusion of alginate droplets using a 5.7 kV fixed-voltage power supply showed that there is a direct relationship between the electrode distance and the bead diameter. For example, at 10 cm electrode distance, the bead diameter was 1500 μm, while at 2 cm it decreased to 800 μm. The greatest effect on bead diameter was observed between 2 and 6 cm electrode distance. Although there was overlap in bead sizes between 6, 8, and 10 cm electrode distances, there was a significant difference (i.e., no overlap in SD) between bead sizes at 2 and 6 cm electrode distances. An inverse relationship between needle size and microbead diameter was

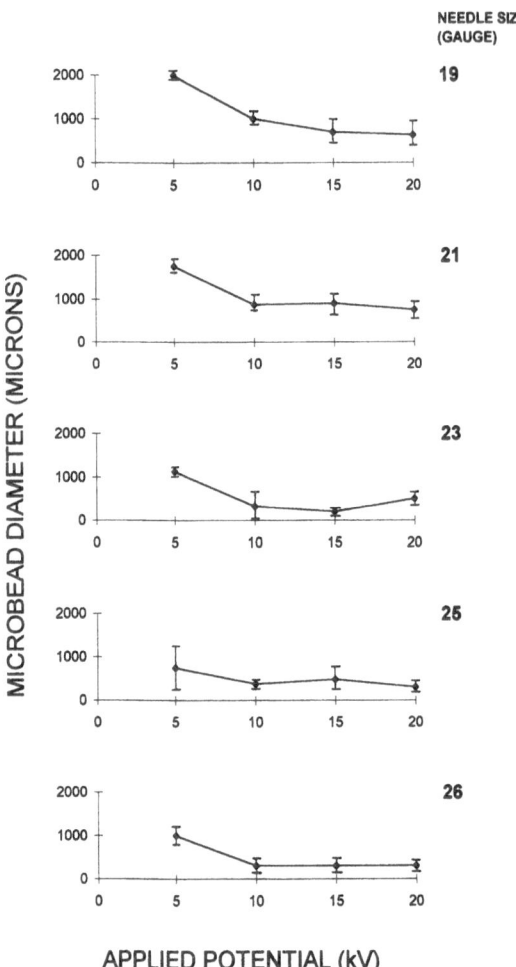

FIGURE 2.8. Effect of needle size on microbead diameter, using a variable high-voltage power supply (±SD) (Goosen et al 1997).

observed. Aside from the 23 G needle, there was a significant difference between bead sizes produced by all needles (i.e., no SD overlap).

As the needle size decreased from 19 to 26 G, the bead size decreased from 1400 to 400 μm, respectively. These results support previous work reported by Bugarski et al (1994). The present investigation showed that the alginate concentration does not appear to be important, because there are overlapping SD intervals for all data points. The bead diameter was found to be 800 μm at both 1% and 3% alginate concentration.

Looking more closely at the effect of needle size on bead diameter, as a function of applied potential (Figure 2.8), we see that the decrease in microbead

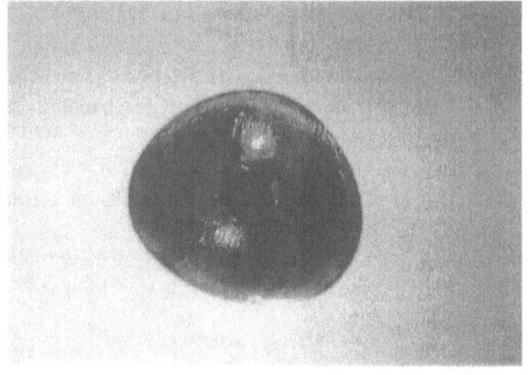

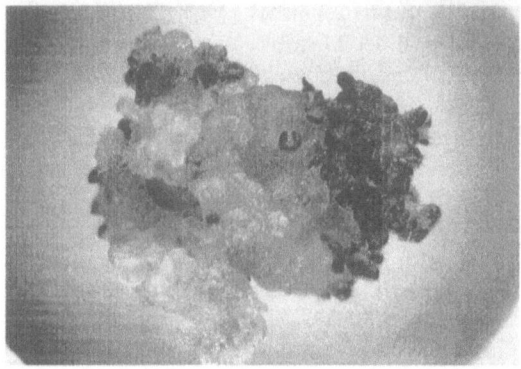

FIGURE 2.9. Callus cells from carnation leaf encapsulated using 4% (w/v) sodium alginate over a 1–month culture period on agar gel. Top: time—zero; middle: 2 weeks; bottom: 1 month.

size was greatest between 5 and 10 kV for all needle gauges tested. For example, when the applied potential was increased from 5 to 10 kV, the microbead diameter decreased from 2000 to 1000 μm and from 1000 to 250 μm for the 19 and 26 G needles, respectively. The smallest bead, 200 μm, was produced with a 26 G needle at 20 kV.

Immobilized callus cells from carnation leaves retained viability as observed by cell growth and plantlet formation (Figure 2.9). In a related study, Shigeta (1995) was able to germinate and grow en-

capsulated somatic embryos of carrot using a 1% sodium alginate solution, as compared to a 2% alginate solution used in the present investigation. The main findings of the present experiment, however, indicate that somatic tissue can be electrostatically extruded and aseptically cultured while retaining viability.

Concluding Remarks

To gain better insight into mass transfer problems in encapsulated cell systems requires not only experimental investigations but also mathematical modeling studies. Furthermore, the ability to develop successful and well-understood systems will necessitate close collaboration among scientists with different areas of expertise such as engineering, microbiology, biochemistry, and medicine. Over the next decade as our knowledge of encapsulated cell systems increases, we can expect to see many new areas of application.

Acknowledgments. The financial support of the Natural Science and Engineering Research Council of Canada, and of Sultan Qaboos University, College of Agriculture (Grant AGBIOR 9505 to Mattheus Goosen) is gratefully acknowledged. The plant cell immobilization study was performed by H. Al-Hajry, S. Al-Maskari, and L. Al-Kharousi (SQU). The assistance of Dr. O. El-Mardi in the plant tissue culture is also acknowledged. *Published with the approval of the College of Agriculture, Sultan Qaboos University as paper number 130897.*

References

Bailey, JE and Ollis, DF (eds). (1986). In: Biochemical Engineering Fundamentals 2nd edition. McGraw-Hill Book Co., NY, Chapter 7.

Bugarski B, Smith J, Wu J, and Goosen MFA. 1993. Methods of Animal Cell Immobilization Using Electrostatic Droplet Generation. *Biotechnol. Tech. 6* (9), 677–682.

Bugarski B, Li Q, Goosen MFA, Poncelet D, Neufeld R, and Vunjak G. 1994. Electrostatic Droplet Generation: Mechanism of Polymer Droplet Formation. *AIChE. J. 440* (6), 1026–1031.

Chang, HN and Moo-Young, M. 1988. *Appl. Microbiol. Biotechnol. 29,* 107.

Colton, CK. 1996. Engineering Challenges in Cell-Encapsulation Technology. *TIBTECH*. Vol. 14.

Deckwer, WD, Burckhart, R, and Zoll, R. 1974. Mixing and Mass Transfer in Tall Bubble Columns. *Chemical Engineering Science. 29*, 2177–2188.

De Vos P, De Haan B, Pater J, and Van Schilfgaarde R. 1996. Association Between Capsule Diameter, Adequacy of Encapsulation, and Survival of Microencapsulated Rat Islet Allografts. *Transplantation. 62*, 893–899, No. 7.

Frame KK and Hu WS. 1988. *Biotechnol. Bioeng. 32*, 1061.

Gan K-H, Geus WP Bakker W, Lamer CBHW, and Heijerman, HGM. 1996. In: Vitro Dissolution Profiles of Enteric-Coated Microsphere/Microtablet Pancreatin Preparations at Different pH Values. *Aliment. Pharmacol. Ther. 10*, 771–775.

Goosen MFA, O'Shea GM, Gharapetian MM, and Sun AM. 1986. Immobilization of Living Cells in Biocompatible Semipermeable Microcapsules: Biomedical and Potential Biochemical Engineering Applications. In: Polymers in Medicine. Chiellini E. (ed.) Plenum Publishing, New York, p. 235.

Goosen MFA, Mahmoud ESE, Al-Ghafri AS, Al-Hajry HA, Al-Sinani YS, and Bugarski B. 1996. Immobilization of Cells Using Electrostatic Droplet Generation. In: Methods in Molecular Biology: Immobilization Enzymes and Cells. Bickerstaff G. (ed). Human Press, Totow, NJ.

Goosen MFA, Al-Ghafri AS, El-Mardi O, Al-Belushi MIJ, Al-Hajry HA, Mahmoud ESE, and Consolacion V. 1997. Electrostatic Droplet Generation for Encapsulation of Somatic Tissue: Assessment of High-Voltage Power Supply. *Biotechnology Progress. 13*(4), 497–502.

Gosmann, B and Rehn, HJ. 1988. *App. Microbiol. Biotechnol. 29*, 554.

Heath C and Belfort G. 1987. Immobilization of Suspended Mammalian Cells: Analysis of Hollow Fiber and Microcapsule Bioreactors. *Advances in Biochemical Engineering/Biotechnology 34*, 1–31.

King GA, Daugulis AJ, Faulkner P, and Goosen MFA. 1987. Alginate-Polylysine Microcapsules of Controlled Membrane Molecular Weight Cut-off for Mammalian Cell Culture Engineering. *Biotechnology Progress. 3*, 231–240.

Klein J, Stock J, and Vorlop DK. 1993. Pore Size and Properties of Spherical Calcium Alginate Biocatalysts. *Eur. J. Appl. Microb. Biotechnol. 18*, 86.

Lanza RP, Hayes JL, and Chick WL. 1996. Encapsulated Cell Technology. *Nature Biotechnology. 14*.

Linek V, Vacek V, and Benes P. 1987. A Critical Review and Experimental Verification of the Correct Use of the Dynamic Method for the Determination of Oxygen Transfer in Aerated Agitated Vessels to Water, Electrolyte Solutions and Viscous Liquids. *Chemical Engineering Journals. 34*, 11–34.

McCabe WL and Smith JC. 1985. Units of Chemical Engineering. McGraw-Hill, New York, 602.

Morgensen AO and Vieth WR. 1973. Mass transfer and biochemical reaction with semipermeable microcapsules. *Biotechnol. Bioeng. 15*, 467–481.

Moo-Young M and Blanch JW. 1983. Kinetics and Transport Phenomena in Biological Reactor Design. *Biochemical Engineering*.

Sharp NA, Daugulis AJ, and Goosen MFA. 1998. Hydrodynamic and Mass Transfer Studies in an External-Loop Air-Lift Bioreactor for Immobilized Animal Cell Culture. *Applied Biochemistry and Biotechnology. 73*(1), 59–77.

Shigeta J. 1995. Germination and Growth of Encapsulated Somatic Embryos of Carrot for Mass Propagation. *Biotechnol. Tech. 10* (9), 771–776.

Siegel MH and Merchuk JC. 1988. Mass Transfer in a Rectangular Air-Lift Reactor: Effects of Geometry and Gas Circulation. *Biotechnology and Bioengineering. 32*, 1128–1137.

Stegemann JP and Sefton MV. 1996. Video Analysis of Submerged Jet Microencapsulation Using HRMA-MMA. *Canadian Journal of Chemical Engineering. 74*, 518–525.

Svec F and Gemeiner P. 1995. Engineering Aspects of Carriers for Immobilized Biocatalysts. *Biotechnology and Genetic Engineering Reviews. 13*.

Takizawa S, Aravinthan V, and Fujita K. 1996. Nitrogen Removal from Domestic Wastewater Using Immobilized Bacteria. *Wat. Sci. Tech. 34* (1–2) 431–440.

Teng W-L, Liu Y-J, Tsai V-C, and Soong T-S. 1994. Somatic Embryogenesis of Carrot in Bioreactor Culture Systems. *Hort. Sci. 29*(11) 1349–1352.

Van't Riet K. 1979. Review of Measuring Methods and Results in Nonviscous Gas-Liquid Mass Transfer in Stirred Vessels. *Ind. Eng. Chem. Process Des. Dev. 18*, 357–364.

Yao K, Peng T, Xu M, and Yuan C. 1994. pH-Dependent Hydrolysis and Drug Release of Chitosan/Polyether Interpenetrating Polymer Network Hydrogel. *Polymer Intl. 34*, 213–219.

Yuet PK, Kwok W, Harris TJ, and Goosen MFA. Mathematical Modeling of Protein Diffusion and Cell Growth. In: Fundamentals of Animal Cell Encapsulation and Immobilization. Goosen MFA. (Ed.) 1993). CRC Press, Boca Raton, Florida. p. 79–112

Yuet PK, Harris J, and Goosen MFA. 1995. Mathematical Modelling of Immobilized Animal Cell Growth. *J. Artif. Cells, Blood Substitutes and Immobilization Biotechno. 23*, 1, 109–133.

3
Polymer Membranes for Cell Encapsulation

Taylor G. Wang

Introduction

Encapsulation of living cells for the treatment of hormone deficiency disease in human beings, such as diabetes, has shown great potential.[1-3] The principle of immunoprotection of protein-secreting cells has two great advantages: (1) cell transplantation without the need for immunosuppression and its accompanying side effects, and (2) transplantation of cells from nonhuman species (xenograft) to overcome the limited supply of donor cells.

To achieve this promise, a capsule must be developed to satisfy a set of stringent and dichotomous requirements. The capsule membrane, to serve as an immunoisolation device, must be able to keep immune system away from the living cells, yet allow the nutrients, oxygen, and proteins to pass through without much impedance. The membrane must be biocompatible to the host and to the cells it encloses. It also needs to be strong enough to survive handling, transplantation, and the hostile environment inside the human body.

The membranes of choice today are polymer systems of both natural and synthetic origin. Polymers offer a large variety of physical, chemical, and structural properties that make them highly favorable for various biomedical applications. When the polymeric material is, as in this case, intended for the immunoprotection of living cells, the optimization of the polymer system demands the ability to meet the following requirements: (1) processing: mechanical stability, thickness, homogeneity, geometry, purification, and sterilization, (2) biological: biocompatibility, biodegradability nontoxicity, retrievability, and permeability adjustment.

All these requirements determine the number of polymer candidates that can be utilized in various types of immunoisolation devices. They are represented by the membranes in the form of hollow fibers (copolymer of acrylonitrile and vinylchloride[4]), tubular diffusion chambers (polyacrylates[5] and sodium alginate[6]), vascularized perfusion devices (polyacrylates[7]), and microcapsules (polyelectrolyte complexes,[8] agarose,[9] and copolymer of hydroxyethyl methacrylate-co-methyl methacrylate[10]). Among them, however, microcapsules seem to exhibit superior insulin response and long-term islet viability and function.[11] The microcapsule system of sodium alginate and poly-L-lysine, originated by Lim and Sun,[8] has demonstrated diabetes reversal in animals, and in a small-scale clinical trial.[12-14] However, the inability to adjust capsule parameters independently (e.g., mechanical strength or permeability) has limited the success of this system.[12,15] Other approaches utilizing different principles of capsule formation, such as temperature-induced sol-gel transition,[9] solvent extraction, and precipitation,[10] suffer similar limitations.

Theoretical Model

Immunoisolation

The effectiveness of immunoisolation of a polymer capsule is closely tied to its membrane's permeability, or more precisely, to the pore size of the capsule membrane. It has been suggested, and argued forcefully, that with proper selection of a cut-off pore

Immunoisolation of Living Cells

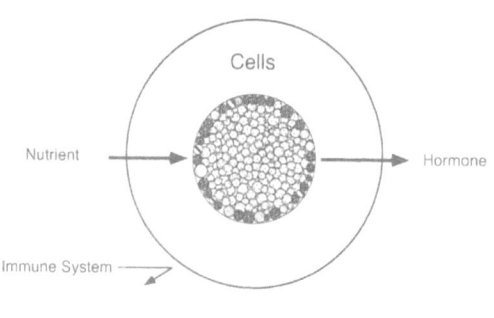

FIGURE 3.1. Gatekeeper model. Components of the immune system are excluded from entering the capsule, yet the nutrients, oxygen, and protein are allowed to pass through without much impedance.

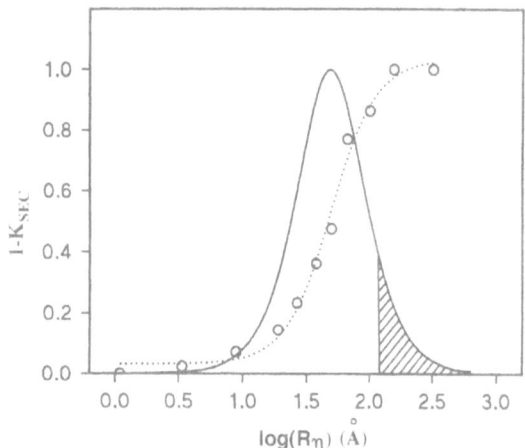

FIGURE 3.2. Pore size distribution of capsule membrane. Apparent pore size distribution (PSD) of SA/CS/PMCG microcapsules as estimated on the basis of dextran viscosity radius R_η is shown. The SA/CS/PMCG capsules were prepared from a mixture of SA and CS; the polycation solution contained PMCG, $CaCl_2$, and NaCl (1 min reaction time). The exclusion limit of this capsule, as determined by SEC with dextran, was 230 kD. The dashed area represents the pores larger than the cut-off limit. The dotted line is the radius of IgG.

size, the polymer membrane can serve as a *gatekeeper* (Figure 3.1), keeping the immune system out, and allowing oxygen and hormones to pass through without much impedance.

Most of the current polymer capsule designs are based on this principle;[8,12,15-16] however, there is a fundamental flaw in this model. Pore sizes of polymer membranes are not homogeneous, by nature, but rather consist of a broad spectrum of sizes, with a very long tail. This problem was first noted by CK Colton as early as 1991,[3] but was largely ignored by capsule designers. On the basis of our dextran permeability measurements, the apparent pore size distribution of the capsular membrane has a cut-off at 117 Å (230 kD),[17-19] as shown in Figure 3.2. The shaded area of Figure 3.2 represents the pore size tail. However, if we replot Figure 3.2 as the surface area distribution versus pore sizes, (Figure 3.3), the tail is no longer a tail. Ignoring this may have contributed to the inconsistent conclusions in the literature.[9,20-26] For the capsule to provide total immunoprotection for the living cells, the polymer membrane must not have pores larger than the antibody complement component.[3] This suggests that unless a gatekeeper capsule has a step function cut-off less than 300 (Å) in pore diameter and is defect-free, the effectiveness of the capsule as an immunoisolation device will be compromised, and less than optimal capsules will be used in human clinical trials.

A new model incorporating the pore size distribution for immunoisolation of living cells has been proposed.[27] In this new model, the capsule wall is thicker and the pores are bigger than in the gatekeeper model. Figure 3.4 is the schematic drawing of the new model (notice that the wall is considerably thicker than in Figure 3.1). The large pores will allow the nutrients and proteins, as well as immune system to freely enter the membrane. However, the small pores inside the membrane will prevent or delay most of the immune system from passing all the way through to the inner volume of the capsule where the cells reside. The capsule membrane serves more as an immune system barrier than as a gatekeeper.

In this calculation, we will focus on preventing the antibody-mediated attack on the living cells. We assume that the immune system density is very diluted, and that antibodies and complement components pass through the immunoisolation membrane as a result of "random walk motion." For the sake of simplicity, we shall study the motion in one dimension, and assume that the probability of going for-

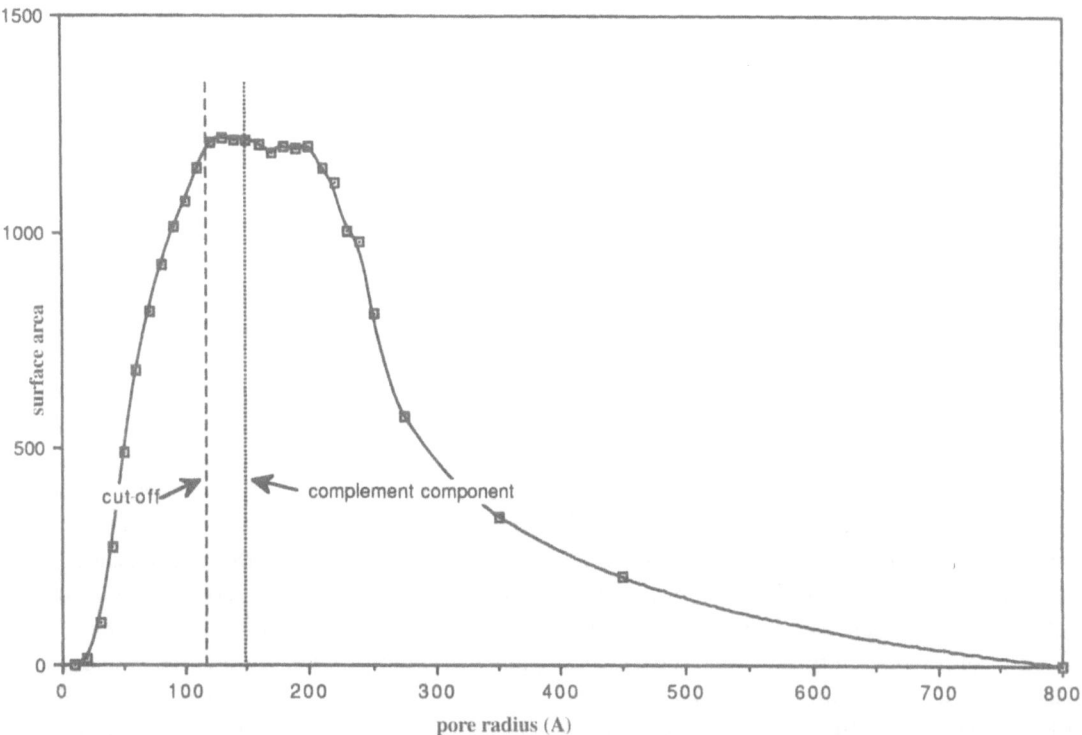

FIGURE 3.3. Surface area distribution of capsule membrane versus pore sizes. The surface area distribution curve of capsule membrane is derived by multiplying the surface area of each pore by the apparent pore size distribution curve above. The dashed line is the cut-off pore size, and the dotted line is the radius of antibody complement component. The solid line is to guide the eye, and represents no theoretical component.

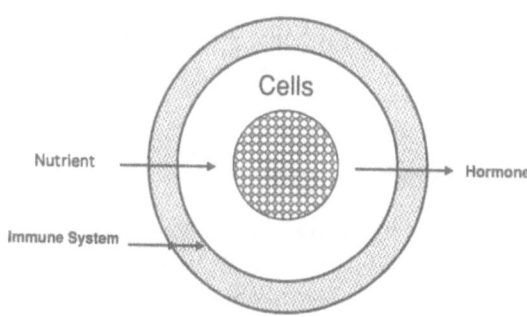

FIGURE 3.4. Barrier model. In the barrier model, the capsule wall is thicker and the pores are bigger than in the barrier model. This allows the immune system to freely enter the membrane. However, there are sufficient small pores inside the membrane to act as traps, to prevent or delay most of the immune system from passing all the way through to the inner volume of the capsule where the islet resides.

ward or backward is the same. After a total of N such steps, the probability of finding an immune system at a position m steps away from the origin is

$$P_N(m) = \frac{N!}{\left(\frac{N+m}{2}\right)! \left(\frac{N-m}{2}\right)!} \left(\frac{1}{2}\right)^N \quad (1)$$

where $m = n_f - n_b$, and $n_f + n_b = N$.

n_f represents the step forward, n_b the step backward. This is a Gaussian distribution curve centered at zero. The root-mean-square displacement, $\sqrt{(\delta x)^2}$, of the immune system after N steps is found to be

$$\sqrt{(\delta x)^2} = \sqrt{(\delta m)^2}\, R = \sqrt{N}\, R \quad (2)$$

where $\delta x = x - \bar{x}$ and $\delta m = m - \bar{m}$. R is the average step size and can be approximated to be the average pore size. If we assume that the immune system is originally located at the surface of the capsule, and that the root-mean-square distance that it needs to travel is the membrane thickness, d, then Equation (2) can be rewritten as

$$\frac{d^2}{R^2} = N. \qquad (3)$$

We can now write the entrapment time, Γ, required for the immune system component to pass through a membrane with a thickness d as

$$\Gamma \sim \frac{d^2}{R^2} f\tau. \qquad (4)$$

Using the equipartition theorem, the random walk motion in three dimensions can be written as

$$\Gamma \simeq 3\frac{d^2}{R^2} f\tau \qquad (5)$$

Where the f is the ratio between surface areas of small pores and large pores, and τ is the time that the immune system can be trapped by the small pores in each collision. The value f can be determined quantitatively from the pore surface area distribution curve of Figure 3.3, and the value τ for different immune system can be estimated qualitatively from the dextrane size exclusion chromatography (SEC) retention time.

Using antibody complement component as an example, the estimated entrapment time for the SA-PLL-SA gatekeeper capsule (110 kD, 2 μm wall thickness) is in the order of days. This calculated value is in agreement with many of the experimental observations on how long the gatekeeper capsule can survive after being transplanted into presensitized animal models, such as NOD mice. More important, this model might suggest that the current barrier model will have limited success in treating Type I diabetic disease in human beings.

Mass Transport

To keep the islets viable inside an immunoisolation capsule, nutrients and oxygen must be allowed to diffuse in, and the insulin to diffuse out. This is a convective diffusive process. Mass transport across a random network structure is an extremely complicated system. Any attempt to undertake a rigorous calculation of this system is far beyond the scope of this paper. Here, the author tries to examine the order of magnitude of the dependence of capsule membrane parameters on the steady state mass transport rate across a capsule membrane.

The diffusion mass transport flux, Q, across the membrane is proportional to the concentration gradient and membrane porosity, as shown below:

$$Q \sim \frac{D}{d}\Delta c\, A_{\text{eff}} \qquad (6)$$

where D is the diffusion coefficient, d is the thickness of the membrane, ΔC is the mass concentration difference, and A_{eff} is the effective capsule surface area that participates in mass transport. The diffusion of biomaterial in liquids occurs under a nonslip condition; the D is proportional to the temperature, T, and inversely proportional to the fluid viscosity, η, and particle radius, r. This is usually called *Stokes* velocity. It represents the diffusion of large spherical biomaterials in an open system. To account for the effect of capsule membrane porosity on mass transport, we need to calculate A_{eff}. We simplify the problem by modeling the polymer membrane as a bundle of interwoven pipes. These pipes have inner diameters, R, which approximate different pore sizes. This reduces the capsule surface area that participates in diffusion to NR^2. N is the total number of pipes per unit surface area. The mass flow inside a pipe has a velocity profile of R^2. Thus, the effective diffusive mass flow passing per unit of time through any cross-section of pipe is proportional to the fourth power of the radius of the pipe. Equation (6) can then be rewritten as

$$Q \sim \frac{R^4 T}{r\eta d}\Delta c\, N. \qquad (7)$$

Equation (7) is crude, and at best phenomenological. However, it might provide some valuable insight into capsule design optimization.

Mechanical Strength

Many of the immunoisolation devices failed inside of animal and/or human hosts because of capsule breakage. It exposes the unprotected living cells to the host's immune system. At best, the cells are destroyed; at worst, an immunoreaction can be triggered. Therefore, if the immunoisolation devices are to be used for the long-term cure of hormone deficiency diseases in human beings, the mechanical strength of capsules must be sufficient to survive the constant pressure exerted on them. Let's examine the case of a spherical capsule with a radius of R. Under an unidirectional pressure, P_x, the shape of the capsule distorts to compensate for the pressure, as follows:

$$P_x = Yd\left(\frac{2}{R_2} - \frac{2}{R_1}\right) \qquad (8)$$

where Y is Young's modulus, d is the thickness, and R_1 and R_2 are major and minor axes of the deformed capsule. If the distortion is small, we can approximate $R_1 = R + \delta r$, and $R_2 = R - \delta r$, and Equation (8) can be rewritten as

$$\frac{\delta r}{R} = \frac{R}{4Yd}P_x. \qquad (9)$$

Equation (9) shows the advantages of smaller capsules and thicker membranes have for resisting deformation, $\delta r/R$, under pressure, P_x.

Microencapsulation versus Macroencapsulation

Debates on the pros and cons of macroencapsulation and microencapsulation are ongoing. Here, we will examine the capsules from the point of view of mass transport. For the same volume and geometry, the total surface area of macrocapsules, (S_{ma}), and microcapsules, (S_{mi}), can be shown as follows:

$$\frac{S_{mi}}{S_{ma}} = \frac{d_{ma}}{d_{mi}} \qquad (10)$$

where d_{ma} and d_{mi} are the dimensions of the macro- and microcapsules. We have shown that the total surface area is inversely proportional to the size of the capsule. Therefore, the macrocapsules will have less capsule surface area to participant in mass transport than the microcapsules.

Capsule Shapes

We investigated the shape of two adjoining media separated by a membrane. Under no external force, using Laplace's formula, we can write the equation for equlibrium as:

$$\frac{1}{R_1} + \frac{1}{R_2} = \text{Constant} \qquad (11)$$

where R_1 and R_2 are the principal radii of curvature at a given point of surface. Thus, the conditions of Equation (11) mean that, under no external force,

the most stable shape of a capsule is spherical and the surface area is minimal.

Experiments

The mathematical models of Equations 5, 7, 9, 10, and 11 show the diverse requirements for designing a capsule that is suitable for immunoisolation. These requirements cannot be met without the ability to adjust all parameters independently. To test our barrier model, a new capsule has been developed. Over 1000 combinations of polyanions and polycations were studied in our search for polymer candidates that would be suitable for barrier capsules. The combination of sodium alginate (SA), cellulose sulfate (CS), poly(methylene-co-guanidine) (PMCG), calcium chloride ($CaCl_2$), and sodium chloride (NaCl)[18-19,28-30] was found to be most promising. The polyelectrolyte complex has distinct advantages. It is very simple and inexpensive. It can be processed under extremely mild physiological conditions. This multicomponent system allows us to adjust the capsule's mechanical properties, permeability,[17] and surface properties[31] independently, as well as the capsule size, for retrieval if needed. A detailed description of the capsule formation process and information on the composition of the solutions are given in the papers cited.[17-19]

Capsule Studies

Biocompatibility

Prior to transplantation of encapsulated islets into animals, extensive biocompatibility testing of empty capsules was performed. Capsules were transplanted into the peritoneal cavity of C57 and NOD mice. We found the biocompatibility of the capsules inside normal C57 mice was not very sensitive to minor polycation concentration changes (Figure 3.5A); however, the NOD mice model had a different response altogether. We found that a small change in the concentration of polycation can alter the capsules inside the NOD mice from clumped to free-floating to breakage (Figure 3.5B). The effect of autoimmune disease on the biocompatibility of the capsule is self-evident here.

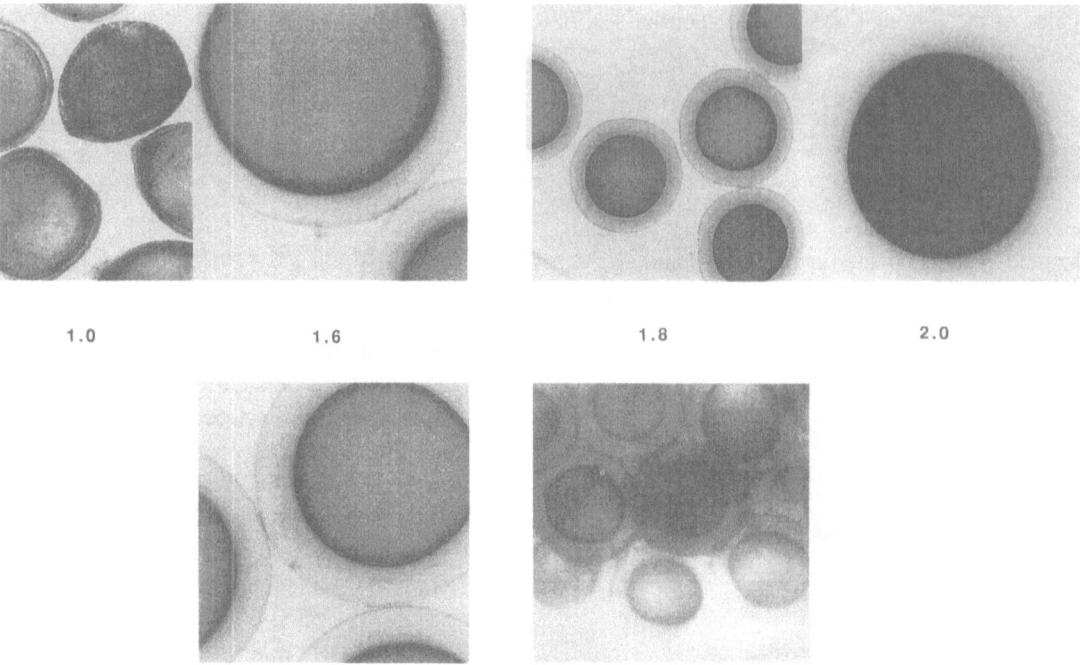

1.0   1.6   1.8   2.0

FIGURE 3.5. Biocompatibility of empty capsules in C57 mice and NOD mice. These capsules were prepared with identical processing steps, but with different PMCG concentrations. [TOP], Capsules retrieved from Normal C57 mice with PMCG concentration percentages of 1.0, 1.6, 1.8, and 2.0. [BOTTOM], Capsules retrieved from Normal NOD mice with PMCG concentration percentages of 1.6 and 1.8. (The 1.0 and 2.0 capsules could not be retrieved from the animals.)

Capsule Permeability

Measurement of capsule permeability was improved by utilizing two complementary methods: (1) size exclusion chromatography (SEC) with dextran molecular weight standards, and (2) a newly developed method to assess the permeability of a series of biologically relevant proteins, using encapsulated protein A-sepharose (PAS).[17] By combining these methodologies to measure permeability and component concentration manipulations to control capsule permeability, a series of capsules with a range of permeabilities (40–230 kD, based on dextran exclusion measurement) was developed and characterized.

The apparent pore size of the capsular membrane was determined by SEC, which measures the exclusion of dextran solutes from the column packed with microcapsules.[17] Using neutral polysaccha-ride molecular weight standards, it is possible to evaluate the membrane properties under conditions in which solute diffusion is controlled only by its molecular dimension. On the basis of the measured values of solute size exclusion coefficients (K_{SEC}) and the known size of solute molecules, one can estimate the membrane pore size distribution (PSD) (Figure 3.2). The capsular membrane excluded dextran molecules with molecular weight above 230 kD, which corresponds to $R_\eta \geqslant 117$ Å (shaded area in Figure 3.2).

Nutrients, Oxygen, and Hormone Transport

In our animal studies with the chemically-induced diabetic mice (STZ), we have noticed the development of central necrosis of the islets retrieved from the mice (Figure 3.6). We have transplanted different sets of islet-carrying capsules (230 kD, 80 μm,

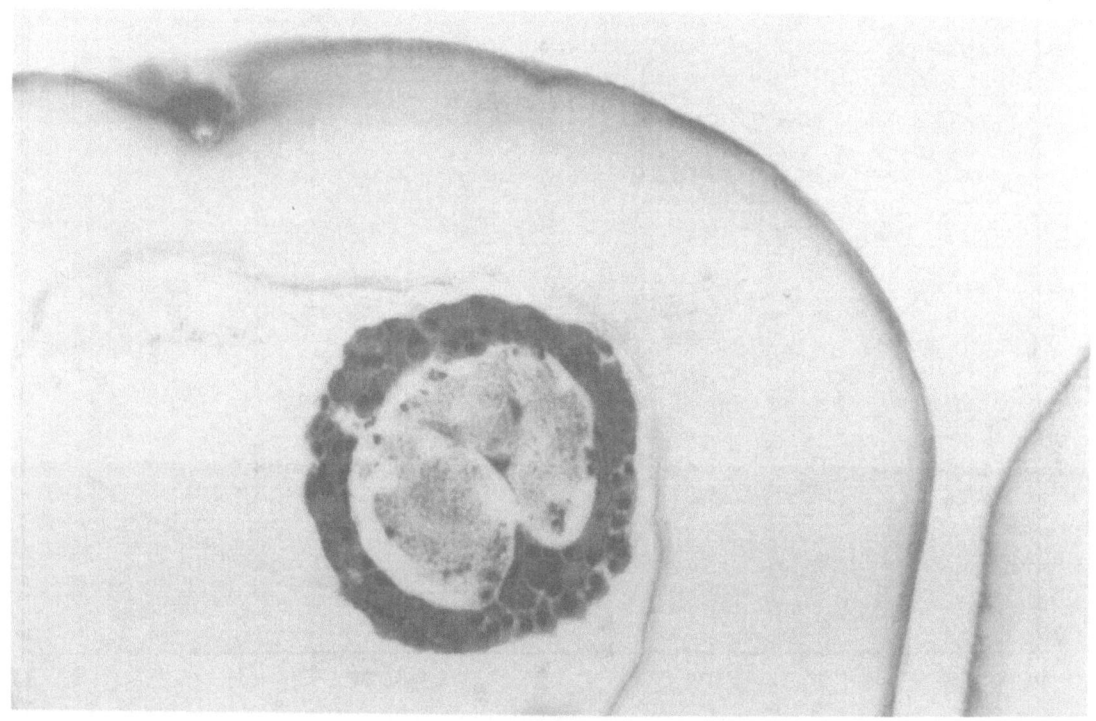

FIGURE 3.6. Effect of nutrient deficiency. Encapsulated islet retrieved from a normal C57 mouse after 4 weeks. The capsule has a wall thickness of 80 μm, with cut-off of 230 kD. Central necrosis is evident. The failure has probably resulted from nutrient deficiency.

and 350 kD, 80 μm; 350 kD, 40 μm, and 80 μm, respectively) into the same animals to study the effect of pore size on nutrient transport. The capsules were retrieved at given intervals, separated and perifused. The insulin responses of the islets encapsulated in different pore size capsules are compared in Figure 3.7. The advantages of encapsulating islets in capsules with larger pore sizes, and with a thinner membrane, are clearly shown.

Diameter Effect

Sets of different diameter capsules (0.8 mm and 0.5 mm, respectively), with the same 80 μm wall thickness, were transplanted into the same animals to study the effects on nutrient transport. The encapsulated islets were retrieved at given intervals, separated, and perifused. The insulin responses of cells encapsulated in capsules of different diameters

were found to be the same within the limits of measurement error (data not shown). This suggests that nutrient transport is strongly affected by the presence of a membrane, but not as much as by the size of the capsule.

Animal Trials

Encapsulated rat islets readily reversed chemically-induced diabetes in mice (strepzotocin-induced), and islet function was maintained for at least 4–6 months (Figure 3.8A). Similar studies were performed with diabetic NOD female mice, and diabetes was reversed for up to 180 days. (Figure 3.8B) This length of immunoprotection is in reasonable agreement with the calculated value suggested by Equation (5). Studies that transplant encapsulated rat islets into additional NOD mice are under way. The encapsulated islets, transplanted into the peritoneal

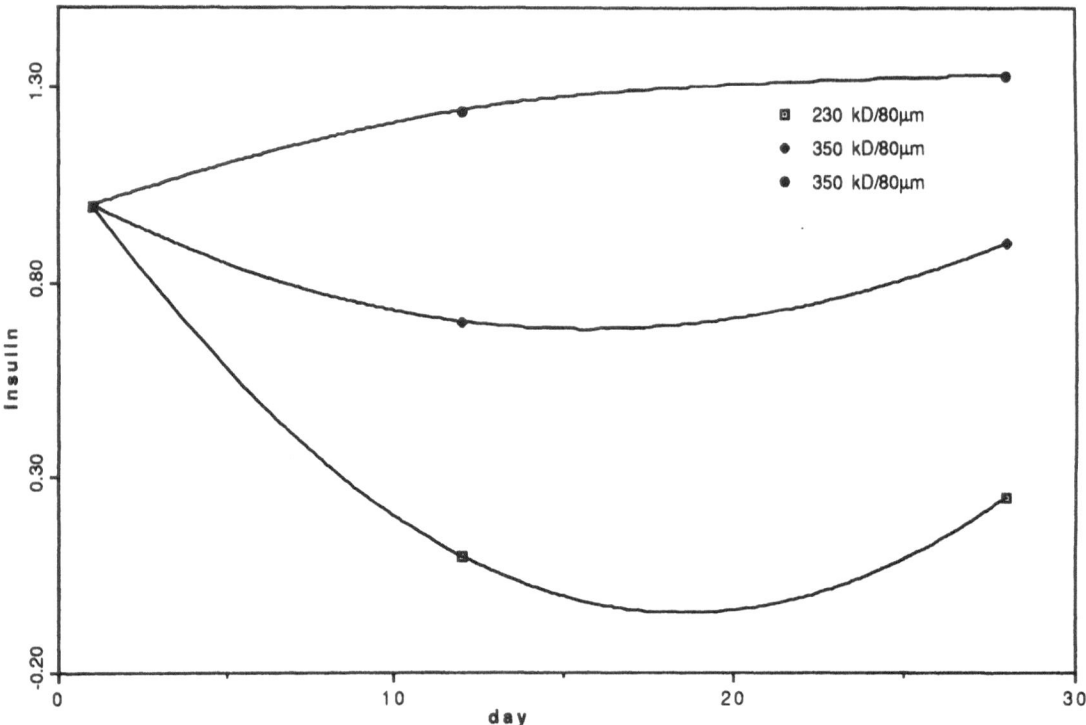

FIGURE 3.7. Insulin response of the islets. Islets encapsulated in two different pore size capsules (230 kD and 350 kD) were transplanted into the same C57 mice. The islets were retrieved, separated, and perifused at 2–week and 4–week intervals.

cavity, are free-floating and quite biocompatible, and show no significant fibrosis histologically or tissue reaction to capsules containing islets. Encapsulated islets have long-term function, as assessed by their ability to secrete insulin following glucose stimulation. The eventual failure of encapsulated islets transplanted into animals with chemically-induced diabetes did not result from capsule rupture or from an immune attack. Capsules retrieved from such animals were free-floating, and their surface was free of fibrosis. Thus, the failure probably results from islet death within the capsule, either from a nutrient deficiency or from islet toxicity caused by soluble immune factors (cytokines). In contrast, the failure of encapsulated islets transplanted into NOD mice may have resulted from an immune or autoimmune attack. Capsules retrieved from such animals were clumped, and they demonstrated marked fibrosis around the capsule surface. Experiments to determine whether this process can be prevented by altering capsule features such as permeability, wall thickness, and surface characteristics are now possible, and large animal studies are under way, with encouraging preliminary results.

Discussion

Each year several hundred thousand Americans die of body organ failure. Organ transplantations have achieved remarkable success in allowing some patients to survive, but the source of human organs is limited. In a given year, only a small portion of patients can benefit from this method of treatment. Immunoisolation of living cells for use as bioartificial organs in human beings has the potential to fill the gap. Encapsulated living cells can be transplanted into human beings without the need for immunosuppression and its accompanying side effects; this process allows cells from nonhuman species to be used, thereby overcoming the limited supply of human cells available for encapsulation.

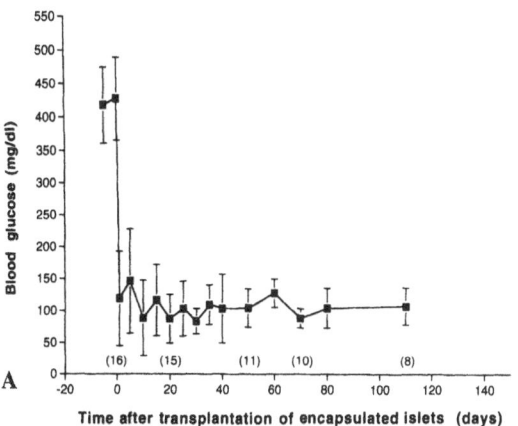

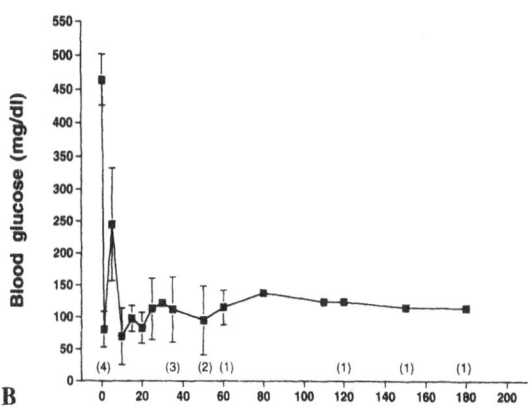

FIGURE 3.8. Transplantation of encapsulated rat islets reverses diabetes. **A**, Approximately 1000 encapsulated rat islets (0.8 mm capsule with 0.1 mm wall thickness and exclusion limit of 230 kD) were transplanted intraperitoneally into diabetic C57/B16 mice (strepzotocin-induced). The mean and the standard deviation of blood sugars of nondiabetic animals (measured using a One-Touch glucometer) were plotted against the number of days following transplantation. Of the 16 diabetic mice that received transplants, the number of nondiabetic animals at selected time points is shown in parentheses. Of the eight animals with normal blood sugars 110 days following transplantation, two later became diabetic, while six were still normoglycemic 300 days following transplantation.

B, Encapsulated rat islets were transplanted into the peritoneal cavity of female NOD mice that had developed spontaneous diabetes by the age of 20 weeks. Of the four diabetic mice that received transplants, the number of nondiabetic animals at selected time points is shown in parentheses; one that was normoglycemic for 6 months following transplantation became diabetic at 241 days.

Polymer membranes have been the choice to date, for encapsulating the living cells for immunoisolation. Recent data has shown some success with this approach in animal models. The approach is based on the premise that a membrane can protect living cells by keeping the host immune system from penetrating them, and yet allow the entrance and exit of substances required for cell survival and function. The concept of such a gate-keeper is rather straightforward, and many immunoisolation capsule designs are based on this principle; however, there is a fundamental flaw in this model. The pore size distribution of capsular membranes is not homogeneous, but rather shows a broad spectrum, with bore radii from a few angstroms to a few hundred angstroms. This heterogenicity in pore size has limited the success of reversing diabetes in human transplantation.

Recently, a new physical picture of the mechanism of immunoisolation has been presented. In the barrier model, with larger pore size and thicker membrane, immune system is free to enter the membrane but is entrapped or delayed from leaving. Limited animal studies have been encouraging. Systematic studies to determine optimal capsule design for human transplantation are now under way.

Experimental Protocol

Assessment of Capsule Permeability

To assess capsule permeability to immunologically relevant proteins such as IgG, protein A-sepharose (PAS, Sigma Chemicals, St. Louis, MO) was encapsulated. The empty capsules and capsules containing ~18 μl of packed PAS were equilibrated in phosphate-buffered saline (PBS) with 0.2% tween 20 and 0.2% bovine serum albumin. Iodinated IgG (approximately 10,000 cpm, New England Nuclear, Wilmington, DE) was added and aliquots removed at selected time points over 24 hours. After removal at each time point, the encapsulated PAS was washed five times with 4 ml of PBS. The amount of IgG that had entered the capsule was quantitated in a gamma counter. Unencapsulated or free PAS served as a positive control. This method of permeability assessment can be adapted for measuring entry of almost any protein, since the only requirements are an antibody against the protein of interest (in this case the PAS would be preincubated with the antibody before encapsulation) and the ability to radiolabel the protein.

Assessment of Insulin Secretion

Insulin secretion by encapsulated rat islets was evaluated in a cell perifusion apparatus with a flow rate of 1 ml/minute with RPMI 1640 with 0.1% BSA as a perfusate. Encapsulated islets were perifused with 2 mM glucose for 30 minutes and the column flowthrough discarded. Three-minute samples of perfusate were collected during a 30-minute perifusion of 2 mM glucose, a 30 minute perifusion of 20 mM glucose + 0.045 mM IBMX, and a 60-minute perifusion of 2 mM glucose. Samples were assayed in duplicate for insulin using Coat-a-Count kits (Diagnostic Products Corporation, Los Angeles, CA) with a rat insulin standard. The amount of insulin secreted was normalized for the number of islets.

Animal Studies

Pancreatic islets were isolated from male Sprague-Dawley rats (250–275 g, Harlan, Indianapolis, IN). Briefly, the pancreatic duct was inflated with a solution of Hank's Balanced Saline Solution (HBSS) containing collagenase (Boehringer-Mannheim, Collagenase P, Indianapolis, IN). Groups of three pancreases were digested in 2 mg collagenase/pancreas in HBSS for 6–13 minutes at 37°C, using a wrist-action shaker. The digestion was stopped by the addition of cold HBSS with 10% Newborn Calf Serum, and the material was shaken vigorously for 10–15 seconds. The digested material was washed three times with cold HBSS and filtered through a wire mesh cell strainer to remove undigested material. Pancreatic islets were separated using a Histopaque-1077 gradient and stored in University of Wisconsin storage solution for 19–24 hours before encapsulation. Encapsulated rat islets were transplanted into the peritoneal cavity of diabetic

mice (either strepzotocin-induced or NOD mice). Diabetes was induced in normal C57/B16 mice by the intraperitoneal injection of strepzotocin (200 mg/kg) 3–7 days prior to transplantation of encapsulated islets. NOD mice were purchased from Taconic Laboratories (Germantown, NY) and bred in microisolator cages with autoclaved bedding and water. Between 18 and 24 weeks of age, 50–75% of females from this colony became diabetic. When diabetes was present for 2–4 weeks, female NOD mice received a transplantation of encapsulated islets. Approximately 1000 encapsulated rat islets (0.8 mm capsule with 0.1 mm wall thickness and an exclusion limit of 230 kD), in a packed capsule volume of 0.2–0.5 ml, were sterilely transplanted intraperitoneally into metofane-anesthetized mice. The blood sugar of the mice was measured using a one-touch glucometer, using blood obtained from the retroorbital plexus or a tail vein.

Acknowledgments. These studies were supported by grants from NASA, Vanderbilt University School of Engineering, and the Vanderbilt Diabetes Research and Training Center (NIH DK20593). The author wishes to express his gratitude for the valuable contribution provided by Igor lacik, Ph.D., Polymer Institute, Slovak Academy of Sciences. The author wishes to thank AV Anilkumar, Ph.D., and Paul LeMaster, M.S., of Vanderbilt University, and AC Powers, M.D., of Vanderbilt University/VA Hospitals for stimulating discussions and critical reading of the manuscript.

References

1. Chang TM. 1992. Hybrid artificial cells: microencapsulation of living cells. ASAIO Journal 38:128–130.
2. Lacy PE. 1995. Treating diabetes with transplanted cells. Scientific American 273:50–58.
3. Colton CK, Avgoustiniatos ES. 1991. Bioengineering in development of the hybrid artificial pancreas. J Biomechanical Engineering 113:152–170.
4. Lacy PE, Hegre OD, Gerasimidi-Vazeou A, Gentile FT, Dione KE. 1991. Maintenance of normaglycemia in diabetic mice by subcutaneous xenografts of encapsulated islets. Science 254:1782–1784.
5. Lanza RP, Butler DH, Borland KM, Staruk JE, Faustman DL, Solomon BA, Muller TE, Rupp RG, Maki T, Monaco AP, Chick WL. 1991. Xenotransplantation of canine, bovine, and porcine islets in di-

abetic rats without immunosuppression. Proc Natl Acad Sci USA 88:11100–11104.

6. Lanza RP, Kühtreiben WM, Ecker D, Staruk JE, Chick WL. 1995. Xenotransplantation of porcine and bovine islets without immunosuppression using uncoated alginate microspheres. Transplantation 59:1377–1384.

7. Maki T, Lodge JPA, Carretta M, Ohzato H, Borland KM, Sullivan SJ, Staruk J, Muller TE, Solomon BA, Chick WL, Monaco AP. 1993. Treatment of severe diabetes mellitus for more than one year using a vascularized hybrid artificial pancreas. Transplantation 55:713–718.

8. Lim F, Sun AM. 1980. Microencapsulated islets as bioartificial endocrine pancreas. Science 210:908–910.

9. Iwata H, Takagi T, Kobayashi K, Oka T, Tsuki T, Ito F. 1994. Strategy for developing microbeads applicable to islet xenotransplantation into a spontaneous diabetic NOD mouse. J Biomed Mat Res 28:1201–1207.

10. Crooks CA, Douglas JA, Broughton RL, Sefton MV. 1990. Microencapsulation of mammalian cells in a HEMA-MMA copolymer: effects on capsule morphology and permeability. J Biomed Mat Res 24:1241–1262.

11. Mikos AG, Papadaki MD, Kouvroukoglou S, Ishaoug SL, Thomson RC, 1994. Mini-review: Islet transplantation to create a bioartificial pancreas. Biotechnol Bioeng 43:673–677.

12. Soon-Shiong P, Feldman E, Nelson R, Heintz R, Yao Q, Yao Z, et al. 1993. Long-term reversal of diabetes by the injection of immunoprotected islets. Proc Nat Acad Sci USA 90:5843–5847.

13. Sun Y, Ma X, Zhou D, Vacek I, Sun AM. 1996. Normalization of diabetes in spontaneously diabetic cynolologus monkeys by xenografts of microencapsulated porcine islets without immunosuppression. J Clin Invest 98:1417–1422.

14. Soon-Shiong P, Heintz RE, Merideth N, Yao QX, Yao Z, Zheng T, et al. 1994. Insulin independence in a Type 1 diabetic patient after encapsulated islet transplantation. Lancet 343:950–951.

15. De Vos P, Wolters GHJ, Fritschy WM, van Schilfgaarde R. 1993. Obstacles in the application of microencapsulation in islet transplantation. Int J Artificial Organs 16:205–212.

16. Goosen MFA. 1994. In: Immunoisolation of pancreatic islets. Lanza RP, Chick WL, editors. Austin, TX: RG Landes Co. p 21–44.

17. Brissova M, Lacik I, Powers AC, Anilkumar AV, Wang TG. 1998. Control and measurement of permeability for design of microcapsule cell delivery system. J Biomed Mat Res 39:61–70.

18. Wang T, Lacik I, Brissova M, Anilkumar AV, Prokop A, Hunkeler D, Green R, Shahrokhi K, Powers AC. 1997. An encapsulation system for the immunoisolation of pancreatic islets, Nature-Biotechnology 15:358–362.

19. Lacik I, Brissova M, Anilkumar AV, Powers AC, Wang TG. 1998. New capsule tailored properties for the encapsulation of living cells. J Biomed Mat Res 39:52–60.

20. Fan MY, Lum Z, Levesque L, Tai IT, Sun AM. 1990. Reversal of diabetes in BB rats by transplantation of encapsulated rat islets. Diabetes 39:519–522.

21. Lum ZP, Tai I, Krestow M, Norton J, Vacek I, Sun AM. 1991. Prolonged reversal of the diabetic state in NOD mice by xenograft of microencapsulated rat islets. Diabetes 40:1511–1516.

22. Weber CJ, Zabinski S, Koschitzky T, Wicker L, Rajotte R, D'Agati V, Peterson L, Norton J, Reemtsma K. 1990. The role of CD4+ helper T cells in the destruction of microencapsulated islet xenografts in NOD mice. Transplantation 49:396–104.

23. Durquy S, Chicheportiche D, Capron F, Boitard C, Reach G. 1990. Comparative study of microencapsulated rat islets implanted in different diabetic models in mice. Horm Metab Res 25:209–213.

24. Wijsman J, Atkinson P, Mazaheri R, Garcia B, Paul T, Vose J, O'Shea G, Stiller C. 1992. Histological and immunopathological analysis of recovered encapsulated allogenic islets from transplanted diabetic BB/W rats. Transplantation 54:588–592.

25. Cole DR, Waterfall M, McIntyre M, Baird JD. 1992. Microencapsulated islet grafts in the BB/E rat: a possible role for cytokines in graft failure. Diabetalogia 35:231–237.

26. Lanza RP, Sullivan SJ, Chick W. 1992. Islet transplantation with immunoisolation. Diabetes 41:1503–1510.

27. Wang TG. 1998 New technologies for artificial cells. Artificial Organs 22. No. 1:68–74.

28. Hunkeler D, Prokop A, Powers A, Haralson M, DiMari S, Wang TG. 1997. A screening of polymers as biomaterial for cell encapsulation. Feature Article of Polymer News 22:232–240.

29. Prokop A, Hunkeler D, DiMari S, Haralson MA, Wang TG. 1998. Water soluble polymer for immunoisolation I: Complex coacervation and cytotoxicity. Advances in Polymer Science 136:1–51.

30. Prokop A, Hunkeler D, DiMari S, Haralson MA, Wang TG. 1998. Water soluble polymer for immunoisolation II: Evaluation of multicomponent microencapsulation systems. Advances in Polymer Science 136:53–73.

31. Xu K, Hercules D, Lacik I, Wang TG. 1998. Atomic force microscopy used for the surface characterization of microcapsule immunoisolation devices. J Biomed Mat Res 41:461–467.

4
Biocompatible Encapsulation Materials: Fundamentals and Application

Ulrich Zimmermann, Christian Hasse, Mathias Rothmund, and Willem Kühtreiber

Despite many advances in transplantation technology, immune rejection of allograft and xenograft tissue remains a problem. Patients still face a lifetime of immunosuppression therapy to inhibit this natural response. An alternative approach is immunoisolated transplantation in which exogenous tissue is encapsulated into semipermeable, artificial membranes. Thus, the tissue is physically isolated from contact with immune effector cells and humoral factors (Lanza et al 1992a Lanza and Chick 1997, Colton 1995). This new technology is highly advantageous, because in principle it does not induce an immune response, thus avoiding immunosuppressant therapy (Colton 1995, Lanza et al 1992a, Lanza and Chick 1997, Lim and Sun 1980). The development of biocompatible materials that can be used in encapsulation of tissue for therapeutic purposes is a rapidly growing field of modern biotechnology. However, most basic materials used for cell encapsulation were designed and developed for various industrial purposes and are usually not commercially available at the high purity and chemical standardization required for medical use (Greco 1994, Hubbell 1995). Impurities may contribute to device failures as well as to undesirable host responses to many implants.

In the past, a whole range of different polymeric materials has been employed to encapsulate cells or tissues for implantation. These materials can be broadly classified as natural polymers and (semi-) synthetic polymers.

Natural Polymers

In the light of therapeutic relevance, natural polymers have significant advantages. These polymers, as well as their degradation products, are usually not toxic or carcinogenic, nor do they produce adverse reproductive and developmental effects (Hubbell 1995). In most cases, entrapment of living cells in matrices made from natural polymers can be accomplished under physiological conditions (Colton 1995, Hubbell 1995, Greco 1994, Lanza et al 1992a, Lanza and Chick 1997, Lim and Sun 1980). However, many commercially available natural polymers display variable biocompatibility, and certain properties may vary considerably among products of different manufacturers or even show large lot-to-lot fluctuations.

Most impurities present in natural polymers originate from their biological source. Common contaminants are proteins, complex carbohydrates, fatty acids and phospholipids, bacterial products (e.g., lipopolysaccharides), and secondary products such as toxins, polyphenols, etc. (De Vos et al 1993, Greco 1994, Skjaek-Braek et al 1989, Sun et al 1996, Zimmermann et al 1992). Many of these compounds exhibit biological activities, such as immunostimulation or cytotoxicity that would prevent their use in implantation, e.g. immunostimulation or cytotoxicity (Klöck et al 1994, Zimmermann et al 1992). The nature and the content of these impurities may vary with the polymer in question, as well as with the supplier of the raw polymer, and even among different lots of the polymer (Zimmermann et al 1992, Klöck et al 1994). Therefore, the prerequisites for the use of natural polymers in implantation are reliable purification procedures and reproducible assay protocols that predict their in vivo performance.

Natural polymers that form semipermeable membranes for cell encapsulation mainly include polysaccharides and/or polyamino acids. Most of these systems have been reviewed elsewhere (Colton 1995, Colton 1996, Galletti 1992, Lanza et al 1992a, Lanza and Chick 1997, Morris 1996). In this review, for reasons of clearness and comprehensibility in discussing the complex problems associated with the use of insufficiently characterized, unpurified raw materials, we will focus on alginates, the material most frequently used for encapsulation of cells and tissues.

Alginates

Alginates are anionic polysaccharides that are extracted from brown seaweeds. They are composed of variable homopolymeric regions of D-mannuronic acid and L-guluronic acid interspaced with regions of alternating blocks (Guiseley 1989). Alginates have hydrogel-forming properties with many di-, tri-, or multivalent cations (e.g., Ca^{2+}, Ba^{2+}, or Fe^{3+}). Semipermeable alginate gels can be prepared under physiological conditions (Guiseley 1989, Sandford 1992).

Alginates find widespread uses as additives in a variety of pharmaceutical products (Sandford 1992). Alginate-based wound dressings, hemostatic agents, and swabs have been used in human therapy for many years (Sandford 1992), and there are only a few reports on complications (Borolla et al 1985, Matthew et al 1993, Odell et al 1994). From clinical experience it might be concluded that alginates should be sufficiently biocompatible for use in implantation and tissue engineering.

However, when alginates are employed as encapsulation material for islets of Langerhans and other tissues, graft failure as a result of fibrosis is a major problem. This has been demonstrated to be more severe in spontaneously diabetic animal models (e.g., BB rats). Mitogenic stimulation of lymphocytes, induction of proinflammatory cytokines,

foreign body reaction, and fibrosis have been reported for most commercial alginates (Clayton et al 1991, Clayton et al 1993, Cole et al 1992, De Vos et al 1993, De Vos et al 1996, De Vos et al 1997, Espevik et al 1993, Fritschy et al 1994, Gotfredsen et al 1990, Jahr et al 1997, Klöck et al 1994, Mazaheri et al 1991, Otterlei et al 1991, Otterlei et al 1993, Pueyo et al 1993, Wijsman et al 1992, Zimmermann et al 1992). Other authors, however, found little or no fibrotic reaction around alginate implants (Clayton et al 1991, De Vos et al 1996).

The biocompatibility of alginates, especially of those alginates containing higher amounts of mannuronic acid, has been examined by several groups. The properties of alginate capsules are highly dependent on the algin composition, in particular the relative content of mannuronic acid to guluronic acid (Guiseley 1989, Sandford 1992). Some groups found that immunostimulatory activities were caused by those alginates with high mannuronic acid content, while immunosuppressive activities were caused by alginates with high guluronic acid content (Espevik et al 1993, Jahr et al 1997, Otterlei et al 1991, Otterlei et al 1993). It was thus concluded that mannuronic acid oligomers would provoke cytokine release from monocytes via a receptor-mediated mechanism. This in vitro induction of proinflammatory cytokines in human monocytes induced by algin preparations in vitro was more or less specific for high-M products. Induction could be inhibited by agents specifically preventing the binding of bacterial lipopolysaccharides (LPS) to CD14, a glycophosphatidylinositol-anchored membrane protein of monocytes and neutrophils. CD14 is responsible for the cellular response to LPS (Gegner et al 1995). It was concluded that poly-M-fragments bind either directly or via a transport molecule (Gegner et al 1995) to this specific receptor (CD14).

Bacterial lipopolysaccharides are major constituents of the cell walls of Gram-negative bacteria and thus a common contaminant in preparations of biological macromolecules (Gegner et al 1995). The release of LPS into the human circulatory system results in the activation of leukocytes and of endothelial cells. This, in turn, results in the release of cytokines, nitrous oxide, and tissue factors. Such responses to LPS or lipotechoic acid cause a clinical syndrome in human beings known as septic shock, which has a high mortality rate (Gegner et al 1995). The CD14 molecule, a glycosylphosphatidylinositol-anchored membrane protein, has been characterized on monocytes, and to a lesser extent on neutrophils. CD14 is a receptor for LPS and the serum LPS-binding protein (LBP) complex. However, several studies indicate that CD14 does not directly mediate intracellular signaling, and evidence exists for an independent pathway for the LPS-induced activation of cells. Additional membrane components are possibly involved in the direct recognition of LPS and these may in turn cooperate with CD14 activation. It has been proposed that as-yet-unidentified proteins are required to explain the signal-transducing properties of LPS (Gegner et al 1995).

Since the mechanism of signal transduction mediated by the CD14 molecule is presently not fully understood (Gegner et al 1995), the significance of the results describing the interaction of polymannuronate with CD14 remains questionable. In addition, the data presented using mannuronate-oligomers are difficult to judge: The in vitro results depended on special depolymerization and purification procedures for the various fractions enriched in each uronic acid monomer. The poly-M and poly-G fragments initially used for these studies were contaminated with lipopolysaccharides (Espevik et al 1993). Furthermore, they were poorly characterized with respect to further contamination with LPS fragments, peptidoglycans, β-1,3,glucans, or other stimulators of TNF-α release in monocytes. In addition, the procedures employed to isolate the poly-mannuronate blocks will result in the formation of terminal 4,5–unsaturated mannuronate derivatives. Similar unsaturated uronic acids have been shown to be potent haptenic structures (Hascall et al 1995).

Further data presented by other authors (Clayton et al 1991, Clayton et al 1993, De Vos et al 1993, De Vos et al 1997, Dorian et al 1995, Klöck et al 1994, Klöck et al 1997, Pfeffermann et al 1996, Sun et al 1996, Zimmermann et al 1992) also did not corroborate the hypothesis that mannuronate-rich alginates generally were potent inducers of monocytes. We analyzed several low-endotoxin alginate samples in different monocyte and macrophage cell systems (Klöck et al 1997, Pfeffermann et al 1996). In contrast to the above, we were unable to establish reliable correlation between fibrogenic response in the BB/OK rat and the release of proinflammatory cytokines in vitro. Rather, even the samples with the lowest capacity for cytokine release from the tested cell systems were still fibrogenic in vivo (Klöck et al 1997, Pfeffermann et al 1996).

Our results, as well as the data published by other groups (Clayton et al 1991, Clayton et al 1993, De Vos et al 1993, De Vos et al 1997, Dorian et al 1995, Sun et al 1996), support the view that mannuronate residues in alginates are not the main reason for their fibrogenic properties. Several in vitro studies have shown that those alginate samples with the best biocompatibility (e.g., the lowest fibrogenic response) are those samples with the highest content of mannuronate. Therefore, other mechanisms responsible for the fibrogenic activity of alginates have to be considered.

Production of Implantation-Grade Alginates

Systematic studies have shown that commercial alginates are contaminated with varying amounts of many different mitogens (Zimmermann et al 1992). These numerous mitogenic contaminants associated with commercial alginates could be detected and quantitatively removed by using the free-flow electrophoresis technique (Zimmermann et al 1992). Separation was based on differences in the electrophoretic mobilities of these compounds as compared to the electrophoretic mobility of the highly charged G-, M-, and MG-blocks.

Free-flow electrophoretic analysis was chosen because even diluted alginate solutions are very viscous and thus difficult to fractionate by other techniques. In free-flow electrophoresis the mixture that is to be separated is injected as a fine stream into a solution flowing perpendicular to the lines of force of an electric field (Zeiller et al 1975). Electrically charged molecules are deflected from the direction of flow at an angle determined by a combination of the flow velocity and the electrophoretic mobility of the molecules. Thus, components with different electrophoretic mobility move in different directions and can be separately collected with a fraction collector after leaving the separation chamber (Zeiller et al 1975). Separations of macromolecules and even of particles can be carried out in a highly reproducible fashion over long experimental periods. The prerequisite for such separations is the maintenance of stationary and absolutely constant running conditions, as well as the homogenous removal of heat induced by the electric field, to prevent the formation of turbulences and thermoconvection as a result of temperature differences (Zeiller et al 1975). Since free-flow electrophoresis has a continuous mode of working, it is not only possible to obtain a high throughput, but also to carry out a rapid change of samples while maintaining constant separation conditions. Thus, the method is ideally suited for the automation of analytical test series, e.g. for quality control (Zimmermann et al 1992, Zeiller et al 1975).

At least 10–20 bioactive, immunostimulating fractions were characterized by such free-flow electrophoretic separations of commercial alginates (Zimmermann et al 1992). The number and distribution of these impurities varied with the source of the alginate and (in some cases) within different batches of the same product.

The mitogenic substances contaminating the fibrogenic alginates were not identified. However, many of the common polysaccharides present in seaweed (e.g., fucoidan, laminarin, carrageenan, etc.; Fleury and Lahaye 1993) are mitogenic and can act costimulatory to other mitogens in splenocyte preparations (Ogata et al 1991, Ohno et al 1995, Okazaki et al 1995). When these compounds were removed from crude alginates by free-flow electrophoresis, the resulting material was no longer fibrogenic after cross-linkage with Ba^{2+} ions and implantation into the peritoneal cavity of rodents (Zimmermann et al 1992) and did not stimulate the mitogenic proliferation of murine splenocytes in vitro.

On the basis of the analytical data obtained by free-flow electrophoresis, we also developed a protocol that allowed us to produce algins free of adverse contaminants on a laboratory scale (Klöck et al 1994, Klöck et al 1997). Briefly, Ba^{2+}-crosslinked alginate beads are prepared by the same procedure as used for cell encapsulation. The beads are then extracted with weak acids, citrate, and ethanol. They are then redissolved in alkaline, EDTA-containing solutions and dialysed extensively. The final product is precipitated with ethanol. The free-flow electrophoretic profiles of such purified alginates revealed no indications for

a significant modification of the surface charge density of the alginate molecule during the purification steps (Klöck et al 1997). The degree of purification was further tested by fluorescence spectroscopy. By this method, polyphenolic impurities present in most conventional alginates can be easily detected (Skjaek-Braek et al 1989). The analysis of the purified alginates by means of fluorescence spectroscopy showed the continued presence of residues that exhibited emission at 445 nm. However, the relative fluorescence intensities of the purified samples were decreased by at least 83% as compared to conventionally extracted alginate (Klöck et al 1997). The biocompatibility test in vitro and in vivo showed that these impurities did not initiate a foreign body reaction. As shown by our earlier data and the data of several other authors, purified alginate product showed no mitogenic activity towards murine lymphocytes in vitro (Gacesa and Russell 1990, Klöck et al 1992, Klöck et al 1997, Pfeffermann et al 1996, Zimmermann et al 1992). In accordance with this finding, the endotoxin content of the purified alginate, as determined using the Limulus lysate assay, was reduced to the level of the solvent (Klöck et al 1992, Klöck et al 1997). In addition, the lymphocytes fully retained their ability to respond to lipopolysaccharides in the presence of purified alginate.

These in vitro data, suggesting that mitogenic contaminants and endotoxins were apparently removed from the raw alginates, were confirmed by further implantation studies. The diabetes-prone BB rat exhibits elevated macrophage activity (Cole et al 1992, Gotfredsen et al 1990, Mazaheri et al 1991, Wijsman et al 1992, Hosszufalusi et al 1992, Rothe et al 1990). However, even in this stringent model, barium alginate capsules retrieved from the kidney capsule were only covered by a very thin fibrotic layer (in contrast to raw alginate; Klöck et al 1997). Barium alginates are preferable for implantation and other biomedical and biotechnological applications (see "Application of Biocompatible Barium Alginate Capsules"). Capsules can be produced in a simple one-step procedure (Geisen et al 1990, Zekorn et al 1992), and there are no "technical imperfections" of the semipermeable membrane, as have been reported as a major factor in implant failure (De Vos et al 1994, De Vos et al 1996). The slight tissue reactions in animals that have received purified Ba^{2+}-alginate capsules could also arise from the implantation process itself and may represent responses to the physical presence of foreign bodies and to specific chemical properties of the alginates.

The significance of these results is further augmented by the animal model used for these studies. Compared to other laboratory rat strains (e.g., Lewis or Wistar), the diabetes-prone BB/OK rat is the most sensitive rat model for assessing the biocompatibility of implanted alginate biomaterials (Clayton et al 1993, Cole et al 1992, De Vos et al 1993, Gotfredsen et al 1990, Hosszufalusi et al 1992, Mazaheri et al 1991, Rothe et al 1990, Wijsman et al 1992). The BB rat has increased proportions of splenic macrophages as compared to other rat strains (Hosszufalusi et al 1992) and, in addition, BB macrophages exhibit TNF-α hypersecretion upon challenge with endotoxin (Rothe et al 1990).

The accuracy of such biocompatibility assays in vivo is further dependent on the site of implantation (Clayton et al 1993, De Vos et al 1993, Greco 1994). After intraperitoneal transplantation, selection for nonreactive capsules occurs during the harvesting of the beads (Clayton et al 1993, De Vos et al 1993). This has hampered previous studies on the biocompatibility of alginate implants. The freely floating, retrievable capsules are not representative of those capsules eliciting the foreign body reaction. The latter are usually overgrown by a fibrotic capsule and strongly adherent to the peritoneal wall. To solve this problem of nonrepresentative retrieval of the implanted material, we implanted alginate capsules under the kidney capsule of BB rats. In contrast to peritoneal implants, the microencapsulated islets implanted beneath the kidney capsule remained localized and allowed the assessment of all implanted capsules (Klöck et al 1997, Pfeffermann et al 1996).

The purification protocol described above (Klöck et al 1994, Klöck et al 1997) has been successfully applied to the purification of alginates regardless of their source and/or uronic acid composition. This is in contrast to other protocols (De Vos et al 1997, Dorian et al 1995), which worked for one specific, commercial product. Biocompatible alginates containing even high amounts of mannuronic acids can be produced on an industrial scale in reproducible quality. This is of exceptional practical relevance since the composition of alginate is crucial for any application of this material

(Guiseley 1989, Sandford 1992). Gels made from alginates composed mainly of guluronic acids (high G) are more brittle (Kelco 1988), and exhibit a very high permeability towards antibodies, as compared to the higher elasticity and lower permeability of gels made from alginates rich in mannuronic acids (Kelco 1988, Klein et al 1983, Steward and Swaisgood 1993). In addition, the high viscosity of high-G alginates makes any immobilization process very difficult, at least for higher polymer concentrations (Kelco 1988). Thus, from the biotechnological point of view, alginates rich in mannuronic acids are the material of choice. Consequently, these highly purified alginates should be chosen for the purpose of clinical application.

Quality Control of Alginates In Vitro

In principle, biocompatibility assays could be performed in vitro and/or in vivo. For practical as well as for ethical reasons, well-designed in vitro experiments should be conducted prior to the final animal work. All too often, experimental animals are unnecessarily used out of convenience (Bruck 1982).

Commercially available alginates represent a family of substances that originate from different algal species. They are extracted by different procedures and are manufactured at different qualities, which may vary even among batches of the same product (Guiseley 1989, Klöck et al 1994, Zimmermann et al 1992). In the light of a more general application for alginates as implantation materials, the need for standardized and reliable procedures for evaluation of the biocompatibility of purified alginate preparations is undisputed.

Systematic studies by our group have shown that the mitogenic activity of raw and purified alginates can be assayed with murine splenic lymphocyte cultures (Klöck et al 1994, Zimmermann et al 1992). In comparison to fibrogenic alginate samples, the mitogenic activity of purified alginate was reduced in these assays to the level of the control. The algin-treated lymphocytes were still able to respond normally to mitogens such as lipopolysaccharides. However, the mitogenic assays used in these early studies were not practical for a routine quality control because partly purified products still induced fibrosis but their mitogenic response was difficult to measure.

Tests that are based on the costimulatory effect of fibrogenic alginates on LPS-induced mitogenic responses of murine splenocytes are considerably more sensitive as compared to the lymphocyte-based tests discussed above (Klöck et all 1997, Pfeffermann et al 1996.) For example, increased proliferation was observed when splenocytes were cultured in the presence of both LPS and alginate.

This costimulatory effect correlates well with the occurrence of fibrosis in the BB/OK rat (Klöck et al 1997, Pfeffermann et al 1996). Fibrogenic alginates always showed a strong costimulatory activity towards the LPS-induced mitogenic proliferation of splenocytes prepared from Balb-c mice. In contrast, nonfibrogenic samples did not influence the response of splenocytes to LPS. Thus, the costimulatory assay is likely to detect those contaminants in the alginate responsible for their fibrogenic properties.

Alginate Gel Capsules

Gels made of calcium alginate have been widely used for the immobilization of sensitive cells (e.g., Brodelius and Vandamme 1987, Colton 1995, Guiseley 1989, Lanza et al 1992a, Lanza and Chick 1997, Sandford 1992, Schnabl and Zimmermann 1989). In biotechnology and biomedicine, immobilized cells are preferred to free cells for the production of metabolites. This is mainly because they are easier to handle, and because the gel matrix also provides the cells with a protective barrier against infection and mechanical stress. However, these calcium alginate gels are not stable in vivo (Brodelius and Vandamme 1987, Schnabl and Zimmermann 1989). Lim and Sun (1980) proposed to coat calcium alginate beads with polycations such as poly-L-lysine (PLL). By simple suspension of the negatively charged alginate beads in a positively charged PLL solution, a 10 to 30 μm PLL membrane will be formed around the beads. In order to improve biocompatibility, another alginate layer is applied to cover the polyamino acids located on the surface of the microcapsules. Citrate treatment is frequently used for the liquefaction of PLL-coated alginate beads. This treatment often damages some microcapsules and also markedly increases the volume of the enclosed alginate. Chitosan (a polyglucosamine) has been proposed to replace PLL (Maysinger et al 1996),

but most research is still based on PLL, especially for encapsulation of animal cells and artificial organs. Despite its widespread use, this system of polyamino acid and polyanionic polysaccharide has had limited success. This is largely due to the cytotoxicity of the polycationic substances used, complement activation induced by polycationic surfaces, and the limited mechanical stability of polycationic membranes (De Vos et al 1993, Clayton et al 1993, Lanza et al 1992).

In order to overcome the technological limitations of polycation-alginate capsules, we have developed an alternate, one-step method for immunoisolation of allogenic or xenogenic islets in barium alginate gel spheres (Geisen et al 1990, Gröhn et al 1994, Hasse et al 1996, Hasse et al 1997a, Hasse et al 1997b, Zekorn et al 1992). Cross-linkage of alginates with barium ions produces a much more resistant matrix than with calcium ions (Brodelius and Vandamme 1987). Barium alginate gels are stable in vitro as well as in vivo for many months (Gröhn et al 1994, Geisen et al 1990, Hasse et al 1996, Hasse et al 1997a, Hasse et al 1997b, Zekorn et al 1992). In contrast to cross-linkage with Ca^{2+} ions, cross-linkage with Ba^{2+} ions leads to gels that are chemically very stable under both in vitro and in vivo conditions. Provided that the encapsulation process is carefully controlled, rodent and porcine islets encapsulated in barium gels do not show any impaired metabolic reactions even after culture in vitro for several months. In contrast to the results obtained with alginate-polycation microcapsules, complement activation by barium alginate capsules was not detected. Several studies discussed in the next section, demonstrated that barium alginate encapsulated islets were able to restore normoglycemia in different animal models for long periods of time.

Application of Biocompatible Barium Alginate Capsules

Diabetes Mellitus

Barium alginate microcapsules have been successfully used for the entrapment of murine, porcine, and human islets of Langerhans (Geisen et al 1990, Zekorn et al 1992). In glucose perifusion challenges, rat islets microencapsulated in barium alginate showed a biphasic insulin release with a short delay of the first phase (Geisen et al 1990, Zekorn et al 1992). During static glucose challenge, the insulin release ranged from 40% to 70% as compared to unencapsulated controls. In diabetic mice, transplantation of barium-alginate-bead encapsulated rat islets resulted in a nonfasting normoglycemia that lasted at least until day 70 (Geisen et al 1990, Zekorn et al 1992). In the same model, transplantation of encapsulated porcine islets resulted in 6 of 10 recipients remaining normoglycemic until day 70, while three animals were still normoglycemic on day 100. Three weeks after transplantation of barium-alginate-encapsulated rat and porcine islets, intraperitoneal glucose tolerance tests were performed. These tests revealed rapid glucose assimilation. Histological examination demonstrated well-preserved islets at the end of the experiments.

Microcapsule diameter and total implantable capsule volume are the major limiting factors for the clinical application of microencapsulated tissue (De Vos et al 1993, Clayton et al 1993, Gröhn et al 1994). The drawbacks of large-diameter capsules are reduced efficiencies in glucose and insulin transfer (Colton 1995). It has been reported by several groups that the maximum diffusion distance for oxygen and nutrient supply that can be tolerated by islets is about 200 μm (see De Vos et al 1993).

Recently, methods for the production of alginate-polylysine-alginate capsules with a diameter of approximately 250–400 μm have been reported (e.g., Sun et al 1996). However, the islets were suspended in the alginate at a density of around 3000 islets per ml (Sun et al 1996). This relatively low density would result in a total capsule (graft) volume of about 300 ml, if 1,000,000 islets were to be implanted in a human being (Clayton et al 1993, De Vos et al 1993). An increased density of islets (>3000/ml) during the encapsulation process, however, results in incompletely covered tissue for many capsules as well as mechanical imperfections (De Vos et al 1996).

Another disadvantage of alginate-polylysine-alginate capsules is their high internal colloid osmotic pressure. Therefore, these capsules have to be at least nearly spherical in shape, and only spherical islets can be conformally coated by alginate-polylysine-alginate capsules. In contrast, conformal coating of islets irrespective of their shape and/or

size can be achieved using the barium alginate technique (Gröhn et al 1994). Small islets (diameter <80–150 μm) were encapsulated in spherical capsules with a total diameter of less than 200 μm. Larger islets (diameter > 100–150 μm) were conformally coated by a thin layer of barium alginate approximately 10–50 μm thick. Colorimetric measurements of the respiratory metabolism of encapsulated islets indicated that this thin layer of barium alginate was able to sustain cells in vitro. These data also demonstrated that neither the barium ions nor the encapsulation procedure (shear forces, washing steps) impaired the islet metabolism.

To further reduce the graft volume, we developed a simple density gradient centrifugation protocol to separate empty microcapsules from capsules containing islets (Gröhn et al 1994). The removal of empty capsules reduced the total implant volume to about two times the volume of the free islets. In addition to the removal of empty capsules, this method also selects for encapsulated islets with a large islet volume as compared to the whole capsule volume. Large capsules containing small islets remained in the fractions containing the empty capsules. Because of the large capsule volume resulting from conventional encapsulation techniques, the peritoneal cavity has been so far the only feasible site for the implantation of microencapsulated islets (Gröhn et al 1994). It has been shown that capsules placed in the peritoneal cavity are prone to much more severe reactions against implants, and that this implant site requires a higher islet mass for normalizing blood glucose in the recipient (De Vos et al 1993). The much smaller volume of implants made possible by the density gradient purification of barium alginate capsules should enable the use of these capsules for transplantation at sites other than the peritoneal cavity, such as the kidney capsule, omentum pouches, spleen or the portal vein of the liver.

Hypoparathyroidism

Clinical hypoparathyroidism is one of the most difficult of all endocrine disorders to treat due to the complexity of parathyroid hormone interactions. Therefore, causative treatment of this disorder by allo- or xenotransplantation of parathyroid tissue is highly desirable. The number of patients who would benefit from parathyroid transplantation has been estimated to be at least 1500 patients per year for Germany alone (Hasse et al 1996).

Individuals affected by permanent hypoparathyroidism face reduced quality of life; however, hypoparathyroidism is rarely a life-threatening condition. Therefore, systemic immunosuppression for recipients of allotransplants is not justified. Immunoisolated transplantation of allogenic or xenogenic parathyroid tissue, without the need for immunosuppression, would be a promising therapeutic approach.

By combining improved tissue culturing techniques with the use of a specially adapted microencapsulation procedure, we recently succeeded in iso-, allo-, and xenotransplantation of microencapsulated parathyroid tissue in experimental hypoparathyroidism in an animal model without systemic immunosuppression (Hasse et al 1996, 1997a, 1997b)

Six months after isotransplantation of microencapsulated parathyroid tissue in parathyroidectomized DA rats, all animals that had received parathyroid tissue were normocalcemic (Hasse et al 1996, 1997a). Following extirpation of the transplant beads from these animals, calcium and PTH dropped to post-PTX levels and matched those of the negative controls. Throughout the entire experiment, PTH and calcium concentrations were always concordant. Correspondingly, in those animals that had received parathyroid tissue implants, histology revealed vital parathyroid tissue inside intact microcapsules. The implants were surrounded by a thin layer of fibroblasts that apparently was functionally irrelevant. Its thickness measured 7 ± 1 µm when purified alginate had been used. (Hasse et al 1996). Microcapsules manufactured of purified alginate, with or without parathyroid tissue, exhibited a significantly reduced pericapsular infiltration of fibroblast as compared to microcapsules from raw alginate (Hasse et al 1996). This would explain the apparently better long-term performance in vivo of the tissue encapsulated with the purified alginate, and thus may have important clinical implications.

In view of the potential clinical use of this method, human parathyroid tissue has been microencapsulated and transplanted over the highest immunological barrier (Hasse et al 1997b). In a controlled, long-term animal study in parathyroidectomized DA rats, the effect of microencapsulation on the survival of xenotransplanted human parathyoid tissue was evaluated over 30 weeks (native and microencapsulated parathyroid tissue each = 40 rats). Functionally, human parathyroid tissue was able to replace that of the rat (Hasse et al 1997b). All animals that had received microencapsulated parathyroid tissue were normocalcemic for 16 weeks. At the end of the study 27 out of 40 rats still maintained normocalcemia. In contrast, serum calcium concentrations dropped to postparathyroidectomy levels within 4 weeks in those animals that had received unencapsulated tissue only. Histologic evaluation of the explanted, functionally successful xenografts showed vital parathyroid tissue inside intact microcapsules, surrounded by a thin layer of fibroblasts. Non-viable fibrotic remnants were demonstrated in animals with nonencapsulated parathyroid tissue. Thus, we have established the feasibility of microencapsulation of human parathyroid tissue, and the preservation of viability over long periods in vivo even when xenotransplanted. In combination with an improved tissue culture method, transplantation of human parathyroid tissue into rats and maintenance of its physiological function was reproducibly achieved without postoperative systemic immunosuppression over the highest transplantation barrier. This may be a crucial step toward the first clinical application of this method (Hasse et al 1997b, 1997c, Chapekar 1996).

Synthetic Polymers

Various synthetic or semisynthetic polymers have been used to encapsulate living cells (e.g., Brodelius Vandamme 1987, Colton 1995, Colton 1996, Clayton et al 1993, Hubbell 1995, Lanza et al 1992, Lanza and Chick 1997, Morris 1996). Islets of Langerhans and other endocrine tissues have been microencapsuled in membranes formed of polyacrylamide-coated agarose, polyvinylalcohol, cellulose sulfate, water insoluble polymethacrylate, polyethylene glycol and derivatives, and PAN-hydrogels (for recent reviews, see Clayton et al 1993, Colton 1996, Hubbell 1995, Lanza and Chick 1997, Morris 1996). Unfortunately, none of these protocols proved to be clinically successful, mostly resulting in fibrosis around the implant.

Recently, screening of no fewer than 1350 combinations of polymers for forming capsules by interfacial coacervation (Wang et al 1997) was reported. However, data on long-term biocompatibility of the optimized polymer blend composed of alginate, cellulose sulfate, and polymethyleneguanidine in sensitive animal models are still lacking.

An alternative approach to the microencapsulation of cells for transplantation is the use of macrocapsules, in which the membrane is usually shaped as a capillary tube or fiber. The materials tried so far include cellulosic membranes, polyvinylchloride, polyamines, cellulose nitrate, cellulose triacetate, and PAN-PVC copolymers (Altman et al 1986, Clayton et al 1993, Colton 1995, Colton 1996, Hubbell 1995, Lanza et al 1992a, Lanza and Chick 1997, Lanza et al 1992b, Morris 1996, Theodorou et al 1979, Zekorn et al 1989). Common problems of such macrocapsular diffusion chambers are mechanical fragility, severe diffusion limitations (Colton 1995), fibrosis, abscess formation, and adhesion (Lanza and Chick 1997), as discussed in chapter 15.

Biocompatibility of Synthetic Polymers

In principle, synthetic polymers can be designed according to specific needs and reproducibly manufactured. However, the use of synthetic polymers is limited because of the unphysiological conditions that are typically necessary for the encapsulation process and/or the potential risks associated with reactive monomers and/or impurities (Brodelius Vandamme 1987, Lodge and Monaco 1994, Silver and Doillon 1989). Synthetic polymers usually are formed by reactive monomers and/or prepolymers. Such components may be highly toxic or carcinogenic, or may produce adverse reproductive and developmental effects (Greco 1994, Hubbell 1995). In addition, synthetic polymer blends usually contain low molecular tweight compounds, such as monomers, oligomers, plasticizers, and catalysts (Greco 1994). If the capsule membrane is formed *in situ* by polymerization of monomers or oligomers, or by interfacial coacervation of prepolymers, these contaminants may

severely damage the encapsulated cells and will also considerably contaminate the final product. It will be very difficult to remove these compounds from the final device, since the encapsulated cells will only tolerate extraction under physiological conditions which typically will not allow for the removal of such contaminants.

The main risk associated with the use of synthetic polymers in implantation is the potential of these materials to release leachable substances when used in implantation (Brodelius Vandamme 1987, International Organization for Standardization 1992, 1997, Lodge and Monaco 1994, Silver and Doillon 1989). Generally, the high molecular weight polymeric materials used for the encapsulation of cells contain low molecular weight components, such as (reactive) monomers, oligomers, and catalysts, which can leach out into the body (Brodelius Vandamme 1987, Lodge and Monaco 1994, Silver and Doillon 1989). Any new synthetic polymeric materials used in therapeutic cell encapsulation may therefore present toxic risks upon long-term, intimate contact with or within the body (International Organization for Standardization 1992, 1997). Therefore, an important requirement of the preclinical toxicology testing of a device is to determine the potential toxicity of these leachable chemicals as they appear in the final device. Chemicals that are recovered by extraction of the implant material, when appropriate, should be used in animal studies after they have been separated, quantified, and identified (Brodelius Vandamme 1987, International Organization for Standardization 1992, 1997, Lodge and Monaco 1994, Silver and Doillon 1989). In addition, the primary concern with any implanted device is its long-term potential to cause cancer (Silver and Doillon 1989). This potential may arise not only from chemical leachables and degradation products of the device, but also from the physical effects of the device at the implanted site. Therefore, adequate long-term studies with implantation of device materials should be conducted to evaluate the carcinogenic potential of polymeric implants.

When selecting the appropriate tests for the biological evaluation of a medical device, one must consider the chemical characteristics of the device materials and the nature, degree, frequency, and duration of the exposure of the device to the body (International Organization for Standardization 1992, 1997). The ISO 10993–1 lists suggested short-term and long-term biological tests. Short-term tests include irritation, sensitization assay, cytotoxicity, acute systemic toxicity, hemocompatibility/hemolysis, pyrogenicity (material-mediated), implantation, and mutagenicity (genotoxicity) tests, as well as pharmacokinetics studies. Long-term tests include subchronic toxicity, chronic toxicity, carcinogenesis bioassay, reproductive, and developmental toxicity. These guidelines may also be used in selecting appropriate tests for the evaluation of encapsulated implants (Chapekar 1996).

Even after purification to homogeneity, the materials that constitute implantable capsules may generate degradation products in the biological environment, and in the body these products may behave differently from bulk materials (Silver and Doillon 1989). According to ISO 10993, degradation studies should be considered if (1) the device is designed to be bioresorbable, or (2) the device is intended to be implanted for a long period, and the potential for significant biodegradation exists (International Organization for Standardization 1992, 1997).

For well-described and clinically accepted degradation products (especially in the case of natural polymers, e.g., alginates) no further investigation may be necessary (Chapekar 1996, International Organization for Standardization 1997). In view of regulatory concerns and safety issues for clinical applications, this higher level of safety is a major advantage of natural polymers such as alginates, as compared to synthetic polymers.

Outlook

As shown above, natural polymers present significant advantages for applications such as implantation. However, the condition for success is efficient purification of the raw materials. Recent advances in alginate purification and successful attempts to reduce the size of the immunoisolated graft are major steps towards making encapsulated cell transplants a viable therapeutic proposition. In the progression toward clinical acceptance, it must be shown that the chemistry and structure of com-

pounds used in any new grafts are safe, and that their biocompatibility and performance are satisfactory without degradation or weakening of the device. Devices made from natural (and therefore toxicologically safe) polymers have been used with significant success in animal—and also in the first small-scale clinical—studies (Hasse et al. 1997).

References

Altman JJ, Houlbert D, Callard P, McMillan P, Solomon BA, Rosen J, Galletti PM. 1986. Long-term plasma glucose normalization in experimental diabetic rats with macroencapsulated implants of benign human insulinomas. Diabetes 35:625–633.

Borolla BJ, Bjorenson JE, Austin BP, Gerstein H. 1985. J Endodont 11:75–83.

Brodelius P, Vandamme EJ. 1987. Immobilized cell systems. In: Biotechnology (Rehm HJ, Reed G, eds.) Vol 7a, Verlag Chemie, Weinheim, p 405–464.

Bruck SD. 1982. Medical applications of polymeric materials. Med Progr Technol 9:1–16.

Chapekar MS. 1996. Regulatory concerns in the development of biologic-biomaterial combinations. J Biomed Mat Res 33:199–203.

Clayton HA, London NJ, Colloby PS, Bell PR, James RF. 1991. The effect of capsule composition on the biocompatibility of alginate-poly-l-lysine capsules. J Microencapsul 8:221–233.

Clayton HA, James RF, London NJ. 1993. Islet microencapsulation: A review. Acta Diabetol 30:181–189.

Cole DR, Waterfall M, McIntyre M, Baird, JD. 1992. Microencapsulated islet grafts in the BB/E rat: a possible role for cytokines in graft failure. Diabetologia 35:231–237.

Colton CK. 1995. Implantable biohybrid artificial organs. Cell Transplantation 4:415–436.

Colton CK. 1996. Engineering challenges in cell-encapsulation technology. Trends Biotechnol 14:158–162.

De Vos P, Wolters GH, Fritschy WM, Van Schilfgaarde R. 1993. Obstacles in the application of microencapsulation in islet transplantation. Int J Artif Organs 16:205–212.

De Vos P, Wolters GH, van Schilfgaarde R. 1994. Possible relationship between fibrotic overgrowth of alginate-polylysine-alginate microencapsulated pancreatic islets and the microcapsule integrity. Transplant Proc 26:782–783.

De Vos P, De Haan BJ, Pater J, Van Schilfgaarde R. 1996. Association between capsule diameter, adequacy of encapsulation, and survival of microencapsulated rat islet allografts. Transplantation 62:893–899.

De Vos P, De Haan DJ, Wolters GHJ, Strubbe JH, Van Schilfgaarde R. 1997. Improved biocompatibility by limited graft survival after purification of alginate for microencapsulation of pancreatic islets. Diabetologia 40:262–270.

Dorian RE, Cochrum KC, Jemtrud SA. 1995. Non-fibrogenic high mannuronate alginate coated transplants, process for their manufacture, and methods for their use. US Patent 5,429,821.

Espevik T, Otterlei M, Skjak-Braek G, Ryan L, Wright SD, Sundan A. 1993. The involvement of CD14 in stimulation of cytokine production by uronic acid polymers. Eur J Immunol 23:255–261.

Fleury N, Lahaye M. 1993. Studies on byproducts from the industrial extraction of alginate. J Appl Phycol 5:605–611.

Fritschy WM, De Vos P, Groen H, Klatter FA, Pasma A, Wolters GH, Van Schilfgaarde R. 1994. The capsular overgrowth on microencapsulated pancreatic islet grafts in streptozotocin and autoimmune diabetic rats. Transpl Int 7:264–271.

Gacesa P, Russell NJ. (eds.) 1990. Pseudomonas infections and alginates. Chapmann & Hall, New York, p 35.

Galletti PM. 1992. Bioartificial organs. Artif Organs 16:55–60.

Gegner JA, Ulevitch RJ, Tobias PS. 1995. Lipopolysaccharide (LPS) signal transduction and clearance— Dual roles for LPS binding protein and membrane CD14. J Biol Chem 270:5320–5325.

Geisen K, Deutschländer H, Gorbach S, Klenke C, Zimmermann U. 1990. Function of barium-alginate microencapsulated xenogenic islets in different diabetic mouse models. In: Frontiers in diabetes research (Shafrir E, ed.). Lessons from animal diabetes III, 8:142–148.

Gotfredsen CF, Steward MG, O'Shea GM, Vose JH, Horn T, Moody AJ. 1990. The fate of transplanted encapsulated islets in spontaneously diabetic BB/WOR rats. Diabetes Research 15:157–163.

Greco, RS (ed.). 1994. Implantation biology, the host response and biomedical devices. CRC Press, Boca Raton, Florida.

Gröhn P, Klöck G, Zimmermann U, Horcher A, Zekorn T, Brandhorst A, Brandhorst T, Federlin K. 1994. Large scale production of barium alginate microcapsules for immunoisolation. Clin Exp Endocrinol 102:380–387.

Guiseley KB. 1989. Chemical and physical properties of algal polysaccharides used for cell immobilization. Enzyme Microb Technol 11:706–715.

Hascall VC, Midura RJ, Sorrell JM, Plaas AH. 1995. Immunology of chondroitin-dermatan sulfate. Adv Exp Med Biol 376:205–216.

Hasse C, Klöck G, Zielke A, Schlosser A, Barth P, Zimmermann U, Rodtmund M. 1996. Transplantation of parathyroid tissue in experimental hypoparathyroidism: In vitro and in vivo function of parathyroid tissue microencapsulated with a novel alginate. Int J Artif Organs 19:735–741.

Hasse C, Zielke A, Klöck G, Schlosser A, Zimmermann U, Rodtmund M. (1997a). Isotransplantation of microencapsulated parathyroid tissue in rats. Exp Clin Endocrinol Diabetes 105:53–56.

Hasse C, Klöck G, Bartsch D, Zielke A, Zimmermann U, Rodtmund M. 1997b. First successful xenotransplantation of microencapsulated human parathyroid tissue in experimental hypoparathyroidism: Long term in vivo function without immunosuppression. J. Microcapsulation 14:617–626.

Hasse C, Klöck G, Schlosser A, Zimmermann U, Rodtmund M 1997c. Parathyroid allotransplantation without immunosuppression. The Lancet 350, 1296.

Hosszufalusi N, Chan E, Granger G, Charles MA. 1992. Quantitative analysis comparing all major spleen cell phenotypes in BB and normal rats: Autoimmune imbalance and double negative T cells associated with resistant, prone and diabetic animals. J Autoimmun 5:305–318.

Hubbell JA. 1995. Biomaterials in tissue engineering. Bio/Technology 13:565–576.

International Organization for Standardization: ISO - 109931–1. (1992). Biological evaluation of medical devices—Part 1: Guidance on selection of tests.

International Organization for Standardization: ISO 10993–16 (1997). Biological evaluation of medical devices—Part 16: Toxicokinetic study design for degradation products and leachables.

Jahr TG, Ryan L, Sundan A, Lichenstein HS, Skjaek-Braek G, Espevik T. 1997. Induction of tumor necrosis factor production from monocytes stimulated with mannuronic acid polymer, and involvement of lipopolysaccharide binding protein, CD14 and bactericidal/permeability increasing factor. Infect Immun 65:89–94.

Kelco. 1988. Algin: Hydrophilic derivatives of alginic acid for scientific water control. Kelco Div of Merck and Co, San Diego, p 51.

Klein J, Stock J, Vorlop KD. 1983. Pore size and properties of spherical Ca alginate biocatalysts. Eur J Microb Biotechnol 18:86–91.

Klöck G, Frank H, Houben R, Zekorn T, Horcher A, Siebers U, Wöhrle M, Federlin K, Zimmermann U. 1994. Production of purified alginates suitable for use in immunoisolated transplantation. Applied Microbiol Biotechnol 40:639–643.

Klöck G, Pfeffermann A, Ryser C, Gröhn P, Kuttler B, Hahn HJ, Zimmermann U. 1997. Biocompatibility of mannuronic-acid rich alginates. Biomaterials 18:707–713.

Lanza RP, Sullivan SJ, Chick WL. 1992a. Perspectives in diabetes. Islet transplantation with immunoisolation. Diabetes 41:1503–1510.

Lanza RP, Borland KM, Staruk JE, Appel C, Solomon BA, Chick WL. 1992b. Transplantation of encapsulated canine islets into spontaneousy diabetic BB rats without immunosuppression. Endocrinology 131:637–42.

Lanza RP, Chick WL. 1997. Transplantation of encapsulated cells and tissues. Surgery 21:1–9.

Lim F, Sun AM. 1980. Microencapsulated islets as a bioartificial pancreas. Science 210:908–910.

Lodge JP, Monaco AP. 1994. Artificial Organs. In: Implantation biology, the host response and biomedical devices (Greco, RS, ed.). Boca Raton, CRC Press, Florida; p. 347–361.

Matthew IR, Brownwe RM, Frame JW, Millar BG. 1993. Tissue response to a haemostatic alginate wound dressing in tooth extraction pockets. Br J Oral Max Surg 31:165–169.

Maysinger D, Berezovskaya O, Fedoroff S. 1996. The hematopoietic cytokine colony stimulating factor 1 is also a growth factor in the CNS: (II). Microencapsulated CSF-1 and LM-10 cells as delivery systems. Exp Neurol 141:47–56.

Mazaheri R, Atkinson P, Stiller C, Dupre J, Vose J, O'Shea G. 1991. Transplantation of encapsulated allogeneic islets into diabetic BB/W rats. Effects of immunosuppression. Transplantation 51:750–754.

Morris PJ. 1996. Immunoprotection of therapeutic cell transplants by encapsulation. Trends Biotechnol 14:163–167.

Odell EW, Oades P, Lombardi T. 1994. symptomatic foreign body reaction to haemostatic alginate. Br J Oral Max Surg 32:178–179.

Ogata M, Yoshida SI, Kamochi M, Shigematzu A, Mitsugishi Y. 1991. Enhancement of LPS triggered tumor necrosis factor production in mice by carrageenan pretreatment. Infect Immun 59:679–683.

Ohno N, Asada N, Adachi Y, Yadomae T. 1995. Enhancement of LPS triggered TNF alpha production by 1,3–β-glucans in mice. Biol Pharm Bull 18:126–133.

Okazaki M, Adachi Y, Ohno N, Yadomae T. 1995. Structure activity relationship of β-1,3–glycans in the induction of cytokine production from macrophages in vitro. Biol Pharm Bull 1320–1327.

Otterlei M, Ostgaard K, Skjaek-Braek G, Ryan L, Smidsrod O, Soon-Shiong P, Espevik T. 1991. Induction of cytokine production from human monocytes stimulated with alginate. J Immunotherapy 10:286–291.

Otterlei M, Sundan A, Skjaek-Braek G, Ryan L, Smidsrod O, Espevik T. 1993. Similar mechanisms of action

of defined polysaccharides and lipopolysaccharides: characterization of binding and tumor necrosis factor alpha induction. Infect Immun 61:1917–1925.

Pfeffermann A, Klöck G, Gröhn P, Kuttler B, Hahn H-J, Zimmermann U. 1996. Assay procedures for the assessment of the biocompatibility of alginate implants. Cell Engineering 4:167–173.

Pueyo ME, Darquy S, Capron F, Reach G. 1993. In vitro activation of human macrophages by alginate polylysine microcapsules. J Biomat Sci 5:197–203.

Rothe H, Fehsel K, Kolb H. 1990. Tumor necrosis factor alpha production is upregulated in diabetes prone BB rats. Diabetologia 33:573–575.

Sandford PA. 1992. High purity chitosan and alginate: preparation, analysis and applications. Front Carb Res 2:250–269.

Schnabl H, Zimmermann U. 1989. Immobilization of plant protoplasts. In: Plant protoplasts and genetic engeneering (Bajay YPS, ed.) Vol 1, Springer Verlag, New York, p 63–96.

Silver F, Doillon C. 1989. Biocompatibility. Interactions of biological and implantable materials. Vol. 1. Polymers. VCH Publishers, New York.

Skjaek-Braek G, Murano E, Poaletti S. 1989. Alginate as immobilization material II: Determination of polyphenolic contaminants by fluorescence spectroscopy and evaluation of methods for their removal. Biotechnol Bioeng 33:90–94.

Stewart WW, Swaisgood HE. 1993. Characterization of calcium alginate pore diameter by size-exclusion chromatography using protein standards. Enzyme Microb Technol 15:922–927.

Sun Y, Ma X, Zhou D, Vacec I, Sun A. 1996. Normalization of diabetes in spontaneously diabetic cynomologous monkeys by xenografts of microencapsulated porcine islets without immunosuppression. Clin Invest 98:1417–1422.

Theodorou NA, Vrbova H, Tyhurst M, Howell SL. 1979. An assessment of diffusion chambers for use in pancreatic islet cell transplantation. Transplantation 27:350–52.

Wang T, Lacik I, Brissova M, Anilkumar AV, Prokop A, Hunkeler D, Green D, Shahrokhi K, Powers AC. 1997. An encapsulation system for the immunoisolation of pancreatic islets. Nature Biotechnol 15:358–362.

Wijsman J, Atkinson P, Mazaheri R, Garcia B, Paul T, Vose J, O'Shea G, Stiller C. 1992. Histological and immunopathological analysis of recovered encapsulated allogeneic islets from transplanted diabetic BB/W rats. Transplantation 54:588–592.

Zeiller K, Loser R, Pascher G, Hannig K. 1975. Free flow electrophoresis. Hoppe-Seyler's Z Physiol Chem 356:1225–1244.

Zekorn T, Siebers U, Filip L, Mauer K, Schmitt U, Bretzel RG, Federlin K. 1989. Bioartificial pancreas: the use of different hollow fibers as a diffusion chamber. Transplant Proc 21:2748–2750.

Zekorn T, Horcher A, Siebers U, Schnettler R, Klöck G, Hering B, Zimmermann U, Bretzel RG, Federlin K. 1992. Barium-cross-linked alginate beads: a simple one-step method for successful immuno-isolated transplantation of islets of Langerhans. Acta Diabetologia 29:99–106.

Zimmermann U, Klöck G, Federlin K, Hannig K, Kowalski M, Brezel RG, Horcher A, Entenmann H, Siebers U, Zekorn T. 1992. Production of mitogen contamination free alginates with variable ratios of mannuronic to guluronic acid by free flow electrophoresis. Electrophoresis 13:269–274.

5

Modulation of Membrane Permeability

Augustine O. Okhamafe and Mattheus F. A. Goosen

Introduction

A majority of mammalian cells, unlike bacterial cells, are anchorage-dependent (Young 1993), a fact that has encouraged the development of immobilization systems. One such, the adsorption or attachment technique (van Wezel 1967), whereby cells are adhered to the surface of microcarriers, suffers several drawbacks. Not only are the strength and stability of the adhesive bonds influenced by formulation and other factors, but the cells are exposed to the abrasion arising from the hydrodynamic forces in an agitated culture system. Although this is not a problem in bacterial culture, where cell walls are thick and protective, considerable cellular damage occurs in animal cells, which are fragile and highly vulnerable to shear forces (Croughan et al 1987). Furthermore, in cell transplantation therapy, this approach does not afford the cells the necessary immunoprotection. Another approach that has been investigated is cell entrapment in a polymer gel or porous matrix. Although this method enhances greatly the available surface, and hence the cell density of the system, the problems of vulnerability to shear and immune assault are at best only partially mitigated. An additional complication is that the gel structure may actually inhibit metabolic exchanges that are vital and, indeed critical, for cell survivability and viability. Encapsulation currently enjoys the widest acceptance and application, and has been used to modify both adsorbed cell and gel-entrapped cell systems for greater efficiency and effectiveness (Okhamafe and Goosen 1993, Young et al 1989).

Encapsulated Cell Systems

A typical encapsulated cell system consists of viable cells or tiny bits of tissue in a core enclosed within a biocompatible membrane that permits ready passage of nutritional factors (oxygen, nutrients, etc.) into the capsule interior, and allows waste products of metabolism, hormones, and bioactive/therapeutic factors to pass in the opposite direction. Two broad types of capsules can be differentiated: (1) *macrocapsules* (both intravascular and extravascular), otherwise known as hollow fibers, are 0.5–10 cm in length with a diameter in the range 0.5–6 mm (Gentile et al 1995, Goosen 1993). They consist of bundles of capillary hollow fibers or a single length of large-diameter fiber with the cells lodged on the exterior surface, and the whole system is sealed within a plastic cylinder with two ports. Blood flow is through the fiber lumen. Sometimes, they are flat with the shape of a disc; (2) *microcapsules* are spherical entities with a fluid core and a diameter usually in the range 0.3–2 mm (Goosen 1987, Gentile et al 1995).

The overall therapeutic objective of cell encapsulation is to provide controlled, efficient, and effective treatment of severe degenerative and disabling disorders in human beings. Most of these disease conditions are attributable to deficient or subnormal metabolic and secretory cell functions. Examples abound: multiple sclerosis, amyotrophic lateral sclerosis, diabetes, blood disorders, Parkinson's disease, hemophilia, Alzheimer's disease, and hepatic failure (Aebischer et al 1995, 1991a, b, de-la-Rubin et al 1996, Gentile et al 1995). The basic presentation of an encapsulated cell system shows a reservoir of discrete cells or tiny bits of tissue bounded by a permselective membrane.

Membrane Function: Issues, Perspectives, and Considerations

The capsule membrane plays a critical role in achieving a successful cell system. It should be nontoxic and biocompatible, demonstrate good mechanical and chemical stability, and show controlled and predictable permeability. Consequently, critical issues pertaining to the barrier and other properties of the membrane will now be briefly addressed.

Immunoisolation. A major prerequisite for successful cell transplantation is that the membrane should effectively isolate (shield) the encapsulated cells from immune system components of the host, and thus minimize transplant rejection. The isolatory mechanism is size exclusion, i.e., immune elements (cells, complex macromolecules, etc.) larger than a particular size are prevented from gaining access to the encapsulated cells. Attainment of total membrane impermeability is unrealistic, and indeed undesirable, since transmembrane metabolic exchange is an absolute requirement. However, it has been suggested that modulation of the permselectivity of the membrane to exclude immunogenic components of molecular weight larger than 1.6×10^5 is a realistic objective (Okhamafe and Goosen 1993).

The prophylactic approach to preventing cell/tissue rejection, i.e., pharmacologic immunosuppression, has achieved only partial success, principally because it is useful in allograft (intraspecies cell/tissue) transplantation but not in xenograft (interspecies) transplantion. Consequently, immunoisolation holds better promise against immune assault in cell/tissue transplantation.

Biocompatibility. A major consideration in encapsulated cell transplantation is the biocompatibility of the membrane. The poor in vivo performance of the early encapsulated islet transplants (Lim and Sun 1980, Sun et al 1983) was due to the incompatibility of the polyethyleneimine membrane, which apparently produced an inflammatory response. Further modification of the microcapsule membrane yielded transplants with greatly improved in vivo results due to better biocompatibility (O'Shea et al

1984). Immunogenic reaction engenders the growth of fibrous tissue around the capsule membrane with the result that transmembrane pathways are occluded, transport of nutritional requirements to the cells is severely impeded, and ultimately cell activity and life span are drastically reduced. A related problem in macrocapsules (vascular devices) is a thrombogenic reaction in the fiber lumen leading to clot formation (Goosen 1993).

Product release and retention. An encapsulated cell/tissue transplant is primarily required to produce therapeutic proteins/peptides and other endogenous chemicals for modulated release into the host system (Aebischer et al 1995, Okhamafe and Goosen 1993). The membrane, unless carefully designed and prepared, will not only ward off the immunoglobulins of the immune defense system of the transplant recipient but could also impede the outward transport of the generated protein from the capsule core. The result would be subnormal transplant function or even outright failure. A careful balance in membrane permeation is necessary so that immunoisolation does not hinder protein release.

On the other hand, the use of microencapsulation to produce peptides and endogenous chemicals in vitro, e.g., Encapsel® (brand of monoclonal antibodies), requires that the bioactive protein/peptide products generated by the encapsulated cells remain trapped within the capsule in order to achieve prepurification and high concentration of the products (Jarvis and Grdina 1983, Posillico 1986). This can only be realized if the capsule is impermeable to the peptide product while readily permitting metabolic exchanges.

Strength and integrity. Membrane permselectivity will be greatly impaired if defects, or even rupture, manifest on the membrane at the time of formulation or afterward as a result of environmentally induced stress or "aging." This exposes the encapsulated cells to attack by elements of the immune system and even outright destruction. Furthermore, weaker membranes tend to be less compact and, therefore, more permeable.

Chang (1993) and coworkers have highlighted a phenomenon that may compromise membrane integrity. The author observed that during the encapsulation of hepatocytes with the poly-L-lysine/alginate system, a few of the cells were embedded in the membrane matrix, resulting in thin and poorly formed sites on the membrane. This not only adversely affected membrane permeability but also elicited acute cell-mediated host reaction that led to membrane penetration and subsequent destruction of the encapsulated cells by components of the immune system.

Transport of metabolic factors. The cells and tissues of the body, under normal circumstances, receive an adequate supply of metabolic requirements via the vascular system by a convective mechanism. In encapsulated systems, it is essential that the membrane freely permit the inflow of anabolic factors (oxygen, glucose, etc.) and the outflow of waste products (carbon dioxide, lactic acid, etc.). The mechanism of transport depends on the membrane type. Three main types of membranes may be differentiated:

(1) *Nonporous membranes.* These are usually hydrophobic. Here, the solute forces its way through the membrane via spaces between polymer segments. These spaces are successively formed as a result of polymer chain mobility, and the overall diffusion process involves partitioning of the solute into the membrane matrix and out on the other side. Richards (1985) has presented relationships to quantify solute flux for three membrane geometries: slab (disc), cylinder (hollow fiber), and sphere (microcapsule). They are outlined in Figure 5.1. Polydimethylsiloxane (PDMS) and ethylene-vinylacetate copolymers (EVAC) are typical examples of nonporous membrane polymers.

(2) *Microporous membranes.* The solute diffuses through discrete liquid-filled pores in the membrane by a partition mechanism. The size of the pores ranges from as small as 1.5 nm to several microns (Gentile et al 1995, Richards 1985), and the properties of the liquid in them exert a great influence on the magnitude of solute flux. Richards (1985) proposed the following equation to describe the solute transport rate, J, or dQ/dt:

$$J = \frac{dQ}{dt} = \frac{\epsilon\, D_L K_{L/S}\, \Delta C}{T \iota_p} \qquad (1)$$

where ϵ, T and ι_p are the porosity, tortuosity, and thickness of the membrane, respec-

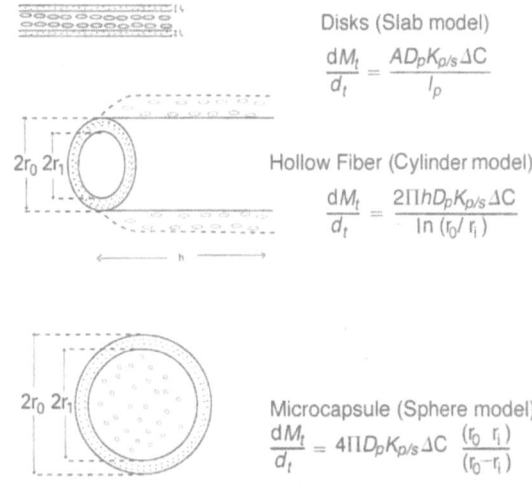

Disks (Slab model)
$$\frac{dM_t}{d_t} = \frac{A D_p K_{p/s} \Delta C}{l_p}$$

Hollow Fiber (Cylinder model)
$$\frac{dM_t}{d_t} = \frac{2 \Pi h D_p K_{p/s} \Delta C}{\ln (r_0 / r_i)}$$

Microcapsule (Sphere model)
$$\frac{dM_t}{d_t} = 4 \Pi D_p K_{p/s} \Delta C \frac{(r_0\ r_i)}{(r_0 - r_i)}$$

FIGURE 5.1. Solute diffusion models for capsule membrane (where dM_1/dt = steady rate of transport at time, t; Dp = diffusion coefficient; K_P^S = partition coefficient of solute between the polymer and the liquid in the capsule core; C = solute concentration gradient across the capsule membrane; l_p = thickness of the membrane; A = the total surface area of the disk [minus the edge effects]; r_o = the outer radius of the device; r_1 = radius of the core; and h = the length of the hollow fiber).

tively, D_L is the diffusion coefficient of the solute in the liquid filling the pores, and $K_{L/S}$ is the partition coefficient of the diffusant between that liquid and the one from which the solute diffuses. Examples include those prepared with reticulated thermoplastics such as acrylonitrile vinyl chloride copolymer (PAN-PVC), polyurethane, and polypropylene (Gentile et al 1995, Zondervan et al 1992).

(3) *Hydrogels.* These polymers are hydrophilic and show a high tendency to take up large amounts of water, which causes them to swell. The absorbed water constitutes an aqueous continuous phase through which the diffusant is transported. Solute permeability is greatly influenced by the solubility of the diffusant in the aqueous phase. The permeation behavior of hydrogel membranes is believed to be intermediate between those of nonporous and porous membranes (Richards 1985). Examples of hydrogel membranes include

those prepared from PLL, chitosan, alginates, methacrylates, and polyacrylates (King et al 1987, Okhamafe et al 1996, O'Shea et al 1984, Ronel et al 1983, Sefton et al 1987).

Composite Membranes

The need for composite or multiple membranes in permeation control is well recognized (Gentile et al 1995, Goosen et al 1990, King et al 1987, Okhamafe and Goosen 1993), although there is still a question regarding whether, in all or some cases, they are true composites or actually multilayer membranes (Okhamafe and Goosen 1993).

Crank (1975) provided mathematical expressions to facilitate the evaluation of permeation phenomena in composite systems. They are based on the assumption that the decrease in solute concentration across the composite equals the sum of the fall in concentration across the individual membranes forming the composite. The composite membrane is composed of n membranes of thickness $l_1, l_2, \ldots l_n$, with diffusion coefficient $D_1, D_2, \ldots D_n$. Since the rate of transfer, F, through the constituent membranes is the same, the total drop in solute concentration can be represented as

$$\frac{Fl_1}{D_1} + \frac{Fl_2}{D_2} + \ldots + \frac{Fl_n}{D_n}$$
$$= (R_1 + R_2 + \ldots + R_n) F \qquad (2)$$

Where $R = l/D$, R being the resistance to diffusion. For composite membranes of different shapes, the following relationships have been further derived (Crank 1975):

(i) Slab: $\dfrac{l_1}{P_1} + \dfrac{l_2}{P_2} + \ldots + \dfrac{l_n}{P_n} = \dfrac{1}{P}$ $\qquad (3)$

(ii) Cylinder:

$$\frac{l_n(R_1 / R_0)}{P_1} + \frac{l_n(R_2 / R_1)}{P_2} + \ldots + \frac{l_n(R_n / R_{n-1})}{P_n}$$
$$= \frac{l_n(R_n / R_0)}{P} \qquad (4)$$

(iii) Sphere:

$$\frac{(1 / R_0) - (1 / R_1)}{P_1} + \frac{(1 / R_1) - (1 / R_2)}{P_2} +$$

$$\ldots + \frac{(1 / R_{n-1}) - (1 / R_n)}{P_n}$$
$$= \frac{(1 / R_0) - (1 / R_n)}{P} \qquad (5)$$

The above relationships are based on the assumption that the solubility coefficient, S, is 1.0. Therefore, since $P = DS$, then $P = D$.

Permeation Control Techniques

In practical terms, several factors could be manipulated to control membrane transport. For the purpose of this write-up, these are placed in three major categories.

Membrane factors. Membrane characteristics and behavior are influenced by several factors. These include:

(1) *Polymer molecular weight.* In general, polymer permeability decreases as the molecular weight of nonionic polymers increases. Intrinsic factors such as the tightness of the polymeric chain packing, the rigidity of the polymer chains, and the degree of polymer crystallinity are known to exert varying influences on solute diffusivity across polymer membranes (Okhamafe and York 1987, O'Neill 1980, Rogers 1976). It is believed that as polymer molecular weight increases, the degree of molecular chain packing, and consequently the network of interchain bonding and polymer chain stiffness, also rises. Thus, the free volume (and hence the diffusion channels) of the amorphous fraction decreases, and so also diffusivity.

However, permeation studies on polymers used in cell encapsulation indicate that the relationship between polymer molecular weight and solute diffusivity elucidated above for nonionic polymers does not hold for polyelectrolytes. Table 5.1 summarizes the findings of various workers using either PLL or chitosan as the membrane polymer. A careful analysis of the conflicting findings suggests that polymer membrane diffusivity, and hence permeability, appears to

TABLE 5.1. Effect of membrane polymer molecular weight on capsule permeability and strength.

Membrane polymer	Permeant	Effect of polymer mol. wt. increase on:		Reference
		Permeability	Strength	
PLL	BSA	↑		Goosen et al 1985
PLL	BSA	↑	↓	King et al 1987
Chitosan	BSA	↑		Shioya and Rha 1989
Chitosan		↑		Shioya and Rha 1989
Chitosan	BSA	no change	↓	McKnight 1987
Chitosan	BSA	↓		Polk 1990

Note: BSA = Bovine serum albumin
↑ = increase
↓ = decrease

be mostly dependent on the extent of ionic interaction between the cationic membrane and the anionic core. Decrease in polymer molecular weight means reduced polymer chain length and complexity as well as lower solution viscosity. Therefore, the membrane solution is afforded a greater capacity to penetrate the capsule core and react by forming thicker, more compact membranes with fewer defects. Thus, polyelectrolytes differ from nonionic polymers, in that a fall in molecular weight decreases membrane permeability, provided molecular weight does not fall below a critical level.

(2) *Strength.* The physical strength of a capsule membrane is a function of several factors. Rowe (1976, 1980) observed that the mechanical strength of hydroxypropyl methylcellulose (HPMC) films rose while their flexibility and incidence of physical defects declined as polymer molecular weight increased. Sato (1980) proposed that cracks or physical defects will occur in a polymeric membrane if the internal stress, P, is greater than or equal to the tensile strength, σ, of the membrane, i.e.,

$$P = \sigma \qquad (8)$$

An effective approach to minimizing internal stress is to plasticize the membrane polymers with noncytotoxic compounds such as the polyethylene glycols and citric acid (Okhamafe and York 1985a, 1988). Applying previously established relationships, Okhamafe and York (1985b) showed that the toughness of a membrane, and hence its stress crack resistance, is dependent on its strength and flexibility. Internal flaws or incipient cracks in the membrane constitute stress locations that may subsequently propagate, leading to physical defects such as rupture.

The significance of the above lies in the fact that defects such as rupture and cracks severely impair the immunoprotective capacity of the membrane, and dislocate transmembrane transport processes. Even where defects are not manifest at the time of capsule production, some encapsulated systems, on incubation in a culture medium, show volume expansion due to liquefication and swelling of the capsule core (Goosen et al 1985, King et al 1987, Yoshioka et al 1990). If the membrane is not sufficiently tough or resistant to the exerted stress, cracks, and even rupture may ensue. Furthermore, the swelling-induced stretching of the membrane could widen its diffusion pores and pathways, thus leading to increased solute diffusivity. For thermoplastic polymers, Gentile et al (1995) have noted that increasing membrane composition and structure through the application of techniques that increase the isoreticulated structure of UF and MF membranes can enhance tensile strength.

So also does a reduction in membrane porosity.

The works of King et al (1987) and Mc-Knight (1990) demonstrated that the higher the molecular weight of PLL and chitosan, respectively, the lower the strength, flexibility, and thickness of the membrane and the higher the incidence of physical defects. Increase in molecular weight (which translates to longer polymer chain length) reduces the capacity of the membrane polymer to penetrate and interact with the core polymer.

(3) *Thickness.* Membrane thickness is closely linked to membrane strength and polymer molecular weight, as already outlined. Any measures taken to enhance membrane strength also increase thickness. Since a direct correlation exists between membrane strength and thickness in encapsulated cell systems, efforts to control these factors often exert a profound impact on membrane permeability.

(4) *Chemical structure.* Alteration of the chemical structure of an existing membrane polymer is perhaps one of the least favored approaches to the control of membrane permeability. For one thing, the outcome of such an effort is often very uncertain. Even where the desired membrane permeation characteristics are achieved, the problems of cytotoxicity and biocompatibility may emerge. Furthermore, because the polymer is now an entirely new material, the potential cost of carrying out all the tests for implantable systems required by the relevant regulatory authorities, as well as the attendant delays, generally makes the approach unattractive. Although the success rate appears low, Gharapetian et al (1987) have reported that modulation of polyacrylate membrane permeability is possible by modifying the copolymer structure.

(5) *Formation of membrane laminates.* There is hardly any doubt now that formation of one or more membranes (or, as it seems in some cases, a multilayer membrane) around a capsule provides an effective tool for modifying membrane transport properties.

King et al (1987) achieved a threefold increase in intracapsular monoclonal antibody production in a PLL-alginate system consisting of two PLL membranes (molecular weight of 2.0×10^5 and 2.2×10^4), compared with a capsule system with a single membrane. Encapsulated cell systems employing composite membranes have been reportedly used to treat Type I diabetes (Gentile et al 1995).

Chang (1993) and his coworkers have demonstrated that using the standard PLL-alginate method to encapsulate hepatocytes leaves a few of the cells trapped within the membrane matrix, with the result that the membrane is thin and malformed at the points of entrapment. When such capsules were implanted peritoneally in mice, a severe cell-mediated immune response resulted not only in the formation of fibrous tissue around the capsule but also in penetration of the membrane at the protruding sites by elements of the immune system. Although these authors also devised a novel technique whereby the cells are initially entrapped in alginate microspheres, which are then embedded in larger alginate beads and finally enclosed within a PLL membrane, it seems that application of multiple membranes could also solve this problem.

Process factors. Modulation of the encapsulation process affords greater flexibility than selective application of membrane factors in membrane permeability control. The more important process factors will now be briefly addressed.

(1) *Membrane polymer solution concentration.* Several studies have examined the effect of polymer solution concentration on the membrane characteristics of encapsulated systems. Goosen et al (1985) and King et al (1987) found that, although membrane thickness remained unchanged when the PLL solution concentration used for preparing PLL-alginate capsules was increased, membrane strength rose while permeability decreased. These and other findings are summarized in Table 5.2.

It would appear that for most polyelectrolytes, a rise in mechanical strength and a

TABLE 5.2. Effect of polymer solution concentration on membrane permeability, strength, and thickness.

Membrane polymer	Permeant	Effect of solution concentration increase on:			Reference
		Permeability	Strength	Thickness	
PLL			↑	no change	Goosen et al (1995)
PLL	Carbonic anhydrase	↓	↓	↑	King et al (1987)
Chitosan	BSA	↓ (small)			Kim and Rha (1989)
Chitosan	BSA	↓			Polk (1990)
Alginate			↓		Daly and Knorr (1988)
Polyacrylate			↑		Sefton et al (1987)

Note: BSA = Bovine serum albumin
  ↑ = increase
  ↓ = decrease

fall in membrane permeation accompany an increase in membrane polymer concentration. This is due to the greater number of interactions at the membrane/core interface, which produces more compact membranes. This is, however, subject to a threshold solution viscosity beyond which penetration of, and interaction with, the core is inhibited.

(2) *Reaction time.* The reaction time denotes the period when the membrane solution is allowed to react or interact with the capsule core before the capsules are separated from the polymer solution and washed. This factor has been shown to be of great importance in the PLL-alginate system, where prolongation of reaction time from 3 to 40 min produced thicker and more compact membranes with correspondingly lower permeability (King et al 1987). Apparently, the higher the reaction time, the greater the opportunity afforded the membrane solution to penetrate and interact with the core. However, reaction time has no influence on chitosan membrane permeability in the chitosan-alginate system (Shioya and Rha 1989, Polk 1990). On the other hand, the alginate membranes of alginate-chitosan capsules exhibited improved strength when reaction time was increased (Daly and Knorr 1988), but the authors did not report any permeation data.

(3) *pH.* This is a handy tool for modulating polyelectrolyte membranes. The cationic polymer, chitosan, appears to have attracted most attention. This polymer is soluble at low pH, but its solubility decreases as alkaline pH is approached. Furthermore, the micropores of chitosan membranes assume minimum dimensions at low pH, but at about pH 6 and higher, repulsive forces cause the pores to open up, and consequently, membrane permeation rises (Kim and Rha 1989, Okhamafe et al 1996). Polk (1990) reported lower membrane strength and flexibility as chitosan pH was lowered from 5.5, but the work of Kim and Rha (1989) shows that this effect notwithstanding, membrane permeability decreased as pH was lowered from 6.0 to 3.2. Ongoing studies in our laboratories further confirm that the permeation characteristics of chitosan membranes can be modulated by varying the polymer solution pH.

(4) *Additives.* In polymer science, the use of additives is undoubtedly one of the most convenient techniques for modulating the permeation and mechanical properties of polymeric membranes. Additives, however, have enjoyed only limited applications in encapsulated cell studies, presumably because of uncertainties regarding their biocompatibility and effect on cell viability.

Two main types of additives have attracted the attention of investigators. First are selected ionic additives, which can be employed to modify the pore dimensions and frequency, as well as the charge density of polyelectrolyte membranes. Sodium chloride was found to enhance the permeability of chitosan membranes to BSA, but the effect was greater for the low molecular weight chitosan than for the higher molecular weight type (Shioya and Rha 1989, Kim and Rha 1989). The effect is thought to be largely due to the anion (Cl^-), which neutralises some of the cationic charge on chitosan, thus limiting its capacity to interact with the anionic core to form compact, thick, and strong membranes. Another probable factor is that reducing the cationic charge density in the chitosan membranes actually facilitates protein transport across the membrane because of lower repulsive forces along the diffusion pores and pathways. The second type of additive is plasticizers. Plasticization results in the severance of the polymer chain segments of a membrane as the plasticizer molecules become sandwiched between the segments. Consequently, polymer chain or segmental mobility increases, which manifests as larger and/or more diffusion pathways as well as increased flexibility (Dechesne et al 1982, Okhamafe and York 1983, 1985a, 1980, Okhamafe and Iwebor 1987, Porter 1982, Sakellariou et al 1986, Rowe et al 1984). Plasticization has generally been found to cause increased permeability (Okhamafe and York 1983) while minimizing coating/membrane defects (Okhamafe and York

1985b). The polyols—polyethylene glycols, propylene glycol, glycerol, etc.—are suitable plasticizers for cellulosic polymers (Okhamafe et al 1983, Sakellariou et al 1986) and are generally biocompatible and noncytotoxic.

A summary of the influences of certain additives on capsule membrane permeability and mechanical properties is listed in Table 5.3.

Miscellaneous factors. There are several other factors that may be exploited to modify membrane permeability. For example, Shioya and Rha (1989) have observed that purifying chitosan and CMC solutions by dialysis lowers membrane permeability. Similarly, Goosen et al (1985) showed that using an alginate of higher purity as the core in PLL-alginate microcapsules yielded more spherical and less defective capsule membranes. Furthermore, capsule swelling ranging up to 200% has been reported for some encapsulated cell systems (Goosen et al 1985, Gharapetian et al 1987, King et al 1987, McKnight 1987, Yoshioka et al 1990). The distended membrane manifested large diffusion pathways and hence increased permeation.

Concluding Remarks

Membrane permeability control is a key requirement in the design of encapsulated cell systems. Cell viability and survival, and hence the therapeu-

TABLE 5.3. Influence of additives on the permeability and mechanical properties of capsule membranes.

Membrane polymer	Additive	Permeant	Effect of additive on:		Reference
			Permeability	Strength	
Chitosan	NaCl	BSA	↑		Shioya and Rha 1989
	Na^{2}Po^4	BSA	↑		Shioya and Rha 1989
Chitosan	NaCl	BSA	↑		Kim and Rha 1989
Chitosan	CaCl			↑	
Alginate	Glucose			↑	Daly and Knorr 1988
Methacrylates	PEG 200		↑		Crooks et al 1990
	Glyerol		↑		Crooks et al 1990
Alginate	Trizma base buffer	BBT	↑	↑	Knorr and Daly 1988
Alginate	Borax			↑	Knorr and Daly 1988

Note: BSA = Bovine serum albumin
     BBT = Brilliant Blue Tartrazine
     ↑ = increase
     ↓ = decrease

tic success of encapulated cell transplants, depend, in large measure, on the capacity of the permselective membrane to afford immunoisolation and adequate transmembrane traffic in nutritional requirements and catabolites. A number of tools are available to the formulator, several of which have been addressed in this chapter, for modulating membrane transport, but the technique or approach adopted should be governed by the type of encapsulated system as well as by the therapeutic objectives and end-use requirements.

References

Aebischer P, Tresco PA, Winn SR, Green LA, Jaeger CB. 1991a. Long-term cross-species brain transplantation of a polymer encapsulated dopamine secreting cell line. Exp Neurol 111:269–275.

Aebischer P, Wahlberg K, Jaeger CB, 1991b. Macroencapsulation of dopamine secreting cells by co-extrusion with an organic polymer solution. Biomaterials 12:50–56.

Aebischer P, Tan S, Deglon N, Heyd B, Zurn A, Baetga E, Sagot Y, Kato A. 1995. Encapsulation of neutrotrophic factor-releasing cells for the treatment of neurodegenerative diseases. Restor Neurol Neurosci 8:65–66.

Chang TMS. 1993. Living cells and microorganisms immobilized by microencapsulation inside artificial cells. In: Goosen MFA, editor. Fundamentals of animal cell encapsulation and immobilization. Boca Raton, Florida: CRC Press. pp 183–196.

Crank, J. 1975. The Mathematics of diffusion. Oxford: Clarendon Press.

Crooks CA, Douglas JA, Broughton RL, Sefton MV. 1990. Microencapsulation of mammalian cells in HEMA-MMA copolymer: effects on capsule morphology and permeability. J Biomed Mater Res 24:1241–1247.

Croughan MS, Hamel J-F, Wang DIC. 1987. Hydrodynamic effects on animal cells grown in microcarrier cultures. Biotechnol Bioeng 29:130–135.

Daly MM, Knorr D. 1988. Chitosan-alginate complex coacevate capsules: effect of calcium chloride, plasticizers, and polyelectrolytes on mechanical stability. Biotechnol Prog 4:76–83.

Dechesne JP, Delporte JP, Jaminet Fr, Venturas K. 1982. Influence of plasticizers on properties of applied films of Eudrajit L30D. J Pharm Belg 37:283–286.

de la Rubin J, Carral A, Montes H, Urquijo JJ, Sanz GF, Sanz MA. 1996. Successful treatment of hepatic veno-occlusive disease in a peripheral blood progenitor cell transplant patient with a transjugular intrahepatic port systemic stent-shunt. Haematologica 81:536–539.

Gentile FT, Doherty EJ, Rein DH, Shoichet MS, Winn SH. 1995. Polymer science for macroencapsulation of cells for central nervous system transplantation. Reactive polymers 25:207–227.

Gharapetian H, Maleki M, O'Shea GM, Carpenter RC, Sun AM. 1987. Polyacrylate microcapsules for cell encapsulation: effects of copolymer structure on membrane properties. Biotechnol Bioeng 30:775–779.

Goosen MFA. 1987. Insulin delivery system and the encapsulation of cells for medical and industrial use. CRC Crit Rev Biocompat 3:1–24.

Goosen MFA. 1993. Toxicity, biocompatibility, and mass transfer effects in immobilized cell systems. In: Goosen MFA, editor. Fundamentals of animal cell encapsulation and immobilization. Boca Raton, Florida: CRC Press. pp 55–78.

Goosen MFA, King GA, Daugulis AJ, Faulkner P. 1990. Multiple membrane encapsulation. U.S. Patent 4,942,129.

Goosen MFA, O'Shea GM, Gharapetian HM, Chou S, Sun AM. 1985. Optimization of microencapsulation parameters: semipermeable microcapsules as a bioartificial pancreas. Biotech Bioeng 27:146–150.

Jarvis AP, Grdina TA. 1983. Production of biologicals from microencapsulated cells. Biotechniques 1:24–29.

Kim SK, Rha C. 1989. Transmembrane permeation of proteins in chitosan capsules. In: Skjaek-Braek G, Anthosen T, Sandford P, editors. Chitin and chitosan. Sources, chemistry, biochemistry, physical properties and applications. London: Elsevier. pp 635–642.

King GA, Daugulis AJ, Faulkner P, Goosen MFA. 1987. Alginate-polylysine microcapsules of controlled membrane molecular weight cut off for mammalian cell culture engineering. Biotechnol Progr 3(4):231–239.

Knorr D, Daly M. 1988. Mechanics and diffusional changes observed in multilayer chitosan/alginate coacervate capsules. Process Biochem 23:48–50.

Lim F, Sun AM. 1980. Microencapsulated islets as bioartificial endocrine pancreas. Science 210:908–911.

McKnight CA. 1987. Chemical modification of chitosan for the microencapsulation of mammalian cells. M.Sc Thesis. Queen's University, Kingston, Canada.

Okhamafe AO, Amsden B. Chu W. Goosen MFA. 1996. Modulation of protein release from chitosan-alginate microcapsules using the pH-sensitive polymer hydroxypropyl methylcellulose acetate succinate. J Microencap 13:497–508.

Okhamafe AO, Goosen MFA. 1993. Control of membrane permeability in microcapsules. In: Goosen MFA, editor. Fundamentals of animal cell encapsulation and immobilization. Boca Raton, Florida: CRC Press. pp 55–78.

Okhamafe AO, Iwebor HU. 1987. Moisture permeability mechanisms of some aqueous-based tablet film coatings containing soluble additives. Pharmazie 42:611–613.

Okhamafe AO, York P. 1985a. Interaction phenomena in some aqueous-based tablet coating polymer systems. Pharm Res 2:19–23.

Okhamafe AO, York P. 1985b. Stress crack resistance of some pigments and unpigmented tablet film coating systems. J Pharm Pharmacol 37:449–454.

Okhamafe AO, York P. 1987. Interaction phenomena in pharmaceutical film coatings and testing methods. Int J Pharm 39:1–21.

Okhamafe AO, York P. 1988. Studies of interaction phenomena in aqueous-based film coatings containing soluble additives using thermal analysis techniques. J Pharm Sci 77:435–444.

O'Neill WP. 1980. Membrane systems. In: Kydonieus AF, editor Controlled release technologies: methods, theory and applications. Vol. 2. Boca Raton, Florida: CRC Press. Chap 4.

O'Shea GM, Goosen MFA, Sun AM. 1984. Prolonged survival of transplanted islets of Langerhans encapsulated in a biocompatible membrane. BBA 804:113–117.

Polk AE. 1990. Development of chitosan-alginate microcapsules for the oral delivery of vaccines in aquaculture. M.Sc. Thesis, Queen's University, Kingston, Canada.

Porter SC. 1982. The practical significance of the permeability and mechanical properties of polymer films used for the coating of pharmaceutical dosage forms. Int J Pharm Technol Prod Manuf 3:21–25.

Posillico EG. 1986. Microencapsulation technology for large-scale antibody production. Bio-Technology 4:114–117.

Richards JH. 1985. The role of polymer permeability in the control of drug release. In: Comyn J, editor. Polymer permeability. London: Elsevier, pp 217–267.

Rogers CE. 1976. Structural factors governing controlled release. In: Paul DR, Harris FW, editors. ACS Symposium Series 33. Washington, D.C.: American Chemical Society. pp 15–25.

Ronel SH, D'Andrea MJ, Hashiguchi H, Klomp GF, Dobelle WH. 1983. Macroporous hydrogel membranes for a hybrid artificial pancreas. I. Synthesis and chamber fabrication. J Biomed Mater Res 17:855–863.

Rowe RC. 1976. The effect of molecular weight on the properties of films prepared from hydroxypropyl methylcellulose. Pharm Acta Helv 51:330–334.

Rowe RC, Forse SF. 1980. The effect of polymer molecular weight on the incidence of film cracking and splitting on film-coated tablets. J Pharm Pharmacol 32:583–584.

Rowe RC, Kotaras AD, White EFT. 1984. An evaluation of the plasticizing efficiency of the dialkyl phthalates in ethylcellulose films using the torsional braid pendulum. Int J Pharm 22:57–62.

Sakellariou P, Rowe RC, White EFT. 1986. An evaluation of the interaction and plasticizing efficiency of the polyethylene glycols in ethylcellulose and hydroxypropyl methylcellulose films using torsional braid pendulum. Int J Pharm 3:55–64.

Sato, K. 1980. The internal stress of coating films, Prog. Org. Coat., 8, 143–148.

Sefton MV, Blysniuk J, Broughton RL, Dawson RM, Sugamori ME. 1987. Microencapsulation of mammalian cells in a water-insoluble polyacrylate by co-extrusion and interfacial precipitation. Biotechnol Bioeng 29:1135–1143.

Shioya T, Rha C. 1989. Transmembrane permeability of chitosan/carboxymethyl cellulose capsule. In: Skjaek-Braek G, Anthosen T, Sandford P, editors. Chitin and chitosan. Sources, chemistry, biochemistry, physical properties and applications. London: Elsevier. pp 627–634.

Sun AM, O'Shea GM, Goosen MFA. 1983. Injectable microencapsulated islet cells as a bioartificial pancreas. Appl Biochem Biotechnol 9:87–95.

van Wezel AL. (1967). Growth of cell strains and primary cells on microcarriers in homogenous culture. Nature (London) 216:64–69.

Yoshioka T, Hirano R, Shioya T, Kako M. 1990. Encapsulation of mammalian cell with chitosan-CMC capsule. Biotechnol Bioeng 35:66–72.

Young DV. (1993). Inverted microcarriers: using microencapsulation to grow anchorage-dependent cells. In: Goosen MFA, editor. Fundamentals of animal cell encapsulation and immobilization. Boca Raton, Florida: CRC Press. pp 243–266.

Young DV, Dobbels S, King L, Deer F, Gillies SD. 1989. Inverted microcarriers: Using microencapsulation to grow anchorage-dependent cells in suspension. Biopharm 2:34–41.

Zondervan GJ, Hoppen HJ, Pennings AJ, Fristchy W, Wolters G, van Schilfgaarde R. 1992. Biomaterials 13:136–144.

6
Biocompatibility Issues

Paul de Vos and Reinout van Schilfgaarde

Introduction

The possibility of transplanting cells in immuno-protective membranes for organ replacement dates back to 1933. At that time, Bisceglie (Bisceglie 1933) replaced the endogenous pancreas by transplanting insulin-producing tissue encapsulated in a semipermeable but immunoprotective membrane to study the effects of absence of vascularization on the survival of tissues. It took, however, until 1943 before the concept of immunoisolation was established by Algire (Algire 1943), who recognized that graft failure was delayed by encapsulating allo- and xenogenic tissues before transplantation. Algire was also the first to illustrate the importance of biocompatibility since he found that graft failure was always accompanied by cellular overgrowth of the membranes. At present, transplantation of cells in immunoisolating devices is under study for the treatment of a wide variety of diseases, including hemophilia B (Liu et al 1993), anemia (Koo and Chang 1993), dwarfism (Chang et al 1993), kidney (Cieslinski and Humes 1994) and liver (Uludag and Sefton 1993) failure, pituitary (Colton 1995) and central nervous system insufficiencies (Aebischer et al 1994), and diabetes mellitus (Lim and Sun 1980). When discussing biocompatibility aspects in this chapter, we will mainly focus on immunoisolation of pancreatic islets for the treatment of diabetes mellitus. We make this restriction for three reasons. The first is that most of the biocompatibility-related problems have been discovered in the application of immunoprotected insulin-producing cells, or the so-called bioartificial pancreas, probably as a conse-quence of the large number of groups involved in this type of research. Second, our own experience with biocompatibility research of immunoisolating capsules is mostly restricted to immunoprotected insulin-producing tissue. And third, most of the considerations regarding the biocompatibility of the bioartificial pancreas are also pertinent to other applications of immunoisolating devices.

Intravascular and Extravascular Approaches

There are several basically different approaches for the transplantation of immunoisolated islets. One major distinction is between intravascular and extravascular devices (Figure 6.1). The intravascular device is the oldest concept and derives from the principle of perfusion chambers (Chick et al 1975). The device is usually composed of a microporous tube with blood flow through its lumen and with a housing on its outside containing the implanted tissue. The device is implanted by vascular anasto-moses to the vessels of the host. The concept of extravascular devices does not require vascular anastomoses, since it is based on the principle of diffusion chambers (Scharp et al 1984). As with intravascular devices, the extravascular devices have initially been designed as macrocapsules with large numbers of islets enveloped together in one immunoisolating membrane, and thus contained together in one compartment. The geometry may be planar in the form of a flat, circular double layer or tube-like as a so-called hollow fiber (Scharp et al

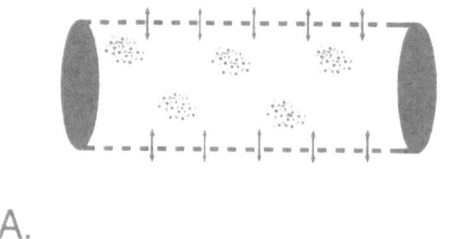

A.

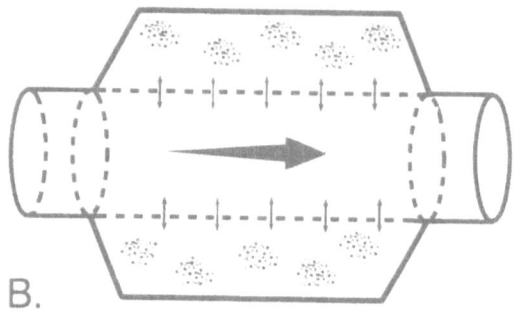

B.

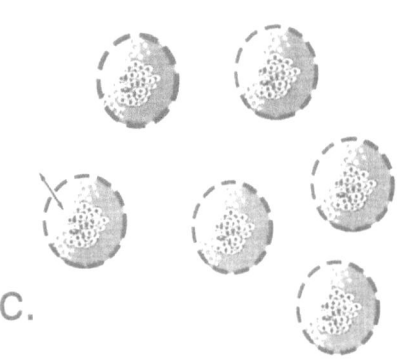

C.

FIGURE 6.1. Immunoisolating devices. A, Immunoisolation in extravascular macrocapsules. Groups of islets are contained in tube- or, alternately, disc-shaped hollow devices. B, Immunoisolation in intravascular devices. Groups of islets are seeded on the outside of artificial capillaries and in close contact with the bloodstream as the devices are implanted as an arterial or arteriovenous conduit. C, Immunoisolation in microcapsules, i.e., individual encapsulation of the islets.

1984). An alternative form of extravascular encapsulation is microencapsulation, which involves the packing of single islets in their own individual, spherical microcapsules.

The type and significance of biocompatibility-related problems vary not only with the chosen principle, i.e., the intravascular or the extravascular technique, but also with the geometry of the device (Woodward 1982), the implantation site (Siebers et al 1990), and the materials applied (Zekorn et al 1994). Since these factors vary considerably among devices, it is not feasible to give a complete overview of all devices and their individual biocompatibility problems. Therefore, in the following sections we have chosen to review the biocompatibility issues of a few well-characterized intravascular and extravascular devices, as an illustration of the problems that may occur. At the same time, we fully recognize that devices may already have been developed in which such problems have been solved.

Biocompatibility and Immunoisolating Devices

Biocompatibility is usually defined as the ability of a biomaterial to perform with an appropriate host response in a "specific application" (Williams 1987). With fully artificial organs such as artificial hips, knees, or middle ears, this definition is easy to interpret. It is, however, far from simple to interpret with bioartificial systems such as the encapsulated cell technique. With extravascular devices it is assumed that a fully biocompatible system is achieved with membranes that elicit no or not more than a minimal foreign body reaction, since overgrowth on the surface of the membrane interferes with optimal diffusion of nutrients and metabolites (De Vos et al 1993, Fritschy et al 1994, Soon-Shiong et al 1991, Zimmermann et al 1992). With intravascular devices the overgrowth of the membrane is considered to be of minor importance. Here, a permanently patent tubular membrane without thrombotic or other adherent layers is required for optimal biocompatibility. Irrespective whether the approach is intravascular or extravascular, the host response to the biomaterial is generally considered as the only and single response causing biocompatibility problems. However, with immunoisolating devices there is interaction not only between the biomaterial and the tissues of the exterior host environment but also between the biomaterial and the encapsulated donor tissue. Recently, some studies have shown that encapsulation directly interferes with the function and viability of the encapsulated tissue (Chang et al 1994, De Vos et al 1997b, Shimi et al 1991). Although this aspect is not covered by the current definition of biocompatibility, it should be considered a true biocompatibility issue since long-term survival of the tissue is required for this "specific application." Therefore, this aspect of biocompatibility will be taken into consideration in the following discussion of biocompatibility issues.

Intravascular Devices

The most intensively studied intravascular device is the modified diffusion chamber of Chick et al (Chick et al 1975). It is technically advanced and has been tested extensively in small (Sun et al 1977) as well as in large animals (Maki et al 1993, Sun et al 1977). The original device was composed of a number of small-diameter artificial capillaries contained by one large-diameter tube. The artificial capillaries were composed of polyacrylonitrile and polyvinylchloride copolymer (PAN-PVC) ultrafiltration capillaries (Colton 1995), and the remaining lumen of the large-diameter tube, i.e., the outside of the artificial capillaries, was loaded with hormone-producing cells. The design permits for close contact between the islets and blood, which are separated only by the microporous walls of the capillaries. These devices were found to induce normoglycemia in diabetic rats (Sun et al 1977), dogs (Maki et al 1993), and monkeys (Sun et al 1977) but required systemic anticoagulation. The duration of this normoglycemia was usually restricted to several hours, and successes of a somewhat longer duration were exceptional. Clotting of the blood in the lumen of these small-diameter artificial capillaries proved to be a major obstacle, in spite of anticoagulant medication in massive doses. This thrombus formation was an early sign of insufficient biocompatibility, and has led to the use of tubular membranes with larger diameters, in the hope of minimizing or eliminating clot formation in the absence

of systemic anticoagulation. The present device is composed of a single coiled tubular membrane with an internal diameter of 5–6 mm. The membrane is somewhat modified but still composed of PAN-PVC with a nominal molecular weight cut-off of 50 kD. The membranes are asymmetric with a dense but thin skin which, at the luminal side, is covered by an open sponge-like matrix. The skin provides solute rejection (i.e., solutes neither pass nor adhere), and hydraulic permeability properties, while the matrix contributes to mechanical strength and to permeability for low molecular weight substances. The membrane is enveloped coaxially by a second, nonpermeable, acrylic housing, containing the islets. The membranes cannot be sutured, and are therefore connected to PTFE grafts for anastomosing the device to the recipient's vessels, for instance the iliac vessels. This approach was found to be rather successful, since these devices, implanted as high-flow arteriovenous fistulas, could remain patent for periods of 7 weeks in the absence of systemic anticoagulant therapy (Galletti et al 1981). This success is in part explained by the high flow rates through the device, which prevent adhesion of cells to the membranes or collection of those cells in the immediate vicinity (Colton and Avgoustiniatos 1991). However, high-flow arteriovenous fistulas are not without risk, and much longer patency rates are required for effective applicability. Obviously, more thromboresistant materials are required for this type of device.

Allo- and xenogenic islets in the wide-bore devices were successfully transplanted to diabetic dogs (Lanza and Chick 1997, Maki et al 1995, Maki et al 1996), but the efforts to improve the blood-compatibility have probably interfered with the efficacy of the device as an implantation site for islets. This view is derived from the following observations. First, two devices per recipient instead of one were required to achieve adequate secretion capacity while maintaining the same numbers of islets per device (Lanza et al 1992b). Furthermore, it has not been possible to load the space between the membranes and the housing with an islet-tissue density higher than 5–10% of the volume (Colton 1995), in spite of the fact that the large lumen is exposed to arterial blood with optimal concentrations of nutrients and oxygen. It is quite plausible that the high flow rates through the device, which are required to keep the device patent, do not allow sufficient exchange of glucose, insulin, and nutrients to permit long-term survival and adequate function of the islets.

There are also indications that the materials applied in this kind of device are not only thrombogenic but also insufficiently compatible with long-term functional survival of the islets. For example, the PTFE as used for vascular anastomosis of the device has been shown to induce IL-1 production by macrophages (Krause et al 1990), a cytokine that is lethal for islets. It is quite plausible that IL-1 causes loss of high numbers of islets during the period between implantation and complete integration of the prothesis, since macrophages are usually the first cells to invade the implant (Remes and Williams 1991). This is another explanation for the fact that so many islets divided over two devices are required for maintaining normoglycemia in dogs.

Although the intravascular devices have shown some degree of success, the problems mentioned above must be solved if clinical application is to be considered. Even then, the complications associated with any type of vascular prosthetic surgery remain a serious threat, such as thrombosis, either primary or secondary to intimal hyperplasia at the venous anastomosis, defects of the device, or infection. This is a major drawback for wide application in large numbers of diabetic patients, since any alternative to conventional insulin treatment should preferably carry no additional risk.

Extravascular Devices

The surgical risks are much lower with extravascular than with intravascular devices. Biocompatibility problems are usually deleterious to the function of the encapsulated tissue only, and have no or only minimal risk for the recipient. The relative safety is an important advantage of extravascular over intravascular devices, but the problems at the level of tissue-material interactions are of similar complexity. These biocompatibility problems are usually related to toxicity and activation of nonspecific foreign body reactions, resulting in fibrotic overgrowth with subsequent necrosis of the encapsulated tissue. In the following sections we will discuss the biocompatibility issues for macrocapsules and microcapsules separately.

Macrocapsules

Macrocapsules usually have the geometry of planar membranes or hollow fibers. They can be implanted with minimal surgery in different sites such as the peritoneal cavity (Archer et al 1980, Lanza et al 1994), the subcutaneous site (Lacy et al 1991, Scharp et al 1994), or the renal capsule (Siebers et al 1990). Also, they can be readily retrieved, and they can be produced in a relatively simple way.

Many different biomaterials have been applied to the production of macrocapsules (Table 6.1). The hollow-fiber geometry is usually preferred over the planar membranes for its smaller foreign body response (Woodward 1982). Most studies on hollow fibers use fibers made of PAN-PVC (Colton and Avgoustiniatos 1991, Lanza et al 1992b), similar to those used in intravascular devices. They have been produced with a smooth, fenestrated outer skin with the same spongy matrix as applied in the intravascular devices, but now this matrix covers the outside rather than the luminal side of the device. The design with the smooth outer skin proved to be the most successful, since it provokes

TABLE 6.1. Main components of capsules proposed for macroencapsulation of pancreatic islets.

Membrane	Geometry	Proposed by
nitro-cellulose acetate planar	Algire 1943	
polycarbonate	planar	Strautz 1970
2-hydroxyethyl methacrylate (HEMA)	planar	Klomp et al 1979
polyacrylonitrile and polyvinylchloride	tubular	Archer et al 1980
cellulose	tubular	Zekorn et al 1989
polyamide	tubular	Zekorn et al 1989
acrylonitrile and sodium-methallyl sulfonate (AN69)	tubular	Kessler et al 1991
alginate-based	tubular	Lanza et al 1995b

much less fibrosis than the rough fenestrated surface, which allows host tissue to grow into the spongy matrix. Many modifications of this concept have been proposed in order to further improve the biocompatibility. One of these was the coating ("grafting") of the membranes with poly(ethyleneoxide) to reduce protein adsorption (Shoichet et al 1994).

Initial studies with macrocapsules were not very successful. This was not so much the consequence of fibrotic overgrowth as of the aggregation of the encapsulated tissue into large clumps (Lacy et al 1991). Extensive necrosis occurred in the center of the clumps as a result of diffusion limitations for nutrients. This problem was readily solved by preventing contact between the encapsulated tissue elements through permanent solitude immobilization in a matrix such as collagen (Metrakos et al 1994), Ca-alginate (Lacy et al 1991), or chitosan (Colton 1995) before encapsulation. The appropriate matrix for a certain cell type is usually selected empirically. It is probably determined by the charge the cells can manage, e.g., alginate is negatively charged, while chitosan is positively charged. This example illustrates the interactions of the biomaterial and the donor tissue, which, obviously, are quite important for long-term function of the donor tissue. This biocompatibility issue is often ignored.

Islets are usually immobilized in alginate before encapsulation in PAN-PVC fibers. The islet density is kept quite low and never exceeds 5–10% of the volume fraction since it has been found that viability is substantially reduced with higher densities. When transplanted in BB-rats, islet-containing fibers were found to induce normoglycemia for up to 8 months, but glucose tolerance remained disturbed and decreased rapidly with time, in spite of prolonged normoglycemia (Lanza et al 1992a). A major factor is that hollow fibers, as a consequence of their shape, tend to break when forced to bend under physiological stress (Colton 1995, Lanza et al 1992b). Another factor is the low number of islets implanted, which may be insufficient for achieving long-term graft survival. The use of higher numbers, however, is impractical, since enormous lengths of fibers would be required as a consequence of the low seeding density of the membranes. A modification is to use tubes with a wider lumen of several millimeters. This, however, is associated with a substantial increase in diffusion distance, which enhances rather than reduces problems like necrosis as a consequence of insufficient nutrient supply, and accumulation of waste materials. Some success with these devices has been reported in diabetic rats, but extreme amounts of islets, i.e., 30,000 islets/kg, were required to maintain normoglycemia for only a few months (Lanza et al 1992b). In addition, the membranes still provoke a foreign body reaction that results in overgrowth by a thin but avascular fibrotic cellular infiltrate after intraperitoneal implantation, which implies further limitations in diffusion capacity and in life span of the islets.

During the past few years, an interesting trend has been the growing number of groups applying hydrogels for macroencapsulation. Hydrogels provide a number of features that are advantageous for the biocompatibility of the membranes. First, as a consequence of the hydrophilic nature of the material, there is a low or zero interfacial tension with surrounding fluids and tissues, which minimizes the protein adsorption and cell adhesion. Furthermore, the soft and pliable features of the gel reduce the mechanical or frictional irritations to surrounding tissue. And, finally, they provide a high degree of permeability for low molecular weight nutrients and metabolites, which is required for the optimal functioning of living cells.

Many hydrogels have been applied, such as gels prepared of alginate (Lanza et al 1995a, Lanza et al 1996), HEMA (Klomp et al 1979, Sefton 1993), and a copolymer of acrylonitrile and sodium-methallyl sulfonate, AN69 (Kessler 1992). Primary attention has been focused on the hydrogel membrane AN69, which was shown to be hemocompatible (Sevastianov and Tseytkina 1984) and to induce only minimal fibrosis in the peritoneal cavity of rats (Kessler et al 1997). Initial experiments with this membrane were not very successful, not so much because of fibrosis as, mainly, because of the low permeability of the membrane for insulin (Kessler et al 1997). Recently, Kessler et al (Kessler et al 1997) have solved this problem quite elegantly by providing the membrane with a more hydrophobic surface through Corona discharge, a technique inducing the formation of radicals and ionic interactions between the membrane constituents and air molecules. Fewer molecules adhered to the surface of the membrane, which improved not only the permeability for insulin but also the long-term biocompatibility of the membranes. One year after implantation in rats, only a few macrophages were found stuck to the membrane's surface. Unfortunately, data on the efficacy of the membrane in the bioartificial pancreas are not yet available.

Microencapsulation

Microcapsules provide a number of advantages over macrocapsules. They are durable and difficult to mechanically disrupt and have a spherical configuration associated with a better surface/volume ratio than the disk or tube geometry, which offers the major advantage of better diffusion. During recent years, normoglycemia has been reported after transplantation of alginate-PLL microencapsulated allo- and xenogenic islets in chemically-induced and in autoimmune diabetic rodents (O'Shea and Sun 1986, Wijsman et al 1992, Fritschy et al 1994), dogs (Soon Shiong et al 1992, Wahoff et al 1994), monkeys (Sun et al 1996, Zhou et al 1994) and, recently, even in a human diabetic patient (Soon Shiong et al 1994). However, normoglycemia was always of limited duration and reportedly varied from several days to several months, although according to one report it lasted 2 years (O'Shea et al 1984).

Microcapsules are almost exclusively produced from hydrogels, obviously for the abovementioned arguments. Table 6.2 lists the most popular biomaterials applied for microencapsulation. A major distinction exists between water-soluble polymers such as alginates and water-insoluble polymers such as HEMA-MMA. Water-insoluble polymers are preferred by a number of groups (Sefton 1993, Wells et al 1993), since it is assumed that they are more stable than water-soluble polymers after implantation. However, a major obstacle in using water-insoluble

polymers for encapsulation of cells is the use of an organic solvent, which usually interferes with cellular function (Sefton 1993, Fritschy et al 1991).

We, like others (Smidsrød and Skjåk-Bræk 1990, Soon Shiong et al 1992, Sun et al 1996), have concentrated on alginate-based capsules, since they have been found, repeatedly, not to interfere with cellular function, and have been shown to be stable for years in small and large animals, and recently also in humans, in spite of their solubility in aqueous solutions. The most commonly used alginate-based capsule is the alginate-polylysine system. The technique is based on entrapment of individual islets in an alginate droplet, which is transformed into a rigid bead by chelification in a divalent cation-rich, mostly Ca^{2+}, solution. Alginate-molecules are composed of mannuronic (M) and guluronic (G) acids and can be connected by Ca^{2+} through binding of consecutive blocks of G-molecules on each of the two molecules. After chelification, the beads are provided with a polylysine membrane by simply suspending the beads in a poly-L-lysine (PLL) solution. During this step, PLL binds to mixed sequences of G and M in the alginate molecules (Bystricky et al 1990). This induces the formation of complexes at the capsule surface, consisting of α-helical PLL surrounded by superhelically orientated polysaccharide chains (Bystricky et al 1990, Bystricky et al 1991). The presence of these complexes decreases the porosity of the capsule. By varying the molecular weight and the concentration of the polylysine and the incubation time, we can modulate the porosity of the capsule membrane (King et al 1987, Vandenbossche et al 1993). Usually, a 10–minute incubation in a solution of 0.1% polylysine with a molecular weight of 22 kD is sufficient to provide the beads with an immunoprotective membrane, but it should be emphasized that the binding of polylysine de-

TABLE 6.2. Main components of capsules proposed for microencapsulation of pancreatic islets.

Main component of the capsule	Source	Initially proposed by
Alginate	Alga	Lim and Sun 1980
Chitosan	Alga	Zielinski and Aebischer 1994
Agarose	Alga	Iwata et al 1989
Poly(hydroxyethylmetacrylate-methyl methacrylate)(HEMA-MMA)	Synthetic	Dawson et al 1987
Copolymers of acrylonitrile (AN69)	Synthetic	Kessler et al 1991

pends not only the incubation time and the molecular weight of the polylysine but also on the type and concentration of alginate applied (De Vos et al 1993), as well as on the incubation temperature (Vandenbossche et al 1993). In a final step, to provide biocompatibility, the capsules are suspended in a solution of alginate or other negatively charged molecules (De Vos et al 1993) to bind all positively charged polylysine residues still present at the capsule surface.

Although many see the procedure as technically simple, it is far from simple to produce a microcapsule that is smooth and mechanically stable, which are the basic characteristics of a biocompatible capsule. In our laboratory we found that high- rather than low-viscosity alginates produce smooth beads with no obvious tails or strains. A modification was to use alginate with an intermediate rather than low G-concentration since we found that low-G alginates swell after implantation, with subsequent breakage of the PLL membrane and overgrowth of the capsules. The stability is also determined by the PLL step, since shorter incubation periods, lower PLL concentrations, lower PLL molecular weight than mentioned above, and incubation at 4°C were associated with an increase in the number of capsules with broken membranes. Finally, we avoid the so-called liquefication of the inner core of the capsule with EGTA or citrate (Fritschy et al 1991), since we have never found any effect of such a treatment on the capsule core, and since we observed that many capsules lose their integrity during the treatment. But tissue reactions continued to interfere with success, in spite of all the improvements in the stability of the capsules. We found that other factors, not directly related to the production procedure as such, determine the tissue reaction that evidenced insufficient biocompatibility. These factors are discussed in the following sections.

Usually, graft failure is interpreted to be the consequence of insufficient biocompatibility of the capsule components, resulting in progressive fibrotic overgrowth of the capsules (Fritschy et al 1994, Soon Shiong et al 1991, Zimmermann et al 1992) and subsequent necrosis of the islets. Consequently, there have been many efforts to solve this problem, but there is still dispute as to which factor is causing the bioincompatibility. Mannuronic acid has been identified as the major initiator of the for-

eign body reaction by some (Soon Shiong et al 1991), while others found guluronic acid to be associated with more severe fibrotic overgrowth (Clayton et al 1991, De Vos et al 1997a). Also, PLL has been suggested as the major initiator of the fibrotic reaction. It should be mentioned that most of these studies were performed with crude alginates, which are known to contain contaminants that may provoke an inflammatory reaction. Consequently, in some studies, the reactions against contaminants are falsely interpreted as reactions against capsule components.

A pertinent factor in the biocompatibility of alginate-PLL capsules is the purity of the alginate applied. We have developed a purification procedure for alginates which, briefly, is composed of a number of filtration, precipitation, and extraction steps (De Vos et al 1997b). Purification substantially improves the biocompatibility, as shown by using purified instead of crude alginate in implantation studies of empty capsules as well as of isogenic and allogenic islet-containing capsules in the abdominal cavity of rats. Empty capsules remained nonadherent and freely floating in the abdominal cavity up to 1 year after implantation, while the percentage of capsules showing overgrowth remained usually well below 10%. Islet allografts encapsulated with purified alginate induced normoglycemia in streptozotocin diabetic recipients. Graft function, however, was limited to periods of time varying from 6 to 20 weeks. Graft failure was not caused by fibrotic overgrowth, since this was usually below 10%. Nor was it caused by insufficient immunoprotection, since similar results were obtained with isografts encapsulated with purified alginate (Figure 6.2), both in terms of success rates as well as percentages of fibrotic overgrowth (Table 6.3).

These studies clearly demonstrate that biocompatibility is substantially improved by applying purified alginate. However, they also demonstrate two other critical problems. First, the duration of graft function remains limited in spite of optimal biocompatibility and virtual absence of fibrotic overgrowth. Second, fibrotic overgrowth is always found in a small portion of the capsules, in spite of using purified alginate. This latter observation indicates that biocompatibility of the microcapsules is influenced not only by the chemical composition and purity of the materials applied, but also by individual imperfections of a more physical nature.

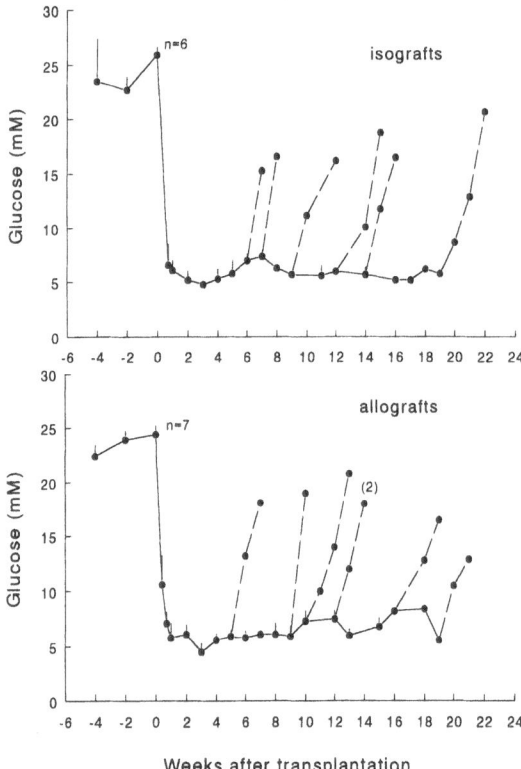

FIGURE 6.2. Nonfasting blood glucose concentrations in streptozotocin diabetic AO-rats after implantation of microencapsulated isografts (top) and allografts (bottom). The broken lines indicate animals returning to hyperglycemia. Values represent mean ± SEM.

To investigate whether such imperfections have any significance and if they can be held responsible for the small but very present portion of islet-containing capsules with overgrowth, we have designed a lectin binding assay by selecting a suitable lectin that combines a high affinity for pancreatic islets with a molecular weight that is prohibitive for diffusing into the microcapsules (De Vos et al 1996a, 1996b). This assay enables us to identify imperfect capsules by specifically and individually labeling those islets that have been inadequately encapsulated. The assay has been applied in several studies with encapsulated islets both in vitro and in vivo, resulting in some interesting observations. Reducing the capsule's diameter from 800 to 500 μm was associated with an increase of inadequately encapsulated islets of 7 to 25%. This in vitro observation was subsequently confirmed to have significant in vivo consequences, since the transplant success rate was 5/5 with 800–μm capsules but only 1/7 with 500–μm capsules (De Vos et al 1996a).

Since the production of small capsules is a major aim in order to improve the functional results of microencapsulation, the above observations have stimulated us to search for a means to reduce the formation of inadequate capsules. One pertinent factor is the composition of the alginate applied, since, in vitro, the number of incompletely encapsulated islets was found to be lower with a high-G than with the usually applied intermediate-G alginates. However, when the composition of the alginate was tested in vivo, we found that fibrotic overgrowth was quite severe when alginate with a high guluronic acid content was applied, while capsules composed of alginate with an intermediate guluronic acid content were found to remain nonadherent and freely floating in the peritoneal cavity. This difference was mainly caused by inadequate binding of polylysine to the high guluronic acid alginate (De Vos et al 1997a). Thus, we should reinforce the interaction of PLL with high-G alginate or search for polycations that interact more effectively with high-G than PLL does, since this type of alginate appears to be advantageous for the process of microencapsulation (De Vos et al 1996b).

Our observations on the bioincompatibility of high-G capsules do not corroborate the findings of Soon Shiong et al (Soon Shiong et al 1991), who

TABLE 6.3. Weeks of graft survival, implanted islet volume, recovery rates, and percentage of capsules with overgrowth, after implantation of microencapsulated rat islet iso- and allografts (Reproduced from De Vos et al 1997b with permission of the publisher).

Encapsulated Transplant	n	Graft survival (weeks)	Implanted islet volume (μl)	Percentage of recovery	Percentage of overgrowth
Isograft	6	11.7 ± 1.8	11.8 ± 0.8	82.0 ± 8.1	10.7 ± 1.9
Allograft	10	12.0 ± 1.6	11.5 ± 0.4	86.1 ± 4.7	10.5 ± 3.6

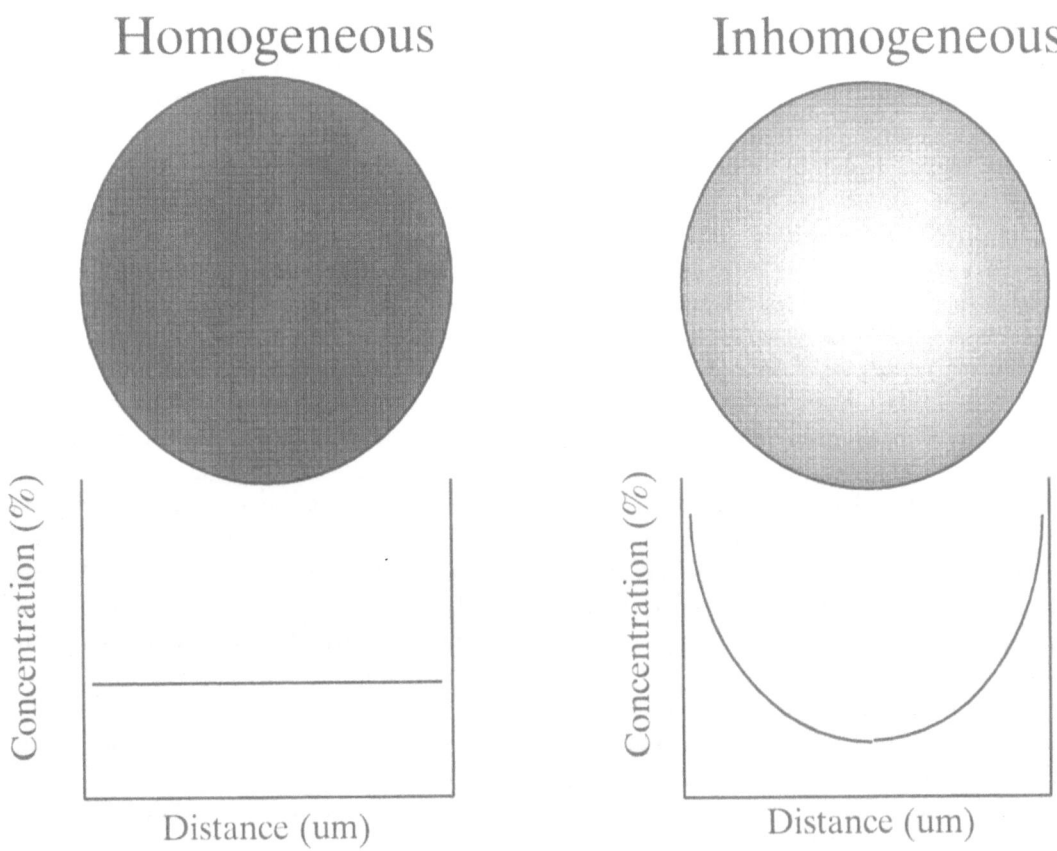

FIGURE 6.3. Alginate concentration profiles in homogeneous (left) and inhomogeneous (right) alginate beads. Inhomogeneous beads provide a reinforcement of the diffusion limitations for immune-active components.

found high-G capsules to be more biocompatible than capsules prepared with alginate with low-G content. This discrepancy should probably be explained by different characteristics of the capsule types used. Soon-Shiong's group (Thu et al 1996a, 1996b) used alginate-PLL capsules with an inhomogeneous alginate core (Figure 6.3), i.e., higher alginate concentrations at the periphery than in the center of the capsules, while we used alginate-PLL capsules with a homogeneous capsule core. The surface of the capsules with an inhomogeneous core has a higher charge density than does the surface of homogeneous capsules (Thu et al 1996b). Consequently, the efficacy of PLL binding is different between the two capsule types, which implies better PLL binding and, thus, a better degree of biocompatibility with inhomogeneous capsules.

As follows from the above, seemingly minor modifications in the encapsulation procedure may

have an important impact on the capsule's biocompatibility and thus on the outcome of the graft. In spite of this recently gained knowledge, the abovementioned factors are poorly standardized. As a consequence, there are many different alginate-PLL procedures, each resulting in capsules with different porosities and with specific chemical and mechanical characteristics which, regretfully, are rarely documented. Obviously, these differences contribute to the enormous variations in reported success rates of encapsulated islet allo- and xenografts.

In spite of differences in the procedures applied, indefinite graft survival has never been demonstrated, which clearly suggests the presence of a fundamental problem. With islet-containing microcapsules it has been assumed that indefinite survival would be achieved with capsules that elicit a minimal foreign body reaction (Fritschy et al 1994,

Soon Shiong et al 1991, Zimmermann et al 1992). However, even when the foreign body reaction affected only an insignificant number of capsules (De Vos et al 1997b), long-term survival of the encapsulated islet graft was not achieved. This phenomenon of graft failure in the absence of overgrowth of the capsules has been reported before (Lum et al 1992, Sun et al 1993) and is usually explained by exhaustion of the graft as a consequence of a too high glycemic load on an insufficient number of transplanted encapsulated islets. In our studies, however, neither the volume of the graft nor the glycemic load on the islets was a causative factor for the graft failure, since with transplantation of a sufficient islet volume of 10 μl, i.e., the equivalent volume of the islets present in the pancreas of a control rat, we found the functional mass of the graft to decrease rapidly not only in diabetic but also in pretransplant normoglycemic recipients. Our results instead suggest that the long-term graft survival is limited by interference of the capsule with the blood supply of the encapsulated tissue, and consequently by interference with optimal nutrition and/or supply of growth factors to the encapsulated tissue. Although this problem is not covered by the current definition, it should be considered to be a biocompatibility problem since long-term survival of the tissue is required for this "specific application" but, regretfully, is not compatible with the presence of the biomaterial around the tissue.

Concluding Remarks

It is clear that tremendous advances have been made in immunoisolating pancreatic islets during the last few decades. Nevertheless, the present status of this technology does not yet qualify for widespread clinical application in the treatment of IDDM as a convincing alternative to insulin treatment. This view is corroborated by the results of the restricted experience obtained in man (Brunetti et al 1991, Scharp et al 1994, Soon Shiong et al 1994).

Optimal biocompatibility of the bioartificial pancreas remains difficult to interpret, but present insights emphasize that we should focus not only on the interaction between the host tissue and the biomaterial but also on the interaction between the donor tissue and the biomaterial. A major chal-

lenge is to find or create techniques to accomplish a higher degree of biocompatibility for bioartificial pancreases, since most available techniques that successfully enhance the biocompatibility of fully artificial organs do not allow for the survival of living tissue.

Although the tissue-material interaction is important, a major factor that explains the insufficient functional performance of an encapsulated islet graft is the diffusion barrier between the islet and the bloodstream. This barrier is composed of the capsule as such in combination with the large distance between the islet tissue and the bloodstream when the graft is placed in the unmodified peritoneal cavity. We can safely assume that this absence of (neo)vascularization interferes with long-term graft survival since effective isolation by the capsule is, apparently, accompanied by isolation from factors required for maintaining tissue viability. This is clearly illustrated by our finding that the survival time of encapsulated islet grafts is limited, even in the absence of fibrotic overgrowth and in the absence of histologic evidence of rejection (De Vos et al 1997b). Thus, even in the presence of full biocompatibility, additional conditions must be met.

Obviously, conceivable approaches are to find or create a transplantation site that offers a more direct contact between the capsule and the recipient's bloodstream. This appears mandatory in order to improve the functional performance and survival to a level and duration acceptable for setting up immunoisolated islet transplantation trials in humans.

References

Aebischer P, Goddard M, Signore AP, Timpson RL. 1994. Functional recovery in hemiparkinsonian primates transplanted with polymer-encapsulated PC12 cells. Exp Neurol 126:151–158.

Algire GH. 1943. An adaption of the transparent chamber technique to the mouse. J Natl Cancer Inst 4:1–11.

Archer J, Kaye R, Mutter G. 1980. Control of streptozotocin diabetes in Chinese hamsters by cultured mouse islet cells without immunosuppression: a preliminary report. J Surg Res 28:77–85.

Bisceglie VV. 1933. Über die antineoplastische Immunität. Krebsforsch 40:141–158.

Brunetti P, Basta G, Faloerni A, Calcinaro F, Pietropaolo M, Calafiore R. 1991. Immunoprotection of pancreatic islet grafts within artificial microcapsules. Int J Artif Organs 14:789–791.

Bystricky S, Malovikova A, Sticzay T. 1990. Interaction of alginate and pectins with cationic polypeptides. Carbohydr Res 13:283–294.

Bystricky S, Malovikova A, Sticzay T. 1991. Interaction of acid polysaccharides with polylysine enantiomers, conformation probe in solution. Carbohydrate Polymers 15:299–308.

Chang PL, Shen N, Westcott AJ. 1993. Delivery of recombinant gene products with microencapsulated cells in vivo. Hum Gene Ther 4:433–440.

Chang PL, Hortelano G, Tse M, Awrey DE. 1994. Growth of recombinant fibroblasts in alginate microcapsules. Biotech Bioeng 43:925–933.

Chick WL, Like AA, Lauris V. 1975. Beta cell culture on synthetic capillaries: an artificial endocrine pancreas. Science 187:847–849.

Cieslinski DA, Humes HD. 1994. Tissue engineering of a bioartificial kidney. Biotechnol Bioeng 43:678–681.

Clayton HA, London NJM, Colloby PS, Bell PRF, James RFL. 1991. The effect of capsule composition on the biocompatibility of alginate-poly-l-lysine capsules. J Microencapsulation 8:221–233.

Colton CK. 1995. Implantable biohybrid artificial organs. Cell Transplantation 4:415–436.

Colton CK, Avgoustiniatos ES. 1991. Bioengineering in development of the hybrid artificial pancreas. J Biomech Eng 113:152–170.

Dawson RM, Broughton RL, Stevenson WT, Sefton MV. 1987. Microencapsulation of CHO cells in a hydroxyethyl methacrylate-methyl methacrylate copolymer. Biomaterials 8:360–366.

De Vos P, Wolters GH, Fritschy WM, Van Schilfgaarde R. 1993. Obstacles in the application of microencapsulation in islet transplantation. Int J Artif Organs 16:205–212.

De Vos P, De Haan BJ, Pater J, Van Schilfgaarde R. 1996a. Association between capsule diameter, adequacy of encapsulation, and survival of microencapsulated rat islet allografts. Transplantation 62:893–899.

De Vos P, De Haan BJ, Wolters GHJ, Van Schilfgaarde R. 1996b. Factors influencing the adequacy of microencapsulation of rat pancreatic islets. Transplantation 62:888–893.

De Vos P, De Haan B, Van Schilfgaarde R. 1997a. Effect of the alginate composition on the biocompatibility of alginate-polylysine microcapsules. Biomaterials 18:273–278.

De Vos P, De Haan BJ, Wolters GHJ, Strubbe JH, Van Schilfgaarde R. 1997b. Improved biocompatibility but limited graft survival after purification of alginate for microencapsulation of pancreatic islets. Diabetologia 40:262–270.

Fritschy WM, Wolters GH, Van Schilfgaarde R. 1991. Effect of alginate-polylysine-alginate microencapsulation on in vitro insulin release from rat pancreatic islets. Diabetes 40:37–43.

Fritschy WM, De Vos P, Groen H, Klatter FA, Pasma A, Wolters GH, Van Schilfgaarde R. 1994. The capsular overgrowth on microencapsulated pancreatic islet grafts in streptozotocin and autoimmune diabetic rats. Transpl Int 7:264–271.

Galletti PM, Trudell LA, Panol G, Richardson PD, Whittemore A. 1981. Feasibility of small bore AV shunts for hybrid artificial organs in nonheparinized beagle dogs. Trans Am Soc Artif Intern Organs 27:185–187.

Iwata H, Amemiya H, Matsuda T, Takano H, Hayashi R, Akutsu T. 1989. Evaluation of microencapsulated islets in agarose gel as bioartificial pancreas by studies of hormone secretion in culture and by xenotransplantation. Diabetes 38 Suppl 1:224–225.

Kessler L, Pinget M, Aprahamian M, Dejardin P, Damge C. 1991. In vitro and in vivo studies of the properties of an artificial membrane for pancreatic islet encapsulation. Horm Metab Res 23:312–317.

Kessler L. 1992. Diffusion properties of an artificial membrane used for Langerhans islets encapsulation: an in vitro test. Biomaterials 13:44–49.

Kessler L, Legeay G, West R, Belcourt A, Pinget M. 1997. Physicochemical and biological studies of corona-treated artificial membranes used for pancreatic islets encapsulation: Mechanism of diffusion and interface modification. J Biomed Mater Res 34:235–245.

King GA, Daugulis AJ, Faulkner P, Goosen MFA. 1987. Alginate-polylysine microcapsules of controlled membrane molecular weight cutoff for mammalian cell culture engineering. Biotechnol Prog 3:231–240.

Klomp GF, Ronel SH, Hashiguchi H, D'Andrea M, Dobelle WH. 1979. Hydrogels for encapsulation of pancreatic islet cells. Trans Am Soc Artif Intern Organs 25:74–76.

Koo J, Chang TSM. 1993. Secretion of erythropoietin from microencapsulated rat kidney cells. Int J Artif Organs 16:557–560.

Krause TJ, Robertson FM, Liesch JB, Wasserman AJ, Grecos RS. 1990. Differential production of IL-1 on the surface of biomaterials. Arch Surg 125:1158–1160.

Lacy PE, Hegre OD, Gerasimidi Vazeou A, Gentile FT, Dionne KE. 1991. Maintenance of normoglycemia in diabetic mice by subcutaneous xenografts of encapsulated islets. Science 254:1782–1784.

Lanza RP, Borland KM, Staruk JE, Appel MC, Solomon BA, Chick WL. 1992a. Transplantation of encapsulated canine islets into spontaneously diabetic BB/Wor rats without immunosuppression. Endocrinology 131:637–642.

Lanza RP, Sullivan SJ, Chick WL. 1992b. Perspectives in diabetes. Islet transplantation with immunoisolation. Diabetes 41:1503–1510.

Lanza RP, Beyer AM, Chick WL. 1994. Xenogeneic humoral responses to islets transplanted in biohybrid diffusion chambers. Transplantation 57:1371–1375.

Lanza RP, Ecker DM, Kühtreiber WM, Marsh JP, Chick WL. 1995a. Simple and inexpensive method for transplanting xenogeneic cells and tissues into rats using alginate gel spheres. Transplant Proc 27:3322.

Lanza RP, Kühtreiber WM, Ecker DM, Marsh JP, Chick WL. 1995b. Successful bovine islet xenografts in rodents and dogs using injectable microreactors. Transplant Proc 27:3211.

Lanza RP, Kühtreiber WM, Ecker DM, Marsh JP, Staruk JE, Chick WL. 1996. A simple method for xenotransplanting cells and tissues into rats using uncoated alginate microreactors. Transplant Proc 28:835.

Lanza RP, Chick WL. 1997. Immunoisolation: At a turning point. Immunology Today 18:135–139.

Lim F, Sun AM. 1980. Microencapsulated islets as bioartificial endocrine pancreas. Science 210:908–910.

Liu HW, Ofosu FA, Chang PL. 1993. Expression of human factor IX by microencapsulated recombinant fibroblasts. Hum Gene Ther 4:291–301.

Lum ZP, Krestow M, Tai IT, Vacek I, Sun AM. 1992. Xenografts of rat islets into diabetic mice. Transplantation 53:1180–1183.

Maki T, Lodge JP, Carretta M, Ohzato H, Borland KM, Sullivan SJ, Staruk J, Muller TE, Solomon BA, Chick WL. 1993. Treatment of severe diabetes mellitus for more than one year using a vascularized hybrid artificial pancreas. Transplantation 55:713–717.

Maki T, Mullon CJ, Solomon BA, Monaco AP. 1995. Novel delivery of pancreatic islet cells to treat insulin-dependent diabetes mellitus. Clin Pharmacokinet 28:471–482.

Maki T, Otsu I, O'Neil JJ, Dunleavy K, Mullon CJP, Solomon BA, Monaco AP. 1996. Treatment of diabetes by xenogeneic islets without immunosuppression— Use of a vascularized bioartificial pancreas. Diabetes 45:342–347.

Metrakos P, Yuan S, Qi SJ, Duguid WP, Rosenberg L. 1994. Collagen gel matrix promotes islet cell proliferation. Transplant Proc 26:3349–3350.

O'Shea GM, Goosen MFA, Sun AM. 1984. Prolonged survival of transplanted islets of Langerhans encapsulated in a biocompatible membrane. Biochim Biophys Acta 804:133–136.

O'Shea GM, Sun AM. 1986. Encapsulation of rat islets of Langerhans prolongs xenograft survival in diabetic mice. Diabetes 35:943–946.

Remes A, Williams DF. 1991. Immune response in biocompatibility. Biomaterials 13:731–743.

Scharp DW, Mason NS, Sparks RE. 1984. Islet immunoisolation: the use of hybrid artificial organs to prevent islet tissue rejection. World J Surg 8:221–229.

Scharp DW, Swanson CJ, Olack BJ, Latta PP, Hegre OD, Doherty EJ, Gentile FT, Flavin KS, Ansara MF, Lacy PE. 1994. Protection of encapsulated human islets implanted without immunosuppression in patients with Type I or Type II diabetes and in nondiabetic control subjects. Diabetes 43:1167–1170.

Sefton MV. 1993. The good, the bad and the obvious: 1993 Clemson Award for Basic Research—Keynote Lecture. Biomaterials 14:1127–1134.

Sevastianov VI, Tseytkina EA. 1984. The activation of the complement system by polymer materials and their blood compatibility. J Biomed Mater Res 18:969–978.

Shimi SM, Hopwood D, Newman EL, Cuschieri A. 1991. Microencapsulation of human cells: its effect on growth of normal and tumour cells in vitro. Br J Cancer 63:675–680.

Shoichet MS, Winn SR, Athavale S, Harris JM, Gentile FT. 1994. Poly(ethylene oxide)-grafted thermoplastic membranes for use as cellular hybrid bio-artificial organs in the central nervous system. Biotech Bioeng 43:563–572.

Siebers U, Zekorn T, Bretzel RG, Planck H, Renardy M, Zschocke P, Federlin K. 1990. Histocompatibility of semipermeable membranes for implantable diffusion devices (bioartificial pancreas). Transplant Proc 22:834–835.

Smidsrød O, Skjåk-Bræk G. 1990. Alginate as immobilization matrix for cells. Trends in Biotechnology 8:71–78.

Soon Shiong P, Feldman E, Nelson R, Komtebedde J, Smidsrød O, Skjåk-Bræk G, Espevik T, Heintz R, Lee M. 1992. Successful reversal of spontaneous diabetes in dogs by intraperitoneal microencapsulated islets. Transplantation 54:769–774.

Soon Shiong P, Heintz RE, Merideth N, Yao QX, Yao Z, Zheng T, Murphy M, Moloney MK, Schmehl M, Harris M. 1994. Insulin independence in a type 1 diabetic patient after encapsulated islet transplantation. Lancet 343:950–951.

Soon Shiong P, Otterlei M, Skjåk-Bræk G, Smidsrød O, Heintz R, Lanza RP, Espevik T. 1991. An immunological basis for the fibrotic reaction to implanted microcapsules. Transplant Proc 23:758–759.

Strautz RL. 1970. Studies of hereditary-obese mice (obob) after implantation of pancreatic islets in Millipore filter capsules. Diabetologia 6:306–312.

Sun AM, Parisius W, Healy GM, Vacek I, Macmorine HG. 1977. The use, in diabetic rats and monkeys, of artificial capillary units containing cultured islets of Langerhans (artificial endocrine pancreas). Diabetes 26:1136–1139.

Sun YL, Ma X, Zhou D, Vacek I, Sun AM. 1993. Porcine pancreatic islets: isolation, microencapsulation, and xenotransplantation. Artif Organs 17:727–733.

Sun YL, Ma XJ, Zhou DB, Vacek I, Sun AM. 1996. Normalization of diabetes in spontaneously diabetic cynomologus monkeys by xenografts of microencapsulated porcine islets without immunosuppression. J Clin Invest 98:1417–1422.

Thu B, Bruheim P, Espevik T, Smidsrød O, Soon Shiong P, Skjåk-Bræk G. 1996a. Alginate polycation microcapsules. I. Interaction between alginate and polycation. Biomaterials 17:1031–1040.

Thu B, Bruheim P, Espevik T, Smidsrød O, Soon Shiong P, Skjåk-Bræk G. 1996b. Alginate polycation microcapsules II. Some functional properties. Biomaterials 17:1069–1079.

Uludag H, Sefton MV. 1993. Microencapsulated human hepatoma (HepG2) cells: in vitro growth and protein release. J Biomed Mater Res 27:1213–1224.

Vandenbossche GMR, Van Oostveld P, Demeester J, Remon JP. 1993. The molecular weight cut-off of microcapsules is determined by the reaction between alginate and polylysine. Biotechnol Bioeng 42:381–386.

Wahoff DC, Stephanian E, Gores PF, Soon Shiong P, Hower C, Lloveras JK, Sutherland DER. 1994. Intraperitoneal transplantation of microencapsulated canine islet allografts with short-term, low-dose cyclosporine for treatment of pancreatectomy-induced diabetes in dogs. Transplant Proc 26:804.

Wells GD, Fisher MM, Sefton MV. 1993. Microencapsulation of viable hepatocytes in HEMA-MMA microcapsules: a preliminary study. Biomaterials 14:615–620.

Wijsman J, Atkison P, Mazaheri R, Garcia B, Paul T, Vose J, O'Shea GM, Stiller C. 1992. Histological and immunopathological analysis of recovered encapsulated allogeneic islets from transplanted diabetic BB/W rats. Transplantation 54:588–592.

Williams DF. 1987. Summary and definitions. In: Progress in biomedical engineering: Definition in biomaterials (4). Amsterdam: Elsevier Science Publisher BV, pp. 66–71.

Woodward SC. 1982. How fibroblasts and giant cells encapsulate implants: considerations in design of glucose sensors. Diabetes Care 5:278–281.

Zekorn T, Siebers U, Filip L, Mauer K, Schmitt U, Bretzel RG, Federlin K. 1989. Bioartificial pancreas: the use of different hollow fibers as a diffusion chamber. Transplant Proc 21:2748–2750.

Zekorn T, Siebers U, Horcher A, Federlin K. 1994. Experimental and clinical islet transplantation; Bioartificial pancreas; Satellite Symposium on the occasion of the 29th Annual Meeting of the German Diabetes Society; Berlin, May 11, 1994. Exp Clin Endocrinol 102:A1–A8.

Zhou D, Sun YL, Vacek I, Ma P, Sun AM. 1994. Normalization of diabetes in cynomologus monkeys by xenotransplantation of microencapsulated porcine islets. Transplant Proc 26:1091.

Zielinski BA, Aebischer P. 1994. Chitosan as a matrix for mammalian cell encapsulation. Biomaterials 15:1049–1056.

Zimmermann U, Klöck G, Federlin K, Hannig K, Kowalski M, Bretzel RG, Horcher A, Entenmann H, Sieber U, Zekorn T. 1992. Production of mitogen-contamination free alginates with variable ratios of mannuronic acid to guluronic acid by free flow electrophoresis. Electrophoresis 13:269–274.

Part II
Encapsulation Systems

Section 1
Microencapsulation

7
Calcium Alginate

Hua Yang and James R. Wright, Jr.

Introduction

The concept of using semipermeable capsules for delivery of therapeutic biological reagents was pioneered by T. M. S. Chang over 30 years ago (Chang 1964). Over the next 15 years, a number of immunoisolation systems were developed in an attempt to protect transplanted tissues or cells from host immune rejection; most of these were hollow fibers that were loaded with isolated pancreatic islets and then transplanted into diabetic rodents as bioartificial pancreas devices (Chick 1977, Tze 1976). In 1980, Lim and Sun successfully applied an alginate-polylysine microencapsulation system to islet transplantation in a rodent model (Lim and Sun 1980). This report attracted more attention to the concept of microencapsulation, especially the alginate-polylysine system. During the last 15 years, much effort has been devoted to the study of this system (reviewed by Lanza and Chick 1997). Most of the procedures for producing these microcapsules involve extruding a mixture of cells and sodium alginate into a divalent cation solution to form water-insoluble gel droplets. The negatively charged gel droplets are then coated with positively charged polymers, such as poly-L-lysine (PLL), through ionic interaction. The primary function of the coating is to form a strong complex membrane that reduces and controls the permeability of the alginate gel sphere. Recently, Lanza et al (1995a) reported prolongation of porcine and bovine islet xenograft survival in diabetic mice without immunosuppression, using uncoated alginate gel spheres—that is, alginate droplets that have undergone gelation in calcium chloride but have not been coated with a synthetic PLL membrane. Sub-

sequently, we applied this method to an even more discordant xenograft model, fish islets transplanted into rodents. In our laboratory, fish islet xenograft survival was significantly prolonged in both diabetic mouse and rat recipients without immunosuppression (Yang et al 1996, 1997b). Further studies revealed that uncoated alginate encapsulation, in combination with immunosuppression, permitted long-term fish islet xenograft survival in diabetic mice and rats (Yang et al 1996, Yang et al 1997b), as well as long-term islet allograft survival in spontaneously diabetic dogs (Lanza et al 1995b). It was surprising that uncoated alginate gel spheres could markedly prolong islet xenograft survival, because data clearly indicate that uncoated alginate gel has sufficient porosity to permit antibodies and complement to enter (Lanza et al 1995a, Martinsen et al 1992, Tanaka et al 1984). It is clear that the immunoprotective effect could not be explained solely by the capsule's mechanical barrier (i.e., porosity of capsule versus molecular weight of diffusents) and that more complex mechanism(s), such as biological and diffusion modulation effects by the encapsulation materials, must also play significant roles. In this chapter, we will focus on technological advances and insights that have been gained through the study of uncoated calcium alginate gel encapsulation systems.

Chemistry and Gel-Forming Kinetics

Alginates are naturally occurring polysaccharides that can be found in all species of brown algae (McNeely and Pettitt 1973) and in some species of bacteria (Morris 1986). Chemically, alginates are

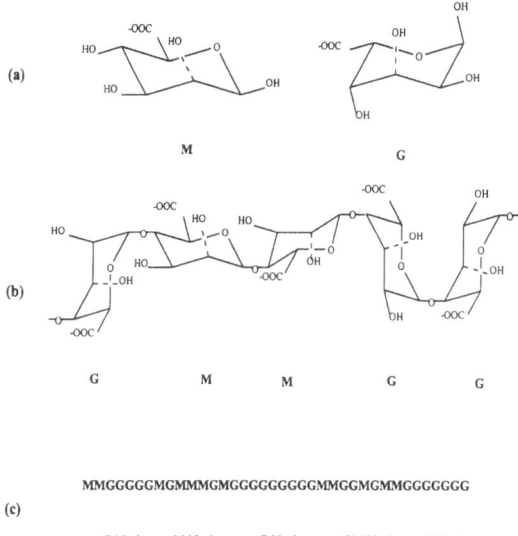

FIGURE 7.1. Structure of alginate: (a) monomer conformation (M: mannuronate, G: guluronate); (b) chain conformation; and (c) schematic alginate chain sequence.

linear copolymers of 1,4–linked α-L-guluronic acid (G) and β-D-mannuronic acid (M) arranged as chains in alternating blockwise patterns (Figure 7.1). Depending on the source, their molecules can be composed of three block types: polymannuronate (MM), polyguluronate (GG), and mixed blocks (MG). Thus, alginates from different sources have different chemical composition and molecular size (Martinsen et al 1989, Smidsrod and Draget 1997). For example, sodium alginate extracted from stipes of *Laminaria hyperborea* (Pronova LVG, MW = 277,000), commercially available from Pronova Biopolymer (Drammen, Norway), is rich in guluronic acid (high G), while sodium alginate isolated from *Macrocystis porifera* (MW = 210,000), commercially available from Kelco division of Merck (San Diego, CA), is high in Mannuronic acid (high M).

Numerous studies have demonstrated that the chemical structure (especially monomeric composition) and molecular size (sequential structure), as well as the gel-forming kinetics of an alginate, will have significant impact on several of its functional properties, including porosity (i.e., diffusion coefficiency), swelling behavior, stability, biodegradability, and gel strength and, even more important, on the gel's immunological characteristics and biocompatibility. Martinsen et al (1989, 1992) and Thu et al (1996a, 1996b) reported a series of stud-

ies on alginate gels. They demonstrated that alginate gel beads made from high-G composition and longer G-blocks have higher mechanical strength, lower shrinkage, more stability towards monovalent cations, and higher porosity. They also demonstrated that the diffusion coefficient decreases with increasing alginate concentration and M/G ratio. In addition, the diffusion process also depends upon the gel-forming methods utilized. When the gel is formed in the presence or absence of a charged osmolyte (e.g., NaCl), the alginate gel sphere will become homogeneous (i.e., isotropic distribution of alginate within the sphere) or inhomogeneous (i.e., exhibiting a higher alginate concentration at the surface of the gel sphere), respectively. Diffusion of intrepid bovine serum albumin (BSA) from homogeneous spheres is much faster than from inhomogeneous spheres (Tanaka et al 1984). Because the alginate gel network is negatively charged, the pH will also influence the diffusion of charged substrates and products; thus, the rate of diffusion of BSA from alginate spheres increases with increasing pH as a result of the increased negative charge on the protein. The temperature dependence of the diffusion coefficient follows Arrhenius's law.

Soon-Shiong et al (1991, 1992) and Otterlei et al (1991) reported that mannuronic acid (M) monomer within the alginate molecule stimulates interleukin-1 (IL-1) and tumor necrosis factor (TNF) production and plays an important role in the induction of fibroplasia. They further demonstrated that a formula low in mannuronic acid content evokes only a minimal cytokine response and thus is more biocompatible.

Because of its mild gelling and good biocompatibility, alginate gel has long been used in the food and pharmaceutical industries (McNeely and Pettitt 1973). As more studies have documented its favorable characteristics, there has been increased interest in using alginate in a wide range of biomedical fields in recent years (reviewed by Smidsrod and Skjak-Braek 1990); these include controlled-release systems for conventional drugs (Bodmeier and Paertakul 1989, Downs et al 1992, Prisant et al 1992), proteins (Shimi et al 1991), hormones (Mumper et al 1994) somatic gene therapy (Marshall 1995), and oral vaccine (Bowersock et al 1996, Pier 1991). It has also been used as supporting matrix for cell proliferation such as is used in rebuilding cartilage (Paige et al 1996) and in wound healing (Kneafsey 1996, Piacquadio and Nelson 1992,

Wilson 1996). It has even been used to encapsulate embryos for in vitro fertilization (Adaniya et al 1993). However, the most intense efforts have related to its superior performance in immunoisolation systems for cell transplantation.

Uncoated Alginate Gel Encapsulation Technique

In our laboratory, anatomically discrete islet tissue harvested from the teleost fish *Oreochromis niloticus* (tilapia) has been used as a valuable donor source for xenotransplantation studies (Wright et al 1992, 1994; Wright and Yang 1997, Yang et al 1995), and we are currently genetically engineering tilapia to have islets that will secrete "humanized" insulin. In the future, these may provide an important source of donor islets for clinical use

(MacKenzie 1996). Using this piscine model, we tested uncoated alginate encapsulated fish islet graft survival in several different strains of diabetic mouse and rat recipients (Yang et al 1996, 1997b).

Figure 7.2 shows the encapsulation procedure as carried out in our laboratory. LVG sodium alginate (Batch No. 901–282–05) and UP-LVG sodium alginate (Batch No. 304–281–05), both purchased from Pronova Biopolymer (Drammen, Norway), were used as basic encapsulation materials. Our encapsulation procedure is as described previously (Yang et al 1997b) and is based on the method described by Lanza et al (1995b). The encapsulation procedure requires three steps:

The first step is the formation of an islet-alginate suspension. Fish islet (Brockmann body) fragments were collected (Yang and Wright 1995) and mixed with 1.5% LVG sodium alginate solution. Islet tissue from 1000 g of donor fish was

(a)

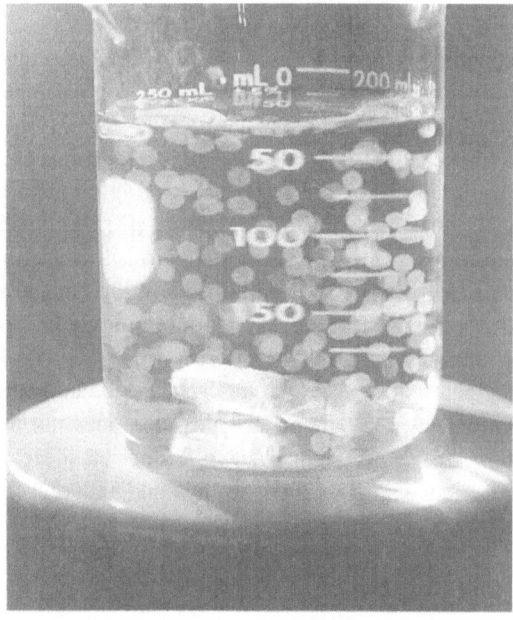

(b)

FIGURE 7.2. Encapsulation process: Islet fragments suspended in 1.5% LVG sodium alginate saline solution are extruded through a 16 G intravenous catheter (Critikon, Tampa, FL) into 1.5% $CaCl_2$ solution (a) to form islet-containing calcium alginate gel spheres (b).

suspended in 1 ml of sodium alginate. This volume of fish islet tissue is functionally equivalent to roughly 600–650 handpicked rodent islets or several thousand human or porcine islet equivalents (Yang et al 1997a).

The second step is the formation of islet-containing calcium alginate gel spheres. The islet-alginate mixture was aspirated into a syringe and then extruded through a 16–gauge intravenous catheter (Critikon, Tampa, FL) into a beaker containing 1.5% $CaCl_2$ solution while stirred with a stirring bar (Figure 7.2). Two different types of capsules were made: spheres, measuring 4.5 mm in diameter (Figure 7.3a) and noodles, measuring about 4.5 mm in diameter and 3 cm in length (Figure 7.3b). The noodle type capsules were initially used because of easy retrieval. However, after 50 days of i.p. implantation in streptozotocin STZ-diabetic nude mice, many of the noodles were found broken into segments of different lengths and, therefore did not offer any significant advantage. Thus, we abandoned noodle-shaped devices in favor of the spheres.

The third step, fine coating, is optional. After gelation, the gel spheres or noodles were washed with normal saline and resuspended in a beaker containing 0.7% UP (ultrapure) LVG sodium alginate solution for 5 minutes while slowly stirred with a stirring bar. Finally, the capsules were washed with Hanks' balanced salt solution and cultured overnight (37°C, 5% CO_2) in CMRL-1066 culture medium with 10% fetal calf serum, 100 U/ml penicillin, and 100 μg/ml streptomycin sulfate before transplantation. In our initial studies, this fine coating was designed to coat islet-containing LVG alginate gel spheres with more purified UP LVG alginate gel to improve the capsule's biocompatibility. Later, we found that this step does not appear to make a significant difference in the capsule's in vivo biocompatibility, and thus can be omitted.

Uncoated Alginate Gel for Islet Xenotransplantation

In our laboratory (Yang et al 1997b), uncoated alginate gel encapsulation markedly prolonged mean graft survival time (mGST) in STZ-diabetic balb/c

mice, with or without adjuvant immunosuppression. Unencapsulated fish islets transplanted i.p. into euthymic balb/c mice uniformly failed within 5 days. The mGST in balb/c mice was significantly prolonged to >27 days by alginate gel sphere encapsulation (p < 0.005), with 1 of 7 mice achieving long-term function. Cyclosporin A (CsA) and 15–deoxyspergualin (DSG) treatments minimally prolonged fish islet graft survival in the absence of encapsulation (p < 0.05, and n.s.); however, when combined with encapsulation, mGSTs were markedly prolonged relative to both control (p < 0.001) and drug-only treatment groups (p < 0.005). Long-term function was achieved in 4 of 6 balb/c recipients receiving 10 mg/kg/day CsA; two of the mice in this group rejected after 39 and 45 days. Daily i.p. injections of 2.5 mg/kg DSG, in combination with encapsulation, resulted in uniform >50 days graft function in balb/c recipients. We also confirmed biocompatibility by transplanting encapsulated fish islets into STZ-diabetic nude mice. Fish islets encapsulated in uncoated alginate gel spheres and noodles uniformly survived >50 or >100 days in STZ-diabetic nude mice. In all instances of long-term graft function in balb/c or nude mice, graft dependency was confirmed by graft retrieval from each recipient's peritoneal cavity 50 to 100 days posttransplantation. Histological examination of the islet grafts showed abundant viable, well-granulated islet tissue (Figures 7.4a and 7.4b) as well as some necrotic, calcified islet tissue in all grafts functioning >50 or >100 days; lesser amounts of islet tissue were found in capsules retrieved from those recipients that did not function long-term. Inflammatory/fibroblastic tissue response to the capsules was minimal to absent, macroscopically (Figure 7.3) and microscopically.

Uncoated alginate encapsulation also prolonged tilapia islet GST in rats. Unencapsulated fish islets transplanted i.p. failed to function in STZ-diabetic Lewis rats, even when immunosuppressed with 2.5 mg/kg/day DSG. Alginate-encapsulated fish islet grafts had a mGST of 13 days in untreated Lewis rats and was prolonged to >46 days with 3/6 rats achieving >50 day survival in Lewis rat recipients receiving DSG (2.5 mg/kg/d). Encapsulation also prolonged mGST to 27 days in Lewis rats receiving 10 mg/kg/day CsA treatment. Histological examination of those grafts retrieved after 50 days of

(a)

(b)

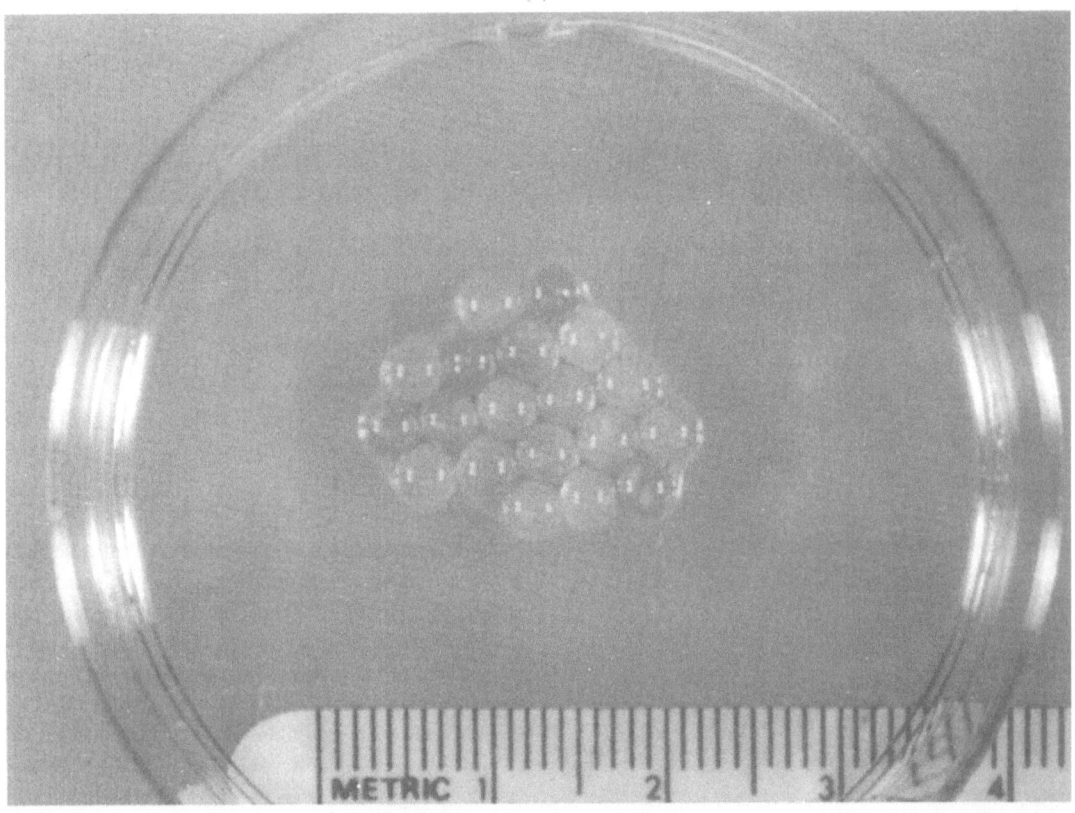

FIGURE 7.3. Encapsulated fish islet grafts recovered from diabetic recipient mice at >50 days posttransplantation. (a) spheres; (b) noodles. Reprinted with the permission of Williams & Wilkins from Yang et al. 1997. Transplantation 64:28–32.

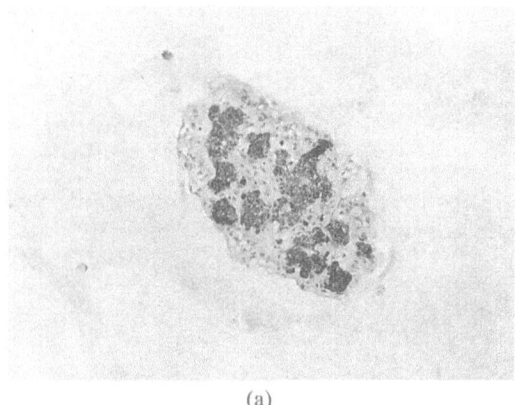

(a)

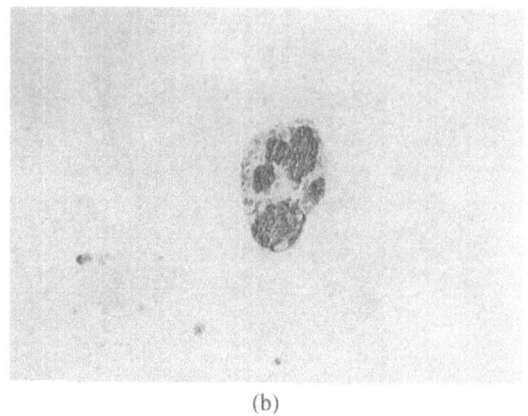

(b)

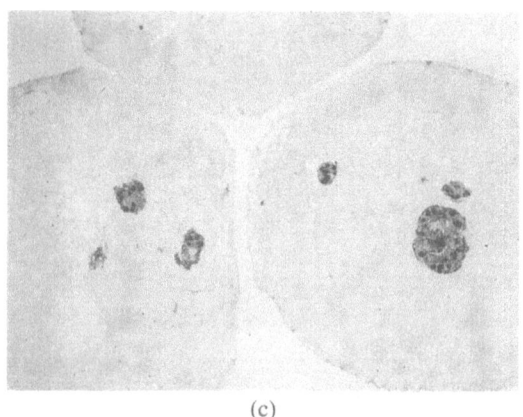

(c)

FIGURE 7.4. Histological sections of encapsulated fish islet grafts removed from (a) a diabetic nude mouse at day 55; (b) a diabetic balb/c mouse receiving DSG treatment at day 50; and (c) a diabetic Lewis rat receiving DSG treatment at day 50 posttransplantation. The islet tissue, stained with immunoperoxidase for tilapia insulin, is viable and contains both beta and non-beta cells (a,b,c), and there is no fibroblastic/inflammatory response to the capsule (c) [original magnification (a) 100×; (b) 100×; and (c) 40×]. Figures (a) and (b) reprinted with the permission of Williams & Wilkins from Yang et al. 1997. Transplantation 64:28–32.

implantation confirmed viable, granulated islet tissue without any significant inflammatory/fibroblastic tissue response (Figure 7.4c). However, in general, fewer viable islets were retrieved from rat recipients compared to mouse recipients after the same periods of implantation.

In 1995, Lanza et al (1995a) reported that uncoated alginate microspheres (800–900 μm in diameter) containing porcine or bovine islets reversed hyperglycemia in (STZ)-diabetic C57 mice for at least 30 days and 43 days respectively without any immunosuppression, while nonencapsulated grafts survived less than 4 days. Lanza et al (1995b) subsequently reported that uncoated alginate encapsulation prolonged porcine islet graft survival in an STZ-diabetic Lewis rat to 13 days. When 10–20 mg/kg/day CsA treatment was added, both porcine and bovine islet graft survival times were prolonged to >100 days in Lewis rat recipients, while nonencapsulated controls experienced primary nonfunction.

Uncoated Alginate Gel for Islet Allotransplantation

In addition to these studies transplanting islet xenografts into rodents, Lanza and Chick (1997) have performed some preliminary studies using a large animal model. They demonstrated that uncoated alginate encapsulated canine islet allografts, when combined with low-dose CsA treatments, completely supplanted exogenous insulin therapy in spontaneously diabetic dogs from at least 60 days to more than 175 days.

Size of the Sphere

Using a porcine-to-rat islet xenotransplantation model, Lanza et al (1995b) tested the viability of islet grafts immobilized in uncoated alginate gel spheres of varying diameters ranging from 800 to

3700 μm. After 2 weeks of i.p. implantation, no islets survived in uncoated alginate gel spheres smaller than 1600 μm in diameter, as confirmed by both loss of blood glucose control and histological examination of islet grafts. However, graft viability was up to 85% for islets immobilized in 3700–μm-diameter gel spheres.

In our laboratory, we further tested the effect of the size of uncoated alginate gel spheres using our fish-to-balb/c mouse islet xenotransplantation model without immunosuppression. A group of 10 mice that received tilapia islet grafts immobilized in 4.5–mm-diameter gel spheres (45 spheres/ml gel) had a mean graft survival time (mGST) of >27 ± 13 days. Another group of 5 balb/c mice received islet grafts immobilized in 6–mm-diameter gel spheres (15 spheres/ml gel); mGST was reduced to 13 ± 2 days. Similar results were obtained in diabetic nude mouse recipients, suggesting that graft failure was due to nonimmune-mediated mechanisms. Islet grafts embedded in 6–mm-diameter gel spheres had a mGST of 15 ± 6.2 days, while those embedded in 4.5–mm-diameter spheres functioned uniformly >50 days or >100 days. These results suggest that oxygen diffusion limits alginate gel sphere size for even tilapia islets, which are known to be markedly more resistant to hypoxia than mammalian islets (Wright et al 1997). It seems likely that 3.7–4.5 mm represents the upper limit of alginate gel sphere diameter for most applications.

Strain and Species Differences

Lanza's laboratory and our laboratory have examined the immunoprotective effect of uncoated alginate gel spheres, using various donor and recipient species. Data reported by Lanza et al (1995a) indicated that uncoated alginate spheres containing porcine or bovine islets consistently provided long-term normoglycemia after transplantation into diabetic C57BL/6J mice. When an STZ-diabetic Lewis rat was used as a recipient, porcine islet grafts failed at 13 days after implantation. However, long-term survival of porcine and bovine islets in Lewis rats, as well as long-term survival of canine islets in spontaneously diabetic dogs, could be achieved with CsA treatments. Uncoated alginate spheres failed to maintain bovine islet graft sur-

vival in diabetic dogs, even when triple immunosuppressive therapy (CsA, azathioprine, and prednisone) was applied (Lanza and Chick 1997).

In our laboratory, using islets harvested from tilapia fish as donor islets, we examined the islet graft survival times in different strains of diabetic mice and in rats after immobilization in uncoated alginate gel spheres. Our data indicated that, although alginate gel spheres had good biocompatibility in both mouse and rat recipients, graft survival times differ significantly between rat and mouse recipients and even between different strains of recipient mice. The mGST of alginate-encapsulated tilapia islet grafts was >49 days in STZ-diabetic C57 BL/6J mouse recipients, with 5 of 6 mice remaining normoglycemic over 50 days, while mGST in balb/c mice was >27 days with only one mouse remaining normoglycemic over 50 days (t = 4.35, df = 11, p < 0.005). Mean GST in Lewis rats was only 13 days. Our data suggest that the immunoprotective effects of uncoated alginate gel spheres vary between recipient strains and species.

Biocompatibility and Biodegradation

For any encapsulation material, good biocompatibility towards both the encapsulated cells and the host is of vital importance. One of the most crucial problems facing almost all encapsulation devices is host "foreign-body" reaction (i.e., fibrosis), resulting in low diffusion efficiency for nutrients and oxygen. Alginate gel appears to be nontoxic to imbedded cells; several investigators have shown that imbedded islets maintain good morphology, insulin content, and glucose responsiveness during long-term *in vitro* culture (Falorni et al 1996, Fraser et al 1995, Lanza et al 1991). However, numerous studies using uncharacterized commercial alginates to produce alginate-polylysine capsules resulted in inconsistent outcomes, mainly owing to their different levels of fibroblastic overgrowth. Later, Soon-Shiong et al (1991, 1992) found that alginate gel of low M content evoked a minimal cytokine response and, thus, reduced the fibrotic reaction. In our laboratory, purified sodium alginate LVG from Pronova (i.e., low M) resulted in

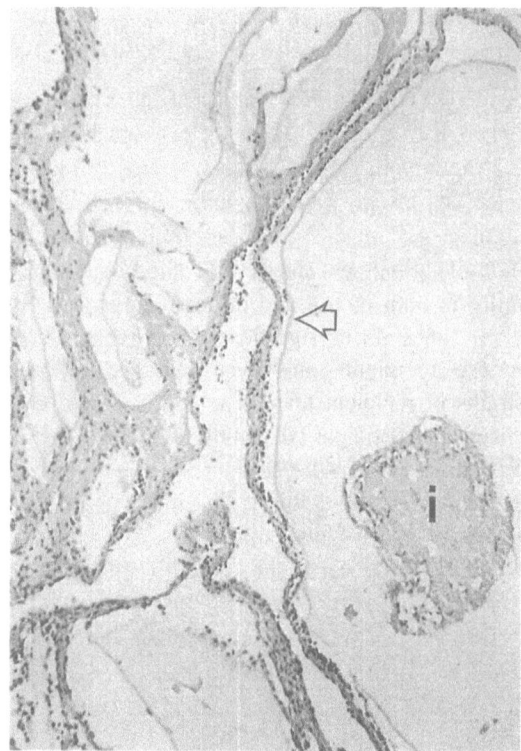

FIGURE 7.5. Histological section showing fibrotic reaction and mild inflammatory response at 20 days post-transplantation to fish islet grafts encapsulated using a poorly characterized alginate. The arrow shows the edge of an alginate gel sphere containing a necrotic islet (i). Several other capsules (left) collapsed during histologic processing (HE, original magnification 40×).

very good biocompatibility in rodents. Fifty to 100 days postimplantation, more than 80% of uncoated alginate gel spheres containing fish islets were found free floating in the abdominal cavity in both diabetic mouse and rat recipients. Microscopic examination revealed mininal-to-absent inflammatory/fibroblastic reaction to the capsules (Figure 7.4). Less than 20% of the beads were imbedded in connective tissue, and even these were encased in mostly adipose tissue with minimal microscopic fibroblastic overgrowth (Yang et al 1997b). These findings were in sharp contrast to our preliminary experiments using other uncharacterized batches of alginate, which promoted strong fibrotic overgrowth (Figure 7.5). Lanza et al (1995a, b) also reported mostly "intact and freely floating" islet-

containing gel spheres in their recipients. Our results support findings by others (Martinsen et al 1989, 1992; Thu et al 1996 a, b) that alginate gel's in vivo biocompatibility depends on its monomeric composition and sequential arrangements.

A recent paper suggests another approach to improve alginate biocompatibility. De Vos et al (1997) recently reported that purification of the alginate to remove impurities improves the biocompatibility of alginate-polylysine microcapsules. Nevertheless, graft survival was still limited, probably as a consequence of the lack of blood supply to encapsulated islets.

Given the limited functional longevity of encapsulated cells, it appears that periodic replacement or reimplantation is inevitable for future applications of immunoisolation systems for cell transplantation. This raises the issues of retrievability and biodegradation. In our rodent experiment, alginate gel had very good biocompatibility. If similar biocompatibility can be achieved after intraperitoneal transplantation in humans, we can expect that the majority of gel spheres could be laparoscopically removed by lavage. Although alginate gel spheres were found to be mostly intact macroscopically after long-term i.p. implantation in small and large animals (Lanza et al 1995a, b; Lanza and Chick 1997, Yang et al 1997b), microscopic chipping and heterophagocytosis were observed (Lanza et al 1995). Several studies have indicated that alginate gel is biodegradable. Al-Shamkhani and Duncan (1995) studied the clearance rate of sodium alginate solution in rats after systemic administration. They found that 24 hours after intravenous administration, the low molecular weight fraction (<48 kD, lower than renal threshold for alginate polymer) of the injected polymer was excreted in the urine, while the larger polymer fraction remained in the circulation and did not readily accumulate in any of the tissues. Almost all of the intraperitoneal dose was transferred from peritoneal cavity to the blood within 24 hours. Several other studies suggested that intraperitoneal and subcutaneous alginate polymer are transferred to the bloodstream through the lymphatic system (Flessner et al 1985, Seymour et al 1987, Tomlinson 1986). From these studies, we may expect that i.p. alginate spheres left over from laparoscopic lavage eventually will be degraded by chipping and heterophagocytosis, transferred to the bloodstream, and excreted in the urine. However, any long-term immunological consequences remain to be addressed.

Protective Mechanism(s)

Uncoated alginate gel encapsulation can achieve marked prolongation of concordant and discordant islet graft survival in several animal models. This is surprising because data clearly indicate that uncoated alginate gel has high porosity. Uptake studies with bovine thyroglobulin (Mw = 669 kD) performed by Lanza et al (1995a) suggest that calcium alginate gel is permeable to molecules with a molecular weight of >600 kD. It is expected that IgG (Mw = 154 kD), various proteins of the complement system (Mw = 24–570 kD), and cytokines (Mw = 10–30 kD) would have free access to the encapsulated grafts. One hypothesis is that the negatively charged molecular network in the alginate gel may restrict the entry of negatively charged immunoglobulins and complements, even though the gel has large pores. On the other hand, the negatively charged gel network may permit entry but also affect the physiologic state of the immunoglobulins and complements through electrostatic interactions, and thus interfere with their action cascade. That means, in addition to the expected dependency upon porosity, the diffusion as well as physiological function of proteins in alginate gel spheres appears to be interfered with by ionic interactions with the gel. There are several reports that support this hypothesis. Tanaka et al (1984) and Yuet et al (1993) found that negatively charged large proteins diffuse rapidly out of alginate gel beads. However, diffusion of the same proteins into the gel beads appeared to be restricted. In another experiment, Thu et al (1996b) found that lysozyme molecules, which are positively charged, are rapidly taken up by alginate gel beads. This suggests a marked effect of the ionic interaction between gel matrix and proteins and could also explain the finding that larger gel spheres provide better protection than smaller gel spheres, since larger spheres have more negatively charged core area. Zekorn et al (1990) suggested that nonspecific coatings of hollow fiber membranes with a molecular cut-off of 50 kD by negatively charged serum proteins may protect the islets inside of the hollow fibers from damage induced by the high dose IL-1 (Mw = 17 kD).

When discussing possible protective mechanisms, we should keep in mind that the mechanism of xenograft rejection of islets or other cell grafts between discordant species is very poorly understood, even when these cell grafts are unencapsulated (Hering, 1992; Wright and Yang, 1997) and that the role of xenoreactive natural antibodies in islet xenograft rejection is still controversial. It is possible that the remarkable protective effect imparted by uncoated alginate gel spheres reflects the significant roles of cell-mediated immunity and direct cellular contact in the process of xenograft rejection.

Future

Microencapsulation offers some distinct advantages, such as greater surface-to-volume ratio, ease of implantation, and ease of replacement, while avoiding several problems common to other types of encapsulation devices. Given alginate's mild gelling condition and superior biocompatibility, alginate-based microcapsules represent one of the easiest and safest immunoisolation systems available today. However, there are still many unanswered questions related to the mechanism(s) responsible for their protective effects, and it remains to be determined whether alginates will prove equally biocompatible in large animals and, ultimately, in humans. Based on preconceived notions derived from over 15 years of research using poly-L-lysine coated alginate microcapsules, the observation that uncoated alginate gel spheres protect either concordant or discordant islet grafts from rejection is an enigma. However, this unexpected observation will undoubtedly precipitate a myriad of studies that will help us better understand the immunoprotective effects of alginate gels and will permit further improvements in immunoisolation systems in the near future.

References

Adaniya GK, Roblero L, Rawlins RG, Miller IF, Quigg JM, Zaneveld LJD. 1993. First pregnancies and live births from transfer of sodium alginate encapsulated embryos in a rodent model. Fertil Steril 59:652–655.

Al-Shamkhani A, Duncan R. 1995. Radioiodination of alginate via covalently-bound tyrosinamide allows monitoring of its fate in vivo. J Bioact Compat Polymers 10:4–13.

Bodmeier R, Paertakul O. 1989. Spherical agglomerates of water insoluble drugs. J Pharm Sci 78:964–967.

Bowersock TL, HogenEsch H, Suckow M, Porter RE, Jackson R, Park H, Park K. 1996. Oral vaccination with alginate microsphere systems. J Controlled Release 39:209–220.

Chang TMS. 1964. Semipermeable microcapsules. Science 146:524–525.

Chick WL, Perna J, Lauris W, Low D, Galletti PM, Whittemore A, Like A, Colton CK, Lysaght M. 1977. Artificial pancreas using living beta cells: effects on glucose homeostasis in diabetic rats. Science 197:780–782.

Downs EC, Robertson NE, Riss IL, Plunkett ML. 1992. Calcium alginate beads as a slow-release system for delivering angiogenic molecules in vivo and in vitro. J Cell Physiol 152:422–429.

Falorni A, Basta G, Santeusanio F, Brunetti P, Calafiore R. 1996. Culture maintenance of isolated adult porcine pancreatic islets in three-dimensional gel matrices: morphologic and functional results. Pancreas 12:221–229

Flessner MF, Dedrick RL, Schultz JS. 1985. Exchange of macromolecules between peritoneal cavity and plasma. Amer J Physiol 248:H15–25.

Fraser RB, MacAulay MA, Wright JR Jr, Sun AM, Rowden G. 1995. Migration of macrophage-like cells within encapsulated islets of Langerhans maintained in tissue culture. Cell Transplant 4:529–534.

Hering BJ. 1992. Islet xenotransplantation. In: Ricordi C, ed. Pancreatic islet cell transplantation: 1892–1992—one century of transplantation for diabetes. Austin, Texas: RG Landes Company. p. 313–335.

Jankovsky M, Vasakova L. 1996. Immobilization in alginate gels. Vet Med Praha 41:159–164.

Kneafsey B, O'Shaughnessy M, Condon KC. 1996. The use of calcium alginate dressings in deep hand burns. Burns 22:40–43.

Lanza RP, Butler DH, Borland KM, Staruk JE, Faustman DL, Solomon BA, Muller TE, Rupp RG, Maki T, Monaco AP, Chick WL. 1991. Xenotransplantation of canine, bovine, and porcine islets in diabetic rats without immunosuppression. Proc Natl Acad Sci USA 88:11100–11104.

Lanza RP, Chick WL. 1997. Transplantation of encapsulated cells and tissues. Surgery 121:1–9.

Lanza RP, Kühtreiber WM, Ecker D, Staruk JE, Chick WL. 1995a. Xenotransplantation of porcine and bovine islets without immunosuppression using uncoated alginate microspheres. Transplantation 59:1377–1384.

Lanza RP, Kühtreiber WM, Ecker D, Staruk JE, Marsh J, Chick WL. 1995b. A simple method for transplanting discordant islets into rats using alginate gel spheres. Transplantation 59:1486–1487.

Lim F, Sun AM. 1980. Microencapsulated islets as bioartifical endocrine pancreas. Science 210:908–10.

Mackenzie D. 1996. Doctors farm fish for insulin. New Scientist 152(2056):20.

Marshall E. 1995. Gene therapy's growing pains. Science 269:1050–1055.

Martinsen A, Skjak-Braek G, Smidsrod O. 1989. Alginate as immobilization material: I. Correlation between chemical and physical properties of alginate gel beads. Biotechnol Bioeng 33:79–89.

Martinsen A, Storro I, Skjak-Braek G. 1992. Alginate as immobilization material: III. Diffusional properties. Biotechnol Bioeng 39:186–194.

McNeely WH, Pettitt DJ. 1973. Algin. In: Whistler RL, ed. Industrial gums—polysaccharides and their derivatives. New York, Academic Press. p 49–81.

Morris VJ. 1986. Gelation of polysaccharides. In: Mitchell JR and Ledward DA, ed. Functional properties of food macromolecules. New York, Elsevier Applied Science. p 121–128.

Mumper RJ, Hoffman AS, Poulakkainen PA, Bouchard LS, Gombotz WR. 1994. Calcium-alginate beads for the oral delivery of transforming growth factor-β_1 (TGF-β_1): stabilisation of TGF-β_1 by the addition of polyacrylic acid within acid-treated beads. J Controlled Rel 30:241–251.

Otterlei M, Espevik T, Ostgaard K, Skjak-Braek G, Soon-Shiong P, Smidsrod O. 1991. Induction of cytokine production from human monocytes stimulated with alginate. J Immunother 10:286–291.

Paige KT, Cima LG, Yaremchuk MJ, Schloo BL, Vacanti JP, Vacanti CA. 1996. De novo cartilage generation using calcium alginate-chondrocyte constructs. Plast Reconstr Surg 97:168–178.

Piacquadio D, Nelson DB. 1992. Alginates. A "new" dressing alternative. J Dermatol Surg Oncol 18:992–995.

Pier GB. 1991. Vaccine potential of Pseudomonas aeruginosa mucoid exopolysaccharide (alginate). Antibiot Chemother 44:134–142.

Prisant LM, Bottini B, DiPiro JT, Carr AA. 1992. Novel drug-delivery systems for hypertension. Am J Med 93:459–559.

Seymour LW, Duncan R, Strohalm J, Kopecek J. 1987. Effect of molecular weight (Mw) of N-(2–hydroxypropyl) methacrylamide copolymers on body distribution and rate of excretion after subcutaneous, intraperitoneal, and intravenous administration to rats. J Biomed Mater Res 21:1341–58.

Shimi SM, Newman EL, Hopwood D, Cushieri A. 1991. Semi-permeable microcapsules for cell culture: ultrastructural characterization. J Microencapsul 8:307–316.

Smidsrod O, Skjak-Braek G. 1990. Alginate as immobilization matrix for cells. TIBTECH 8(3):71–78.

Smidsrod O, Draget KI. Alginate gelation technologies. 1997. Special Publications of the Royal Society of Chemistry. 192:279–294.

Soon-Shiong P, Otterlie M, Skjak-Braek G, Smidsrod O, Heintz R, Lanza RP, Espevik T. 1991. An immunological basis for the fibrotic reaction to implanted microcapsules. Transplant Proc 23:758–759.

Soon-Shiong P, Feldman E, Nelson R, Komtebedde J, Smidsrod O, Skjak-Braek G, Espevik T, Heintz R, Lee M. 1992. Successful reversal of spontaneous diabetes in dogs by intraperitoneal microencapsulated islets. Transplantation 5:769–774.

Soon-Shiong P, Feldman E, Nelson R, Heintz R, Yao Q, Yao Z, Zheng T, Merideth N, Skjak-Braek G, Espevik T, Smidsrod O, Sandford P. 1993. Long-term reversal of diabetes by the injection of immunoprotected islets. Proc Natl Acad Sci USA 90:5843–5847.

Tanaka H, Matsumura M, Veliky IA. 1984. Diffusion characteristics of substrates in Ca-alginate gel beads. Biotechnol Bioeng 26:53–58.

Thu B, Bruheim P, Espevik T, Smidsrod O, Soon-Shiong P, Skjak-Braek G. 1996a. Alginate polycation microcapsules. 1. Interaction between alginate and polycation. Biomaterials 17:1031–1040.

Thu B, Bruheim P, Espevik T, Smidsrod O, Soon-Shiong P, Skjak-Braek G. 1996b. Alginate polycation microcapsules. 2. Some functional properties. Biomaterials 17:1069–1079.

Tomlinson E. 1986. Site-specific drug carriers. Eng-Med 15:197–202.

Tze WJ, Wong FC, Chen LM, O'Young S. 1976. Implantable artificial endocrine pancreas unit used to restore normoglycemia in the diabetic rat. Nature 264:466–467.

Wilson PR. 1996. Dressed to heal: new options for graft site dressing. Australas J Dermatol 37:157–8.

Wright JR Jr, Polvi S, MacLean H. 1992. Experimental transplantation using principal islets of teleost fish (Brockmann bodies): Long-term function of tilapia islet tissue in diabetic nude mice. Diabetes 41:1528–1532.

Wright JR Jr. 1994. Procurement of fish islets (Brockmann bodies). In: Lanza RP, Chick WL, ed. Pancreatic islet transplantation series. Vol. 1. Procurement of pancreatic islets. Austin, Texas: RG Landes. p 123–133.

Wright JR Jr, Yang H. 1997 Tilapia Brockmann bodies: An inexpensive, simple model for discordant islet xenotransplantation. Ann Transplant. 2(3):72–76.

Wright JR Jr, Yang H, Dooley KC. 1998. Tilapia—A source of hypoxia-resistant islets for encapsulation. Cell Transplant. 7:299–307.

Yang H, O'Hali W, Kearns H, Wright JR Jr. 1996. Reversal of diabetes in rodents by encapsulated fish islets (abstract). Cell Transplant 5:5S-2.

Yang H, Wright JR Jr. 1995. A method for mass harvesting islets (Brockmann bodies) from teleost fish. Cell Transplant 4:621–628.

Yang H, Dickson BC, O'Hali W, Kearns H, Wright JR Jr. 1997a. Functional comparison of mouse, rat, and fish islet grafts transplanted into diabetic nude mice. Gen Comp Endocrinol 106:384–388.

Yang H, O'Hali W, Kearns H, Wright JR Jr. 1997b. Long-term function of fish islet xenografts in mice by alginate encapsulation. Transplantation 64:28–32.

Yuet PK, Kwok WY, Harries TJ, Goosen MFA. 1993. Mathematical modelling of protein diffusion and cell growth in microcapsules. In: Goosen MFA, ed. Fundamentals in animal cell encapsulation and immobilization. Boca Raton, Florida, CRC Press. p 79–111.

Zekorn T, Siebers U, Bretzel RG, Renardy M, Planck H, Zgchocke P, Federlin K. 1990. Protection of islets of Langerhans from interleukin-1 toxicity by artificial membranes. Transplantation 50:391–394.

8
Immunoprotection of Islets of Langerhans by Microencapsulation in Barium Alginate Beads

Tobias D.C. Zekorn and Reinhard G. Bretzel

Introduction

Today the fate of diabetic patients is determined by the development of secondary complications such as blindness, neuropathy, and end-stage renal failure. The DCCT study[1] confirmed a positive correlation of intensified insulin treatment with a reduced development of these complications. Further improvement of metabolic control could be achieved by transplantation of insulin-producing islet cells. However, a principal obstacle for this method is side effects due to chronic application of the required immunosuppression, i.e., cyclosporine A-induced hypertension and islet toxicity,[2] that could be more harmful than the disease itself. Therefore, successful transplantation of islets entrapped in immunoprotective membranes (bioartificial endocrine pancreas = BAEP) would circumvent these problems and also allow transplantation of islet tissue from other species (i.e., pig) or of engineered cells. Moreover, after successful application of this principle to diabetes, it can be applied to other diseases such as phenylketonuria, pain, adrenal and neuroendocrine deficiency syndromes, hematologic diseases, etc.

A major task in BAEP research is to achieve biocompatibility. In the case of bioartificial pancreas, this requires that (1) the recipient's reaction does not hinder diffusional transport and (2) there is no release of substances such as interleukin-1[3] or nitric oxide by the recipient's immune system that can diffuse through the membrane and may affect survival and adequate function of the encapsulated islets. In addition, soluble substances released by the entrapped islets should not provoke an inflammatory or immunological reaction in the recipient.

Technology

Alginate

Since the early 1970s alginates have been used for the immobilization of cells.[4] The structure of calcium-alginate gels was described in principle as early as 1965[5] and later in more detail by others.[6] The size of the beads depends on a variety of factors such as viscosity, concentration, and type of alginate. It can be controlled by a coaxial air jet in a range of 200 to 5000 microns.[7] The production of smaller beads needs more sophisticated techniques.[8]

Already in 1980[9] alginates were introduced to surround single islets with an immunoprotective membrane. One reason for the use of this polymer from extracted seaweed is to obtain polymerization under physiological conditions.[4] Alginates are anionic polysaccharides composed of homopolymeric regions of β-D-mannuronic (M-block) and a-L-guluronic (G-block) acids interspaced with regions of mixed sequence. Because of their affinity to di-, tri-, and multivalent cations (calcium, barium, polylysine), alginates have hydrogel-forming properties.[5,6] Alginate is a natural polymer containing many impurities after the extraction process. A variety of mitogenic fractions within raw alginates has been described[10] that could be removed by free-flow electrophoresis or other techniques[11] without changing the material's in vitro characteristics. Otterlei et al[12] and Espevik et al[13] reported an in vitro activation of monocytes by soluble M-blocks. In contrast, Mai et al[14] found an immunosuppressive effect of Na-alginate related to increasing molecular size. Barium-complexed purified alginate did not induce lymphocyte proliferation in vitro and induced a negligible foreign body reaction after implantation of empty beads into rodents[15] only. However, the choice of the animal model has a major impact on the histological results of in vivo biocompatibility tests. When empty beads were implanted into dogs or pigs, a major reaction could be found directed toward the implants after only 3 weeks.[16] But, there are not only differences between species but also between strains or colonies: In BB/OK rats microbeads from purified material induce a remarkable reaction, whereas they do not when implanted into BB/Gi rats. To date, there are no reports of a predictable reaction against empty beads in human volunteers.

Today, a major concern about the use of alginate for microencapsulation is still the molecular variability of a natural product, which results in a wide, unpredictable range of physicochemical properties of the complexed material and a large intercapsular variation. This variation has been more precisely described for materials other than alginate.[17] In our hands a composite of polyethylenimine and barium-complexed alginate resulted in tighter capsules. However, we also observed an increased fibrotic reaction after transplantation of encapsulated islets, finally resulting in an early loss of graft function. Previously performed in vitro viability tests and implantation of empty capsules revealed promising results. Nevertheless, there are some indications that blending of different materials may overcome some of the addressed problems of alginates.[18]

Barium Complexation

Kierstan and Bucke[19] substituted calcium for the initially recommended complexation with aluminum ions. This technology is still used today in many areas.[20,21] However, in principle, alginates can be complexed by any polyvalent cation. Haug[22] described the differences in affinity of cations to alginates:

$$Pb > Cu > Cd > Ba > Sr$$
$$> Ca > Co, Ni, Zn > Mn.$$

The mechanical rigidity reflected the degree of affinity.[23] A stabilization of calcium-alginate beads could be achieved by supplementation with other multivalent cations such as Ti^{2+} (see Reference 24)

or Al^{2+} (see Reference 25). Alginates also form strong complexes with polycations such as chitosan,[26] polyamino acids,[9] polyethylenimine,[27] or polyacrylamide.[28] A complexation with polyamino acids can be strengthened by the addition of isocyanate.[29,30] Unfortunately, many of these methods cannot be applied to living cells.

Because calcium-complexed alginate beads are not as stable as others and the use of polycations may cause other difficulties, such as a lack of biocompatibility, we developed in our laboratory a complex with barium ions.[31] This is a simple, one-step technique that also provides an almost complete gelling of the alginate core. On the bead's surface, only alginate is exposed, and the complexation is strong enough to avoid any toxic effects to animals due, theoretically, to a release of small amounts of the cation.[32] Barium alginate beads are monocomponent microcapsules that provide complete encapsulation of islets without many disadvantages of other techniques, i.e., capsule rupture due to swelling of the core.

Encapsulation Procedure

In our experiments islets were encapsulated in barium-alginate beads made from purified alginate. Alginate (2% w/v, purified Manugel GHB, Kelco, Hamburg, Germany) was dissolved in 0.9% saline (pH 7.5). After we suspended the islets in alginate solution (2000 islets/ml), droplets were formed by a homemade nozzle with a surrounding air stream. Droplets became gel beads by falling into a precipitation bath containing 20 mM $BaCl_2$ in 10 mM isotonic MOPS-buffer (morpholinopropanesulfonic acid; Sigma, St. Louis, U.S.A.; 1 ml alginate in a minimum of 50 ml precipitation bath). After several (3–4×) intensive washings in saline (10 mM MOPS-buffer), the islet-containing beads (diameter 600 to 700 μm) were cultured overnight at 37°C in regular culture medium (i.e., RPMI 1640 + 10% heat-inactivated fetal calf serum).

Variations of this procedure can easily be made by changing the following variables: alginate viscosity, M-G-ratio, barium chloride concentration, temperature, alginate flow rate, air jet, and droplet generator. The use of purified material is recommended; however, an improperly performed purification may alter the material's physicochemical properties.

Several groups have tried to reduce capsule size in order to improve diffusional properties. Using barium for alginate gelling, we were able to reduce the size of a microcapsule to a 10 μm film conformally coating each single islet.[33] However, we often found some hints of local instabilities in this coat. De Vos and colleagues[34] demonstrated with alginate-polylysine capsules that smaller capsules often exhibit inadequate or particularly incomplete encapsulation. There is evidence that these findings may be a result of local inhomogeneities within alginate and consecutive irregularities in its complexation.

Biocompatibility and Immunology

In the case of bioartificial pancreas, the evaluation of biocompatibility requires consideration of three different aspects:

1. impact of the encapsulation procedure and materials on islet survival and function
2. reaction of a recipient towards implanted artificial material
3. interaction between host and encapsulated islets

Impact of the Encapsulation Procedure and Material on Islet Survival and Function

Neither conditions during the encapsulation procedure nor the resulting capsule itself should alter the islet. This requires an adjustment of the complexing agents to near physiological conditions. The material should not interact with the B-cell's glucose-sensing or pattern of insulin release (i.e., sulfonurea-like activity). After overnight culture, encapsulated rat islets respond to a glucose stimulus in a typical biphasic pattern, indicating a preservation of the mechanisms for rapid insulin release and biosynthesis.[31] In the case of the preparation of alginate-polylysine capsules, Fritschy et al[35] found a remarkable decrease in acute insulin release if calcium alginate has been resolved during the process by citrate instead of EGTA. Besides these data, very little is known, especially for long-term material-islet interactions. In principle, a prolongation of cell survival could be assumed if cells were embedded within a matrix.[20] Another aspect is the reduction of the supply of nutrients to the entrapped, nonvascularized tissue by the diffusion properties of the capsule, i.e., the encapsulation material itself and its adjusted molecular cut-off. This obstacle may have a major impact on long-term islet performance.[32,36]

Reaction of a Recipient Toward the Implanted Material

Function of a bioartificial pancreas requires a reaction of the recipient that does not result in a diffusion barrier finally leading to cell death. All attempts to evaluate biocompatibility in vitro[12,15] would show some lack of predictability in in vivo experiments. This is especially true for experiments testing liquid alginate, but not the gelled form used for in vivo experiments. Therefore, implantation experiments are necessary to evaluate the reaction of a recipient model towards the implanted device. The majority of experiments for testing a bioartificial pancreas have been performed in rodents,[37,38] and there are only a few reports on systematic experiments in large animal models.

The choice of an animal model should reflect the human situation. In diabetes and islet transplantation research, the diabetic BB-rat is accepted as one animal model of human autoimmune diabetes.[39,40] Therefore, many studies on biocompatibility have been performed in different breeding lines of these animals. However, it is still questionable whether this (these) model(s) represent(s) the human situation, given the BB-rat's immunological properties (macrophage activity, lymphopenia, etc.).

Most authors[38,40,41,42] have described lack of pericapsular fibrosis as "biocompatibility." However, local irritation of the environment by the operation procedure, the device itself, an antigen released from the device, etc., provokes an inflammatory infiltrate and may induce the release of substances[43] that are known to be toxic to encapsulated islets, e.g., interleukin-1 and NO, because these small inflammatory molecules are not excluded by an alginate-based microdevice.[32] Therefore, histological examination of intraperitoneally implanted capsules requires not only removal of the capsules by lavage but also a careful investigation of the peritoneal tissue. In our experience, even sham operations in rats, with injection of culture medium without any protein content, resulted in a slightly activated peritoneum for at least several weeks.

Three weeks after intraperitoneal implantation in Lewis and BB/Gi rats, barium-alginate beads (implanted volume: 2–3 ml; for encapsulation procedure, see I.3) were almost completely free from any fibrotic overgrowth.[16] Compared to this, the reaction was slightly increased after 15 weeks in Lewis rats. Besides this negligible fibrotic overgrowth, at both times we found a mild inflammatory infiltration within the peritoneal fatty tissue (biopsy). This infiltration was located not only in areas of microcapsules but also in areas without any microcapsules, indicating a persistent peritoneal reaction to implanted, free-floating beads.

To investigate the reaction towards beads that are not free-floating within the peritoneal cavity but localized, we implanted large and small barium-alginate beads beneath the kidney capsule of BB/OK rats and German landrace pigs, respectively. After 3 weeks, the implants were removed and examined histologically (for details, see 16).

In rats, large beads (600 μm Ø) were surrounded by a 5–10 cell layer wall of fibrotic tissue. Planimetric evaluation of serial sections (reactive area per μm surface) confirmed this qualitative impression: large beads provoke a significantly intense reaction compared to small beads. In pigs, there was a similar reaction. However, the space between single beads was filled with scar tissue. Moreover, pigs evidenced many giant cells close to the bead's surface. A pig-specific reactivity is likely since the appearance of giant cells is typical for a variety of granulomatosus inflammatory reactions in this animal model.

Additional experiments in dogs with implantation of beads beneath the renal capsule revealed a similar reaction as described in pigs. However, in contrast to the results from pigs, only very few giant cells could be observed.

From these experiments we conclude that implantation of empty capsules in humans will not result in a complete lack of reactivity. Different material *and* recipient-related factors influence the process, finally resulting in a slight infiltrate or severe fibrosis. Careful investigation and control of procedures and materials may lead to a material that provokes only a minor reaction that does not hinder the device's intended function. The heterogeneity of patients may lead to the necessity of using at least an initial immunosuppression to control the nonspecific reaction against a bioartificial pancreas. Using technologies other than barium-alginate-based microcapsules, several groups have shown that this reaction may be reduced by antiphlogistic treatment.[37,40]

Interaction Between Host and Encapsulated Islets

The aim of using a membrane is to separate the recipient's immune system from its target. For this idea the term "*immunoisolation*" has been introduced. However, there is strong evidence that there is not a strict immunological "isolation" of encapsulated islets, and there are even some antigens, such as insulin, that should be released by the devices. In rats, we observed an increasing fibrotic overgrowth and inflammatory reaction of the peritoneal fatty tissue when using microencapsulated allografts as opposed to syngeneic grafts.[44] This reaction was even a more intense with xenografts.[16] By means of mixed lymphocyte islet reaction, our group has demonstrated a lymphoid activation by encapsulated islets[15] in the same donor-recipient combination (rat-to-mouse) in which we had previously found, in one third of the recipients, a long-lasting normalization of random blood glucose.[31] Also, in the case of polylysine-alginate capsules, we[45] and others[46] have described a similar immunological recognition of encapsulated cells by the recipient. To date, there is no systematic study available demonstrating the impact of a reduced molecular cut-off on the immunological interaction between encapsulated graft and host. The factors responsible for recognition and secondary damage of the encapsulated graft still have to be identified. They are not necessarily released by endocrine cells but may also be released by other cells such as endothelial cells, macrophages, lymphocytes, etc. Upon application of an immunoaltering protocol, we observed a remarkable improvement of allograft function[47] in rats. Interestingly, this effect was more evident if islets were kept for 2 weeks in a low-temperature culture, free-floating rather than encapsulated.[48] Unfortunately, we were not able to transfer these results from an allogeneic to a xenogeneic donor-recipient combination utilizing porcine islets (Figure 8.1). In diabetic mice, however, microencapsulated porcine islets resulted, in

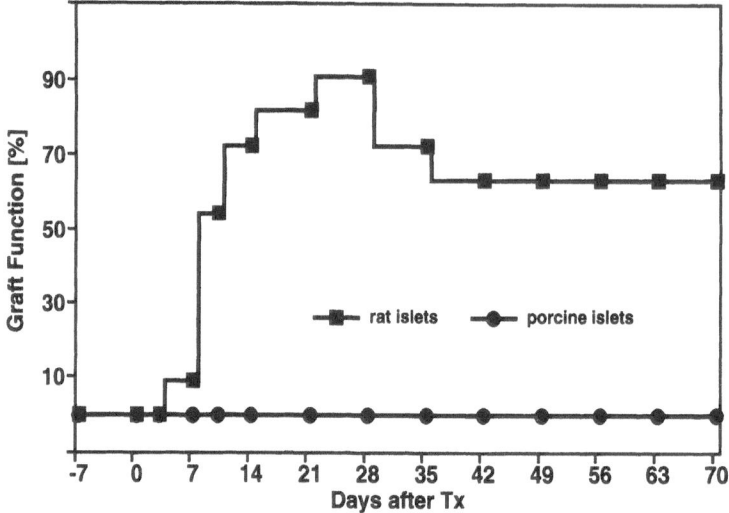

FIGURE 8.1. Graft function (random blood glucose >200 mg/dl) after allogeneic (3500 Wistar rat islets, n = 11) and xenogeneic (6000–7000 porcine, n = 6) transplantation of islets into streptozotocin-diabetic Lewis rats. Before transplantation, microencapsulated islets were kept in 22°C culture during 14 days for immunoalteration.

30% of the recipents, in a long-lasting graft function,[31] while nonencapsulated islets did not show any evidence for graft function.

Function and Secondary Complications

The idea behind a bioartificial pancreas is to transplant mechanically immunoprotected insulin-secreting cells into a diabetic organism. The devices have to react to a glucose challenge in a distinct manner with an appropriate release of insulin. This requires a comparable physiology between recipient and insulin-producing tissue/cells. A principal disadvantage of all kinds of micro- and macrodevices is the diffusional transport through membranes that leads to a prolonged reaction time, compared to vascularized grafts.

Using barium-alginate beads for immunoprotection, we were able to demonstrate allogeneic graft function for at least for 3 months in 70% of the transplanted rats. To exclude any immunological interaction we performed syngeneic transplantation experiments in rats and were able to keep glucose metabolism almost normalized at least for 6 months.[32]

Besides dispensing with exogenous insulin, transplantation of microencapsulated insulin-producing tissue has to establish a glucose regulation that minimizes or even abolishes the risk of diabetic late complications. Preliminary results from our experiments give evidence that the onset of retinopathy may be prevented after successful transplantation of microencapsulated rat islets.

Summary and Conclusion

Microencapsulation of islets in barium-alginate beads for transplantation in insulin-deficient diabetes is a simple one-step procedure resulting in a monocomponent microcapsule that provides some degree of immunoprotection. However, a certain persistent immunological interaction between encapsulated graft and recipient has been proven both in vitro and in vivo. Encapsulated islets that are transplanted intraperitoneally and remain nonvascularized are able to correct the recipient's diabetic state sufficiently for at least several months. Further investigations are being directed to the production of defined, purified, and homogeneous alginate that is complexed under controlled conditions. Moreover, production of hydrogels with a low molecular cut-off will most likely require a composite material. In contrast, besides technical aspects, other issues such as biocompatibility and performance of the encapsulated, nonvascularized, dystopically transplanted islet[34] remain to be evaluated critically in the development of a bioartificial pancreas. It has been conclusively demonstrated by numerous groups that such a therapeutic approach is possible in principle. However, further metabolic and immunological studies are necessary to define mechanisms that lead to the loss of an immunoprotected graft before this principle of treatment can be applied to treatment of diabetic patients.

References

1. The diabetes control and complications trial research group. 1993. The effect of intensive treatment of diabetes on the development and progression of long-term complications in insulin-dependent diabetes mellitus. NEJM 329: 977–986.
2. Alejandro R, Feldman EC, Bloom AD, Kenyon NS. 1989. Effects of cyclosporin on insulin and C-peptide secretion in healthy beagles. Diabetes 38: 698–703.
3. Mandrup-Poulsen T, Bendtzen K, Nerup J, Egeberg J, Nielsen JH. 1986. Mechanisms of pancreatic islet cell destruction. Allergy 41: 250–259.
4. Hackel U, Klein J, Megenet R, Wagner F. 1975. Immobilization of microbial cells in polymeric matrices. Eur. J. Appl. Microbiol. Biotechnol. 1: 291–293.
5. Baardseth E. 1965. Localization and structure of alginate gels. In: Young EG, McLachlan JL, editors. Proceedings of the fifth international seaweed symposium. Oxford: Pergamon Press; p. 19–28.
6. Skjak-Braek G, Martinsen A. Applications of some algal polysaccharides in biotechnology. 1991. In: Guiry MD, Blunden G, editors. Seaweed resources in Europe: uses and potentials. John Wiley & Sons Ltd.; p. 219–257.
7. Klein J, Vorlop KD, Eng H, Kluge HM, Washausen P. 1979. Procedures for polymer entrapment of whole cells. Deutsche Gesellschaft für chemisches Apparatewesen. Monographien. 84: 274–276.
8. Gröhn P, Klöck G, Schmitt J, Zimmermann U, Horcher A, Bretzel RG, Hering BJ, Brandhorst D, Brandhorst H, Zekorn T, Federlin K. 1994. Large-scale production of Ba2+-alginate-coated islets of Langerhans for immunoisolation. Exp. Clin. Endocrinol. 102: 380–387.
9. Lim F, Sun A. 1980. Microencapsulated islets as bioartificial endocrine pancreas. Science 210: 908–910.
10. Zimmermann U, Klöck G, Federlin K, Hannig K, Kowalski M, Bretzel RG, Horcher A, Entenmann H, Siebers U, Zekorn T. 1992. Production of mitogen-contamination free alginates with variable ratios of mannuronic acid to guluronic acid by free flow electrophoresis. Electrophoresis 13: 269–274.
11. Klöck G, Frank H, Houben R, Zekorn T, Horcher A, Siebers U, Woehrle M, Federlin K, Zimmermann U. 1994. Production of purified alginates suitable for use in immunoisolated transplantation. Appl. Microbiol. Biotechnol. 40: 638–643.
12. Otterlei M, Ostgaard K, Skjak-Braek G, Smidsrod O, Soon-Shiong P, Espevik T. 1991. Induction of cytokine production from human monocytes stimulated with alginate. J. Immunother. 10: 286–291.
13. Espevik T, Otterlei M, Skjak-Braek G, Ryan L, Wright SD, Sundan A. 1993. The involvement of CD 14 in stimulation of cytokine production by uronic polymers. Eur. J. Immunology 23: 255–261.
14. Mai GT, Seow WK, Pier GB, McCormack JG, Thong YH. 1993. Suppression of lymphocyte and neutrophil functions by Pseudomonas aeruginosa mucoid exopolysaccharide (alginate): reversal by physiochemical, alginase, and specific monoclonal antibody treatments. Infection and Immunity 61: 559–564.
15. Zekorn T, Klöck G, Horcher A, Siebers U, Woehrle M, Kowalski M, Arnold MW, Federlin K, Bretzel RG, Zimmermann U. 1992. Lymphoid activation by different crude alginates and the effect of purification. Transpl. Proc. 24: 2952–2953.
16. Zekorn TDC, Horcher A, Mellert J, Siebers U, Altug T, Emre A, Hahn H-J, Federlin K. 1996. Biocompatibility and immunology in encapsulation of islets of Langerhans (bioartificial pancreas). Int. J. Artif. Organs 19: 251–257.
17. Uludag H, Hwang JR, Sefton MJ. 1995. Microencapsulated human hepatoma cells: capsule-to-capsule variations in protein secretion and permeability. J. Control. Release 33(2): 273–283.
18. Wang T, Lacik I, Brissiva M, Anilkumar AV, Prokop A, Hunkeler D, Green R, Shahrokhi K, Powers AC. 1997. A new generation capsule and encapsulation system for immunoisolation of pancreatic cells. Nature Biotechnology 15: 358–362.
19. Kierstan M, Bucke C. 1977. Immobilization of microbial cells, subcellular organelles, and enzymes in calcium alginate gels. Biotechnol. Bioeng. 19: 387–397.
20. Cheetham PSJ, Garrett C, Clark J. 1985. Isomaltose production using immobilized cells. Biotechnol. Bioeng. 27: 471–480.
21. Duff RG. 1985. Microencapsulation technology: A novel method for monoclonal antibody production. Trends Biotechnol. 3: 167–170.
22. Haug A. 1964. Composition and properties of alginates. Rep. Norw. Inst. Seaweed Res. 30: 25–45.
23. Smidsrød O, Haug A. 1972. Properties of poly(1,4–hexuronates) in gel state. II. Comparison

of gels of different chemical composition. Acta Chem. Scand. 26: 79–88.

24. Burns MA, Kvesitadze GI, Graves DJ. 1985. Dried calcium alginate/magnetite spheres: A new support for chromatographic separations and enzyme immobilization. Biotechnol. Bioeng. 27: 137–145.

25. Rochefort WE, Rehg T, Chau PC. 1986. Trivalent cation stabilization of alginate gel for cell immobilization. Biotechnol. Lett. 8: 115–120.

26. Rha CK. 1984. Chitosan as a biomaterial. In: Colwell RR, Pariser ER, Sinskey AJ, editors. Biotechnology in marine sciences. New York: Wiley; p. 177–189.

27. Veliky IA, Williams RE. 1981. The production of ethanol by Saccharomyces cerevisiae immobilized in polycation-stabilized calcium alginate gels. Biotechnol. Lett. 33: 275–280.

28. Rosevar A. Improvements in composite materials. European Patent 048109A2.

29. Kitajima M, Sekiguchi W, Kondo A. 1971. A modification of red blood cells by isocyanates. Bull Chem. Soc. Jap. 44: 139–143.

30. Sato T, Chiba T, Yoshinaga K, Kitajima M, Tershima M. 1988. Improvement of in vivo stability of alginate-polylysine capsules. Tohuku J. Exp. Med. 155: 271–274.

31. Zekorn TDC, Horcher A, Siebers U, Schnettler R, Hering B, Zimmermann U, Bretzel RG, Federlin K. 1992. Barium-cross-linked alginate beads: A simple, one-step-method for successful immuno isolated transplantation of islets of Langerhans. Acta Diabetologica 29: 99–106.

32. Zekorn T. unpublished data.

33. Zekorn T, Siebers U, Horcher A, Schnettler R, Zimmermann U, Bretzel RG, Federlin K. 1992. Alginate coating of islets of Langerhans: In vitro studies on a new method for immuno isolated transplantation. Acta Diabetologica 29: 41–45.

34. De Vos P, De Haan B, Pater J, Van Schilfgaarde R. 1996. Association between capsule diameter, adequacy of encapsulation, and survival of microencapsulated rat islet allografts. Transplantation 62: 893–899.

35. Fritschy WM, Wolters GHJ, van Schilfgaarde R. 1991. Effect of alginate-polylysine-alginate microencapsulation on in vitro insulin release from rat pancreatic islets. Diabetes 40: 37–43.

36. De Vos P, Nieuwenhuizen A, De Haan BJ, van Schilfgaarde R. 1997. An imbalance between beta-cell birth and beta-cell death contributes to failure of encapsulated islet grafts. Diabetologia 40 (Suppl. 1): A 124.

37. Christenson L, Aebischer P, McMillan P, Galletti PM. 1989. Tissue reaction to intraperitoneal polymer implants: Species difference and effects of corticoid and doxyrubicin. J. Biomed. Mater. Res. 23: 705–718.

38. Cole DR, Waterfall M, McIntyre M, Baird JD. 1992. Microencapsulated islet grafts in the BB/E rat: a possible role for cytokines in graft failure. Diabetologia 35: 231–237.

39. Woehrle M, Markman JF, Silvers WK, Barker CF, Naji A (1986) Transplantation of cultured pancreatic islets to BB rats. Surgery 10: 334–341.

40. Wijsman J, Atkinson P, Mazaheri R, Garcia B, Paul T, Vose J, O'Shea G, Stiller C. 1992. Histological and immunopathological analysis of recovered encapsulated allogeneic islets from transplanted diabetic BB/W rats. Transplantation 54: 588–592.

41. Siebers U, Sturm R, Planck H, Zschocke P, Bretzel RG, Zekorn T, Federlin K. 1990. Morphological studies on biocompatibility of membranes for immunoisolated islet transplantation. Horm. Metabol. Res. Suppl. 25: 206–208.

42. Darqui S, Chicheportiche D, Capron F, Boitard C, Reach G. 1990. Comparative study of microencapsulated rat islets implanted in different diabetic models in mice. In: Federlin K, Bretzel RG, Hering BJ, editors. Methods in islet transplantation research. Horm. Metabol. Res. Suppl. 25: 209–213.

43. Kröncke K-D, Funda J, Berschick B, Kolb H, Kolb-Bachofen V. 1991. Macrophage cytotoxicity towards isolated rat islet cells: Neither lysis nor its protection by nicotinamide are beta-cell specific. Diabetologia 24: 232–238.

44. Horcher A, Zekorn T, Siebers U, Klöck G, Frank, H, Houben R, Bretzel RG, Zimmermann U, Federlin K. 1994. Transplantation of microencapsulated islets in rats: Evidence for induction of fibrotic overgrowth by islet allo-antigens released from microcapsules. Transpl. Proc. Transpl. Proc. 26(2): 784–786.

45. Zekorn TDC, Endl U, Horcher A, Siebers U, Bretzel RG, Federlin K. 1995. Evidence for antigen-release induced cellular immune response against alginate-polylysine encapsulated islets. Xenotransplantation 2: 116–119.

46. Vandenbossche GMR, Bracke ME, Cuvelier CA, Bortier HE, Marbel MM, Remon J-P. 1993. Host reaction against alginate-polylysine microcapsules containing living cells. J. Pharm. Pharmacol. 45: 121–125.

47. Zekorn T, Horcher A, Siebers U, Federlin K, Bretzel RG. 1999. Synergistic effect of microencapsulation and immunoalteration on islet allograft survival in bioartificial pancreas. J. Mol. Med. 77:193–198.

48. Horcher A, Siebers U, Bretzel RG, Federlin K, Zekorn T. 1995. Transplantation of microencapsulated islets in rats: Influence of low temperature culture before and after the encapsulation procedure on graft function. Transpl. Proc. 27: 3232–3233.

9
Agarose

Hiroo Iwata and Yoshito Ikada

Introduction

Microencapsulation is the envelopment of small solid particles, liquids, or gas bubbles with coating. Many methods and materials for preparing microcapsules have been developed and employed in cosmetics, pharmaceutics, and agricultural products. A core material is encapsulated through phase separation of polymer solutions, *in situ* interfacial polymerization, spray drying, or other techniques. Since most of the techniques used include processes in which the core material is exposed to organic solvents, high temperature, or chemical reactions, they are not applicable to the microencapsulation of mammalian cells, which are so fragile that they are easily destroyed when exposed to such radical processes. Few methods that can successfully microencapsulate mammalian cells without damaging them have been reported. Polysaccharides, such as alginic acid (Lim and Sun 1980), agarose (Iwata et al 1989) and chitosan (Matthew et al 1993), have been used for microencapsulation of mammalian cells, because insoluble membranes or firm hydrogels can be formed from aqueous solutions of these polysaccharides by bringing the solutions into contact with polyvalent cationic metal ions or polyelectrolytes, and by lowering the temperature of the solutions.

This chapter gives an overview of research on microencapsulation of mammalian cells, especially islets of Langerhans (islets), in agarose gel and their application as a bioartificial pancreas (BAP).

Agarose

Agar, obtained from seaweed, *Gelidium* and *Gracilaria*, is a mixture of agarose and agaropectin with the average agarose content 55 ~ 65% (Meer, 1980). Agarose is a gelling agent predominantly composed of repeating units of alternating β-D-galactopyranosyl and 3, 6-anhydro-α-L-galactopyranosyl coupled through 1 → 3 (Figure 9.1). Agaropectin has a similar structure with addition of 4–10% ester sulfate and D-glucuronic acid. Upon heating and cooling of its aqueous solution, agarose forms a thermally reversible gel that is hard and rather brittle, but not rubber-like in its mechanical behavior. Above the melting point of the gel, agarose chains are believed to exist in solution as random coils, while cooling of the agarose solution leads to formation of a three-dimensional network in which double helices form junction zones between the chains (Clark and Ross-Murphy 1987). Agarose is unique among polysaccharides, because gelation occurs at temperatures far below the gel-melting temperature (Dumitriu et al, 1996). Most agarose solutions congeal around 40°C to form a firm gel but do not melt below 80°C. This large hysteresis is advantageous in microencapsulation of cells. Physical properties of the agarose gel and temperatures of the gelation and melting depend on the source of seaweed from which it is extracted. They can also be altered by chemical modifications. Agarose with the gelling temperature lower than 37°C is suitable for microencapsulation of mammalian cells.

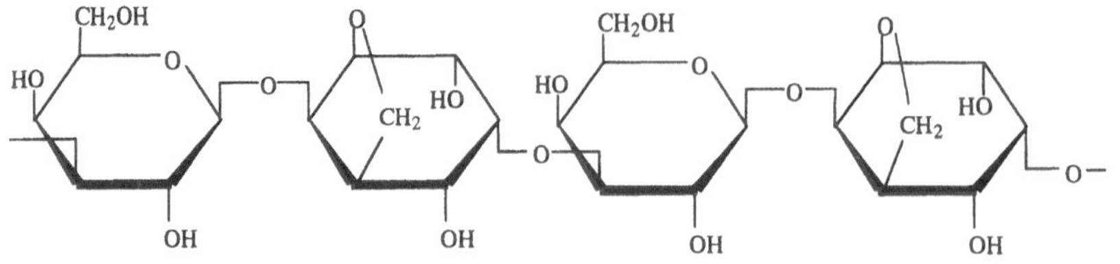

FIGURE 9.1. Molecular structure of agarose.

Pure agarose gels are fairly stable at room temperature, and few microorganisms metabolize agarose or produce enzymes that degrade it. It is generally thought that agarose cannot be metabolized by mammals. As will be mentioned below, implanted agarose microcapsules stay intact in a mouse peritoneal cavity for more than 200 days and are therefore stable enough to be applied for at least several hundred days of implantation into the body.

Agar, a mixture of at least two polysaccharides, agarose and agaropectin, is approved for food use and is in the GRAS (Generally Recognized As Safe) list under the Food and Drug Act in the United States of America. Investigations by Frohberg and coworkers (Frohberg et al 1969) disclosed that intraperitoneal daily injection of 0.2 ml of 1% agar solution from the 11th to the 15th day of gestation in NMRI mice produced no fetotoxic effects.

Agarose has many advantages over other polysaccharides as a material to microencapsulate mammalian cells, as it is easily handled and nontoxic, and can form a firm, stable gel.

Microencapsulation of Mammalian Cells

Mammalian cell technology has attracted considerable interest because of the capacity of animal cell cultures to synthesize or transform complex compounds such as antibodies, viral vaccines, immunochemicals, hormones, and enzymes. Agarose was first applied to microencapsulation of mammalian cells by Mosbach et al (Scheire et al 1983, Nilsson et al 1983). They have demonstrated that entrapment in the gel gives increased stability to normally fragile cells, enables high cell densities within the microcapsules, and makes such preparations suitable for continuous operation. However,

their work has not been extended very much, although this technology was applicable in preparation of bioartificial organs, that is, artificial organs prepared by a combination of viable cells and biomaterials. Islets, clusters of cells in the pancreas that produce insulin, were microencapsulated in agarose hydrogel to prepare a BAP (Gin et al 1987, Iwata et al 1985), while microencapsulated hepatocytes were examined to prepare a liver assist device (Enosawa et al 1995).

Two procedures used to microencapsulate islets in agarose hydrogel are illustrated in Figure 9.2 and briefly explained below.

Procedure I (Iwata et al 1992, Nilsson et al 1983): As schematically shown in Figure 9.2 (a), a required amount of low gelling temperature agarose (0.15 g for 5% hydrogel) is suspended in 3 ml of serum-free Eagle's minimum essential medium (MEM) solution in a 50 ml glass centrifuge tube. The suspension is autoclaved to dissolve the agarose and sterilize the solution, which is then mixed well, cooled, and kept at 37°C for 10 min. Several thousand islets are suspended in approximately 50 μl of MEM and mixed with the agarose solution. Twenty ml of paraffin oil already sterilized by autoclaving and kept at 37°C is added to the glass tube, which is agitated manually until agarose solution droplets of 500 to 800 μm in diameter are formed. To induce gelation of the agarose droplets, the glass tube is immersed in an ice bath for 5 min under gentle agitation. Thirty milliliters of ice-cold Hanks' balanced salt solution is added, followed by centrifugation at 800 $\times$ g for 20 min. The oil phase is removed by suction, and the resulting microcapsules are washed several times with Hanks' solution. A phase-contrast micrograph of microencapsulated hamster islets is shown in Figure 9.3. An islet composed of several thousand endocrine cells is seen in the core part of each microcapsule. The layer from the surface of

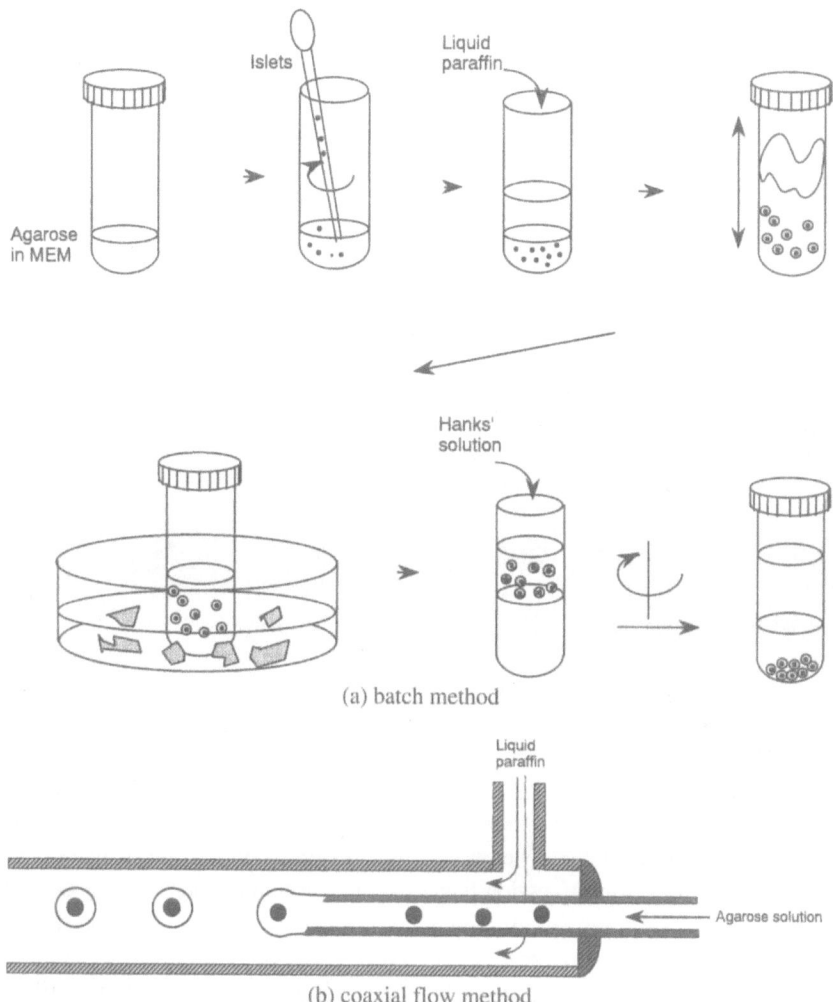

FIGURE 9.2. Two procedures for microencapsulation of mammalian cells in agarose hydrogel. (a) batch method, (b) coaxial flow method.

the islet to the outer surface of the microcapsule is composed of a uniform agarose hydrogel membrane. This probably gives a certain mechanical strength to the microcapsule.

Procedure II (Gin et al 1987): Microencapsulation is done using a system of two coaxial nozzles, shown in Figure 9.2 (b). The needle of a syringe (nozzle I) filled with islet suspension in agarose solution is placed inside a slightly larger plastic capillary tube (nozzle II) through which paraffin oil is pumped. The two media are forced through the tubes using a peristaltic pump connected to a T-piece. The flow rate of each medium is adjusted to retrieve the agarose solution as droplets by the paraffin oil flow and to form perfectly spherical agarose microcapsules. Their diameter is a function of the internal diameter of nozzle I and the flow rate of the agarose solution. The end of the tube is cooled in a water bath at 4°C to allow the droplets to set to gel. They are collected in an aqueous medium to rapidly remove the hydrophobic medium.

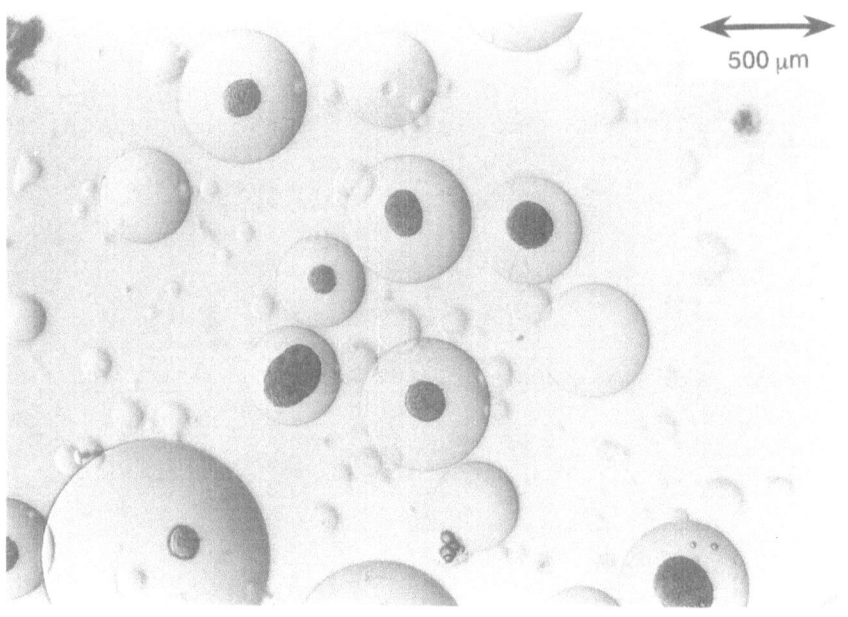

FIGURE 9.3. Microencapsulated hamster islets in 5% agarose hydrogel.

Bioartificial Pancreas

More than 8 years have already passed since the first series of successful clinical trials of islet transplantation with high-dose immunosuppression therapy. Complications caused by intensive immunosupressive therapy still remain to be overcome. An attractive solution to this problem is BAP composed of islets enclosed in a semipermeable membrane to repel components of the host immune system that cause rejection, but to permit the passage of oxygen, nutrients, glucose, and insulin between encapsulated living islet cells and the host. The transplanted islets, isolated from the host immune system by the semipermeable membrane, are expected to survive and be able to control glucose metabolism for a long period without administration of immunosuppresive drugs.

In Vitro Culture
of Microencapsulated Islets

In vitro culture of microencapsulated islets has been employed to examine their functions and preservation. Microencapsulated hamster islets in agarose hydrogel maintained their insulin-release ability for more than 100 days in culture media supplemented with and without serum, while the native nonencapsulated islets gradually lost their insulin-release ability with excessive proliferation of contaminated fibroblasts in culture medium with serum or disintegrated during several days of culturing in serum-free medium (Iwata et al 1991). Human islets are more fragile than those of hamster. Mintz et al (Brendel et al 1994) attempted a combination of microencapsulation of islets in agarose hydrogel and in vitro culture to preserve human islets. The absolute number and volume of islets retrieved from the hydrogel were significantly higher than those in the conventional free-floating media after 2 and 4 weeks of culture. They also evaluated the effectiveness of islets for reversal of insulin-dependent diabetes mellitus by transplantation of the islets under the kidney capsule of diabetic nude mice. Only the islets embedded in the agarose hydrogel were consistently able to induce normoglycemia in diabetic recipients after 14 days of culture.

Seven to ten thousand islet equivalents (IEQ, average diameter 150 μm) per kilogram of recipient body weight are considered to be the critical mass needed to induce normoglycemia (Hering et al 1996). Several donors are required for consistently obtaining this critical amount. Cryopreservation and free-floating suspension cultures are utilized as the main islet preservation methods, but both of them result in a substantial loss of islet viability. Preservation of islets in agarose gel matrix by in vitro culturing may be of benefit for accumulation of functionally competent islets, thus facilitating the implementation of clinical protocols that utilize freshly isolated islets from multiple donors, without the need for cryopreservation.

Semipermeable membranes with the cut-off molecular weight of 50 kD, that is, an intermediate molecular weight between insulin (6 kD) and immunoglobulin IgG (160 kD), have been preferentially used for preparing BAPs. However, unknown components with higher molecular weights than 50 kD that might be fed from the blood and/or body fluid might be needed for islet survival. However, the study mentioned above (Iwata et al 1991) demonstrated that microencapsulated islets maintained insulin release in the range of $250 \sim 400$ μU/islet/day for more than 100 days, even in serum-free RPMI 1640 medium. This fact shows that islets can survive and preserve the ability to release insulin in a medium containing only low molecular weight chemicals, such as amino acids, vitamins, salts, and glucose. The highest molecular weight component in RPMI 1640 is vitamin B12, with the molecular weight of 1355 De. In BAPs, insulin with the molecular weight of 6 kD should freely permeate through the membrane into the host to control glucose metabolism. The molecular weights of components of RPMI 1640 and insulin indicate that a membrane with a cut-off molecular weight larger than 6 kD is applicable to BAP if its permeability is high enough.

Graft Rejection

An allograft is a graft between genetically different individuals within one species, while a xenograft is a graft between individuals from different species (Janeway and Travers 1997, Paul 1991). The usual sequence of allograft rejection in unsensitized recipients is less rapid than that of allograft rejection in sensitized recipients and that of xenograft rejection, and is known as first set rejection. Histocompatibility antigens presented on the surface of grafted cells induce an immune response in graft recipients who lack these antigens. The T cell immunity directed against the alloanti-gens probably results from their similarity to self major histocompatibility complex molecules. The activation of cellular immunity by the T cells is the predominant cause of first set allograft rejection.

T cells might actually have a weaker affinity for disparate xenoantigens since xenoantigens are more foreign than alloantigens, and hence T cell responses to xenontigens are generally weaker than to alloantigens. The humoral immunity resulting from immunoglobulins and complement proteins dominates the rejection process of xenografts. Many animals have natural antibodies that react with cells and tissues from other species, although they are totally unrelated to those species and have not been previously exposed to xenogeneic tissues. Even when the natural antibody levels are low or undetectable in some donor-recipient combinations, recipients of xenografts are induced to form new antibodies against the xenoantigens, which are important components of the humoral response to xenografts. Both types of antibodies are significant obstacles.

In summary, cell-mediated immunity plays a major role in first set allograft rejections, whereas humoral immunity induces major damage to xenografts. Thus, the properties required for semipermeable membranes used in BAP depend on the source of islets, allo- and xenogeneic.

Allograft

The effect of microencapsulation on allo-islet survivals was studied using streptozotocin-induced diabetic (STZ-diabetic) mice and nonobese diabetic (NOD) mice as recipients (Iwata 1992). The STZ-diabetic BALB/c mice intraperitoneally received 1500 C57BL/6 mice islets microencapsulated into hydrogel with 5% agarose concentration. In control experiments, 1500 native islets without microencapsulation were also intraperitoneally transplanted. The recipients with native islets demonstrated less than 11−day normoglycemic periods, but all of the STZ-induced diabetic BALB/c mice receiving microencapsulated islets maintained normoglycemia indefinitely. Agarose microcapsules with originally spherical shape could be retrieved intact from the recipient at 235 postoperative days. There are arguments about the significance of infinite normoglycemia because the possibility of B cells or B-cell tumor generation in STZ-induced diabetic animals

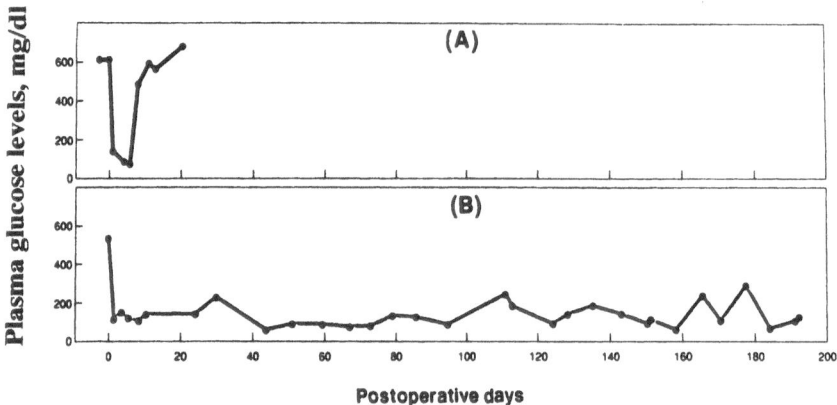

FIGURE 9.4. Nonfasting plasma glucose levels of recipient NOD mice after intraperitoneal implantation of 1500 islets. (A), native islets, (B), microencapsulated islets.

cannot be ruled out (Richardt et al 1984). NOD mice in which diabetes is induced by autoimmune reactions are a better choice for evaluating the function of BAP. Functional graft survival periods in NOD mice were 46, 83, and 94 days for 3 of 5 recipients, and the grafts in the remaining 2 mice were functioning until the animals were sacrificed at 102 and 192 postoperative days. Plasma glucose level changes of the NOD mice that showed the longest normoglycemic period are shown in Figure 9.4, along with those of a recipient of native islets. The microcapsule made of 5% agarose can effectively prolong allo-islet functioning in diabetic mice. The agarose microcapsule must act as a mechanical barrier, prohibiting contact between the islets and the recipient immunocompetent cells that have a major role in first set allograft rejection, thus effectively immunoisolating the allo-islets.

Xenograft

Long-term reversals of diabetes in the recipients of microencapsulated allo-islets are encouraging, but allogeneic donors—human donors—of pancreatic tissue are limited in clinical practice. The use of animal islets may overcome the shortage of human donors.

Agarose Microcapsule

Performance of BAPs containing xeno-islets has been studied by intraperitoneal implantation of 1000 hamster islets in 5% agarose microcapsules

into STZ-diabetic BALB/c mice (Iwata et al 1994a). Immunoisolative effectiveness of the 5% agarose microcapsule was limited in xenotransplantation because the grafts in 12 of 16 recipients lost function within 20 days after implantation. The survival periods of allo- and xenografts reflect the difference in the mechanism of graft rejection between allo- and xenotransplantation.

As mentioned above, humoral immunity is suspected to play a major role in the rejection of the xenografts. Therefore, permeation of IgG through the microcapsules was assessed, using Protein A-Sepharose, which is able to selectively bind IgG (Iwata et al 1995). The microcapsules containing the Protein A-Sepharose were placed in human serum. IgG, which diffused through the microcapsules and attached to the Protein A-Sepharose, was visualized by immunostaining. After 5–days incubation, the Protein A-Sepharose in 5% agarose microcapsules was deeply stained, clearly indicating that the 5% agarose hydrogel could not prevent IgG penetration. However, antibody binding to the antigens on the islet cell surface alone cannot do lethal damage to the islet cells. For destruction of islet cells, complement proteins must be activated by immune complexes on the cell surface. To examine the protective effect of the microcapsule, against the complement attack, permeation of active complement proteins through the microcapsule was assessed using sheep erythrocytes coated with the antibody against Forssman antigen (EA) (Morikawa et al 1996). Following encapsulation in agarose microcapsules, EA was incubated in

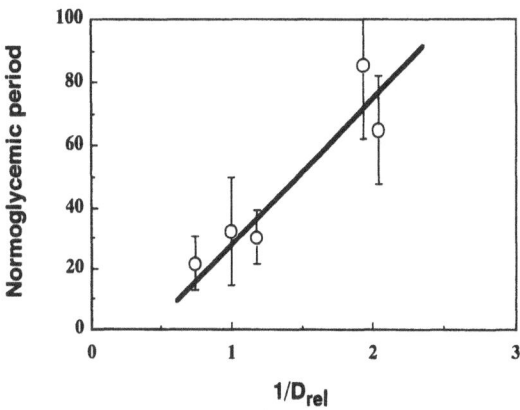

FIGURE 9.5. Dependence of xeno-islet graft survivals on the diffusion coefficients of IgG in agarose hydrogels. D_{rel} is the diffusion coefficient of IgG in different concentrations of agarose, relative to the diffusion coefficient in 5% agarose hydrogel.

human serum, and the percentage of hemolysis was calculated from the hemoglobin released into the serum. More than 50% of EA in the agarose microcapsule was lysed within a 10–hr incubation. Permeation of active complement proteins could not be prevented by the 5% agarose microcapsule either. The limited immunoisolative effect of the 5% agarose microcapsule observed in xenotransplantation can be explained in terms of the permeation of antibodies and complement proteins though the agarose hydrogel.

The network of the agarose hydrogel can be rendered denser by increasing the concentration of agarose to further restrict the diffusion of antibodies and complement proteins through the microcapsule. To demonstrate this immunoisolative effect of the high concentration agarose (Iwata et al 1994a), the graft survival period is plotted in Figure 9.5 against the relative diffusion coefficient, D_{rel}, which is the diffusion coefficient in the tested hydrogel divided by the diffusion coefficient in the 5% concentration hydrogel. The graft survival period is inversely proportional to the relative diffusion coefficient of IgG through agarose hydrogels. This suggests that the graft survival was prolonged with the decreasing permeability of high molecular weight proteins through the microcapsule.

Jain et al reported the use of agarose macrocapsules to prevent xenogeneic islet graft rejection

(Jain et al 1996). They enclosed rat islets in macrocapsules (6000 ~ 8000 μm) made with various combinations of agarose, collagen, and Gelfoam, and transplanted them intraperitoneally in STZ-diabetic mice. The encapsulated xenografts maintained normoglycemia for more than 170 days. They did not mention the mechanism of prevention of rejection and prolongation of graft survival by macroencapsulation, but it seems probable that passing of antibodies and complement proteins through the thick hydrogel layer of the macrocapsule until they reached islet cells took a long time. This might contribute to the long-term survival of the xeno-islets in the macrocapsule.

Microcapsule Containing a Complement Activator

The complement system must be activated by antigen-antibody complexes or other pathways on the cell surface for destruction of the graft. There are two pathways for complement activation. One of them is the classical complement pathway initiated by the Cl_q component of Cl, which binds with a high affinity to the Fc region of immunoglobulins in antigen-antibody complexes. The other is the alternative complement pathway activated nonimmunologically by a variety of activators, including polysaccharides, bacteria, and damaged tissues. C3 represents the common component of both the classical and the alternative complement pathways, and thus depletion of C3 disarms both the pathways. Pretreatment of the recipient by injection of cobra venom factor, which activates complements resulting in C3 exhaustion (Cochrane et al 1970), could significantly prolong the survival of vascularized xenografts (Adachi et al 1987). These facts suggest that a membrane containing a complement activator that functions like cobra venom factor will effectively consume the cytolytic complement activity and could be expected to protect xeno-islets from humoral immunities. To confirm this assumption, microcapsules containing a complement activator were prepared (Morikawa et al 1996, Takagi et al 1994). It was reported that sulfated polyanions, such as dextran sulfate, were strong activators of the alternative pathway to deplete C3 activity (Bitter-Suermann et al 1981, Burger et al 1975). As a complement activator, poly(styrene sulfonic acid) (PSSa) was selected,

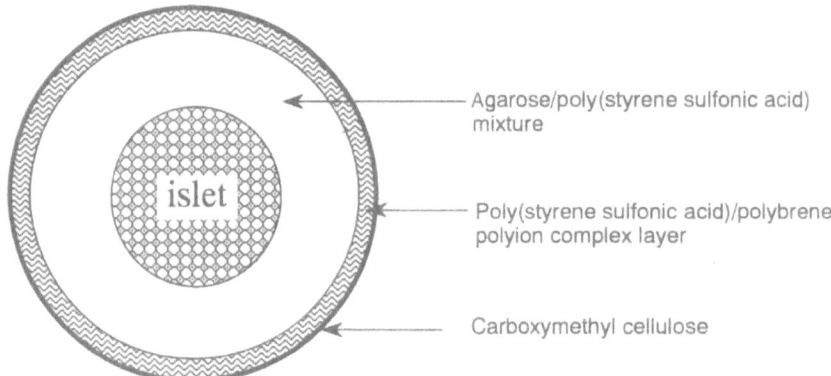

FIGURE 9.6. Schematic representation of the agarose/poly(styrene sulfonic acid) (PSSa) microcapsule.

because addition of 1×10^{-2} g of PSSa to 100 ml of serum was enough to completely exhaust the hemolytic complement activity, and PSSA is miscible with agarose. The structure of the microcapsule containing PSSa and the molecular structure of the polymers used are illustrated schematically in Figures 9.6 and 9.7, respectively. The layer from the surface of the islet to the outer surface of the microcapsule is composed of an agarose/PSSa mixed gel in which the PSSa is dissolved in the molecular network of the agarose hydrogel. A polyion complex layer was formed between the polyanionic PSSa and the polycationic polybrene at the periphery of the microcapsule, to confine the PSSa molecules within the microcapsule. The surface of the microcapsule was further coated with carboxymethyl cellulose to give tissue compatibility to the microcapsule. Hamster islets enclosed in 5% agarose/10% PSSa microcapsules were implanted into STZ-diabetic BALB/c mice. All the mice receiving 1000 microencapsulated hamster islets demonstrated more than 50 days normoglycemia, as shown in Figure 9.8. Thus, the microcapsule containing sulfated polyanion could effectively protect the xeno-islets from rejection.

There is still ambiguity in the interaction mechanism of sulfated polyanions with the complement system. Bitter-Suermann et al (Bitter-Suermann et al 1981) reported that sulfated polyanions were capable of reversibly binding factor H, leading to an extensive activation, thus consumption, of C3 and factor B, because the regulatory function of factor H was blocked in the fluid-phase C3b-dependent amplification system of the alternative pathway.

Meri and Pangburn (1990) reported that heparin and low molecular weight dextran sulfate inhibited lysis of cells through enhancing binding of H to activator-bound C3b. These two studies are contradictory concerning the mechanisms of inhibiting cell lysis, but suggest that the microcapsule containing sulfated polyanions can protect the entrapped xenogeneic cells from humoral immunity. However, Montdargent et al (1991) reported that consumption of complement protein by PSSa was not through true activation, but through complement protein adsorption. Koistinen (1993) denied that the inhibitory effect of heparin and dextran sulfate on the alternative pathway was due to the augmenting of the function of factor H, as proposed by Meri and Pangburn. These studies suggest that the microcapsule containing PSSa has only a limited protective effect on xenografts because PSSa will lose the protective effect after full adsorption of complement proteins. However, the microcapsule containing PSSa effectively prolonged xeno-islet survival for a long period of time, as demonstrated above. On human cells and tissues, polyanions such as sialoglycoproteins and glycosaminoglycans would be recognized by factor H, and the increased affinity for C3b would result in rapid C3b inactivation (Carreno et al 1989). These facts strongly suggest that an addition of sulfated polyanions to microcapsules is promising for developing microcapsules applicable to islet xenografts, although more detailed and well-organized studies are needed to make clear the mechanism by which sulfated polyanions affect the complement functions and to select a suitable polyanion that can ef-

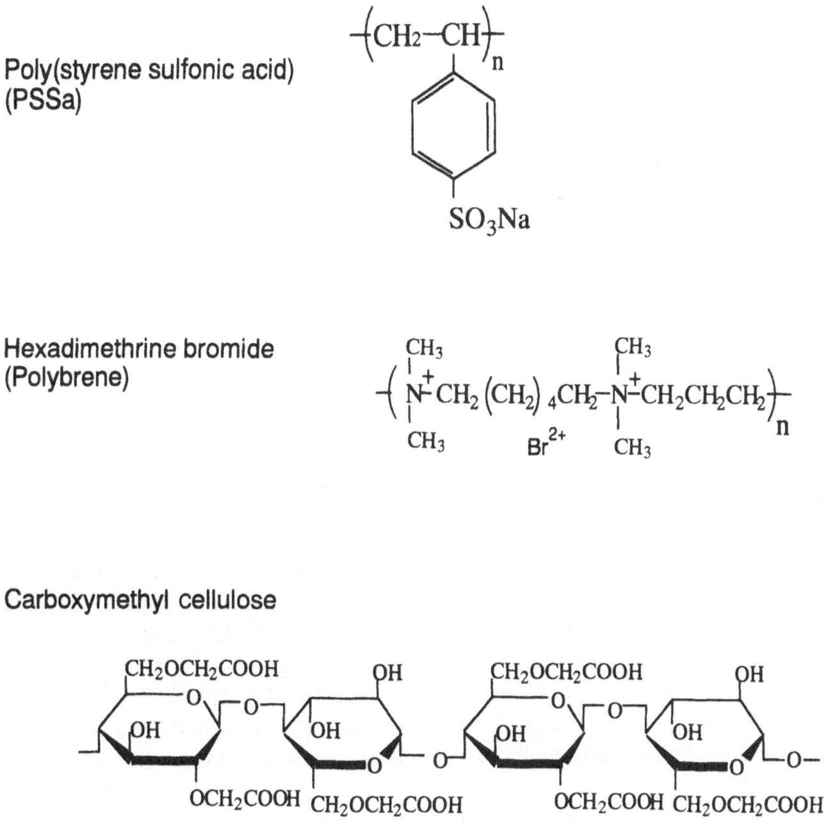

Poly(styrene sulfonic acid)
(PSSa)

$$\left(CH_2{-}CH \right)_n$$

SO_3Na

Hexadimethrine bromide
(Polybrene)

$$\left(\overset{CH_3}{\underset{CH_3}{N^+}}CH_2(CH_2)_4CH_2{-}\overset{CH_3}{\underset{CH_3}{N^+}}CH_2CH_2CH_2 \right)_n$$

Br^{2+}

Carboxymethyl cellulose

CH_2OCH_2COOH OH CH_2OCH_2COOH OH

OH OH OH OH

OCH_2COOH CH_2OCH_2COOH OCH_2COOH CH_2OCH_2COOH

FIGURE 9.7. Molecular structure of polymers used for preparing agarose/PSSa microcapsules.

fectively regulate the complement functions for preparation of microcapsules applicable to clinical xeno-islet transplantation.

Conclusion

Agarose is easily handled and nontoxic and can form a firm gel that is stable in a mammalian body. Mammalian cells can be easily microencapsulated in the agarose hydrogel without any sophisticated apparatus. The agarose microcapsule acts as a mechanical barrier, prohibiting contact between the islets and the immunocompetent cells, and thus can effectively protect allo-islets against rejection. In xenotransplantation, the agarose microcapsule cannot protect xeno-islets by itself, but an addition of PSSa to microcapsules is able to reduce the complement cytolytic activity, effectively prolonging the xeno-islet functional survival. Agarose has many advantages over other polysaccharides for preparation of BAP.

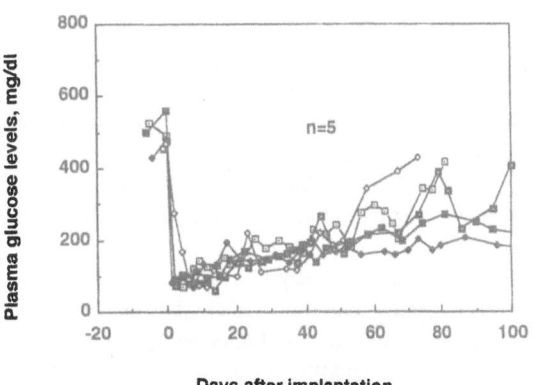

FIGURE 9.8. Nonfasting plasma glucose levels of STZ-diabetic BALB/c mice after intraperitoneal implantation of 1000 islets enclosed in 5% agarose/10% PSSa microcapsule.

References

Adachi H, Rosengard BR, Hutchins GM. Hall TS, Baumgartner WA, Borkon AM, Reitz BA. 1987. Effect of cyclosporine, aspirin, and cobra venom factor on discordant cardiac xenograft survival in rats. Transplant Proc 19:1145–1148.

Bitter-Suermann D, Burger R, Hadding U. 1981. Activation of the alternative pathway of complement: efficient fluid-phase amplification by blockade of the regulatory complement protein β 1H through sulfated polyanions. Eur J Immunol 11:291–295.

Brendel MD, Kong SS, Alejandro R, Mintz DH. 1994. TI: Improved functional survival of human islets of Langerhans in three-dimensional matrix culture. Cell-Transplant 3:427–435.

Burger R, Hadding U, Schorlemmer HU, Brade V, Bitter-Suermann D. 1975. Dextran sulphate: a synthetic activator of C3 via the alternative pathway, Immunology 29:549–554.

Carreno MP, Labarre D, Maillet F, Jozefowics M, Kazatchkine MD. 1989. Regulation of the human alternative complement pathway: Formation of a ternary complex between factor H, surface-bound C3b and chemical groups on nonactivating surfaces. Eur. J. Immunol. 19:2145–2150.

Clark AH, Ross-Murphy SB. 1987: Structural and mechanical properties of biopolymer gel. Adv Polym Sci 83:57–207.

Cochrane CG, Muller-Eberhard HJ, Aikin BS. 1970. Depletion of plasma complement in vivo by a protein of cobra-venom: its effect on various immunologic reactions. J Immunol 105:55–69.

Dumitriu S, Vidal PF, Chornet E. 1996. Hydrogels based on polysaccharides. In: Dumitriu S, editor. Polysaccharides in medical applications. New York: Marcel Dekker, pp. 125–241.

Enosawa S, Suzuki S, Kakefuda T, Amemiya H, Iwata H, Ito N. 1995. Examination of ammonia removal activity by agarose-encapsulated rat hepatocytes and in vivo estimation of ammonia toxicity in the rat model. Jpn J Artif Organs 24:736–739.

Frohberg H, Oettel H, Zeller H. 1969. On the mechanism of the fetotoxic effect of tragacanth. Arch Toxikol 25:268–295.

Gin H, Dupuy B, Baquey C, Ducassou D, Aubertin J. 1987. Agarose encapsulation of islets of Langerhans: reduced toxicity in vitro. J Microencapsul 4:239–242.

Hering BJ, Brendel MD, Schultz AO, Schultz B, Bretzel RG, editors. 1996. International islet transplant registry, Third Medical Department, Center of Internal Medicine Justus-Liebig-University of Giessen.

Iwata H, Amemiya H, Matsuda T, Matsuo Y, Takano H, Akutsu T. 1985. Investigation of biomedical materials for hybrid artificial organs. Jpn J Artif Organs 14:864–867.

Iwata H, Amemiya H, Matsuda T, Takano H, Hayashi R, Akutsu T, 1989. Evaluation of microencapsulated islets in agarose gels as bioartificial pancreas by studies of hormone secretion in culture and xenotransplantation. Diabetes 38(Suppl. 1):224–225.

Iwata H, Amemiya H, Matsuda T, Msuda M, Takano H, Akutsu T. 1991. Examination of molecular weight cut-off of a semipermeable membrane for a bioartificial pancreas: islets cultured in chemically defined media. Artif Organs Today 1:229–236.

Iwata H, Takagi T, Amemiya H, Shimizu H, Ymashita K, Kobayashi K, Akutsu T. 1992. Agarose for a bioartificial pancreas. J Biomed Mater Res 26:967–977.

Iwata H, Kobayashi K, Takagi T, Oka T, Yang H, Amemiya H, Tsuji T, Ito F. 1994a. Feasibility of agarose microbeads with xenogeneic islets as a bioartificial pancreas. J Biomed Mater Res 28:1003–1011.

Iwata H, Morikawa N, Fujii T, Takagi T, Samejima T, Ikada Y. 1995. Does immunoisolation need to prevent the passage of antibodies and complements? Transplant Proc 27:3224–3226.

Janeway CA, Travers P. 1997. In: Transplant rejection: responses to alloantigens. Immunobiology: the immune system in health and disease—third ed. Current Biology Ltd./Garland Publishing Inc. 12:19–12:42.

Jain K, Yang H, Asina SK, Patel SG, Desai J, Diehl C, Stenzel K, Smith BH, Rubin AL. 1996. Long-term preservation of islets of Langerhans in hydrophilic macrobeads. Transplantation 61:532–536.

Koistinen V. 1993. Effects of sulphated polyanions on functions of complement factor H. Molecular Immunology 30:113–118.

Lim F, Sun AM. 1980. Microencapsulated islets as bioartificial endocrine pancreas. Science 210:908–910.

Matthew HW, Salley SO, Peterson WD, Klein MD. 1993. Complex coacervate microcapsules for mammalian cell culture and artificial organ development. Biotechnol Prog 9:510–519.

Meer W. 1980. Agar. In: Davidson RL, editor. Handbook of water-soluble gums and resins. v: McGraw-Hill, Inc. Ch. 7, pp. 1–19.

Meri S, Pangburn MK. 1990. Discrimination between activators and nonactivators of the alternative pathway of complement: Regulation via a sialic acid/polyanion binding site on factor H. Proc Natl Acad Sci USA 87:3982–3986.

Montdargent B, Labarre D, Jozefowicz M. 1991. Interactions of functionalized polystyrene derivatives with the complement system in human serum. J Biomater Sci Polymer Edn 2:25–35.

Morikawa N, Iwata H, Fujii T, Ikada Y. 1996. An immuno-isolative membrane capable of consuming cytolytic complement proteins. J Biomater Sci Polymer Edn 8:225–236.

Nilsson K, Scheirer W, Merten OW, Ostberg L, Liehl E, Katinger HW, Mosbach K. 1983. Entrapment of animal cells for production of monoclonal antibodies and other biomolecules. Nature 302:629–630.

Paul LC. 1991. Mechanism of humoral xenograft rejection. In: Cooper DKC, Kemp E, Reemtsma K, White DJG, editors. Xenotransplantation, Springer-Verlag, pp. 47–67.

Richardt M, Menden A, Bretzel RG, Federlin K. 1984. Islet transplantation in experimental diabetes of the rat. VIII. B-cell restoration following islet transplantation. Preliminary results. Horm Metabol Res 16:551–552.

Scheirer W, Nilsson K, Merten OW, Katinger HW, Mosbach K. 1984. Entrapment of animal cells for the production of biomolecules such as monoclonal antibodies. Develop Biol Standard 55:155–161.

Takagi T, Iwata H, Tashiro H, Tsuji T, Ito F. 1994. Development of a novel microbead applicable to xenogeneic islet transplantation. J Control Release 31:283–291.

10
Poly(ethylene glycol)

Amarpreet S. Sawhney

Introduction

The demands placed on a material to be used for microencapsulation are stringent. The material should (1) be stable in the physiological environment over several years, (2) not engender any cytotoxicity, (3) be permselective so as to be immunoprotective and yet allow nutrient and metabolite access, and (4) be biocompatible so as not to elicit an inflammatory or fibrotic response from the host. Of the wide range of materials used, ionically coacervated microcapsules of alginate and poly(l-lysine) (PLL) have shown promise. O'Shea and Sun (1986) demonstrated rat islet survival times of 2 to 3 months, and occasionally of 1 year, in mice, using alginate/PLL/alginate trilayered microcapsules. The xenografts apparently failed as a result of overgrowth with fibroblast-like and macrophage-like cells upon the microcapsules. This cellular overgrowth is due to a nonspecific foreign body reaction elicited by the microcapsules and is by no means restricted to alginate/PLL/alginate microcapsules. Roberts et al, using HEMA-MMA copolymers for microencapsulation, also reported seeing up to a 10 μm thick layer of cellular overgrowth after 4 weeks in vivo (Roberts et al 1991).

Cell adhesion is preceded by the adsorption of proteins on the surface of the implanted material. Almost all surfaces adsorb proteins to some extent. Protein adsorption is mediated by hydrophobic interactions as well as by coulombic interactions (McMahon et al 1990). Contact of cells with adsorbed proteins that mediate cell adhesion, such as fibronectin or vitronectin, leads to cell attachment and spreading (O'Shea et al 1984). Protein-repellent surfaces have been created by the immobilization of poly(ethylene glycol) (PEG) or poly(ethylene oxide) (PEO). PEG and PEO are used synonymously in this chapter, and they differ only in that PEG has two terminal hydroxyls, whereas PEO has one. Steric repulsion has been suggested to be responsible for the repulsive force generated between two surfaces adsorbed with PEG chains in aqueous solvents (O'Shea and Sun 1986). Protein adsorption is found to be very low for surfaces consisting of PEG chains with a degree of polymerization above 100 (Andrade and Hlady 1986). The decrease in protein adsorption is attributed to osmotic pressure and elastic restoring forces generated by the PEG chains when they interact with a protein molecule (Buck and Horwitz 1987). The protein-repellent character is a function of the PEG chain length and surface density (Luckham and Klein 1985). Andrade et al (1987) have determined that high surface density is of greater importance than long chain length.

Sawhney and Hubbell (1992) have described the synthesis and use of a graft copolymer of PLL and methoxy poly(ethylene glycol) to enhance the biocompatibility of alginate-PLL microcapsules. Sawhney et al have also reported on an *in situ* polymerization process that uses macromonomers based on PEG to encapsulate islets of Langerhans (Pathak et al 1992). These polymerizations can be carried out either in a bulk fashion, whereby the photoinitiator is dispersed uniformly through the macromonomer solution, or in a surface-conforming or "interfacial" way so as to give thin hydrogel barriers around islets. This chapter will briefly describe these three methods whereby PEG is used either to modify the

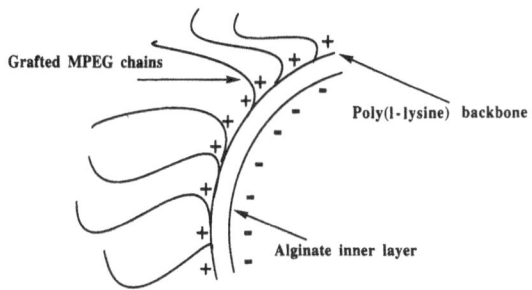

FIGURE 10.1. Cutaway schematic view of a PLL-g-MPEG sphere. The nonionic MPEG chains are expected to extend into the surrounding aqueous environment and cover up the ionically coacervated microsphere surface.

surface of microcapsules or to encapsulate tissue with a biocompatible hydrogel coating such that the coating is formed either in a "bulk" or an "interfacial" fashion.

Use of PEG-Based Copolymers for Surface Modification

The presence of PEG at the interface can mask the underlying highly ionically charged surface of the alginate-PLL-based microcapsules by extending outwards into the surrounding aqueous environment to sterically hinder the approach of approaching macromolecules. In order to achieve this, graft copolymers of PEG and polycations such as PLL have been synthesized. The use of these graft copolymers instead of the homopolymeric PLL enables the PLL to adsorb along the surface of the microsphere to form the wall of the microsphere by ionic coacervation, while allowing the grafted PEG chains to extend into the surrounding aqueous medium. A schematic cutaway view of the microsphere wall formed in this fashion is shown in Figure 10.1. This graft copolymer has been shown to have properties of reduced protein adsorption, complement binding, and cell adhesion *in vitro.*

Synthesis and Microencapsulation

The graft copolymer was synthesized by end-activating MPEG ($\propto$-methoxy, ω-hydroxy polyethylene glycol, MW 5000), using 1,1 carbonyldiimidazole (CDI; Aldrich) in a dry organic solvent. The end-activated MPEG can in turn be coupled to

amine groups present along the PLL backbone in a borate buffer at an elevated pH. An optimal degree of MPEG grafting is needed, whereby the interaction of the PLL component with cells and tissue is reduced, but its interaction with alginate remains strong enough to form stable permselective capsules. The resulting polymer can be purified by dialysis against a bicarbonate-buffered saline. The reaction scheme for activation of the MPEG and the grafting of the activated MPEG onto the primary amines present along the PLL backbone is shown below.

End-activation of the MPEG can be confirmed by elemental analysis, and successful grafting to the PLL backbone can be confirmed using FTIR by observing the presence of a typical ether stretch of ($-CH_2-O-CH_2-$) at 1110 cm^{-1}. The ^1H NMR spectrum for PLL-g-MPEG can also be used to follow the formation of a graft copolymer by the detection of a small new peak due to the terminal methylene of the lysine side chain appearing as a shoulder on the $-CH_2-NH_2-$proton peak of the PLL side chains at 2.9 ppm. The grafted MPEG chains resulted in a substantial increase in the molecular weight of the resulting polycation. For instance, the PLL-g-MPEG used, with 17 chains of MPEG (5000 g/mol), had a molecular weight of about 105,000 (calculation based on NMR data), as compared to the original PLL molecular weight of 20,100.

Microcapsules can be formed that exhibit a PEG-rich phase on their surface by simply washing the alginate-PLL-alginate microcapsules with a solution of 0.3% (on a PLL basis) PLL-g-MPEG over a 20-min period. The microcapsule fabrication procedure for the alginate-PLL-alginate microcapsules described by O'Shea and Sun (1986) can essentially be used for fabricating the alginate-PLL microcapsules. The presence of PEG on the surface can be verified by surface analysis techniques such as ESCA, by following the decrease in nitrogen atomic concentration on the surface as a result of the presence of the PEG.

Protein Adsorption and Cell Adhesion In Vitro

PLL-g-MPEG has been optimally synthesized so as to be capable of coacervating with alginate because of its polycationic nature, but at the same time to be protein-repellent in nature as a result of the grafted MPEG chains (Sawhney et al 1992). The presence of the grafted MPEG was seen to reduce protein adsorption as expected. Adsorption of bovine serum albumin and human fibrinogen was seen to be decreased on surfaces coated with the PLL-g-MPEG compared to alginate-PLL surfaces. Additionally, the adhesion of human foreskin fibroblasts cultured in vitro on PLL-g-MPEG surfaces was also dramatically reduced over that of alginate-PLL surfaces. These phenomena of reduced protein adsorption and consequent cell adhesion were attributed to the "masking" of the ionically coacervated microcapsule surface with the grafted MPEG chains.

Fibroblasts adhered well and spread over alginate/PLL surfaces after a 4-hr incubation. In contrast, the alginate/PLL-g-MPEG surfaces showed less than 5% of the cell attachment seen on the alginate-PLL surfaces. None of the attached cells was well spread or appeared rounded (Sawhney et al 1992).

Permeability

The presence of the grafted MPEG chains can interfere with the ionic coacervation process by the same mechanism of steric hindrance that is a desirable attribute of these graft copolymers in minimizing protein adsorption. It is expected that the affinity of the graft copolymer for the underlying alginate will be attenuated over that of the original PLL chains. In order to determine the extent and impact of this phenomenon on the permeability of microcapsules, bilayered (alginate-PLL or alginate-PLL-g-MPEG) capsules were fabricated as previously described with PLL or PLL-g-MPEG having

grafted MPEG (MW 5000). The diffusion of radiolabeled albumin as a marker through the microcapsule membrane was used to study the permeability differences between different types of microcapsules.

The radiolabeled albumin (^{125}I-labeled BSA) was added to a 1.4% w/v solution of alginate (Kelco HV) in saline. This solution was then used to form the calcium alginate microspheres. Other steps for the formation of the microcapsule membrane were as described before and have been detailed elsewhere (Sawhney 1992). After the final saline wash, each type of microcapsule was incubated in a citrate solution. The microcapsules were periodically agitated and allowed to settle, and the supernatant was assayed for released albumin using a γ-scintillation counter. The attenuated affinity of the graft copolymer for the alginate was evidenced by the increased permeability to bovine serum albumin in trilayered alginate-PLL-g-MPEG-alginate microcapsules over that in trilayered alginate-PLL-alginate microcapsules. These capsules were also unstable. However, by decoupling the surface modification step from the microcapsule wall fabrication step, pentalayered microcapsules could be fabricated that had a basic trilayered structure of alginate-PLL-alginate and were then further surface modified by the PLL-g-MPEG copolymer. These microcapsules were seen to retain both the desirable permeability property of the alginate-PLL-alginate microcapsules and the biocompatibility property added by the PLL-g-MPEG (Sawhney 1992).

Complement Binding

The activation of the complement cascade can lead to macrophage activation, and subsequent stimulation of fibroblast activity, by the secretion of chemotactic factors. Complement activation through the alternate pathway proceeds by the immobilization of iC3b on the surface of any foreign surface. The bound iC3b acts to stimulate the further activation of the complement activation by a feedback loop. The decrease of nonspecific protein adsorption on MEPG-containing surfaces can thus be expected to also reduce the adsorption of iC3b on the microcapsule surface. The extent of iC3b immobilization on the surface of microcapsules can thus be used as a measure for complement binding and activation.

A radioimmunoassay for iC3b was used to determine the relative amounts of immobilized iC3b on microcapsules having alginate-PLL or alginate-PLL-g-MPEG surfaces. The specific binding of anti-iC3b was seen to be significantly lower on the microcapsules with PLL-g-MPEG. This indicates that the presence of MPEG on the surface of the microcapsules may make them less interactive with the complement and hence with the immune system.

Implantation and Recovery

Microcapsules exhibiting either PLL or the MPEG-g-PLL were implanted intraperitoneally into mice in order to assess the in vivo biocompatibility. Two indices of biocompatibility in vivo were measured, the recruitment of macrophages into the peritoneal fluids as measured by a count of suspended cells obtained from a peritoneal lavage, and the attachment of cells to the microcapsule wall as seen microscopically and histologically.

A sharp reduction in nonattached cell counts in the peritoneal fluids for PLL-g-MPEG-containing microcapsules explanted after 4 days was seen relative to control microcapsules that possessed a PLL-rich surface. The inflammatory reaction tended to subside once the microcapsule implants were encased in a fibrous envelope, and thus the trilayered control microcapsules did not show a significantly higher nonattached cell count over saline vehicle control levels after 1 month.

The microcapsules having a PLL-g-MPEG-rich surface were almost totally free of adherent cells after 4 days, while the control microcapsules, having a PLL-rich surface were overgrown with multiple layers of cells. Relatively few free trilayered control microcapsules were recovered at the end of 1 month. A large number were found adherent to the intestines and entangled in the omental tissue. The microcapsules found attached to peritoneal organs and tissue showed even greater fibrosis than the free-floating microcapsules. This can be seen in Figure 10.2.

PLL provides a strongly cell-adhesive surface which, if not shielded by the MPEG chains, can result in severe overgrowth within a few days. This was seen to be the case in both the 4-day and the 1-month studies in vivo. The alginate/PLL-g-MPEG coacervated surfaces were quite cell-nonadhesive.

Because of the methoxy end group on the MPEG chain, the MPEG is anchored at only one point along its entire chain length. In an aqueous environment, the pendent MPEG chains are extended and have a high mobility. The nonionic MPEG chains apparently shield the ionically charged coacervated surface of the underlying microcapsule from the biological environment.

Histologically, several layers of cells can be seen upon the trilayered capsule wall after 1 month intraperitoneally. The tissue response to these microcapsules was typical of fibrosis of an implant. A number of macrophages along with several spindle-shaped fibroblasts were evident. Microcapsules fabricated with the grafted MPEG chains were free of fibrous reaction, and the microcapsule surface was free of any adherent tissue after 1 month implantation intraperitoneally.

The biocompatible and permselective microcapsules exhibiting an MPEG surface, described herein, appear to be stable and retain their biocompatibility for at least 1 month, with no indication of incipient failure. The fibrotic and inflammatory response to these capsules appears to have been reduced by the MPEG as a result of reduced protein adsorption and complement binding. These microcapsules show a definite improvement in biocompatibility over the alginate/PLL/alginate capsule, and may represent an important step toward obtaining long-term survival of xenograft tissue for human therapy.

PEG-Based Macromonomers for "Bulk" Microencapsulation

Few methods for polymer synthesis have been carried out in direct contact with living tissues (Dupuy et al 1988, Ishihara and Nakabayashi 1990, Kobayashi et al 1991, Koehnlein and Lemperle 1969, Potts and Petrou 1991, Ronis et al 1984) or organisms (Tanaka et al 1977). Cytotoxicity in such syntheses can be associated with exposure to toxic monomers (Koehnlein and Lemperle 1969, Yashon et al 1966), heating during polymerization, dehydration, use of organic solvents, and prolonged exposure to short wavelength ultraviolet light. An *in situ* photopolymerization process has been developed (Sawhney et al 1992) that allows the noncytotoxic polymerization of macromonomers on living tissue. Since water-soluble species are used, the toxicity associated with organic solvents can be

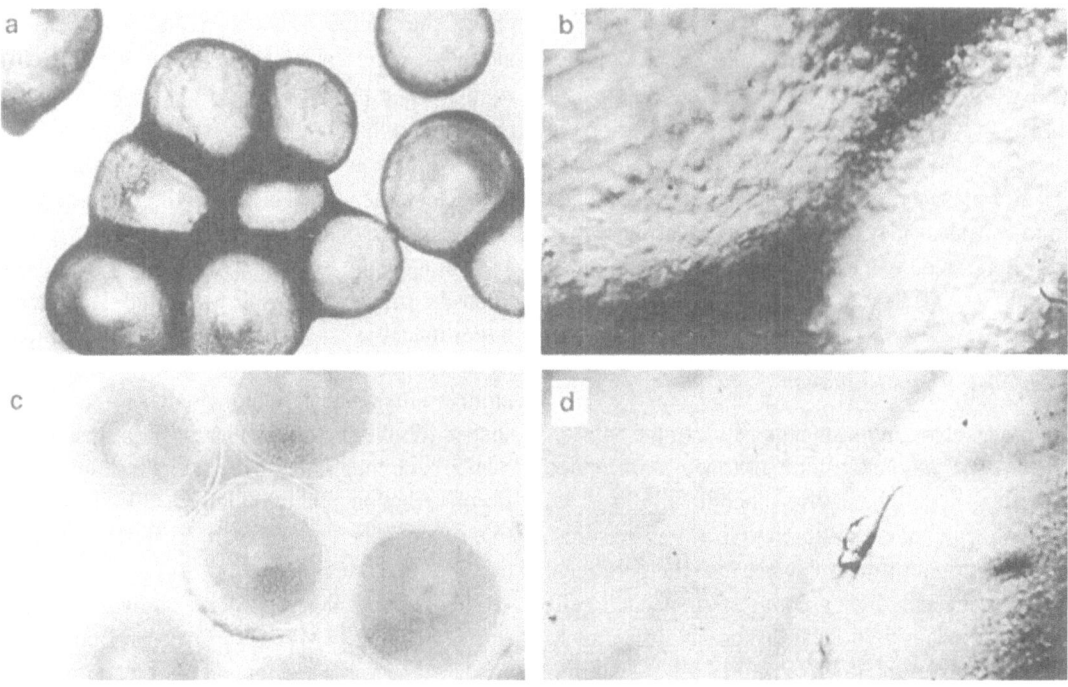

FIGURE 10.2. Photomicrographs of microcapsules explanted from peritoneal cavity of mice after 1-month period. (A) 50× phase contrast view of trilayered control alginate/PLL/alginate microcapsules after showing extensive fibrous overgrowth and clumping, (B) 400× Hoffman modulation contrast surface view of trilayered control microcapsules, (C) 50× phase contrast and (D) 400× Hoffman modulation contrast views of alginate/PLL/alginate/PLL-g-MPEG/alginate microcapsules. Used with permission of Elsevier Science Limited.

prevented. The use of macromonomers reduces monomer-related toxicity for two reasons. First, the macromonomers cannot readily diffuse across cell membranes, and changes in osmotic pressure are minimal, thus the macromonomers are not likely to interfere with intracellular processes. Second, the number of acrylate groups undergoing polymerization is drastically reduced when a macromonomer is used to achieve a cross-linked network, compared to using a monomer. This in turn reduces the heat of polymerization so as to have essentially no perceptible increase in temperature in the microenvironment of the tissue being encapsulated. The use of visible-light-based photoinitiating systems can also markedly reduce the toxicity that can be associated with the use of ultraviolet (UV) light, since tissue is essentially transparent to the blue-green light used in these systems, compared to the strong absorption of UV light by proteins and nucleic acids.

PEG was selected as the backbone molecule for synthesis of the macromonomers for two reasons.

First, the biocompatibility of PEG in biological environments is well documented, as already mentioned. Thus, it was considered that a hydrogel that was composed primarily of PEG would provide a suitably compatible biomaterial for microencapsulation. Second, the hydrophilic nature of the PEG backbone and the hydrophobic nature of the acrylate end groups enabled the formation of micellar structures in an aqueous environment. This micellar nature greatly enhances the rate of polymerization by speeding up the rate of free radical propagation reactions compared to termination reactions. This characteristic allows the rapid polymerization and facile encapsulation of cells and tissue.

Synthesis and Microencapsulation

PEG multiacrylates were synthesized by reaction of tetrahydroxy PEG (MW 18,500) with acryloyl chloride (4-fold excess based on PEG hydroxyl end groups), using triethyl amine (equimolar with acryloyl chloride) as a proton acceptor in a dry benzene reflux under a dry nitrogen atmosphere. The resulting product was filtered and precipitated with hexane. It was redissolved in dichloromethane and reprecipitated in an excess of anhydrous ether, filtered, and dried under vacuum at 60°C. The PEG was purchased from Polysciences and used for the multiacrylate synthesis. The structure of the tetrahydroxylated PEG is shown in Figure 10.3.

A solution (23% w/w in 10 mM HEPES buffered saline) of the PEG multiacrylate (PEG-MA) was mixed with a cell suspension along with ethyl eosin (0.5 mM) and triethanolamine (100 mM) as the photosensitizer/electron donor initiating system and 1 ml/ml of N-vinyl pyrrolidinone. The rate of polymerization was followed by measuring the length of a spike produced by the gelation of macromer within a square cuvette following penetration of an argon ion laser beam (500 mW, beam diameter 3 mm, emission max. = 514 nm). The mass of gel produced was observed to be an exponential function of irradiation time (Sawhney et al 1992). Using this information it was estimated that a sphere 500 μm in diameter could be polymerized in about 100 ms at this illumination intensity (6 W/cm^2). A coextrusion apparatus was used to form microspheres 0.5–0.8 mm in diameter by extruding the liquid macromer into droplets and subsequently gelling them by exposure to laser light.

In Vitro and In Vivo Properties

In order to assess the effect of the polymerization process on the viability of cells and tissue, the viability of the encapsulated cells was measured by a trypan blue exclusion assay (Sigma). Human foreskin fibroblasts (HFF), Chinese hamster ovary cells (CHO-K1), and beta cell insuloma cells (RiN5F) were encapsulated by the technique described above. Using the dye exclusion test, all types of cells were found to be viable (more than 95%) after encapsulation; viability of control cells without encapsulation was also 95%. This illustrates the nontoxicity of the encapsulation process. The laser light is not absorbed by cells in the absence of an exogenous, cell-binding chromophore (Karu 1990). There is no significant heat of polymerization because of the nature, size, and dilution of the macromers used. These polymerizations can proceed extremely rapidly in oxygen-containing aqueous environments at physiological pH.

Microspherical gels made from the PEG-MA precursor were seen to be stable for over one year in phosphate-buffered saline. Gels made from the PEG-MA precursor contained 91.4 ± 0.9% water, and the fraction of macromer undergoing gelation was seen to be 80.3 ± 0.9%. Disc-shaped gels made from this precursor were implanted subcutaneously in 4-week-old rats and were monitored for calcification and mechanical deterioration. These gels were seen not to calcify or lose tensile strength significantly when compared to gels of the same composition that had not been implanted (control) over an 8-week implantation period (Sawhney 1992).

Gels with the proper formulation were found to be capable of being immunoprotective, as indicated by protein diffusion. These hydrogels were seen to allow the slow diffusion of bovine serum albumin (BSA) (MW 67,000), but were impermeable to immunoglobulin G (IgG) (MW approx. 150,000) and fibrinogen (Fg) (MW approx. 350,000).

PEG-Based Macromonomers for "Interfacial" Microencapsulation

Numerous approaches have been developed for the microencapsulation of tissue particles and cell aggregates. Lim and Sun formed permselective membranes by coacervating water-soluble polycations, such as polylysine, upon the surface of spherical gel particles of calcium alginate (Lim and Sun 1980). These membranes were used to correct diabetes in mice with transplanted rat islets, although some limitations in biocompatibility and fibrous encapsulation were noted (O'Shea and Sun 1986). Dawson et al (1987) used a phase inversion process to form spherical polyacrylate membranes around

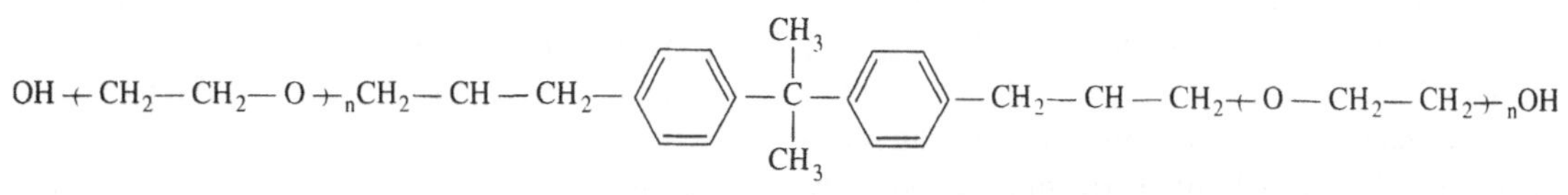

$$OH + CH_2 - CH_2 - O +_n CH_2 - CH - CH_2 - \langle\rangle - \underset{\underset{CH_3}{|}}{\overset{\overset{CH_3}{|}}{C}} - \langle\rangle - CH_2 - CH - CH_2 +O - CH_2 - CH_2 +_n OH$$

FIGURE 10.3. Structure of the tetrahydroxylated PEG.

cell suspensions. Dupuy et al (1988) used reactive approaches to polymerize acrylamide around cell suspensions entrapped within spherical gel particles of agarose. In each of the above approaches, the size of the membrane capsule was determined by the hydrodynamics of particle formation, rather than by the size of the cell aggregate. All macro- and microencapsulation techniques attempted with various materials to date have some important drawbacks that can limit the practical application of these approaches in the development of microencapsulated biohybrid organs. One of these limitations is the size of the microcapsule that can be fabricated and the stagnant fluid environment between the tissue and the microcapsule wall. This stagnant fluid environment creates diffusional limitations for nutrients, oxygen, and metabolic byproducts. This diffusional resistance has been postulated to be responsible for necrosis of encapsulated cells (Fan et al 1990). The larger size of the microcapsule compared to the tissue aggregate that is to be transplanted also leads to inefficiencies in transplant volumes. Thus, huge volumes of microcapsules will need to be implanted, for example, to restore normoglycemia in a Type I diabetic.

Another limitation has been the encapsulation efficiency. This is especially true for encapsulation of tissue aggregates such as islets of Langerhans. Separation of "empty" microcapsules from those containing islets is difficult. The presence of these empty microcapsules can lead to further unnecessary increase in implant volumes. An "interfacial" polymerization approach to forming conformal PEG hydrogel barriers, only tens of microns thick, upon tissues has been described (Sawhney et al 1994). This route desmonstrated high tissue compatibility, even at the outermost cellular layers of the treated tissues.

As a model system to investigate biological barriers formed by interfacial photopolymerization, the formation of hydrogel membrane coatings upon the surfaces of islets of Langerhans was examined. Immunoisolating barriers should prohibit contact with immune cells and proteins, while permitting adequate nutrient and hormone permeation (Lacy et al 1991). The interfacial approach sought to address the first-mentioned limitation, that of size; the membrane geometry would be defined by the size and shape of the cell aggregate to be microencapsulated. Interfacial polymerization can possibly provide such control since the surface of the islet would define the inner surface of the membrane, and the conditions of photopolymerization would define its thickness.

Islet Microencapsulation

Rat islets were exposed to a solution of eosin Y (Aldrich), which resulted in the adsorption of the dye to the islets. The islets were then washed thrice with HEPES buffered saline to remove excess dye, and were settled each time by centrifugation. The synthesis of the PEG-MA acrylated macromer has been described previously. The macromer solution consisted of 23% w/w sterile solution of PEG-MA, 30 μl/ml of triethanolamine (Aldrich), and 1 μl/ml of N-vinyl pyrrolidinone (Aldrich). After the final wash the macromer solution was added, and the islets were suspended by gentle agitation in a transparent centrifuge tube. This suspension was exposed to light from an argon ion laser (emitting at 514 nm; American Laser) for 40 seconds at an irradiance of 70 mW/cm^2. The islets were washed free of nongelled macromer, using one wash with saline and two with tissue culture medium.

A thin conformal coating of the PEG-based hydrogel was seen to form on the surface of the islets. Since eosin Y was in contact with the macromer-triethanolamine solution only at the islet-solution interface, polymerization was limited to that interface. This can be seen in Figure 10.4.

The final thickness of the hydrogel membrane coatings could be controlled by changing exposure time as well as irradiation intensity. Various factors such as irradiation intensity, time, formulation ingredients, etc., can influence the photopolymerization process. Lyman et al (1996) have described the effect of these factors on the resulting conformal hydrogel in a similar system. For example, the thickness of the hydrogel barrier may be increased by prolonging the exposure to light, by exposure to higher intensity light, or by exposure at higher concentrations of the macromolecular precursor in solution. Some alterations in PEG functionality or molecular weight may be necessary in future work to provide identical immunoprotectivity in these interfacially polymerized gels as in the bulk polymerized ones. For example, these very thin membranes may require a lower molecular weight pre-

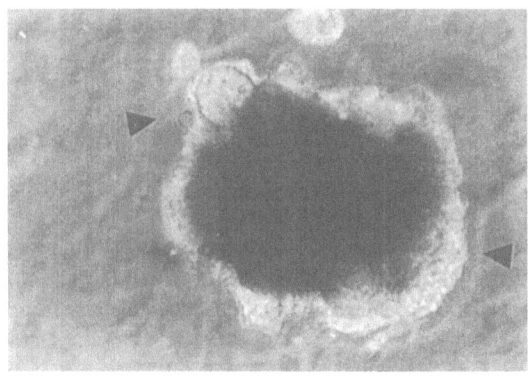

FIGURE 10.4. Rat islet conformally coated with a poly(ethylene glycol) multiacrylate hydrogel. Phase contrast microscopy was used, and the hydrogel barrier, which is highly swollen and consists of in excess of 90% water, is barely visible, as indicated with arrowheads. A representative islet is shown. Magnification, 200×. The surface of these microcapsules is almost free of adherent cells. Used with permission of John Wiley and Sons, Inc.

cursor to effectively prevent immunoglobulin permeation. Further investigation is also necessary to determine intra- and interislet variability in coating thickness.

In Vitro Evaluation of Encapsulated Islets

Islets that had been treated to form the conformal barrier, as well as islets that had been maintained in culture, were fixed and examined using transmission electron microscopy (TEM). Samples for TEM were fixed with glutaraldehyde and postfixed with 2% osmium tetroxide in phosphate-buffered saline, dehydrated with a graded ethanol series, and embedded in Spurr's medium. Thin sections were cut on an ultramicrotome and stained with uranyl acetate and lead citrate. In both cases, a representative β-cell on the surface of the islet was visualized. The cell membrane appeared intact and equivalent between treated and control samples, in spite of photopolymerization directly upon it. β-cells in both treated and control islets were well granulated with densely staining insulin granules, a characteristic ultrastructural morphological feature of islet β-cells.

An assay utilizing 2 mM calcein-AM and 4 mM ethidium homodimer solutions (Molecular Probes)

was used to determine islet viability. Islets were examined using an inverted stage fluorescence microscope (Leitz Fluovert) with a fluorescein filter set. The live cells within the islets were seen to fluoresce green, while dead cells fluoresced red. The projected area of each islet observed was determined using an image processing system (Argus, Hammatsu), and the fraction of the islet area that was seen to be viable was determined. Nonviable cells even in the center of the islets were visible by this method. The viability of islets after interfacial polymerization of the PEG gel on the surface of the islet was not found to be significantly different from that of control islets that had been maintained in culture.

Static glucose stimulation was performed on islets that had been treated to form the PEG conformal coatings and on islets that had been maintained in culture over the same period, to determine continued islet function and response. The islets were exposed to a sequence of low, high, and low levels of glucose in a serial experiment (low glucose [0.5 mg/ml glucose] or high glucose [3.0 mg/ml glucose]). The supernatant solutions were analyzed for insulin concentration with a radioimmunoassay kit, using rat insulin as a standard (Amersham). The islets with the PEG gel coating showed rapid insulin secretion in response to high glucose challenge, and a slow decline in insulin secretion in response to the step change back to the low-glucose environment (Sawhney 1992). The islets that had been cultured for the same time period with no treatment did not show a significantly different normalized insulin secretory response for all stimulation.

The interfacial polymerization approach permits the custom treatment of tissues with conformal protective coatings 10–50 μm thick. These barriers may be useful in interrupting immune and cell adhesive recognition events between the body and a treated tissue. Such modifications can be made without gross alterations in the tissue, such as loss of cellular viability, normal ultrastructural morphology, or normal function. Further investigation is required to demonstrate a lack of fine alterations, such as modifications at the level of the glycoproteins in the cell's glycocalyx. Longer-term investigation will also be necessary to examine the stability of the barriers and, in the case of the present application, the long-term impact on islet function.

Conclusion

PEG lends itself easily to structural modification through its end groups, for both surface modification and macromer synthesis purposes. PEG has been and will continue to be an important material for microcapsule surface modification as well as for encapsulation. *In situ* polymerization techniques described herein may make it possible to custom-tailor immunoprotective membrane barriers around tissue.

References

Andrade JD, Hlady V. 1986. Protein adsorption and materials biocompatibility: a tutorial review and suggested hypotheses. Adv Polym Sci 79:1–63.

Andrade JD, Nagoaka S, Cooper S, Okano T, Kim SW. 1987. Surfaces and blood biocompatibility. Current hypothesis. ASAIO Trans 10:75–76.

Buck CA, Horwitz AF. 1987. Cell surface receptors for extracellular matrix molecules. Ann Rev Cell Biol 3:179–205.

Dawson RM, Broughton RL, Stevenson WTK, Sefton MV. 1987. Microencapsulation of CHO cells in a hydroxyethyl methacrylate-methyl methacrylate copolymer. Biomaterials 8:360–366.

Dupuy B, Gin H, Baquey C, Ducassou D. 1988. In vitro polymerization of a microencapsulating medium around living cells. J Biomed Mater Res 22:1061–1070.

Fan M-Y, Lum Z-P, Fu X-W, Levesque L, Tai IT, Sun AM. 1990. Reversal of diabetes in BB rats by transplantation of encapsulated pancreatic islets. Diabetes 39:519–522.

Ishihara K, Nakabayashi N. 1990. Adhesive bone cement both to bone and metals: 4–META in MMA initiated with tri-n-butyl borane. J Biomed Mater Res 23:1475–1482.

Karu TI. 1990. Effects of visible radiation on cultured cells. Photochem and Photobiology 53(6):1089.

Kobayashi H, Hyon S-H, Ikada Y. 1991. Water-curable and biodegradable prepolymers. J Biomed Mater Res 25:1481–1494.

Koehnlein HE, Lemperle G. 1969. Experimental studies with a new gelatin-resorcin-formaldehyde glue. Surgery 66:377–382.

Lacy PE, Hegre OD, Gerasimidi-Vazeou A. 1991. Maintenance of normoglycemia in diabetic mice by subcutaneous xenografts of encapsulated islets. Science 254:1782–1784.

Lim F, Sun AM. 1980. Microencapsulated islets as a bioartificial pancreas. Science 210:908–910.

Luckham P, Klein J. 1985. Interactions between smooth solid surfaces in solutions of adsorbing and nonadsorbing polymers in good solvent conditions. Macromolecules 18:721.

Lyman MD, Melanson, D, Sawhney, AS. 1996. Characterization of the formation of interfacially photopolymerized thin hydrogels in contact with arterial tissue. Biomaterials 17:359–364.

McMahon J, Schmid S, Weislow O, Stinson S, Camalier R, Gulakowski R, Shoemaker R, Kiser R, Dykes D, Harrison S, Mayo J, Boyd MJ. 1990. Feasibility of cellular microencapsulation technology for evaluation of anti-human immunodeficiency virus drugs in vivo. J Natl Cancer Inst 82:1761.

O'Shea GM, Goosen MFA, Sun AM. 1984. Prolonged survival of transplanted islets of Langerhans encapsulated in a biocompatible membrane. Biochim Biophys Acta 804:133–136.

O'Shea GM, Sun AM. 1986. Encapsulation of rat islets of Langerhans prolongs xenograft survival in diabetic mice. Diabetes 35:943–946.

Pathak CP, Sawhney AS, Hubbell JA. 1992. Rapid photopolymerization of immunoprotective gels in contact with cells and tissue. J Am Chem Soc 114:8311–8312.

Potts TV, Petrou A. 1991. Argon laser initiated resin photopolymerization for the filling of root canals in human teeth. Lasers Surg Med 11:257–262.

Roberts T, deBoni T, Sefton MV. 1991. Microencapsulation of dopamine secreting cells (PC12) in a HEMA-MMA copolymer. Trans Soc Biomater 14:157.

Ronis ML, Harvick JD, Fung R, Dellavecchia M. 1984. Review of cyanoacrylate tissue glues with emphasis on their otorhinolaryngological applications. Laryngoscope 94:210–213.

Sawhney AS. 1992. Biocompatible microspheres and microcapsules for animal tissue encapsulation and transplantation. Dissertation. University of Texas at Austin, Austin, Texas.

Sawhney AS, Hubbell JA. 1992. Poly(ethylene oxide)-graft-poly(l-lysine) copolymers to enhance the biocompatibility of poly(l-lysine)-alginate microcapsule membranes. Biomaterials 13:863–870.

Sawhney AS, Pathak CP, Hubbell JA. 1994. Modification of Langerhans surfaces with immunoprotective poly(ethylene glycol) coatings. Biotech Bioeng 44:383–386.

Tanaka A, Yasuhara S, Osumi M, Fukui S. 1977. Immobilization of yeast microbodies by inclusion with photocrosslinkable resins. Eur J Biochem 80:193–197.

Yashon D, Jane JA, Gordon MC. 1966. Effects of methyl-2–cyanoacrylate adhesives on the somatic vessels and the central nervous system of animals. J Neurosurg 25:883–888.

11

Long-Term Survival of Poly-L-Lysine-Alginate Microencapsulated Islet Xenografts in Spontaneously Diabetic NOD Mice

Collin J. Weber, Judith A. Kapp, Mary K. Hagler, Susan Safley, John T. Chryssochoos, and Elliot L. Chaikof

Introduction and Background

Problems with Current Therapy for Diabetes

Each year, approximately 15,000 new cases of insulin-dependent diabetes mellitus (IDDM) are diagnosed in the United States. Diabetes-related health care costs are staggering, and the life expectancy of a teenager is reduced by 30 years from the onset of IDDM. For many patients with IDDM, exogenous insulin therapy is not adequate to prevent the complications of the disease. The Diabetes Control and Complications Trial (DCCT) found that intensive insulin therapy delayed the onset and slowed progression of retinopathy, nephropathy, and neuropathy in patients with IDDM.[1] Unfortunately, intensive insulin therapy is not appropriate for many IDDM patients, and even with careful monitoring, DCCT patients had increased episodes of severe hypoglycemia, compared to conventionally treated patients.[1] Thus, results of the DCCT support the rationale for transplantation of insulin-producing cells. Several investigators are using genetic engineering to introduce the insulin gene into cells such as bone marrow stem cells from a diabetic patient that could be transplanted back into the same patient. However, the physiological regulation of insulin production, processing, and secretion is so complex that gene therapy remains highly experimental. Pancreas or islet transplantation is a more feasible approach for the near future.

Limitations and Risks of Pancreas Allotransplantation

Human pancreas allografts have been effective in normalizing blood glucose and stabilizing diabetic nephropathy. Unfortunately, there are significant risks associated with this procedure, including graft rejection, bleeding, abscess formation, and the side effects of immunosuppression. The first-year mortality of human pancreatic allografts remains high in many centers (approximately 10%), immunosuppression is required, and human pancreases are in short supply. Therefore, the number of clinical whole-organ pancreatic transplants being done worldwide is extremely limited.[2,3] Clearly, therapeutic alternatives to pancreas transplantation for IDDM are needed.

The Rationale for Pancreatic Islet Xenografts

Islet transplantation is an attractive therapy for patients with IDDM because problems related to exocrine pancreas grafting (bleeding, abscess, pancreatitis, etc.) can be avoided. Since the inception of islet transplant experiments, it has been hoped that islets would supply insulin more homeostatically than exogenous insulin can, and that "near-normal" modulation of carbohydrate metabolism might prevent or stabilize the secondary complications of IDDM.[4] A large number of experimental studies, done over the last two decades, support the feasibility of islet transplantation in humans.[4] Limited numbers of human islet isografts and rare, successful human islet allografts have confirmed that isolated islets can reverse human diabetes.[2] However, almost all allografts of human islets have failed, as a result of rejection,[3] and both availability and yield of human islets are extremely limited. Therapeutic islet transplants for large numbers of patients with IDDM almost certainly will require the use of donor islets harvested from animals (xenografts) or genetically engineered islet cell lines.[2,4]

Preventing Islet Xenograft Rejection Requires Immunoisolation

In spite of significant progress in our understanding of xeno-antigen recognition and xenograft rejection, there is still no safe immunosuppressive regimen that can prevent destruction of widely unrelated, "discordant" (e.g., pig-to-man) xenografts.[5,6] By contrast, certain porous mechanical devices have been shown to protect xenogeneic cell grafts, such as islets.[4,7,8] Both intra- and extravascular devices are under development. However, potential clinical complications, such as bleeding, coagulation, and bioincompatibility mitigate against the use of most currently available devices in diabetic patients.[7,9] For example, acrylic-copolymer hollow fibers, placed subcutaneously, have been shown to maintain viability of human islet allografts for up to 2 weeks (50 islets per 1.5 cm fiber with 65,000 MW permeability).[10] However, to implant 500,000 islets would require >15 meters of these hollow fibers for one patient, which is not clinically feasible.

Microencapsulation Is Currently the Optimal Immunoisolation Technique

One of the most promising islet envelopment methods is the polyamino acid-alginate microcapsule.[11] A large number of studies have shown that intraperitoneal xenografts of encapsulated rat, dog, pig, or human islets into streptozotocin-diabetic mice or rats can normalize blood glucose for >100 days.[11-13] In addition, long-term normalization of hyperglycemia by microencapsulated canine islet allografts and pig islet xenografts, and early function of one human islet allograft have been reported.[14-17]

Microcapsules have the three-dimensional design that optimizes volume-to-surface-area,[18,19] and they may actually enhance islet viability.[20,21] It has been postulated by several investigators that microcapsules, like other bioartificial membrane devices, promote survival of xenogeneic and allogenic islets by: (1) preventing or minimizing release of donor antigen(s), thereby reducing host sensitization, and/or (2) preventing or reducing exposure of the islets to host effector mechanisms (i.e., contact with cytotoxic T-cells, macrophages, antibodies, and complement).[7,22-24]

Exciting new approaches to modify xenogeneic donors include genetic manipulations of donor pigs to (1) reduce human-anti-pig naturally occurring antibody toxicity by producing transgenic pigs expressing human decay accelerating factor (DAF) to inhibit complement-mediated lysis by antibody,[25] (2) inhibit glycosylation (α-GAL), which is a major target of natural human anti-pig antibodies,[6,26] and (3) to "knock out" pig MHC antigens.[27,28] However, the MHC is only one of many classes of antigens that differ between humans and pigs. Since host APC can process and express foreign antigens of all types, other immunogenic proteins produced by pig islets would be expected to elicit human-anti-pig rejection responses. Thus, it is likely that even with the advent of genetically engineered donor (e.g., pig) islet cells, mechanical devices such as microcapsules will remain an important tool for protecting xenogeneic cells and providing delivery of these cells in an optimal, physiologic manner. Microcapsules will minimize sensitization of the host toward the graft and protect the graft from antibodies and direct cytotoxicity by T cells and macrophages.

Improvements in Microcapsule Design

We have characterized an improved formulation of poly-l-lysine-alginate microencapsulation that involves forming a "double-wall" membrane and increasing alginate concentration.[29,30] This "double-wall" microcapsule is more durable than conventional microcapsules; it reduces the likelihood of islets protruding from the capsule surface. Our technique of preparation of "double-wall" microcapsules is described below. Furthermore, we have found that increasing alginate from 1.85% to 2.0% reduces membrane permeability to approximately 100,000.

kD. This capsule excludes IgG (unlike conventionally designed capsules, which allow passage of IgG and 148,000 kD fluoresceinated dextran).[11,29] These improved microcapsules have resulted in prolonged and occasionally indefinite survival of rat islet xenografts in NOD mice, with no host immunosuppression required.

The technique of microencapsulation currently employed by our laboratory utilizes alginate (beta-D-mannpyroanospyluronic acid and alpha-l-fulopyranosyluornic acid) (Kelco low viscosity, bacterial count <25/gm; Kelco Div. Merck, San Diego, CA) in aqeuous phase, under conditions that are physiological with regard to pH and temperature (24°C).[11,12,29,31,32]

Briefly, isolated islets are suspended 1:10 (v/v) in 2.0% sodium alginate in 0.9% saline. Droplets containing islets in alginate are produced by extrusion (1.7 ml/min) through a 22 gauge air-jet needle (air flow 5 L min). Droplets fall 2 cm into a 20 ml beaker containing 10 ml of 1.1% $CaCl_2$ in 0.9% saline, pH 7.1. Gelled droplets are decanted and transferred to a 50 ml centrifuge tube, filled completely with 1.1% $CaCl_2$ for each 2–4 ml of microcapsules, and the tube is rotated gently, end-over-end, one revolution each 10 seconds for 10 minutes. Microcapsules are allowed to settle, supernatant is aspirated, and then microcapsules are washed in 0.5% $CaCl_2$ followed by 0.28% $CaCl_2$ (in 0.9% saline, pH 7.1). After a final wash in 0.9% saline, poly-l-lysine (MW 18,000) 0.5 mg/ml in saline is added (to fill the tube), and the tube is rotated for 6 minutes. Microcapsules are allowed to settle, and then are washed in 0.1% CHES in saline, pH 8.2, followed by another wash in 0.9% saline. Next, 0.2% sodium alginate is added, and the microcapsules are rotated again for 4 minutes. Thereafter, alginate is aspirated and discarded and microcapsules are washed in 0.9% saline. A final wash in 55 mM sodium citrate is used to solubilize any alginate not reacted with poly-l-lysine, and thereafter, capsules may be implanted or placed in petri dish culture. This technique produces "single-wall" microcapsules, which are similar to those reported from several other laboratories.[14,17,22,33-36]

We have found that "single-wall" capsules are fragile. In order to make microcapsules more resilient, we have added several additional steps to the protocol prior to the final sodium citrate step. In essence, reincubation of capsules with both poly-l-lysine and dilute alginate are done, to prepare "double-wall" microcapsules. Specifically, additional poly-l-lysine (0.5 mg/ml in saline) is added. Microcapsules are rotated for an additional 6 minutes, then washed again in 0.1% CHES (in 0.9% saline, pH 8.2) and then in saline, following which they are reincubated on the rotator, in dilute 0.2% sodium alginate for 4 minutes Microcapsules are then washed in 0.9% saline, followed by addition of 55 mM sodium citrate in saline, pH 7.4, in which the microcapsules are rotated for an additional 6 minutes. Finally, microcapsules are washed three times in 0.9% saline, and then transferred to conventional tissue culture medium.

"Single-wall" microcapsules are translucent, measuring approximately 700 microns, while "double-wall" microcapsules are slightly smaller (approx. 500 microns) and are somewhat opaque. When implanted in NODs, neither "single-wall" nor "double-wall" (empty) capsules excited a cellular reaction if intact. After 6 months in vivo "double-wall" microcapsules demonstrated better capsule integrity and greater capsule wall thickness than was the case with "single-wall" microcapsules.

More recently, we have made another modification in capsule methodology, namely, "pretreating" (PT) islets in dilute alginate, prior to "double-wall" encapsulation. We developed this pretreatment technique in order to minimize defective microcapsules with islets embedded within the microcapsule membrane, a difficulty reported by de Vos[33,34] and Chang.[35] Briefly, the pretreatment technique is as follows. Cells are washed in Hank's balanced salt solution (HBSS) and allowed to settle, and excess HBSS is removed, leaving a pellet of cells and approximately 0.2 cc of HBSS. Thereafter, 1.5 ml of 2% alginate and 13.5 ml normal saline (NS) are added to produce a final concentration of 0.2% alginate. Cells in this liquid are then rotated for 10 minutes to mix. Thereafter, they are centrifuged to pellet the islets, at 2000 rpm for 3 minutes. Supernatant is removed and cells washed three times in normal saline and spun to pellet as above three times. Thereafter, the cell pellet is resuspended in 0.2 cc NS, and then 15 ml of 1.1% calcium chloride is added and incubated by rotating for 10 minutes. Thereafter, the cells are washed twice in normal saline.

This procedure constitutes a precoating of cells with alginate and calcium chloride, which results in

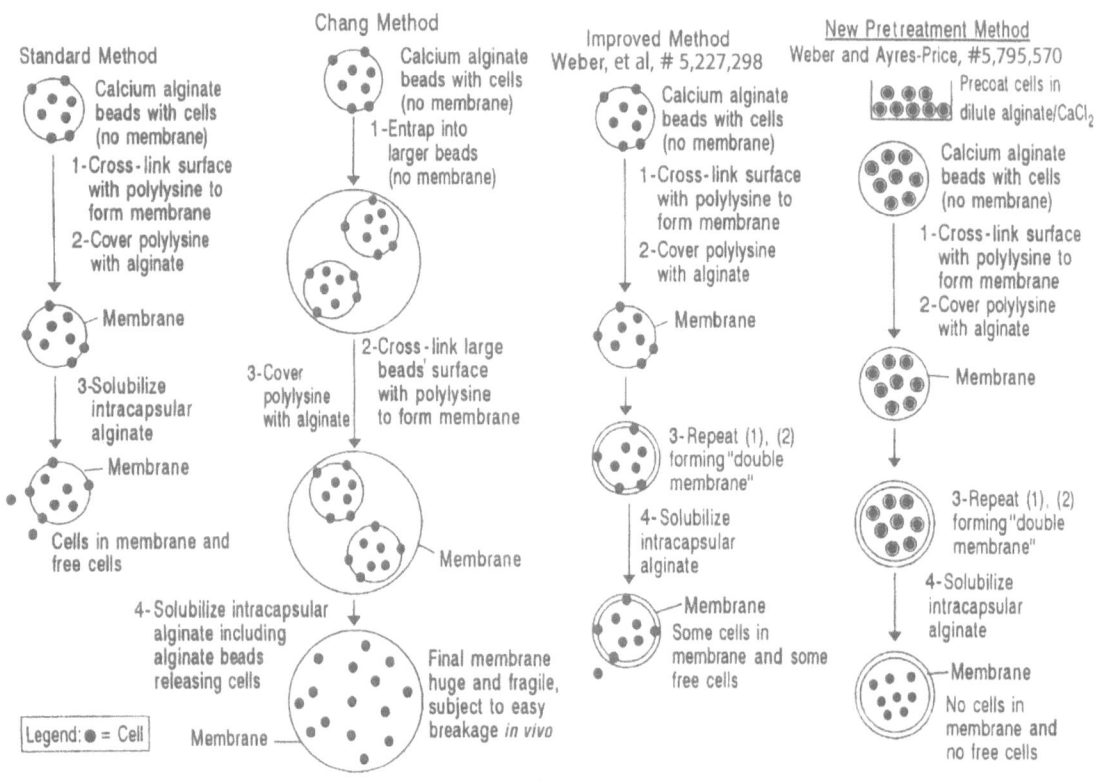

FIGURE 11.1. Diagram of various methods of poly-l-lysine-alginate microencapsulation.

a thin layer of cross-linked alginate surrounding each cell or tissue. Thereafter, these precoated cells are processed to form conventional double-wall microcapsules. We have found that this technique is more effective than the reencapsulation method of Chang,[35] the large final outer capsules of which are fragile and unstable (see Figure 11.1). This maneuver has reduced the incidence of defective capsules having islets incorporated within capsule membranes from 30% to 3–5%, as assessed by phase and confocal microscopy and by FITC-labeled islet-specific lectin staining (see Figure 11.2).

With these refinements in microcapsule preparation, we have observed significantly increased encapsulated islet xenograft survival, compared to prior "double-wall" (DW), 1.85% alginate capsules. Rat-to-NOD encapsulated islet survival increased from 19 ± 13 days for DW 1.85% to 86 ± 24 days for DW 2.0% ($p < .001$) (one graft, biopsy-proven, functioned >400 days). These maneuvers have also resulted in prolonged survival of rabbit-to-NOD islet xenografts (from 19 ± 3 days

for DW 2.0% capsules to 31 ± 6 days for PT DW 2.0% capsules) ($p < .02$) (see Table 11.1). Improvements in encapsulated islet xenograft survival may reflect reduced donor antigen release (exposure), or prevention of entry of host IgG.

We postulated that microencapsulated islet xenograft survival would be influenced by microcapsule permeability. We have found recently that microcapsule permeability may be altered by increasing or decreasing the concentration of PLL in the microcapsule formula. Red blood cells were encapsulated in alginate via our air-jet system and then incubated with various polyamino acids, including PLL. The RBCs were then lysed, and hemoglobin (MW 64,500) efflux was measured spectrophotometrically at 480 nm as a function of time alongside a concurrent control. The permeability coefficient was calculated according to the following formula: $(2.303 * Cf * Vt * S)/(Ci * At)$, where Ci and Cf are the initial and final hemoglobin concentrations, respectively, Vt and At are the total volumes and areas of capsules, respectively, and $S =$

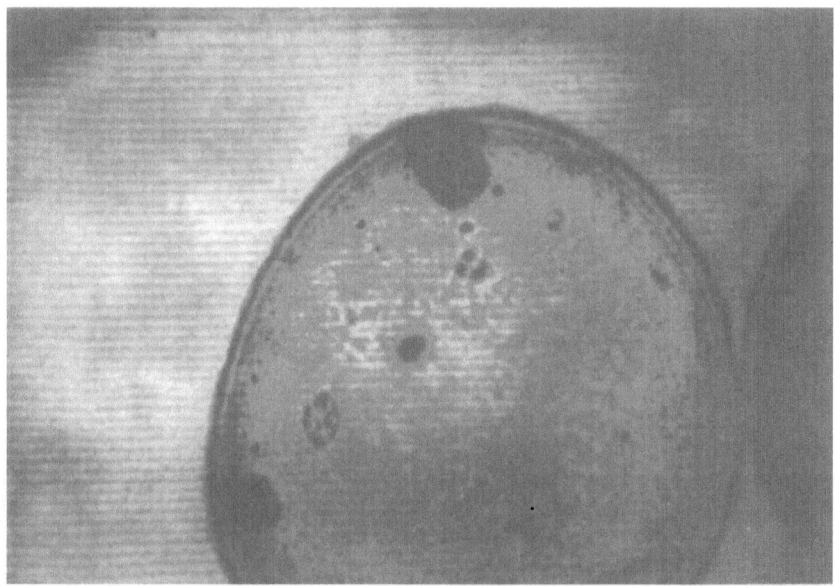

FIGURE 11.2. A, Confocal microscopy of 2.0% "double-wall" poly-l-lysine microcapsule, prepared without islet "pretreatment." Note islet partially embedded within microcapsule wall. Approx. 250×. (*Continued*)

slope of 1n $(Ct - Cf)/(Ci - Ct)$.[36] PLL substitutions (poly-l-ornithine, alanine, asparate, and histidine) did not result in suitable capsules. PLL molecular weight alterations did not affect permeability. In our studies, PLL *concentration* was the most critical factor in altering capsule diffusion. These observations are similar to the recent findings of other investigators.[36] There was a thirteenfold decrease in hemoglobin efflux occurring in capsules that had a fourfold increase in PLL (see Figure 11.3).

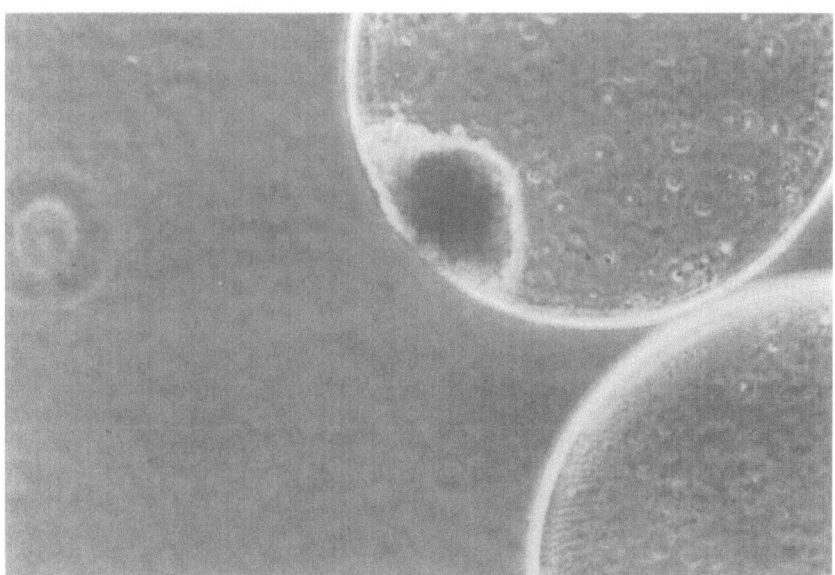

FIGURE 11.2. (*Continued*) B, Same technique as Figure 11.2A, phase contrast microscopy, showing nonpretreated islet incorporated into microcapsule membrane. Approx. 250×.

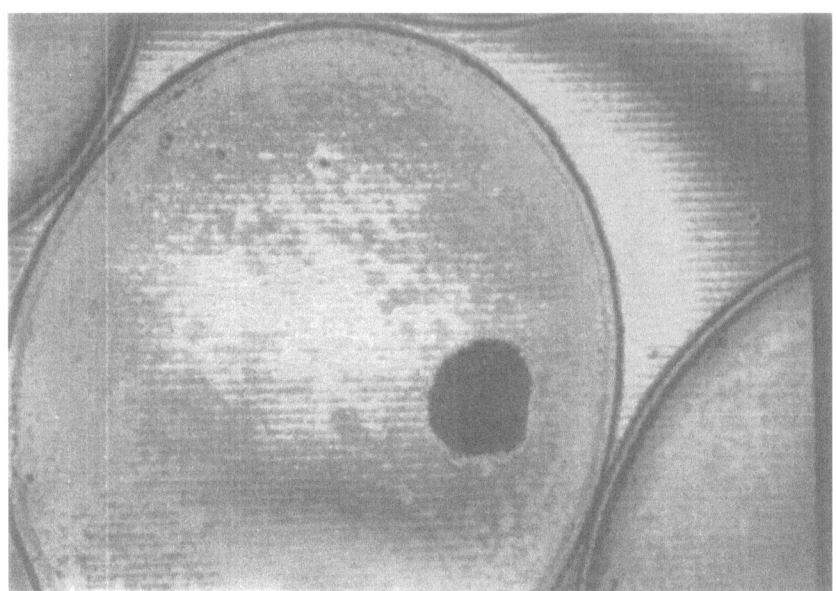

FIGURE 11.2. (*Continued*) C, Same technique as in Figure 11.2A, but with islet "pretreatment." Note that islet is well within microcapsule and is *not* incorporated within microcapsule membrane.

TABLE 11.1. Effect of microcapsule design on islet xenograft survival in diabetic mice.

Donor recipient	Site	Capsule type	Graft survival (days)(BG<250.mg/dl)		
			(N)	X ± SEM	NOD cellular reaction (0−4+)
RAT-NOD mouse	I.P.	SW, 1.85	10	13 ± 1*	4+
RAT-NOD mouse	I.P.	DW, 1.85	9	19 ± 3*‡‡	4+
RAT-NOD mouse	S	(-)	5	5 ± 0.5	4+
RAT-B10. mouse	I.P.	DW, 1.85	5	132 ± 17‡	0
RAT-B10. mouse	S	(-)	5	10 ± 0.6‡	4+
RAT-NOD mouse	I.P.	DW, 2.0#	5	86 ± 24##	1+
RAT-NOD mouse	I.P.	PT, DW, 2.0#	7	126 ± 47##	1+
RABBIT-NOD mouse	I.P.	DW, 2.0#	5	19 ± 3	4+
RABBIT-NOD mouse	I.P.	PT, DW, 2.0#	4	31 ± 6@	4+

$* = p < .001$ vs. intrasplenic contr.
‡‡ $= p < .01$ vs. SW
‡ $= P < .001$ vs. NODs
$= p < .002$ vs. DW, 1.85%
\# = excludes IgG
I.P. = intraperitoneal microcapsules
S = nonencapsulated islets, intrasplenic
@ $= p < .02$ vs. DW, 2.0%
SW, 1.85 = "single-wall," 1.85% alginate
DW, 1.85 = "double-wall," 1.85% alginate
DW, 2.0 = "double-wall," 2.0% alginate
PT, DW, 2.0 = islet pretreatment, "double-wall," 2% alginate

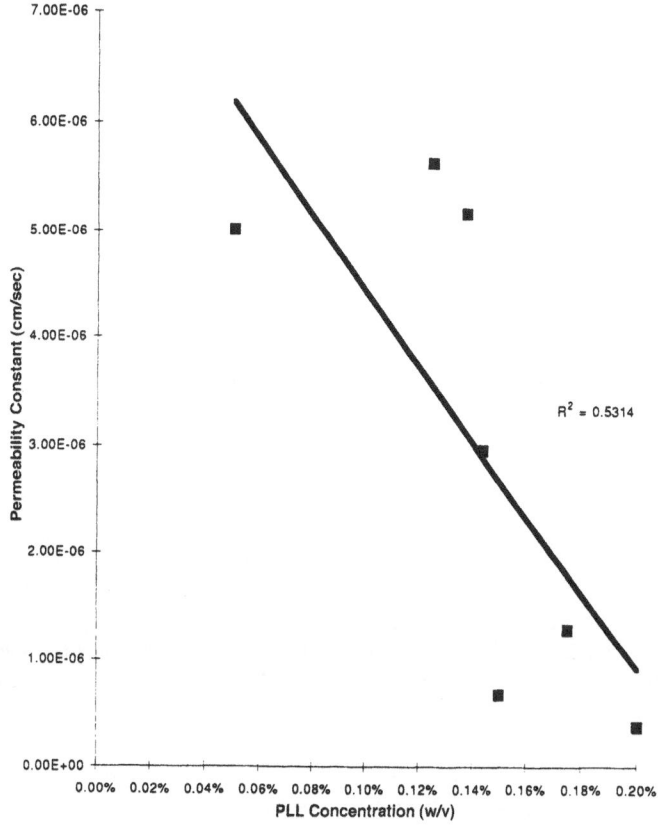

FIGURE 11.3. Correlation between PLL concentration and microcapsule permeability to hemoglobin (MW 64,500 kD).

Microcapsule Biocompatibility

Our empty microcapsules are biocompatible in rodents. We have recovered intact empty microcapsules, free of host reaction, from the peritoneal cavity of NOD mice and Lewis rats after >180 days, in multiple experiments.[11,12,29,37,38] Similar observations of the biocompatibility of empty microcapsules have been made by others.[14,17,19,39-44] We have found a limited cellular reaction to fractured microcapsules and capsule fragments, but uniform absence of cellular reactions to intact empty capsules (Figure 11.4).

Not all investigators have found that empty alginate-poly-l-lysine microcapsules were biocompatible.[34,45-50] Instead, some have reported that the fibrotic reactions surrounding empty microcapsules were similar to those observed around microcapsules containing islets. It is possible that impurities in reagents, such as contamination with endotoxin, may be responsible for reactions against some formulations of microcapsules. Standardized reagents for microcapsule preparation will aid in interpretations of data on functional characteristics of microcapsules formulated by different methods. Assurance of biocompatibility of empty microcapsules is a prerequisite for their use as protective membranes for islet xenografts.

Obviously, standardized reagents for microencapsulation would be advantageous if human trials are contemplated. However, we have not encountered problems with biocompatibility in our murine experiments. From a practical and expedient point of view, we have found it quite acceptable to select Kelco, low viscosity, low bacterial count alginate, and screen each batch with empty capsules in NOD recipients for 2–4 months prior to use with islets.

The Optimal Source of Xenogeneic Islets

The optimal source of xenogeneic islets for clinical use remains controversial. Islets have been isolated from subhuman primates and xenografted into immunosuppressed, diabetic rodents, with short-term

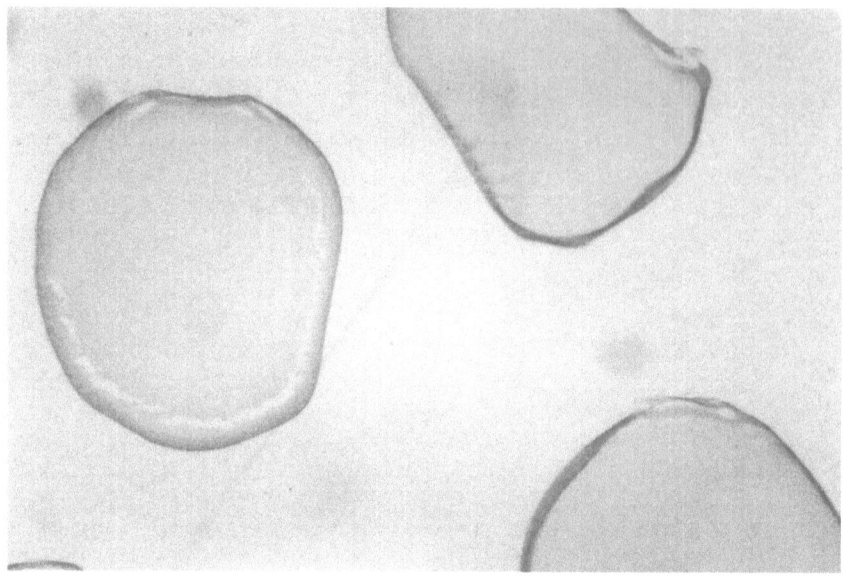

FIGURE 11.4. Empty "double-wall" microcapsules, biopsied from NOD peritoneal cavity after 6 months. Note lack of NOD reactive cells, uniform capsule size, and relatively thick capsule wall. H. & E., approx. 160×.

reversal of diabetes.[51] However, there are significant ethical issues surrounding the use of primates as a source of islets for human diabetes. Other promising sources are pig, cow, dog, and rabbit islets, all of which function remarkably well (i.e., maintaining normoglycemia) in diabetic rodents until transplant rejection occurs.[12,20,21,52-56] Long-term human, bovine, canine, and pig islet xenograft survival has been documented in nude mice and rats, suggesting that sufficient islet-specific growth factors are present in xenogeneic recipients to permit the xenografts' metabolic survival.[4,20,54,56-59]

Recent development of techniques for large-scale isolation of islets from neonatal pigs[53,60] and increased success in genetic modification of pigs[25,28] makes pig donors the most promising xenogeneic source for use in human diabetes.

Large-Scale Neonatal Pig Islet Isolation

In collaboration with Drs. Korbutt and Rajotte,[20,53,56,60] we have developed a reproducible method for isolation of large numbers of functionally viable islets from neonatal pig donors. With this technique, 30,000–100,000 islets may be obtained from each donor pig. Neonatal pig islet cells continue to secrete insulin in vitro after microen-

capsulation.[56,60] These neonatal pig islets are actually dispersed neonatal pig pancreatic cells that reaggregate to form "islet"-like spheroids with approximately 5–10% beta cells, which is significantly higher than the 1–2% beta cell concentration in the adult pig pancreas.

In order to assess the viability of neonatal "islets" in vivo, we have grafted them into immunoincompetent NOD-Scid mice. Recently, the Scid mutation has been back-crossed onto the NOD background, resulting in immunodeficient NOD-Scid mice.[61-64] These mice are homozygous for the Scid mutation, which results in an inability to rearrange T-cell receptor and immunoglobulin genes.[61,62] Consequently, these mice lack T and B lymphocytes. NOD-Scid mice do not develop diabetes spontaneously, but they may be rendered diabetic with multiple low-dose streptozotocin (MLD-SZN).[62-64] NOD-Scid mice express NOD MHC genes and other genes that are required for development of diabetes, upon transfer of lymphocytes from diabetic NOD mice.

To document the long-term functional activity of neonatal pig islets, we grafted them into NOD-Scid mice rendered diabetic with streptozotocin (SZN). Both nonencapsulated intrasplenic and microencapsulated intraperitoneal neonatal pig

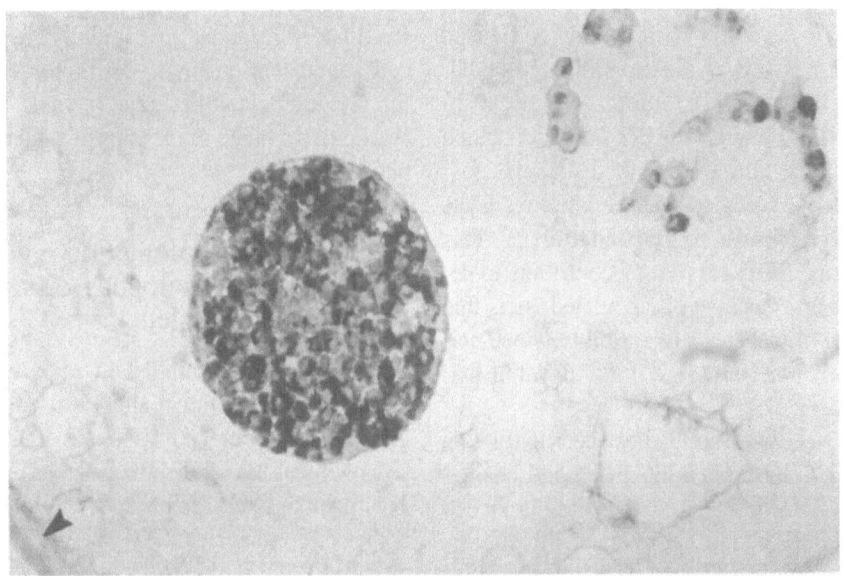

FIGURE 11.5. Neonatal pig islet in microcapsule, biopsied day #189 from SZN-diabetic NOD-SCID mouse. Anti-insulin immunohistochemistry, showing intensely insulin-positive beta cells, occupying approx. 80% of islet. Approx. 400×. Arrow points to inner surface of microcapsule membrane.

islets normalized hyperglycemia for >100 days. Biopsies of neonatal pig islets within microcapsules for up to 189 days following intraperitoneal xenotransplantation into immunoincompetent, streptozotocin-diabetic NOD-Scid mice revealed increased numbers of intensely insulin-positive islet cells (Figure 11.5). Thus, neonatal pig islets have an added advantage over adult islets, in that they appear to differentiate and proliferate within microcapsules.

These data demonstrate that in the absence of an immunological attack, neonatal pig islets survive and function physiologically in xenogeneic recipients for prolonged periods. In addition, NOD-Scid mice transplanted with encapsulated neonatal pig islets provide a convenient, in vivo model to identify the cells and factors responsible for rejection, by adoptive transfer of cells and sera from NOD mice.

The use of xenogeneic tissues such as pig islets in human transplantation has raised concerns about the potential risks of transmission of recognized zoonotic pathogens and other unknown xenogeneic agents to the recipient.[65,66] For example, porcine endogenous retrovirus has been shown to be capable of infecting human cells in vitro.[66] However,

there is no evidence of transmission to humans. Historically, porcine skin grafts were used as human burn wound dressings clinically for many years without serious sequelae. Recently, fetal pig neurons have been xenografted into the brains of patients with Parkinson's disease without evident viral infection.[67] While this is an important issue, the clinical need for organs (hearts, livers, kidneys) is so acute in this country that the development of effective protocols for generation of safe porcine organs may be anticipated, with islet xenografts being a beneficiary of these developments.

Spontaneously Diabetic NOD Mice Are the Best Model of Human IDDM

NOD mice are the most appropriate small animal model for studying the feasibility of islet xenotransplants because their disease resembles human IDDM in several ways. NOD mice develop diabetes spontaneously, beginning at approximately 12 weeks of age. Macrophage, dendritic cell, and lymphocytic infiltration of islets can be detected as early as 4 weeks of age and precedes overt hyperglycemia.[68-71] Diabetes in NOD mice is T-lymphocyte-dependent,[68-69] and it is associated with the expression of certain MHC Class II genes.[72-74] We have found that both NOD and (SZN-diabetic) B10.H-2^{g7} (expressing the NOD-MHC-linked disease allele) mice rejected encapsulated rat islets, while NOD.H-2^b mice, which express all of the non-MHC-linked diabetes susceptibility genes, accepted encapsulated rat islets for >100 days (similar to B10 controls).[75] This suggests that the NOD-MHC may contribute to destructive responses against encapsulated islets that are distinct from diabetes susceptibility, since neither B10.H-2^{g7} nor NOD.H-2^b mice develop diabetes spontaneously.[11,75]

Cytotoxic T cells and antibodies specific for beta cells and insulin have been identified and characterized from NOD mice.[69,76-78] Loss of tolerance to islet antigens in NOD mice correlates with appearance of Th1 immune responses, particularly toward glutamic acid decarboxylase (GAD), a factor that has been reported to be a primary autoantigen in human IDDM.[79] The disease can be induced in nondiabetic, syngeneic mice by transfer of CD8$^+$ and CD4$^+$ T cells or T-cell clones from diabetic NOD mice.[69,78,80-82] Moreover, inhibition of NOD macrophages or CD4$^+$ T lymphocytes, or treatment with anti-Class II monoclonal antibodies prevents or delays diabetes onset in NOD mice.[83] It has been suggested that helper T cells function to activate CD8$^+$ cells, which damage beta cells by direct cytotoxic attack in NOD mice. However, some recent studies have suggested that the killing

of beta cells may be indirect, from a nonspecific inflammatory response that initially involves CD4$^+$ cells, but that also includes infiltrating macrophages that release cytokines and oxygen free-radicals (particularly nitric oxide), all of which are known beta cell toxins.[84-91]

Microcapsules Prolong Survival and Function of Isologous, Allogeneic, and Xenogeneic Islets

Spontaneously diabetic NOD mice reject unprotected islet xenografts, allografts, and even isografts,[12,55,92-94] and conventional immunosuppressive regimens have little effect on this reaction.[54,94-97] Treatment of NOD recipients with monoclonal antibodies directed against CD4$^+$ helper T lymphocytes or with FK506 prolongs islet graft function (from 5 to 25 days),[12,92,94,97] but does not result in long-term islet graft survival in NOD mice.

We have found that microencapsulation prolongs the functional survival of islet xenografts in NOD mice, when compared to survival of unencapsulated islets injected into the spleen.[12,38,54,55] The same is true for islet allografts and isografts into NOD mice[12,54,55] (see Table 11.2). NOD rejection of encapsulated islet xenografts is accompanied by an intense cellular reaction, composed primarily of macrophages and lymphocytes, which entraps islet-containing microcapsules and results in recurrence of hyperglycemia within (2–4) weeks[12,38]

TABLE 11.2. Beneficial effect of "double-wall" microencapsulation on survival of islet iso-, allo-, and xenografts in NOD mice.

Donor	Recipient	Technique	(N)	Surv (days)@
NOD	NOD	CAP/I.P.	4	44 ± 7*
NOD	NOD	Splenic	3	6, 7, 7
Balb	NOD	CAP/I.P.	4	73 ± 31*
Balb	NOD	Splenic	2	5, 5
Lewis rat	NOD	CAP/I.P.	8	79 ± 15*
Lewis rat	NOD	Splenic	9	19 ± 3
Dog	NOD	CAP/I.P.	3	14 ± 4*
Dog	NOD	Splenic	2	0, 0
Rabbit	NOD	CAP/I.P.	7	20 ± 2*
Rabbit	NOD	Splenic	2	5, 6
Neonatal pig	NOD	Cap/I.P.	8	27 ± 13*
Neonatal pig	NOD	Splenic	3	6 ± 1

*p < .01 vs. splenic: @ = Mean ± SEM / CAP/I.P. = microencapsulated islet graft to peritoneal cavity;
Splenic = nonencapsulated islets grafted beneath splenic capsule.

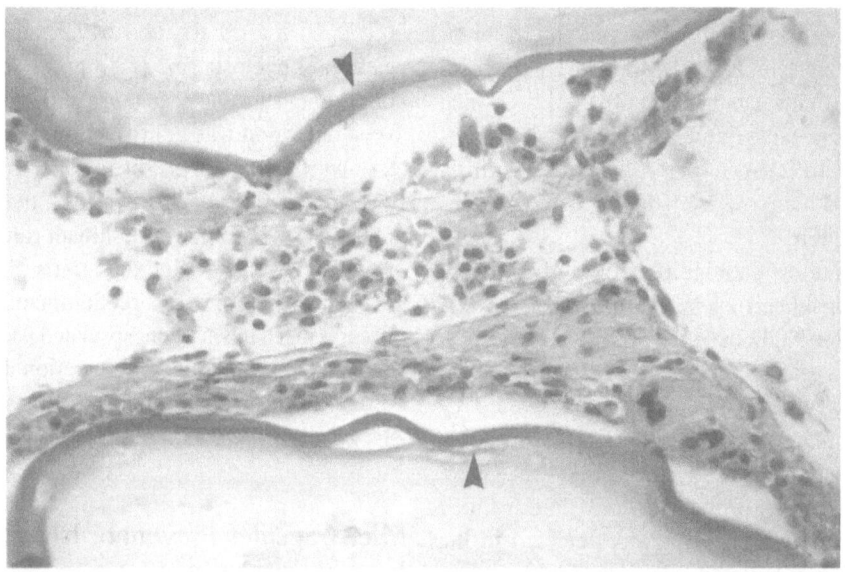

FIGURE 11.6. Pericapsular NOD cellular reaction to encapsulated neonatal pig islets. Note peritoneal NOD cells between two adjacent microcapsules. Biopsy done at time of graft rejection, day #20. H & E 400×. Arrows point to inner aspect of each capsule.

(see Figure 11.6). Rejection of encapsulated islets by NOD mice is not completely understood. However, the observation that rejection of encapsulated islets rarely occurs in mice with drug-induced (SZN) diabetes suggests that preexisting islet-specific autoimmunity may be involved in islet graft destruction. Similar problems are to be expected in transplanting xenogeneic islets into humans with IDDM.

Antigraft antibodies, macrophages, T lymphocytes, and cytokines all may be involved in encapsulated islet xenograft rejection. Halle[22] and Darquy[39] reported that microcapsules protected donor islets from host immunoglobulins, specifically human anti-islet antibodies and complement effects in vitro, Although antibodies and complement components are too large (>150,000. kD) to enter conventional poly-l-lysine microcapsules, they may still be instrumental in rejection. Antibodies may combine with shed donor antigens to form complexes that could bind to FcR of macrophages in vivo (in the peritoneal cavity), inducing cytokine release that ultimately could cause destruction of encapsulated islets.[6,98] Complement could facilitate bind-

ing of complexes to macrophages via the C3b receptor or by the release of chemotactic peptides that could increase the number of macrophages.[98]

We proposed involvement of NOD T lymphocytes in rejection of encapsulated islet xenografts on the basis of findings that treatment of NOD mice with antibodies directed against CD4+ helper T cells prolonged function of encapsulated islet xenografts in NOD mice.[11,12] These findings are similar to observations of others[5,6,99–102] and suggest that CD4+ T cells play a central role in rejection of xenogeneic islet grafts.

Cytokine Messenger RNA (mRNA) in Encapsulated Islet Xenograft Biopsies

Macrophages/monocytes are the predominant cells that adhere to encapsulated islet allografts and xenografts in NOD mice.[11,12,38] Cytokines known to be products of macrophages, including IL-1 and TNFα,[46,84,103] may be involved in destruction of encapsulated islets. For example, both IL-1 and TNFα have been reported to reduce insulin secretion and cause progressive damage of islet cells in vitro.[80,84–86,103,104] Cytokine-mediated injury might occur directly or indirectly, by activation of an intraperitoneal inflammatory response.[7,11] IL-1 induces nitric oxide synthase (NOS),[85,86,105] which generates nitric oxide (NO), causing injury to mitochondria and to DNA in beta cells.[85,86,105] Furthermore, this pathway of islet damage may be enhanced by TNFα.[103,106]

To elucidate the pathogenesis of NOD destruction of encapsulated islets, we extracted mRNA from recipient NOD peritoneal cells harvested at the time of graft rejection, and expression of mRNA for several lymphokines was studied by RT-PCR.[55] Integrity of RNA samples was assessed by inspection of Northern blots that were hybridized with the probe for the 3′ untranslated region of beta actin.[107] IL-4 was detected in the majority of encapsulated islet xenografts undergoing rejection by NOD mice. IL-10 expression was variable. IL-2 was detected during autoimmune destruction of NOD isografts but rarely in rejection of xenografts.[55] O'Connell et al[108,109] demonstrated that IL-2 messenger RNA is also detected in biopsies of rejected allogeneic islets. These data suggest that the primary T-cell response in rejecting encapsulated islet xenografts may be "Th2–like" and that it differs from the T-cell response to allogeneic islets. Increased Th2, relative to Th1[110,111] activity is distinct from the known NOD Th1 anti-islet immune response.[79,112] The Th2 response is characteristic of evoked antibody responses to foreign antigens, and suggests that humoral reactions may be important to rejection of encapsulated islet xenografts.[107] Therefore, strategies designed to abrogate Th2 responses may prolong encapsulated islet xenograft survival.

Depletion of CD4+, but Not CD8+, T Cells Prolongs Encapsulated Islet Xenografts in NOD Mice Indefinitely

Because encapsulation reduces but does not prevent rejection of xenogeneic islets, we have tried various strategies to further reduce the immune response to encapsulated islets. For example, treatment of NOD recipients of encapsulated islet xenografts with monoclonal anti-CD4+ antibody (GK1.5) resulted in prolonged survival of xenografts.[12] On the other hand, treatment with monoclonal anti-CD8 antibody 53.6.7 (100 μg i.p., day 5 and then twice weekly) or cyclosporine (30 mg/kg, s.c., daily) did not prolong survival.[54] In these experiments, CD8+ T-cell depletion was confirmed by flow cytometry of NOD spleen and peritoneal cells. These data are consistent with prior observations, that CD4+ (but not CD8+) T cells play a dominant role in rejection of nonencapsulated islet xenografts.[5,12,99–101] They also are consistent with a predominantly Th2 NOD rejection mechanism of encapsulated islet xenografts, which is suggested by demonstration that IL-4, but not IL-2, message is found in cells associated with islet xenograft rejection.[55]

Microcapsules Prevent or Delay Host Sensitization

To clarify the mechanism of long-term microcapsule protection of xenogeneic rat islets, we have performed preliminary experiments in which paired diabetic NODs were pretreated with Hank's balanced salt solution (HBSS), Lewis rat islets (200 i.p.), or Lewis rat splenocytes (10^6 i.p.). Encapsulated Lewis islets were xenografted into presensitized and HBSS-treated control NODs 14 days later. As shown in Figure 11.7, both donor islet and splenocyte pretreatment resulted in rapid graft rejection, while nonpresensitized NODs accepted encapsulated islet xenografts for > 100 days. These data suggest that a major function of microcapsules is to prevent host sensitization, rather than to protect grafts from the effector arm of the response. Thus, maneuvers that reduce islet immunogenicity may be synergistic with islet encapsulation.

Model of NOD Recognition/Reaction of Microencapsulated Islet Xenografts

On the basis of our experimental data, we have developed a model to describe the mechanisms that we think may be involved in the rejection of encapsulated, xenogeneic islets by autoimmune NOD mice (Figure 11.8). Xenogeneic islet antigens are released by the occasional broken capsule or by donor proteins less than 100,000 MW (AgX) that diffuse out of the capsule (1) and are endocytosed

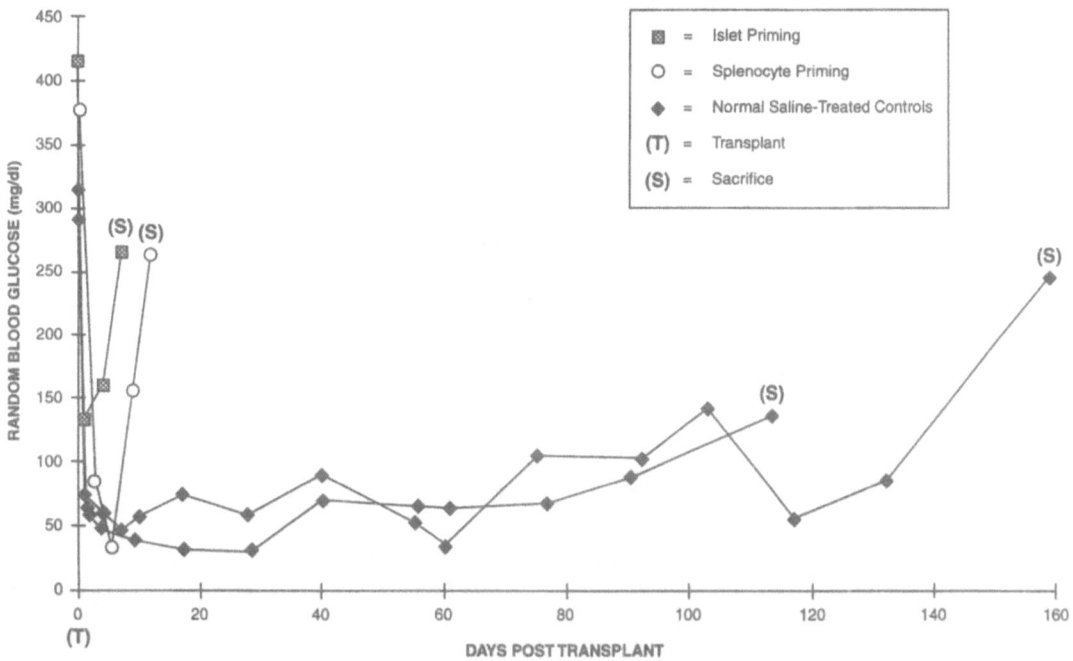

FIGURE 11.7. Effect of Lewis rat splenocyte priming on Lewis-rat-to-NOD encapsulated islet xenograft survival.

by dendritic cells. Dendritic cells process proteins via the MHC Class II pathway and present peptide X complexed with Class II and costimulatory molecules to CD4$^+$ T cells (2). In the presence of appropriate cytokines, CD4$^+$ T cells are activated and develop into Th2 cells that express CD40L. B cells, with surface IgM that binds AgX, endocytose and process it into peptides that bind MHC Class II, which are expressed on the surface of B cells. T cells specific for peptide X bind B cells, and the interaction of CD40 with CD40L causes activation of B cells. Activated B cells mature into plasma cells under the direction of Th2 lymphokines (3). Plasma cells secrete specific antibody that forms complexes with AgX (4). Antibodies cannot directly damage the encapsulated islets because they are too large to enter the capsules. However, antibodies in the peritoneal cavity could form complexes with antigens shed or secreted from the capsules. Such antigen-antibody

complexes efficiently bind to FcR expressed on the surface of peritoneal macrophages. Binding of complexes to FcR activates macrophages to secrete a variety of mediators (5), including IL-1, TNFα, and nitric oxide (NO),[113] all of which have toxic effects on islets, and all of which are small enough to cross the double wall of the capsules. Potentially, the effector arm could be further augmented by the activation of complement (C) by antigen-antibody complexes. C3b bound to the complexes enhances the activation of macrophages by increasing the binding of the complexes via the C3b receptor,[114] and small peptides such as C3a released during C activation induce local inflammatory responses, thereby attracting more macrophages into the peritoneal cavity.[115] Furthermore, preexisting autoantibodies to islet antigens might augment this immunogenicity of islet antigens by forming complexes that can bind to and activate APC.

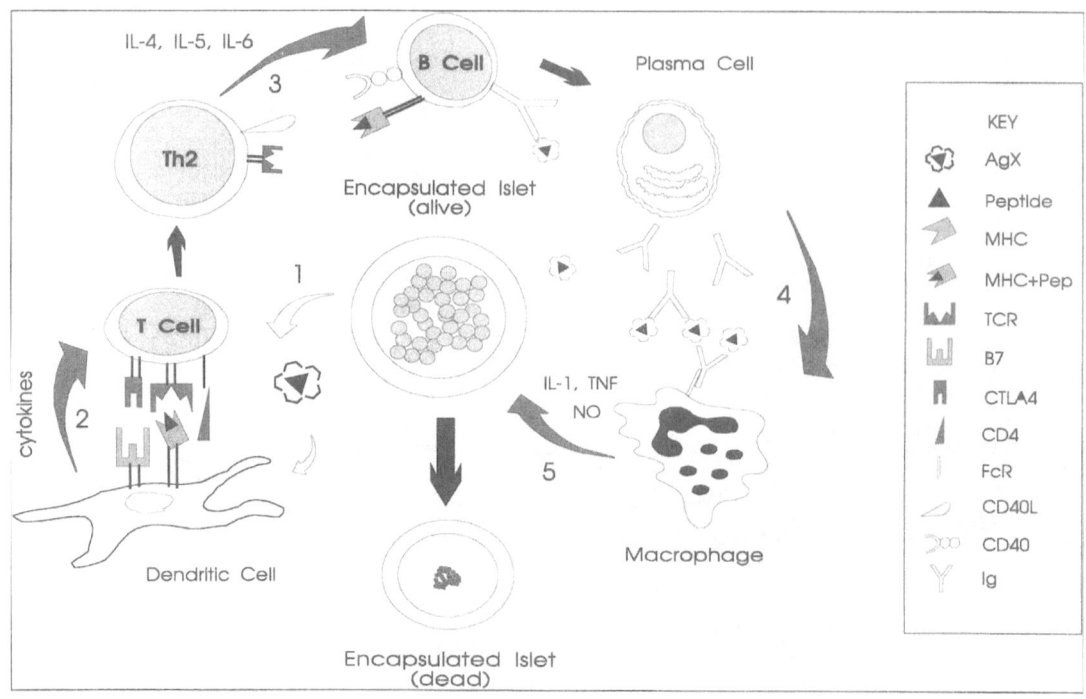

FIGURE 11.8. Potential responses involved in rejection of encapsulated, xenogeneic islets by NOD recipients. AgX = donor antigen; MHC = major histocompatibility complex; Pep = peptide; TCR = T-Cell receptor; FCR = Fc receptor; Ig = immunoglobulin.

CTLA4lg Blocks Costimulatory Interactions

Several studies, in vitro and in vivo, have shown that foreign molecules that interact with the T-cell receptor (peptides, specific antibodies, mitogens) fail on their own to stimulate naive T cells to proliferate,[111,116] and may induce antigen-specific anergy. An additional (costimulatory) signal is required, and it is delivered by APCs. In mice, one such costimulatory pathway involves the interaction of the T-cell surface antigen CD28 with either one of two ligands, B7–1 and B7–2, on the APCs.[117-123] Once this full interaction of T cells and APCs occurs, however, subsequent reexposure of T cells to peptide, mitogen, etc. may result in proliferation in the absence of costimulation.[120,124,125]

CTLA-4 is a cell surface protein that is closely related to CD28; however, unlike CD28, CTLA-4 is expressed only on activated T cells. It has been suggested that CTLA4 may modulate functions of CD28.[123,124,126,127] CTLA4lg is a recombinant soluble fusion protein, combining the extracellular binding domain of the CTLA4 molecule with the constant region of the IgG_1, gene. Administration of CTLA4lg to mice induced antigen-specific unresponsiveness in murine lupus and encephalomyelitis models[122,128] and long-term acceptance of murine cardiac allografts.[129-135] In addition, it induced tolerance to human islets in SZN-diabetic mice.[57]

CTLA4lg Treatment of NOD Recipients of Microencapsulated Neonatal Pig and Adult Rabbit Islets Results in Prolonged Islet Xenograft Survival

The most significant obstacle to islet xenotransplantation in human IDDM is the lack of an effective immunosuppressive regimen to prevent cross-

TABLE 11.3. Survival of microencapsulated (MC) adult rabbit and neonatal pig islets in NOD mice: effects of NOD treatment with CTLA4lg.

Donor	Technique	Rx	(N)	Graft survival (days)(BG<250 mg/dl)	
				$x \pm$ S.E.	Days
Rabbit	MC/IP	None	7	20 ± 2	12, 16, 18, 18, 20, 28, 28
Rabbit	MC/IP	CTLA4-lg[a]	8	108 ± 24*	37, 43, 47, 58, 148, 151, 173, 205
Rabbit	MC/IP	CTLA4-lg[b]	4	70 ± 8*	48, 66, 81, 83
Rabbit	Splenic	CTLA-lg[a]	3	6 ± 1*	5, 6, 6
Rabbit	Splenic	None	2	—	5, 6
Neonatal pig	MC/IP	None	8	27 ± 36	9, 10, 12, 12, 14, 18, 23, 118[s]
Neonatal pig	MC/IP	CTLA4-lg[c]	10	115 ± 19*	61, 74[s], 80[s], 101[s], 108, 137[s], 161[s], 266[s]
Neonatal pig	MC/IP	CTLA4-lg[c]	10	103 ± 28	12, 32, 55, 63, 75, 78, 83, 103, 239, 287
Neonatal pig	Splenic	CTLA4-lg[d,c]	3	5 ± 1	4, 5, 5
Neonatal pig	Splenic	None	3	6 ± 1	5, 6, 7

IP = intraperitoneal
CTLA4lg, 200 mcg I.P., QOD
* = p > .04 vs. MC alone
MC = microcapsules

(a) × 92 days
(b) × 14 days
(c) × 21 days
(d) = mutant CTLA 4–Ig which does not bind complement.
(s) = sacrifice for biopsy

species graft rejection.[5,6,25–28] It has been reported that human islets will survive long-term in mice with chemically-induced (SZN) diabetes, either when treated with anti-CD4 antibody[59] or CTLA4lg (a high-affinity fusion protein that blocks CD28–B7 interactions),[57] or when treated by exposure to purified high-affinity anti-HLA F(ab)$_2$.[136] Unfortunately, none of these regimens alone has been effective in prolonging survival of xenogeneic islet grafts in animals with spontaneous, autoimmune diabetes (NOD mice).[12,54,56,100] Thus, rejection of xenogeneic islets can be overcome by various immunosuppressive strategies, providing the recipient does not express autoimmune diabetes. There is considerable evidence that xenorecognition (unlike allorecognition) occurs primarily via the "indirect" antigen presentation pathway, by which host APCs present peptides scavenged from extracellular (donor) proteins to host helper T cells.[5,137–139] Our recent report, that host MHC is critical to NOD rejection of encapsulated islet xenografts,[75] and our prior finding, that helper T cells are essential for this response,[12] are both consistent with this "indirect" pathway.

We have found that a combination of microencapsulation of donor neonatal pig or adult rabbit islets with treatment of recipient NOD mice with murine CTLA4lg results in prompt and long-term reversal of spontaneous diabetes, allowing these xenogeneic islets to survive and function in NOD mice for 60–200 days[54,56] (see Table 11.3, Figures 11.9 and 11.10). These data demonstrate synergy between CTLA4lg treatment and islet encapsulation, since neither encapsulation nor CTLA4lg alone was effective. We conclude that interference with the B7/CD28 costimulatory pathway by recipient treatment with CTLA4lg synergizes with encapsulation to prolong survival of discordant islet xenografts in diabetic NOD mice. The data support the hypothesis that NOD antigen-presenting cells (APC) process soluble, shed donor antigens and activate NOD T cells via the "indirect" antigen presentation pathway. We believe that this approach holds great promise for treatment of patients with diabetes since human CTLA4lg is available, neonatal pig islets are plentiful, and our microcapsules are durable and biocompatible.

These data demonstrate that combination therapy markedly prolongs islet xenograft survival in autoimmune recipients, whereas none of the therapies given alone does so. Anti-CD4 antibodies and CTLA4lg both synergize with encapsulation to permit prolonged to indefinite survival of neonatal pig islets. Thus, additional combinations of therapies will likely be even more effective in the induction of graft acceptance. Our studies provide evidence that xenogeneic islet grafts may be feasible for treatment of IDDM in the foreseeable future.

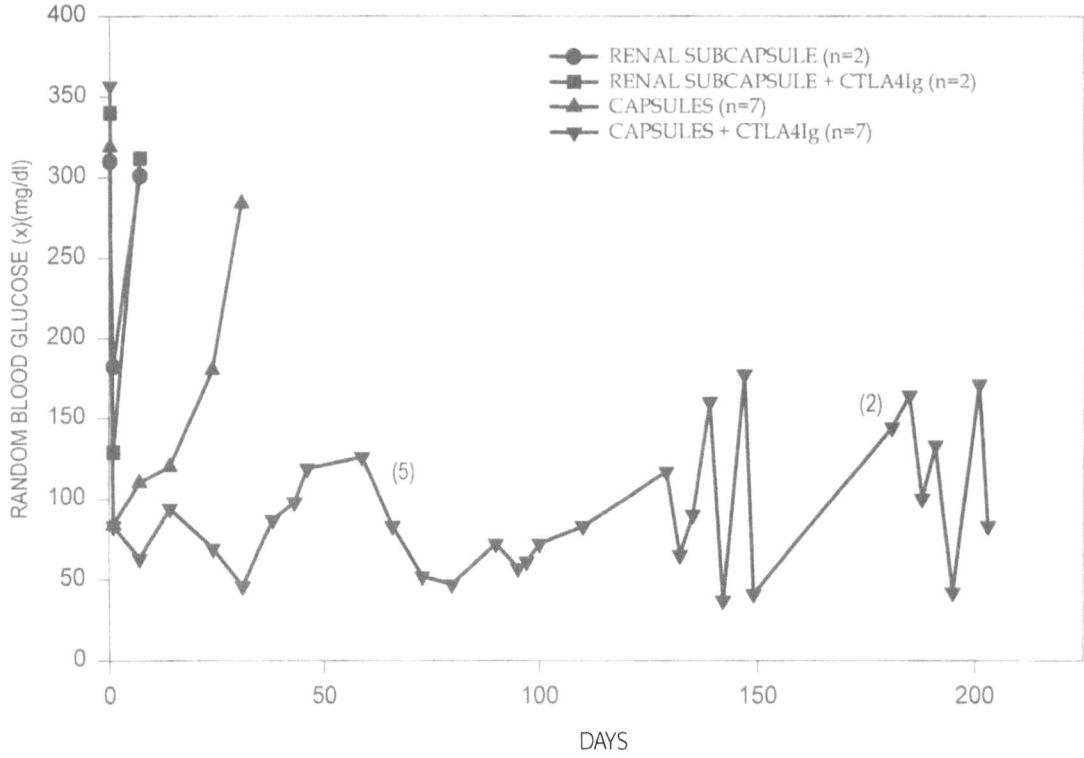

FIGURE 11.9. Intraperitoneal microencapsulated adult rabbit islet xenografts into diabetic NOD mice: effects of encapsulation and CTLA4lg.

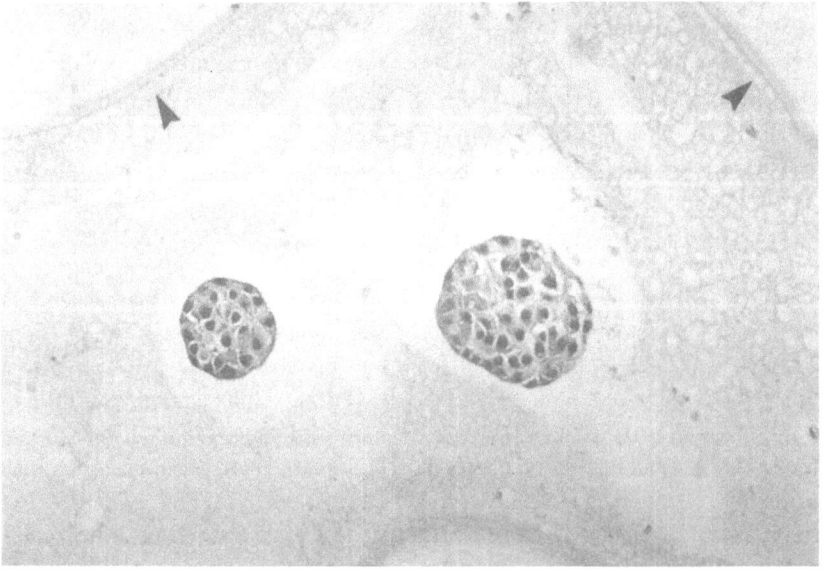

FIGURE 11.10. Biopsy of long-term functioning encapsulated rabbit islets from diabetic CTLA4lg-treated NOD mouse. Note absence of host NOD peritoneal cells. Arrows point toward inner aspect of microcapsule membrane. H & E Approx. 400×.

References

1. The Diabetes Control and Complications Trials Research Group. 1993. The effect of intensive treatment of diabetes on the development and progression of long-term complications in insulin-dependent diabetes mellitus. NEJM 329:977–86.
2. Warnock G, Rajotte R. 1992. Human pancreatic islet transplantation. Transplantation Reviews 6:195–208.
3. Remuzzi G, Ruggenenti P, Mauer S. 1994. Pancreas and kidney/pancreas transplants: experimental medicine or real improvement? Lancet 343:27–31.
4. Lacy P. Status of islet cell transplantation. 1993. Diabetes Review 1:76–92.
5. Auchincloss H, Sachs DH. Xenogeneic transplantation. 1998. Ann. Rev. Immunol. 16:433–470.
6. Parker W, Saadi S, Lin SS, Holzknecht ZE, Bustos M, Platt JL. 1996. Transplantation of discordant xenografts: a challenge revisited. Immunology Today 17:373–8.
7. Colton C, Avgoustiniatos E. 1991. Bioengineering in development of the hybrid artificial pancreas. J Biochem Eng 113:152–70.
8. Colton CK. 1992. The engineering of xenogeneic islet transplantation by immunoisolation. Diab Nutr Metab 5:145–9.
9. Lanza R, Sullivan S, Chick W. 1992. Islet transplantation with immunoisolation. Diabetes 41:1503–10.
10. Scharp D, Swanson C, Olack B, et al. 1994. Protection of encapsulated human islets implanted without immunosuppression in patients with Type I or Type II diabetes and in nondiabetic control subjects. Diabetes 43:1167–70.
11. Weber C, Reemtsma K. 1994. Microencapsulation in small animals-II: Xenografts. In: Pancreatic islet transplantation Series: Vol III: immunoisolation of pancreatic islets. Lanza R, Chick W, eds. Austin, Texas: R. Landes, 59–79.
12. Weber C, Zabinski S, Koschitzky T, et al. 1990. The role of CD4+ helper T cells in destruction of microencapsulated islet xenografts in NOD mice. Transplantation 49:396–404.
13. Siebers U, Zekorn T, Horcher A, et al. 1994. Microencapsulated transplantation of allogeneic islets into specifically presensitized recipients. Transpl Proc 26:787–8.
14. Soon-Shiong P, Feldman E, Nelson R, et al. 1992. Successful reversal of spontaneous diabetes in dogs by intraperitoneal microencapsulated islets. Transplantation 54:769–74.
15. Soon-Shiong P, Heintz RE, Merideth N, et al. 1994. Insulin independence in a type 1 diabetic patient after encapsulated islet transplantation. Lancet 343:950–1.
16. Zhou D, Sun Y, Vacek I, Ma P, Sun A. 1994. Normalization of diabetes in cynomolgus monkeys by xenotransplantation of microencapsulated porcine islets. Transp Proc 26:1091–2.
17. Sun Y, Ma X, Zhou D, Vacek I, Sun A. 1996. Normalization of diabetes in spontaneously diabetic cynomologus monkeys by xenografts of microencapsulated porcine islets without immunosuppression. J Clin Invest 98:1417–22.
18. Iwata H, Kobayashi K, Takagi T, et al. 1994. Feasibility of agarose microbeads with xenogeneic islets as a bioartificial pancreas. Journal of Biomedical Materials Research 28:1003–11.
19. Wang T, Lacik I, Brissova M, et al. 1996. An encapsulation system for the immunoisolation of pancreatic islets. Nature Biotechnology 15:358–62.
20. Korbutt GS, Ao Z, Warnock GL, Flashner M, Rajotte RV. 1995. Successful reversal of diabetes in nude mice by transplantation of microencapsulated porcine neonatal islet cell aggregates. Transplantation Proceedings 27:3212–2.
21. Ao Z, Korbutt GS, Warnock GL, Flashner M, Colby CB, Rajotte RV. 1995. Microencapsulation enhances canine islet survival during long-term culture. Transplantation Proceedings 27:3350–0.
22. Halle J, Bourassa S, Leblond F, et al. 1993. Protection of islets of Langerhans from antibodies by microencapsulation with alginate-poly-l-lysine membranes. Transplantation 44:350–4.
23. Iwata H, Morikawa N, Fujii T, Takagi T, Samejima T, Ikada Y. 1995. Does immunoisolation need to prevent the passage of antibodies and complements? Transplantation Proceedings 27:3224–6.
24. Zekorn T, Endl U, Horcher A, Siebers U, Bretzel RG, Federlin K. 1995. Mixed lymphocyte islet culture for assessment of immunoprotection by islet microencapsulation. Transpl Proc 27:3362–3.
25. Rosengard AM, Cary N, Horsley J, et al. 1995. Endothelial expression of human decay accelerating factor in transgenic pig tissue: a potential approach for human complement inactivation in discordant xenografts. Transplant Proc 27:326–6.
26. Cairns T, Lee J, Goldberg L, et al. 1995. Inhibition of the pig to human xenograft reaction using soluble GALx1–3GAL and GALx1–3GALB1–4GIcNAc. Transpl 60:1202–7.
27. Bravery CA, Batten P, Yacoub MH, Rose ML. 1995. Direct recognition of SLA- and HLA-like class II antigens on porcine endothelium by human T cells results in T cell activation and release of interleukin-2. Transpl 60:1024–33.
28. Seebach JD, Yamada K, McMorrow IS, Sachs DH, DerSimonian H. 1996. Xenogeneic human anti-pig cytotoxicity mediated by activated natural killer cells. Xenotransplantation 3:188–97.

29. Weber C, Costanzo M, Krekun S, D'Agati V. 1993. Causes of destruction of microencapsulated islet grafts: Characteristics of a 'double-wall' poly-l-lysine—alginate microcapsule. Diabetes, Nutrition and Metabolism 1:167–71.

30. Weber C, Norton J, Reemtsma K. 1993. Method for microencapsulation of cells or tissue, United States Patent #5,227,298: July 13, 1993;(Abstract).

31. Weber C, Costanzo M, Zabinski S, et al. 1992. Xenografts of microencapsulated rat, canine, porcine, and human islets into streptozotocin (SZN)- and spontaneously diabetic NOD mice. In: Pancreatic islet transplantation. Ricordi C, ed. Austin, Texas: R.G. Landes, 177–90.

32. Weber C, Zabinski S, Norton J, Koschitzky T, D'Agati V, Reemtsma K. 1989. The future role of microencapsulation in xenotransplantation. In: Xenograft 25. Hardy M, ed. Amsterdam: Elsevier, 297–308.

33. de Vos P, Wolters G, Van Schilfgaarde R. 1994. Possible relationship between fibrotic overgrowth of alginate-polylysine-alginate microencapsulated pancreatic islets and the microcapsule integrity. Transp Proc 26:782–3.

34. de Vos BJ, de Haan GHJ, Wolters R, Van Schilfgaarde R. 1996. Factors influencing the adequacy of microencapsulation of rat pancreatic islets. Transplantation 62:888–93.

35. Chang T. 1992. Artificial cells in immobilization biotechnology. Art Cells & Immob Biotech 20:1121–43.

36. Vandenbossche G, Van Oostveldt P, Demeester J, Remon J. 1993. The molecular weight cut-off of microcapsules is determined by the reaction between alginate and polylysine. Biotechnology and Bioengineering 42:381–6.

37. Weber C, Krekun S, Koschitzky S, et al. 1991. Prolonged functional survival of rat-to-NOD mouse islet xenografts by ultraviolet-B (UV-B) irradiation plus microencapsulation of donor islets. Transplantation Proceedings 23:764–6.

38. Weber C, Price J, Costanzo M, Becker A, Stall A. 1994. NOD mouse peritoneal cellular response to poly-l-lysine-alginate microencapsulated rat islets. Transplantation Proceedings 26:1116–9.

39. Darquy S, Reach G. 1985. Immunoisolation of pancreatic B cells by microencapsulation. Diabetologia 28:776–80.

40. Zekorn T, Siebers U, Horcher A, et al. 1992. Alginate coating of islets of Langerhans: in vitro studies on a new method for microencapsulation for immuno-isolated transplantation. Acta Diabetol 29:41–5.

41. Calafiore R. 1992. The large-scale microencapsulation of isolated and purified human islets of Langer-hans. In: Pancreatic islet cell transplantation. Ricordi C, ed. Austin, Texas: R. Landes, 207–14.

42. Goosen M, O'Shea G, Gharapetian H, Chou S, Sun A. 1985. Optimization of microencapsulation parameters: Semipermeable microcapsules as a bioartificial pancreas. Biotechnology and Bioengineering 27:146–50.

43. Horcher A, Zekorn T, Siebers U, et al. 1994. Transplantation of microencapsulated islets in rats: Evidence for induction of fibrotic overgrowth by islet alloantigens released from microcapsules. Transpl Proc 26:784–6.

44. Lanza R, Kuhtreiber W, Ecker D, Staruk J, Chick W. 1995. Xenotransplantation of porcine and bovine islets without immunosuppression using uncoated alginate microspheres. Transplantation 59:1377–84.

45. Wijsman J, Atkinson P, Mazheri R. 1992. Histological and immunopathological analysis of recovered encapsulated allogeneic islets from transplanted diabetic BB/W rats. Transplantation 54:588–92.

46. Cole D, Waterfall M, McIntyre M, Baird J. 1992. Microencapsulated islet grafts in the BB/E rat: A possible role for cytokines in graft failure. Diabetologia 35:231–7.

47. Clayton H, London N, Bell P, Jams R. 1992. The transplantation of encapsulated islets of Langerhans into the peritoneal cavity of the biobreeding rat. Transplantation 54:558–60.

48. Zekorn T, Klock G, Horcher A, et al. 1992. Lymphoid activation by different crude alginates and the effect of purification. Transpl Proc 24:2952–3.

49. Soon-Shiong P, Oterlie M, Skjak-Braek G, et al. 1991. An immunologic basis for the fibrotic reaction to implanted microcapsules. Transpl Proc 23:758–9.

50. Mazaheri R, Atkinson P, Stiller C, Dupre J, Vose J, O'Shea G. 1991. Transplantation of encapsulated allogeneic islets into diabetic BB/W rats: Effects of immunosuppression. Transplantation 51:750–4.

51. Weber C, Hardy M, Rivera S, et al. 1986. Diabetic mouse bioassay for functional and immunologic human and primate islet xenograft survival. Transplantation Proceedings 18:823–8.

52. Giannarelli R, Marchetti P, Villani G, et al. 1994. Preparation of pure, viable porcine and bovine islets by a simple method. Transplantation Proceedings 26:630–1.

53. Korbutt GS, Ao Z, Warnock GL, Rajotte RV. 1995. Large-scale isolation of viable porcine neonatal islet cell (NIC) aggregates. Transplantation Proceedings 27:3267–7.

54. Weber CJ, Hagler MK, Chryssochoos CP, et al. 1996. CTLA4–Ig prolongs survival of microencapsulated rabbit islet xenografts in spontaneously diabetic NOD mice. Transplantation Proceedings 28:821–3.

55. Weber CJ, Hagler M, Konieczny B, et al. 1995. Encapsulated islet iso-, allo-, and xenografts in diabetic NOD mice. Transpl Proc 27:3308–11.

56. Weber CJ, Hagler MK, Chryssochoos JC, et al. 1997. CTLA4lg prolongs survival of microencapsulated neonatal porcine islet xenografts in diabetic NOD mice. Cell Transplantation 6:505–8.

57. Lenschow D, Zeng Y, Thistlethwaite J, et al. 1992. Long-term survival of xenogeneic pancreatic islet grafts induced by CTLA4lg. Science 257:789–95.

58. Falqui L, Finke E, Carel J, Scharp D, Lacy P. 1991. Marked prolongation of human islet xenograft survival (human-to-mouse) by low temperature culture and temporary immunosuppression with human and mouse antilymphocyte sera. Transplantation 51:1322–5.

59. Ricordi C, Lacy P, Sterbenz K, Davie J. 1987. Low-temperature culture of human islets or in vivo treatment with L3T4 antibody produces a marked prolongation of islet (human-to-mouse). Transplantation 44:465–8.

60. Korbutt GS, Elliot JE, Ao Z, Smith DK, Warnock GL, Rajotte RV. 1996. Large scale isolation, growth, and function of porcine neonatal islet cells. J Clin Invest 97:2119–29.

61. Christianson S, Shultz L, Leiter E. 1993. Adoptive transfer of diabetes into immunodeficient NOD-Scid/Scid mice. Diabetes 42:44–5.

62. Gerling I, Friedman H, Greiner D, Shultz L, Leither E. 1994. Multiple low-dose streptozotocin-induced diabetes in NOD-Scid/Scid mice in the absence of functional lymphocytes. Diabetes 43:433–40.

63. Rohane P, Shimada A, Kim D, et al. 1995. Islet-infiltrating lymphocytes from prediabetic NOD mice rapidly transfer diabetes to NOD-scid/scid mice. Diabetes 44:550–4.

64. Shultz L, Schweitzer P, Christian S, et al. 1995. Multiple defects in innate and adaptive immunologic function in NOD/LtSz-*scid* mice. J of immunology 154:180–91.

65. Chapman LE, Folks TM, Salomon DR, Patterson AP, Eggerman TE, Noguchi PD. 1995. Xenotransplantation and xenogeneic infections. New England J Med 333:1498–501.

66. Patience C, Takeuchi Y, Weiss R. 1997. Infection of human cells by an endogenous retrovirus of pigs. Nature Medicine 3:282–6.

67. Deacon T, Shumacher J, Dinsmore J, et al. 1997. Histological evidence of fetal pig neural cell survival after transplantation into a patient with Parkinson's disease. Nature Medicine 3:350(Abstract).

68. Jarpe A, Hickman M, Anderson J. 1990. Flow cytometric enumeration of mononuclear cell populations infiltrating the islets of Langerhans in prediabetic NOD mice: Development of model of autoimmune insulitis for Type I diabetes. Regional Immunology 3:305–17.

69. Miller B, Appel M, O'Neil J, Wicker L. 1988. Both the Lyt-2+ and L3T4+ T cell subsets are required for the transfer of diabetes in nonobese diabetic mice. J Immunol 140:52–8.

70. Haskins K, Portas M, Bergman B, Lafferty K, Bradley B. 1989. Pancreatic islet-specific T-cell clones from nonobese diabetic mice. PNAS 86:8000–4.

71. Jansen A, Homo-Delarche F, Hooijkaas H, Leenen P, Dardenne M, Drexhage H. 1994. Immunohistochemical characterization of monocytes, macrophages and dendritic cells involved in the initiation of the insulitis and B-cell destruction in NOD mice. Diabetes 43:667–74.

72. Lipes M, Rosenzweig A, Tan K, et al. 1993. Progression to diabetes in nonobese diabetic (NOD) mice with transgenic T cell response. Science 259:1165–9.

73. Gelber C, Pabrosky L, Singer S, et al. 1994. Isolation of nonobese diabetic mouse T-cells that recognize novel autoantigens involved in the early events of diabetes. Diabetes 43:33–9.

74. Podolin P, Pressey A, DeLarato N, Fischer P, Peterson L, Wicker L. 1993. I-E+ nonobese diabetic mice develop insulitis and diabetes. J Exp Med 178:793–803.

75. Weber C, Tanna A, Costanzo M, Price J, Peterson L, Wicker L. 1994. Effects of host genetic background on survival of rat → mouse islet xenografts. Transplantation Proceedings 26:1186–8.

76. Haskins K, McDuffie M. 1990. Acceleration of diabetes in young NOD mice with a CD4+ islet-specific T cell clone. Science 249:1433–6.

77. Bergman B, Haskins K. 1994. Islet-specific T-cell clones from the NOD mouse respond to B-granule antigen. Diabetes 43:197–203.

78. Peterson J, Pike B, McDuffie M, Haskins K. 1994. Islet-specific T-cell clones transfer diabetes to nonobese diabetic (NOD) F$_1$ mice. J Immunol 153:2800–6.

79. Tisch R, Yang X, Singer S, Liblau R, Fugger L, McDevitt H. 1993. Immune response to glutamic acid decarboxylase correlates with insulitis in nonobese diabetic mice. Nature 366:72–5.

80. Dylan D, Gill R, Schloot N, Wegmann D. 1995. Epitope specificity, cytokine production profile and diabetogenic activity of insulin-specific T cell clones isolated from NOD mice. Eur J Immunology 25:1062.

81. Peterson JD, Haskins K. 1996. Transfer of diabetes in the NOD-*scid* mouse by CD4 T-cell clones: differential requirement for CD8 T-cells. Diabetes 45:328–36.

82. Utsugi T, Yoon JW, Park BJ, et al. 1996. Major histocompatibility complex class I-restricted infiltration

and destruction of pancreatic islets by NOD mouse-derived B-cell cytotoxic CD8$^+$ T-cell clones in vivo. Diabetes 45:1121–31.

83. Boitard D, Bendelac A, Richard M, Carnaud C, Bach J. 1988. Prevention of diabetes in nonobese diabetic mice by anti-l-A monoclonal antibodies: Transfer of protection by splenic T cells. PNAS 85:9719–23.

84. Rabinovitch A, Sumoski W, Rajotte R, Warnock G. 1990. Cytotoxic effects of cytokines on human pancreatic islet cells in monolayer culture. J of Clinical Endocrinology and Metabolism 71:152–6.

85. Bergmann L, Kroncke K, Suschek D, Kolb H, Kolb-Bachofen V. 1992. Cytotoxic action of IL-1B against pancreatic islets is mediated via nitric oxide formation and is inhibited by N^G-monomethyl-L-arginine. FEBS Letter 299:103–6.

86. Xenos E, Stevens R, Gores P, et al. 1993. IL-1 induced inhibition of B-cell function is mediated through nitric oxide. Transpl Proc 25:994–4.

87. Vara E, Arias-Diaz J, Garcia C, et al. 1995. Production of TNF, IL-1, IL-6 and nitric oxide by isolated human islets. Transplantation Proceedings 27:3367–71.

88. Hao L, Wang Y, Gill RG, La Rosa FG, Talmage DW, Lafferty KJ. 1990. Role of lymphokine in islet allograft rejection. Transplantation 49:609–14.

89. Jahr H, Bretzel RG, Wacker T, et al. 1995. Toxic effects of superoxide, hydrogen peroxide, and nitric oxide on human and pig islets. Transplantation Proceedings 27:3220–1.

90. Simeonovic C, Wilson J, Ramsay A. 1995. Role of IL-4 in pig proislet xenograft rejection in mice. Transplantation Proceedings 27:3571–1.

91. Allison J, Oxbrown L, Miller F. 1994. Consequences of *in situ* production of IL-2 for islet cell death. Internation Immunol 6:541–9.

92. Mandel T, Koulmanda M, Loudovaris T, Bacelj A. 1989. Islet grafts in NOD mice: A comparison of iso-, allo- and pig xenografts. Transplant Proceedings 21:3813–4.

93. Ricker A, Stockberger S, Halban P, Eisenbarth G, Bonner-Weir S. 1986. Hyperimmune response to microencapsulated xenogeneic tissue in non obese diabetic mice. In: The immunology of diabetes mellitus. Jaworski M, ed. Amsterdam: Elsevier, 193–200.

94. Akita K, Ogawa M, Mandel T. 1994. Effect of FK506 and anti-CD4 therapy on fetal pig pancreas xenografts and host lymphoid cells in NOD/Lt, CBA, and BALB/c mice. Cell Transplantation 3:61–73.

95. Ricker A, Bhatia V, Bonner-Weir S, Eisenbarth G. 1989. Microencapsulated xenogeneic islet grafts in NOD mouse: Dexamethasone and inflammatory response. Diabetologia 32:53.

96. Lafferty K. 1988. Circumventing rejection of islet grafts: An overview. In: *Transplantation of the endocrine pancreas in diabetes mellitus*. Van Schilfgaarde R, Hardy M, eds. Amsterdam: Elsevier, 279–91.

97. Lafferty KJ, Hao L. 1994. Approaches to the prevention of immune destruction of transplanted pancreatic islets. Transpl Proc 26:399–400.

98. Baldwin W, Pruitt S, Brauer R, Daha M, Sanfilippo F. 1995. Complement in organ transplantation. Transplantation 59:797–808.

99. Pierson R, Winn H, Russell P, Auchincloss H. 1989. CD-4 positive lymphocytes play a dominant role in murine xenogeneic responses. Transplantation Proceedings 21:519–21.

100. Loudovaris T, Charlton B, Mandel T. 1992. The role of T cells in the destruction of xenografts within cell-impermeable membranes. Transplantation Proceedings 24:2938–9.

101. Gill R, Wolf L, Daniel D, Coulombe M. 1994. CD4+ T cells are both necessary and sufficient for islet xenograft rejection. Transp Proc 26:1203–4.

102. Ahn M, Chang E, Chur G, Barker CF, Markmann JF. 1995. Cellular requirements for pancreatic islet xenograft rejection. Transplantation Proceedings 27:3302–3.

103. Campbell I, Iscaro A, Harrison L. 1988. Interferon gamma and tumor necrosis factor alpha: cytotoxicity to murine islets of Langerhans. J Immunol 141:1325–9.

104. Rabinovitch A. 1994. Immunoregulatory and cytokine imbalances in the pathogenesis of IDDM. Diabetes 43:613–21.

105. Andersen H, Jorgensen K, Egeberg J, Mandrup-Poulsen T, Nerup J. 1994. Nicotinamide prevents interleukin-I effects on accumulated insulin release and nitric oxide production in rat islets of Langerhans. Diabetes 43:770–7.

106. Mandrup-Poulsen T, Bendtzen K, Dinarello C, Nerup J. 1987. Human tumor necrosis factor potentiates interleukin-1 mediated rate of pancreatic B-cell cytotoxicity. J Immunol 139:4077–82.

107. Takeuchi T, Lowry R, Konoieczny B. 1992. Heart allografts in murine systems. Transplantation 53:1281–94.

108. O'Connell P, Pacheoco-Silva A, Nickerson P, et al. 1993. Unmodified pancreatic islet allograft rejection results in preferential expression of certain T cell activation transcripts. J Immunol 150:1093–104.

109. Nickerson P, Pacheco-Silva A, O'Connell P, Steurer W, Kelly V, Strom T. 1993. Analysis of cytokine transcripts in pancreatic islet cell allografts during rejection and tolerance induction. Transplantation Proceedings 25:984–5.

110. Lowry R, Takeuchi T. 1994. The Th1, Th2 paradigm and transplantation tolerance. R Landes, Austin, Texas, (in press).

111. Janeway C, Bottomly K. 1994. Signals and signs for lymphocyte responses. Cell 76:275–85.

112. Kaufman D, Clare-Salzler M, Tian J. 1993. Spontaneous loss of T-cell tolerance to glutamic acid decarboxylase in murine insulin-dependent diabetes. Nature 365:69–72.

113. Ke Y, Li Y, Kapp JA. 1995. Ovalbumin injected with complete Freund's adjuvant stimulates cytolytic responses. Eur J Immunol 25:549–53.

114. Krych M, Atkinson J, Holers V. 1992. Complement receptors. Curr Opin Immunol 4:8–13.

115. Frank M, Fries L. 1991. The role of complement in inflammation and phagocytosis. Immunol Today 12:322–6.

116. Jenkins M. 1994. The ups and downs of T cell costimulation. Immunity 1:443–6.

117. Liu Y, Jones B, Brady W, Janeway C, Linsley P. 1992. Co-stimulation of murine CD4 T cell growth: cooperation between B7 and heat-stable antigen. Eur J Immunol 22:2855–9.

118. Harding F, McArthur J, Gross J, Raulet D, Allison J. 1992. CD28–mediated signalling co-stimulates murine T cells and prevents induction of anergy in T-cell clones. Nature 356:607–9.

119. Galvin F, Freeman G, Razi-Wolf Z, et al. 1992. Murine B7 antigen provides a sufficient costimulatory signal for antigen-specific and MHC-restricted T cell activation. J Immunol 149:3802–8.

120. Linsley PS, Brady W, Urnes M, Grosmaire LS, Damle NK, Ledbetter JA. 1991. CTLA-4 is a second receptor for the B cell activation antigen B7. J Exp Med 74:561–9.

121. Hathcock K, Laszio G, Dickler H, Bradshaw J, Linsley P, Hodes R. 1993. Identification of an alternative CTLA-4 ligand costimulatory for T cell activation. Science 262:905–7.

122. Perrin P, Scott D, Quigley L, et al. 1995. Role of B7: CD28/CTLA-4 in the induction of chronic relapsing experimental allergic encephalomyelitis. J of Immunology 154:1481–90.

123. Freeman GJ, Boussiotis VA, Anumanthan A, et al. 1995. B7–1 and B7–2 do not deliver identical costimulatory signals, since B7–2 but not B7–1 preferentially costimulates the initial production of IL-4. Immunity 2:523–32.

124. Larsen CP, Ritchie SC, Hendrix R, et al. 1994. Regulation of immunostimulatory function and costimulatory molecule (B7–1 and B7–2) expression on murine dendritic cells. J Immunol 152:5208–19.

125. Hathcock KS, Laszlo G, Pucillo C, Linsley P, Hodes RJ. 1994. Comparative analysis of B7–1 and B7–2 costimulatory ligands: Expression and function. J Exp Med 180:631–40.

126. Gimmi C, Freeman G, Gribben J, Gray G, Nadler L. 1993. Human T-cell clonal anergy is induced by antigen presentation in the absence of B7 costimulation. Immunology 90:6586–90.

127. Guerder S, Meyerhoff J, Flavell R. 1994. The role of the T cell costimulator B7–1 in autoimmunity and the induction and maintenance of tolerance of peripheral antigen. Immunity 1:155–66.

128. Finck B, Linsley P, Wofsy D. 1994. Treatment of murine lupus with CTLA4Ig. Science 265:1225–7.

129. Lin H, Bolling S, Linsley P, et al. 1993. Long-term acceptance of major histocompatibility complex mismatched cardiac allografts induced by CTLA4Ig plus donor-specific transfusion. J Exp Med 178:1801–6.

130. Bolling S, Turka L, Wei R, Linsley P, Thompson C, Lin H. 1994. Inhibition of B7–induced CD28 T-cell activation with CTLA4Ig prevents cardiac allograft rejection: evidence for costimulation. Transplantation 413–5.

131. Pearson T, Alexander D, Winn K, Linsley P, Lowry R, Larsen C. 1994. Transplantation of tolerance induced by CTLA4–Ig. Transplantation 57:1701–6.

132. Steurer W, Nickerson PW, Steele AW, Steiger J, Zheng XX, Strom TB. 1995. Ex vivo coating of islet cell allografts with murine CTLA4/Fc promotes graft tolerance. J Immunol 155:1165–74.

133. Bolling SF, Lin H, Turka LA. 1996. The time course of CTLA4Ig effect on cardiac allograft rejection. J Surg Res 63:320–3.

134. Pearson TC, Alexander DZ, Hendrix R, et al. 1996. CTLA4–Ig plus bone marrow induces long-term allograft survival and donor-specific unresponsiveness in the murine model. Transplantation 61:997–1004.

135. Larsen CP, Elwood ET, Alexander DZ, et al. 1996. Long-term acceptance of skin and cardiac allografts after blocking CD40 and CD28 pathways. Nature 381:434–8.

136. Faustman D, Coe C. 1991. Prevention of xenograft rejection by masking donor HLA class I antigens. Science 252:1700–2.

137. Platt J, Back F. 1991. The barrier to xenotransplantation. Transplantation 52:937–47.

138. Moses R, Winn H, Auchincloss H. 1992. Evidence that multiple defects in cell-surface molecule interactions across species differences are responsible for diminished xenogeneic T cell responses. Transplantation 53:203–9.

139. Moses R, Pierson R, Winn H, Auchincloss H. 1990. Xenogeneic proliferation and lymphokine production are dependent on CD4+ helper T cells and self antigen-presenting cells in the mouse. J Exp Med 172:567–75.

12
Alginate/Poly-L-Ornithine Microcapsules for Pancreatic Islet Cell Immunoprotection

Riccardo Calafiore and Giuseppe Basta

Foreword

Over the past few decades several laboratory procedures have been developed for immobilization and encapsulation of either animal or plant cellular products, or cells/tissues for different applications. Encapsulation techniques usually reflected the physical-chemical properties associated with the microcapsule's membrane, depending upon its specific employment. In fact, use of microcapsules for drug or metabolite time-controlled release or delivery poses a number of technical problems that substantially differ from those that arise when the membranes are to be rigorously immunoselective so as to protect the enveloped cell/tissue graft (TX) from the host's immune response. This being the case, it seems critical that microcapsules keep their constituent physical-chemical structure as immutable as possible throughout the post-TX period. Moreover, encapsulated cell TX addresses those recipients that may require replacement of a function that has been lost, thus introducing additional variables. For instance, in the specific case of insulin-dependent diabetes mellitus (IDDM), it is not sufficient that microcapsules serve as a biocompatible and competent physical immunobarrier for the implanted islets, but it is also necessary that the membranes be specially engineered so as not to interfere with normal β-cell insulin secretory kinetics, which are indispensable for the prevention of blood glucose spikes. In other words, here it is not only a matter of immunoprotecting cells that release a certain product, but also, and rather, of preserving an extremely complex sensing apparatus that is finely tuned to maintain the glucose homeostasis. Of course, as a further variable, IDDM is an autoimmune disorder, which means that an extra immunoprotection may be required to avoid recurrence of the disease.

In summary, the chemical formulation and the engineering of the microcapsules play a major role in providing these artificial membranes with either biocompatibility, immunobarrier competence, or long-term stability, so as to fulfill requirements for the application of this technology to biomedical as well as other fields.

Microencapsulation for Pancreatic Islet TX Immunoprotection

Background

In spite of the recent introduction of highly purified and pharmacologically flexible insulin preparations, administered to IDDM patients according to intensive treatment regimens, either conventionally or by open-loop artificial delivery systems, to achieve an optimized blood glucose control[1], the inability of such a therapeutic approach to prevent both the onset of secondary complications of IDDM and the elevated incidence of acute, severe hypoglycemic episodes is becoming increasingly apparent.[2] In fact, subcutaneously injected exogenous insulin could never even approximate the pulsatile insulin secretory patterns of normal islet β-cells. On the contrary, TX of the endocrine pancreas, comprising either whole organs[3] or isolated and purified islet cells,[4] could theoretically

TABLE 12.1. Potential advantages of isolated islets over whole/segmental pancreas.

1. Low mass (1–2% of the total pancreas)
2. Minor surgery
3. Access to immunomodulatory intervention
4. Access to cryogenic storage
5. Access to immunoisolation within permselective artificial membranes

correct hyperglycemia to a state very close to the physiological model. In particular, islet cells seem to hold potential advantages over whole pancreata (Table 12.1).

Now, after more than 30 years of study,[5] it has been clearly demonstrated that islet cell TX may induce, in the best conditions, reversal of hyperglycemia together with exogenous insulin withdrawal in either animal models of, or patients with, IDDM. However, success so far achieved in lower mammals has not been matched by that accomplished in higher mammals with IDDM, with special regard to humans. In the latter, at this juncture, long-standing (>1 year) full remission of hyperglycemia, together with insulin independence, has been obtained in as little as 10% of the total grafted individuals,[6] clearly signifying that there are still many problems to be solved prior to even thinking of islet cell TX as a substantial strategy for the cure of IDDM. The two key points that play a major role in the process of transplanting islets in IDDM recipients, namely (1) provision of an adequate viable and functional islet cell mass, and (2) adequate protection from islet graft directed immune response have been only partially fulfilled, with many issues still waiting for an answer. In particular, as problems of obtaining high mammalian islet cell yield and purity are gradually being overcome,[7,8] the islet-graft-related immune problems continue to represent a major obstacle to the success of this approach. At this juncture, the only islet immunoprotection strategy that has been widely pursued in humans, has consisted of general, pharmacologic immunosuppression of the host. Unfortunately, this approach, with its severe, life-threatening side effects, imposed, for ethical reasons, the necessity of combining islet with major organ (i.e., kidney or liver) grafts, in IDDM patients affected by end-stage renal or liver disease, and also virtually prevented younger subjects, solely affected by IDDM, from participating in these clinical trials. Therefore, although general immunosuppression can reasonably be employed for organ grafts, which are indispensable for the patients' survival, this is certainly not the case for islet cells, whose functional activity can be replaced, though imperfectly, as mentioned elsewhere, by sq. exogenous insulin injections. In light of the above reviewed concerns, potential alternative solutions that do not imply any recourse to general immunosuppression for islet graft immunoprotection should be pursued and implemented (Table 12.2). Of these, the only strategy that, after substantial study in lower animal models, has been initially used in higher mammals, including humans, consists of islet TX immunoisolation within highly biocompatible and immunoselective biomembranes.[9] Although there are different approaches to immunoisolate an islet TX within a physical barrier (Table 12.3), the principle consists of preventing access by both humoral and cellular mediators of the immune system, while allowing for biochemical exchange between the enveloped tissue and the host. Aside from intra- or extravascular TX sites, which may apply to both

TABLE 12.2. Alternate strategies to circumvent islet-graft-directed immune destruction.

1. Induction of immune unresponsiveness
  • immunomodulation (successful only in rodents)
  • microchimerism: donor bone marrow and islet cell graft-tolerogenic effect on Class II APCs
2. Islet graft immunoisolation

TABLE 12.3. Device types for islet graft immunoisolation.

1. Macrodevices
   a. Vascular
      (A-V shunt prostheses)
   b. Extravascular
      • tubular ("straws")
      • planar
2. Microcapsules
   a. Alginate-based (large, minimal volume)
   b. Other polymer-based

TABLE 12.5. Islet transplantation: Immunoisolation.

Macrodevices (hollow fibers, "straws," vascular prosthesis)	
PROS	CONS
• No immunosuppression	• High morbidity
• Ease of implantation	• Low biocompatibility
• Retrievability	• Low tissue loading
• Xenotransplantation	• Large size

macrodevices[10] and microcapsules,[11] and despite the fact that both approaches have shown, so far, preliminary—although partial—evidence of function upon TX into diabetic higher mammal recipients, pros and cons associated with the two procedures have begun to be identified (Tables 12.4 and 12.5). Some potential advantages of microcapsules over macrodevices could consist, in particular, of the fact that the former, unlike the latter, envelop each individual islet, thereby offering better opportunities for islet TX immunoprotection, should disruption or breakage of any membrane occur. Moreover, microcapsules, unlike macrodevices, occupy far smaller volume, with associated benefits for the TX process, and do not share with the latter the low tissue loading capability,[12] which may represent a major pitfall for the TX functional performance.

General Technical Principles

Preparation of microcapsules, embodying intact islets or small cell clusters, consists of a delicate, stepwise technical procedure that, by combining appropriate chemical polymer blends with particular physical conditions, results in tissue entrapment within translucent, homogeneous, and morphologically intact artificial membranes, with no compromise of either the viability or the physiological competence of islet cells. Although microcapsules'

TABLE 12.4. Immunoisolation Microcapsules.

Pros
• Provide immunoprotection to individual islets
• Prolong *in vitro* maintenance of islets
• Minimize fibroblast growth in grafted tissue
Cons
• Inadequacy of implant site
• Biopolymer availability

chemical composition may differ among preparative methods, the common, critical criteria that must be complied with consist of the membrane's physical intactness, selective filtration/permeability properties, and long-term stability, regardless of the formulation employed. It is, in fact, evident that even a minimal discontinuation of the capsule's wall would irreversibly damage the competence of the micromembrane as an immunobarrier.

Since 1987, we have favored, in our laboratory the principle of using alginate (AG)/polyamino acid (PA) complex for fabrication of our microcapsules, with major, multiple-level adjustments of the original, previously described method.[11] In particular, as reported in greater detail elsewhere, we, unlike all the other laboratories engaged in AG/PA microencapsulation research, have uniquely employed poly-l-ornithine (PLO) in place of poly-l-lysine (PLL).[13] Moreover, we have been using at least two major, different approaches for preparation of our AG/PLO microcapsules,[14] depending on the capsules' final desired physical configuration and size. We have been able to remain with the original chemical recipe, consisting of the AG/PLO sandwich, for membrane preparation, although we had to vary the reagent's stoichiometric molar ratios when making conventional size microcapsules (CSM) and minimal volume microcapsules (MVC).[15,16]

Atomizing Method (CSM)— University of Perugia

Briefly, in our procedure, a suspension composed of Na-AG and islet cells at a 30:1 ratio was thoroughly mixed by magnetic stirring, and subsequently extruded through a single or multiple jet-head microdroplet generator, fabricated in our facility, where air at variable flow rates in combination with mechanical pressure, sprayed the atom-

FIGURE 12.1. Freshly prepared porcine-islet-containing CSM, under inverted phase microscopy.

ized suspension into a 1.5% CaCl$_2$ collection solution. Upon contact, islet-containing Na-AG microdroplets turned into Ca-AG gel microspheres. After being washed several times with saline, the gel beads were covered with two layers of PLO, at different concentrations, and finally with a Na-AG outer, single layer. The final microcapsules looked very uniform and well defined, each one containing one individual or two islets, under inverted phase microscopy examination (Figure 12.1). Their average diameter measured 700–800 μm, and their molecular weight cut-off (MWCO), as assessed by transmembrane fluorescein-conjugated (FITC)-dextran fluxes, ranged from approximately 70 to 100 kD. Key points of the procedure were the following:

- Na-AG purity grade and concentration, as well as full removal of endotoxins
- AG/islet suspension extrusion velocity
- air flow rate
- distance between jet-head device's tip and CaCl$_2$ collection bath
- Na-AG viscosity
- PLO concentration

Histological examination of either empty or islet-containing CSMs was compatible with the presence of an intact, multilayered membrane. This finding was substantiated by ultrastructural examination conducted with both scanning/transmission electron and laser confocal microscopy.

Nonatomizing Method (CM)— University of Perugia

As a major difference from the above-described procedure, the nonatomizing method for microcapsule fabrication, generated in our laboratory, does not involve any mechanical extrusion of the islet/AG suspension through an air-flow device. Rather, it is a matter of casting AG on each individual islet by a special chemical procedure, with the final result that microcapsules tightly adhere to the islets with no wasted, dead space left in between. Importantly, the nonatomizing procedure is totally harmless to the islets since it involves no traumatizing preparative steps.

Briefly, our method is based on the principle of an aqueous two-phase, ficoll-poly-ethylene glycol-AG emulsification process. The emulsion comprised islet-containing AG microdroplets, which immediately turned into Ca-AG gel microbeads upon reaction on CaCl$_2$. Because of the different fabrication process, coherent microcapsules (CM) were not

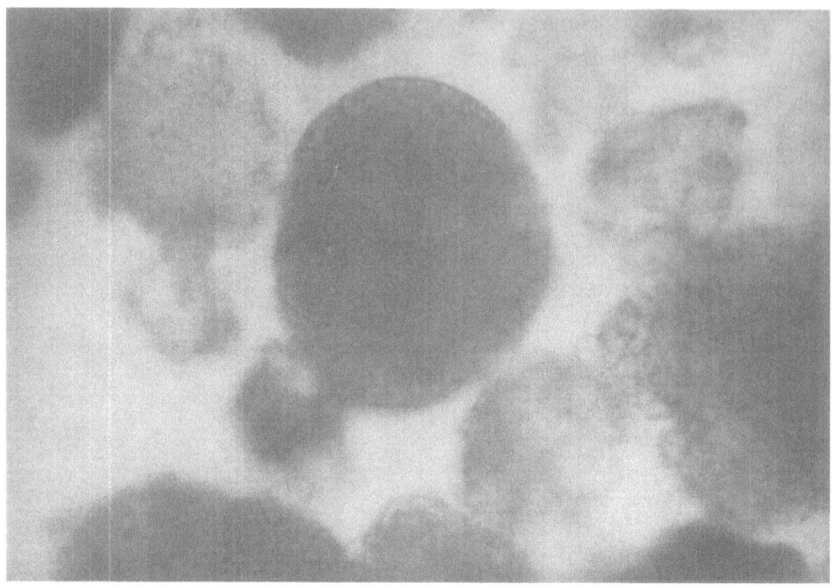

FIGURE 12.2. Freshly prepared rat-islet-containing CM under inverted phase microscopy

associated with a spherical shape, but rather constituted thin hydrogel films, tightly conforming to, and thus reproducing, the physical configuration of each individual islet (Figure 12.2). Appropriate titration of PLO as well as outer AG coating solutions permitted us to provide CMs, like CSMs, with a permselective membrane with no shrinkage or collapse[17] of the gel cast, a phenomenon that would be observed with inappropriate PLO and AG concentrations.

General Properties of AG/PLO CSMs and CMs

1. *Retention of encapsulated in vitro islet cell viability and function.* Neither CSM nor CM prototypes demonstrated adverse effects on *in vitro* islet viability (Figure 12.3a,b) or responsiveness to glucose (Figure 12.4) during or after the preparative process, as compared with unencapsulated control islets.

2. *Immunobarrier competence.* The AG/PLO membrane was extensively proven in our system to prevent both access to Ig and mononuclear cell activation in mixed islet/mononuclear cell co-cultures,[18] thereby confirming the protection afforded by this formulation against major hu-

moral and cellular mediators of the immune system. Both CSM and CM prototypes were associated with full survival of the enveloped islets upon TX, into allo- or xenogeneic hosts, thus proving their in vivo immunoprotective capacity.

3. *Biocompatibility.* When empty, neither CSMs nor CMs ever provoked, upon grafting into either rodent or higher mammal hosts, any significant inflammatory cell reaction, throughout several months of TX, thus confirming full biological acceptance of these membranes. *Ad hoc* studies in which high-grade purified AG preparations were compared to unpurified ones clearly documented that complete endotoxin removal greatly contributed to preventing the microcapsules' overgrowth with inflammatory cell tissue after fixation and staining with hematoxilin-eosin (H&E) (Figure 12.5a,b).

4. *Long-term durability.* AG/PLO microcapsules were shown to maintain their original morphological integrity and multilayered structure, upon intraperitoneal graft in rodents (Figure 12.6), or within a special vascular chamber in dogs (Figure 12.7), for up to 1 year or several months of TX, respectively.

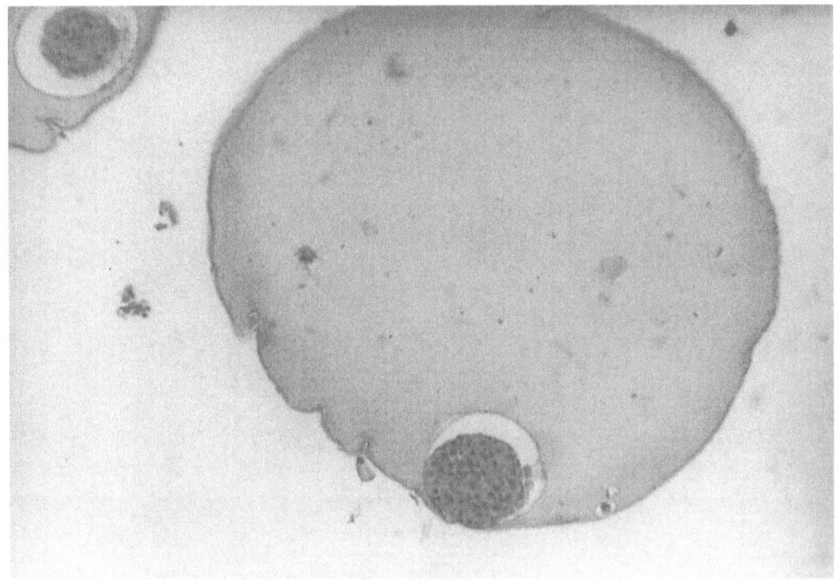

(a)

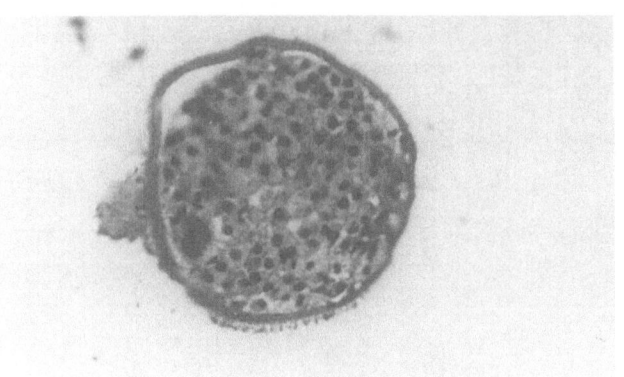

(b)

FIGURE 12.3. a, Rat-islet-containing CSM upon fixation and staining with H&E; b, Rat-islet-containing CM upon fixation and staining with H&E.

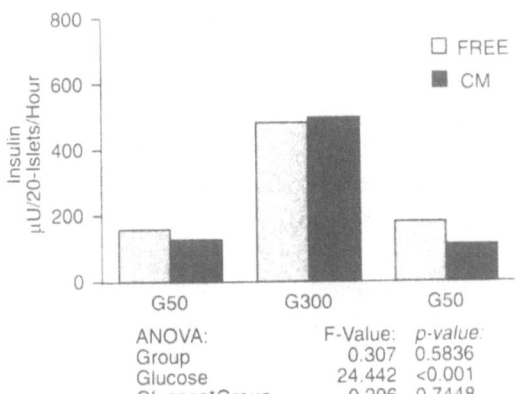

In vitro Static Incubation of Free vs Coherent Microcapsules Rat Islets

ANOVA:	F-Value:	p-value:
Group	0.307	0.5836
Glucose	24.442	<0.001
Glucose*Group	0.296	0.7448

FIGURE 12.4. In vitro insulin release from encapsulated rat islets upon static incubation with glucose.

Specific Properties of AG/PLO CSMs and CMs

We uniquely selected PLO instead of PLL, as a PA constituent of the outer gel coating membrane for a number of reasons, summarized as follows:

1. *Chemical composition.* PLO differs from PLL only at the level of the hydrocarbon side chain, which lacks a methyl group. We observed that this small structural difference may result in an AG/PLO complex that has physical-chemical advantages over AG/PLL (Figure 12.8).

2. *Morphology.* AG/PLO at equimolar reagent concentrations, makes sturdier, more uniform, and better-defined microcapsules than AG/PLL, under inverted phase microscopy examination. Upon histologic fixation, sectioning, and staining with H&E, AG/PLO membranes shows a

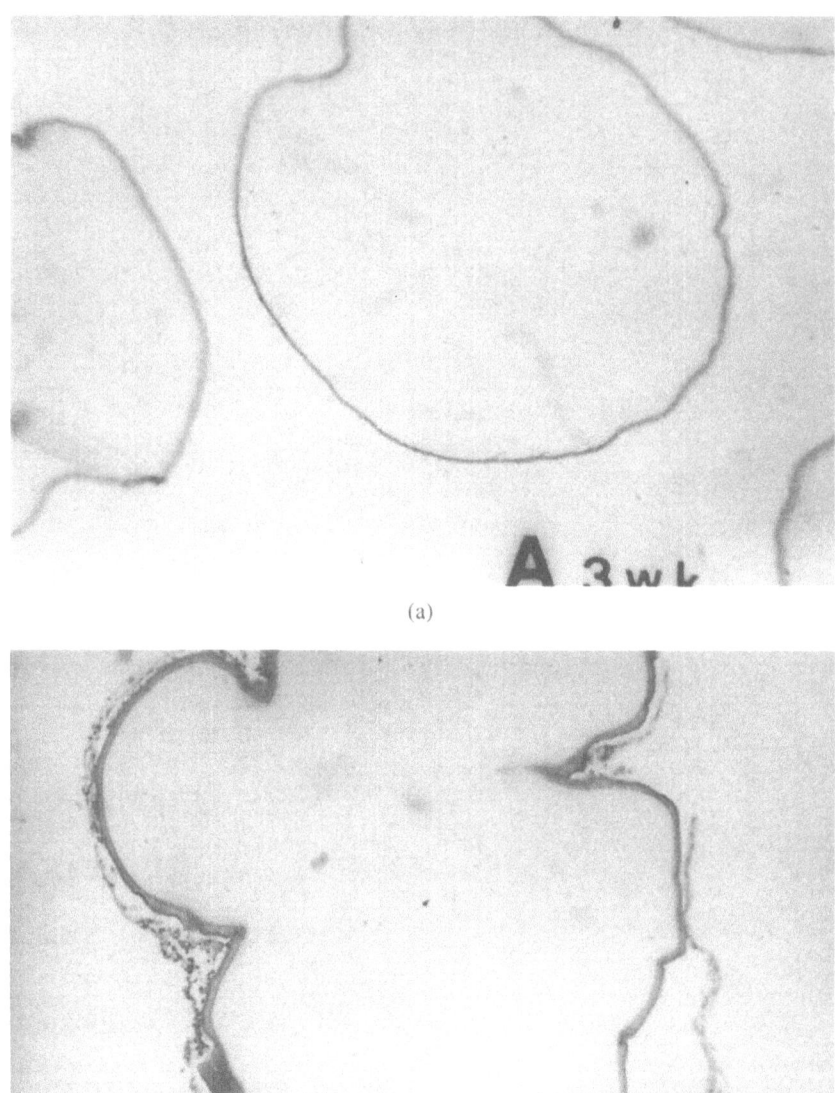

(a)

(b)

FIGURE 12.5. Empty CSM formulated with purified, endotoxin-free (a) or unpurified (b) AG, upon explant from normal rats, fixation, and staining with H&E.

clearer intactness of the multilayers composing the capsule's wall, as compared to AG/PLL.

3. *Selective permeability.* In order to assess whether AG/PLO and AG/PLL membranes, fabricated with equimolar reagents, would be associated with comparable MWCO properties, we evaluated them in two distinct series of experiments:

a. Experiments:

(1) AG/PLO and AG/PLL transmembrane fluxes of either FITC- or C_{14}-radiolabeled dextran, at different molecular weights

(2) The AG/PLO and AG/PLL membranes' permeability to monoclonal immunoglobulins (MoAb) produced by encapsulated hybridoma cell lines (Figure 12.9).

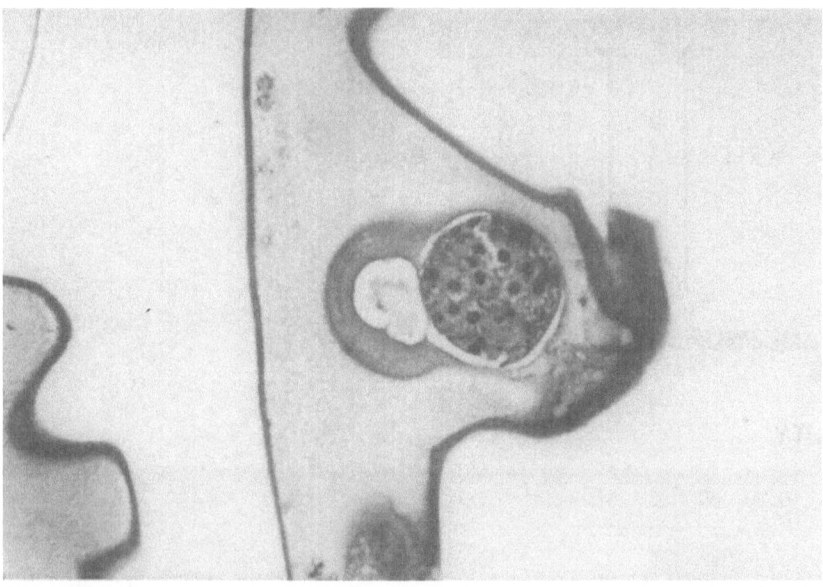

FIGURE 12.6. Viable rat-islet-containing CSM retrieved from remitter NOD mice at 1 year of TX (fixation and staining with H&E).

b. Results:

(1) AG/PLO membrane clearly showed better selectivity properties as compared to AG/PLL, for all three dextran molecular weights tested (Figure 12.10).

(2) No leakage of MoAb was detected through AG/PLO, unlike AG/PLL, which allowed some transmembrane MoAb leakage, as assessed by nephelometry. These results convinced us that from a strict physical-chemical

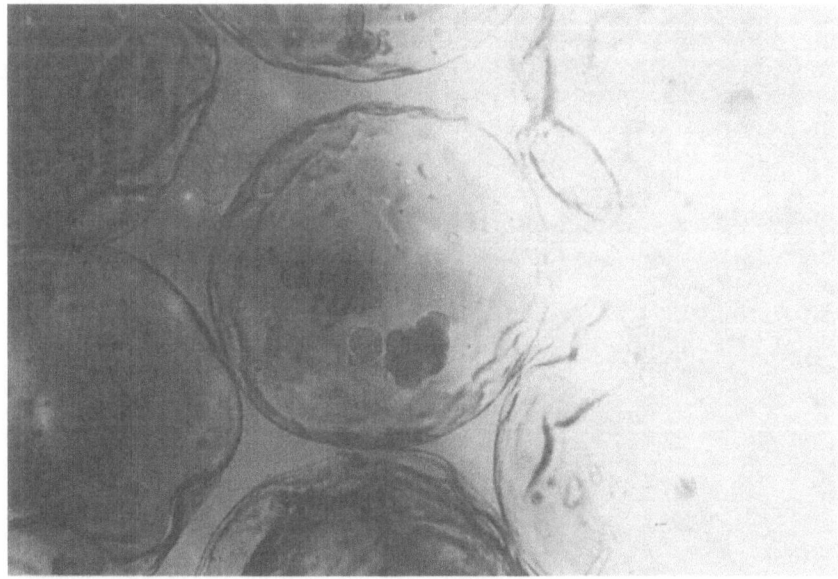

FIGURE 12.7. Viable porcine islet containing CSM embodied in vascular device, retrieved from remitter dogs (inverted phase microscopy examination).

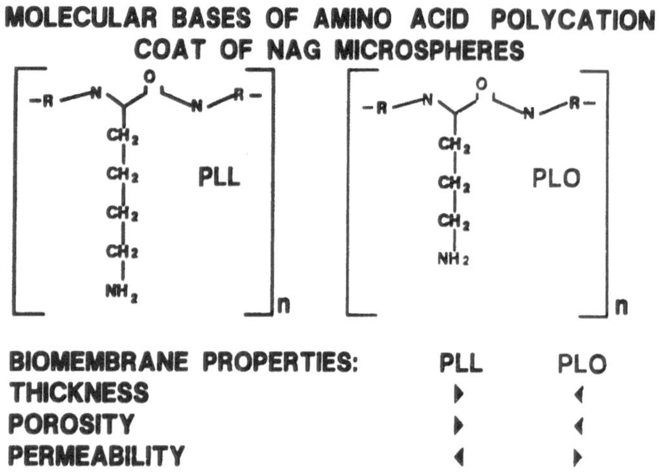

MOLECULAR BASES OF AMINO ACID POLYCATION COAT OF NAG MICROSPHERES

FIGURE 12.8. Chemical formulations and properties of PLO vs. PLL.

PLL

PLO

BIOMEMBRANE PROPERTIES:	PLL	PLO
THICKNESS	▶	◀
POROSITY	▶	◀
PERMEABILITY	◀	▶

point of view, AG/PLO would perform better than AG/PLL in formulating membranes for islet TX immunoisolation.

4. *Biocompatibility.* We compared biologic acceptance of empty AG/PLO vs. AG/PLL microcapsules, after 30 days of intraperitoneal TX in normal rats. Although microcapsules prepared with either formulation did not show any significant signs of inflammatory cell overgrowth, AG/PLO microcapsules resulted in attraction of fewer fibroblasts or other cell types than did AG/PLL, upon histologic examination.

The in vitro and in vivo studies conducted in our laboratory led us to select AG/PLO membranes for all our subsequent in vitro immunological and functional studies and in vivo hyperglycemia-correction studies by TX into diabetic recipients.

FIGURE 12.9. Hybridoma-cell-line-containing AG/PLO CSM.

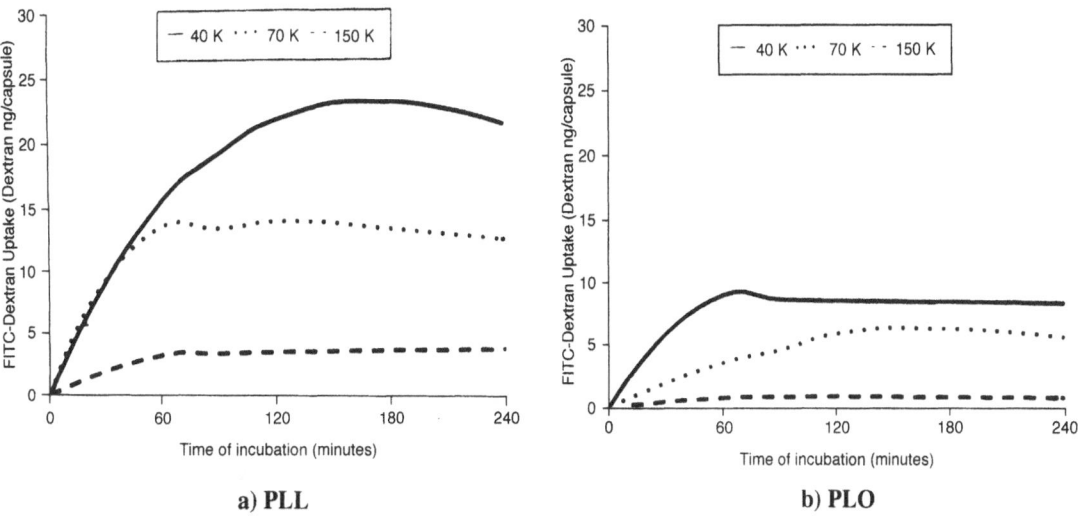

a) PLL   b) PLO

FIGURE 12.10. FITC-dextran fluxes across AG/PLL (a) vs. AG/PLO (b) CSM.

Summary of In Vivo TX Trials of Islet-Containing AG/PLO Microcapsules in Diabetic Recipients

1. *Rodents.* Full remission of hyperglycemia was achieved in 100% of rodents with either spontaneous (NOD mice, BB rats) or streptozotocin-induced diabetes mellitus receiving allo- or xeno/discordant or nondiscordant AG/PLO encapsulated islet TX. These conditions were sustained, in both basal and postabsorptive conditions, for extraordinarily long periods of time, up to a maximum 1 year of post-TX follow-up. Tissue explantation from the animals that were still euglycemic, irrespective of remission length, was associated with very viable islet cells enveloped within clean microcapsules; on the contrary, when microcapsules were retrieved from the animals that had failed after the initial remission, necrotic islet debris was found to be embodied within microcapsules surrounded by a thick layer of inflammatory cell reaction. Whether time-related, intracapsular islet cell death, possibly due to insufficient oxygen/nutrient supply, or to other unknown, immune or nonimmune intrinsic mechanisms occurring at the TX site, accounted for the finite life span of these grafts, will hopefully be unfolded by further study.

2. *Canines.* In our system, intraperitoneal grafts of either allo- or xenogeneic porcine-islet-containing AG/PLO microcapsules into nonimmunosuppressed dogs with either spontaneous, insulin-dependent, or pancreatectomy-induced diabetes mellitus, generally provided controversial results. Using CSMs that contained an average 80,000 islets, within 50 ml per dog, we obtained only partial and transient remission of hyperglycemia in the diabetic recipients, regardless of the TX barrier (allo- vs. xenogeneic). Upon laparotomy, we found that in these animals the grafted microcapsules had formed visible clusters, often attached to the mesenteric organs' surface. Fixation and H&E staining of the retrieved material was associated with intracapsular necrotic debris and pericapsular inflammatory cell reaction. Only about 30% of the total grafted microcapsules, which appeared to be freely floating in the peritoneal cavity, were devoid of any inflammatory infiltration and contained still viable islets at 90 days after TX (Figure 12.11). In order to establish whether these findings were in any way related to the host's immune response, we implanted autologous canine islets in CSMs into dogs that had undergone total pancreatectomy. Again, these animals did not require any insulin supplementation within 30 days of TX, when peritoneal biopsies were associated with freely floating microcapsules containing viable and functional islets. However, insulin dependence, which recurred at 60

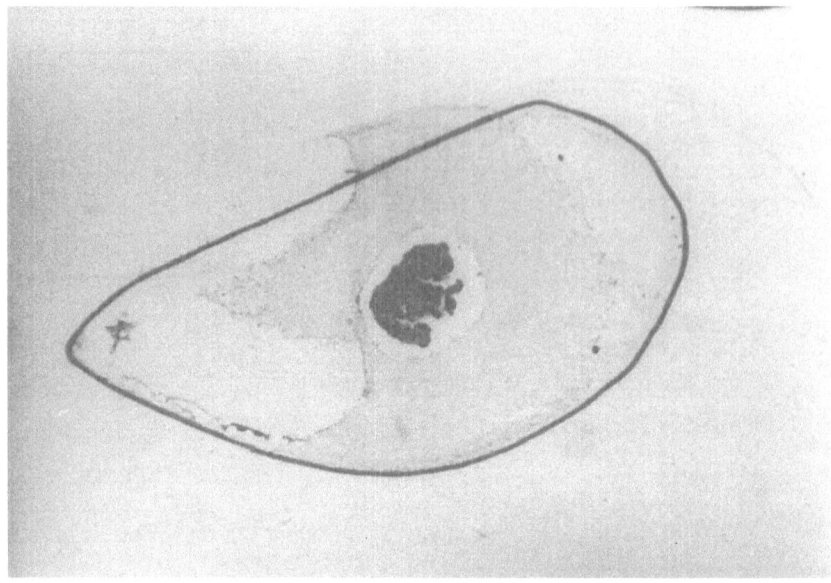

FIGURE 12.11. Viable-porcine-islet-containing CSM retrieved from the peritoneal cavity of remitter dogs at 90 days post-TX.

days after TX, reflecting the presence of subliminal islet cell mass, was most likely related to fibrotic overgrowth of microcapsules and intracapsular islet cell death. This was a demonstration that the TX outcome was not necessarily affected by the host's immune reactivity.[13]

In an attempt to improve on these results, we used an alternate TX site and loaded our islet-containing CSMs in special coaxial vascular devices, directly anastomosed with blood vessels, serving as A-V shunts or arterial bypasses (Figure 12.12). With the vascular chambers, we avoided the peritoneal, site-related "capsule clustering" phenomenon and achieved a better system performance that resulted, for the first time, in full remission of hyperglycemia in one of five diabetic dogs receiving xenogeneic encapsulated islets as a vascular TX.[19]

3. *Humans.* Results achieved in canines with vascular TX of encapsulated islets encouraged us to initiate, in 1989, pilot clinical trials in which human islets, upon envelopment in CSMs, were embodied in coaxial vascular chambers and grafted as arterial bypasses into nonimmunosuppressed, insulin-dependent diabetic patients who would have required surgery for vascular disease any-

way.[20] Although TX metabolic function was documented in two patients who underwent successful surgery, it only resulted in reduction, not withdrawal, of daily exogenous insulin supplementation. We believe that grafted islet mass probably was insufficient, in this instance, because of restrictions in the devices' CSM loading capability.

Outlook

From these studies it emerged that CSMs, while able per se to constitute an immunoprotective shield for islet grafts, did not generate consistent, long-term metabolic TX function. It remains unclear, at this juncture, whether microcapsules' immunobarrier competence is the only issue involved with islet graft limited functional life span, or whether other aspects should be considered, beginning with the islets' biologic survival since islets are grafted heterotopically across major histocompatibility barriers. As far as immunoisolatory microcapsules are concerned, a major problem was associated with the CSMs' individual excessive size (600–800 μm), which brought the final TX volume up to 120–180 ml/patient. Such a huge

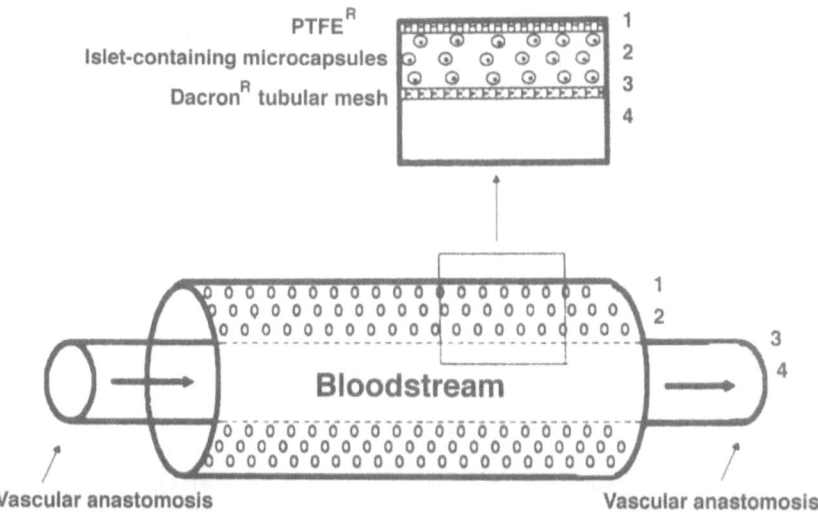

FIGURE 12.12. Schematic representation of coaxial vascular prosthesis for embodiment of microencapsulated islets.

mass would provoke, per se, an inflammatory cell reaction, especially upon TX in a demanding site such as the peritoneal cavity, where oxygen/nutrient supply and phagocyte activity are likely to be inadequate and intense, respectively. The large volume of CSMs also hampered appropriate loading of the vascular chambers, thereby creating the risk of grafting an inadequate islet cell functional mass. For these reasons, we addressed all our research efforts to trying to minimizing the AG/PLO microcapsules' size, so as to widen the potential TX site umbrella.[21] The newly generated CMs offer the unique opportunity to access graft sites so far interdicted to CSMs, with special regard to organs or tissues situated in the mesenteric area. These sites could bring tangible advantages in terms of either facilitated gas/nutrient supply or improved insulin secretory kinetics. Higher mammal in vivo trials, actually in progress in our laboratory, with this latest generation of AG/PLO microcapsules, will reveal whether these new membrane prototypes can fulfill the requirements for an effective and safe islet graft immunoprotection for IDDM therapy.[22]

References

1. The Diabetes Control and Complications Trial Research Group. 1994. The effect of intensive treatment of diabetes on the development of long-term complications in insulin-dependent diabetes mellitus. N Engl J Med 329:976–986.

2. The Diabetes Control and Complications Trial Research Group. 1996. The absence of a glycemic threshold for the development of long-term complications. The perspective of the Diabetes Control and Complications Trial Research Group. Diabetes 45:1289–1298.

3. Sutherland DER. 1996. The case for pancreas transplantation. Diabete Metab. 22: 132–138.

4. Secchi A, Socci C, Maffi P, Taglietti MV, Falqui L, Bertuzzi F, De Nittis P, Piemonti L, Scopsi L, Di Carlo V, Pozza G. 1997. Islet transplantation in IDDM patients. Diabetologia 39:225–231.

5. Lacy E, Kostianovsky M. 1967. Method for the isolation of intact islets of Langerhans from the rat pancreas. Diabetes 16:35–39.

6. International Islet Transplantation Registry. 1997. Newsletter on 15th Artificial Insulin Delivery Systems Pancreas and Islet Transplantation European Study Group Meeting, Igls, Austria.

7. Basta G, Falorni A, Osticioli L, Brunetti P, Calafiore R. 1995. Method for mass retrieval, morphologic and functional characterization of adult porcine islets of Langerhans. A potential nonhuman pancreatic tissue for xenotransplantation in insulin-dependent diabetes mellitus. J Inv Med 43:555–566.

8. Marchetti P, Giannarelli R, Cosimi S, Masiello P, Coppini A, Viacava P, Navalesi R. 1995. Massive isolation, morphologic and functional characterization, and xenotransplantation of bovine pancreatic islets. Diabetes 44:375–381.

9. Lanza RP, Sullivan SJ, Chick WL. 1990. Islet transplantation with immunoisolation. Diabetes 41:1503–1510.

10. Maki T, Otsu I, O'Neill JJ, Dunleavy K, Mullon CJP, Solomon BA, Monaco AP. 1996. Treatment of diabetes by xenogeneic islets without immunosuppression: use of a vascularized artificial pancreas. Diabetes 45:342–347.

11. Lim F, Sun AM. 1980. Microencapsulated islets as bioartificial endocrine pancreas. Science 210:908–910.

12. Lacy PE, Hegre OD, Gerasimidi-Vazeou A, Gentile FT, Dionne KE. 1991. Maintenance of normoglycemia in diabetic mice by subcutaneous xenografts of encapsulated islets. Science 254:1782–1784.

13. Calafiore R, Basta G. 1995. Microencapsulation of pancreatic islets: Theoretical principles, technologies and practice. In: Ricordi C, editor. Methods in cell transplantation. New York: Landes, p. 587–609.

14. Brunetti P, Basta G, Falorni A, Calcinaro F, Pietropaolo M, Calafiore R. 1991. Immunoprotection of pancreatic islet grafts within artificial microcapsules. Int J Artif Organs 14:789–791.

15. Basta G, Osticioli L, Rossodivita ME, Sarchielli P, Tortoioli C, Brunetti P, Calafiore R. 1995. Method for fabrication of coherent microcapsules: a new potential immunoisolatory barrier for pancreatic islet transplantation. Diab Nutr Metab 8:105–112.

16. Calafiore R, Basta G, Osticioli L, Luca G, Tortoioli C, Brunetti P. 1995. Coherent microcapsules for pancreatic islet transplantation: A new potential approach for bioartificial pancreas. Trans Proc 28:822–823.

17. Basta G, Rossodivita ME, Osticioli L, Luca G, Brunetti P, Calafiore R. 1996. Ultrastructural examination of pancreatic islet containing alginate/polyaminoacidic coherent microcapsules. J Submicr Cytol Pathol 28(2):209–213.

18. Calafiore R, Basta G, Sarchielli P, Luca G, Tortoioli C, Brunetti P. 1996. A rapid qualitative method to assess in vitro immunobarrier competence of pancreatic islet containing alginate/polyaminoacidic microcapsules. Acta Diabetol 33:150–153.

19. Calafiore R, Basta G, Falorni A, Ciabattoni P, Brotzu G, Cortesini R, Brunetti P. 1992. Intravascular transplantation of microencapsulated islets in diabetic dogs. Trans Proc 24:935–936.

20. Calafiore R, Basta G, Falorni A, Brotzu G, Alfani D, Cortesini R, Brunetti P. 1991. Vascular graft of microencapsulated human pancreatic islets in nonimmunosuppressed diabetic recipients. Diab Nutr Metab 4:45–48.

21. Calafiore R. 1997. Perspectives in pancreatic and islet cell transplantation for the therapy of IDDM. Diabetes Care 20(5):889–896.

22. Calafiore R, Basta G, Boselli C, Bufalari A, Giustozzi GM, Luca G, Tortoioli C, Brunetti P. 1997. Effects of alginate/polyaminoacidic coherent microcapsule transplantation in adult pigs. Trans Proc 29:2126–2127.

13
Chitosan

SungKoo Kim, JoonHo Choi, Estela A. Balmaceda, and ChoKyun Rha

Introduction

Chitosan is poly-β(1→4)-2-amino-2-deoxy-β-D-glucan (Figure 13.1), deacetylated chitin. Chitin is widely found in crustacean shells, fungi, insects, annelids, mollusks, coelenterates, etc. (Muzzarelli 1990). Chitosan is produced commercially from chitin, the major component of crustacean shells, most often from shrimp-processing waste. Chitosan has positively charged ionic groups that function uniquely among all polysaccharides available today (Hirano 1989, Sandford 1989) and therefore has great potential for a variety of applications.

Chitosan is partially deacetylated chitin, poly-acetoglucosamine with charged amine groups on carbon number 2 of glucose units that constitute the cellulose backbone. Chitosan is soluble in acidic aqueous solution. Chitosan assumes conformations ranging from a quasi-globular state to a random coil, depending on the solution conditions (Kienzle-Sterzer et al 1982a, Rodriguez-Sanchez et al 1982). The hydrodynamic volume of chitosan varies with solution conditions, as shown by intrinsic viscosities in Figure 13.2 (Kienzle-Sterzer, 1984). The Mark-Houwink exponent, a, of chitosan is given in Table 13.1. Studies report chain stiffness parameter of 0.04–0.1 (Kienzle-Sterzer et al 1982a, 1982d, Kienzle-Sterzer et al 1984, Terbojevich et al 1986) and a characteristic ratio of 15 (kienzle-Sterzer et al 1984).

The mechanical properties of chitosan film also indicate that chitosan is sensitive to the solution conditions (Kienzle-Sterzer et al 1982b, 1982c, Lang et al 1982, Rinaudo and Domard 1989, Rodriguez-Sanchez et al 1982). Considerable ef-forts on the study of chitosan are evident in the literature. Table 13.2 summarizes some general properties of chitosan qualitatively described in the literature.

Chitosan can be described in terms of its polymer characteristics (molecular weight and molecular weight distributions, degree of deacetylation, viscosity, solubility, impurity, etc.), general physical characteristics (density, color, particle size, etc.), and biological activities (toxicity, microbial contamination, immunopotential effect, therapeutic properties, etc.) (Knapczyk et al 1989). Chitosan and chitin can be subjected to a wide range of chemical and physical modifications to generate unique properties and to expand applications. The vast array of available chemical, enzymatic, and genetic engineering techniques for modification are described in the literature (Gorin and Barreto-Berger 1983, Kenne and Linderberg 1983, Yalpani 1988).

Chitosan is generated by alkaline deacetylation of chitin. The degree of deacetylation controls the solubility and is a useful category for classification of chitosan. N-acetylation of chitosan leads to fully N-acetylated chitin (Nishi et al 1979). Chitosan derivatives that have been produced include carboxyacyl-chitosan (Kurita 1986), succinylated chitosan (Yamaguchi et al 1981), O-hydroxyethyl-chitosan (Yamaguchi et al 1981), hydroxypropyl-chitosan (Maresch et al 1989), and carboxymethyl-chitosan (Hayes 1986, Davies et al 1989, Rinaudo et al 1992). Sulfation, sulfonation, cyanoethylation, and graft polymerization can also be employed for modification of chitosan (Hayashi 1993, Kurita and Inoue 1989).

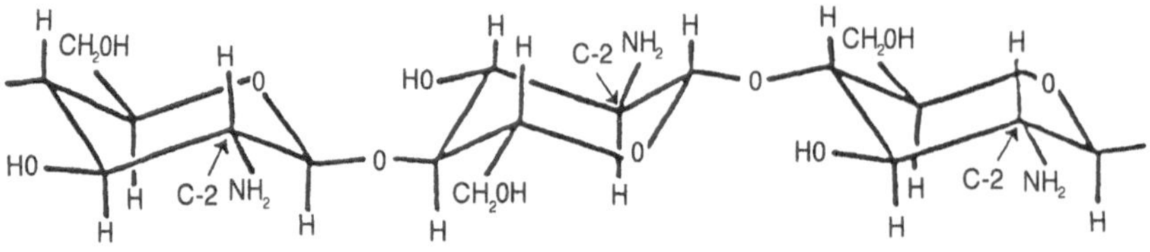

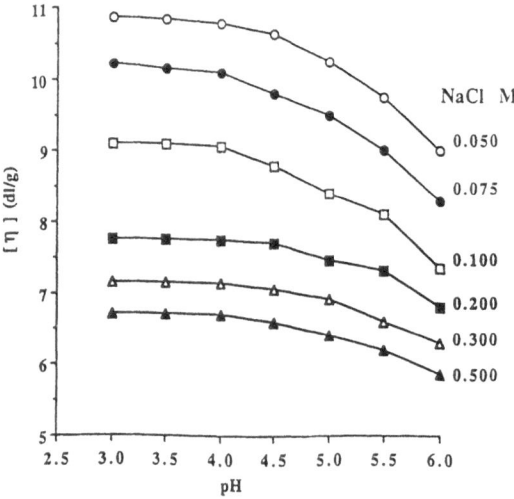

FIGURE 13.1. Chemical structure of chitosan.

FIGURE 13.2. Intrinsic viscosity of chitosan (in acetic acid) (Kienzle-Sterzer 1984).

TABLE 13.1. Mark-Houwink exponent*, a, of chitosan.

Solvent	a	References
Trifluoroacetic acid	0.30	Berkovich et al 1980
Acetic acid 1%, NaCl 2.8%	0.15	Berkovich et al 1980
Acetic acid 1%, LiCl 2%	0.19	Berkovich et al 1980
Acetic acid 0.2M, NaCl 1M, Urea 4M	0.71	Muzzarelli 1977
Acetic acid 0.1M, NaCl 0.2M	0.93	Robert and Domszy 1982
Acetic acid 0.2M, Sodium acetate 0.1M		Wang et al 1991
Degree of deacetylation = 69%	1.12	
Degree of deacetylation = 84%	0.96	
Degree of deacetylation = 91%	0.88	
Degree of deacetylation = 100%	0.81	

* $[\eta] = K \, Mw^a$

where a < 0.5: spherical globular shape

      0.5 < a < 0.8: nondraining random coil

      0.8 < a < 1.2: free-draining

TABLE 13.2. General properties of chitosan.

1. Chemical Properties
   • Linear polyamine (poly-D-glucosamine)
   • Reactive amino groups
   • Available reactive hydroxyl groups
   • Chelates many transitional metal ions
   • Strong affinity with polyanions
2. Solution Properties
   • Soluble at pH < 6.5, insoluble at pH > 6.5
   • Viscous solutions
   • Shear thinning
   • Gel-forming with polycations
   • Soluble in alcohol-water mixtures
3. Biological Properties
   • Biocompatible
   • Biodegradable
   • Nontoxic
   • Therapeutic

Chitosan is modified in such manner as to alter its properties to improve its functional performance.

Chitosan and its derivatives are known to display functional properties desirable for a wide spectrum of commercial applications (Austin et al 1981). These biological properties, including bioadhesion, biocompatibility, biodegradability and nontoxicity, are directly related to the physical properties, and consequently to the chemical structure, of chitosan and its derivatives. The behavior of chitosan as a polyelectrolyte is dictated by its overall molecular conformation, which is effected by the degree of hydrogen bonding or electrostatic repulsion between neighboring chain segments. The physical properties depend on degree of acetylation, charge density, charged groups, and molecular weight of chitosan. Chitosan often complexes with metal ions and polyanions, and interacts with or adheres readily to biologicals, including natural polymers, negatively charged mucopolysaccharides, proteins, membranes and cells. The modification of the physical properties of chitosan is aimed at designing stable solutions, gels, powders, films, and membranes for practical applications.

General Applications

Chitin is one of the two most abundant naturally occurring polymers; thus, availability of the raw material imposes no problem for the industrial production of chitosan. Chitosan can be prepared in a variety of forms (powder, solution, gel, film/membrane, fiber, bead, capsules), facilitating potential applications.

Chitosan has been evaluated for a variety of therapeutic applications, including wound-healing, hemostatic, antimicrobial, osteoconductive, cholesterol-lowering, and drug delivery agents (Sandford 1989). Chitosan is also used for skin and hair care, utilizing the ability of chitosan to form films that adhere to skin or hair, to thicken gels and emulsions, and to act as a moisturizer (Gross et al 1979, Gross et al 1980, Lang and Clausen 1989, Leuba et al 1991, Sandford 1989).

Chitosan has been approved by the United States Environmental Protection Agency as a seed treatment/growth regulator and as a pesticide/nematocide (Federal Register 1988 a–d, Federal Register 1989, Toda and Matsuda 1993). Chitosan also has been approved by the Association of American Feed Control Officials (1986) as a flocculant in recovering proteinaceous wastes for animal feed. Experimental use of chitosan as a controlled-release agent for pesticides and herbicides continues. Chitosan can be added to croplands to stimulate natural microbes that provide protection to certain crops (Sandford 1989, Weiner 1992). Spraying of chitosan solutions is also known to increase the yield of some agricultural crops.

Because it is nontoxic, biocompatible, and available in a variety of forms, chitosan is conveniently used as a matrix for immobilizing enzymes and cells, for chromatographic supports, and for the recovery of proteins (Mooyoung et al 1992, Moriguchi et al 1989, Senstad and Mattiasson 1989, Uragami 1991). The large-scale application of chitosan is as a nontoxic cationic flocculant in the treatment of industrial wastes (sewage, sludge, slaughterhouse, brewery, canning, and gravel wastes, etc.) and as a chelating agent for toxic (heavy and radioactive) metals. Chitosan has been approved for metal chelation in water treatment, as a selective adsorbent and as a coagulant aid for potable water (Weiner 1992). Chitosan was approved as a food additive by the Japanese Ministry of Health and Welfare in 1983 (Jap. Nat. Additive List, 1983). Studies on chitosan as a dietary fiber and on its ability to protect foods stimulated applications of chitosan in wine and juice clarification, as a protective fruit coating, etc. (Weiner 1992).

In summary, chitosan is an abundant natural polymer, biodegradable, biocompatible, bioadhesive, nontoxic, and less antigenic, with reactive

TABLE 13.3. Therapeutic applications of chitosan and its derivatives.

Activity	Application
Wound healing	Wound healing
	Wound dressing
	Wound filling
Cell binding	Hemostatic
	Antibacterial
	Antifungal
	Antitumor
	Spermicidal
Osteoconductive	Bone tissue generation
	Bone filling
Immunological	Macrophage activation
	Induction of cytokine production
	Cell function
	Host defense response
Hypocholesterolemic	Lipid trapping/binding
	Dietary fiber and supplement
Miscellaneous	Surgical glove powder
	Contact lenses
	Dialysis membrane
	Anticoagulant
	Antigastritis
	Antibilirubinemia
	Dental surgical exclusion
Drug release	Entrapment
	Encapsulation
	Covalent linking

amine groups with $\beta(1\rightarrow4)$ glycosidic linkages that endow the contour conformation and the conformational flexibility of the molecule in solution, properties that are responsible for its biological functions and engineering amenability. Therefore, chitosan is likely to have a potentially important role in therapeutics and encapsulations.

Therapeutic Functions of Chitosan

A variety of therapeutic applications of chitosan and its derivatives has been reported in the literature, and some of these are listed in Table 13.3. More important applications are summarized below.

Wound Healing

Skin lesions, especially those associated with the loss of tissue, promote a number of reactions leading to full recovery. The spontaneous healing process, however, does not always lead to morpho-

functional normality in the injured areas. In addition, the presence of pathologies, such as diabetes and hypertension, leads to incomplete and prolonged healing reaction, and spontaneous healing frequently results in the formation of anomalous connective tissue that has a fibrous appearance, reduced mechanical resistance, and a tendency to retraction (Muzzarelli et al 1989). In such cases, guided healing may be mediated by favoring physiological reconstruction of the skin through a replacement-like process, rather than allowing the scar tissue formation that often results (Biagini et al 1989, 1992). Chitosan is a candidate for such therapeutic use.

The improvement or accelerating effect of chitosan and its derivatives in wound healing is related to several possible mechanisms such as blood coagulation, abnormal fibroblastic reactivity, bactericide, and cellular activation. Soft pads containing freeze-dried chitosan, when administered to patients undergoing plastic surgery, showed organized tissue formation both at the dermal and epidermal level, indicating that chitosan may promote healing in plastic surgery (Biagini et al 1991). Wound healing with minimal scar formation and the promotion of normal tissue generation was possible with the use of chitosan solution or solid chitosan (Malette and Quigley 1985). The chitosan, placed in contact with the tissue wound, prevented bleeding and fibroblast formation, and proliferation of fibroblasts (Bartone and Adickes 1988, Diegelmann et al 1996).

Chitosan-based wound-dressing materials have been designed as a chitosan-gelatin complex (Sparkes and Murray 1986), as microfungal fibers (Sagar et al 1990), and as chitosan-glycerol-water gels (Jackson 1987). Wound-filling compositions are also proposed for open wounds such as ulcers, using chitosan and compatible hydrocolloid materials (Mosbey 1990). An in vitro model for human skin was used to determine the effect of heparin-chitosan complexes on the epithelialization of full-thickness wounds (Kratz et al 1997). The heparin-chitosan complex stimulated skin growth on 90% of the wounds, compared to the control. Further animal studies indicate that chitosan stimulates wound healing by reducing the growth of macrophages, thus reducing angiogenesis, fibroplasia, and connective tissue deposition (Diegelmann et al 1996). Animal tests indicate that open wounds treated with chitosan showed a significant decrease in fibrosis, compared to control; at the same time, no adverse effects on the wound-healing process were observed (Bartone and Adickes 1988). Chitin and chitosan significantly decreased the growth of inflammatory cells and suppressed the formation of rete ridges in the wounds (Okamoto et al 1995).

Chitosan can generate tough, water-absorbent, biocompatible films. These films can be formed directly on open wound sites by application of an aqueous solution of chitosan. The solution provides a cool and pleasant soothing effect when applied to the open wounds, allows excellent oxygen permeability, and prevents oxygen deprivation of injured tissues. Chitosan films are claimed to inhibit infection and pain, maintain adequate water retention, and prevent escape of wound secretions, and they are naturally degraded by body enzymes, eliminating the need for removal. In most injuries (and especially burns), removing the wound dressing can cause damage to the injury site and pain. Artificial skin produced by chitin fibers in the form of an unwoven fabric may have an advantage, and has been successfully used for such cases (Kifune 1992). Chitosan-collagen complexes have been prepared for potential application in the regeneration and healing of skin burns (Berthod et al 1994, Shahabeddin et al 1990). The multilayer wound-covering materials proposed comprise a supporting layer, a moisture permeation controlling layer (Koide et al 1995), and a chitosan foil for wound sealing (Kaessmann and Haak 1997).

Cell Binding

Deacylated chitin, or chitosan, as a polycation aggressively binds to a variety of mammalian and microbial cells, and is an effective cellular agglutinating agent (Evan and Kent 1962, Katchalsky et al 1959). The binding property of chitosan offers a variety of biomedical functions, i.e., hemostatic, bacteriostatic, fungistatic, and spermicidal. The mechanism of cell binding is known to be due to ionic interactions between the positive charges on chitosan and the negative charges on the cell surface. Normally, the repulsive force between red blood cells is maintained as a result of the high net negative charge on the cell membranes, predominantly due to the presence of neuraminic acid residues (Eylar et al 1962, Sapoznikov et al 1984). The re-

moval of this acid by neuraminidase enzyme results in a decrease of the high net negative charge of the red blood cells and diminishes gelling with chitosan. Therefore, the flocculation, gel formation, or binding of red blood cells is caused by the electrostatic interactions of the positive amine groups of chitosan with receptors containing neuraminic acid residues on the cell surface.

Hemostatic

Chitosan forms a coagulum with red blood cells, prevents rebleeding, and promotes smooth muscle reformation (Malette et al 1983, Rao and Sharma 1997). The hemostatic activity of chitosan has been shown to act independently of the normal blood-clotting cascade and is dependent upon the cellular agglutination of the red blood cell component of blood, which is due to ionic interactions, as in the mechanism discussed above. The agglutination of red blood cells by polycations is dependent both on polymer structure or conformation and on molecular weight. Chitosan with molecular weight of 35,000 was only able to produce a loose coagulum in heparinized blood, while chitosan with molecular weight of 600,000 or above produced a firm coagulum (Olsen et al 1989). Forms of chitosan used in hemostatic wound dressing (Errede et al 1983) and in mucosal hemostatic agents have been patented (Ito et al 1985).

Bacteriostatic and Fungistatic

Antimicrobial activities have been explored using chitosan and its derivatives for bacterial, fungal, and yeast strains (Evan and Kent 1962, Lazarenko et al 1986, Malette et al 1983, Muzzarelli et al 1990, Papineau et al 1991, Ralston et al 1963, Segal and Lehre 1987, Seo et al 1992, Smith 1984, Svensson et al 1986, Tanigawa et al 1992, Uchida et al 1989, Yalpani et al 1992). These activities of chitosan and its derivatives provide the advantages in wound healing, food preservation (Ueno et al 1996), bacteria growth inhibition in myocyte culture (Malette et al 1986), and antimicrobial effect in personal care products (Nelson et al 1991) as well as in clinical and hospital wares.

Spermicidal

The binding of chitosan with sperm cells is directly related to its spermicidal activity and is used to kill or inactivate mammalian spermatozoa in the uterine cavity (Smith 1984).

Anticancer and Antitumor

The binding of chitosan with cancer cells is responsible for its anticancer activity (Evan and Kent 1962, Sirica and Woodman 1971).

Immunological

Chitosan and its derivatives can activate a variety of cell types, and this cell-activating property could be responsible for their immunomodulating properties. These immunomodulating properties may be utilized in therapeutic applications. In particular, chitosan can activate peritoneal exudate cells (Suzuki et al 1985) and macrophages (Nishimura et al 1986a), and induce thymocyte adherence to macrophages (Sapoznikov et al 1984). The cell-activating properties and immunoadjuvant activities initiate host defense responses that can prevent microbial infection as well as inhibit tumor formation (Nishimura et al 1984, Nishimura et al 1986b). Chitosan also activates some macrophage functions, including superoxide production and cytokine production, which influence other cell functions such as lymphocyte activity, fibroblast function, and endothelial cell function (Nishimura et al 1986c, Otterlei et al 1992).

Osteoconductive

Chitosan has a marked osteogenic effect in the endochondral long bones and in all elements of the adjacent bone structure (Borah et al 1992). A chitosan derivative (methylpyrrolidinone chitosan) also stimulates intramembraneous bone formation by chelating the osteoblast-like precursors, followed by tissue mineralization (Muzzarelli et al 1993a). Pastes made from chitosan and tri-calcium phosphates are effective in stimulating bone tissue formation (Ito et al 1994, Kawakami et al 1992, Mattioli-Belmonte et al 1995). A bone-filling material using chitosan as an osteoinduction substance was patented by Ito (1997).

Hypocholesterolemic

Chitosan shows a significant hypocholesterolemic activity in various experimental animals. Chitosan may be effective in controlling the absorption of fats and cholesterol in the diet (Vahouny et al 1983) and in lowering cholesterol and triglyceride absorption (Hirano et al 1990, Ikeda et al 1989, Kobayashi et al 1979, Sugano et al 1980, Sugano et al 1988, Vahouny et al 1983). Prefeeding rats with 5% chitosan reduces the absorption of cholesterol by about 66% and the absorption of oleic acid by 60%. The extensively hydrolyzed chitosan, glucosamine oligomer, does not exhibit a cholesterol-lowering activity, therefore it is likely that a certain minimum molecular size is required to exert the hypocholesterolemic effect. In fact, the lower limit of the molecular weight required for the cholesterol-lowering action of chitosan has been estimated to be about 7000 (Sugano et al 1992). The hypocholesterolemic potential of chitosan may be related to its solubility since the mode of chitosan activity is attributed to its soluble dietary-fiber-like behavior in the intestine (Nauss et al 1980, Sugano et al 1988).

Chitosan is attracted to lipids and has the ability to significantly prevent fat adsorption in the digestive tract. It is proposed that chitosan dissolves in the stomach and converts to a gel that traps fat, thereby preventing its absorption and subsequent storage (Kanauchi et al 1995). Addition of 5% chitosan reduced fat digestibility by about 50% compared to the control diet without added fibers when tested in rats, while at the same time no significant change in the fatty acid composition was detected in the feces (Deuchi et al 1994).

Chitosan and cross-linked chitosan have been patented in the form of (1) a nonabsorbable lipid binder for use as a food additive or as a pharmaceutical preparation (Furda 1980), (2) a dietary lipid digestion-absorption inhibitory agent (Kanauchi and Deuchi 1995, Kanauchi et al 1997), and (3) an antihypercholesterolemic agent (Scopelianos and Wilmington 1989).

Miscellaneous

Chitosan provides a diverse spectrum of uses in the pharmaceutical and biotechnology areas. Chitosan has been used as surgical glove powder (Casey 1977), for contact lenses (Allan et al 1985, Markey et al 1989), as an anticoagulant (Hirano et al 1985, Muzzarelli et al 1986), as an antigastritis agent (Ferro 1991), as an antibilirubinemia agent (Nagyvary 1982), as a hemodialysis membrane (Amiji 1995, Amiji 1996), and for dental surgical (Muzzarelli et al 1993b, Sapelli et al 1986) and other dental applications (Shibasaki et al 1994a, Shibasaki et al 1994b). Chitosan suture and related surgical aids are being used. The ability of chitosan to form sulfate esters, which are nonthrombogenic, appears to make it a promising candidate for prosthetic structural devices of any shape or size. Therefore, chitosan could serve as a replacement for bone, cartilage, arteries, veins, and musculofascial prostheses.

Chitosan for Therapeutic Delivery

Chitosan has been examined for its potential use in controlled drug delivery systems (Muzzarelli 1983, Nagai et al 1984). Its polymeric and cationic nature, with potentially reactive functional groups, offers unique possibilities for utilization in controlled-release technologies. Chitosan can be used in solution form (Illum et al 1994), in microgranulation systems (Hou et al 1985), in sustained release matrices (Nagai et al 1984, Thakur and Jain 1986), in erodible matrices (Pangburn et al 1984), in controlled-release gel systems (Miyazaki et al 1981), and as a sustained-release vehicle in wet granulation (Kawashima et al 1985). Many patents have been issued for methods related to drug release and delivery systems, such as oral composition (Komiyama et al 1985), controlled drug release (Cardinal et al 1990, Hashimoto et al 1995, Illum 1997, Makino et al 1989, Thakur and Jain 1988, Tsuru and Masuno 1992, Yamada et al 1996, Yen and Reed 1995), delivery vehicles (Gallo and Hassan 1992, Henderson et al 1997), and medication (Collombel et al 1992, Homme and Rostaing 1997, Sekigawa and Onda 1993). Figure 13.3 gives schematic representations of various forms of therapeutic delivery systems that can be prepared with chitosan.

Solutions

Chitosan solutions containing insulin for nasal delivery enhance insulin adsorption across the mu-

cosa of rat and sheep (Illum et al 1994), deliver clinically relevant levels of insulin, and cause minimal membrane and cellular damage (Aspden et al 1996). The mucociliary transport rates (MTR) of chitosan solution, examined on human nasal tissue, is dependent on the amount and molecular weight of chitosan (Aspden et al 1997). The effect of the chemical composition and molecular weight of chitosan on epithelial cells, epithelial permeability, and toxicity was studied at the cellular level using Caco-2 monolayers and hydrophilic marker molecules of varying molecular weights (Schipper et al 1997) and using model human intestinal epithelial cells (Schipper et al 1996). Positive charges on chitosan induce the reorganization of tightly associated proteins, followed by enhanced transport through the cellular pathway. Types of chitosan with a low degree of acetylation are effective absorption enhancers at low and high molecular weights (Schipper et al 1996).

Granules (Tablets)

Tablet drug release matrices have been prepared from chitosan (Brine 1989), chitosonium malate (Akbuga 1993) or mixtures of chitosan with carbomer (Nigalaye et al 1990), hyaluronic acid (Takayama et al 1990), and alginic acid (Miyasaki et al 1994). The drug release rates were dependent on the polymer mixing ratio, solubility, and molecular size of the drugs. These tablet formulations could provide bioavailability comparable to that of a commercial oral mucosal adhesive tablet.

Gel (Matrix)

Chitosan gels or matrices of various forms are used to entrap therapeutics for sustained release. Chitosan formulations were used for encapsulations in the forms of porous beads (Cardinal et al 1990, Moriguchi, et al 1989, Mosbach and Nilsson 1987), granular porous chitosan (Kawamura et al 1989), polycationic substance (Gallo and Hassan 1992, Illum 1996), beads (Freeman 1996, Mosbach and Nilsson 1987), polyionic hydrogels (Dumitriu et al 1997), and gel mixtures (Collombel et al 1992, Lorenz and Lee 1995).

Chitosan gel microspheres for entrapping pharmaceuticals have been prepared by complexing with sodium tripolyphosphate (Aydin and Akbuga

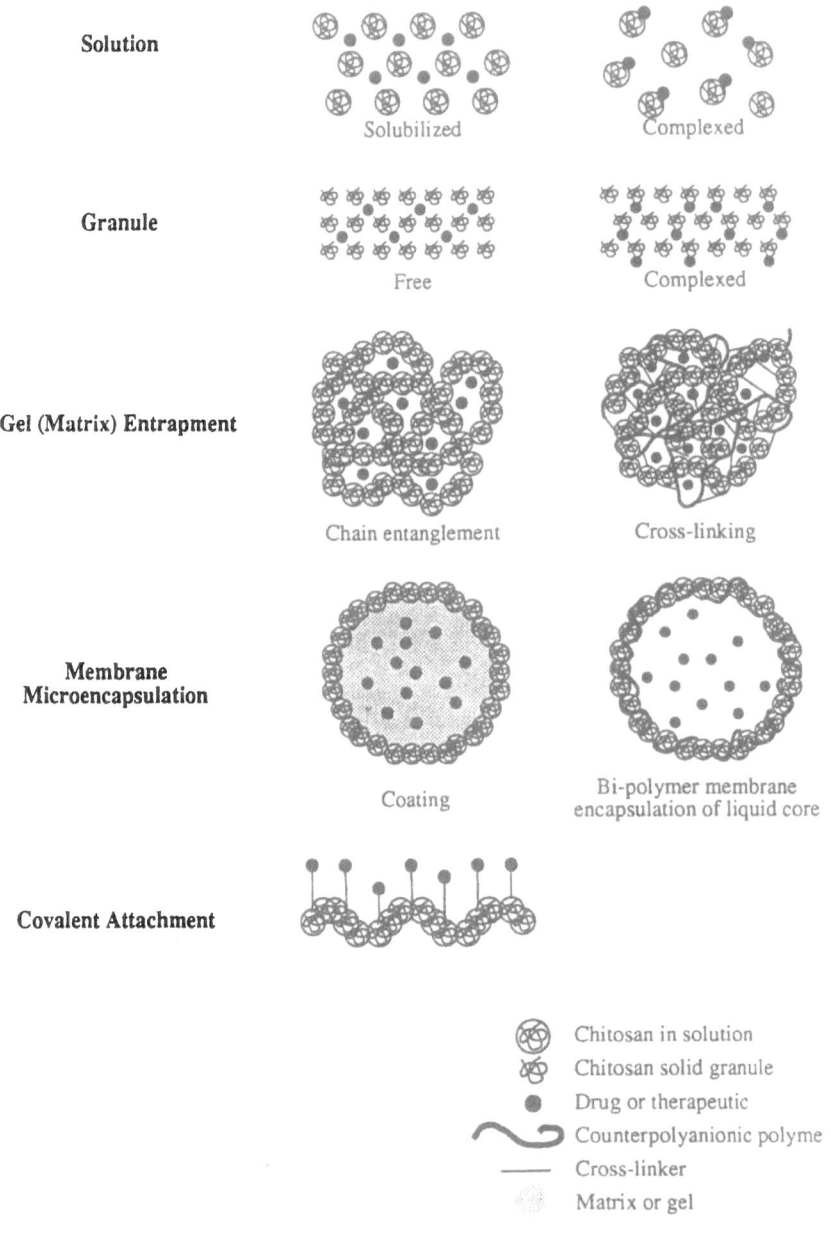

Solution

Solubilized

Complexed

Granule

Free

Complexed

Gel (Matrix) Entrapment

Chain entanglement

Cross-linking

Membrane Microencapsulation

Coating

Bi-polymer membrane encapsulation of liquid core

Covalent Attachment

Chitosan in solution
Chitosan solid granule
Drug or therapeutic
Counterpolyanionic polyme
Cross-linker
Matrix or gel

FIGURE 13.3. Chitosan for encapsulations and therapeutics.

1996, Sezer and Akbuga 1995, Shiraishi et al 1993), sodium sulfate (Berthold et al 1996), and citric acid (Orienti et al 1996). The cross-linked chitosan microspheres have also been formulated via glutaraldehyde and other cross-linking agents (Akbuga and Bergisadi 1996, Filipovic Grcic et al 1996, Jameela et al 1994, Jameela and Jayakrish-nan 1995, Jameela et al 1996, Thanoo et al 1992). Another form of drug delivery system, magnetic cross-linked chitosan microspheres has been reported (Hassan et al 1992, Hassan and Gallo 1993).

The drug release characteristics and kinetics in chitosan microspheres were studied for several different drugs. The release kinetics depend on chitosan concentration in case of cisplatin (Nishioka et al 1990), indomethacin (Orienti et al 1996), ampicillin (Chandy and Sharma 1993), and piroxicam (Sezer and Akbuga 1995). The drug release rate of cisplatin from chitosan microspheres appears to be controlled by the dissolution and diffusion of the drug from the chitosan matrix (Wang et al 1996). The drug release rate also depends on the drug-polymer ratio with an anti-inflammatory drug (Berthold et al 1996), the drying method in preparing the microsphere (Sezer and Akbuga 1995), and the pH of the release medium with indomethacin (Miyazaki et al 1988, Munjeri et al 1997, Orienti et al 1996) and sulphamethoxazole (Munjeri et al 1997). Release rates of indomethacin decrease with the increasing molecular weight of chitosan (optimum = 25,000), and it also depends on the dispersion of the drug and on the porosity, tortousity, and surface area of microspheres (Shiraishi et al 1993). The release rate of nifedipine from the chitosan matrix is slower for microbeads than for microgranules (Chandy and Sharma 1992).

Implantation of cross-linked chitosan microspheres containing antineoplastic agent in the skeletal muscles is well tolerated by the living tissue with no significant degradation (Jameela and Jayakrishnan 1995). The glutaraldehyde cross-linked chitosan microsphere minimizes the drug toxicity and maximizes the therapeutic efficacy (Jameela et al 1996). Drug release characteristics from cross-linked chitosan microspheres are affected by initial drug (5–fluorouracil) concentration, type and concentration of chitosan, and glutaraldehyde concentration (Akbuga and Bergisadi 1996). Release rates from cross-linked chitosan microspheres with nifedipine (Filipovic et al 1996) and theophylline, aspirin, or griseofulvin (Thanoo et al 1992) are dependent on cross-linking density, particle size, and initial drug loading. The drug loading efficiency exceeded at least 70% (Filipovic et al 1996, Thanoo et al 1992). Cross-linked chitosan microspheres can be used as drug carriers for biological molecules that are sensitive to organic solvents, pH, temperature, and ultrasound (Jameela et al 1994).

Film/Membrane

Diffusion through pores of membranes and partition mechanisms limit the transport of hydrophobic drugs through cross-linked chitosan membranes; thus, the release rates of bioactive materials could be controlled using cross-linking variations that determine the membrane permeability (Thacharodi and Rao 1993, Thacharodi and Rao 1995). Chitosan films and microbeads containing steroid drugs were examined for potential applications in contraceptive and novel drug delivery (Chandy and Sharma 1991).

Coating

Film-forming properties make chitosan a desirable candidate for use in coating (Homme and Rostaing 1997, Saunders and Pegg 1993, Sekigawa and Onda 1993, Yamada et al 1995), covering materials (Kaessmann and Haak 1997, Koide et al 1995), and membrane generation (Arai and Akiya 1978, Mima et al 1976, Mooyoung et al 1992, Uragami 1991).

Chitosan-coated liposomes show increased adhesion to the intestinal walls, compared to control liposomes, which show minimal adhesion (Takeuchi et al 1994). Chitosan-coated liposomes containing insulin effectively decrease the blood glucose levels as a result of mucoadhesion that is dependent on chitosan concentration (Takeuchi et al 1996). The potential of chitosan as a support in artificial organs was evaluated with bioartificial liver support systems coated with cross-linked chitosan. The chitosan support system retains its structure and releases lactate dehydrogenase much more slowly than the support prepared with collagen. Hepatocytes attached to the chitosan system also show higher urea synthesis activity than those attached to the collagen-coated surface (Kawase et al 1997).

Encapsulation with Bi-Polymer Membrane

The process of encapsulation, initially forming a transient of alginate-calcium gel particles, then reacting with a polycation, in this case polylysine, to form surface film, and subsequently liquefying the gel by removing calcium ions by washing procedures to make a capsule containing a liquid core was reported (Lim 1983) and is known to have been used in commercial mammalian cell culture processes. Unfortunately, this process is not very practical because of the many steps involved, which are tedious and difficult to control. Furthermore, there are inherent problems associated with this encapsulation process because the anionic groups of alginate are not available to form a bi-polymer membrane, since some of them are already engaged in calcium linkages to form the gel. In addition, the alginate in the capsule, even though liquefied in the process, can revert to a gel when calcium or other divalent ions become available, as is the case in cell culture or physiological systems. Thus, a liquid core in the microcapsule is not possible.

Chitosan makes it possible to form bi-polymer membranes instantaneously encapsulating a liquid core, thereby eliminating all these problems (Kim 1992, Kim and Rha 1989a,b, Rha and Rodriguez-Sanchez 1988a,b). Microcapsules containing a liquid core were formed by ionic interactions of chitosan and counterpolyions (Jarvis 1989, Kim and Rha 1989b, Rha and Rodriguez-Sanchez, 1988a,b, Shioya and Rha 1989). The encapsulation procedure is shown in Figure 13.4 (Kim 1992). This simple one-step encapsulation process is commercially used for cell culture, has been proposed as a method for immunoisolation in artificial organs, and is known to be used in encapsulating β-cells for artificial pancreas for insulin therapy (Gupta et al 1993, Kim and Rha 1989a). Since with this encapsulation procedure the pore size of the bipolymer membrane can be controlled so that it is sufficiently large to allow diffusion of nutrients and products, and at the same time small enough to prevent diffusion of the immunoglobulin, cell-based artificial organs for transplantation can be engineered.

Various methods for trapping or encapsulation of bioactive materials, the inclusion of oil emulsions and/or microbial and mammalian cells (Dumitriu et al 1997, Freeman 1996, Jarvis 1989, Magdassi et al 1996, Mooyoung et al 1992, Mosbach and Nilsson 1987) and the formation of semipermeable membranes encapsulating cells, and the concepts of artificial cells (Chang 1964, 1972, 1992, Chang et al 1971, Goosen 1992), cell culture and artificial organs (Gupta et al 1993, Kim and Rha 1989a), and biorectors (Hu et al 1997) have been considered and developed with chitosan and its derivatives.

Encapsulated rabbit hepatocytes with chitosan mixed with carboxymethylcellulose support the survival and growth of liver endothelial cells (Matthew et al 1993). Endothelial cell growth factors (ECGF) encapsulated into chitosan-albumin microspheres and fibers promote significant neo-

vascularization, whereas the direct injection of ECGF into animals did not show neovascularization during the same time period (Elcin et al 1996). High chitosan viscosity promotes the slow release of cytokine into cerebral cortex lesions, both from encapsulated cytokine and from microencapsulated fibroblast-like cells producing colony stimulating factor (CSF-1) (Maysinger et al 1996).

Covalent Attachment

Covalently binding a drug to chitosan is reported as an alternate drug delivery method. 5−fluorouracil covalently attached to chitosan through hexamethelene via urea bonds exhibits strong survival effects against leukemia, in addition to growth inhibitory effects on fibrosarcoma and hepatoma with decreased toxicity (Ouchi et al 1991).

Structure-Property-Function Relationships in Chitosan

The general rule that the chemical structure of a polymer dictates its intrinsic or physical properties and that its functional performance in practice is the expression of these properties under process or use conditions is also true for chitosan. The chemical structure, glucose monomers joined by $\beta(1{\to}4)$ linkage, permits a certain bond angle, contour conformation, and persistent length and chain flexibility to a chitosan molecule, leading to a random coil globular chain conformation that is not very extended.

Given the conformation and potentially active amine groups of chitosan, the possibility for ultramolecular structure synthesis for a matrix or membrane becomes apparent. Taking chitosan with globular conformation as the primary polymer component of the membrane, a positively charged anionic polymer with extended chain conformation, e.g., alginate, can be used as a secondary or connecting polymer that allows intermolecular electrostatic interactions to take place between the two polymers. The extended polymer acts as the glue between the chitosan molecules to create a membrane or network matrix (Figure 13.5).

According to the structural model, shown in Figure 13.5, the pore size of the biopolymer membrane can be estimated from the intrinsic viscosity, which indicates the molecular volume of the chi-

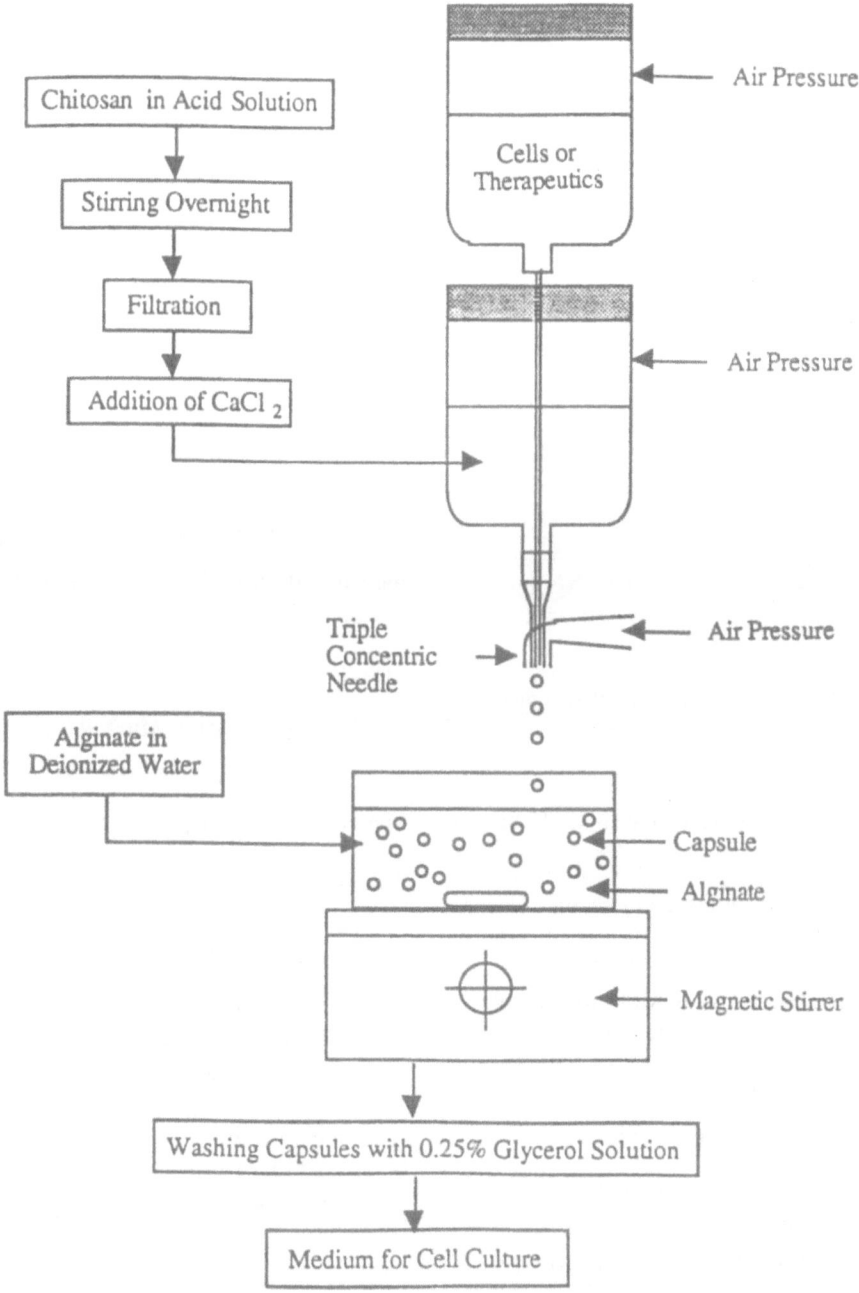

FIGURE 13.4. Encapsulation procedure (Kim 1992).

tosan in solution, and from the overlapping parameter. Assuming that chitosan has a spherical shape
and that the chain is distributed evenly throughout
the molecular domain, as shown in Figure 13.6, the
intrachain distance is obtained by stretching out the
whole chain length of the chitosan molecule and
evenly distributing the total hydrated volume of
chitosan along the fully stretched chain. Then, the
distance equivalent of two radii gives the intrachain distance of the chitosan molecule. The intrachain distance can be calculated with experimentally obtained values of hydrated volume, degree of

A Primary Polymer (Chitosan)

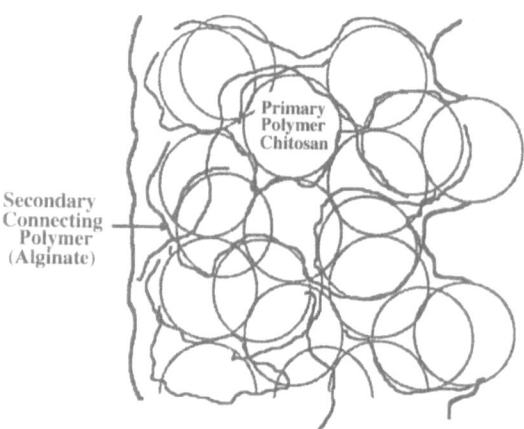

Secondary
Connecting
Polymer
(Alginate)

Primary
Polymer
Chitosan

FIGURE 13.5. Schematic model of bi-polymer membrane (Kim 1992).

polymerization, and molecular weight of chitosan, as follows:

Hydrated volume of a chitosan molecule: (1)

Hydrated volume of a chitosan molecule

$$= \frac{\text{Intrinsic Viscosity} \times \text{Molecular Weight}}{\text{Avogadro's Number}}$$

$$= [\eta] \times Mc / A_{vo}$$

Extended total chain length of chitosan: (2)

With the degree of polymerization (DP) and monomer dimension

$$DP = \frac{\text{Molecular weight of chitosan}}{\text{Molecular weight of glucosamine unit}}$$

$$= \frac{Mc}{Mu}$$

Extended chain length

$$= DP \times \frac{\text{unit cellobiose length}}{2}$$

cellobiose is composed of 2 glucoses linked by $\beta(1{\rightarrow}4)$ glycosidic linkage

$$= 4.75 \text{ Å} \times Mc / Mu$$

Cross-sectional area of total hydrated volume along the extended chain: (3)

Cross-sectional area (A)

$$= \frac{\text{Hydrated volume of a chitosan molecule}}{\text{Extended chain length}}$$

$$= \frac{[\eta] \times Mc}{A_{vo}} + \left(4.75 \text{ Å} \times \frac{Mc}{Mu} \right)$$

$$= \frac{[\eta] \times Mu}{4.75 \text{ Å} \times A_{vo}}$$

Intrachain distance (ID):   (4)

Intrachain distance is the diameter of the cross-sectional area less the diameter of unhydrated chitosan chain (assuming the chain diameter = 6.5 Å).

Intrachain distance = (Cross-sectional area $\times 4/\pi)^{1/2}$ − Chain diameter

Therefore, the intrachain distance is expressed as:

$$ID = \left(\frac{[\eta] \times Mu \times 4}{A_{vo} \times 4.75 \text{ Å} \times \pi} \right)^{1/2} - d_c$$

Therefore, $ID = c\,[\eta]^{1/2} - d_c$

where, ID = intrachain distance (Å)
  $[\eta]$ = intrinsic viscosity (dl/g)
  Mc = molecular weight of chitosan
  A_{vo} = Avogadro's number
  Mu = molecular weight of glucose unit
  c = constant
  d_c = chain diameter.

The intrachain distance (ID) can be considered as the pore size (P) of the capsule membrane when chitosan concentration (Cc) is less than the critical concentration (Ccr).

$$P = ID, \text{ when } Cc , Ccr.$$

However, when the chitosan concentration is greater than the critical concentration at which the zero shear rate viscosity reaches infinity (empirical expression: $Ccr = 4/[\eta]$), the pore size of the capsule can be calculated by dividing intrachain distance by the overlapping parameter.

$$P = \frac{ID}{OP} = \frac{ID}{(Cc / Ccr)} = \frac{ID}{[Cc / (4 / 100\,[\eta])]},$$

$$\text{when } Cc > Ccr.$$

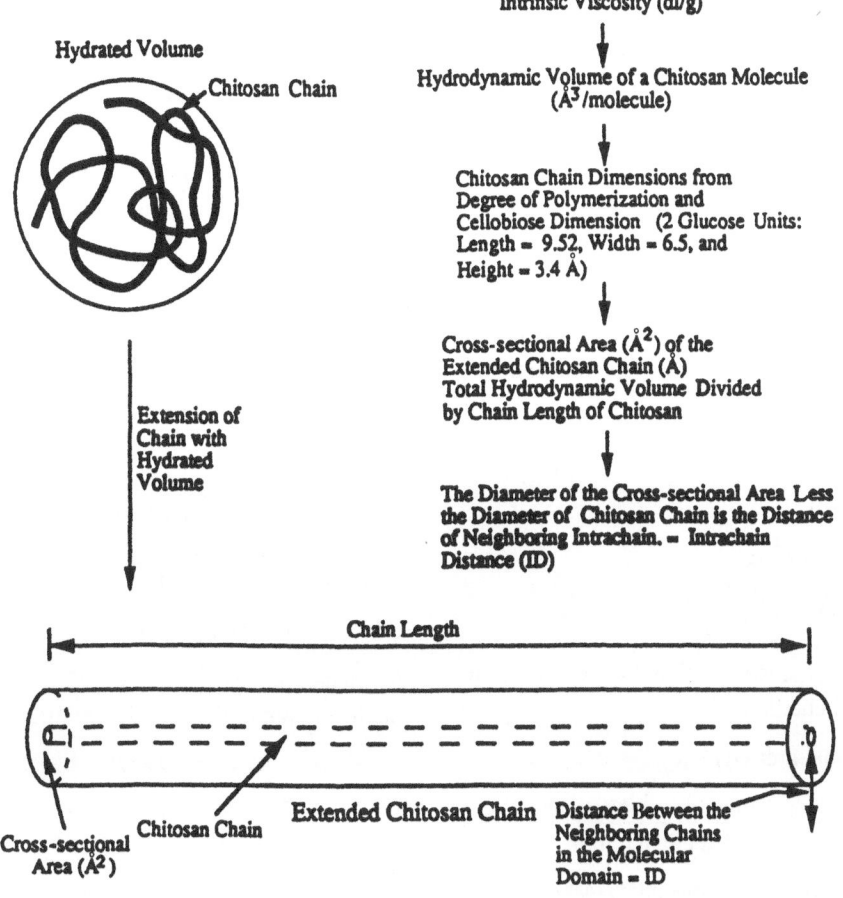

FIGURE 13.6. Schematic for the estimation of intrachain distance.

Therefore, using the expression for intrachain distance obtained above, P becomes

$$P = \frac{c[\eta]^{1/2} - d_c}{Cc \times [\eta] / 4} = \frac{4(c[\eta]^{1/2} - d_c)}{Cc \times [\eta]}.$$

Then, if $c[\eta]^{1/2} - d_c \approx c[\eta]^{1/2}$, the equation becomes

$$P = \frac{4 c[\eta]^{1/2}}{Cc \times [\eta]} = \frac{c'}{Cc \times [\eta]^{1/2}}$$

where, P = pore size (Å)
  ID = intrachain distance (Å)
  OP = overlapping parameter
  Cc = chitosan concentration (g/ml)
  Ccr = critical concentration (g/ml)
  c' = constant
  d_c = chain diameter (Å).

Intrachain distance within the molecular domain increases with increasing intrinsic viscosity. Therefore, the pore size of the bi-polymer membrane increases with the increase in intrinsic viscosity when the concentration of chitosan is lower than critical concentration. However, when the chitosan concentration is higher than the critical concentration, the pore size of the bi-polymer may not increase proportionally with intrinsic viscosity as a result of the overlapping of the molecular domains of polymers.

The intrachain distance and pore size of the membrane is estimated as follows for the case in which 0.5% chitosan solution, having intrinsic viscosity $[\eta]$ = 8.1 dl/g and molecular weight of 1.5 $\times 10^5$, is used to form the bi-polymer membrane.

Hydrated volume of a chitosan molecule: (5)

Hydrated volume

$$= [\eta] \times Mc / A_{vo}$$

$$= 8.1 \text{ dl} / \text{g} \times 100 \text{ ml} / \text{dl} \times 1.5 \times 10^5 \text{ g} / \text{mole}$$

$$\times 10^{24} \text{Å}^3 / \text{ml} \times 1 / (10^{24})$$

$$= 1.2 \times 10^8 \text{ Å}^3 / \text{molecule}$$

Extended chain length of chitosan: (6)

With the degree of polymerization (DP) and glucose dimension

$DP = Mc/Mu$

$= 1.5 \times 10^5 \text{ g}/178 \text{ g}$

$= 842$

Glucose dimension

$= 4.75 \text{ Å} \times 6.5 \text{ Å} \times 3.4 \text{ Å}$; a glucose unit with $\beta(1\rightarrow4)$ linkage based on the bond angle

Chain length

$= DP \times$ Unit length of glucose (with $\beta(1\rightarrow4)$ linkage)

$= 842 \times 4.75 \text{ Å}$

$= 4000 \text{ Å}$

Cross-sectional area of total hydrated volume along the extended chain: (7)

Cross-sectional area (A)

$= \dfrac{\text{Hydrated volume of a chitosan molecule}}{\text{Extended chain length}}$

$= 1.2 \times 10^8 \text{ Å}^3 / \text{Molecule} / 4000 \text{ Å}$

$= 3.0 \times 10^4 \text{ Å}^2$

Intrachain distance (ID):   (8)

Intrachain distance is the diameter of the cross-sectional area of a rodlet with the hydrated volume of chitosan molecule less the diameter of the unhydrated chitosan chain (assuming the diameter of the unhydrated chitosan chain is 6.5 Å).

Intrachain distance

$= (\text{Cross-sectional area} \times 4/\pi)^{1/2}$
$\quad - \text{Chain diameter}$

$= (3.0 \times 10^4 \text{ Å}^2 \times 4/\pi)^{1/2} - 6.5 \text{ Å}$

$= 193 \text{ Å}$

Pore size (P):   (9)

Pore size (P) can be obtained by considering the overlapping parameter of the chitosan solution, when 0.5% chitosan solution is used for the membrane formation.

$$\text{Pore size } (P) = \frac{\text{Intrachain distance } (ID)}{\text{Overlapping parameter } (OP)}$$

Overlapping parameter (OP)

$$= \frac{\text{Chitosan concentration used for the formation of membrane } (Cc)}{\text{Critical concentration } (Ccr)}$$

Critical Concentration (Ccr)

$= 4 / [\eta]$

$= 4 / (8.1 \text{ dl} / \text{g} \times 100 \text{ ml} / \text{dl})$

$= 4.9 \times 10^{-3} \text{g} / \text{ml}$

$= 0.49\%$

Overlapping parameter with final concentration of chitosan solution used for the bi-polymer membrane formation (0.5%)

$OP = 0.50 / 0.49 = 1.02$

Estimated pore size (P) of 0.50% chitosan used for the formation of the bi-polymer membrane

$= ID / OP$

$= 193 \text{ Å} / 1.02$

$= 189 \text{ Å}$

Using the literature, intrinsic viscosities of chitosan, interchain distance, and pore sizes are estimated by the same procedures and assumptions, and are listed in Table 13.4 (Kim 1992).

The pore sizes of bi-polymer membranes formed at a variety of intrinsic viscosities and concentrations of chitosan solution were calculated (Table 13.4) and plotted (Figure 13.7). The pore sizes thus estimated from the above calculations ranged from 60 to 300 Å for the practical values of intrinsic viscosities of 1 to 18 dl/g and chitosan concentrations of 0.1% to 1.0%. In fact, the pore sizes estimated were confirmed to be correct by the diffusion experiments performed with model proteins of known dimensions (Kim 1992).

Therefore, this indicates that microcapsules with bi-polymer membranes of desirable pore size can be designed and engineered with chitosan. Similarly, it would be possible to design and engineer the therapeutics and microencapsulations for desirable specific functions on the basis of the structure-property-function relationship.

TABLE 13.4. Estimated pore size of the chitosan-alginate membrane (estimated* from the intrinsic viscosity) (Kim 1992).

Molecular weight	Solvent	[n], dl/g	$ID^{(1)}$, Å	Pore size$^{(2)}$, Å	Reference
1.7×10^4	Trifluoroacetic acid	3.6	125	125 (Ccr > 0.5%)	Berkovich et al 1980
1.7×10^4	1% Acetic acid, 2.8% NaCl	4.7	144	144 (Ccr > 0.5%)	Berkovich et al 1980
1.7×10^5	1% Acetic acid, 2.8% NaCl	6.4	169	169 (Ccr > 0.5%)	Berkovich et al 1980
1.5×10^5	1% Acetic acid, 2% LiCl	6.7	174	174 (Ccr > 0.5%)	Shioya et al 1989
1.7×10^5	Trifluoroacetic acid	7.0	177	177 (Ccr > 0.5%)	Berkovich et al 1980
13×10^5	pH 2.5, 0.2M NaCl	7.7	186	186 (Ccr > 0.5%)	Rodriguez-Sanchez et al 1982
1.5×10^5	1% Glutamic acid 2% LiCl	8.1	191	189	Shioya et al 1989
1.3×10^5	pH 2.5, 0.1 M NaCl	8.5	196	185	Rodriguez-Sanchez et al 1982
7.2×10^5	1% Acetic acid, 2% LiCl	9.0	202	180	Shioya et al 1989
8.9×10^5	1% Acetic acid, 2% LiCl	9.3	205	177	Shioya et al 1989
8.9×10^6	1% Acetic acid, 2% LiCl	10.4	218	167	Shioya et al 1989
2.8×10^6	1% Acetic acid, 2% LiCl	11.5	229	159	Shioya et al 1989
7.2×10^5	1% Glutamic acid, 2% LiCl	12.1	235	155	Shioya et al 1989

(1) Intrachain distance
(2) Pore size = ID when Cc < Ccr, Pore size = ID/OP when Cc < Ccr
* 0.5% chitosan solution

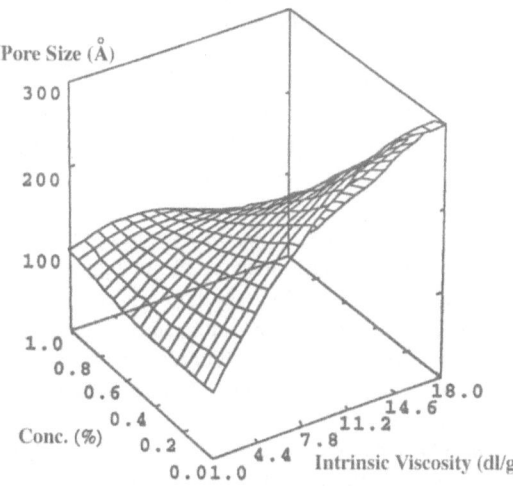

FIGURE 13.7. Estimated pore size of bi-polymer capsule membrane.

Conclusions

Chitosan with its $\beta(1{\to}4)$ glycosidic linkage is endowed with a chain conformation that stearically supports ionic intermolecular interactions that amine groups on C2 of glucose are capable of. These structural characteristics conductive to intermolecular interactions are responsible for the direct therapeutic functions of chitosan in broad areas of application. Chitosan's chemical structure translate into properties that are amenable to controlled ultramolecular structure or network matrix synthesis; that is, the modulation of various physical properties to engineer functional parameters such as pore size of membrane, degrees of swelling, gel density, or structure reactivity/stability is possible with chitosan. Therefore, chitosan has inherent advantages for application in therapeutics and microencapsulation.

Chitosan can be produced readily and is possibly the most effective cationic carbohydrate polymer that can be biosynthesized: Glucose is the most abundant, least costly, and best understood sugar monomer; β(1→4) glycosidic linkage provides the most extended and most flexible chain conformation among all glucans; and its amine groups have the most effective cationic function. Chitosan has demonstrated physiological compatibility with living tissues. It would be difficult to engineer another polysaccharide that could surpass chitosan in its variety of functions and applications for therapeutics and encapsulation.

References

AAFCO. 1986. Official Publication of the Association of American Feed Control Officials Inc. 177.

Akbuga J. 1993. The effect of the physiochemical properties of a drug on its release from chitosonium malate matrix tablets. Int J Pharm (Amsterdam) 100: 257–261.

Akbuga J, Bergisadi N. 1996. 5–Fluorouracil-loaded chitosan microspheres: Preparation and release characteristics. J Microencapsul 13: 161–168.

Allan GG, Altman LC, Bensinger RE, Ghosh DK, Hirabayashi Y, Neogi AN, Neogi S. 1985. Biomedical application of chitin and chitosan. In: Chitin chitosan and related enzymes. Zikakis, JP, ed. New York: Academic Press Inc. 119–134.

Amiji MM. 1995. Permeability and blood compatibility properties of chitosan-PEO blend membranes for haemodialysis. Biomaterials 16: 593–599.

Amiji MM. 1996. Surface modification of chitosan membranes by complexation-interpenetration of anionic polysaccharides for improved blood compatibility in haemodialysis. J Biomat Sci Polym Edition 8: 281–298.

Arai S, Akiya F. 1978. Desalination reverse osmotic membranes and their preparation. US Patent 4111810.

Aspden TJ, Illum L, Skaugrud O. 1996. Chitosan as a nasal delivery system: Evaluation of insulin absorption enhancement and effect on nasal membrane integrity using rat models. Eur J Pharm Sci 4: 23–31.

Aspden TJ, Mason JDT, Jones NS, Lowe J, Skaugrud O, Illum L. 1997. Chitosan as a nasal delivery system: The effect of chitosan solutions on in vitro and in vivo mucociliary transport rates in human turbinates and volunteers. J Pharm Sci 86: 509–513.

Austin PR, Brine CJ, Castle JE, Zikatis JP. 1981. Chitin: New facets of research. Science 212: 729–753.

Aydin Z, Akbuga J. 1996. Gelled spheres containing salmon calcitonin prepared from chitosan and tripolyphosphate solution by ionotropic gelation. Int J Pharm (Amsterdam) 131: 101–103.

Bartone FF, Adickes ED. 1988. Chitosan effects on wound healing in urogenital tissue: Preliminary report. J Urology 140: 1134–1137.

Berkovich LA, Timofeyeva MP, Tsyurupa MP, Davankov VA. 1980. Hydrodynamic and conformational parameters of chitosan. Polym Sci USSR 22: 2009–2018.

Berthod F, Saintigny G, Chretien F, Hayek D, Collombel C, Damour O. 1994. Optimization of thickness pore size and mechanical properties of a biomaterial designed for deep burn coverage. Clin Materials 15: 259–265.

Berthold A, Cremer K, Kreuter JJ. 1996. Preparation and characterization of chitosan microspheres as drug carrier for prednisolone sodium phosphate as model for anti-inflammatory drugs. J Controlled Release 39: 17–25.

Biagini G, Pugnaloni A, Frongia G, Gazzanelli G, Lough C, Muzzarell RAA. 1989. N-carboxymethyl chitosan induces neovascularization. In: Chitin and chitosan. Skjåk-Bræk, G, Anthonsen, T, and Sandford, P, eds. New York: Elsevier Applied Science 671–678.

Biagini G, Bertani A, Muzzarelli R, Damadei A, Benedetto G, Belligolli A, Riccotti G, Zucchini C, Rizzoli C. 1991. Wound management with N-carboxybutyl chitosan. Biomaterials 12: 281–286.

Biagini G, Muzzarell RAA, Giardino R, Castaldini C. 1992. Biological materials for wound healing. In: Advances in chitin and chitosan. Brine, CJ, Sandford, P, and Zikakis, JP, eds. New York: Elsevier Applied Science 16–24.

Borah G, Scott G, Wortham K. 1992. Bone induction by chitosan in endochondral bones of the extremities. In Advances in chitin and chitosan. Brine, CJ, Sandford, P, and Zikakis, JP, eds. New York: Elsevier Applied Science 47–53.

Brine CJ. 1989. Controlled release pharmaceutical applications of chitosan. In: Chitin and chitosan. Skjåk-Bræk, G, Anthonsen, T, and Sandford, P, eds. New York: Elsevier Applied Science 679–691.

Cardinal JR, Curatolo WJ, Ebert CD. 1990. Chitosan compositions for controlled and prolonged release of macromolecules. US Patent 4895724.

Casey, DJ. 1977. Chitin derived surgical glove powder. US Patent 4064564.

Chandy T, Sharma CP. 1991. Biodegradable chitosan matrix for the controlled release of steroids. Biomaterials Artificial Cells Immobilization Biotechnol 19: 745–760.

Chandy T, Sharma CP. 1992. Chitosan beads and granules for oral sustained delivery of nifedipine in-vitro studies. Biomaterials 13: 949–952.

Chandy T, Sharma CP. 1993. Chitosan matrix for oral sustained delivery of ampicillin. Biomaterials 14: 939–944.

Chang TMS. 1964. Semipermeable microcapsules. Science 146: 524–525.

Chang TMS, MacIntosh FC, Mason SG. 1971. Encapsulated hydrophilic compositions and methods of making them. Canadian Patent 873815.

Chang TMS. 1972. Artificial Cells. Springfield: Thomas Publisher.

Chang TMS. 1992. Recent advances in artificial cells with emphasis on biotechnological and medical approaches based on microencapsule. In: Microcapsules and nanoparticles in medicine and pharmacology. Donbrow, M, ed. Boca Raton: CRC Press. 323–339.

Collombel C, Damour O, Gagnieu C, Poinsignon F, Echinard C, Marichy J. 1992. Biomaterials with a base of mixtures of collagen chitosan and glycosaminoglycans process for preparing them and their application in human medicine. US Patent 5166187.

Davies DH, Elson CM, Hayes ER. 1989. NO-carboxymethyl chitosan a new water soluble chitin derivative. In: Chitin and chitosan Skjåk-Bræk, G, Anthonsen, T, and Sandford, P, eds. New York: Elsevier Applied Science 467–472.

Deuchi K, Kanauchi O, Imasato Y, Kobayashi E. 1994. Decreasing effect of chitosan on the apparent fat digestibility by rats fed on a high-fat diet. Biosci Biotechnol Biochem 58: 1613–1616.

Diegelmann RF, Dunn JD, Lindblad WJ, Cohen IK. 1996. Analysis of the effects of chitosan on inflammation angiogenesis fibroplasia and collagen deposition in polyvinyl alcohol sponge implants in rat wounds. Wound Repair Regeneration 4: 48–52.

Dumitriu S, Kahane I, Guttmann H. 1997. Supported polyionic hydrogels containing biologically active material. US Patent 5648252.

Elcin YM, Dixit V, Gitnick G. 1996. Controlled release of endothelial growth factor from chitosan-albumin microspheres for localized angiogenesis: in vitro and in vivo studies. Artificial Cells Blood Substitute Immobilization Biotechnol 24: 257–271.

Errede LA, Stoecz JD, Winter GD. 1983. Composite wound dressing. US Patent 4373519.

Evan EE, Kent SP. 1962. The use of basic polysaccharides in histochemistry and cytochemistry: IV. Precipitation and agglutination of biological materials by Aspergillus polysaccharide and deacetylated chitin. J Histochem Cytochem 10: 24–28.

Eylar EH, Madoff MA, Brody OV, Oncley JL. 1962. The contribution of sialic acid to the surface of the erythrocyte. J Biol Chem 237: 1992–2000.

Federal Register. 1988a. 53: 10428 Mar 30.

Federal Register. 1988b. 53: 17191 May 16.

Federal Register. 1988c. 53: 3075 Feb 3.

Federal Register. 1988d. 53: 4713 Feb 17.

Federal Register. 1989. 53: 11949 Mar 23.

Ferro A. 1991. Appetite moderating and anti-gastritis composition. US Patent 4999341.

Filipovic GJ, Becirevic LM, Skalko N, Jalsenjak I. 1996. Chitosan microspheres of nifedipine and nifedipine-cyclodextrin inclusion complexes. Int J Pharm (Amsterdam) 135: 183–190.

Freeman A. 1996. Method for entrapment of active materials in chitosan. US Patent 5489401.

Furda I. 1980. Nonabsorbable lipid binder. US Patent 4223023.

Gallo JM, Hassan EE. 1992. Receptor-mediated delivery system. US Patent 5129877.

Goosen M. 1992. Fundamentals of animal cell encapsulation and immobilization. Boca Raton: CRC Press.

Gorin PAJ, Barreto-Berger E. 1983. The chemistry of polysaccharides of fungi and lichens. In: Polysaccharides. Aspinall, GO, ed. New York: Academic Press 2: 365–409.

Gross P, Konrad E, Mager H. 1979. Hair setting lotion containing a chitosan derivative. US Patent 4134412.

Gross P, Konrad E, Mager H. 1980. Hair shampoo and conditioning lotion. US Patent 4202881.

Gupta S, Kim SK, Vemuru RP, Aragona E, Cperneni PR, Burk RD, Rha CK. 1993. Hepatocyte transplantation: An alternative system for evaluating cell survival and immune isolation. Int J Artificial Organs 16: 3155–3165.

Hashimoto M, Otagiri M, Imai T. 1995. Drug composition. US Patent 5474989.

Hassan EE, Parish RC, Galla JM. 1992. Optimized formulation of magnetic chitosan microspheres containing the anticancer agent oxantrone. Pharm Res (New York) 9: 390–397.

Hassan EE, Gallo JM. 1993. Targeting anticancer drugs to the brain. I: Enhanced brain delivery of oxantrone following administration in magnetic cationic microspheres. J Drug Target 1: 7–14.

Hayashi J. 1993. Process for preparing a sulfonated chitosan. US Patent 5229504.

Hayes ER. 1986. NO-carboxymethyl chitosan and preparative method therefor. US Patent 4619995.

Henderson SE, Curran DT. Elson CM. 1997. Negatively charged chitosan derivative semiochemical delivery system. US Patent 5645844.

Hirano S, Tanaka Y, Hasegawa M, Tobetto K, Nishioka A. 1985. Effect of sulfated derivatives of chitosan on some blood coagulant factors. Carbohydr Res 137: 205–215.

Hirano S. 1989. Production and application of chitin and chitosan in Japan. In: Chitin and chitosan. Skjåk-Bræk, G, Anthonsen, T, and Sandford, P, eds. New York: Elsevier Applied Science 37–43.

Hirano S, Itakura C. Seino H, Akiyama Y, Nonaka I, Kanbara N, Kawakami T. 1990. Chitosan as an ingredient for domestic animal feeds. J Agric Food Chem 38: 1214–1217.

Homme C, Rostaing JF. 1997. Chitosan-based nutrient or medicinal compositions for administration to ruminants. US Patent 5616339.

Hou WM, Miyazaki S, Takada M, Komai T. 1985. Sustained release of indomethacin from chitosan granules. Chem Pharm Bull 33: 3986–3992.

Hu WS, Cerra FB, Nyberg SL, Scholz MT, Shatford RA. 1997. Bioreactor device with application as a bioartificial liver. US Patent 5605835.

Ikeda I, Tomari Y, Sugano M. 1989. Interrelated effects of dietary fiber and fat on lymphatic cholesterol and triglyceride absorption in rats. J Nutr 119: 1383–1387.

Illum L, Farraj NF, Davis SS. 1994. Chitosan as a novel nasal delivery system. Pharm Res (New York) 11: 1186–1189.

Illum L. 1996. Systemic drug delivery compositions comprising a polycationic substance. US Patent 5554388.

Illum L. 1997. Composition for nasal administration. US Patent 5629011.

Ito H, Komiyama N, Sano H, Mandai H. 1985. Pharmaceutical bandages. JP Patent kokai 60–142927.

Ito M, Miyazaki A, Yamaagishi T, Yagsasaki H, Hashem A, Oshida Y. 1994. Experimental development of a chitosan-bonded beta-tricalcium phosphate bone filling paste. Biomedical Materials Engineer 4: 439–449.

Ito M. 1997. Osteoinduction substance method of manufacturing the same and bone filling material including the same. US Patent 5618339.

Jackson DS. 1987. Chitosan-glycerol-water gel. US Patent 4659700.

Jameela SR, Misra A and Jayakrishnan A. 1994. Cross-linked chitosan microspheres as carriers for prolonged delivery of macromolecular drugs. J Biomaterials Sci Polym Edition 6: 621–632.

Jameela SR, Jayakrishnan A. 1995. Glutaraldehyde cross-linked chitosan microspheres as a long acting biodegradable drug delivery vehicle: Studies on the in vitro release of mitoxantrone and in vivo degradation of microspheres in rat muscle. Biomaterials 16: 769–775.

Jameela SR, Latha PG, Subramoniam A, Jayakrishnan A. 1996. Antitumor activity of mitoxantrone-loaded chitosan microspheres against Ehrlich ascites carcinoma. J Pharm Pharmcol 48: 685–688.

Japanese Natural Additive List Ministry of Health and Welfare Food Chemistry Report. 1983. Ser No 32.

Jarvis AP. 1989. Microencapsulation with polymers. US Patent 4803168.

Kaessmann HJ, Haak KW. 1997. Chitosan foil for wound sealing and process for its preparation. US Patent 5597581.

Kanauchi O, Deuchi K. 1995. Dietary lipid digestion-absorption inhibitory agents and ingesta US Patent 5453282.

Kanauchi O, Deuchi K, Imasato Y, Shizukuishi M, Kobayashi E. 1995. Mechanism for the inhibition of fat digestion by chitosan and for the synergistic effect of ascorbate. Biosci Biotechnol Biochem 59: 786–790.

Kanauchi O, Deuchi K, Imasato Y. 1997. Dietary lipid digestion-absorption inhibitory agents and ingestion. US Patent 5654001.

Katchalsky A, Danon D, Nevo A, De Vries A. 1959. Interaction of basic polyelectrolytes with the red blood cell: II. Agglutination of red blood cells by polymeric bases. Biochem Biophys Acta 33: 120–138.

Kawakami T, Antoh M, Hasagawa H, Yamagishi T, Ito M, Eda S. 1992. Experimental study on osteoconductive properties of a chitosan-bonded hydroxyapatite self-hardening paste. Biomaterials 13: 759–763.

Kawamura Y, Tanibe H, Kurahashi I, Seo H, Nakajima S. 1989. Process for producing granular porous chitosan. US Patent 4833237.

Kawashima Y, Lin SL, Kasai A, Handa T, Takenaka H. 1985. Preparation of a prolonged release tablet of aspirin with chitosan. Chem Pharm Bull 33: 2107–2113.

Kawase M, Michibayashi N, Nakashima Y, Kurikawa N, Yagi K, Mizoguchi T. 1997. Application of glutaraldehyde-crosslinked chitosan as a scaffold for hepatocyte attachment. Biol Pharm Bull 20: 708–710.

Kenne L, Linderberg B. 1983. Bacterial polysaccharides In: Polysaccharides. Aspinall, GO, ed. New York: Academic Press 2: 287–363.

Kienzle-Sterzer CA, Rodriguez-Sanchez D, Rha CK. 1982a. Dilute solution behavior of a cationic polyelectrolyte. J Appl Polym Sci 27: 4467–4471.

Kienzle-Sterzer CA, Rodriguez-Sanchez D, Karalekas D, Rha CK. 1982b. Stress relaxation of polyelectrolyte network as affected by ionic strength. Macromolecules 15: 631–635.

Kienzle-Sterzer CA, Rodriguez-Sanchez D, Rha CK. 1982c. Mechanical properties of chitosan film: Effect of solvent acid. Makromol Chem 183: 1353–1359.

Kienzle-Sterzer CA, Rodriguez-Sanchez D, Rha CK. 1982d. Intrinsic viscosities of chitosan solutions as affected by ionic strength. Proc 2nd Int Conf on Chitin Chitosan Sapporo. 26–29.

Kienzle-Sterzer CA. 1984. Transport properties of a cationic polyelectrolyte in dilute and concentrated solutions: Chitoson. Massachusetts Institute of Technology Ph.D. thesis.

Kienzle-Sterzer CA, Rodriguez-Sanchez D, Rha CK. 1984. Solution properties of chitosan: chain conformation. In: Chitin chitosan and related enzymes. Zikakis, JP, ed. New York: Academic Press Inc. 383–393.

Kifune K. 1992. Clinical application of chitin artificial skin (Beschitin W). In: Advances in chitin and chitosan. Brine, CJ, Sandford, P, and Zikakis, JP, eds. New York: Elsevier Applied Science 9–15.

Kim SK, Rha CK. 1989a. Chitosan for the encapsulation of mammalian cell culture. In: Chitin and chitosan. Skjåk-Bræk, G, Anthonsen, T, and Sandford, P, eds. New York: Elsevier Applied Science 617–626.

Kim SK, Rha CK. 1989b. Transmembrane permeation of proteins in chitosan capsules. In: Chitin and chitosan. Skjåk-Bræk, G, Anthonsen, T, and Sandford,

P, eds. New York: Elsevier Applied Science 635–642.

Kim SK. 1992. Modulation of the porosity of bi-polymer membrane in encapsulation. MIT Ph.D. Thesis.

Knapczyk J, Krówczynski L, Krzek J, Brzeski M, Nürnberg E, Schenk D, Struszczyk H. 1989. Requirements of chitosan for pharmaceutical and biomedical application. In: Chitin and chitosan. Skjåk-Bræk, G, Anthonsen, T, and Sandford, P, eds. New York: Elsevier Applied Science 657–663.

Kobayashi T, Otsuka S, Yugari Y. 1979. Effect of chitosan on serum and liver cholesterol levels in cholesterol-fed rats. Nutr Rep Int 19: 327–334.

Koide M, Konishi J, Ikegami K, Osaki K. 1995. Multilayer wound covering materials comprising a supporting layer and a moisture permeation controlling layer and method for their manufacture. US Patent 5395305.

Komiyama N, Itoi H, Sano H. 1985. Oral compositions. US Patent 04512968.

Kratz G, Arnander C, Swedenborg J, Back M, Falk C, Gouda I, Larm O. 1997. Heparin-chitosan complexes stimulate wound healing in human skin. Scand J Plastic Reconstr Surg Hand Surg 31: 119–123.

Kurita K. 1986. Chemical modifications of chitin, chitosan. In: Chitin in nature and technology. Muzzarelli, R, Jeuniaux, C, and Gooday, GW, eds. New York: Plenum Press 287–294.

Kurita K, Inoue S. 1989. Preparation of iodo-chitins and graft copolymerization onto the derivatives. In: Chitin and chitosan. Skjåk-Bræk, G, Anthonsen, T, and Sandford, P, eds. New York: Elsevier Applied Science 365–372.

Lang ER, Kienzle-Sterzer CA, Rodriguez-Sanchez D, Rha CK. 1982. Rheological behavior of a typical random coil polyelectrolyte: chitosan. Proc 2nd Int Conf on Chitin Chitosan Sapporo 34–38.

Lang G, Clausen T. 1989. The use of chitosan in cosmetics. In: Chitin and chitosan. Skjåk-Bræk, G, Anthonsen, T, and Sandford, P, eds. New York: Elsevier Applied Science 51–69.

Lazarenko EN, Baran AA, Medvedev Yu V. 1986. Flotation of bacterial suspensions using flocculants. Kolloidnyi Z 48: 571–573.

Leuba J, Link H, Stoessel P, Viret J. 1991. Cosmetic preparation containing chitosan. US Patent 5057542.

Lim F. 1983. Microcapsules containing viable tissue cells. US Patent 4391900.

Lorenz DH, Lee CC. 1995. Gels formed by the interaction of polyvinylpyrrolidone with chitosan derivatives. US Patent 5420197.

Magdassi S, Mumcuoglu, K, Bach U. 1996. Method of preparing natural-oil-containing emulsions and microcapsules and its uses. US Patent 5518736.

Makino Y, Matugi H, Suzuki Y. 1989. Sustained release preparation. US Patent 4814176.

Malette WG, Quigley HJ, Gaines RD, Johnson ND, Rainer WG. 1983. Chitosan: a new haemostatic. Ann Thorac Surg 36: 55–58.

Malette WG, Quigley HJ. 1985. Method of achieving haemostasis inhibiting fibroplasia and promoting tissue regeneration in a tissue wound. US Patent 4532134.

Malette WG, Quigley HJ. 1986. Method of altering growth and development and suppressing contamination microorganisms in cell or tissue culture. US Patent 4605623.

Maresch G, Clausen T, Lang G. 1989. Hydroxypropylation of chitosan. In: Chitin and chitosan. Skjåk-Bræk, G, Anthonsen, T, and Sandford, P, eds. New York: Elsevier Applied Science 389–396.

Markey ML, Bowman LM, Michael VW. 1989. Contact lenses made of chitosan. In: Chitin and chitosan. Skjåk-Bræk, G, Anthonsen, T, and Sandford, P, eds. New York: Elsevier Applied Science 713–717.

Matthew SW, Salley SO, Peterson WD, Klein MD. 1993. Complex coacervate microcapsules for mammalian cell culture and artificial organ development. Biotech Progr 9: 510–519.

Mattioli-Belmonte, M, Biagini G, Muzzarelli RAA, Castaldini C, Gandolfi MG, Krajewski A, Ravaglioli A, Fini M, Giardino R. 1995. Osteoinduction in the presence of chitosan-coated porous hydroxyapatite. J Bioactive Compatible Polym 10: 249–257.

Maysinger D, Berezovskaya O, Fedoroff S. 1996. The hematopoietic cytokine colony stimulating factor 1 is also a growth factor in the CNS: II. Microencapsulated CSF-1 and LM-10 cells as delivery systems. Experimental Neurology 141: 47–56.

Mima S, Yoshikawa S, Miya M, Toyonaka JA, Ikeda JA. 1976. Hydrophilic polymer membranes of polyvinyl alcohol and chitosan. US Patent 3962158.

Miyazaki S, Ishii K, Nadai T. 1981. The use of chitin and chitosan as drug carriers. Chem Pharm Bull 29: 3067–3069.

Miyazaki S, Yamaguchi H, Yokouchi C, Tadaka M, Hou WM. 1988. Sustained release of indomethacin from chitosan granules in beagle dogs. J Pharm Pharmacol 40: 642–643.

Miyazaki S, Nakayama A, Oda M, Tadaka M, Attwood D. 1994. Chitosan and sodium alginate based bioadhesive tablets for intraoral drug delivery. Biol Pharm Bull 17: 745–747.

Mooyoung M, Bols NC, Overgaard SE, Scharer JM. 1992. Immobilization of biologically active material in capsules prepared from a water-soluble polymer and chitosan acetate. US Patent 5116747.

Moriguchi S, Suzuki H, Watanabe H, Satoh M, Abe M, Iwata Y. 1989. Adsorbent composed of porous beads of chitosan and adsorption method using same. US Patent 4879340.

Mosbach K, Nilsson K. 1987. Method of encapsulating biomaterial in bead polymers. US Patent 4647536.

Mosbey DT. 1990. Wound filling compositions. US Patent 4956350.

Munjeri O, Collett JH, Fell JT. 1997. Hydrogel beads based on amidated pectins for colon-specific drug delivery: The role of chitosan in modifying drug release. J Controlled Release 46: 272–278.

Muzzarelli RAA. 1977. Chitin. New York: Pergamon press.

Muzzarelli RAA. 1983. Chitin and its derivatives: new trends of applied research. Carbohydr Polym 3: 53–75.

Muzzarelli RAA, Tanafani F, Emanuelli M, Chiurazzi E, Piani M. 1986. Sulfated N-carboxymethyl chitosans as blood anticoagulants. In: Chitin in nature and technology. eds, Muzzarelli, R, Jeuniaux, C, and Gooday, GW, eds. New York: Plenum Press 469–476.

Muzzarelli RAA, Biagini G, Pugnaloni A, Filippini O, Baldassarre V, Castaldini C, Rizzoli C. 1989. Reconstruction of periodontal tissue with chitosan. Biomaterials 10: 598–603.

Muzzarelli RAA. 1990. Chitin. In: Encyclopedia of Polymer Science and Engineering 3: 430–441.

Muzzarelli RAA, Tarsi R, Flippini O, Giovanettei E, Biagini G, Varaldo PE. 1990. Antimicrobial properties of N-carboxybutylchitosan. Antimicrobial Agents and Chemotherapy 34: 2019–2023.

Muzzarelli RAA, Zucchini C, Ilari P, Pugnaloni A, Mattioli-Belmonte M, Biagini G, Castaldini C. 1993a. Osteoconductive properties of methylpyrrolidinone chitosan in an animal model. Biomaterials 14: 925–929.

Muzzarelli RAA, Biagini G, Bellardini M, Simonelli L, Castaldini C, Fratto G. 1993b. Osteoconduction exerted by methylpyrrolidine chitosan used in dental surgery. Biomaterials 14: 39–43.

Nagai T, Sawayanagi Y, Nambu N. 1984. Application of chitin and chitosan to pharmaceutical applications. In: Chitin chitosan and related enzymes. Zikakis, JP, ed. New York: Academic Press Inc. 21–39.

Nagyvary JJ. 1982. Method for treating hyperbilirubinemia. US Patent 4363801.

Nauss JL, Thompson JL, Nagyvary JJ. 1980. The binding of micellar lipids to chitosan. Lipids 18: 714–719.

Nelson JD. 1991. Chitosan pyrithione as an antimicrobial agent useful in personal care products. US Patent 5015632.

Nigalaye AG, Adusumilli P, Bolton S. 1990. Investigation of prolonged drug release from matrix formulations of chitosan. Drug Develop Ind Pharm 16: 449–468.

Nishi N, Noguchi J, Tokura S, Shiota H. 1979. Studies on chitin I. Acetylation of chitin. Polymer J 11: 27–32.

Nishimura K, Nishimura S, Nishi N, Tokura S, Azuma I. 1984. Immunological activity of chitin and its derivatives. Vaccine 2: 93–99.

Nishimura K, Nishimura S, Seo H, Nishi N, Tokura S, Azuma I. 1986a. Macrophage activation with multiporous beads prepared from partially deacetylated chitin. J Biomed Mat Res 20: 1359–1272.

Nishimura K, Nishimura S, Nishi N, Tokura S, Azuma I. 1986b. Immunological activity of chitin derivatives. In: Chitin in nature and technology. Muzzarelli, R, Jeuniaux, C, and Gooday, GW, eds. New York: Plenum Press 477–483.

Nishimura K, Ishihara C, Ukei S, Tokura S, Azuma I. 1986c. Stimulation of cytokine production in mice using deacetylated chitin. Vaccine 4: 151–156.

Nishioka Y, Kyotani S, Okamura M, Miyazaki M, Okazaki K, Ohnishi S, Yamamoto Y, Ito K. 1990. Release characteristics of cisplatin chitosan microspheres and effect of containing chitin. Chem Pharm Bull (Tokyo) 38: 2871–2873.

Okamoto Y, Shibazaki K, Minami S, Matsuhashi A, Tanioka SI, Shigemasa Y. 1995. Evaluation of chitin and chitosan on open wound healing in dogs. J Vet Med Sci 57: 851–854.

Olsen R, Schwartzmiller D, Weppner W, Winandy R. 1989. Biomedical applications of chitin and its derivatives. In: Chitin and chitosan. Skjåk-Bræk, G, Anthonsen, T, and Sandford, P, eds. New York: Elsevier Applied Science 813–828.

Orienti I, Aiedeh K, Gianisi E, Bertasi V, Zecchi V. 1996. Indomethacin loaded chitosan microspheres: Correlation between the erosion process and release kinetics. J Microencapsul 13: 463–472.

Otterlei M, Espevik T, Skjak-Braeck G, Smidsrod O. 1992. Diequatorially bound β-14–polyuronates and use of same for cytokine stimulation. US Patent 5169840.

Ouchi T, Banba T, Masuda H, Matsumoto T, Suzuki S, Suzuki M. 1991. Design of chitosan 5FU conjugate exhibiting antitumor activity. J Macromol Sci-Chem A28.

Pangburn SH, Trescony PV, Heller J. 1984. Partially deacetylated chitin: Its use in self-regulated drug delivery system. In: Chitin chitosan and related enzymes. Zikakis, JP, ed. New York: Academic Press Inc. 3–20.

Papineau AM, Hoover DG, Knorr D, Farkas DF. 1991. Antimicrobial effect of water-soluble chitosans with high hydrostatic pressure. Food Biotechnol 5: 45–57.

Ralston GB, Tracey MV, Wrench PM. 1963. The inhibition of fermentation in baker's yeast by chitosan. Biochem Biophys Acta 93: 652–655.

Rao SB, Sharma CP. 1997. Use of chitosan as a biomaterial: Studies on its safety and haemostatic potential. J Biomed Materials Research 34: 21–28.

Rha CK, Rodriguez-Sanchez D. 1988a. Process of encapsulation and encapsulated active material system. US Patent 4744933.

Rha CK, Rodriguez-Sanchez D. 1988b. Encapsulated active material system. US Patent 4749620.

Rinaudo M, Domard A. 1989. Solution properties of chitosan. In: Chitin and chitosan. Skjåk-Bræk, G, Anthonsen, T, and Sandford, P, eds. New York: Elsevier Applied Science 71–86.

Rinaudo M, Dung P, Milas M. 1992. A new and simple method of synthesis of carboxymethylchitosan. In: Advances in chitin and chitosan Brine, CJ, Sandford, P, and Zikakis, JP, eds. New York: Elsevier Applied Science 516–525.

Robert GAF, Domszy JG. 1982. Determination of viscometric constants for chitosan. Int J Biol Macromol 4: 374–377.

Rodriguez-Sanchez D, Kienzle-Sterzer CA, Rha, CK. 1982. Intrinsic viscosity of chitosan solutions as affected by ionic strength. Proc 2nd Int Conf on Chitin Chitosan Sapporo 30–34.

Sagar B, Hamlyn P, Wales D. 1990. Wound dressing. US Patent 4960413.

Sandford PA. 1989. Chitosan: commercial uses and potential applications. In: Chitin and chitosan. Skjåk-Bræk, G, Anthonsen, T, and Sandford, P, eds. New York: Elsevier Applied Science 51–69.

Sapelli PL, Baldassare V, Muzzarelli RAA, Emanuelli M. 1986. Chitosan in dentistry. In: Chitin in nature and technology. Muzzarelli, R, Jeuniaux, C, and Gooday, GW, eds. New York: Plenum Press 507–512.

Sapoznikov AM, Galaktionov VG, Khromykh LM, Senchenkov YeP. 1984. Effect of chitosan on the electrophoretic mobility of thymocytes. Biophysics 29: 300–304.

Saunders MS, Pegg RK. 1993. Reactive chitosan coated articles and test kit for immunoassay. US Patent 5208166.

Schipper NGM, Varum KM, Artursson P. 1996. Chitosan as absorption enhancers for poorly absorbable drugs: 1. Influence of molecular weight and degree of acetylation on drug transport across human intestinal epithelial (Caco-2) cells. Pharm Res (New York) 113: 1686–1692.

Schipper NGM, Olsson S, Hoogstraate JA, Deboer AG, Varum KM, Artursson P. 1997. Chitosan as absorption enhancers for poorly absorbable drugs: 2. Mechanism of absorption enhancement. Pharm Res (New York) 14: 923–929.

Scopelianos AG, Wilmington DEE. 1989. Polymeric aminosaccharides as anti-hypercholesterolemic agents. US Patent 4877775.

Segal E, Lehre N. 1987. Topical pharmaceutical preparations containing chitin soluble extract. US Patent 4701444.

Sekigawa F, Onda Y. 1993. Coated solid medicament form having releasability in large intestine. US Patent 5217720.

Senstad C, Mattiasson B. 1989. Chitosan as a ligand carrier in affinity precipitation. In: Chitin and chitosan. Skjåk-Bræk, G, Anthonsen, T, and Sandford, P, eds. New York: Elsevier Applied Science 589–604.

Seo H, Mitsuhashi K, Tanibe H. 1992. Antibacterial and antifungal fiber blended by chitosan. In: Advances in chitin and chitosan. Brine, CJ, Sandford, P, and Zikakis, JP, eds. New York: Elsevier Applied Science 34–40.

Sezer AD, Akbuga J. 1995. Controlled release of piroxicam from chitosan beads. Int J Pharm (Amsterdam) 121: 113–116.

Shahabeddin L, Berthod F, Damour O, Collombel C. 1990. Characterization of skin reconstructed on a chitosan-crosslinked collagen glycosaminoglycan matrix. Skin Pharmacol 3: 107–114.

Shibasaki K, Sano H. Matsukubo T, Takaesu Y. 1994a. pH response of human dental plaque to chewing gum supplemented with low molecular chitosan. Bull Tokyo Dent Coll 35: 61–66.

Shibasaki K, Sano H. Matsukubo T, Takaesu Y. 1994b. The influences of the buffer capacity of various substances on pH changes in dental plaque. Bull Tokyo Dent Coll 35: 27–32.

Shioya T, Rha CK. 1989. Transmembrane permeability of chitosan/carboxymethylcellulose capsule. In: Chitin and chitosan. Skjåk-Bræk, G, Anthonsen, T, and Sandford, P, eds. New York: Elsevier Applied Science 627–634.

Shiraishi S, Imai T, Otagiri M. 1993. Chitosan gel beads containing indomethacin prepared by polyelectrolyte complex of sodium tripolyphosphate and chitosan. J Controlled Release 25: 217–225.

Sirica AE, Woodman RJ. 1971. Selective aggregation of L1210 leukemia cells by the polycation chitosan. J Nat Cancer Inst 47: 377–388.

Smith RL. 1984. Chitosan as a contraceptive US Patent 4474769.

Sparkes BG, Murray DG. 1986. Chitosan based wound dressing materials. US Patent 4572906.

Sugano M, Fujikawa T. Hiratsuji Y, Nakashima K, Fukunda N, Hasegawa Y. 1980. A novel use of chitosan as a hypocholesterolemic agent in rats. Am J Clin Nutr 33: 787–793.

Sugano M, Watanabe S, Kishi A, Izume M, Ohtakara A. 1988. Hypocholesterolemic action of chitosans with different viscosity in rats. Lipids 23: 187–191.

Sugano M, Yoshida K, Hashimoto M, Enomoto K, Hirano S. 1992. Hypocholesterolemic activity of partially hydrolyzed chitosan in rats. In: Advances in chitin and chitosan. Brine, CJ, Sandford, P, and Zikakis, JP, eds. New York: Elsevier Applied Science 472–478.

Suzuki K, Tokoyo A, Okawa Y, Suzuki S, Suzuki M. 1985. Enhancing effects of N-acetyl-chito-oligosaccharides on the active oxygen-generating and microbial activities of peritoneal exudate cells in mice. Chem Pharm Bull 20: 1359–1372.

Svensson S, Mardh PA, Lindh F, Lonn H, Kallin E, Nilsson B, Mansson O. 1986. Carbohydrate derivatives and compositions thereof for therapeutic or diagnostic use and methods for their use. Int Patent W086/05789.

Takayama K, Hirata M, Machida Y, Masada T, Sannan T, Nagai T. 1990. Effect of interpolymer complex formation on bioadhesive property and drug release phenomenon of compressed tablet consisting of chitosan and sodium hyaluronate. Chem Pharm Bull (Tokyo) 38: 1993–1997.

Takeuchi H, Yamamoto H, Niwa T, Hino T, Kawashima Y. 1994. Mucoadhesion of polymer-coated liposomes to rat intestine in vitro. Chem Pharm Bull (Tokyo) 42: 1954–1956.

Takeuchi H, Yamamoto H, Niwa T, Hino T, Kawashima Y. 1996. Enteral absorption of insulin in rats from mucoadhesive chitosan-coated liposomes. Pharm Res (New York) 13: 896–901.

Tanigawa T, Tanaka Y, Sashiwa H, Saimoto H, Shigemasa Y. 1992. Various biological effects of chitin derivatives. In: Advances in chitin and chitosan. Brine, CJ, Sandford, P, and Zikakis, JP, eds. New York: Elsevier Applied Science 206–215.

Terbojevich M, Cosani A, Scandola M, Fornasa A. 1986. Solution properties and mesophase formation of chitosan. In: Chitin in nature and technology. Muzzarelli, R, Jeuniaux, C, and Gooday, GW, eds. New York: Plenum Press 349–351.

Thacharodi D, Rao KP. 1993. Release of nifedipine through crosslinked chitosan membranes. Int J Pharm (Amsterdam) 96: 33–39.

Thacharodi D, Rao KP. 1995. Development and in vitro evaluation of chitosan-based transdermal drug delivery systems for the controlled delivery of propanolol hydrochloride. Biomaterials 16: 145–148.

Thakur AB, Jain NB. 1986. Application of chitosan in preparation of controlled release tablets. Proc 13th Int Symp Controlled Release of Bioactive Materials 13: 151–152.

Thakur AB, Jain NB. 1988. Controlled release formulation and method. US Patent 4738850.

Thanoo BC, Sunny MC, Jayakrishnan A. 1992. Crosslinked chitosan microspheres preparation and evaluation as a matrix for the controlled release of pharmaceuticals. J Pharm Pharmcol 44: 283–286.

Toda T, Matsuda H. 1993. Antibacterial anti-nematode and/or plant-cell activating composition and chiti-nolytic microorganisms for producing the same. US Patent 5208159.

Tsuru S, Masuno I. 1992. Drug-releaser. US Patent 5164186.

Uchida Y, Izume M, Ohtakara A. 1989. Preparation of chitosan oligomers with purified chitosanase and its application. In: Chitin and chitosan. Skjåk-Bræk, G, Anthonsen, T, and Sandford, P, eds. New York: Elsevier Applied Science 373–382.

Ueno R, Fujita Y, Nagamura Y, Kamino Y, Tabata A. 1996. Preservation of foods by the combined action of a natural antimicrobial agent and separately packaged deoxidizing agent. US Patent 5549919.

Uragami T. 1991. Selective permeable membrane for separation of liquid solution. US Patent 5006255.

Vahouny GV, Satchithanandam S, Cassidy MM, Lightfoot FB, Furda I. 1983. Comparative effects of chitosan and cholestyramine on lymphatic absorption of lipids in the rat. Am J Clin Nutr 38: 278–284.

Wang W, Bo S, Li S, Qin W. 1991. Determination of the Mark-Houwink equation for chitosans with different degrees of deacetylation. Int J Biol Macromol 13: 281–285.

Wang YM, Sato H, Adachi I, Horikoshi I. 1996. Optimization of the formulation design of chitosan microspheres containing cisplatin. J Pharm Sci 85: 1204–1201.

Weiner ML. 1992. An overview of the regulatory status and of the safety of chitin and chitosan as food and pharmaceutical ingredients. In: Advances in chitin and chitosan. Brine, CJ, Sandford, P, and Zikakis, JP, eds. New York: Elsevier Applied Science 663–672.

Yalpani M. (ed) 1988. Polysaccharides: synthesis modifications and structure/property relations. New York: Elsevier Science Publisher BV.

Yalpani M, Johnson F, Robinson LE. 1992. Antimicrobial activity of some chitosan derivatives. In: Advances in chitin and chitosan. Brine, CJ, Sandford, P, and Zikakis, JP, eds. New York: Elsevier Applied Science 543–548.

Yamada A, Wato T, Uchida N, Fujisawa M, Takama S, Inamoto Y. 1996. Systemic drug delivery compositions comprising a polycationic substance. US Patent 5554388.

Yamaguchi R, Arai Y, Itoh T, Hirano S. 1981. Preparation of partially N-succinylated chitosan and their cross-linked gels. Carbohydr Res 88: 172–175.

Yen SF, Reed KW. 1995. Liquid ophthalmic sustained release delivery system. US Patent 5422116.

14
Polyacrylates

Julia E. Babensee and Michael V. Sefton

HEMA-MMA

Why Polyacrylates?

The suitability of a polymer for microencapsulation of cells is determined by its processability (e.g., its viscosity and solubility, especially in solvents that are tolerated by the cells), its permselectivity in the form of the microcapsule wall, and its biocompatibility. The capsule wall must have a high permeability to nutrients and cell-derived products, yet must exclude antibodies, complement components, and inflammatory and immune cells. Since the surface chemistry has a large role in defining the capsule biocompatibility, it too is significant. An understanding of the influence of various factors such as polymer chemistry and microencapsulation conditions is our research objective. We have focused on using polyacrylates for microencapsulation with the view to deepening our understanding of these issues.

The polyacrylate on which our current research is based is a thermoplastic hydroxyethyl methacrylate-methyl methacrylate (HEMA-MMA, ~75 mole % HEMA) copolymer prepared by solution polymerization after careful monomer purification to reduce the cross-linker content.[1] This copolymer is hydrophilic with a ~25–30% (w/w) water uptake[2] consistent with the poly(HEMA) content, but has mechanical strength, toughness, and elasticity imparted by the poly(MMA) component. These properties lead to adequate permeability of the polymer capsules to aqueous solutes for cellular sustenance[3] and sufficient mechanical durability to tolerate normal handling and stresses *in vivo*. The critical requirement of microcapsule biocompatibility was considered likely to be met, given the common use of the homopolymers poly (MMA) and poly (HEMA) in biomedical applications (e.g., bone cement and intraocular and contact lenses, respectively). The water insolubility of the HEMA-MMA polymer provides stability in the aqueous physiological environment but necessitates the use of an organic solvent to prepare the polymer solution. The ultimate success of this material depended on the ability to select a tolerable solvent and to design a gentle encapsulation process.

Microencapsulation Processes

We have prepared microcapsules using three different water-insoluble polyacrylates, namely EU-DRAGIT RL,[4,5,6] copolymers of dimethylaminoethyl methacrylate and methyl methacrylate (DMAEMA-MMA),[7] and most often, HEMA-MMA. Common to all these polymers has been the use of coaxial extrusion of cell suspension and polymer solution, shearing of the capsule droplet, and polymer wall formation by interfacial precipitation upon nonsolvent contact. However, each polymer has different properties, such as its solubility in solvents that would be suitable for microencapsulation with cells (diethyl phthalate or polyethylene glycol), the solution viscosity, and the precipitation characteristics in an appropriate nonsolvent. Therefore, their application to microencapsulation required appropriately selected polymer solvents and nonsolvents and droplet shearing conditions.

The early microencapsulation studies involving commercially available EUDRAGIT RL or the

DMAEMA-MMA (reviewed in[8]) indicated that encapsulation of living cells within water-insoluble polyacrylate polymers was feasible but that improvement of polymer properties was necessary in the areas of biocompatibility and permeability. The expected properties of HEMA-MMA, as supported by the use of the parent monomers (HEMA and MMA) in biomedical applications, suggested its suitability for microencapsulation purposes. Modifications to the coaxial extrusion, interfacial precipitation process used for EUDRAGIT RL were required to accommodate HEMA-MMA.[3,9] Microcapsule preparation using HEMA-MMA has been described in more detail elsewhere.[10]

Large-diameter microcapsules were prepared using a coaxial extrusion submerged jet, interfacial precipitation process.[3] Microcapsules were produced as capsule droplets, consisting of the cellular core surrounded by the polymer solution, by pumping the HEMA-MMA/PEG-200 polymer solution (10%, w/v) and the mammalian cell suspension in their complete tissue culture medium, to the tip of a coaxial needle assembly. The polymer solution flowed through the outer needle, and the cell suspension through the inner needle. The cell suspension was augmented with the viscosity/density enhancer 20% (w/v) Ficoll-400.[11] Each capsule droplet was sheared from the needle assembly as its tip was withdrawn from the hexadecane overlayer, and it then passed through the hexadecane overlayer into the PBS precipitation bath, which contained 100 ppm of the Pluronic surfactant L101. The latter was added to facilitate the passage of the droplet through the hexadecane/PBS interface. In this precipitation bath, the polymer solvent was extracted, leaving behind a polymer wall surrounding the cellular core. Microcapsules produced by this process were spherical, opaque, and uniform in their diameters of 750–900 μm.

A consequence of the biocompatibility of HEMA-MMA is its failure to support the attachment, spreading, and growth of anchorage-dependent cells. The nonadherent HEMA-MMA intracapsule environment was modified by coencapsulating cell attachment and growth substrates such as the commercially available (Collaborative Research, Bedford, MA) Matrigel®[12,13] or hydrated cell-preloaded Cytodex beads[14] for anchorage-dependent cells or cell immobilization matrices such as agarose or chitosan.[15] Although the latter two cell immobi-

lization matrices would not present sites for cellular attachment, their presence would distribute the cells within the capsule core, preventing their aggregation in this nonadherent environment and the consequent necrosis due to nutrient diffusion limitations. The cell attachment and growth substrates also provide these functions. Modifications to the microencapsulation process were made to accommodate these additives in the capsule core solution.

A more recent development has been the submerged-nozzle liquid-jet extrusion process, depicted schematically in Figure 14.1, to produce HEMA-MMA microcapsules of even smaller diameter (300–600 μm), Figure 14.2.[16] In this process, the polymer solution and cell suspension are delivered to the tip of a coaxial needle assembly, as above. However, the needle assembly remains stationary, while the hexadecane (or dodecane) flows uniformly, coaxially to the needle assembly to shear off each capsule droplet. It is this inherently higher capsule-shearing force that produces the smaller capsules. Microcapsules have been prepared with cells suspended in medium containing either Ficoll-400, Matrigel®, or agarose. The hydrophobic shearing liquid was typically hexadecane; for Matrigel® and agarose capsules, however, because of the need for cooling to control gelation, dodecane was used. The flow rate of hexadecane/dodecane affected both the diameter of the resultant capsules and their production rate; the typically used flow rate of 130 ml/min resulted in capsule diameters of ~400 μm at a rate of 96 capsules/min. Capsule droplets were coextruded into the PBS + L101 precipitation bath during a time period lasting up to 1 hour, followed by two fresh PBS (no L101) washing steps of 20 minutes each.

HEMA-MMA Microcapsule Permeability

To provide an environment that is favorable for the maintenance of cellular viability and function, the capsules must be permeable to small molecules (nutrients and metabolites: glucose, oxygen, lactate) and intermediate or large molecules (growth factors, [~13–30 kD], transferrin [80 kD]). Yet following implantation, the capsules must be impermeable to components of the host immune system (IgG antibodies, 150 kD; complement components such as C1q, 410 kD) such that the encapsu-

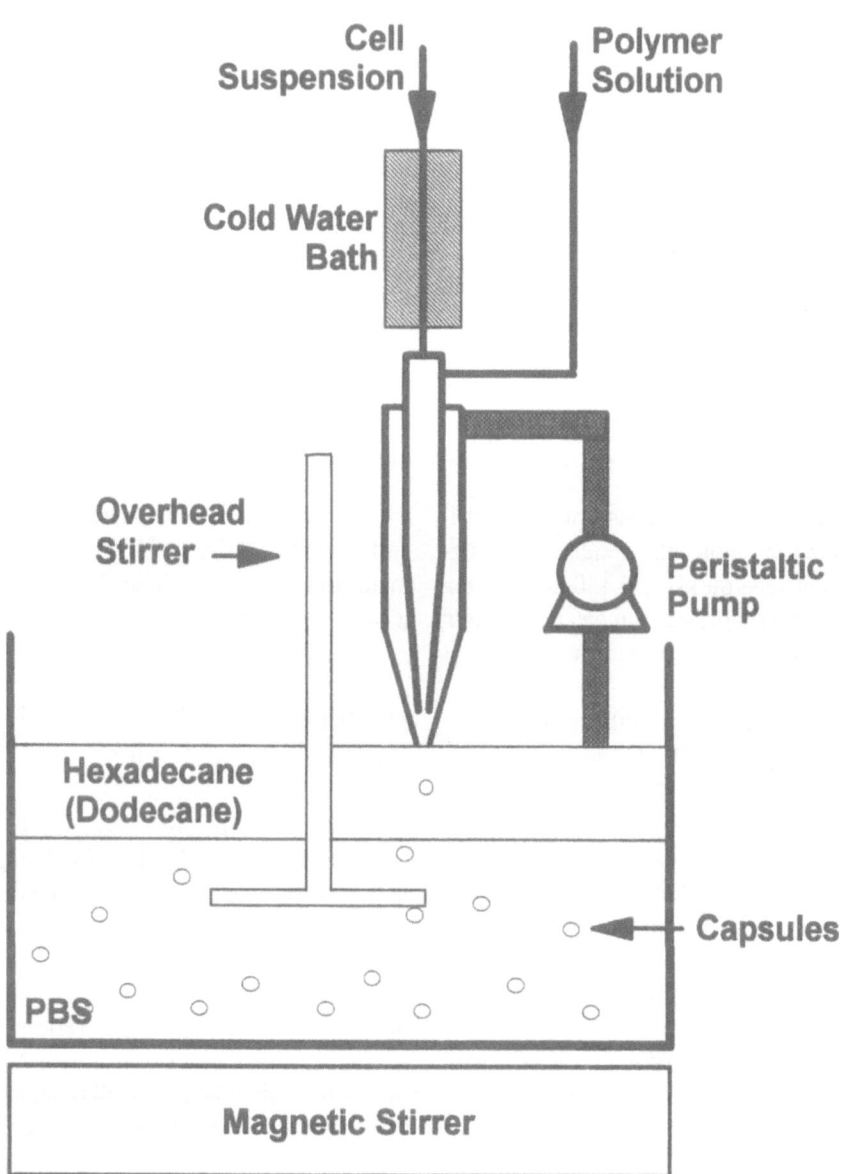

FIGURE 14.1. Schematic drawing of the encapsulation apparatus. Cells and polymer solution are pumped by syringe pumps through a needle assembly. A peristaltic pump recirculates the hexadecane and dodecane. The capsules are kept in suspension by a magnetic overhead stirrer. (Reprinted with the permission of J. Wiley from Reference 16.)

lated cells are not attacked by the host immune system on a molecular level. Furthermore, for a functioning cell-based biomolecule delivery system, the polymer wall must be permeable to the therapeutic product: e.g., dopamine (153 Da), insulin (6000 Da, if a monomer), and growth hormone (23 kDa). We have developed techniques to determine the molecular weight cut-off and the permeability of HEMA-MMA capsules, and to modify these membrane characteristics.

The permeability of HEMA-MMA capsules was estimated by determining the time-dependent release, into an extracapsular PBS sink, of one of several molecules of different molecular weights.[3]

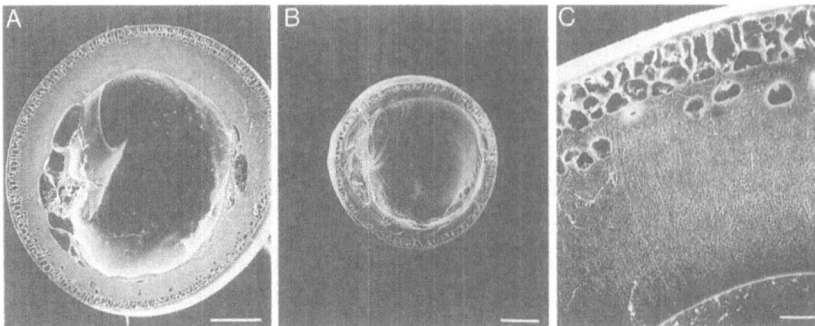

FIGURE 14.2. Scanning electron micrographs of HEMA-MMA microcapsules: A, large-diameter capsule (OD: 660 μm); B, small-diameter capsule (OD: 450 μm); C, wall structure of a large capsule.

There was a decrease in the mass transfer coefficient with increasing molecular weight, and a significant drop between 69 kD and 150 kD (log-log plot). This results suggested a molecular weight cut-off for the membrane that was on the order of 100 kD.

More recent capsule permeability studies have been directed at examining the permeability of capsules at the single-capsule level[17] using horseradish peroxidase (HRP, M_W 40 kD). The enzyme was loaded by diffusion from an external incubation solution, and its release from a single capsule was detected using a sensitive method based on 3,3′,5,5′-tetramethylbenzidine (TMB) as a chromogenic substrate. The capsule-to-capsule variation in HRP permeability was dependent on the care taken in setting up the encapsulation assembly. It was possible to reduce this variability significantly by careful processing. However, reengineering of the co-extrusion nozzle has begun to make the process more reproducible on a regular basis.

Control over microcapsule permeability has been extended by replacing the polymer solvent, PEG-200, with triethylene glycol (TEG, M_w 150 Da), which has a higher diffusivity to enhance the rate of solvent removal upon precipitation. Microcapsules prepared from a 9% (w/v) HEMA-MMA solution in TEG had an increased permeability.[18] An increase in capsule permeability was also noted upon the addition of a pore-forming additive poly(vinyl pyrrolidone) (PVP, M_w 10,000) to the 10% HEMA-MMA in TEG solution at a ratio of 0.3 parts PVP to 1 part HEMA-MMA on a weight basis. This improvement in permeability was at-tributed to a significant increase in the apparent surface pore density in the presence of PVP. It remains to be seen how these changes affect the molecular weight cut-off.

Biological Properties: Specific Examples and Issues

Microencapsulation processes using HEMA-MMA have been developed solely for living cells. Initially, it was necessary to ensure that enough cells survived the encapsulation process and, as appropriate, that they proliferated within the capsule and remained viable long-term. The maintenance of differentiated functions by a variety of encapsulated cells was also studied, since this was the means by which the potentially therapeutically active biomolecule would be produced. Also important has been the effect of the intracapsular environment, determined by the polymer wall on cell arrangement and their consequent viability and function.

Cell Survival and Proliferation

The success of encapsulation is initially dependent on the number of viable cells that are entrapped within the capsule, on an absolute basis or compared to the theoretical number of cells fed to each capsule—the encapsulation efficiency. It may be desirable in certain situations, such as when few cells are encapsulated, that cell proliferation occur within the microcapsule. Such proliferation also indicates a reasonable microenvironment within the capsule core for the cells. The rate of proliferation may, however, be lower than under normal tissue conditions, in part because the environment inside the capsule and the environment in the normal tissue culture dish are different. On the other hand, cells in an unproliferative state may function in a more differentiated manner with enhanced production of the desired product. For other cell types, such as pancreatic islets, which do not proliferate in culture, the initial encapsulation efficiency must be high to provide adequate cell mass within the capsule.

A variety of cell types including the Chinese hamster ovary (CHO) fibroblast,[11] rat pheochromocytoma, PC12,[19] human hepatoma, HepG2,[13] murine fibroblast, L929[20] cell lines, primary rat hepatocytes,[21] and rat islet tissue[22] has been encapsulated within HEMA-MMA microcapsules. These studies have shown that the cells survive the encapsulation procedure, despite the exposure to shear forces and organic solvents/nonsolvents, and grow or function afterwards in vitro for periods from 2 to at least 6 weeks. More recently we have microencapsulated transfected cell lines, specifically mouse fibroblasts (2A-50) that had been engineered to secrete human growth hormone and β-glucuronidase,[23] as a potential application of gene therapy for the delivery of therapeutic proteins using genetically engineered cells.

Generally, only a fraction of the cells delivered to the needle assembly actually become enclosed by the polymer wall. For example, when HepG2 cells were encapsulated with medium augmented with 20% Ficoll-400 (regular capsules), the encapsulation efficiency was ~20–25%.[13] However, inclusion of the cell immobilization matrix Matrigel® in the capsule core increased the encapsulation efficiency for HepG2 cells to ~50%. The encapsulation efficiency was also dependent on the sensitivity of the cells to the encapsulation process and on the quality of the encapsulation run, with better-centered capsules having higher efficiencies as more cells were completely enclosed by the polymer wall.[24]

When anchorage-dependent cells were encapsulated along with an attachment substrate (Matrigel®), cells proliferated over the 2-week period to reach higher cell number values per capsule than in the absence of the attachment matrix.[13] However, while the number of cells per capsule increased, the cellular metabolic activity per capsule did not parallel this increase, but rather decreased slightly after 2 weeks. HepG2 cells encapsulated at the high density within Matrigel® in small-diameter capsules (~400 μm) maintained a constant level of metabolic activity over 3 weeks, similar to what was described above for the large-diameter capsules, but not decreasing after 2 weeks as for large-diameter capsules.[25]

The picture that is emerging is that cells grow to fill the capsule space, but this growth is limited by the nutrient supply and intracapsule and intercellular diffusion gradients (see p. 178). They may maintain a roughly constant cell number and cell activity level thereafter, provided that the intracapsule environment is suitable: e.g., anchorage-dependent cells have the necessary attachment substrate. The rate of proliferation is likely less than that in normal tissue culture, as is the rate of metabolic activity on a per cell basis (as exemplified by MTT conversion). How long this steady cell number/activity is maintained is not known, but MTT conversion has been observed for several months after encapsulation. Interestingly, cells are not found outside of the capsule, except on rare occasions, suggesting that the capsule wall is strong enough to limit proliferation to the intracapsular space.

Protein Release: HepG2 Cells

For the ultimate application of encapsulated cells as a bioartifical organ or physiologically controlled biomolecule delivery system, the cells must not only survive the encapsulation process and remain viable, but they must also express their differentiated functions. The ability of the HEMA-MMA microcapsule to support the functional state of the cells has been assessed by the quantification of encapsulated cell-derived biomolecule release into the extracapsular milieu.

As a protein release example, human hepatoma, HepG2, cells, were used as a model for hepatocytes.[13] Four plasma proteins that span the molecular weight range of interest (α_1-acid glycoprotein, AG, Mw 42 kD; α_1-antitrypsin, AT, Mw 52 kD; haptoglobin, Hap, Mw 98 kD; and fibrinogen, Fbg, Mw 340 kD) were released into the surrounding medium by HepG2 cells from an aliquot of ~100

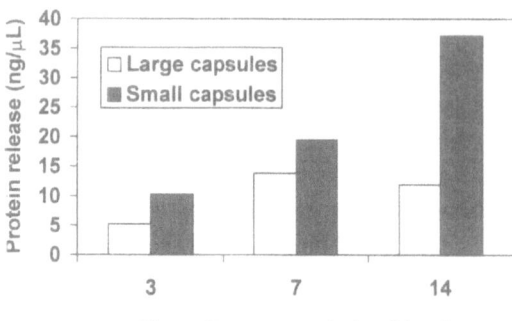

FIGURE 14.3. Comparison of acid glycoprotein release by small (~400 μm) and large (~900 μm) HEMA-MMA microcapsules containing HepG2 cells (initial encapsulation density: 5×10^6 cells/ml) in Matrigel®. Large-diameter capsules prepared by withdrawing nozzle assembly from stagnant hexadecane;[13] small-diameter capsules prepared as in Figure 14.1 by pumping dodecane past a stationary nozzle. Capsules compared on the basis of equal intracapsule space: 11.3 μL/100 large capsules, 2.83 μL/200 small capsules. (Reprinted with the permission of the American Chemical Society from J. E. Babensee and M. V. Sefton. 1997. Controlled Drug Delivery: Challenges and Strategies, K Park, ed.)

Matrigel® capsules. Protein release curves over a 2-week period showed higher amounts of secreted protein, paralleling the increase in the number of cells per capsule (see above); significantly more protein was secreted at day 7, as compared to day 3. The amount of fibrinogen released, relative to α_1-acid glycoprotein, was lower for encapsulated cells than for control unencapsulated cells within Matrigel® in tissue culture, which was consistent with a sieving effect of the polymer membrane on the higher molecular weight protein, Fbg, relative to the smaller AG for large (~800 μm) diameter capsules.[13]

HepG2 cells within small-diameter capsules of ~400 μm also showed a significantly lower Fbg:AG secretion rate ratio as compared to unencapsulated cells, just as with the aforementioned large-diameter capsules of ~800 μm.[13] A comparison of the amount of protein released from small- and large-diameter capsules containing HepG2, on the basis of the internal capsule volume, is shown in Figure 14.3. At all time points after encapsulation, especially after 14 days, there was more acid glycoprotein secreted by small microcapsules than

by large ones, when compared on the basis of equivalent core volume. This was attributed to a more efficient use of internal capsule volume for the smaller capsules and a lower degree of cellular necrosis at the center of the core of the smaller capsules.

Intracapsule Environment

Diffusion limitations associated with the capsule (related to the capsule's wall and size) affect not only the transport of products, such as insulin, but also the transport of metabolites (nutrients and waste products). The corresponding changes in the intracapsular environment (e.g., oxygen, nutrient, or metabolite concentrations) and the presence of products from dying or dead cells further influence cell behavior. These are expressed both through their morphology or three-dimensional arrangement and through protein secretion. The three-dimensional arrangement further influences the transport limitations, because the intracapsular or intercellular space may be rate-limiting. The changes in cell arrangement and morphology are illustrated by anchorage-dependent cells, such as CHO cells[24] and HepG2 cells,[12] and anchorage-independent PC12[19] cells placed into the nonadherent environment of the HEMA-MMA capsule. These have been found to grow as spherical or elliptical aggregates, of varying sizes, within intracapsular polymer pockets. This was consistent with several reports on unencapsulated cell (e.g., hepatocyte) culture under nonadherent conditions.[26] In the absence of a substrate for the cells to attach to and grow on, as in a normal tissue culture situation, cells attach to each other or to cell-secreted extracellular matrix components such as laminin, fibronectin, and collagen, in order to arrange themselves into a spheroidal shape. When HepG2 cells (without coencapsulated Matrigel®) were arranged within large aggregates (~300 μm × 100 μm) in capsules after ~3–7 days in vitro, the cells at the center of these aggregates became necrotic, while cells in an outer shell, 2 to 5 cells in thickness (~30 μm) remained viable. Similar outer shells of viable cells surrounding necrotic cores were observed within aggregates of encapsulated PC12 cells after about 1 week in vitro. This morphology was maintained throughout the 6-week observation period. Transmission electron microscopy (TEM) observations of PC12 cell spheroids within HEMA-MMA capsules showed the presence of dopamine granules in the healthy cells of the outer shell.

The environment within the HEMA-MMA capsule core may be altered by the coencapsulation of extracellular components such as a cell attachment substrate, Matrigel®, or the coencapsulation of an immobilization matrix, such as agarose or chitosan. Whereas the former provides a substrate for cell attachment (which HEMA-MMA does not), the latter two may simply improve the distribution of cells within the capsule, and minimize intercellular diffusion limitations. Since Matrigel® is derived from a mouse sarcoma, its stability and biocompatibility in vivo may be questionable because contact with inflammatory cell-derived proteolytic enzymes, such as collagenase or elastase, could release Matrigel® fragments whose antigenicity (low molecular weight and xenogenicity) might exacerbate any inflammatory reaction. These concerns will likely preclude the use of Matrigel® in vivo despite its positive cellular effects in vitro. The issue of xenogenicity may be circumvented by using host-species-specific extracellular matrix components such as collagen derived from human umbilical cords, which is at present commercially available. On the other hand, agarose, as a cell immobilization matrix in the capsule core, might elicit less of an immune and inflammatory response (if complement activation is not a problem).

Future Directions

We have reached the stage where microparticles containing immobilized or encapsulated cells can be produced with synthetic polymers (e.g., HEMA-MMA), using water or organic solvents. Cell survival during the process and the preservation of cell viability or function afterwards, at least in vitro, have been demonstrated. There is much to do to optimize the processes that are used, to improve their reproducibility, to be able to control permeability or molecular weight cut-off, or simply to reduce the capsule diameter. These are primarily technical problems that will resolved as the processes are scaled up and as further understanding of the processes is achieved.

There is much to do as well in understanding how the capsule wall affects the encapsulated cell

behavior. The complexities of the interrelationships among wall permeability, extracellular matrix, cell density, etc., are beginning to be recognized, but we are still a long way from knowing how to design an encapsulated cell system. We have learned not to expect cell behavior that is comparable to what is seen in conventional tissue culture—the three-dimensional arrangement of cells inside a capsule creates a unique microenvironment and may even be more physiological than that on a polystyrene dish. We are at the stage where any cell can be encapsulated, and, provided that the conditions are reasonable, its differentiated function can be retained in vitro.

On the other hand, we are only beginning to understand what happens in vivo. After implantation, the tissue reaction, the implant site, the presence of cytokines, the release of antigenic proteins or cell debris, the presence of adsorbed protein, and the capsule chemistry and durability all intervene to affect the encapsulated cell behavior, its ability to secrete the desired product, and the duration of viability. Although it will take considerable effort to sort out these interrelationships, the therapeutic benefit of cell transplantation warrants the effort.

Acknowledgments. We acknowledge the financial support of the Medical Research Council of Canada, the Natural Sciences and Engineering Research Council, The Canadian Parkinson's Foundation, and the Canadian Liver Foundation, the latter in providing a fellowship to JEB. We also acknowledge the assistance of Mr. Vlad Horvath, and the advice of Professor U. De Boni.

References

1. Stevenson WTK, Evangelista RA, Broughton RL, Sefton MV. 1987. Preparation and characterization of thermoplastic polymers from hydroxyalkyl methacrylates. J. Appl. Polm. Sci. 34: 65–83.
2. Stevenson WTK, Sefton MV. 1988. The equilibrium water content of some thermoplastic hydroxyalkyl methacrylates. J. Appl. Poly. Sci. 36: 1541–1553.
3. Crooks CA, Douglas JA, Broughton RL, Sefton MV. 1990. Microencapsulation of mammalian cells in a HEMA-MMA copolymer. Effects on capsule morphology and permeability. J. Biomed. Mater. Res. 24: 1241–1262.

4. Boag AH, Sefton MV. 1987. Microencapsulation of human fibroblasts in a water insoluble polyacrylate. Biotech. Bioeng. 30: 954–962.

5. Sugamori ME, Sefton MV. 1989. Microencapsulation of pancreatic islets in a water insoluble polyacrylate (EUDRAGIT RL), Trans. ASAIO 35: 791–799.

6. Broughton RL, Sefton MV. 1989. Effect of capsule permeability on growth of CHO cells in EUDRAGIT RL microcapsules: Use of FITC dextran as a marker of capsule quality. Biomaterials 10: 462–465.

7. Mallabone CL, Crooks CA, Sefton MV. 1989. Microencapsulation of human diploid fibroblasts in cationic polyacrylates. Biomaterials 10: 380–386.

8. Sefton MV, Stevenson WTK. 1993. Microencapsulation of live animal cells using polyacrylates. Adv. Poly. Sci. 107: 145–197.

9. Dawson RM, Broughton RL, Stevenson WTK, Sefton M.V. 1987. Microencapsulation of CHO cells in a hydroxyethyl methacrylate-methyl methacrylate copolymer. Biomaterials 8: 360–366.

10. Babensee JE., Sefton MV. 1996. Microparticulates for Cell Delivery. In: Microparticulates—preparation characterization and application to medicine. Cohen S., Bernstein H., eds. New York: Marcel Dekker 477–519.

11. Uludag H, Sefton MV. 1993. Metabolic activity and proliferation of CHO cells in hydroxyethyl-methacrylate (HEMA-MMA) microcapsules. Cell Transplantation 2: 175–182.

12. Babensee JE, De Boni U, Sefton MV. 1992. Morphological assessment of hepatoma cells (HepG2) microencapsulated in a HEMA-MMA copolymer with and without Matrigel. J. Biomed. Mater. Res. 26: 1401–1408.

13. Uludag H, Sefton MV. 1993. Microencapsulated human hepatoma cells: In vitro growth and protein release. J. Biomed. Mater. Res. 27: 1213–1224.

14. Patterson WJ. 1992. Microencapsulation of growth factor-secreting cells on microcarriers. Masters of Applied Science, Department of Chemical Engineering and Applied Chemistry, University of Toronto.

15. De Castro A. 1994. HEMA-MMA microencapsulated agarose, chitosan or Matrigel® as a PC12 cell immobilization matrix. Bachelor of Applied Science Thesis, Department of Chemical Engineering and Applied Chemistry, University of Toronto.

16. Uludag H, Horvath V, Black JP, Sefton MV. 1994. Viability and protein secretion from human he-patoma (HepG2) cells encapsulated in 400 μm polyacrylate microcapsules by submerged nozzle–liquid jet extrusion. Biotech. Bioeng. 44: 1199–1204.

17. Uludag H, Hwang JR, Sefton MV. 1995. Microencapsulated human hepatoma (HepG2) cells: Capsule-to-capsule variations in protein secretion and permeability. J. Cont. Rel. 33: 273–283.

18. Hwang JR, Sefton MV. 1995. The effects of polymer concentration and a pore forming agent (PVP) on HEMA-MMA microcapsule structure and permeability. Journal of Membrane Science 108: 257–268.

19. Roberts T, De Boni U, Sefton MV. 1996. Dopamine secretion by PC12 cells microencapsulated in a hydroxyethyl methacrylate-methyl methacrylate copolymer. Biomaterials 17: 267–276.

20. Ung D Y-P. 1993. The effect of cytokines on microencapsulated mammalian model cells. Bachelor of Applied Science Thesis, Department of Chemical Engineering and Applied Chemistry, University of Toronto.

21. Wells GDM, Fisher MM, Sefton MV. 1993. Microencapsulation of viable hepatocytes in HEMA-MMA microcapsules: a preliminary study. Biomaterials 14: 615–620.

22. Sefton MV, Kharlip Lilia. 1994. Insulin release from rat pancreatic islets microencapsulated in a HEMA-MMA polyacrylate, Pancreatic Islet Transplantation Volume III: Immunoisolation of Pancreatic Islets. Lanza RP, Chick WL, eds. Austin, Texas: R.G. Landes Company. p. 107–117.

23. Tse M, Uludag H, Sefton MV, Chang PL. 1996. Secretion of recombinant proteins from hydroxyethyl methacrylate-methyl methacrylate capsules. Biotech. Bioeng. 51: 271–280.

24. Uludag H, Sefton MV. 1992. Metabolic activity of CHO fibroblasts in HEMA-MMA microcapsules. Biotech. Bioeng. 39: 672–678.

25. Uludag H, Horvath V, Black JP, Sefton MV. 1994. Viability and protein secretion from human hepatoma (HepG2) cells encapsulated in 400 μm polyacrylate microcapsules by submerged nozzle–liquid jet extrusion. Biotech. Bioeng. 44: 1199.

26. Koide N, Sakaguchi K, Koide Y, Asano K, Kawagughi M, Matsushima H, Takenami T, Shinji T, Mori M, Tsuji T. 1990. Formation of multicellular spheroids composed of adult rat hepatocytes in dishes with positively charged surfaces and under other nonadherent environments. Exp. Cell Res. 186: 227.

Section 2
Macroencapsulation

15
Diffusion Chambers

Robert P. Lanza

Diffusion chambers include a diversity of planar and tubular configurations, including flat sheets, disks, hollow fibers and wide-bore tubes.[1-5] Although various device designs have been studied for nearly half a century, these experiments have been performed mainly in rodents, and only recently has any progress with the diffusion chamber technique been reported in large animals.[6] In the late 1940s/early 1950s, Algire and coworkers[7-9] first transplanted cells enclosed in Millipore diffusion chambers into mice, in order to study the mechanisms of cellular immune rejection. The viability of the implanted tissue was severely compromised by fibrous overgrowth of the extracapillary-situated chambers. Subsequent studies with islet allografts and xenografts in rodents were equally discouraging.[10-17] Although partial or transient normoglycemia was achieved in some studies, the outer membrane surface was rapidly covered by fibrosis, and islet survival was not observed after a few weeks of transplantation. Furthermore, Theodorou and Howell[18] investigated the in vitro glucose-insulin transfer kinetics of this kind of macroencapsulation device and found that the transit half-time of insulin (almost an hour) was less than optimal for physiologic regulation of blood glucose concentrations.

Neovascularized Diffusion Chambers

Recently, there has been renewed interest in planar diffusion chambers prepared with parallel, flat-sheet microporous membranes.[19-23] These diffusion chambers are similar in design to those that were extensively studied by Algire and others[7-17] several decades ago, except that the laminated surface texture induces a greater degree of neovascularization at the host-membrane interface.[19,20] This induced perimembrane vasculature is intended to furnish the implanted cells with an improved source of oxygen and nutrients. Using these diffusion chambers, Brauker et al[19] have shown that embryonic rat and mouse lung tissues were able to survive and remain differentiated in an allogeneic model system for up to a year. In contrast, xenogeneic tissues died within a few weeks even when implanted with intact membranes.[20] The death of the tissue was accompanied by a severe local inflammatory response around the device. The pore size of the membranes used in these studies was 0.4 μm, and therefore did not prevent large molecular weight molecules, such as IgG, complement, and IgM, from entering into the chambers.

Narrow-Bore Hollow Fibers

A widely studied variation of the diffusion chamber technique involves inclusion of cells inside hollow fibers (Figure 15.1). In 1982, Woodward et al[24] examined the influence of geometry on the occurrence of fibrosis. He demonstrated that a disc-shaped chamber induced a zone of collagen-rich connective tissue between the implant and the vascular system, whereas a hollow fiber elicited only a minimal response. A number of investigations have been carried out to assess the potential of these narrow-bore tubular chambers (<1 mm i.d.) as an artificial pancreas. Others found that islet and

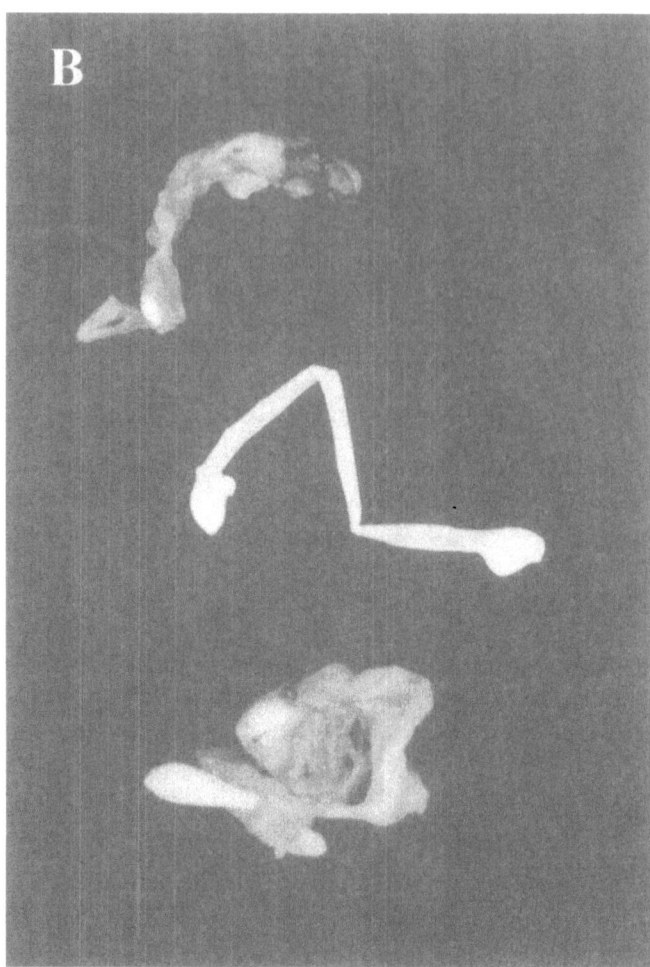

FIGURE 15.1. Hollow fibers before (left) and after (right) implantation into a dog for 3 weeks. Despite over a decade of research, reversal of diabetes in larger animal models has not been attained with narrow-bore hollow fiber device designs.

insulinoma tissue seeded within hollow fibers restored normoglycemia in diabetic rodents when implanted intraperitoneally.[25-27] Histologic analysis of recovered implants, however, revealed a fibrous tissue layer surrounding the membranes. The same type of fiber was also observed to elicit an inflammatory pericapsular response in the pig.[28] This tissue reaction was generally more intense, though qualitatively similar to that seen in the rat, except for lymphoid clusters with giant and pseudo-epithelioid cells that were observed only in the pigs. The reaction consisted of several layers of fibroblasts and collagen with polymorphonuclear leukocytes, macrophages, histiocytes, and small lymphocytes. The fenestrated outer wall of the tubular membrane was always infiltrated by collagen, fibroblasts, and macrophages. To date, reversal of diabetes in larger animal models has not been attained with narrow-bore hollow fiber device designs.[21]

Wide-Bore Tubes

Recent reports[6,29-32] describe a series of experiments using wide-bore tubular chambers with a diameter of >1 to 5 mm. These studies were carried out with polyacrylonitrile-polyvinyl chloride (PAN-

PVC) membranes having a smooth outer skin. Porcine, bovine, and canine islets placed within these chambers restored normoglycemia in streptozotocin (STZ)-induced diabetic rats for more than a year without immunosuppression.[32] Only minimal tissue reactivity was observed. The external membrane surfaces were generally free of fibrotic overgrowth and exhibited only occasional host cell adherence. Encapsulated canine xenografts implanted in spontaneously diabetic BB rats also had the same success, resulting in fasting normoglycemia for periods of several weeks to more than 8 months.[31] Intravenous glucose tolerance test (IVGTT) K-values (decline in glucose levels, percent per minute) after implantation in spontaneously diabetic BB and STZ rats were 2.3 ± 0.4 and 2.6 ± 0.2 to 3.5 ± 0.3, respectively, compared with 3.1 ± 0.1 and 3.3 ± 0.1 for normal control groups. In contrast, the K-values for untreated diabetic rats were <1. Both light and electron microscopy of long-term functioning grafts revealed well-preserved islets, with hormone-producing alpha, beta, and delta cells.

While these wide-bore PAN-PVC membranes solved many of the problems associated with diffusion chambers (e.g., fibrosis, abscess formation, adhesions),[33,34] studies in large animals closer to man will likely be required before clinical trials can be contemplated. Experiments in totally pancreatectomized, severely diabetic dogs have in fact recently been performed.[6] They indicated that canine islet implants can provide long-term correction of hyperglycemia without the use of immunosuppressive and/or anti-inflammatory drugs. Insulin independence was achieved for >10 weeks in dogs with preimplantation insulin requirements of 30–40 units per day (a dosage in the range of what most human patients require). Little or no fibrosis was observed for periods as long as 30 weeks (Figure 15.2).

In view of these encouraging results, a number of unsolved issues critical to the wide-scale clinical success of these devices must be addressed. These include (1) long-term biocompatibility (with risk of fibrosis, peritonitis, intestinal adhesions, and abscess formation), (2) membrane breakage, (3) suitability for retrieval (or, alternately, the use of the peritoneum or other implantation site as a "dumping ground"), (4) further improvements in glycemic control, and (5) potential limitations imposed by the size and geometry of these chambers.

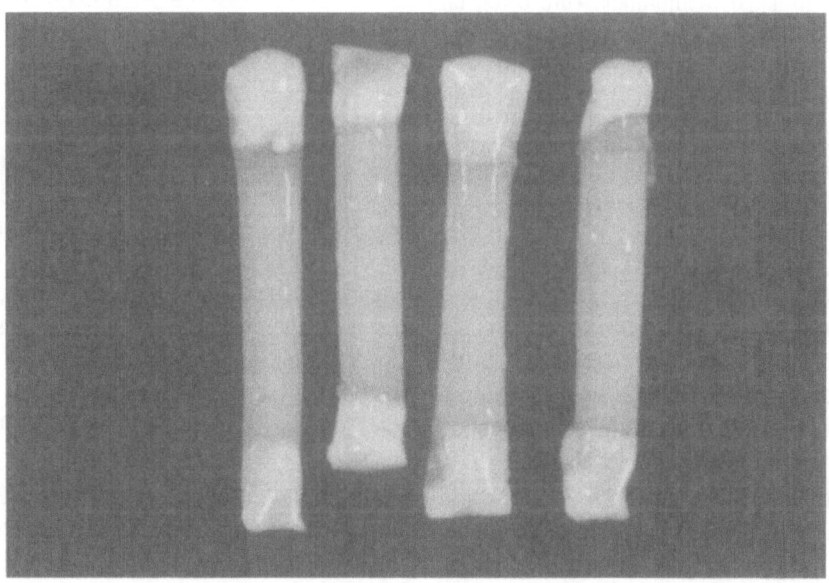

FIGURE 15.2. Wide-bore tubes retrieved from the peritoneal cavity of a diabetic dog 215 days after implantation. Note the absence of fibroencapsulation. Although more durable than hollow fiber device designs, under stress, wide-bore chambers can also bend, leading to fracture of the membrane wall and subsequent destruction of the encapsulated cells.

Biocompatibility

Experiments performed in our laboratory have demonstrated the feasibility of long-term immunoisolation of islets by wide-bore [PAN-PVC] membranes and the long-term biocompatibility of the membrane versus the graft and versus the recipient.[29-32] These data indicate that islet implants can provide correction of hyperglycemia in dogs and rodents for periods of several months to more than a year without the use of immunosuppressive drugs. Diffusion chambers fabricated from wide-bore acrylic membranes showed little or no evidence of an inflammatory response when implanted intraperitoneally in either spontaneously or streptozotocin-induced diabetic rats.[31,32] Complications such as abscess formation or intestinal adhesions, which have been observed with other technologies,[33,34] were not observed with these implants. These studies have recently been extended to implantation in the peritoneal cavity of a large animal, the dog, with surgically induced diabetes.[6] Histological examination of the chambers revealed that they were biocompatible. The outer surface showed only scattered foci of macrophages and lymphocytes. Intactness and sterility of the chambers, however, were crucial factors in the success of the implants. In addition to loss of islet viability, damaged or contaminated membranes were often encapsulated by fibrous tissue that exhibited an interstitial acute and/or chronic inflammatory reaction, and development of granulation tissue was observed. Before testing can be undertaken in diabetic patients, it will be important to determine the cause(s) of this peritoneal tissue reaction.

Membrane Breakage

Most of the transplants described above ultimately failed because of membrane breakage. Under stress, the tubular chambers can bend, leading to fracture of the membrane walls and subsequent destruction of the encapsulated islet tissue. By 5–7 months postimplantation, 80–90% of the membrane chambers in dogs had broken. The tubular membranes used in most of these studies had a wall thickness of only 69–105 μm. The chambers fabricated from these membranes were relatively fragile, and susceptible to breakage. An increase in the membrane wall thickness may minimize this problem.

Methods for Removal and/or Sampling

Because of limitations imposed by islet longevity and resultant requirements for reseeding, membrane chambers will eventually require localization and removal. If necessary, and depending upon the site of implantation, the devices could be removed by laparoscopy. However, surgical excision would be necessary if diffusion chambers were to become fibroencapsulated and/or vascularized. Open surgery, of course, carries risk of infection and would be a more extensive surgical procedure.

Blood Glucose Control

The motivation for islet transplantation is to provide physiologic control of blood glucose concentration. In vitro and in vivo experiments with tubular diffusion chambers have demonstrated only moderately delayed changes (lagtime <10 minutes) in insulin secretion in response to changes in glucose concentration. Perifusion of encapsulated canine islets with glucose elicited an approximately fourfold average increase from the basal insulin secretion.[29] There was a delay of only 7 ± 1 minutes before the insulin concentration in the perfusate began to increase. Although this response is well within a time frame compatible with closed-loop insulin delivery (pharmacokinetic modeling of glucose homeostasis in man suggests that the lag time of the increase in insulin delivery by an artificial pancreas must be <15 minutes to avoid the overexcursion of postprandial blood glucose),[35] a reduction in the volume of the islet cell compartment would further improve the transmission of the glycemic signal from the blood to the islets and of insulin from the islets to the recipient.

In a set of experiments,[6,36] wide-bore chambers were implanted into the peritoneum of six totally pancreatectomized dogs, and the animals monitored for glycemic control by fasting and postprandial blood glucose determinations, and by responses to both intravenous and oral glucose. All of the dogs had varying degrees of reduced insulin requirements for control of fasting blood glucose levels. Implantation of the chambers completely supplanted exogenous insulin therapy in three animals for 51 to >90 days (each of these implants continued to maintain blood glucose control for >20 weeks) (Figures 15.3 and 15.4). The fasting glucose concentrations averaged 81 ± 6 mg/dl for these three animals during the first month. This was lower than the fasting glucose

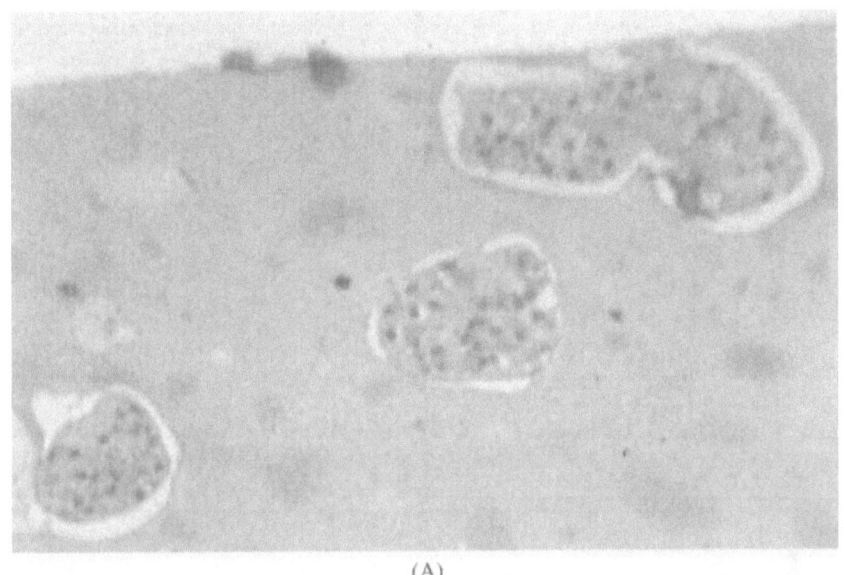

(A)

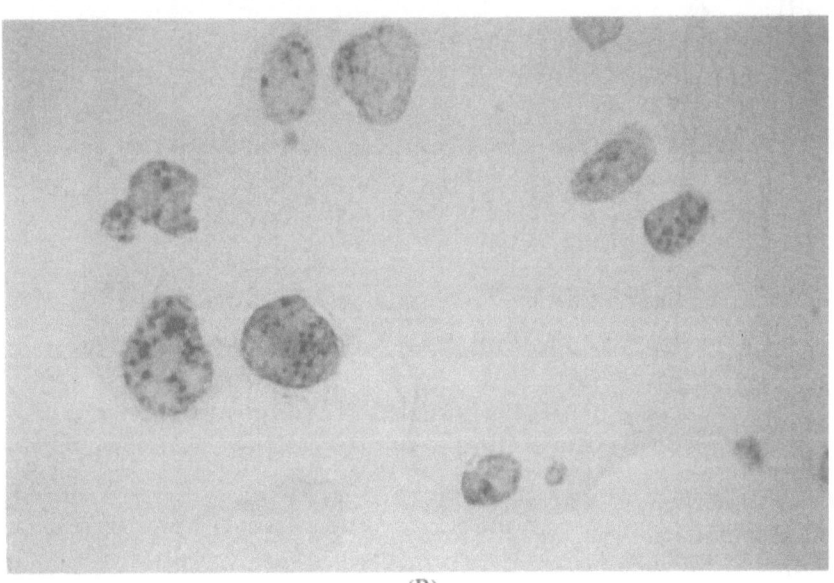

(B)

FIGURE 15.3. (A) Canine islets retrieved from the peritoneal cavity of a diabetic dog 155 days after transplantation inside wide-bore tubes (H & E). (B) Immunoperoxidase stained for insulin, revealing multiple viable hormone-containing beta cells.

levels prior to pancreatectomy, which averaged 91 ± 3 mg/dl. The precise reason for these slightly lower levels is presently unclear.

Diffusion Limitations

Immunoisolated islets lack intimate vascular access, and must be supplied with oxygen and nutrients by diffusion from the nearest blood vessels over distances greater than those normally encountered. In wide-bore membrane chambers, the problem of cell death or dysfunction as a result of oxygen supply limitations, or accumulation of wastes or other cellular products is likely to be more severe. Our observations with 4.8 mm i.d. chambers are consistent with this. Chambers retrieved from the peritoneal cavity of dogs several months after allotransplantation contained a central core of

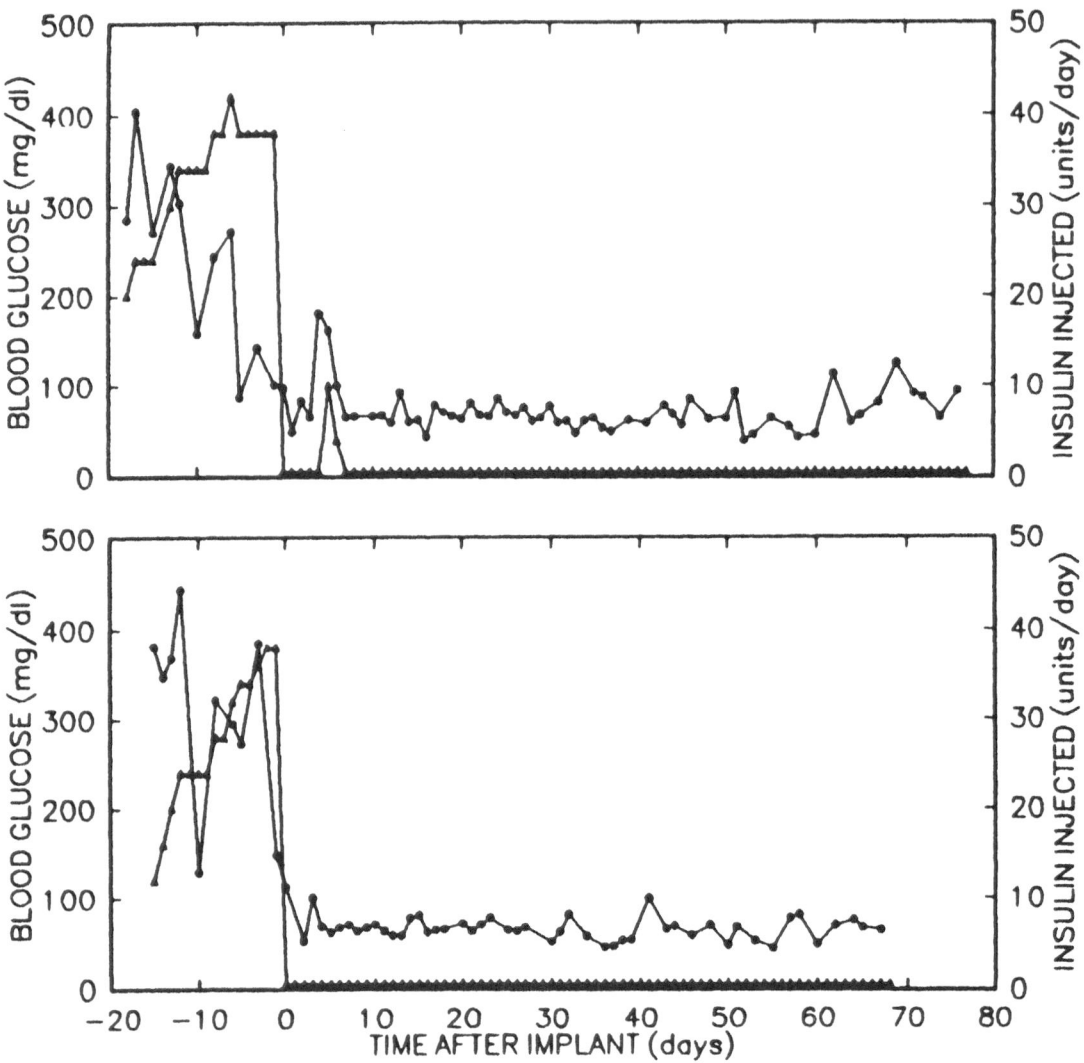

FIGURE 15.4. Reversal of diabetes in dogs, using islets encapsulated in wide-bore tubes. This is the only report of the ability of diffusion chambers to sustain normoglycemia in a large animal model without any immunosuppression. Exogenous insulin requirements (▲) and fasting blood glucose concentrations (●) in two dogs before and after device implantation. Reprinted with permission.[6]

necrotic islets. Only a rim of islets remained viable within approximately 0.5–1 mm of the inner membrane wall. Similar results were obtained with canine islet implants into rats (Figure 15.5). These findings may also explain the surprisingly large number of islets required to achieve blood glucose control. Clearly, careful attention must be paid to the diffusion distances and transport properties of the membranes.

Host Sensitization

Although encapsulation systems serve to block uptake of antibodies and complement, the possibility must be considered that antigens released from the cell compartment could stimulate a host humoral response. Such antibodies could be induced by antigens shed from the cell surface, or by proteins secreted by live cells or liberated after cell death. This

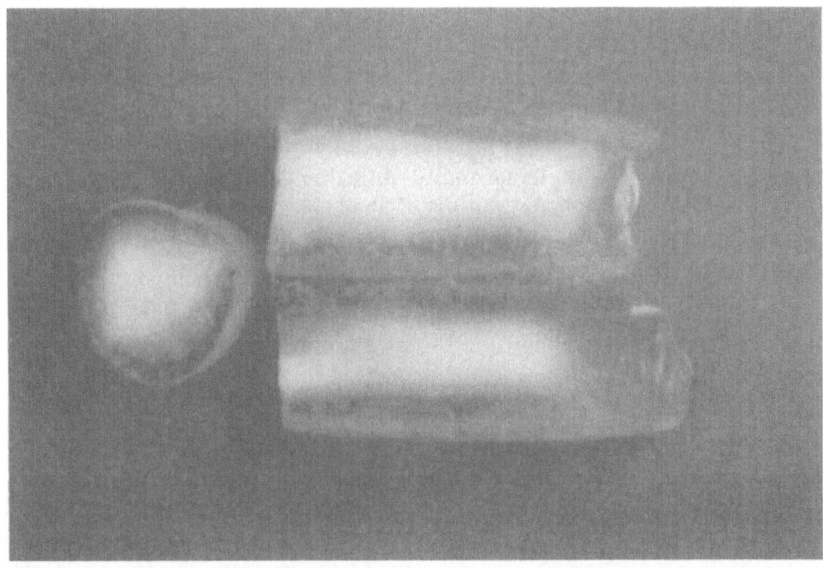

FIGURE 15.5. The problem of cell death as a result of oxygen supply limitations or accumulation of wastes or other cellular products is likely to be more severe in wider-bore diffusion chambers. Our observations with 4.8 mm i.d. tubes are consistent with this. PAN-PVC chambers retrieved from the peritoneal cavity of dogs several months after allotransplantation contained a central core of necrotic islets. Only a rim of islets remained viable within approximately 0.5–1 mm of the inner membrane wall. Similar results were obtained with canine islet implants into rats. The figure shows the islet-containing alginate matrix removed (and sliced open) from a wide-bore diffusion chamber retrieved 2 months after xenotransplantation into a rat. Clearly, careful attention must be paid to the diffusion distances and transport properties of the membranes.

could lead to an allergic response and/or immune complex disease. Studies have been performed that may help to rule out the possibility of serious immunologic sequelae in recipients.[37,38] Although these studies were carried out using islets encapsulated inside wide-bore tubes, the results may be reflective of other cell encapsulation systems as well, including microcapsules, devices anastomosed to the vascular system as AV shunts, hollow fibers, and other types of implantable diffusion chambers.

In these studies, wide-bore chambers containing porcine or bovine islets were implanted intraperitoneally into streptozotocin-induced diabetic rats. Sera were collected at various intervals and tested against isolated canine and porcine islets for tissue specificity and interspecies cross-reactivity by fluorescence immunocytochemistry. No immunofluorescence (or only weak background staining) was obtained when islets were exposed to horse sera, or to sera obtained prior to implantation of devices containing xenogeneic islets. Within 2–6 weeks,

however, the postimplantation sera showed strong immunoreactivity. The antibodies were found to be reactive to multiple tissues, and to possess little or no interspecies cross-reactivity.

The appearance of these xenoantibodies coincided with the appearance of circulating soluble immune complexes. However, none of the respiratory, cutaneous, or gastrointestinal manifestations that are characteristic of an anaphylactic reaction or of the diseases of immediate type hypersensitivity were observed, even following the intraperitoneal injection of additional naked islet tissue. Renal glomeruli did not stain for IgG or C3 in islet recipients. These results suggest that islet cell antigens crossed the membrane and stimulated antibody formation in the host, although they did not appear to cause renal or immune-complex disease during the course of this study.

In theory, the essential feature of immune-complex disease would be the demonstration of immune-complexes in tissues and biological fluids. However, the formation, fate, biologic activities, and pathologic potential of immune complexes depend on the nature of the antigens (size, valence, chemical composition, and magnitude and duration of exposure) and antibodies (affinity and reactivity with phlogogenic mediators) involved, the molar ratio and production rate of the two reactants, and the state of the host's phagocytes, both circulating leukocytes and the reticuloendothelial system.[39] In the experimental serum sickness model, >99% of the immune complexes formed are eliminated by phagocytes, predominantly by the liver's Kupffer cells, leaving <1% of the complexes capable of producing disease.[40] Overload of this system may be causative in precipitating disease.[41-44]

It is important to note that in diseases of immediate type hypersensitivity (or allergies), genetic factors exert an important influence on many specific immunoglobulin responses.[45] In humans, individual haplotypes have been shown to be associated with the magnitude of responsiveness to certain antigens,[46-48] and could predispose certain xenodevice-sensitized patients to an anaphylactic reaction following exposure to other xenogeneic proteins in the form of foods (for instance, the ingestion of pork). However, none of the clinical manifestations that are characteristic of the anaphylactic syndrome, or of the disease of immediate type hypersensitivity was observed in this study, even after reexposure to a significant quantity of islet material.

Clinical Trials

Diffusion chamber technology may be applicable to transplanting a wide variety of primary cells and bioengineered cells into patients. Human clinical trials have already been initiated, not only using allogeneic islets encapsulated in hollow-fiber membrane devices,[49] but also using tubular membrane chambers containing bovine adrenal chromaffin cells for the treatment of chronic pain[50,51] and cells releasing recombinant ciliary neurotrophic factor (CNTF) for the treatment of amyotrophic lateral sclerosis (ALS).[52]

Diabetes

Using PAN-PVC hollow fibers, Scharp et al[49] successfully allotransplanted human islets into patients with both Type I and Type II diabetes. Although the blood glucose concentrations of the patients remained unchanged during the 2-week implantation interval (only a small number of islets [between 150–200 islet equivalents] were implanted into each patient), this study clearly demonstrated that diffusion chambers can protect islets against both allogeneic immune rejection and against the autoimmune component of Type I diabetes. Islet viability was confirmed histologically, and ranged from approximately 70 to 95% at the end of the 2-week implantation period.

Chronic Pain

Preliminary clinical trials have also been initiated in Europe involving end-stage cancer patients suffering from intractable pain.[51] The patients received bovine chromaffin cells encapsulated inside a tethered tubular membrane chamber. The use of chromaffin cells in these studies was based on their ability to secrete natural analgesic substances such as catecholamines, epinephrine, and opioid peptides.[50] The chambers were implanted into either the subarachnoid space using a minimally invasive surgical procedure similar to a spinal tap or into the lateral ventricle of the brain. There were no serious complications, and 7 of 10 patients showed a decrease in narcotic intake and improved pain scores. The FDA has now approved phase I clinical trials in the United States.

ALS

In animal models of chronic degeneration, CNTF has been shown to prevent motor neuron loss.[52] In light of these results, baby hamster kidney (BHK) cells have been genetically engineered to produce CNTF. Encapsulation and subcutaneous implantation of transfected BHK cells in membrane chambers slowed down progressive motor neuropathy and reduced motoneuron death in a mouse model of ALS.[54] Using this approach, phase I clinical trials have been initiated in Europe using cells transfected with the gene for human CNTF. The investigators hope that low doses of CNTF will be effective when delivered to specific sites by implanting the chambers in the lumbar spine area, where the motor neurons degenerate in ALS patients.

References

1. Colton CK 1995. Implantable biohybrid artificial organs. Cell Transplant 4:415–436.
2. Lanza RP, Hayes JL, Chick WL. 1996. Encapsulated cell technology. Nature Biotechnology 14:1107–1111.
3. Mikos AG, Papadaki MG, Kouvroukoglou S, Ishaug SL, Thomson RC. 1994. Mini-review: Islet transplantation to create a bioartificial pancreas. Biotech Bioeng 43:673–677.
4. Zielinski BA, Goddard MB, Lysaght MJ. 1997. Immunoisolation. In: Principles of tissue engineering (Lanza RP, Langer R and Chick WL, eds) pp. 321–326, San Diego: Academic Press.
5. Lanza RL, Chick WL. 1997. Transplantation of encapsulated cells and tissues. Surgery 121:1–9.
6. Lanza RP, Borland KM, Lodge P, et al. 1992. Treatment of severely diabetic, pancreatectomized dogs using a diffusion-based hybrid pancreas. Diabetes 41:886.
7. Algire GH, Legallais FY. 1949. Recent developments in the transplant chamber technique as adapted to the mouse. J Natl Cancer Inst 10:225–253.
8. Algire GH, Weaver JM, Prehn RT. 1954. Growth of cells in vivo in diffusion chambers. I. Survival of homografts in immunized mice. J Natl Cancer Inst 15:493–507.
9. Prehn RT, Weaver JM, Algire GH. 1954. The diffusion-chamber technique applied to a study of the nature of homograft resistance. J Natl Cancer Inst 15:509–17.
10. Garvey JF, Morris PJ, Finch DR, et al. 1979. Experimental pancreas transplantation. Lancet 971–972.
11. Theodorou NA, Vrbova H, Tyhurst M, Howell SL. 1980. Problems in the use of polycarbonate diffusion chambers for syngeneic pancreatic islet transplantation in rats. Diabetologia 18:313–317.
12. Strautz RL. 1970. Studies of hereditary-obese mice (ob/ob) after implantation of pancreatic islets in Millipore filter capsules. Diabetologia 6:306–312.
13. Weber C, Weil R, McIntosh R, et al. 1975. Xenotransplantation of porcine islets into hyperglycemic rats. Surgery 77:208–215.
14. Buschard K. 1975. Cultivation of islets of Langerhans in Millipore chambers in vivo. Horm Metab Res 7:441–442.
15. Jolley WB, Hinshaw DB, Call TW, Alverel LS. 1977. Xenogeneic pancreatic islet transplantation in proteolytic enzyme bonded diffusion chambers in diabetic rats. Transplant Proc 9:363–365.
16. Andersson A. 1979. Survival of pancreatic islet allografts. Lancet 2:585.
17. Valente U, Ferro M, Campisi C, et al. 1981. Islet transplantation by means of artificial membrane chambers in diabetic recipients. Artif Organs 5(Suppl):780–783.
18. Theodorou NA, Howell SL. 1979. An assessment of diffusion chambers for use in pancreatic islet cell transplantation. Transplantation 27:350–352.
19. Brauker JH, Carr-Brendel VE, Martinson LA, et al. 1995. Neovascularization of synthetic membranes directed by membrane microarchitecture. J Biomed Mat Res 29:1517.
20. Brauker J, Martinson LA, Young SK, and Johnson RC. 1996. Local inflammatory response around diffusion chambers containing xenografts. Transplantation 61:1671–1677.
21. Dionne K, Scharp D, Lysaght MJ, Hegre O, Lacy P. 1994 Macroencapsulation of islets for the treatment of diabetes. In: Pancreatic islet transplantation Volume III: Immunoisolation of pancreatic islets (Lanza RP and Chick WL, eds) pp. 119–131, Austin, Texas: RG Landes Company.
22. Suzuki K, Bonner-Weir S, Hollister J, Weir GC. 1996. A method for estimating number and mass of islets transplanted within a membrane device. Cell Transplant 5:613–625.
23. Loudovaris T, Mandel TE, Charlton B. 1996. CD4+ T cell mediated destruction of xenografts within cell-impermeable membranes in the absence of CD8+ T cells and B cells. Transplantation 61:1678–1684.
24. Woodward CL. 1982. How fibroblasts and giant cells encapsulate implants: Considerations in design of glucose sensors. Diabetes Care 5:278.
25. Altman JJ, Houlbert D, Bruzzo F, et al. 1982. Implantation of semipermeable hollow fibers to prevent immune rejection of transplanted pancreatic islets. In: Islet-pancreas-transplantation and artificial pancreas (Federlin K, Pfeiffer E, Raptis S, eds) New York: Thieme-Stratton.
26. Altman JJ, Houlbert D, Callard P, et al. 1986. Long-term plasma glucose normalization in experimental diabetic rats with microencapsulated implants of benign human insulinomas. Diabetes 35:625.
27. Archer J, Kaye R. Matter G. 1980. Control of streptozotocin diabetes in Chinese hamsters by cultured mouse islet cells without immunosuppression. J Surg Res 28:77.
28. Icard P, Penfornis F, Gotheil C, et al 1990. Tissue reaction to implanted bioartificial pancreas in pigs. Transplant Proc 22:724.
29. Lanza RP, Butler DH, Borland KM et al 1991. Xenotransplantation of canine, bovine, and porcine islets in diabetic rats without immunosuppression. Proc Natl Acad Sci USA 88:11100.
30. Lanza RP, Butler DH, Borland KM, et al 1992. Successful xenotransplantation of a diffusion-based biohybrid artificial pancreas: A study using canine, bovine, and porcine islets. Transplant Proc 24:669.
31. Lanza RP, Borland KM. Staruk JE, et al 1992. Transplantation of encapsulated canine islets into

spontaneously diabetic BB/Wor rats without immunosuppression. Endocrinology 131:637.

32. Lanza RP, Beyer AM, Staruk JE, Chick WL. 1993. Biohybrid artificial pancreas: Longterm function of discordant islet xenografts in streptozotocin diabetic rats. Transplantation 56:1067.

33. Theodorou NA, Vrbova H, Tyhurst M, et al 1979. An assessment of diffusion chambers for use in pancreatic islet cell transplantation. Transplantation 27:350.

34. Andersson A. Survival of pancreatic islet allografts 1979. Lancet 2:585.

35. Kraegen EW, Chisholm DJ, MacNamara ME. 1981. Timing of insulin delivery with meals. Horm Metab Res 13:365.

36. Lanza RP, Lodge P, Borland KM, et al 1993. Transplantation of islet allografts using a diffusion-based biohybrid artificial pancreas: long-term studies in diabetic, pancreatectomized dogs. Transplant Proc 25:978.

37. Lanza RP, Beyer AM, Chick WL. 1994. Xenogeneic humoral responses to islets transplanted in biohybrid diffusion chambers. Transplantation 57:1371.

38. Lanza RP, Kühtreiber WM, Beyer A, et al 1994. Humoral response to encapsulated islets. Transplant Proc 26:3346.

39. Theofilopoulos AN, Dixon FJ. 1979. The biology and detection of immune complexes. Adv Immunol 28:89.

40. Wilson CB, Dixon FJ. 1971. Quantification of acute and chronic serum sickness in the rabbit. J Exp Med 134:7S.

41. Benacerraf B, Sebestyen M, Cooper HS. 1959. The clearance of antigen-antibody complexes from the blood by the reticulo-endothelial system. J Immunol 82:131.

42. Cochrane CG, Koffler D. 1973. Immune complex disease in experimental animals and man. Adv Immunol 16:185.

43. Kijlstra A, Van Der Lelij A, Knutson DW, et al 1978. The influence of phagocyte function on glomerular localization of aggregated IgG in rats. Clin Exp Immunol 32:207.

44. Haakenstad AO, Mannik M. 1974. Saturation of the reticuloendothelial system with soluble immune complexes. J Immunol 112:1939.

45. Austen KF. 1991. Diseases of immediate type hypersensitivity. In: Harrison's principles of internal medicine, 12th edition (Wilson JD, Braunwald KJ, Isselbacher KJ, eds) New York: McGraw-Hill, p. 1422.

46. Sasazuki T, Kohno Y, Iwamoto I, et al 1978. Association between an HLA haplotype and low responsiveness to tetanus toxiod in man. Nature 272:359.

47. de Vries RP, Kreeftenberg HG, Loggen HG, et al 1977. In vitro immune responsiveness to vaccinea virus and HLA. N Engl J Med 297:692.

48. Christiansen FT, Hawkins BR, Dawkins RL. 1978. Immune function in ankylosing spondylitics and their relatives: influence of disease and HLA B27. Clin Exp Immunol 33:270.

49. Scharp DW, Swanson CJ, Olack BJ, et al 1994. Protection of encapsulated human islets implanted without immunosuppression in patients with type I or type II diabetes and in nondiabetic control subjects. Diabetes 43:1167–1170.

50. Sagen J. 1996. Chromaffin cell transplantation. In: 1996/97 Yearbook of Cell and Tissue Transplantation (Lanza RP, Chick WL, eds) Dordrecht, The Netherlands: Kluwer Academic Press.

51. Aebischer P, Buchser E, Joseph JM, et al 1994. Transplantation in humans of encapsulated xenogeneic cells without immunosuppression. Transplantation 58:1275–7.

52. Ezzell C. 1995. Tissue engineering and the human body shop: encapsulated-cell transplants enter the clinic. J NIH Research 7:47–51.

53. Sendtner M, Schmalbruch H, Stockli KA, et al 1992. Ciliary neurotrophic factor prevents degeneration of motor neurons in mouse mutant neuropathy. Nature 358:502–4.

54. Sagot Y, Tan SA, Baetge E, et al 1995. Polymer encapsulated cell lines genetically engineered to release ciliary neurotrophic factor can slow down progressive motor neuropathy in the mouse. Eur J Neurosci 7:1313–22.

16
Vascular Devices

Takashi Maki, Anthony P. Monaco, Claudy J.P. Mullon, and Barry A. Solomon

Introduction

Our group at the Beth Israel Deaconess Medical Center (formerly Deaconess Hospital) and a team at Circe Biomedical, Inc. (formerly Grace Biomedical, *WR Grace and Co.-Conn*) have studied the use of a vascularized bioartificial pancreas (VBAP) for treatment of diabetes for the past 6 years.[1-4] We were the first to demonstrate the long-term use of a bioartificial pancreas in large animals, and our data represented the longest survival of discordant xenogeneic islets. In this report we will briefly describe the VBAP device, a large animal diabetes model, experimental results with the use of allogeneic and xenogeneic islets in the device, and plans to initiate a clinical trial of the VBAP containing porcine islets for treatment of patients with early-stage IDDM.

The Vascularlized Bioartificial Pancreas (VBAP)

The VBAP device consists of a single coiled hollow-fiber membrane contained within a disc-shaped acrylic housing that provides a compartment for the islet suspension surrounding the membrane (Fig. 16.1). The specially prepared membrane has a wall thickness of 120 to 140 μm and is fabricated from an acrylic copolymer material using a phase inversion spinning technique.[5] The resulting coiled ultrafiltration membrane tube has a nominal internal diameter of 6.0 mm and a molecular weight cut-off of 70,000 daltons, which excludes effector lym-phocytes and immunoglobulins. The membrane tube is connected to commercially available expanded polytetrafluoroethylene (e-PTFE) vascular grafts. These grafts match the inner diameter of the coiled device membrane to which they are connected. The islets are suspended in 2% agar and seeded into an annular space outside the tubular membrane through a seeding port. The vascular grafts are anastomosed directly to the host blood vessels such that the device acts as an arteriovenous shunt, and arterial blood perfuses the device. Thus, the VBAP has the advantage of a high oxygen tension of the arterial blood exposed to the islets unlike the microcapsules and diffusion chambers, whose islets usually are exposed to a low oxygen tension in the abdominal cavity. The VBAP device also has the advantage of easy retrievability. Replacement of nonfunctioning islets with freshly prepared islets may be feasible with development of a new islet suspension matrix that can be easily retrieved from the device. When the VBAP devices were implanted in normal dogs with daily low-dose (75 mg) aspirin therapy, patency was lost after 9, 16, 21, 26, and 42 months, while three devices were removed patent after 18, 42, and 47 months.[6] This finding is remarkable because the dog is considered to be a hypercoagulable model and therefore a severe test of device patency.[7] The possible limitations of the VBAP are all associated with the creation of an arteriovenous shunt, i.e., the requirement of major surgery, possible vascular thrombosis, and a potential for cardiac stress and diversion of blood from the distal extremity with a peripheral "steal" syndrome caused by a significant volume of shunted blood.

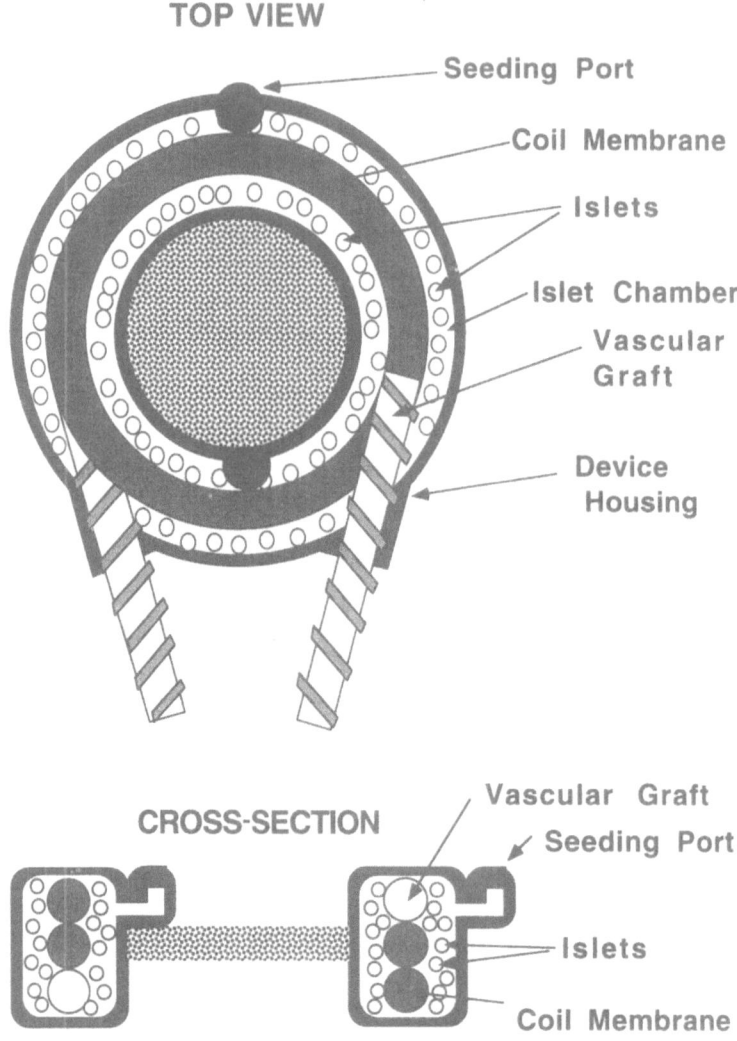

FIGURE 16.1. Schematic diagram of the vascularized bioartificial pancreas (VBAP).

A Large Animal Diabetes Model

VBAP devices were tested in a dog model of diabetes for their functional and therapeutic efficacy. The canine species was chosen for a number of reasons: (1) it is an acceptable and well-established laboratory large animal model, (2) its diet and metabolism resemble those of a human more closely than those of other models, (3) it can be consistently rendered diabetic surgically, avoiding the difficulty and complications associated with chemical induction by alloxan or streptozocin, and (4) it

is more cost efficient than nonhuman primate models. However, the canine species possesses a number of shortfalls; (1) it is hypercoagulable, making hemocompatibility and long-term performance evaluation difficult to carry out, especially in the presence of diabetes, which has been reported to enhance clotting,[8] and (2) pancreatectomy-induced diabetes is severely brittle and associated with loss of both endocrine and exocrine pancreas function.

Dogs were rendered diabetic by a surgical total pancreatectomy (>95%) under general anesthesia.

Following pancreatectomy, multivitamins and pancreatic enzyme were given daily, mixed with food. Induction of hyperglycemia was highly reproducible; of 147 total pancreatectomies, only 3 dogs (2.0%) failed to become diabetic, probably as a result of the presence of ectopic pancreases. Hyperglycemia (>19.4 mmol/L), which usually developed within 2 weeks after pancreatectomy, was brittle in nature with wide day-to-day changes despite a large amount of daily exogenous insulin (30–40 units/day[9]). Diabetic dogs were used for implantation of vascularized bioartificial devices 2–7 weeks after pancreatectomy.

VBAP Containing Allogeneic Canine Islets

Our groups was the first to demonstrate the long-term use of a bioartificial pancreas in large animals. A prototype device (a membrane length of 30 cm and an islet compartment volume of 7.1 ml) was seeded with canine islets and implanted in the lower abdominal cavity 1–3. Because a single device in a recipient dog failed to achieve consistent glycemic control, two devices were implanted in each recipient dog in an effort to increase insulin delivery to the recipient. As shown in Table 16.1, the VBAP devices controlled severe diabetes for up to 1 year with zero or minimal exogenous insulin. No immunosuppression was used at any time, thus confirming the immune exclusive nature of the device. Blood glucose levels after oral or intravenous administration of glucose challenge were improved greatly compared to those in preimplant diabetic dogs. In two dogs, elective removal of the devices after one year led to a rapid increase in fasting blood sugar levels despite an increased exogenous insulin dose. In vitro perfusion of one of the removed devices produced significant insulin output over a 2-week period, indicating the functional viability of islets after 1 year in the device. Histological examination revealed no gross fibrosis

TABLE 16.1. The FBG and exogenous insulin requirement prior to and after implantation of two VBAP devices seeded with allogeneic canine islets.

Animal	Preimplantation[a]		Postimplantation[b]		Period of Function (post-impl day)	Duration of Implant[c] (days)	Cause of Termination
	FBG (mmol/L)	Insulin (units/day)	FBG (mmol/L)	Insulin (units/day)			
PS22	12.3 ± 2.1	18	7.1 ± 0.2	1	1–262	284	Loss of function[d]
PS23	17.1 ± 1.8	30	8.4 ± 0.4	0	1–209	209	Ileus
PS24	11.6 ± 2.6	27	9.3 ± 0.8	8 ± 1	9–42	64	Loss of function
PS26	24.0 ± 2.8	28	10.2 ± 0.5	4 ± 1	1–167	237	Loss of function
PS28	20.0 ± 0.9	23 ± 1	8.5 ± 0.5	1	6–50	51	BW loss[e]
PS30	16.5 ± 4.3	16	7.0 ± 0.4	0	1–27	28	BW loss
PS35	12.3 ± 3.7	30 ± 1	12.5 ± 1.9	13 ± 1	8–26	28	Loss of function
PS36	10.7 ± 4.4	32 ± 1	8.3 ± 0.5	2	1–99	99	BW loss
PS48	6.7 ± 0.3	22	9.0 ± 0.5	4	1–70	90	Thrombosis
PS49	7.8 ± 1.8	21	6.7 ± 0.6	0	1–25	26	BW loss
PS50	11.4 ± 3.4	22	7.7 ± 0.3	1	1–188	191	Loss of function
PS56	7.1 ± 1.5	31 ± 1	8.0 ± 0.2	5	5–373	373	(Elective)[f]
IP1	—	—	7.3 ± 0.2	3	1–370	370	(Elective)

[a] Average FBG and exogenous insulin requirement 1 week before device implantation. All the results are expressed as mean ± SEM.

[b] Average FBG and exogenous insulin requirement during the period of function. The period of function is defined as a period during which time the FBG levels were consistently maintained below 250 mg/dl with less than 50% of preimplantation exogenous insulin dosage. The first and the last day of the period are shown.

[c] The postimplantation day when the device was removed.

[d] Increase of exogenous insulin dose to more than 50% of preimplantation dosage with patient devices was considered as loss of islet function.

[e] The experiment was terminated when a recipient animal lost more than 15% of prepancreatectomy body weight regardless of the cause.

[f] The experiment was terminated electively with a patent and functioning device.

throughout the membrane, except for occasional foci of thin layers of fibrin-like material adhered to the lumenal surface of the membrane. A varying amount of beta cell amylin (islet amyloid polypeptide) was noted in the majority of the islets, suggesting beta cell exhaustion.

VBAP Containing Xenogeneic Porcine Islets

We have extended our studies to the use of islet xenografts in the VBAP[4]. Porcine islets were selected for potential clinical application because donors are readily available, being a food source, and porcine insulin has been used clinically to treat diabetes. Islets were isolated from market weight Yorkshire pigs by a standard collagenase digestion and Ficoll gradient separation method. In a series of islet isolations, a donor pancreas (~90 g) yielded approximately 3×10^5 islet equivalents (IE) (or 10^6 islets of 85 μm in diameter) with >90% purity and viability. When isolated porcine islets (5000 IE/mouse) were transplanted into the renal subcapsular space of streptozocin-induced diabetic nude mice, mice remained normoglycemic for >1 year, with progressive weight gain, high circulating porcine C-peptide values, and rapid glucose clearance rate in intravenous glucose tolerance tests.[10] For the use of porcine islets, the VBAP devices were modified to contain a longer semipermeable membrane coil and/or a larger islet compartment volume than the prototype, in an attempt to increase a total insulin secretory capacity. A single device containing 160,000–430,000 IE per device was implanted into each totally pancreatectomized diabetic recipient. Again no immunosuppression was used.

Our data represent the longest survival of discordant xenogeneic islets and showed that porcine islet function could be supported by the VBAP for over 9 months in pancreatectomized diabetic dogs without immunosuppression. As summarized in Table 16.2, 9 of 17 dogs maintained FBG levels at below 11 mmol/L for 53 to 263 days with a 54–74% reduction in exogenous insulin requirements. Although complete elimination of exogenous insulin therapy was not accomplished, the improvement achieved by a single device with small doses of exogenous insulin is significant consider-

TABLE 16.2. The FBG and exogenous insulin requirement prior to and after implantation of a single VBAP device seeded with xenogeneic porcine islets.

Animal[a]	Pre-Implantation[b]		Post-Implantation[c]		Period of Function (post-impl day)	Plasma Porcine C-Peptide[d] (ng/ml)	Duration of Implant[e] (days)	Cause of Termination[f]
	FBG (mmol/L)	Insulin (units/day)	FBG (mmol/L)	Insulin (units/day)				
GP10	21.0 ± 2.3	35	10.4 ± 0.5	16	13–127	0.12 ± 0.02	144	LF
GP18	26.6 ± 0.8	39	9.9 ± 0.3	12	7–270	0.21 ± 0.02	271	RB
GP19	21.6 ± 1.8	40	11.3 ± 0.3	19	11–229	0.28 ± 0.01	233	Th
GP14	16.4 ± 4.1	31	4.9 ± 0.2	6 ± 1	1–56	0.45 ± 0.08	57	Th
GP23	22.9 ± 0.8	38	8.7 ± 0.5	16 ± 1	13–71	0.49 ± 0.04	90	Th
GP24	22.2 ± 0.9	36	9.9 ± 0.6	8	9–91	0.37 ± 0.06	157	Th
GP25	21.0 ± 0.8	44	12.9 ± 0.6	14 ± 1	9–73	0.47 ± 0.05	84	LF
GP27	26.8 ± 0.7	44	7.0 ± 0.5	13 ± 1	11–64	0.73 ± 0.07	77	Th
GP28	19.1 ± 2.3	35	7.5 ± 0.4	4	1–78	0.55 ± 0.04	> 366[g]	

[a] Total of 17 dogs were implanted with a single porcine islet seeded device. Eight dogs (not shown above) lost devices due to device nonfunction, early thrombosis, and recipient bowel obstruction.

[b] Average FBG and exogenous insulin requirement 1 week before device implantation. All the results are expressed as mean ± SEM.

[c] Average FBG and exogenous insulin requirement during the period of function. The period of function is defined as a period during which time the FBG levels were consistently maintained below 13.9 mmol/L. The first and the last day of the period are shown.

[d] Average plasma porcine C-peptide values during the period of function.

[e] The postimplantation day when the device was removed.

[f] LF, Loss of function; Th, thrombosis, RB, reduced bruit.

[g] Coumadin (6mg) was given daily instead of aspirin after 89 days postimplantation.

ing the very severe and brittle nature of the diabetes induced by total pancreatectomy in the dog. Significant porcine C-peptide detected in the dogs' sera confirmed that the device was producing porcine insulin over the period of implantation. Two dogs achieved good glycemic control for more than 8 months, with improved glucose tolerance tests and a significant time-dependent reduction in hemoglobin A1c levels. Histological evaluation revealed no evidence of immune response (i.e., lymphocytic infiltration). Although a high titer of naturally occurring antiporcine cytotoxic antibody was found in the sera of recipient dogs prior to and after the device implantation, survival of porcine islets contained in the devices was not affected. There was no significant increase in cytotoxic activity of the sera during long-term implantation of the device. Although late vascular thrombosis was seen more frequently in these studies than with the use of allogeneic islets, whether this is attributable to the use of xenogeneic porcine islets and/or to the device modification is not known at present.

Plans for Clinical Phase I/II Trial of VBAP Containing Porcine Islets

There is a group of diabetic patients who experience frequent, severe episodes of hypoglycemia requiring assistance. These patients with "hypoglycemic unawareness" cannot live alone because of the unanticipated and sudden onset of hypoglycemic episodes. Patients who experience sever, injury-related hypoglycemia and hypoglycemic unawareness are not considered candidates for intensive insulin therapy, such as the regimen used in the Diabetes Control and Complications Trial[11] At the present time, a pancreas transplant is the only potential therapy available to this group of patients. Unfortunately, the pancreas transplanted alone in nonuremic diabetic patients has a greater tendency to be rejected, in spite of high-dose immunosuppression, than the pancreas simultaneously transplanted with kidney in diabetic patients with chronic renal failure[12].

Results of our large animal studies provided evidence that the VBAP achieves long-term glycemic control without chronic immunosuppression. The use of porcine islets in the VBAP also alleviates the shortage of human donor organs. The surgical implantation of the VBAP will be considerably less complicated than a pancreas transplant, principally because no provision need be made for the exocrine function of the normal pancreas. These observations have provided reasonable support to justify plans for a clinical phase I/II trial to test the tolerability, safety, and function of the VBAP containing porcine islets in the subset of IDDM patients who suffer from severe hypoglycemic events and whose only therapy today is solitary pancreas transplantation. Initially, six patients will be included in the study, who have a history of severe hypoglycemia despite assiduous attempts at diabetes control by a variety of insulin regimens and nutritional counseling. Patients who have any transplanted organ and who have developed serious vascular, neurological, and nephological complications will be excluded from the study. The VBAP device that will be used in the clinical trial will be the same as that used in the xenograft experiments. The device is 7.2 cm in diameter by 3 cm in height, containing a 45 cm semipermeable membrane tube and a 12 ml islet compartment, and weighs 78 g (empty) or approximately 100 g (all spaces filled). The device will be implanted at the iliac fossa with vascular anastomoses in a fashion similar to renal transplantation. No immunosuppression will be used at any time points. The patients will be given 325 mg/day aspirin for prevention of clotting. Following implantation, the patients will be monitored for changes in the frequency of hypoglycemic events, frequency of insulin injections, and total daily insulin requirements. Frequent monitoring of clinical and laboratory parameters will be done to assess changes in the patients' cardiovascular and pulmonary functions as well as any changes in overall clinical condition. The vascular patency of the device will be carefully monitored. Although this will be a phase I/II study, frequent monitoring of the function of the VBAP will also be done. Reduction of exogenous insulin requirements, less frequent hypoglycemic episodes, and/or the presence of porcine C-peptide will be considered as presumptive evidence of device function. The patients will be followed for as long as the device continues to function. The device will be removed if thrombosis or any other adverse event occurs.

References

1. Maki T, Ubhi CS, Sanchez-Farpon H, Sullivan SJ, Borland K, Muller TE, Solomon BA, Chick WL, Monanco AP. 1991. Successful treatment of diabetes with the biohybrid artificial pancreas in dogs. Transplantation 51:43.

2. Sullivan SJ, Maki T, Borland KM, Mahoney MD, Solomon BA, Muller TE, Monaco AP, Chick WL. 1991. The biohybrid artificial pancreas: Long-term implantation studies in diabetic, pancreatectomized dogs. Science 252:718.

3. Maki T, Lodge JPA, Caretta M, Ohzato H, Borland KM, Sullivan SJ, Staruk J, Solomon BA, Chick WI, Monaco AP. 1993. Treatment of severe diabetes mellitus for more than one year using a vascularized hybrid artificial pancreas. Transplantation 55:713.

4. Maki T, Otsu I, O'Neil JJ, Dunleavy K, Mullon CJP, Solomon BA, Monaco AP. 1996. Treatment of diabetes by xenogeneic islets without immunosuppression: Use of a vascularized bioartificial pancreas. Diabetes 45:342.

5. Michaels AS. 1971. High flow membrane, US parent #3,615,024, October 26.

6. Maki T, Mullon CJP, Solomon BA Monaco AP. 1995. Novel delivery of pancreatic islet cells to treat insulin-dependent diabetes mellitus. Clin Pharmacokinet 28:471.

7. Dewanjee MK, Kapadvanjwala M, Sanchez A, Elson R, Serafini AN, Zilleruelo GE, Sfankianakis GN. 1992. Quantitation of comparative thrombogenicity of dog, pig, and human platelets in a hemodialyzer. ASAIO J 38:88.

8. McMillan DE. 1992. Clotting disorders in diabetes. In: Alberti KGMM, DeFonzo RA, Keen H, Zimmer P, eds. International textbook of diabetes mellitus. Chichester: John Wiley & Sons Ltd., 1447.

9. Maki T, Monanco AP, Mullon CJP, Solomon BA. 1996. Early treatment of diabetes with porcine islets in a bioartificial pancreas. Tissue Engineering 2:299.

10. Maki T, O'Neil JJ, Porter J, Nicholson D, Mullon CJP, Solomon BA, Monanco AP. 1996. Porcine islets for xenotransplantation. Transplantation 62:136.

11. The Diabetes Control and Complications Trial Research Group. 1993. The effect of intensive treatment of diabetes on the development and progression of long-term complications in insulin-dependent diabetes mellitus. New Engl J Med 329:977.

12. Sutherland DER, Gruessner A, Moundry-Munns K. 1992. Analysis of United Network for Organ Sharing (UNOS) United States of America (USA) pancreas transplant registry data according to multiple variables. In: Terasaki P, Cecka JM, eds. Clinical transplants. UCLA Tissue Typing Laboratory, California p 45.

17
Macrocapsules Based on Ultrafiltration and Diffusion

Krystyna Burczak and Yoshito Ikada

The recognition that living islets of Langerhans from one species, enclosed within a synthetic permselective immunoisolating membrane, could be transferred to another organism and remain viable and, at the same time, functionally active (Archer et al 1980) has given great impetus to the development of different bioartificial pancreatic systems (Scharp et al 1984). They are expected to overcome the immune rejection problem which currently limits transplantation of pancreatic tissue for the treatment of insulin-dependent diabetes mellitus.

Similar to microencapsulation, macroencapsulation of islets of Langerhans in large-sized membranes impermeable to components of the recipient's immune system has the potential to solve the problems of poor availability and strong rejection of islets in the treatment of insulin-dependent diabetes with natural organ replacement. Also, macroencapsulation enables us to use xenogeneic tissue, such as porcine islets and insulin-secreting clonal cell lines provided by genetic manipulation techniques, to overcome the problem of the shortage of transplantable human pancreatic tissue. The greatest advantage of devices based on macrocapsules over those based on microcapsules is that the macrocapsules can be readily retrieved from the implanted site if something abnormal has happened in the patient carrying the device. In contrast, it is too difficult to remove all the implanted microcapsules from the patient's body because of the tremendous number of implanted microcapsules.

Two kinds of devices based on macrocapsules have been reported for encapsulation of islets of Langerhans. One is fabricated from long hollow fibers and the other from hydrogel sheets. The solute permeation is effected through ultrafiltration for hollow fibers, while hydrogels allow solutes to permeate by diffusion.

The aim of this chapter is to review the potential, feasibility, and efficacy of the two systems that have been studied for the replacement of the biological function of the pancreas.

Hollow Fiber Macrocapsules

Introduction

The development of bioartificial pancreas (BAP) based on hollow fibers has its origin in the work of Chick et al (1975). They studied in vitro culture of neonatal rat pancreatic beta cells on the external surface of bundled hollow fibers sealed within a chamber, and perfused the cells with tissue culture medium. The encapsulated cells continued to synthesize and release insulin in response to changes in glucose concentration in the perfusion medium. Many later ex vivo and in vivo studies have indicated the high potential of this approach to restore short-term normoglycemia in chemically or surgically induced diabetic animals (Colton and Avgoustiniatos 1991, Lanza et al 1992).

Device Function

The principle of BAP devices fabricated from hollow fibers is the same as with other extravascular systems, that is, to protect islets of Langerhans from immune rejection by the host through their physical isolation by a membrane permeable to glucose and insulin, but impermeable to the humoral

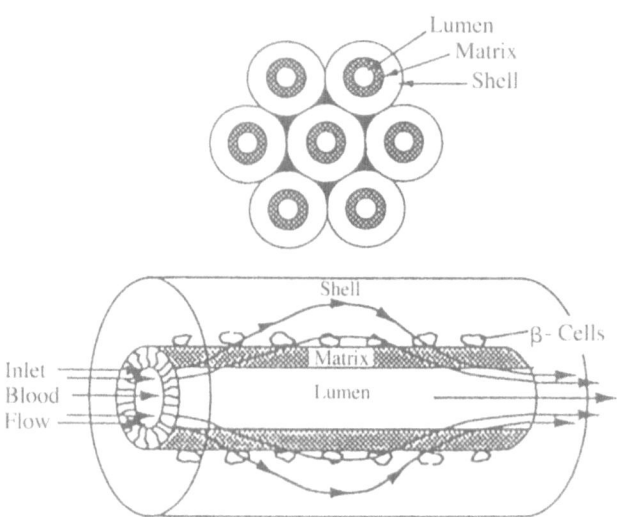

FIGURE 17.1. Bottom panel: Cross-sectional view of a single hollow fiber. Beta cells are located at the matrix-shell interface. Convective flow stream lines show typical recirculation.

Top panel: Schematic of a hexagonal array of hollow fibers.

(Reprinted with the permission of Lippincott-Raven Publishers from Pillarella MR, Zydney AL. 1990 a. Trans Am Soc Artif Intern Organs 36: M716.)

and cellular components of the host's immune system. However, in comparison to systems relying on a pure diffusion process, such as extravascular diffusion chambers, the devices based on ultrafiltration can significantly reduce the overall response time after glucose challenge because of a direct, rapid exchange between the bloodstream and the islet graft across the semipermeable membrane. The blood circulating through the lumen of the intravascular device serves as the culture medium, providing oxygen and nutrients to the immunoisolated tissue.

A BAP based on ultrafiltration consists of either a single hollow-fiber semipermeable membrane or a bundle of parallel fibers (Figure 17.1). The insulin-producing islets can be cultured in the membrane matrix or in the region surrounding the fiber(s) (shell), which is enclosed in a rigid plastic housing. The device is connected to the blood circulation via an arteriovenous (AV) shunt. Blood flow through the membrane lumen is driven by the physiological blood pressure drop between artery and vein, approximately 90 mm Hg. The unidirectional blood flow causes the pressure in the lumen of the first part of the hollow fiber to be greater than in the islet compartment, and an ultrafiltration flux is generated from the bloodstream toward the islet compartment. When the hydrostatic pressure becomes lower in the second half of the fiber than in the islet compartment, an equal flux occurs in the reverse direction, from this compartment toward the bloodstream. The blood is ultrafiltered by the hollow fiber membrane and stimulates the islets to release insulin in response to the glucose concentration in the blood ultrafiltrate. Perfusion experiments with islets of Langerhans enclosed in a Millipore chamber were performed using the blood ultrafiltrate produced in vivo from the blood of normal rats, and they verified that blood ultrafiltrate could acutely support the normal islet beta cell function (Reach et al 1982).

What is expected from the devices based on ultrafiltration is not only a sufficient supply of nutrients, oxygen, and glucose directly from the blood, but also advantageous rapid exposure of islets to any changes in blood glucose concentration so as to take advantage of their glucose-stimulated insulin secretory feature and thereby achieve a physiological feedback control system of blood glucose concentration. If sufficiently fast transmission of glucose signal and outflow of insulin from the islet compartment are provided, the device will function as a closed-loop insulin delivery system, in which insulin secretion by islets can be regulated minute by minute by the concomitant blood glucose concentration, and the secreted insulin will regulate the blood glucose level. Figure 17.2 illustrates a device-function-dependent time scale of the feedback control of blood glucose concentration by the encapsulated islet tissue. In this figure the line G_B represents the change in blood glucose concentration from a basal level G_o, after oral glucose loading. The line G_i represents the change in glucose concentration within the islet compartment, and G_{ti} is a threshold value of the blood glucose level that stimulates the beta cells to secrete insulin under normal condi-

FIGURE 17.2. Schematic representation of the changes in blood glucose concentration (G_B) and in glucose concentration within the islet chamber (G_i) from a basal (postabsorptive) value (G_o). G_{ti} is the glucose concentration that triggers insulin release. G_{tbp} is the blood glucose concentration that is reached when insulin release from the bioartificial pancreas begins to increase; it therefore represents the apparent glucose threshold of the device. Left panel: long lag time in the transmission of the glucose signal from the blood to the islet compartment; the system cannot regulate blood glucose (failure). Right panel: short lag time is a necessary condition for the successful functioning of the system (i.e., G_{tbp} within the physiological range, represented by the hatched area).

(Reprinted with the permission of ISAO Press from Reach G et al. 1984 b. Progress in Artificial Organs: 1983. 766.)

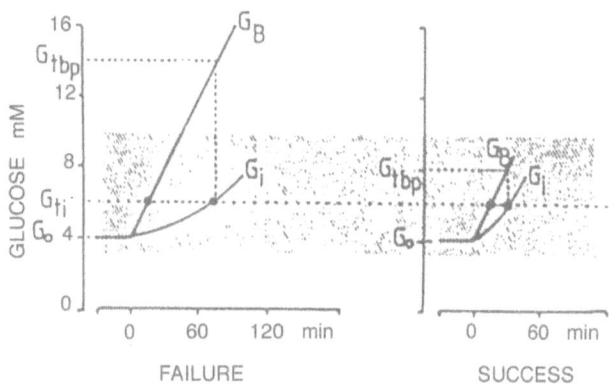

tions. In properly projected devices the lag time, i.e. the time necessary to reach the G_{ti} value after glucose loading, is short enough, and triggering of insulin secretion will occur while the blood glucose concentration G_{tbp} is still in the physiological range.

Mass Transfer and Dynamic Response

Most of the extensive studies with implanted ex vivo and in vivo hollow-fiber devices showed their high feasibility in long-term islet immunoisolation (Colton and Avgoustiniatos 1991). The restoration and maintenance of normoglycemia were observed for short periods of time in chemically or surgically induced diabetic rats, dogs, and monkeys (Chick et al 1977, Sun et al 1977, Tze et al 1980) with the short response time required to attain normoglycemic blood levels. The restoration and maintenance of normoglycemia are related to the size of the diabetic animal and to the way the device is connected to the blood circulation, as either an arteriovenous or arterio-arterio (AA) shunt (Sun et al 1987). In parallel with those extensive experimental studies, theoretical models for glucose-insulin kinetics were elaborated, aiming at analysis of available experimental data and identification of the important parameters influencing the dynamic response as well as guiding the design of intravascular devices that would provide more precise matching of insulin secretion by islets to changes

in glucose concentration. Initially it was assumed that diffusion governs the mass transfer of glucose and consequently the dynamic response of intravascular devices, since some in vitro and in vivo experiments with 2.7 mm i.d. tubular membranes as AV shunts demonstrated a delayed change in glucose-stimulated insulin secretion (Colton and Avgoustiniatos 1991). Theoretical studies on the model of diffusion and secretion in devices allowed the identification of the important parameters influencing their dynamic response, such as diffusion distance in the islet compartment, membrane matrix thickness, and flow rate of circulating medium (Catapano et al 1990, Sparks et al 1982, Sparks et al 1983, Weinless and Colton 1983). Since the work of Colton et al (1980), it has been known that a minute ultrafiltration-reabsorption cycle is present in the closed islet compartment surrounding the hollow fiber membrane. Moreover, the aforementioned effects of diabetic animal size and the way the device is connected to the circulation seemed to indicate that, in addition to diffusion, a convective mode (ultrafiltration) contributes to the overall mass transfer of glucose and insulin (Sun et al 1985). Reach et al (1981) were the first investigators who attempted to make use of convective transport to improve the dynamic response of the projected intravascular BAP. With respect to the model of the hollow-fiber device (Vitafiber Unit, 3 × 50, Amicon Corporation), a

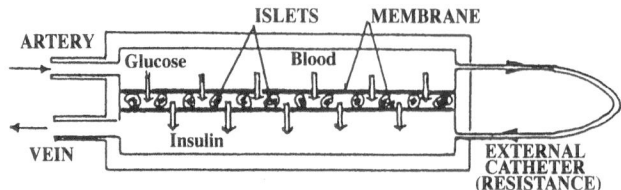

FIGURE 17.3. Scheme of a bioartificial pancreas with U-shaped blood channel.

(Reprinted with the permission of The American Society of Mechanical Engineers from Jaffrin MY et al. 1988. J Biomech Eng 110:3.)

flux of ultrafiltration was indicated to be involved in the mass transfer of glucose from the blood to the islet compartment, in addition to diffusion (Velho and Reach 1989). This ultrafiltration flux proceeded through the first half of the hollow fiber, and then liquid was reabsorbed through the second half of the fiber. In vitro investigations quantitatively proved that ultrafiltration flux was proportional to the flow rate of the medium circulating through the fibers (Velho and Reach 1989). The dependence of glucose transfer across the membrane to the islet compartment on the flow rate of circulating medium was predicted by Reach et al (1984c) and confirmed experimentally by sampling the medium from the islet compartment during the glucose concentration change (Velho and Reach 1989). However, the heterogeneity of glucose concentration inside the islet compartment surrounding the hollow fibers was also shown experimentally. It seems that the device design should be modified to yield glucose-insulin kinetics compatible with the performance of a closed-loop insulin delivery system and to reduce the large volume of the islet compartment (2.6 cm^3 extracapillary volume), which is responsible both for the delay in transmission of the change in glucose concentration from the blood to the islet compartment and for the heterogeneity of the glucose concentration inside the compartment.

A new ultrafiltration device designed by Reach et al (1984d) as a U-shaped BAP is schematically shown in Figure 17.3. The blood channel surrounds the islet chamber, which consists of two flat membranes, and blood circulates first above the upper membrane and then below the bottom membrane in the reverse direction. The ultrafiltration-reabsorption flux in this device crosses the islet compartment as a countercurrent perpendicular to the surface of the

ultrafiltration-immunoisolation membrane. Table 17.1 gives characteristics of the ultrafiltration chamber of the U-shaped BAP. Interestingly, such a design of an ultrafiltration device enabled a significant decrease in the volume of the islet compartment (about 17 times, comparing 0.15 cm^3 to 2.6 cm^3 for the islet compartment in an ultrafiltration hollow fiber device) (Velho and Reach 1989). The main advantages of this U-shaped BAP include: (1) accelerated triggering of insulin secretion in response to glucose concentration, as a result of the decrease in volume of the islet compartment, and (2) shortening of the lag time for insulin release into the circulation. This system involves interesting features (Reach et al 1984d) such as insulin release within 5 min in response to glucose loading, irrespective of glucose stimulus (square-wave or progressive rise simulating the physiological increase in blood glucose level following a meal), in-

TABLE 17.1. Characteristics of the ultrafiltration chamber.

Islet compartment	
width	0.6 cm
length	7.0 cm
height	360 μm
volume	0.15 ml
exchange surface area	$2 \times 4.2 \text{ cm}^2$
Blood channel	
thickness	400 μm
volume	0.7 ml (including the connector)
drop in hydrostatic pressure	40 mm Hg for a 4-ml/min flow rate
AN69S membrane	
hydraulic permeability	$40 \text{ ml/h/m}^2/\text{mm Hg}$
thickness	20 μm

(Reprinted with the permission of The American Diabetes Association, Inc. from Reach G et al. 1984c. Diabetes 33:753)

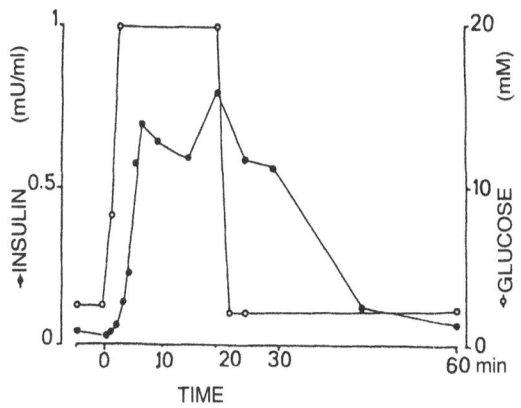

FIGURE 17.4. In vitro evaluation of the U-shaped bioartificial pancreas. Glucose and insulin concentration in the effluent from the chamber during a square-wave glucose stimulation.

(Reprinted with the permission of ISAO Press from Reach G et al. 1984 a. Progress in Artificial Organs-1983. 767.)

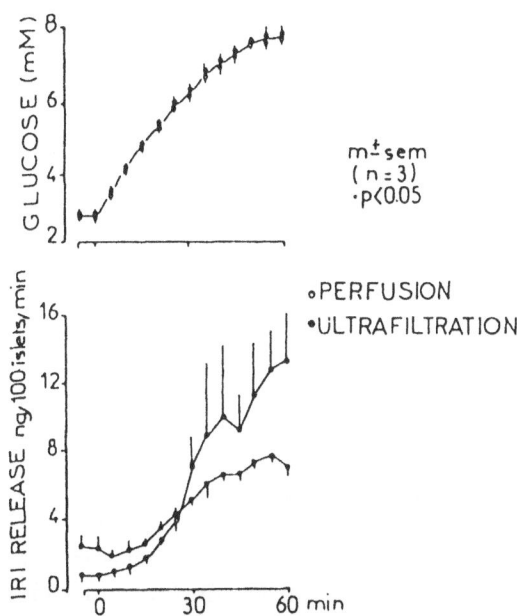

FIGURE 17.5. Insulin release (IRI) from perfused islets (open circles) or from islets placed in the ultrafiltration chamber (closed circles) during 0.1 mM/1/min ramp glucose stimulation. Upper panel: glucose concentration. Lower panel: insulin release from the chamber into the effluent medium.

(Reprinted with the permission of The American Diabetes Association, Inc. from Reach G et al. 1984 d. Diabetes 33:755.)

dependence of the response time on the membrane surface area, and proportionality of the insulin release from the device to the number of islets. The beta cell function was not suppressed by insulin, which was continuously carried away across the adjacent part of the membrane by the ultrafiltration flux (Reach 1984a). The U-shaped device caused a significant rise in concentration of insulin released as early as 5 min after square-wave glucose stimulation, with a maximal response after 10–20 min (Figure 17.4), thus providing much more satisfactory kinetics than those obtained with the ultrafiltration device employing hollow fibers.

A ramp stimulation test was performed in order to mimic a physiological challenge such as oral glucose loading at a blood glucose increase rate of 0.1 mM/1/min (Sorensen et al 1982). This test proved that the kinetics of insulin release from the U-shaped BAP were compatible with closed-loop insulin delivery at a response time of less than 15 min (Theodorou and Howell 1979). Figure 17.5 shows the kinetics of insulin release at a ramp glucose stimulation rate of 0.1 mM/1/min from the islet chamber into the effluent medium. The results of in vitro (Reach et al 1984d) and in vivo (Reach et al 1986) studies performed using normal rats, and those of an ex vivo study using normal dogs

(Lepeintre et al 1990) have demonstrated the high potential of the U-shaped design of BAP, given its ability to respond to glucose loading by acutely increasing insulin production. Evaluation of this U-shaped device under a simulated hemodynamic condition has revealed that there is a distinct relationship between the geometry and the functionality of the device. Thus, the kinetics of insulin release in response to glucose improved when the flow of circulating medium increased, as a result of the elimination of resistance between the two parts of blood channel (Moussy et al 1989). Recent theoretical analysis of ultrafiltration and mass transfer in the U-shaped BAP (Jaffrin et al 1988, Reach and Jaffrin 1990) has led to the conclusion that solute convection is not important under applied experimental conditions, whereas diffusion dominates, and that a rapid response of the system would be achieved by having an islet compartment with a small volume.

Recently, a novel mathematical model was proposed that predicted well the performance of the U-shaped artificial pancreas for both square-wave and ramp glucose stimulations (Mullon and Norton 1990). A spiral-wound device and a 2-channel BAP have recently been described, that operate on a principle similar to that of the U-shaped BAP, combining the diffusive and convective transports to shorten the lag time between insulin release and a rise in blood glucose concentration (Moussy and Moussy 1995, Sarver and Fournier 1990). However, comparison of their designs to those of much simpler U-shaped or hollow-fiber BAPs indicates that their clinical application will meet with difficulties and that they still need technical feasibility studies.

The effects of convective flow on solute transport and insulin release in hollow-fiber BAP were theoretically analyzed by Pillarella and Zydney (1990b). They developed a detailed model for glucose and insulin transfer and insulin secretion in a hollow-fiber BAP on the basis of mass and momentum conservation equations describing flow and transport in the lumen, matrix, and shell. This model accounts for the glucose and insulin diffusion and convection in both the axial and radial directions, the time-dependent insulin secretion kinetics, and the spatial distribution of islets in their compartment. In addition, this model provides an insight into the potential benefits of enhanced solute transport due to convective recirculation and of the nonuniform distribution of insulin-producing cells in the hollow-fiber ultrafiltration device (Pillarella and Zydney 1990a). The model simulations demonstrated that convective recirculation flow could dramatically improve insulin response, allowing the device to accurately realize the biphasic insulin secretion characteristic of the normal physiological response (Pillarella and Zydney 1990b). An in vitro perfusion study by Ramírez et al (1992) validated the predictions of the Pillarella and Zydney model; the solute radial convection and islet spatial distribution can profoundly affect the insulin response of the device to increased glucose concentration, and the insulin release from islets occurred within physiologically acceptable time limits under optimum conditions. Their study offered the first experimental evidence that the BAP design constructed of a straight tube, with minimized space volume of the islet compartment and seeding of

islets at the lumen outlet of hollow fiber membrane, could be suitable for use in humans.

Hydrogel Macrocapsules

Introduction

As mentioned earlier, BAP is designed to enclose the islet tissue within a selectively permeable membrane that immunoisolates the tissue from the recipient. So far, not much attention has been directed to cross-linked networks of synthetic water-soluble polymers (hydrogels) as materials for macroencapsulation and immunoisolation of islets of Langerhans. Klomp et al studied a macroporous hydrogel membrane made from poly(hydroxyethyl methacrylate) (PHEMA) for this purpose (Klomp et al 1983, Ronel et al 1983). This study revealed a minimal extent of tissue reaction when the macrocapsule was implanted into the peritoneal cavity of diabetic rats for 7 weeks, but granulation and fibrosis took place thereafter, with neovascularization up to 12 weeks. The blood glucose level was normalized for more than 6 months (Klomp et al 1981).

Because of their high water content, softness, and pliability, hydrogels are not expected to cause mechanical irritation to tissues in contact with them. Moreover, their diffusive properties can be regulated by cross-linking, which creates a kind of macromolecular sieve that can efficiently prevent the access of undesirable solutes to the macrocapsule interior.

We will describe the results on macrocapsules prepared from poly(vinyl alcohol) (PVA) hydrogel to immunoisolate islet tissue.

Macrocapsule Design

PVA hydrogels are highly water-swellable materials (Burczak et al 1991, 1994), but they cannot be used for the purpose of macroencapsulation unless reinforced with a mechanically strong support. Two designs of PVA hydrogel macrocapsules have been developed: a tube type of mesh-reinforced PVA hydrogel (MRPT) and a bag type of mesh-reinforced PVA hydrogel (MRPB). In both cases, the PVA membrane acts as a selectively permeable membrane having a mean mesh size of 15 nm. An in vitro study on the permeability of the PVA hydrogel membrane with a water content of 97% and a

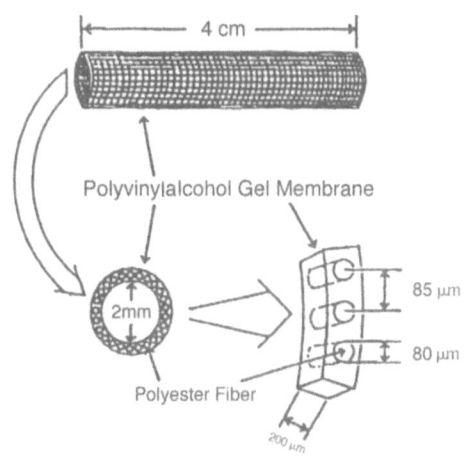

FIGURE 17.6. Structure of mesh-reinforced poly (vinyl alcohol) tube (MRPT).

(Reprinted with the permission of Elsevier Science Inc. from Mitsuo M et al. 1992. Transplant Proc 24:2939.)

thickness of 120 μm demonstrated that glucose, insulin, and nutrients passed through the membrane easily, whereas the passage of IgG was almost entirely blocked, indicating that the membrane could effectively prevent immunoattack on encapsulated islets (Inoue et al 1992). Figure 17.6 shows the structure of MRPT, which is 4 cm in length, 2 mm in inner diameter, 2.5 cm^2 in surface area, and 0.125 cm^3 in volume. The thickness of PVA hydrogel membrane is 200 μm. The MRPB shown in Figure 17.7 is approximately 2 cm in length, 2 cm in width, and 1.5 cm^3 in capacity.

These devices are small in size because they were developed for in vivo experiments using streptozotocin (STZ)-induced diabetic rats. The detailed method for the macrocapsule preparation has been described elsewhere (Aung et al 1993, Inoue et al 1992). Briefly, a poly(ethylene terephthalate) (PET) mesh tube (or bag) was immersed in 3 wt% PVA solution (polymerization degree of 7200) to which 0.08 wt% glutaraldehyde and 0.1 N HCl had been added. After coating with PVA, the tube (or bag) was left for 18 hrs at 100% humidity. After cross-linking, the macrocapsule was washed in boiling water for 30 min. Thus, the macrocapsules are composed of PVA hydrogel made by chemical cross-linking through acetalization and are reinforced with a thin, coarse PET mesh to provide the hydrogel with integrity against stresses appearing

in in vivo environments. The long-term biocompatibility of PVA hydrogel was proved by its insignificant tissue encapsulation after 18 weeks of intraperitoneal implantation and by low IL-1 production from adhered macrophages as compared with other polymers (Inoue et al 1992). The MRPT implanted for 1 month scarcely adhered to the surrounding tissue and was easily retrievable.

In Vitro Glucose-Insulin Kinetics

The results obtained using STZ-induced diabetic rats showed that long-term immunoisolation of allo- (Inoue et al 1992) and xenografts (Gu et al 1994) was possible. However, in order to regulate the physiological glucose level, encapsulated islets must respond to glucose within a short time in a closed-loop insulin delivery system (Kraegen et al 1981). To evaluate the kinetic performance of constructed MRPT, 1000 rat islets encapsulated within MRPT were perfused for 60 min at a rate of 1 ml/min with Krebs-Ringer bicarbonate buffer (KRBB) containing either 3.3 or 16.7 mM glucose. The result of in vitro insulin release from MRPT is shown in Figure 17.8. Insulin release began to increase at 9 ± 3 min after stimulation with glucose and reached a plateau at approximately 40 min, when the glucose concentration in the perfusate increased to 16.7 mM. A static incubation study on MRPT seeded with 250 and 1000 islets showed the maintenance of increased insulin release for 3 hrs after stimulation with 16.7 mM glucose and a clear dependence of the amount of secreted insulin on the number of encapsulated islets. Table 17.2 shows the effect of islet number on the insulin release from MRPT. These results demonstrate that encapsulated islets can respond rapidly to stimulation with 16.7 mM glucose solution. The lag time in insulin release (9 ± 3 min) satisfies the prerequisite that the delay in insulin delivery from BAP should be less than 15 min to avoid the overexcursion of blood glucose after a meal.

Islet Encapsulation

2500 to 3000 islets were isolated from 12-week-old male Lewis rat pancreases and seeded into MRPTs by injection. After being plugged, the MRPTs were implanted into the abdominal cavity of STZ-induced diabetic recipients (male Lewis rats). Nonfasting

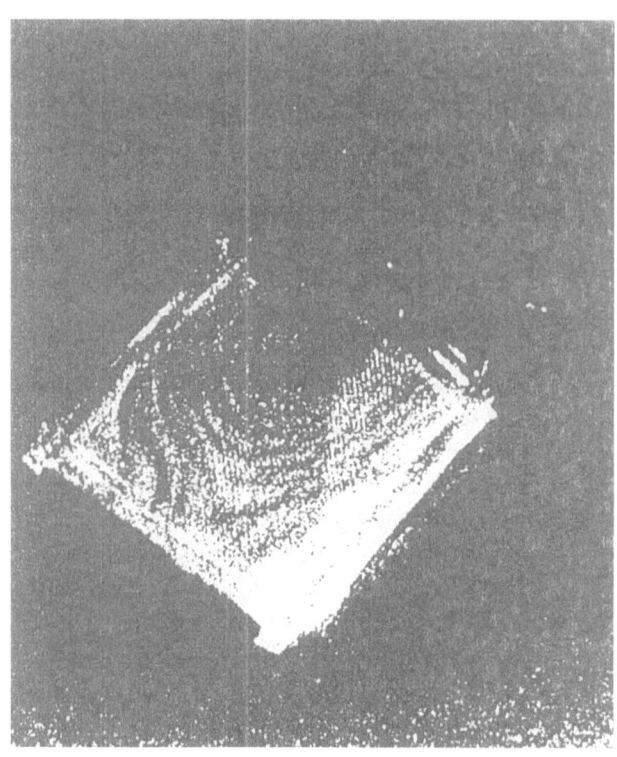

FIGURE 17.7. Mesh-reinforced poly(vinyl alcohol) hydrogel bag (MRPB).

(Reprinted with the permission of Elsevier Science Inc. from Gu YJ et al. 1994. Cell Transplant 3(Suppl 1):S20.)

(NPG) and fasting plasma glucose (FPG) levels were monitored daily before and after transplantation. Intravenous glucose tolerance tests (IVGTT, 0.5 g/kg) were performed at 2 and 4 weeks after MRPT implantation and compared with those of normal, untreated diabetic rats. The FPG levels of MRPT recipients were normalized during 1 month after transplantation (Tx) (pre-Tx: 283 ± 60; post-Tx: 85 ± 21; P < 0.001), with return to hyperglycemic levels after MRPT removal. NFPG levels were normalized for the first 1 week and then elevated (Mitsuo et al 1992). Table 17.3 summarizes

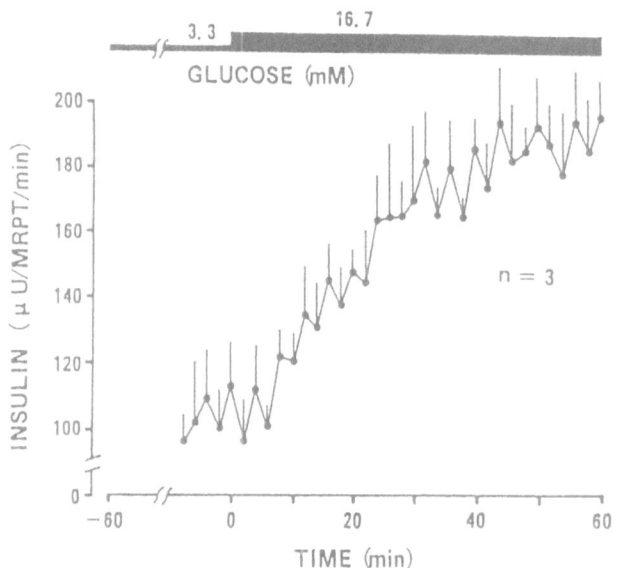

FIGURE 17.8. In vitro insulin release from MRPT containing 1000 islets. The values are mean ± SEM of three separate experiments, performed with duplicate samples.

(Reprinted with the permission of Lippincott-Raven Publishers from Aung T et al. 1993. ASAIO Journal 39:94.)

TABLE 17.2. Effects of number of islets seeded in MRPT on insulin release.

	Static incubation (hr)			
	0–1	1–2	2–3	3–4
Glucose concentration (mmol)	16.7	16.7	16.7	3.3
Insulin release (mU/MRPT/hr)*				
250 islets (n = 4)	0.6 ± 0.2	1.7 ± 0.4	2.1 ± 0.4	2.1 ± 0.
1000 islets (n = 5)	2.4 ± 0.2†	3.9 ± 0.3†	3.6 ± 0.1†	3.4 ± 0.

*Values are mean ± SEM
† p < 0.05 compared with 250 islets
(Reprinted with the permission of Lippincott-Raven Publishers from Aung T et al. 1993 ASAIO Journal 39:95)

the plasma glucose levels and calculated rates of glucose clearance (K values) during the IVGT test.

Although the initial plasma glucose levels of MRPT recipients were high after glucose injection, they declined at a rate comparable to that of glucose clearance in normal rats. The mean K value (%/min) before transplantation was extremely low (0.2 ± 0.1), but increased markedly to 0.9 ± 0.2 at 2 weeks and to 1.1 ± 0.3 at 4 weeks after transplantation. These results show that MRPT implantation resulted in only partial amelioration of carbohydrate metabolism. This may be due to a gradual loss of encapsulated tissue viability since necrosis of islets was observed. However, an unequivocal decline of glucose level was detected at 8 min after glucose loading, indicating that insulin was released from islets and through the PVA hydrogel membrane fairly rapidly.

Earlier allogeneic transplantation of islets enclosed within MRPT (Inoue et al 1992) demonstrated that transplantation reduced the NFBG levels for a significantly longer period of time than intraperitoneal transplantation of free, nonencapsulated islets (Figure 17.9). Approximately 2000 islets, nonentrapped and entrapped in MRPT, were transplanted into the peritoneal cavity of diabetic Wistar rats. The nonfasting serum glucose levels decreased from pre-Tx levels of 440–500 to 100–200 mg/dl. A sustained decrease in nonfasting serum

glucose levels was observed in 1 of 6 recipient diabetic animals until the removal of BAP on the 97th day, and then nonfasting serum glucose levels began to increase gradually to over 500 mg/dl. When approximately 3000 dog islets entrapped in MRPB were transplanted into diabetic rats, no significant decrease in either the nonfasting or fasting blood sugar level was noticed. However, when the islets, before seeding in MRPB, were immobilized in collagen type IV matrix to prevent their aggregation, a significant decrease in fasting blood sugar levels of recipient diabetic rats was observed after xenotransplantation of islets in MRPB (Gu et al 1994). The fasting blood sugar level decreased even at 2 weeks after xenotransplantation. An in vitro incubation study showed that a considerable amount of insulin was still released from the MRPB with islets immobilized in collagen matrix in response to a high glucose concentration (16.7 mM), indicating that the islets encapsulated in MRPB could survive even after xenotransplantation (Gu et al 1994).

Islet Immobilization in Gel Matrices

One of the most serious problems associated with macroencapsulated BAPs is the loss of viability of the transplanted islets, probably because of cluster formation, resulting in the disturbance of oxygen

TABLE 17.3. Changes in plasma glucose levels and K values in IV-GTTs.

	Mean Plasma Glucose (mg/dL)					
	0 min	8 min	15 min	30 min	60 min	K Value (%/min)
Normal rats (n = 7)	72 ± 15	214 ± 39	163 ± 25	136 ± 29	117 ± 18	1.0 ± 0.3
Pre-Tx (n = 4)	379 ± 61	485 ± 71	478 ± 76	453 ± 71	441 ± 73	0.2 ± 0.1
Post-Tx; 2 wk (n = 4)	69 ± 10*	233 ± 29*	211 ± 36*	171 ± 19*	147 ± 37*	0.9 ± 0.2†
Post-Tx; 4 wk (n = 4)	77 ± 4*	467 ± 148*	245 ± 38*	244 ± 14*	160 ± 12*	1.1 ± 0.3†

* $P < .001$, † $P < .01$ vs pre-Tx
(Reprinted with the permission of Elsevier Science Inc. from Mitsuo M et al. 1992 Transplant Proc 24:2940)

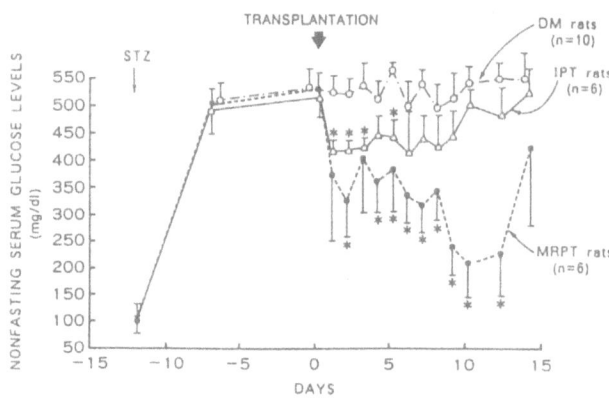

FIGURE 17.9. Comparative study on the change in nonfasting blood glucose levels among DM rats, IPT rats, and MRPT rats. DM, diabetes mellitus; IPT, intraperitoneal transplantation; MRPT, mesh-reinforced poly(vinyl alcohol) tube, $*p < 0.05$; significant difference in glucose levels from those found in DM rats. Each value represents M ± SEM.

(Reprinted with the permission of Lippincott-Raven Publishers from Inoue K et al. 1992 Pancreas 7:566.)

and nutrient transport into the central portion of these clusters (Lacy et al 1991). Histological examination of the transplanted islet tissue seeded in MRPT (Mitsuo et al 1992) and MRPB (Gu et al 1994) showed that the islets enclosed in a small space within the devices tended to aggregate into large clumps. The aggregation caused necrosis of the central part of the islet clumps, severe impairment of islet function, and gradual loss of tissue viability within 1–2 weeks (Mitsuo et al 1992). To prevent this disadvantageous aggregation of islets in MRPT and MRPB devices, and to preserve long-term islet function, islets were first immobilized in gel matrices made from biopolymers. Among them are collagen, hyaluronic acid, sodium alginate, and agarose. Collagen of type IV was found to be effective in preventing the islet aggregation in the bag type of PVA hydrogel chamber (Gu et al 1994). Hyaluronic acid is naturally present in the human body and is believed to be biocompatible with cells. Agarose has been used as a culture medium and has been successfully applied in microencapsulation of islets (Iwata et al 1989). Sodium alginate is also widely employed in microencapsulation of islets. It should be stressed that, in the procedure of sodium alginate gel formation, only treatment by calcium ions was done to maintain the gel state in the MRPTs, not poly-L-lysine coating or solubilization of alginate gel by sodium citrate, which are commonly employed for microencapsulation (Aung et al 1995).

Prior to seeding, 1000 to 2000 islets were suspended in solutions of collagen (type I-A), 0.5 wt % hyaluronic acid, 1 wt % sodium alginate, or 5 to 7 wt % agarose. The islets suspended in RPMI-1640

medium were injected into MRPTs as controls. After 3 weeks of culture in RPMI-1640 medium, the MRPTs were subjected to perfusion to investigate insulin release in response to square-wave glucose stimulation. The tissue enclosed in MRPTs was then retrieved for evaluation by histological examination. The results of these studies showed that, among the biopolymer matrices used, the agarose matrix was the most suitable for immobilizing islets before enclosing in MRPT or MRPB (Aung et al 1995). Islets suspended in 5 wt % agarose gel showed no tendency to aggregate within 3 weeks of in vitro culture after the islet-containing agarose gel was enclosed in MRPT (Aung et al 1994). Perfusion of MRPT containing agarose-immobilized islets showed fairly rapid response of islets to square-wave glucose stimulation, initiating at 9 ± 3 min after the increase in glucose concentration (Aung et al 1994). In contrast to agarose, other biopolymers did not offer efficacious prevention of islet aggregation. Islets immobilized in collagen, hyaluronic acid, and sodium alginate gels showed varying extents of aggregation, with central necrosis similar to that observed for the islets retrieved from the control group.

MIN6 Cell Entrapping in Macrocapsules

Recent advances in gene manipulation technology have provided tools for modifying cellular characteristics and intracellular events. One example is the establishment of the MIN6 cell line, derived by the gene-targeting method from insulinoma cells of a transgenic mouse expressing the large T-antigen of SV40 (Miyazaki et al 1990). MIN6 cells retain

FIGURE 17.10. Perfusion study on MRPT: insulin release from the MRPT began to increase at 8 min and reached a peak approximately 44 min after the glucose concentration in the perfusate rose to 16.7 mM.

(Reprinted with the permission of Elsevier Science Inc. from Hayashi H et al. 1996 a. Cell Transplant 5:S67.)

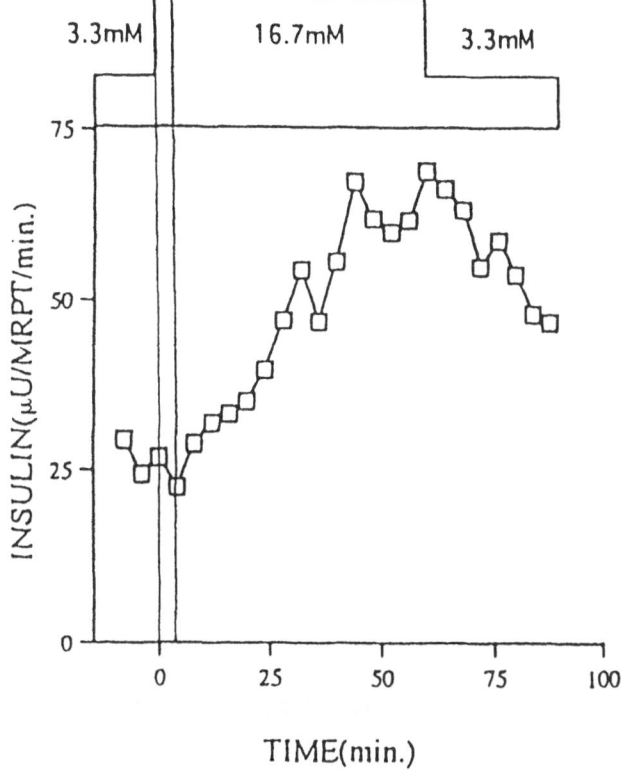

physiological glucose-concentration-dependent insulin secretion similar to that of normal islets (Ishihara et al 1993, Miyazaki et al 1990). Therefore, MIN6 is presumed to be superior to isolated islets in terms of both its potential for insulin secretion and the ability to proliferate in sufficient quantities to meet the demands of patients awaiting pancreas transplantation.

Before seeding in MRPTs or MRPBs, MIN6 cells were immobilized in 5 wt % low-temperature-gelling agarose solution. The number of cells seeded in one MRPB was 1.0×10^7 (Hayashi et al 1995). Glucose stimulation was performed after a 2-week culture of MIN6 cells (seeded in MRPBs) in Dulbecco's modified Eagle's medium (DMEM, 25 mM glucose) with 10% FBS and antibiotics. An approximately twofold increase in insulin secretion was recorded in response to 16.7 mM glucose stimulation as compared to 3.3 mM glucose. The insulin release from the MRPT macrocapsules containing 5.0×10^6 MIN6 cells began to increase at 8 min and reached a peak at approximately 44 min after a rise of glucose concentration in the perfusate to 16.7 mM (Figure 17.10). An approxi-

mately threefold increase in insulin release was observed in response to 16.7 mM glucose (Hayashi et al 1996a).

MRPBs containing 1.0×10^7 or 3.0×10^7 MIN6 cells, embedded in agarose gel, were xenogeneically transplanted into the peritoneal cavity of STZ-induced diabetic Lewis rats (Hayashi et al 1996b, 1996c). The fasting serum glucose levels of transplanted recipients were monitored every week following the intraperitoneal transplantation of MRPBs. As shown in Figure 17.11, the serum glucose levels of the control diabetic animals indicate the persistence of hyperglycemia at serum glucose levels of 281 ± 14 mg/dl (n = 6). On the other hand, the serum glucose levels of diabetic recipients showed a rapid and significant decrease from the pre-Tx level of 366 ± 64 mg/dl to less than 200 mg/dl for 2–3 months after xenotransplantation of 1.0×10^7 MIN6 cells in one MRPB. The mean survival time of the enclosed cells was 77 days. When the macrocapsules containing 3.0×10^7 MIN6 cells were transplanted, the diabetic recipient rats showed a decrease of serum glucose levels from the pre-Tx levels of 231 ± 25 mg/dl to levels

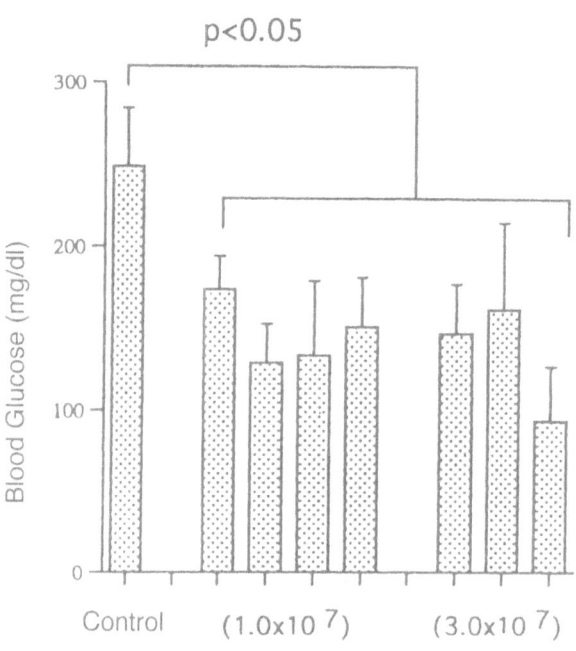

FIGURE 17.11. Effect of the number of enclosed cells on serum glucose levels.

(Reprinted with the permission of Elsevier Science Inc. from Hayashi H et al. 1996 b. Transplant Proc 28:1098.)

below 200 mg/dl in a week, lasting 3–4 months, with a mean cell survival time of 93 days.

Summary

A goal of all the encapsulated systems is to assure the correction of hyperglycemia in a response time shorter than 15 min. In intravascular systems the pressure drop between artery and vein creates an ultrafiltration flux that enhances transmission of the glucose signal to the islet compartment beyond the level caused by diffusion. The insulin released by islets is pulled by backultrafiltration flux into the circulation, effecting amelioration of glucose metabolism. In contrast, this solute drag effect is not present in extravascular systems, where glucose and insulin cross the encapsulating membrane by diffusion alone.

In the peritoneum, which is commonly used as an implantation site for extravascular systems, peritoneal glucose concentration follows blood glucose changes with a 5–10 min delay, and reaches only 60–70 % of that in blood (Velho et al 1989). Thus, the lag time and insulin level of extravascular BAPs will always differ from those of intravascular systems. However, the results of IVGTTs performed at 2 and 4 weeks after transplantation of the islet tissue seeded in MRPTs show that K values of glucose clearance were comparable to those of normal rats, and that the decline in glucose level observed at 8 min after glucose loading was still within an acceptable range, i.e. <15 min. These results suggest that the BAP based on hydrogel macrocapsules is satisfactory in terms of in vivo glucose-insulin kinetics.

It will be possible to design more or less complex intravascular BAPs that perform as potential closed-loop insulin delivery systems. However, their long-term performance will be limited either by lack of hemocompatibility of the membrane in contact with the flowing blood or by the occurrence of thrombosis at the anastomotic site. On the other hand, extravascular systems using hydrogel macrocapsules are technically easier to fabricate, but encapsulation of the device by a collagenous connective tissue will occur to hinder the diffusional transport of nutrients. The decrease in islet viability may also be due to imperfect procedures to provide long-lasting isolation of islets from the pancreas.

Even though there are still several problems to be addressed for BAP development, BAP is a highly promising approach to reducing diabetic complications.

References

Archer J, Kaye R, Mutter G. 1980. Control of Streptozotocin diabetes in Chinese hamsters by cultured mouse islet cells without immunosuppression. J Surg Res 28: 77–85.

Aung T, Kogire M, Inoue K, Fujisato T, Gu YJ, Burczak K, Shinohara S, Mitsuo M, Maetani S, Ikada Y, Tobe T. 1993. Insulin release from bioartificial pancreas using mesh-reinforced polyvinyl alcohol hydrogel tube: in vitro study. ASAIO Journal 39: 93–96.

Aung T, Inoue K, Kogire M, Sumi S, Fujisato T, Gu YJ, Shinohara S, Hayashi H, Doi R, Imamura M, Mitsuo M, Nakai I, Maetani S, Ikada Y. 1994. Improved insulin release from a bioartificial pancreas using mesh-reinforced polyvinyl alcohol hydrogel tube: immobilization of islets in agarose gel. Transplant Proc 26: 790–791.

Aung T, Inoue K, Kogire M, Doi R, Kaji H, Tun T, Hayashi H, Echigo Y, Wada M, Imamura M, Fujisato T, Maetani S, Iwata H, Ikada Y. 1995. Comparison of various gels for immobilization of islets in bioartificial pancreas using a mesh-reinforced polyvinyl alcohol hydrogel tube. Transplant Proc 27: 619–621.

Burczak K, Fujisato T, Hatada M, Ikada Y. 1991. Protein permeation through polymer membranes for hybrid-type artificial pancreas. Proc Jpn Acad 61B: 83–88.

Burczak K, Fujisato T, Hatada M, Ikada Y. 1994. Protein permeation through poly(vinyl alcohol) hydrogel membranes. Biomaterials 15: 231–238.

Catapano G, Iorio G, Drioli E, Lombardi CP, Crucitti F, Doglietto GB, Bellantone M. 1990. Theoretical and experimental analysis of a hybrid bioartificial membrane pancreas: a distributed parameter model taking into account Starling fluxes. J Member Sci 52: 351–378.

Chick WL, Like AA, Lauris V, Galletti PM, Richardson PD, Panol G, Mix TW, Colton CK. 1975. A hybrid artificial pancreas. Trans Am Soc Artif Intern Organs 21: 8–14.

Chick WL, Perna JJ, Lauris V, Law D, Galletti PM, Panol G, Whittemore AD, Like AA, Colton CK, Lysaght MJ. 1977. Artificial pancreas using living beta cells: effects of glucose homeostasis in diabetic rats. Science 197: 780–782.

Colton CK, Solomon BA, Galletti PM, Richardson PD, Takahashi C, Naber SP, Chick WL. 1980. Development of novel semipermeable tubular membranes for a hybrid artificial pancreas. In: Cooper AR, editor. Ultrafiltration membranes and applications. New York: Plenum Publishing, pp. 541–555.

Colton CK, Avgoustiniatos ES. 1991. Bioengineering in development of the hybrid artificial pancreas. J Biomech Eng (Trans ASME) 113: 152–170.

Gu YJ, Inoue K, Shinohara S, Doi R, Kogire M, Aung T, Sumi S, Imamura M, Fujisato T, Maetani S, Ikada Y. 1994. Xenotransplantation of bioartificial pancreas using a mesh-reinforced polyvinyl alcohol bag. Cell Transplant 3 (Suppl 1): S19–S21.

Hayashi H, Inoue K, Aung T, Tun T, Echigo Y, Gu YJ, Shinohara S, Kaji H, Kato M, Imamura M, Maetani S, Morikawa N, Iwata H, Ikada Y, Miyazaki J. 1995. Xenotransplantation of a novel B-cell line (MIN6) in mesh-reinforced polyvinyl alcohol hydrogel bag. Transplant Proc 27: 3358–3361.

Hayashi H, Inoue K, Aung T, Tun T, Gu YJ, Wang WJ, Shinohara S, Kaji H, Doi R, Setoyama H, Kato M, Imamura M, Maetani S, Morikawa N, Iwata H, Ikada Y, Miyazaki J. 1996a. Application of a novel B cell line MIN6 to a mesh-reinforced polyvinyl alcohol hydrogel tube and three-layer agarose microcapsules: an in vitro study. Cell Transplant 5: S65–S69.

Hayashi H, Inoue K, Aung T, Tun T, Wang WJ, Gu YJ, Shinohara S, Echigo Y, Kaji H, Kato M, Setoyama H, Kawakami Y. Imamura M, Morikawa N, Iwata H, Ikada Y, Miyazaki J. 1996b. Prolongation of survival of a xenografted bioartificial pancreas with a mesh-reinforced polyvinyl alcohol hydrogel bag employing a B-cell line (MIN6). Transplant Proc 28: 1097–1098.

Hayashi H, Inoue K, Aung T, Tun T, Wenjing W, Gu YJ, Shinohara S, Echigo Y, Kaji H, Setoyama H, Kawakami Y, Imamura M, Morikawa N, Iwata H, Ikada Y, Miyazaki J. 1996c. Long survival of xenografted bioartificial pancreas with a mesh-reinforced polyvinyl alcohol hydrogel bag employing a B-cell line (MIN6). Transplant Proc 28: 1428–1429.

Inoue K, Fujisato T, Gu YJ, Burczak K, Sumi S, Kogire M, Tobe T, Uchida K. Nakai I, Maetani S, Ikada Y. 1992. Experimental hybrid islet transplantation: application of polyvinyl alcohol membrane for entrapment of islets. Pancreas 7: 562–568.

Ishihara H, Asano T, Tsukuda K, Katagiri H, Inukai K, Anai M, Kikuchi M, Yazaki Y, Miyazaki J-I, Oka Y. 1993. Pancreatic beta cell line MIN6 exhibits characteristics of glucose metabolism and glucose-stimulated insulin secretion similar to those of normal islets. Diabetologia 36: 1139–1145.

Iwata H, Amemiya H, Matsuda T, Takano H, Hayashi R, Akutsu T. 1989. Evaluation of microencapsulated islets in agarose gel as bioartificial pancreas by studies of hormone secretion in culture and by xenotransplantation. Diabetes 38 (Suppl 1): 224–225.

Jaffrin MY, Reach G, Notelet D. 1988. Analysis of ultrafiltration and mass transfer in a bioartificial pancreas. J Biomech Eng (Trans ASME) 110: 1–10.

Klomp GF, Hashiguchi H, Dobelle WH. 1981. Hybrid artificial pancreas with hydrogel membranes. Artif Organs 5(A): 38.

Klomp GF, Hashiguchi H, Ursell PC, Takeda Y, Taguchi T, Dobelle WH. 1983. Macroporous hydrogel membranes for a hybrid artificial pancreas. II. Biocompatibility. J Biomed Mater Res 17: 865–871.

Kraegen EW, Chisholm DJ, MCNamara ME. 1981. Timing of insulin delivery with meals. Horm Metab Res 13: 365–367.

Lacy PE, Hegre OD, Gerasimidi-Vazeou A, Gentile FT, Dionne KE. 1991. Maintenance of normoglycemia in diabetic mice by subcutaneous xenografts of encapsulated islets. Science 254: 1782–1784.

Lanza RP, Sullivan SJ, Monaco AP, Chick WL. 1992. The hybrid artificial pancreas: Diffusion and vascular devices. In: Ricordi C, editor. Pancreatic islet cell transplantation. Austin, Texas: R.G. Landes Company, pp. 223–237.

Lepeintre J, Briandet H, Moussy F, Chicheportiche D, Darquy S, Rouchette J, Imbaud P, Duron JJ, Reach G. 1990. Ex vivo evaluation in normal dogs of insulin released by a bioartificial pancreas containing isolated rat islets of Langerhans. Artif Organs 14: 20–27.

Mitsuo M, Inoue K, Nakai I, Oda T, Gu Y, Shinohara S, Kogire M, Fujisato T, Maetani S, Ikada Y, Tobe T, Oka T. 1992. Efficacy of mesh reinforced polyvinyl-alcohol tube as a novel device for bioartificial pancreas: a functional study of rat islets in vivo. Transplant Proc 24:2939–2940.

Miyazaki J-I, Araki K, Yamato E, Ikegami H, Asano T, Shibasaki Y, Oka Y, Yamamura K-I. 1990. Establishment of a pancreatic B cell line that retains glucose-inducible insulin secretion. Special reference to expression of glucose transporter isoforms. Endocrinology 127:126–132.

Moussy F, Rouchette J, Reach G, Cannon R, Jaffrin MY. 1989. In vitro evaluation of a bioartificial pancreas under various hemodynamic conditions. Artif Organs 13:109–115.

Moussy Y, Moussy F. 1995. Analysis of glucose and insulin mass transfer in a novel 2-channel bioartificial pancreas. Art Cells, Blood Subs, Immob Biotech 23:163–173.

Mullon CJP, Norton CA. 1990. A mathematical analysis of the U-shaped hybrid artificial pancreas. A novel insulin release rate equation. Biomat, Art Cells, Artif Organs 18:43–57.

Pillarella MR, Zydney AL. 1990a. Effect of beta cell distribution on the performance of a bioartificial pancreas. Trans Am Soc Artif Intern Organs 36:M715–M719.

Pillarella MR, Zydney AL. 1990b. Theoretical analysis of the effect of convective flow on solute transport and

insulin release in a hollow fiber bioartificial pancreas. J Biomech Eng 112:220–228.

Ramírez CA, López M, Stephens CL. 1992. In vitro perfusion of hybrid artificial pancreas devices at low flow rates, ASAIO Journal 38:M443–M449.

Reach G, Poussier P, Sausse A, Assan R, Itoh M, Gerich JE. 1981. Functional evaluation of a bioartificial pancreas using isolated islets perifused with blood ultrafiltrate. Diabetes 30:296–301.

Reach G, Poussier P, Sausse A, Assan R, Itoh M, Furman B, Gerich JE. 1982. Use of ultrafiltration (instead of dialysis) to achieve rapid glucose-insulin kinetics in a bioartificial pancreas. Horm Metab Res 12 (Suppl):177–179.

Reach G. 1984a. Bioartificial pancreas. Present state and future prospects. Biomed Biochim Acta 43:569–576.

Reach G, Jaffrin MY, Desjeux J-F. 1984b. Design and in vitro evaluation of a new bioartificial pancreas. In: Atsumi K, Maekawa M, Ota K, editors. Progress in artificial organs-1983. Cleveland: ISAO Press. p 765–768.

Reach G, Jaffrin MY, Vanhoutte C, Desjeux J-F. 1984c. Importance of convective transport in a model of bioartificial pancreas. ASAIO Journal 7:85–90.

Reach G, Jaffrin MY, Desjeux J-F. 1984d. A U-shaped bioartificial pancreas with rapid glucose-insulin kinetics. In vitro evaluation and kinetic modelling. Diabetes 33:752–761.

Reach G, Chenard PS, Darquy S, Lepeintre J, Desjeux J-F, Cannon R, Jaffrin MY. 1986. Bioartificial pancreas: in vivo evaluation in conscious normal rats. In: Nosé Y, Kjellstrand C, Ivanovich P, editors. Progress in artificial organs-1985. Cleveland: ISAO Press. p 621–626.

Reach G, Jaffrin MY. 1990. Kinetic modelling as a tool for the design of a vascular bioartificial pancreas: feedback between modelling and experimental validation. Computer Methods and Programs in Biomedicine 32:277–285.

Ronel SH, D'Andrea MJ, Hashiguchi H, Klomp GF, Dobelle WH. 1983. Macroporous hydrogel membranes for a hybrid artificial pancreas. I. Synthesis and chamber fabrication. J Biomed Mater Res 17:855–864.

Sarver JG, Fournier RL. 1990. Numerical investigation of a novel spiral wound membrane sandwich design for an implantable bioartificial pancreas. Comput Biol Med 20:105–119.

Scharp DW, Mason NS, Sparks RE. 1984. Islet immuno-isolation: the use of hybrid artificial organs to prevent islet tissue rejection. World J Surg 8:221–229.

Sorensen JT, Colton CK, Hillman RS, Soeldner JS. 1982. Use of a physiologic pharmacokinetic model of glucose homeostasis for assessment of performance requirements for improved insulin therapies. Diabetes Care 5:148–157.

Sparks RE, Mason NS, Finley TC, Scharp DW. 1982. Development, testing and modeling of an islet transplantation chamber. Trans Am Soc Artif Intern Organs 28:229–231.

Sparks RE, Mason NS, Finley TC, Scharp DW. 1983. A distributed source-model for the hybrid artificial pancreas. Trans Am Soc Artif Intern Organs 29:460–462.

Sun AM, Parisius W, Healy GM, Vacek I, Macmorine HG. 1977. The use, in diabetic rats and monkeys, of artificial capillary units containing cultured islets of Langerhans (artificial endocrine pancreas). Diabetes 26:1136–1139.

Sun AM, Goosen MFA, O'Shea GM, Gharapetian HM. 1987. Recent advances in the development of a bioartificial pancreas. In: Williams DF, Lyman DJ, editors. Blood compatibility. New York: CRC Series.

Theodorou N, Howell SL. 1979. An assessment of diffusion chambers for use in pancreatic islet cell transplantation. Transplantation 27:350–353.

Tze WJ, Tai J, Wong FC, Davis HR. 1980. Studies with implantable artificial capillary units containing rat islets on diabetic dogs. Diabetologia 19:541–545.

Velho G, Reach G. 1989. Monitoring the kinetics of glucose transfer across a bioreactor membrane with a glucose sensor. Int J Artif Organs 12:539–543.

Velho G, Froguel P, Reach G. 1989. Determination of peritoneal glucose kinetics in rats: Implications for the peritoneal implantation of closed-loop insulin delivery systems. Diabetologia 32:331–336.

Weinless NL, Colton CK. 1983. A theoretical model for insulin secretory dynamics in a hybrid artificial pancreas. Ann NY Acad Sci 413:421–423.

Part III
Applications of Encapsulated Cells

Section 1
Primary Cells

18
Artificial Pancreas

Willem M. Kühtreiber, Robert P. Lanza, and William L. Chick

Introduction

Patients who suffer from advanced cases of diabetes must take daily insulin injections. Milder forms of the disease can be managed by a variety of oral hypoglycemic drugs. While such therapies can restore the average blood glucose levels to normal, the moment-to-moment fine regulation that is performed by the pancreas of a nondiabetic person is very difficult or impossible to achieve. Failure to achieve such true glucose homeostasis leads to the serious secondary complications associated with diabetes, such as diabetic neuropathy, nephropathy, and retinopathy. Indeed, the results of the Diabetes Control and Complications Trial (DCCT) have confirmed that improved blood glucose control by means of intensive insulin therapy markedly reduces these secondary complications (The Diabetes Control and Complications Trial Research Group 1993).

The β-cells of pancreatic islets sense fluctuations in the concentration of blood glucose via complex electrophysiological and biochemical interactions, and adjust their moment-to-moment insulin secretion accordingly (Rajan et al 1990, Robertson et al 1991). The transplantation of islets therefore carries the promise of true physiologic control for the diabetic patient, resulting in a reduction or potential elimination of the secondary complications of diabetes.

In an artificial pancreas, the islets are shielded from the immune system of the host by a selectively permeable barrier. Low-molecular-weight substances, such as nutrients, oxygen, electrolytes, and secretory products can diffuse across the barrier.

However, the molecular weight exclusion limit of the barrier is such that immunocytes, immunoglobulins, certain components of the complement system, and other large immune-effector molecules are excluded. This approach has the potential to be used for allogeneic as well as xenogeneic applications.

Several types of artificial pancreas devices are being developed. These include intravascular devices, diffusion chambers, and microcapsules. This chapter summarizes the results that have been obtained with these three types of devices. The islets inside an artificial pancreas must survive and function while being subjected to low partial oxygen tension and restricted nutrient availability. We will therefore start with a discussion of the physiological characteristics of islets that enable them to function under such conditions.

Isolation and Physiology of Porcine Islets

The supply of human donor material is extremely limited. A large-scale allograft transplantation program with human islets is therefore presently not feasible, and this will remain the case unless it becomes possible to propagate islets or β-cells in vitro while maintaining normal function. Many groups, including our own, are therefore using animal pancreatic islets. The large-scale harvesting of islets is most feasible with domesticated (farm) animals, which already are being slaughtered in large quantities, such as cows or pigs. Porcine islets are considered to be a particularly promising donor source because there is only a difference of 1 amino acid between the structure of human and porcine insulin (van Haeften 1989), and because of the relatively low levels of preformed human xeno-antibodies to pig tissues (Kirkman 1989).

Pancreatic Islet Isolation

Pancreatic islets constitute only 1–2% of the pancreatic tissue. Porcine and canine islets are best isolated from adult animals, whereas bovine islets are prepared from the pancreas of newborn calves less then 2 weeks of age (Kühtreiber et al 1994a). The isolation is performed according to procedures modified from the method of Warnock and Rajotte (Lanza et al 1991, Warnock and Rajotte 1988). Briefly, the pancreas is infused with cold University of Wisconsin organ preservation solution (UW, see Sumimoto et al 1989) and excised. The gland is transported on ice to the laboratory, where a solution of crude collagenase is infused via the pancreatic duct system. The pancreas is then digested and the dissociated tissue washed with cold medium. Islet purification is based on density differences between islets and exocrine tissue and is performed on a discontinuous Ficoll gradient. After density centrifugation, the islets are collected, washed several times, and put in tissue culture in α-MEM or Hams F12 based medium. Freshly isolated islets initially look ragged, but they regain their compact appearance in culture within 24 to 48 hours.

Long-Term Insulin Secretion

An artificial pancreas must be capable of producing large amounts of insulin over extended periods of time, while maintaining adequate glucose responsiveness. We have therefore studied the insulin secretion of porcine islets during long-term culture. These studies are summarized below.

We immobilized porcine islets in agar and maintained them in α-MEM based medium containing 200 mg/dl glucose. Aliquots of medium were collected three times per week and frozen for insulin analysis, using a standard radioimmunoassay. The islets secreted substantial amounts of insulin for extended periods of time at these mildly stimulating glucose concentrations (Kühtreiber et al 1994b). Although the amount of secreted insulin declined

over time, some preparations still secreted approximately 20 μU/islet/day after 1 year in culture (Figure 18.1). Large amounts of insulin were also produced by islets that were immobilized within immunoisolation devices. For example, a biohybrid artificial pancreas seeded with 200,000 EIN porcine islets and maintained in in vitro culture for 3 months, produced 25 to 40 units of insulin per day for the duration of the experiment (Kühtreiber et al 1994b). We have also obtained long-term in vivo functional data for porcine, bovine, and canine islets (Lanza et al 1991, Lanza et al 1993).

In addition to being capable of static insulin secretion under continuous stimulating glucose conditions, islets must respond appropriately to acute glucose stimuli. To determine whether isolated islets retain this capacity, porcine islets in long-term culture were subjected to glucose challenges in a perifusion system (Kühtreiber et al 1994b). In such studies, the islets were perifused at 37°C for 1 hour with medium containing 100 mg/dl glucose. The glucose concentration was then stepped up to 300 mg/dl for 1 hour and stepped back down to 100 mg/dl for an additional hour. Perifusate samples were collected every 2 minutes and analyzed for insulin and glucose.

Good insulin secretory responses to such glucose challenges were obtained for porcine islets that had been in culture for up to 3 months. The insulin secretion at 100 mg/dl glucose was considered to be basal insulin secretion. The islets responded to the glucose step from 100 to 300 mg/dl by increasing insulin secretion by 5.2 ± 1.4 times basal ($n = 10$). Insulin secretion then leveled off to a lower plateau value of 3.2 ± 0.6 times basal ($n = 10$) and returned to near basal when the glucose stimulus was reduced to 100 mg/dl (Kühtreiber et al 1994b).

Insulin Secretion Under Low Oxygen Conditions

The capillary system of pancreatic islets collapses and vanishes after isolation. As a result, islets must rely on diffusion from and to the surrounding medium for their oxygen and nutrient supply, as well as for the removal of metabolic waste products and secreted substances such as insulin. The cells in isolated islets thus encounter diffusion distances that are far greater than those encountered in

FIGURE 18.1. Insulin secretion from porcine islets that were immobilized in agar and cultured in vitro for 1 year. The regression line was derived by exponential curve fitting using Microsoft Excel.

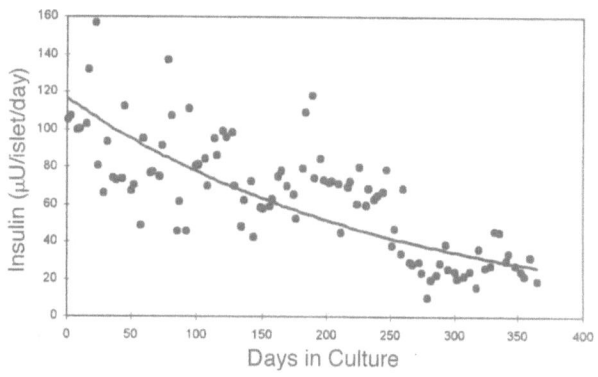

vascularized islets, where each cell is very close to a capillary. For encapsulated islets, diffusion is further inhibited by the permselective nature of the capsule membranes (see chapter 2 by Goosen for information on diffusion theory and oxygen utilization). In addition, most capsules are implanted in the peritoneal cavity, where the partial oxygen tension (pO_2) is reduced as compared to that of arterial blood. It is easy to see that a combination of all these factors imposes limits on the dimensions for both islets and capsules, beyond which the islet cells will not be able to function properly, or will die (Dionne et al 1989, Dionne et al 1993, Kirchgessner et al 1992, Kühtreiber et al 1993, Ohta et al 1990, Schrezenmeir et al 1992).

We measured the pO_2 inside islet-containing diffusion chambers and showed that encapsulated islets within the peritoneal cavity of rats experience an average pO_2 of 38 mm Hg (Kühtreiber et al 1993). We then cultured islets under these low oxygen conditions in an incubator that allowed control of oxygen in addition to CO_2. Control islets were cultured at ambient oxygen (159 mm Hg) in a regular incubator. The glucose concentration in the culture medium was chosen to be mildly stimulatory (200 mg/dl). Periodic media samples were taken and analyzed for insulin. For porcine islets the ratio of insulin secretion at low versus ambient oxygen tension was 0.92 ± 0.27 (mean ± SD; n = 10). For canine islets this ratio was 0.80 ± 0.08 (n = 5), suggesting that canine islets were slightly more affected by low oxygen than were porcine islets. These results suggest that encapsulated islets should be able to function well as an intraperitoneal implant.

Types of Artificial Pancreas Devices

Three types of artificial pancreas systems can be distinguished (Figure 18.2). First, in a vascular system, the islets are contained in a plastic chamber that surrounds a selectivity permeable membrane (Figure 18.2A). The device is implanted as a shunt in the vascular system of the recipient. Second, islets can be placed inside membrane diffusion chambers, which can be implanted in a number of extravascular sites, such as intraperitoneally or subcutaneously (Figure 18.2B). Third, the islets can be encapsulated within microcapsules, which can be simply injected into the implant site (Figure 18.2C). In the following section, the characteristics of these three types of devices are summarized.

Vascular Biohybrid Pancreas

The original vascular biohybrid pancreas consisted of bundles of capillary fibers seeded on their outside surfaces with clusters of pancreatic islet cells (Chick et al 1975). While these devices worked well in vitro, they suffered from rapid clotting when implanted in vivo. This clotting problem was addressed by vascular devices with tubular immunoisolation membranes that had a larger lumen. These devices consist of acrylic housings that contain coiled tubular membranes (diameter 5–6 mm) of polyacrylonitrile-polyvinyl chloride (PAN-PVC). The islet chamber is formed by the space between the tubular membrane and the housing. Because the PAN-PVC membrane cannot be sutured

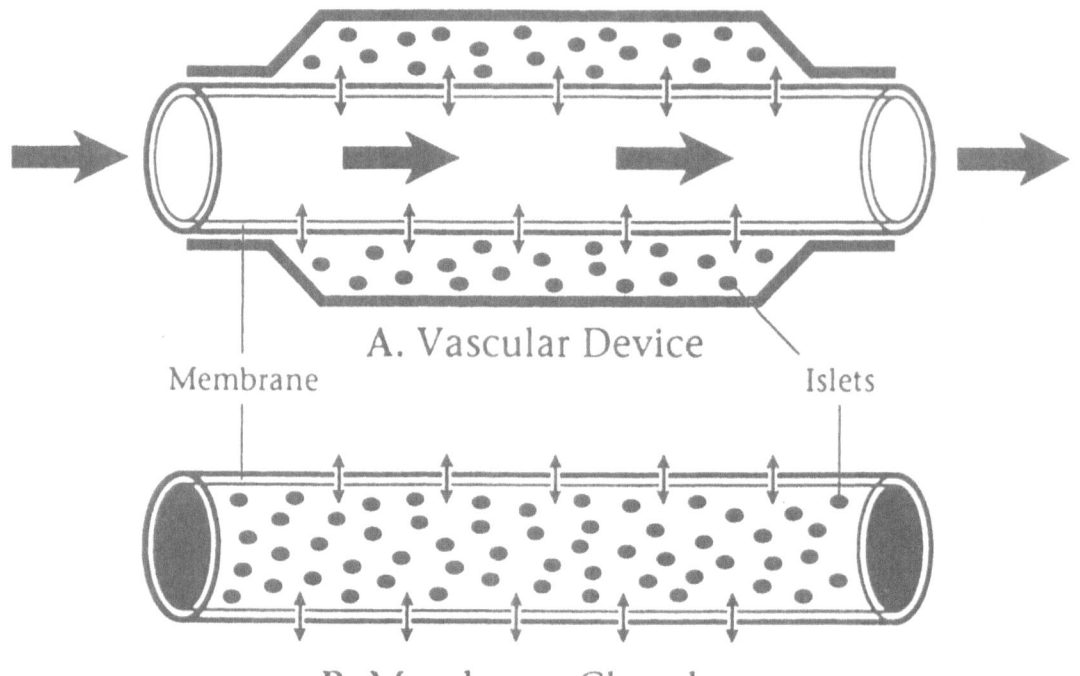

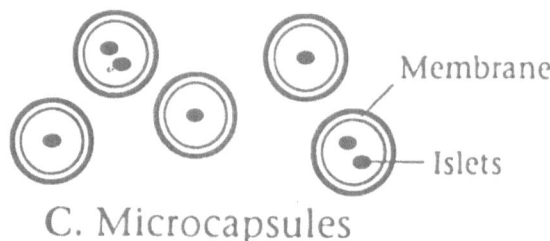

FIGURE 18.2. Three types of artificial pancreas devices can be distinguished. In the vascular biohybrid pancreas, the islets are contained in a plastic chamber that *surrounds* a selectively permeable membrane (A). In membrane diffusion chambers, the islets are placed *within* a PAN-PVC membrane that is closed on both sides (B). In microcapsules, the islets are immobilized within small hydrogel-based beads (C). (Figure reprinted with permission from Lanza et al. 1992. Diabetes 41:1503–1510. Copyright American Diabetes Association, Inc.)

directly, it is connected to polytetrafluoroethylene graft material that is used to suture the device as an anastomosis to the vascular system.

The vascularized biohybrid pancreas has been used for both allografts and xenografts. In allograft applications, the devices were seeded with canine islets and implanted in dogs that were rendered diabetic by total pancreatectomy (Lanza et al 1994,

Sullivan et al 1991). The dogs received no immunosuppressive therapy. The devices significantly decreased the exogenous insulin requirements of these dogs. Out of 17 dogs that received two devices containing canine islets, 11 dogs maintained normal fasting blood glucose concentrations without the need for exogenous insulin for periods ranging from 26 days to over 1 year. The devices were removed from two animals that showed function for over one year. In both cases, the exogenous insulin requirements of the dogs promptly increased by more than 20 units per day, confirming that normoglycemia in these dogs was due to the functioning of the devices. Histologic examination after removal showed that the devices contained viable islets with granulated β-cells and no evidence for infiltration of immunocytes.

More recently, the vascularized biohybrid pancreas has also been used for xenografts (Maki et al 1996). Porcine islets were seeded into the devices and implanted into totally pancreatectomized dogs. No immunosuppression was used. The devices remained functional for 52 to 270 days. Although none of the dogs became completely insulin independent, their insulin requirements were significantly reduced. Nine of 17 dogs maintained fasting blood glucose levels at below 11 mmol/L with a 54–74% reduction in exogenous insulin requirements. Significant levels of porcine C-peptide were detected in the dogs' sera, and there was a time-dependent improvement in glycosylated hemoglobin levels.

The PAN-PVC membrane inside a vascular biohybrid pancreas has a limited surface area. The number of islets that can be in close proximity to the membrane, where the availability of nutrients and oxygen is highest, is thus limited. Adding more islets beyond a certain critical density results in reduced insulin secretion per islet. Therefore, more insulin can only be obtained by simultaneously increasing the number of islets and the membrane surface area. However, lengthening the tubular membrane within the device results in a higher incidence of blood clotting. Since it is thus difficult to increase the amount of insulin that a single unit can provide, two devices will have to be implanted to treat the average Type I diabetic patient with an insulin requirement of approximately 40 units per day. Nevertheless, human clinical trials with such devices containing porcine islets are planned (see chapter 16, "Vascular Devices" by Maki et al).

Diffusion Chambers

Reports from our laboratory describe the use of membrane diffusion chambers manufactured from PAN-PVC tubes with inner diameters of up to 4.8 mm and wall thicknesses of 70 to 100 μm (Lanza et al 1991, 1992a, 1992b, 1992c, 1993). The membranes had a nominal molecular weight exclusion of 50–80 kD. These chambers were implanted in the peritoneal cavity instead of being anastomosed to the vascular system. Porcine, bovine, or canine islets placed in these chambers restored normoglycemia in streptozotocin diabetic rats for more then a year without immunosuppression. The external surfaces of retrieved chambers were generally free of fibrotic overgrowth and exhibited only occasional host cell adherence.

Implantation of encapsulated canine islets in spontaneously diabetic BB/Wor rats had similar results, with fasting normoglycemia for over 8 months. Intravenous glucose tolerance tests (IVGTT) showed that the K rates (decline in blood glucose levels, expressed in percent per minute) were close to values for nondiabetic control animals, increasing from less than 1 for untreated diabetic animals to approximately 3 for animals that were treated with diffusion chambers (Lanza et al 1992b). Light microscopy and electron microscopy on retrieved implants showed islets with normal histology, and well-preserved ultrastructure. Similar diffusion chambers containing canine islets have also been used in totally pancreatectomized diabetic dogs. Although the dogs had insulin requirements of up to 40 units per day, the implants reversed the diabetic state of the animals for over 10 weeks (Lanza et al 1992a). Little or no fibrosis was observed for periods as long as 30 weeks.

Diffusion chambers have a number of limitations that stand in the way of their clinical use. First, the chambers are subject to continuous bending due to movements of the abdominal organs. Over time, this leads to membrane breakage that results in the immune-destruction of the islets inside. The sharp edges of the fractures can cause peritoneal scarring and inflammation. It is possible that an increase in membrane thickness and a decrease of the chamber length could alleviate this breakage problem. Second, the inner diameter of the chambers is large (up to 4.8 mm). As described above, such diameters impose limitations on the amount of oxygen and nutrients that can diffuse into the axial centers of the chambers. As a result, cells located at the axial centers of the wide-bore chambers tend to become necrotic. Indeed, chambers that had been implanted in the peritoneal cavity for several months contained a central necrotic core with only a rim of viable islets within approximately 0.5 mm of the inner membrane wall. Third, glucose perifusion challenges of islets encapsulated in diffusion chambers have shown delays of 7 ± 1 minutes in insulin secretory responses to changes in glucose concentration. These responses are fast enough to prevent postprandial blood glucose overexcursions (Kraegen et al 1981). However, because of the greater surface-to-volume ratio, microencapsulated islets are able to respond in a more physiological fashion (lag time <2 minutes; see below). Fourth, encapsulated pancreatic islets only have a finite life expectancy. It will therefore be necessary to perform periodic booster implants. Since the PAN-PVC copolymer does not biodegrade, it may be necessary to surgically remove old chambers.

Microcapsules

Microcapsules offer potential solutions to the geometry, diffusion, breakage, and retrieval problems of the vascular and diffusion devices already discussed. First, because of their spherical shape, microcapsules have larger surface-to-volume ratios. This translates into improved diffusion characteristics. Second, the microcapsules can simply be injected, for example, into the peritoneal cavity, without the need for surgery. Third, it is possible to adjust the rate of degradation of the microcapsules to correspond to the functional longevity of the islets, eliminating the need to retrieve them. The same may be the case for some of the other biomaterials used to encapsulate islets. If necessary, for example in the event of an acute infection, microcapsules could be removed by peritoneal lavage.

Over the past two decades, a number of systems for the microencapsulation of pancreatic islets have been developed. These include the alginate-poly-l-lysine system developed by Lim and Sun (1980), the alginate-poly-ornithine system of Calafiore and Basta (1994), the barium-alginate system of Zekorn et al (1992), the chitosan system introduced by Rha (1984), the encapsulation of islets in polyacrylate by Sefton et al (1987), the use of agarose by Iwata et al (1989), and the coherent microcapsules of Calafiore et al (1995, 1997). The characteristics of these encapsulation materials, as well as the mi-

croencapsulation methods themselves, are described in separate chapters of this book. Here we describe the experience of our group with microencapsulated islets.

Xenografts in Rats Using Uncoated Alginate Spheres

We have shown that simple, uncoated alginate spheres with a large diameter of 3 to 4 mm can be used to prevent xenograft rejection in the rat with low-dose CsA (Lanza et al 1995b, 1995c). These spheres were made by extruding a mixture of 1.5% sodium alginate (Pronova, Norway) and islets through a 14–16 gauge needle into a 1.5% $CaCl_2$ solution. The calcium ions instantly gelled the alginate by the formation of ionic cross-links. Adult male Lewis rats were made diabetic 10–14 days prior to implantation, with a single injection of streptozotocin into the tail vein. Only rats with a plasma glucose level of >400 mg/dl were used. Rats were anesthetized, and 16,000 islet equivalents of free or encapsulated islets (porcine or bovine) were introduced into the peritoneal cavity. The animals received low-dose cyclosporin A (CsA, 10–20 mg/kg s.q.). The control group consisted of the animals that were implanted with free islets.

The islets in the control animals failed, with euglycemia lasting less then 1 day. In contrast, animals that received encapsulated islets all remained normoglycemic for >100 days (Figure 18.3). Nonfasting plasma glucose levels dropped from preimplantation levels of 530 ± 8 mg/dl to 149 ± 17 mg/dl for bovine and 156 ± 12 mg/dl for porcine islets during the first week. These animals were sacrificed after 100 to 150 days and a necropsy performed. The alginate capsules were found intact and free-floating in the peritoneal cavity. The external surfaces of the spheres were free of fibrosis and host-cell adherence. Immunohistochemical staining of the islets revealed mostly viable islets that were very similar to the day zero control specimens. All three types of islet cells (α, β, and δ cells) were well represented. Histology on the pancreata of the rats showed few or no granulated β-cells, confirming that the results were not due to spontaneous recovery of β-cells.

In a second set of experiments, the role of the capsule diameter in xenograft survival was investi-

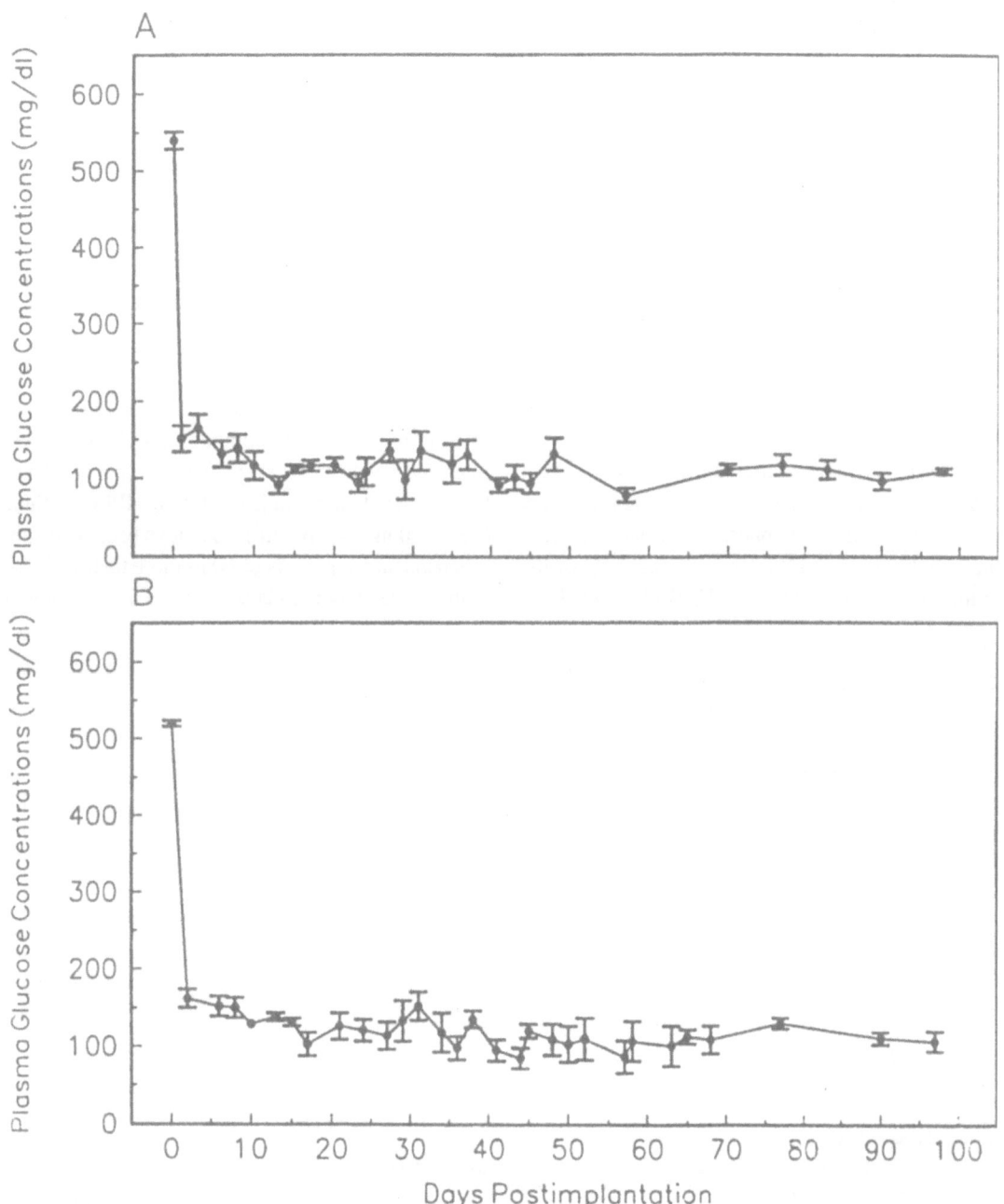

FIGURE 18.3. Mean plasma glucose levels in STZ diabetic rats that received intraperitoneal implants of alginate-encapsulated bovine (A, n = 4) and porcine (B, n = 5) islets. Except for one of the porcine islet implants (which failed at 13 days postimplantation), all of the animals received low-dose CsA (10–20 mg/kg s.q.). (Figure reprinted with permission from Lanza et al, Transplantation 59:1485–1487. Copyright 1995 by Williams & Wilkins.)

224   Willem M. Kühtreiber, Robert P. Lanza, and William L. Chick

gated (Lanza et al 1995b). Porcine islets were immobilized in uncoated alginate capsules with diameters ranging from 0.8 to 3.7 mm and implanted
in the peritoneal cavity of nonimmunosuppressed
streptozotocin induced diabetic rats for 2 weeks.
The viability of the islets in the retrieved capsules
ranged from 0% in spheres with diameters of up to
1.6 mm to 85% in the 3.7 mm spheres. It was thus
clear that the survival of the islets improved with
increasing diameter. However, the protection fur-
nished by capsules with large diameters was only
temporary. Even for the largest capsules, the viability of the enclosed islets dropped to less than
20% after 1 month. When low-dose CsA was administered, long-term survival increased dramatically to in vitro control levels.

The results with uncoated alginate spheres, presented above, are remarkable if one realizes that
these simple, uncoated spheres are readily permeable to antibodies and the large components of the
complement system, such as C1Q. In fact, we have

shown (Lanza et al 1995c) that such spheres are
permeable to molecules as large as bovine thyroglobulin (MW 669 kD) and Blue Dextran (average MW 2600 kD).

Allografts in Dogs Using Uncoated Alginate Microcapsules

We also achieved long-term function of allografts in
spontaneously diabetic dogs, using uncoated alginate microcapsules (Lanza et al 1995d, 1996, 1998).
Canine islets were encapsulated in uncoated alginate
microcapsules (diameter 800–1200 μm) and implanted into the peritoneal cavities of five spontaneously diabetic dogs. Although the dogs received
low-dose Cyclosporin A, by 3 weeks postimplantation, the levels were undetectable by HPLC methods
(i.e., <30 ng CsA per ml). The dogs became insulin
independent for periods of 60 to more then 175 days
with average blood glucose levels of 122 ± 4 mg/dl

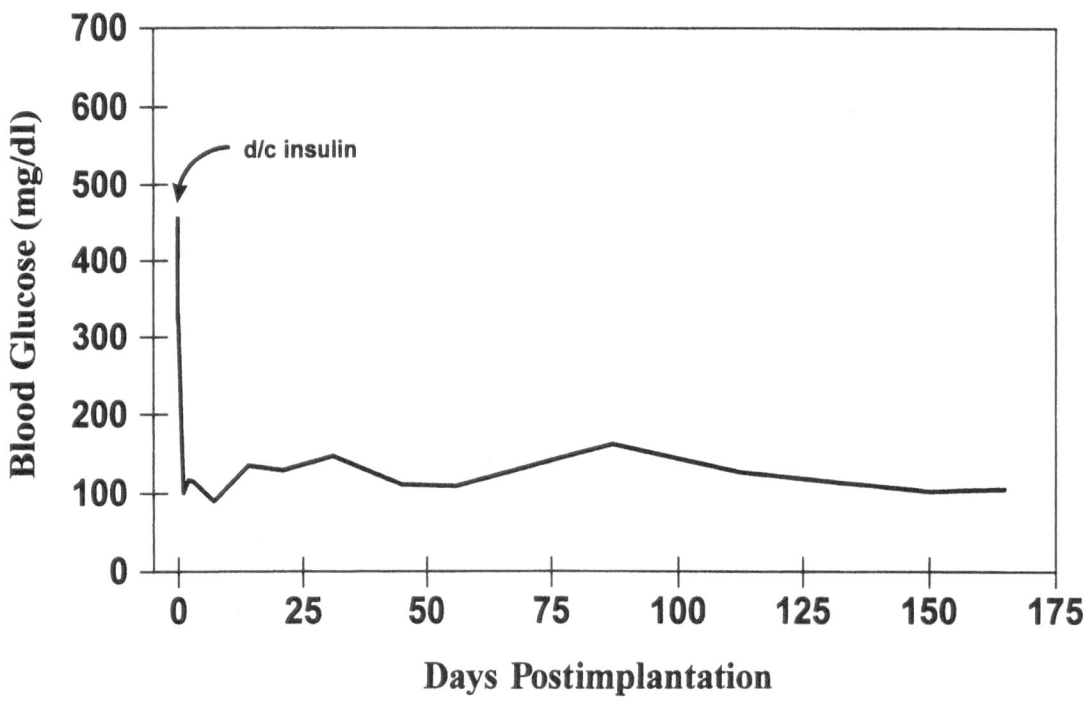

FIGURE 18.4. Fasting blood glucose concentrations of a spontaneously diabetic dog that received an intraperitoneal
implant of alginate encapsulated canine islets.

during the first 2 months (Figure 18.4). Glycosylated hemoglobin levels (Hb_{AlC}) improved from 6.7 ± 0.5% to 4.2 ± 0.2%. Intravenous glucose tolerance tests showed that the glucose clearance rates were also significantly improved.

The above experiments showed that for allografts there was no need for a permselective barrier. Indeed, we were able to achieve results similar to published work that employed PLL-coated microcapsules. For example, Soon-Shiong et al (1993) used PLL-coated microspheres containing canine islets to treat spontaneously diabetic dogs that were receiving a cyclosporin A regimen. The dogs did not require exogenous insulin therapy for periods of 63 to 140 days. Limited success in the use of polyamino acid coatings in the transplantation of encapsulated islets has also been reported by Calafiore et al (1991, 1992, 1994). They encapsulated human pancreatic islets in poly-l-ornithine (PLO) coated alginate microbeads. These beads were then loaded into coaxial vascular chambers and grafted as bypasses into two Type I human diabetics and into three dogs that were rendered diabetic by treatment with alloxan. No immunosuppression was used. Insulin independence was achieved in one out of three dogs (for about 1 month) and transiently (4 days) in one of the two human diabetic patients.

Sun et al also reported on the use of PLL-coated capsules in higher animals (1996). They encapsulated porcine islets in alginate-PLL-alginate capsules and transplanted them intraperitoneally into nine spontaneously diabetic cynomologus monkeys with preimplant insulin requirements of 3–11 units per day. Some monkeys received multiple transplants of 30,000 to 70,000 islets. Seven monkeys became insulin independent for periods ranging from 120 to 804 days, with fasting blood glucose levels in the normoglycemic range. The monkeys had improved glucose clearance rates and decreased hemoglobin A_1C levels within 2 months after implantation. Porcine C-peptide was detected in all monkeys throughout the normoglycemic period. Capsules recovered from two monkeys after 3 months were physically intact, free of fibrosis, and with enclosed islets clearly visible.

As described above, our studies showed that uncoated microcapsules were able to adequately protect allografts in dogs. Therefore, these uncoated reactors may well suffice for the treatment of human diabetes with encapsulated human islets. Unfortunately, the availability of human donor pancreas is too limited for the large-scale application of this method. These simple microcapsules may also be applicable for allografting other primary or genetically engineered human-derived cells, to treat patients in need of hormone or enzyme replacement therapy.

Xenografts Using a New Type of Microcapsule

As illustrated above, we achieved prolonged survival and function of xenografts in rats and of allografts in dogs with the use of uncoated microcapsules. However, these simple microcapsules failed when xenografts were attempted in dogs. We have therefore further improved the immunoprotective properties of the microcapsules by decreasing their molecular permeability, using proprietary methods. Porcine islets were immobilized in these improved microcapsules and implanted I.P. in normal Lewis rats. No immunosuppression was used. Controls consisted of uncoated alginate spheres as described above. Two weeks postimplantation, no surviving islets were found in the controls. In contrast, islets immobilized in the new microcapsules retrieved up to 9 months after implantation were all found in excellent condition, with viabilities at preimplant levels. The improved microcapsules were intact and free floating, with their surfaces mostly free of fibrosis. Immunohistochemical staining of retrieved islets revealed well-granulated, insulin-containing β-cells. Although the periphery of the microcapsules stained intensely for rat IgG, the islet compartment remained negative, showing that immunoglobulins could not enter (Figure 18.5). We have also performed preliminary experiments with this new type of microcapsule in the canine model. We were able to recover microcapsules containing viable bovine islets after implantation into the peritoneal cavity of normal dogs for periods of 4–6 weeks both with and without low-dose immunosuppression (Lanza and Chick 1995a, 1996).

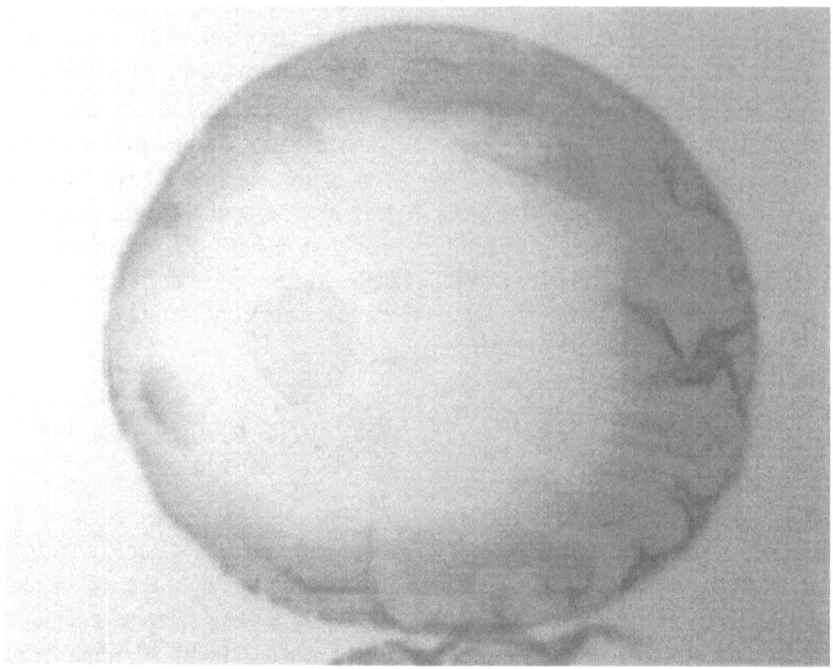

FIGURE 18.5. A microreactor removed from the peritoneum of a dog 26 days after implantation, immunoperoxidase stained for IgG. The center of the reactor was protected from antibody penetration. The faint spot in the clear, IgG-free area is a bovine islet. (Reprinted with permission from Lanza and Chick., 1995. Scientific American Science & Medicine 2:16–25. Copyright Scientific American, Inc.)

References

Calafiore R, Basta G, Falorni A, Brotzu G, Alfani D, Cortesini R, Brunetti P. 1991. Vascular graft of microencapsulated human pancreatic islets in nonimmunosuppressed diabetic recipients. Diab Nutr Metab 4:45–48.

Calafiore R. 1992. Transplantation of microencapsulated pancreatic human islets for therapy of diabetes mellitus. A preliminary report. ASAIO Journal 38:34–37.

Calafiore R, Basta G. 1994. Transplantation of microencapsulated human islets in nonimmunosuppressed diabetic patients. In: Lanza RP, Chick WL (eds). Pancreatic islet transplantation: Volume III. Immunoisolation of pancreatic islets. Austin, Texas: Landes/CRC Press. pp. 93–106.

Calafiore R, Basta G, Osticioli L, Luca G, Tortoioli C, Brunetti P. 1995. Coherent microcapsules for pancreatic islet transplantation: a new potential approach for bioartificial pancreas. Trans Proc 28:812–813.

Calafiore R, Basta G, Broselli C, Bufalari A, Giustozzi GM, Luca G, Tortoioli C, Brunetti P. 1997. Effects of alginate/polyaminoacidic coherent microcapsule transplantation in adult pigs. Trans Proc 29:2126–2127.

Chick WL, Like AA, Lauris V. 1975. Beta cell culture on synthetic capillaries: an artificial endocrine pancreas. Science 187:847–849.

The Diabetes Control and Complications Trial Research Group. 1993. The effect of intensive treatment of diabetes on the development and progression of long-term complications in insulin-dependent diabetes mellitus. N Engl J Med 329:977–986.

Dionne KE, Colton CK, Yarmush ML. 1989. Effect of oxygen on isolated pancreatic tissue. ASAOI Trans 35:739–741

Dionne KE, Colton CK, Yarmush MK. 1993. Effect of hypoxia on insulin secretion by isolated rat and canine islets of Langerhans. Diabetes 42:12–21.

Iwata H, Amemiya H, Matsuda T, Takano H, Hayashi R, Akutsu T. 1989. Evaluation of microencapsulated islets in agarose gel as bioartificial pancreas by studies of hormone secretion in culture and by xenotransplantation. Diabetes 38(Suppl. 1):224–225.

Kirchgessner J, Gerö L, Mueller-Klieser W, Kunz L, Beyer J, Schrezenmeir J. 1992. Oxygen distribution in islet organs: Effect of convection and barium-alginate encapsulation. Diabetologia 35:A188.

Kirkman RL. Of swine and men: organ physiology in different species. In: Hardy MA (ed.). Xenograft 25 Amsterdam. New York Elsevier. 1989, p. 125.

Kraegen EW, Chisholm DJ, MacNamara ME. 1981. Timing of insulin delivery with meals. Horm Metab Res 13:365–367.

Kühtreiber WM, Lanza RP, Beyer AM, Kirkland KS, Chick WL. 1993. Relationship between insulin secretion and oxygen tension in hybrid diffusion chambers. ASAIO J. 39:M247–M251.

Kühtreiber WM, Lanza RP, Chick WL. 1994a. Bovine islet isolation. In: Lanza RP and Chick WL, editors. 1994. Pancreatic islet transplantation: Volume I. Procurement of pancreatic islets. Austin, Texas: R.G. Landes Company. pp. 71–79.

Kühtreiber WM, Lanza RP, Chick WL. 1994b. Secretory function of BioHybrid pancreas devices containing isolated porcine islets. ASAIO J. 40:M789–M792.

Lanza RP, Butler DH, Borland KM, et al. 1991. Xeno-transplantation of canine, bovine, and porcine islets in diabetic rats without immunosuppression. Proc Natl Acad Sci USA 88:11100–11104.

Lanza RP, Borland KM, Lodge P, et al. 1992a. Treatment of severely diabetic, pancreatectomized dogs using a diffusion-based hybrid pancreas. Diabetes 41:886–889.

Lanza RP, Borland KM, Staruk JE, Appel MC, Solomon BA, Chick WL 1992b. Transplantation of encapsulated canine islets into spontaneously diabetic BB/Wor rats without immunosuppression. Endocrinology 131:637–642.

Lanza RP, Sullivan SJ, Chick WL. 1992c. Islet transplantation with immunoisolation. Diabetes 41:1503–1510.

Lanza RP, Beyer AM, Staruk JE, Chick WL. 1993. Biohybrid artificial pancreas: Long term function of discordant islet-xenografts in streptozotocin diabetic rats. Transplantation 56:1067–1072.

Lanza RP, Solomon BA, Monaco AP et al. 1994. Devices implanted as AV shunts. In: Lanza RP, Chick WL (eds). Pancreatic islet transplanation: Volume III. Immunoisolation of pancreatic islets. Austin, Texas: Landes/CRC Press. pp. 157–168.

Lanza RP and Chick WL. 1995a. Encapsulated cell therapy. Scientific American Science and Medicine 2:16–25.

Lanza RP, Ecker D, Kühtreiber WM, Staruk JE, Marsh J, Chick WL. 1995b. A simple method for transplanting discordant islets into rats using alginate gel spheres. Transplantation 59:1485–1487.

Lanza RP, Kühtreiber WM, Ecker D, Staruk JE, Chick WL. 1995c. Xenotransplantation of porcine and bovine islets without immunosuppression using uncoated alginate microspheres. Transplantation 59:1377–1384.

Lanza RP, Kühtreiber WM, Chick WL. 1995d. Encapsulation technologies. Tissue Engineering 1995 1:181–196.

Lanza RP, Chick WL. 1996. Pancreas. In: Lanza RP, Chick WL (eds) 1996/97 Yearbook of cell and tissue transplantation. Dordrecht, The Netherlands: Kluwer Academic Publishers. pp. 253–264.

Lanza RP, Ecker DM, Kühtreiber WM, Marsh JP, Ringeling J, Chick WL. 1999. Transplantation of islets using microencapsulation: Studies in diabetic rodents and dogs. J Molecular Medicine 77:206–210.

Lim F, Sun AM. 1980. Microencapsulated islets as bioartificial endocrine pancreas. Science 210:908–910.

Maki T, Otsu I, O'Neil JJ, Dunleavy K, Mullon CJP, Solomon BA, Monaco AP. 1996. Treatment of diabetes by xenogeneic islets without immunosuppression. Use of a vascularized bioartificial pancreas. Diabetes 45:342–347.

Ohta M, Nelson D, Nelson J, Meglasson MD, Erecinska M. 1990. Oxygen and temperature dependence of stimulated insulin secretion in isolated rat islets of Langerhans. J Biol Chem 265:17525–17532.

Rajan AS, Aguilar-Bryan L, Nelson DA, Yaney GC, Hsu WH, Kunze DL, Boyd III AE. 1990. Ion channels and insulin secretion. Diabetes Care 13:340–363.

Rha C. 1984. Chitosan as a biomaterial. In: Colwell RR, Parish ER and Sinskey AJ (eds) Biotechnology in the marine sciences. New York: Wiley.

Robertson PA, Seaquist ER, Walseth TF. 1991. G Proteins and modulation of insulin secretion. Diabetes 40:1–6.

Schrezenmeir J, Gerö L, Laue C, Kirchgessner J, Müller A, Hüls A, Passman R, Hahn HJ, Kunz L, Mueller-Klieser W, Altman JJ. 1992. The role of oxygen supply in islet transplantation. Transplantation Proceedings 24:2925–2929.

Sefton MV, Dawson RM, Broughton RL, Blysnivk J, Sugamori ME. 1987. Microencapsulation of mammalian cells in a water insoluble polyacrylate by coextrusion and interfacial precipitation. Biotechnol Bioeng 29:1135–1143.

Soon-Shiong P, Feldman E, Nelson R, et al 1993. Long-term reversal of diabetes by the injection of immunoprotected islets. Proc Natl Acad Sci USA 90:5843–5847.

Sullivan SJ, Maki T, Borland KM, Mahoney MD, Solomon BA, Muller TE, Monaco AP, Chick WL. 1991. Biohybrid artificial pancreas: long-term implantation studies in diabetic, pancreatectomized dogs. Science 252:718–721.

Sumimoto R, Jamieson NV, Wake K, Kamada N. 1989. 24-hour rat liver preservation using UW solution and some simplified variants. Transplantation 48:1–5.

Sun Y, MA X, Zhou D, Vacek I, Sun AM. 1996. Normalization of diabetes in spontaneously diabetic Cynomologus monkeys by xenografts of microencapsulated porcine islets without immunosuppression. J Clin Invest 98:1417–1422.

van Haeften TW. 1989. Clinical significance of insulin antibodies in insulin-treated diabetic patients. Diabetes Care 12:641–648.

Warnock GL, Rajotte RV. 1988. Critical mass of purified islets that induce normoglycemia after implantation into diabetic dogs. Diabetes 37:467–470.

Zekorn TDC, Horcher A, Siebers U, Schnettler R, Hering B, Zimmermann U, Bretzel RG, Federlin K. 1992. Barium-cross-linked alginate beads: A simple, one-step-method for successful immuno isolated transplantation of islets of Langerhans. Acta Diabetologica 29:99–106.

19
Microencapsulated Islets in Type I Diabetics: Clinical Experience

Patrick Soon-Shiong

Introduction

Nonencapsulated pancreatic islets, transplanted into Type I diabetic patients, have generally yielded disappointing results because immune rejection has remained a major obstacle to successful treatment, despite the use of high-dose immunosuppressive regimens. Immunosuppressive drugs are necessary for transplantation of organs or tissues directly into human patients, but severe side effects frequently limit their continued use and diminish their therapeutic effectiveness.

Encapsulation circumvents the need for immunosuppressive drugs because the transplanted living cells are surrounded by a semipermeable membrane that protects the cells from the host's immune system. Since encapsulation prevents rejection, investigators have sought ways to reverse insulin-dependent diabetes by a simple injection of immunoprotected insulin-secreting cells without immunosuppression therapy (Darquy and Reach 1985, Lim and Sun 1980, O'Shea et al 1984, Sun et al 1984, Sun and Shea 1985). The idea of using encapsulation to prevent the immune system from gaining unwanted access to cells evolved more than 40 years ago (Prehn et al 1954), when the advantages of diffusion-chamber devices were explored. Since that time, many kinds of immunoisolation systems have been described (Albisser et al 1974, Altman et al 1981, Archer et al 1980, Chick et al 1977, Gates and Lazarus 1977, Reach et al 1981, Strautz 1970, Sun et al 1977, Sullivan et al 1991, Theodoru et al 1979, Tze et al 1976, Tze et al 1980). However, clinical applications of those early devices raised significant challenges, such as their mechanical fragility, their lim-

ited surface area, and (in the case of vascularized devices or chamber systems) their need for a major surgical procedure with the risk of thrombosis and infection.

Microencapsulation technologies obviate many of these difficult technical issues. They also provide safe and simple techniques for implanting immunoisolated cells into various sites of the human body. Previous attempts at implantation of microencapsulated islets were unsuccessful because the early semipermeable membranes did not form a biocompatible and mechanically stable immunoprotective system that could allow sufficient oxygenation of the enclosed insulin-secreting cells and could also provide adequate in vivo kinetics of insulin diffusion in response to a glycemic signal.

Alginate-Based Microcapsules

The reversal of diabetes in rats by alginate-based encapsulated islets was first reported nearly 20 years ago (Lim and Sun 1980). Since that time, successful attempts in large animal models or in Type I diabetic patients have been reported only very recently (Soon-Shiong et al 1992, Soon-Shiong et al 1993, Sullivan et al 1991). The progress of this technology had been limited by a lack of understanding of the fundamental mechanisms affecting biocompatibility of the alginate capsule membrane.

It is now known that some of the factors affecting long-term viability of alginate-based encapsulated islets include: the immunostimulatory capacity of the alginate material, the mechanical integrity of the microcapsule, the chemical stability of the alginate membrane, and the cell-adherence propensity of the outer capsule membrane. These factors are themselves modulated by the sequential structure, chemical composition, and molecular size of the alginate polymer, as well as by the kinetics of the gel-formation process.

Alginate microcapsules provide many important practical advantages for encapsulation of living cells: they have a large surface area in comparison with intravascular devices, they supply enhanced nutrition and oxygen to the cells, they permit precise tailoring of the porosity of their alginate polylysine membranes to allow diffusion of nutrients while simultaneously excluding immunoglobulins, they minimize overall risk of membrane failure by using thousands of small membranes instead of a single large membrane, and they can be injected directly into the peritoneal cavity or implanted with minimal invasive surgery. Microencapsulation within spherical beads, especially alginate-polylysine beads, is currently one of the most effective immunoisolation methods for cell therapy (Thu et al 1996a, Thu et al 1996b). Research and development efforts have brought about continuous improvements to the biocompatibility, mechanical strength, and chemical stability of alginate-polylysine microcapsules. These technical accomplishments have finally made possible the successful therapeutic transplantation of human-insulin-producing cells into a human diabetic patient.

Therapeutic Applications of Alginate Microcapsules

The results reported here describe the first known achievement of long-term insulin independence following an intraperitoneal transplantation of alginate-polylysine-encapsulated islets in a Type I diabetic patient (Soon-Shiong et al 1994). These encouraging results suggest that therapeutic implantation of microencapsulated living cells may someday become the preferred treatment for early-onset diabetes and may thereby prevent the complications of this disease.

Improved glycemic control delays the onset and slows the progression of neuropathy, nephropathy,

and retinopathy in insulin-dependent diabetic patients (Diabetes Control and Complications Trial Research Group 1993). Physicians thus face the challenge of intervening before diabetes leads inexorably to one or more of its secondary complications, such as renal failure, blindness, or coronary and peripheral vascular-occlusive disease. Since conventional insulin therapy has failed to achieve tight glucose control, an alternate treatment for diabetes has been urgently needed: transplantation of encapsulated human islets using a minimally invasive minor surgical procedure, without the risks of high-dose or lifelong immunosuppression.

The shortage of donor human tissue for encapsulation will be overcome eventually by the use of either xenograft mammalian (e.g., porcine) islets or proliferated human islets. Thus, encapsulated islet therapy may be made available to the millions of diabetic patients who could benefit from islet cell transplantation.

Diabetes Reversal in Dogs, Using Microencapsulated Islets

Ten insulin-dependent spontaneously diabetic dogs were enrolled into a blind controlled study of microencapsulated canine islets, in comparison with free unencapsulated islets (Soon-Shiong et al 1992, Soon-Shiong et al 1993). Each dog had an insulin requirement of 1–4 units/kg per day, diabetic K values of 0.6 ± 0.6, and an absence of cumulating C-peptide. The first four animals all received encapsulated islets. The remaining six dogs were randomized to receive either free or encapsulated islets. The veterinarians monitoring the recipients posttransplant were not informed about the choice of therapy.

Pancreatic islets were isolated from canine donors using a standard collagenase digestion technique, and were purified using a two-phase aqueous separation solution (Lanza et al 1990). Encapsulated or free islets were implanted by a simple injection via a 2 cm incision in the abdominal cavity. Serum glucose levels were determined daily following transplant, and euglycemia was achieved within 24 hours (serum glucose falling from 304 ± 117 to 116 ± 72 mg/dl) in all seven encapsulated islet recipients. An intravenous glucose tolerance test

(IVGTT), performed on all recipients at day 14, demonstrated normalization of K values from a pretransplant level of 0.6 ± 0.4 to a posttransplant level of 2.6 ± 0.6. All animals receiving encapsulated islets remained euglycemic, free of the need for exogenous insulin, for a period of 63–172 days, with a median insulin-independence period of 105 days. In contrast, the recipients receiving free islets with cyclosporine rejected their grafts within 7 days of implantation.

Of the seven dogs receiving encapsulated islets in the initial trial (Soon-Shiong et al 1992), five remained available for follow-up for 2 years. In the follow-up study (Soon-Shiong et al 1993), three of these five dogs received a second transplant after loss of insulin independence, although they still had ongoing islet function from the first transplant at the time of the second transplant. The duration of insulin independence from the first transplant (125 $\pm$ 33 days; range, 95–172 days) did not differ significantly from the duration following the second transplant (102 $\pm$ 20 days; range, 83–130 days). One of the five dogs received a third injection of encapsulated islets. This dog demonstrated insulin independence for 172 days after the first transplant, 83 days after the second, and 138 days after the third. The anti-inflammatory dose of cyclosporine A was discontinued 30 days after transplantation in all five dogs. Thus, these recipients were followed for a range of 110 days to more than 600 days on no immunosuppressive agents at all.

Tight glycemic control was achieved in all five dogs receiving encapsulated islets, as shown by significant improvements in hemoglobin Alc levels, in K values, and in C-peptide release. Long-term islet function was also achieved in all five dogs. Grafts survived for 228 to 641 days in the dogs receiving multiple implants, and for more than 726 days in the dog receiving a single transplant. A strict definition of positive C-peptide secretion was chosen as evidence of ongoing graft survival, even though this choice unfavorably skews our data when compared to the definition of graft survival reported by other investigators who used reduction of insulin dose as the measure for ongoing islet function. According to the insulin-dose criterion, ongoing islet survival would be between 619 and 780 days in four of these five dogs.

The potential duration of long-term graft survival following a single injection was highlighted by the dog receiving only a single encapsulated islet transplant. This recipient demonstrated continuous islet function for more than 726 days, as evidenced by ongoing basal C-peptide secretion (0.18 pmol/mL). On the basis of these preclinical data, as well as controlled studies demonstrating absence of acute toxicity or mutagenicity of the purified high-G-alginate material, we initiated the first human clinical trial to assess the safety and efficacy of intraperitoneally transplanted encapsulated human islets in Type I diabetic patients.

Human Clinical Trials with Encapsulated Islets

Governmental regulatory (FDA) and institutional review board (IRB) approvals were obtained to initiate these trials in insulin-dependent diabetic patients with functioning renal grafts, who were currently candidates for whole organ pancreas transplantation. Eligibility for this study was limited to Type I diabetic patients 18–50 years of age with evidence of insulin dependence for more than 5 years, with a functioning stable kidney transplant, with glucagon-stimulated C-peptide levels less than 0.2 ng/mL, and with an absence of severe coronary artery disease.

The first patient (Soon-Shiong et al 1994) to receive encapsulated islets was a 38-year-old white male with insulin-dependent diabetes for 30 years, requiring a mean $\pm$ SE of 0.7 ± 0.01 units insulin/kg/day, amounting to 45 to 50 units of insulin daily. The patient suffered from severe complications of the disease, including a 7-year history of lower extremity peripheral neuropathy (daily sharp shooting pains of the left lower foot with progressive sensory loss), ulcerations of the left foot, retinopathy, left eye vitrectomy, and end-stage renal failure resulting in a living-related kidney transplant. The patient's renal function was stable (serum creatinine 1.0 mg/dl) on low-dose maintenance immunosuppression of cyclosporine and Imuran 50 mg daily. In the 9-month post-islet-transplant period, the patient's cyclosporine A level ranged from 59 to 197 ng/mL.

Encapsulated islets were injected directly into the peritoneal cavity through a 2 cm midline incision. A full therapeutic dose would have been

20,000 islets/kg (based on our preclinical data), but this patient was given a subtherapeutic dose of 10,000 islets/kg at the first procedure, followed by supplemental doses of 5000 islets/kg each at 7 months and at 33 months, as part of a dose escalation study. To date, the patient has had no adverse effects. Despite the significant reduction of insulin, his body weight has increased from 68.1 to 72.7 kg (Table 19.1).

Ongoing insulin secretion from the intraperitoneally transplanted encapsulated islets was demonstrated for more than 58 months. Insulin secretion from the transplanted cells was detected within 24 hours after injection of the initial subclinical dose of 10,000 islets/kg. On the second postoperative day, all exogenous insulin was discontinued. The patient maintained normoglycemia and tolerated breakfast and lunch without requiring any insulin therapy. Occasional hyperglycemic episodes were noted postprandially, so the patient was placed on a minimal dose of insulin to maintain normoglycemia. He maintained stable daily mean blood glucose levels ranging from 129 ± 3 to 150 ± 5 mg/dl, with less lability relative to his pretransplant levels, while on a significantly ($P < 0.001$) reduced dose of insulin of approximately 0.2 units/kg/day for a period of 6 months (Figure 19.1, Table 19.1).

Following the dose escalation of a further 5000 islets/kg in the 7th month, the patient's insulin requirements dropped even further to .07 units/kg/day, and he achieved insulin independence in the 9th month (Figures 19.1 to 19.3, Table 19.1). The patient maintained insulin independence for the entire 30-day period, with a daily mean blood glucose of 134 ± 4.2 mg/dl (114 observations) and an M Value of 1.16. During this period of insulin independence, he demonstrated no evidence of ketosis, and maintained his body weight at 67.3 kg (Figure 19.2).

Fasting proinsulin levels were high (1.24 ng/ml), suggesting stress from the subclinical dose of islets, and it was decided to supplement the patient with NPH insulin on a daily basis. He was thus returned to exogenous insulin therapy and for 58 months demonstrated ongoing islet function with tight glycemic control (Table 19.1). Significantly, in the face of this tight glycemic control, the patient reported no episodes of symptomatic hypoglycemia,

in sharp contrast to the one to two episodes of insulin reaction noted routinely on a weekly basis prior to the transplant.

During the long-term follow-up period, the patient found that there were occasions when he had to remove himself from exogenous insulin in order to maintain normoglycemia. For example, on days 416 and 417, the patient's mean blood glucose levels were 79 and 106 mg/dl while on zero exogenous insulin, confirming long-term islet function from the encapsulated islet transplant (Table 19.2). In the 20th month, his daily mean blood sugar was 120 ± 3.9 mg/dl (120 observations), with a normal M value of -3.97 (Table 19.1). Basal C-peptide secretion increased, concomitant with the drop in insulin requirement, from a pretransplant level of less than 0.1 ng/ml to a posttransplant fasting level of 1.0 ng/ml at the 8th month (Figure 19.4, Table 19.1), confirming sustained insulin secretion from the encapsulated islets.

M value showed impressive improvement at 6 months (0.20), at 20 months (-3.88), and at 31 months (-1.91), compared to the pretransplant level (4.35), indicating significantly less glycemic lability since the transplant (Figure 19.4, Table 19.1). These improvements of glycemic control over an extended time period were corroborated by improvements in both glycosylated serum albumin and glycosylated hemoglobin levels. Glycosylated serum albumin decreased from 10.6% pretransplant to 5.1% at 6 months, 5.1% at 13 months, and 3.4% at 33 months. Glycosylated hemoglobin levels fell from 9.3% pretransplant to 8.1% at 20 months and 7.3% at 48 months (Figure 19.5).

The patient reported subjective improvements in his lower extremity peripheral neuropathy symptoms. The sharp shooting pains in his left lower foot, which occurred constantly on a daily basis pretransplant, abated in the posttransplant period. EMG studies (Figure 19.6) confirmed improvement in axonal nerve function by demonstration of continued bilateral increases of amplitude of peroneal motor latencies at 3 months, 6 months, and 11 months posttransplant (from 45 m.v. to 194 m.v. at 11 months on the left, and from 200 m.v. to 425 m.v. at 11 months on the right). The left foot ulcer, which required approximately 3 months to heal pretransplant, recurred on week 10 posttransplant. This foot ulcer healed completely within 7 days, possibly

(continued on page 238)

TABLE 19.1. Indices demonstrating improved glycemic control and insulin requirement postencapsulated islet transplantation in patient #1. (HR1) 3/19/98

	Pre-Tx	Post-Tx														
		1 mo.	2 mo.	3 mo.	4 mo.	5 mo.	6 mo.	7 mo.	8 mo.	9 mo.	10 mo.	11 mo.	12 mo.	13 mo.	14 mo.	15 mo.
		(1st Tx)							(2nd Tx)							
Blood Glucose Evaluation																
# of observations	60	201	182	154	114	120	117	131	120	114	116	117	120	116	111	65
Daily Mean ± SE (mg/dl)	146 ±7.1	150 ±3.62*	150 ±4.01*	141 ±4.18*	150 ±4.95*	131 ±4.10*	129 ±2.61*	132 ±4.79*	144 ±4.52*	134 ±4.20*	164 ±5.6*	127 ±4.94*	124 ±1.69**	117 ±1.59**	108 ±3.50**	108 ±4.69**
<50 mg/dl episodes (%)	0.00	0.47	0.00	0.00	0.00	0.00	0.00	0.00	0.83	0.00	0.00	0.9	0.00	0.00	1.8	0
>200 mg/dl episodes (%)	11.7	15.42	16.48	4.55	15.79	5.00	1.71	4.57	10.00	6.14	22.41	7.6	0.00	0.86	0.00	0
M-Value (Standard=120 mg/dl)	4.35	3.77	4.83	1.72	3.60	0.53	0.20	0.45	3.04	1.16	6.18	0.35	-0.22	-0.83	-3.97	-2.56
Body Weight (Kg)	68.1	68	68.1	69.1	67.7	69.5	67.3	67.3	66.4	67.3	68.1	67	68.2	69.1	69.1	69.5
Daily Insulin Requirement																
Daily Mean ± SE (Units/kg/day)	0.69 ±0.01	.21 ±0.01**	.21 ±0.01**	.22 ±0.01**	0.27 ±0.00**	.25 ±0.01**	0.25 ±0.01**	0.22 ±0.01**	0.07 ±0.01**	0.00 ±0.00**	0.14 ±0.00**	.31 ±0.00**	.35 ±0.00**	.42 ±0.00**	0.33 ±0.00***	.28 ±0.03**
Metabolic Parameters																
Fasting Pro-Insulin (ng/ml)	<0.04	0.92	—	0.68	0.36	—	0.28	0.80	0.59	1.24	—	0.33	1.06	2.30	1.63	—
Fasting C-Peptide (ng/ml)	0.1	0.70	0.40	0.30	0.60	—	0.6	0.60	1.0	—	—	0.6	0.3	0.2	0.1	0.3
Hemoglobin A1c (% TL HB)	9.3	8.70	8.20	7.40	7.60	—	7.8	7.90	7.8	7.9	—	7.1	8.1	7.9	8.4	—
Glycosylated Albumin (% TL ALB.)	10.6	5.20	4.40	4.40	4.30	—	5.1	5.30	5.1	5.3	—	4.2	5.6	5.1	3.0	—
Renal Function																
Creatinine	1.1	1.0	—	1.2	—	—	1.3	1.0	1.1	1.0	—	1.0	1.0	1.1	1.1	—
BUN	19	17	—	22.5	—	—	25	23	22.1	23	—	22.7	19.8	27.6	27.6	—

(continued)

233

TABLE 19.1. (continued).

	Post-Tx															
	16 mo.	17 mo.	18 mo.	19 mo.	20 mo.	30 mo.	31 mo.	33 mo.	34 mo. (3rd Tx)	35 mo.	36 mo.	43 mo.	44 mo.	48 mo.	55 mo.	58 mo.
Blood Glucose Evaluation																
# of observations	116	120	120	120	120	92	99	160	176	84.00						
Daily Mean ± SE (mg/dl)	119 ±3.7**	118 ±4.0**	127 ±2.8**	128 ±2.8**	120 ±3.9**	127 ±3.03**	114 ±2.00**	121 ±10.8	158 ±40.0	152 ±5.0						
<50 mg/dl episodes (%)	1.72	0.8	0.8	0.00	0.00	0.00	0.00	0.63	0.00	0.00						
>200 mg/dl episodes(%)	2.59	3.33	2.5	2.5	1.67	2.17	0.00	1.88	9.70	15.50						
M-Value (Standard = 120 mg/dl)	−2.21	3.38	−1.02	−0.73	−3.88	−0.20	−1.91	−0.51	5.43	4.50						
Body Weight (Kg)	72.7	72.7	72.7	72.7	72.7	73.2	73.2	70	70	70	72			79.5	76	73
Daily Insulin Requirement																
Daily Mean ± SE (Units/kg/day)	.44 ±0.00**	.51 ±0.00**	.57 ±0.0*	.59 ±0.00**	0.59 ±0.00**	.51 ±0.01**	0.52 ±0.00**	.49 ±0.17**	0.46 ±0.14**	0.51 ±0.02			0.6	0.55	0.55	
Metabolic Parameters																
Fasting Pro-Insulin (ng/ml)			0.45				0.23	0.34	0.45	0.72	0.37	0.87	0.6	0.32	0.7	0.7
Fasting C-Peptide (ng/ml)					<0.1		0.3	0.30	1.2	1.7	1.7	0.6		0.9	0.7	
Hemoglobin A1c (% TL HB)	7.6				8.1		6.4	7.50	6.4	6.9	7.4	6.3	6.8	7.3	7.2	7
Glycosylated Albumin (% TL ALB.)								3.40	4.1	3.9	2.4	3.8	3.1	3.6	3.1	2.9
Renal Function																
Creatinine				1.0	1.1		1.0	0.9	1		1	0.9	1	0.9	0.9	
BUN				29.4	21.20		24	19.00	19	20	18	15	23	18	17	

* P > 0.05, Not significantly different from Pre-Tx

** P < 0.05, Significantly different from Pre-Tx

*** P < 0.001, Significantly different from Pre-Tx

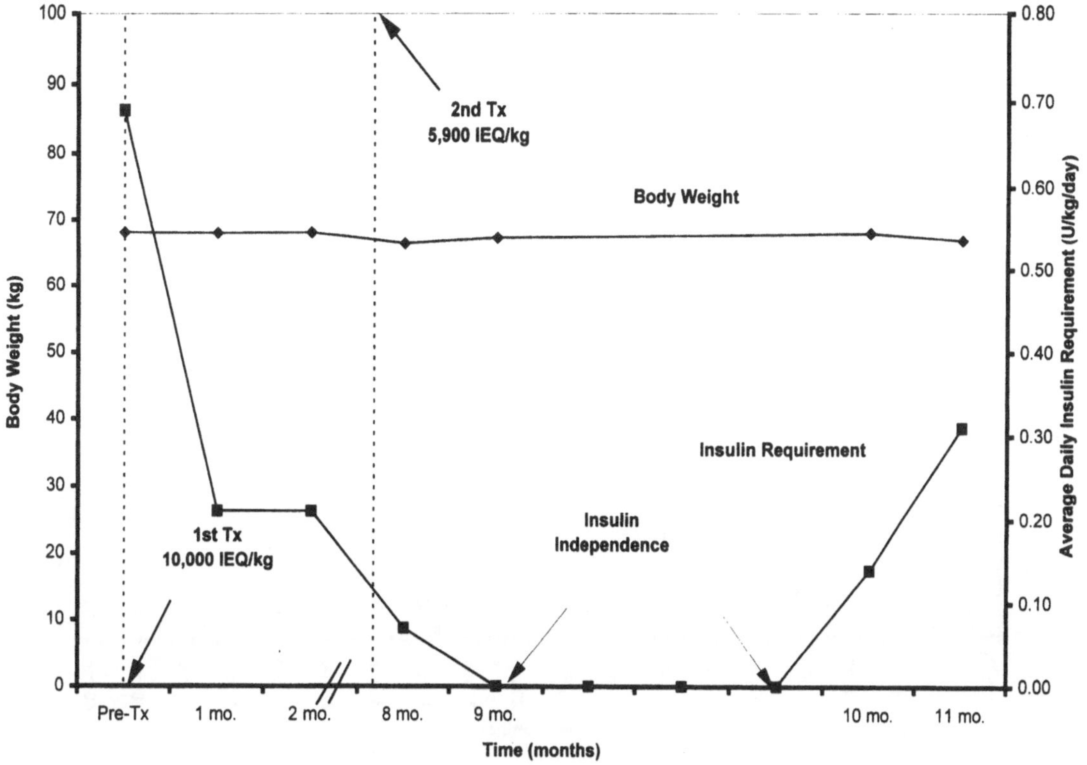

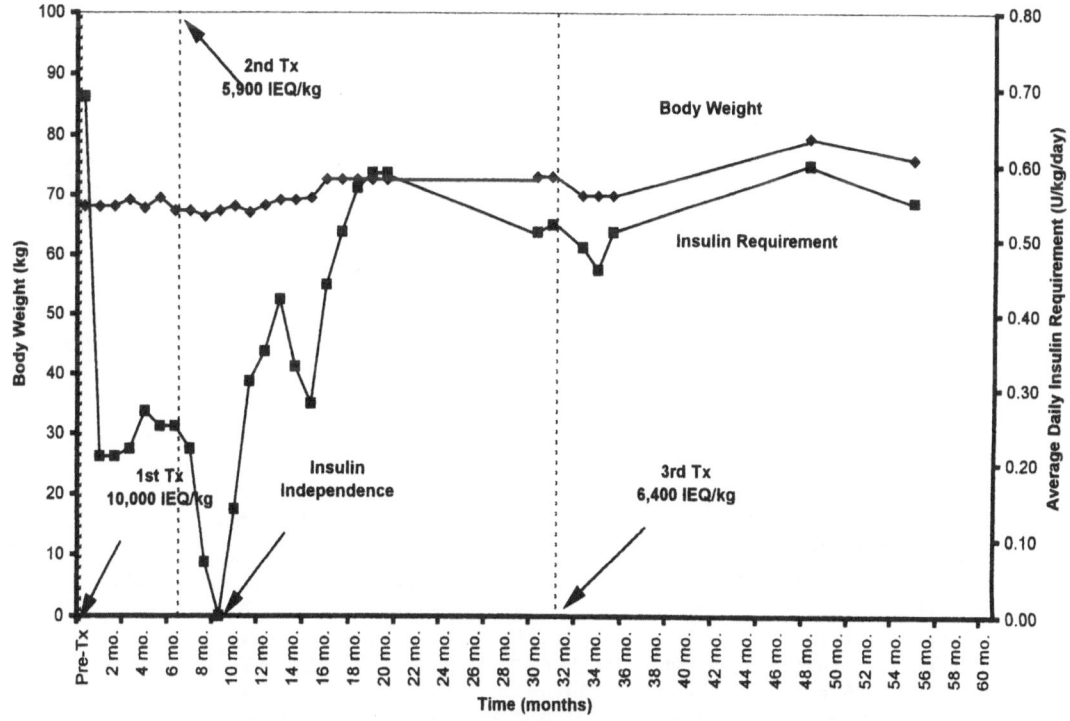

FIGURE 19.1. Improved gylcemic control postencapsulated islet transplantation: body weight and insulin requirement pre-Tx to 11 months post-Tx-1.

FIGURE 19.2. Improved glycemic control postencapsulated islet transplantation: glucose levels and insulin requirements following encapsulated islet transplantation: body weight and insulin requirement pre-Tx to 55 months post-Tx-1.

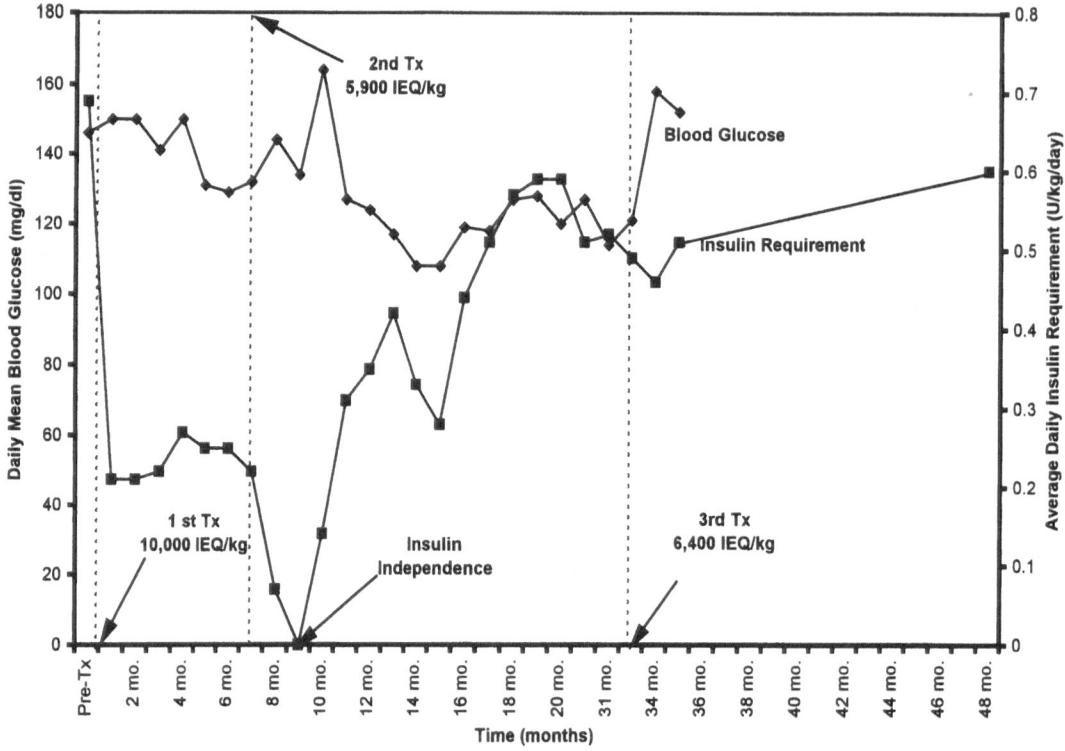

FIGURE 19.3. Improved glycemic control postencapsulated islet transplantation: glucose levels and insulin requirements following encapsulated islet transplantation: blood glucose levels and insulin requirements.

TABLE 19.2. Blood sugar levels on days 416 and 417 while on no insulin therapy, demonstrating long-term islet function.

Time	Day 415	Insulin	Day 416	Insulin	Day 417	Insulin
0700	65 mg%	6 units	49 mg%	0 units	83 mg%	0 units
1300	76 mg%	—	72 mg%	—	120 mg%	—
1900	53 mg%	4 units	90 mg %	—	122 mg%	—
2300	69 mg%	—	102 mg%	—	100 mg%	—
	X = 66 mg%	10 units	X = 79 mg%	0 units	X = 106 mg%	0 units

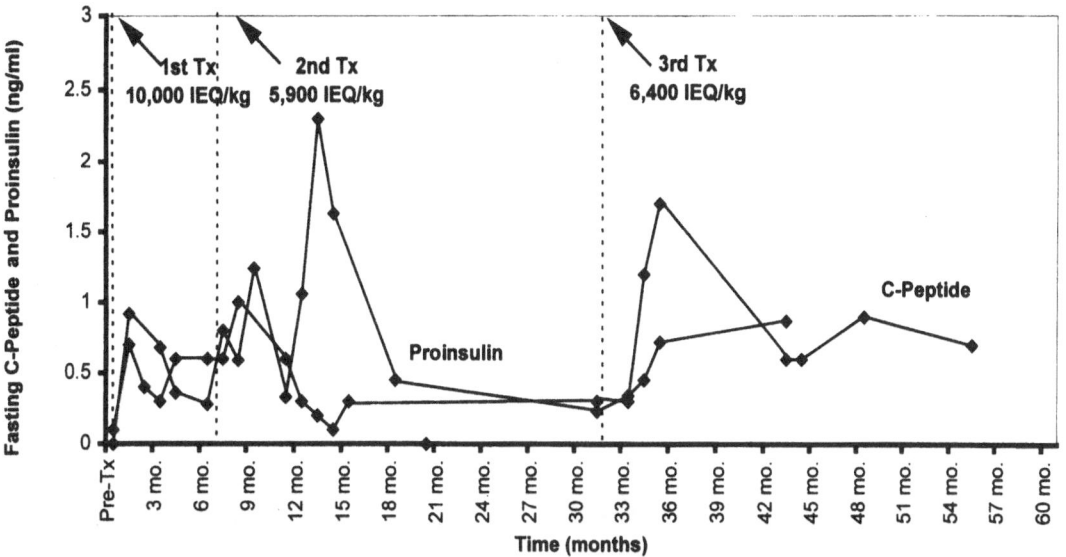

FIGURE 19.4. Islet beta cell function parameters: fasting C-peptide and proinsulin.

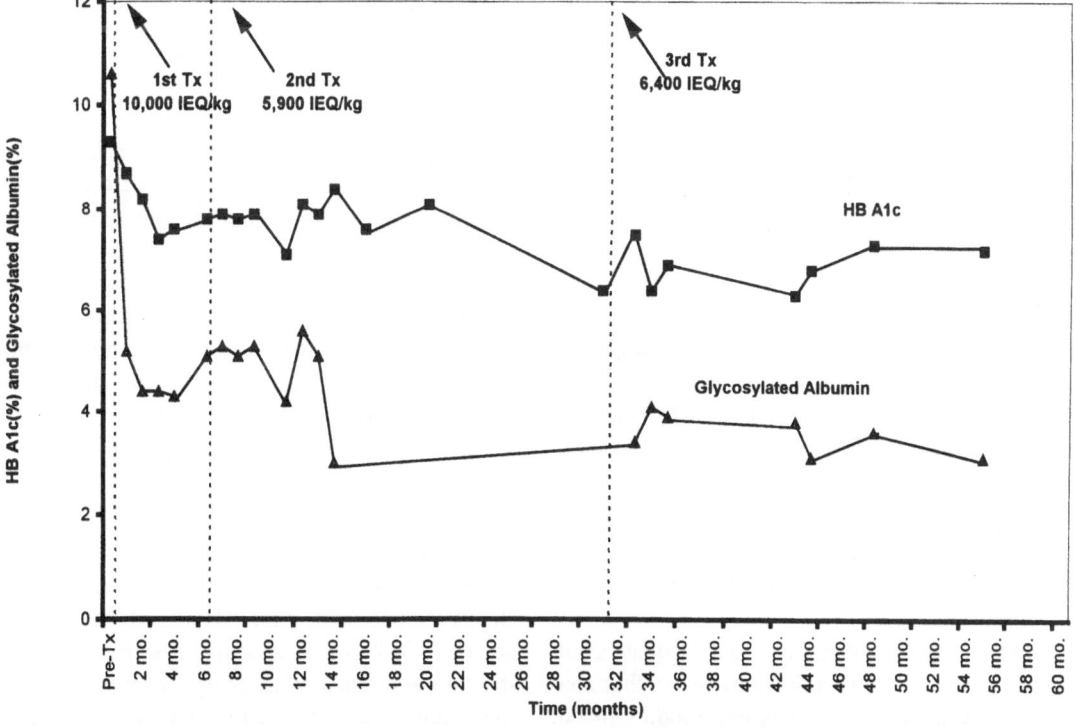

FIGURE 19.5. Glycemic control parameters: hemoglobin A1c (HbA1c) and glycoalbumin.

Improved EMG Result Post Encapsulated Islet Transplantation:
Peroneal Motor Amplitude

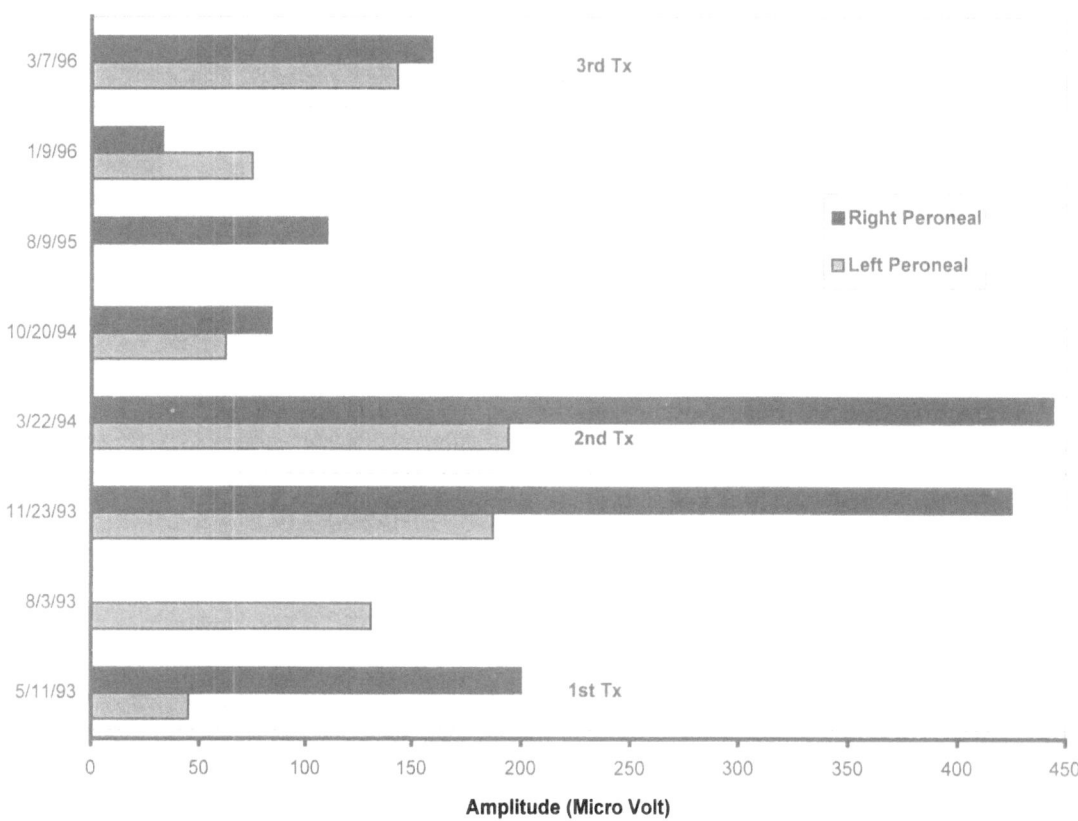

FIGURE 19.6. Improved EMG result postencapsulated islet transplantation: peroneal motor function.

as a consequence of his improved glycemic control. Throughout the 44-month period, the patient's renal function has remained stable with serum creatine levels ranging from 0.9 to 1.1 mg/dl. The patient reported significant improvement in the quality of his life, including an increased energy level, an ability to walk further, and a general feeling of improved health. He has recently secured full-time employment, an accomplishment he was not able to achieve because of his diabetes for almost a decade prior to his transplant.

Prospects for Future Improvements

Mechanical stability of capsules and adequate availability of islets are two important issues that continue to be addressed by ongoing studies in our laboratory. Long-term mechanical stability of the microcapsule formulation has been enhanced by use of a photopolymerizable covalent-cross-linkable from of alginate. These microcapsules have demonstrated biocompatibility as well as successful xenograft transplantation (dog to rat and dog to mouse) without immunosuppression. Possibilities for adequately increasing the source of insulin-secreting cells for treating the large number of patients who may benefit from encapsulated islet transplantation are being explored by the use of porcine islets and also by the proliferation of beta cells.

Acknowledgments. The authors acknowledge R. Heintz, N. Merideth, M. Moloney, M. Murphy, M. Schmehl, Q. Yao, Z. Yao, and T. Zheng for their outstanding assistance in isolating, purifying, and encapsulating islets, and also Q. Yao, Z. Yao, and S. Flechsig for their help in preparing this manuscript.

References

Albisser AM, Leibel BS, Ewart G, Davidovac Z, Botz CK, Zinger W. 1974. An artificial endocrine pancreas. Diabetes 23:389–396.

Altman JJ, Manoux A, Callard P, Houlbert D, Desplanques N, Bruzzo F, Galletti PM. 1981. Successful pancreatic xenografts using semipermeable membranes. Artif Organs (Suppl) 5:776–779.

Archer J, Kaye R, Mutter G. 1980. Control of streptozotocin diabetes in Chinese hamsters by cultured mouse islet cells without immunosuppression: a preliminary report. J Surg Res 28:77–85.

Chick WL, Perna J, Lauris W, Low D, Galletti PM, Whittemore A, Like A, Colton CK, Lysaght M. 1977. Artificial pancreas using living beta cells: Effects on glucose homeostasis in diabetic rats. Science 197:780–782.

Darquy S, Reach G. 1985. Immunoisolation of pancreatic B cells by microencapsulation: An in vitro study. Diabetologia 28:776.

Diabetes Control and Complications Trial Research Group. 1993. The effect of intensive treatment of diabetes on the development and progression of long-term complications in insulin-dependent diabetes mellitus. N Engl J of Med 329:977–986.

Gates RJ, Lazarus NR. 1977. Reversal of streptozotocin-induced diabetes in rats by intraperitoneal implantation of encapsulated neonatal rabbit pancreatic tissue. Lancet II:1257–1259.

Lanza RP, Heintz R, Merideth N, Yao Z, Yao Q, Zavitz D, Chen C, Soon-Shiong P. 1990. Large-scale canine and human islet isolation using a physiological islet purification solution. Diabetes 39(1):309A.

Lim F, Sun AM. 1980. Microencapsulated islets as a bioartificial endocrine pancreas. Science 210:908.

O'Shea GM, Goosen MFA, Sun AM. 1984. Prolonged survival of transplanted islets of Langerhans encapsulated in a biocompatible membrane. Biochim Biophys Acta 804:133.

Prehn RT, Weaver JM, Algire GH. 1954. The diffusion-chamber technique applied to a study of the nature of homograft resistance. J Nat Cancer Inst 15:509–517.

Reach G, Poussier P, Sausse A, Assan R, Itoh M, Furman BL, Gerich JE. 1981. Functional evaluation of a bioartificial pancreas using isolated islets perfused with blood ultrafiltrate. Diabetes 30:296–301.

Soon-Shiong P, Feldman E, Nelson R, Komtebedde J, Smidsrod O, Skjak-Braek G, Espevik T, Heintz R, Lee M. 1992. Successful reversal of spontaneous diabetes in dogs by intraperitoneal microencapsulated islets. Transplantation 54:769–774.

Soon-Shiong P, Feldman E, Nelson R, Heintz R, Yao Q, Yao Z, Zheng T, Merideth N, Skjak-Braek G, Espevik T, Smidsrod O, Sandford P. 1993. Long-term reversal of diabetes by injection of immuno-protected islet cells. Proc. Natl. Acad. Sci. USA 90:5843–5847.

Soon-Shiong P, Heintz R, Merideth N, Yao Qiang X, Yao Z, Zheng T, Murphy M, Moloney M, Schmehl M, Harris M, Mendez R, Mendez R, Sandford P. 1994. Insulin independence in a Type I diabetic patient after encapsulated islet transplantation. Lancet (April 16) 343:950–951.

Strautz RL. 1970. Studies of hereditary-obese mice (obob) after implantation of pancreatic islets in Millipore filter capsules. Diabetologia 6:306–312.

Sullivan SJ, Borland KM, Mahoney MD, Chick WL, et al. 1991. Biohybrid artificial pancreas: Long term implantation studies in diabetic, pancreatectomized dogs. Science 252:718.

Sun AM, O'Shea GM, Goosen MFA. 1984. Injectable microencapsulated islets as a bioartificial endocrine pancreas. Appl Biochem Biotechnol 10:87.

Sun AM, Parisius W, Healy G, Vacek I, Macmorine HG. 1977. The use, in diabetic rats and monkeys, of artificial capillary units containing cultured islets of Langerhans (artificial endocrine pancreas). Diabetes 26:1136–1139.

Sun AM, O'Shea GM. 1985. Microencapsulation of living cells—a long term delivery system. J Controlled Release 2:137.

Theodorou NA, Vrbova H, Tyhurst M, Howell SL. 1979. An assessment of diffusion chambers for use in pancreatic islet cell transplantation. Transplantation 27:350–352.

Thu B, Bruheim T, Smidsrod O, Soon-Shiong P, Skjak-Braek G. 1996a. Alginate polycation microcapsules I. Interaction between alginate and polycation. Biomaterials 17:1031–1040.

Thu B, Bruheim T, Smidsrod O, Soon-Shiong P, Skjak-Braek G. 1996b. Alginate polycation microcapsules II. Some functional properties. Biomaterials 17:1069–1079.

Tze WJ, Tai FC, Davis HR. 1980. Studies with implantable artificial capillary units containing rat islets on diabetic dogs. Diabetologia 19:541–545.

Tze WJ, Wong FC, Chen LM, O'Young S. 1976. Implantable artificial endocrine pancreas unit used to restore normoglycemia in the diabetic rat. Nature 264:466.

Woodward SC. 1982. How fibroblasts and giant cells encapsulate implants: Considerations in design of glucose sensors. Diabetes Care 5:278–281.

20

Transplantation of Microencapsulated Parathyroid Tissue: Clinical Background, Methods, and Current Status of Research

Christian Hasse, Andreas Zielke, Ulrich Zimmermann, and Mathias Rothmund

History of Parathyroid Allotransplantation

While autotransplantation of parathyroid tissue is a well-established clinical procedure (Wells et al 1974), allotransplantation of the parathyroid gland still is at the level of experimental surgery. (See Table 20.1.)

In 1884, the Swiss Moritz Schiff was the first to demonstrate that tetany, occurring after total thyroidectomy, could be reversed by replantation of the thyroid (Welbourn 1990). Six years later, Loeb, who investigated the cramps of isolated muscles submersed in a low-calcium saline solution in vitro, demonstrated that addition of calcium to the medium resolved the spasms—uncovering the relation between hypocalcemia and tetany (Loeb 1900). In 1909, MacCallum and Voegtlein discovered that hypocalcemia after thyroidectomy was caused by parathyroidectomy (MacCallum and Voegtlein 1909). Hanson in 1924 and Collip in 1925 were the first to try to overrule this causality by intramuscular injections of parathyroid extracts to patients with postthyroidectomy tetany. For limited periods of time, they were able to restore normocalcemia in these patients (Collip 1925, Hanson 1924).

In 1972, almost 90 years after Schiff's first report of thyroid replantation, Tochilin performed the first allotransplantation of fetal thyroid and parathyroid glands in humans and documented long-term reversal of hypothyroidism, but only short-term reversal of hypoparathyroidism (Tochilin 1972a–c). Tochilin later combined this treatment with simultaneous transplantation of the pancreas, and achieved similar results (Kiprenskii et al 1990). The first allotransplantation of parathyroid glands with systemic immunosuppression was reported by Groth et al in 1973, in a patient who had previously undergone successful renal transplantation. Since the patient became normocalcemic, with no further need of calcium supplementation, the transplant apparently was functioning; however, the further course of the patient and the period for which normocalcemia was maintained are unknown (Groth et al 1973). The potential long-term function of simultaneous renal- and parathyroid allotransplants with systemic immunosuppression was outlined in a case report of 1979. In this particular case, the parathyroid tissue's function was documented by determination of parathyroid hormone (transplant- versus non-transplant-bearing forearm) as well as histology from a biopsy of the allotransplanted tissue. It remained functioning until complete rejection of both organs 30 months postoperatively (Ross et al 1979). Also in 1979, two children with hypoparathyroidism due to DiGeorge syndrome (see Section 2) received parathyroid allotransplants without immunosuppression. Because of a severe concomitant immune defect, both children died too early to allow for an assessment of allograft function (Wells et al 1979). Kunori et al also completely ignored the ensuing immunologic reaction following transgenous transplantation. They reported a first successful long-term therapy of postoperative hypoparathyroidism by repetitive allotransplantation of fresh and cryoconserved parathyroid tissue particles in 1991, although the patient received initial immunosuppressive medical therapy for the first 2 days postoperatively. For a period of 6 months, their patient reportedly had normal levels of serum calcium and improvement of symptoms with half the substitutive therapy (Kunori et al 1991). One year later, Segerberg et al performed the first successful syngenous transplantation of parathyroid glands in identical twins, one of whom suffered from permanent postoperative hypoparathyroidism (Segerberg et al 1992).

Since then, many in vitro investigations and experimental animal studies have been performed to accomplish parathyroid allotransplantation without the need of postoperative immunosuppression. The earlier assumption, that parathyroid tissue would be less immunogenic than other tissue types, was proved wrong (Wells 1973a and b). In general, two basic lines of research have evolved. Some have tried to transplant parathyroid tissue to regions with apparently little or no immunologic competence. For instance, Fisher et al used the anterior eye chamber, a Chinese group reported transplantation to the later ventricle of the brain, and Dib-Kuri et al 1975 to the scrotum. However, none of the three achieved long-term graft function (Dib-Kuri et al 1975, Fisher et al 1967, Yao et al 1993). Others have aimed to reduce the immunogenicity of the tissue to be transplanted. UV and ionizing irradiation, various tissue culture methods, coating of the graft with Millipore membranes, nude mice in vivo passage, and treating the recipient with antidonor serum have been employed in an attempt to reduce the immunologic response. However, a reproducible increase of in vivo graft survival for more than just several weeks has not yet been accomplished (Gough and Finnimore 1980, Raaf et al 1974, Starling et al 1977, Swan et al 1960). Only Sollinger and coworkers were able to obtain functioning parathyroid transplants in two patients without postoperative immunosuppression for a period of 15 months, following pretreatment of parathyroid tissue with ionizing irradiation and subsequent nude mouse in vivo passage. The course of both patients and whether or not the patients required further substitutive therapy is, however, unclear. Although the authors concluded that their procedure was effective and very promising, there have been no further clinical reports on this combination of techniques (Sollinger et al 1983). Dyna-beads, i.e., magnetic microspheres loaded with specific antibodies, effectively reduce MHC I and MHC II antigen concentrations of single parathyroid cells. Using this technique, a first successful transplantation of single parathyroid cells was reported. With the patients receiving "mild immunosuppression," transplants were functioning for periods of up to 12 weeks, allowing for a reduction of oral calcium and vitamin

D substitution (Anton et al 1995). Recently, a combination of this technique with cell separation and in vitro culture passage yielded hormonally active allotransplants without immunosuppression for periods of 3 months and—in one case—for a period of 1 year (Tolloczko et al 1996). In 1989, Fu and Sun published first short-term in vivo results (12 weeks) of functioning parathyroid allotransplantations without postoperative immunosuppression, implanting parathyroid cells coated by an alginate, a natural substrate used for immunoisolation of cells (Fu and Sun 1989). The technique of microencapsulation has since been continuously enhanced. Several longterm animal experiments have documented longterm survival of syn- and transgenous parathyroid transplants, without any immunosuppressive therapy (Hasse et al 1994, 1996, 1997a–c). Most recently, the first two clinical cases in patients with postoperative hypoparathyroidism receiving successful parathyroid allotransplants without immunosuppression have been reported, utilizing the technique of microencapsulation (Hasse et al 1997d).

Hypoparathyroidism

Definition

Hypoparathyroidism is defined as a reduction of the concentration of intact parathyroid hormone and (total) serum calcium to subnormal levels, caused by impaired or absent parathyroid function (Kruse and Kuhlencordt 1994). Persistent, symptomatic hypoparathyroidism is diagnosed when the parathyroid hormone deficiency syndrome persists for more than 6 months and/or is accompanied by complaints or sequelae commonly associated with hypoparathyroidism.

Etiology and Classification

Symptomatic hypoparathyroidism may occur in a primary and secondary form (Potts 1995). Cases in which the etiology is unknown or cannot be established with certainty are referred to as idiopathic hypoparathyroidism (Ziegler 1976). In principle, all of the different forms of symptomatic hypoparathyroidism may be an indication for parathyroid allotransplantation. Therefore, all of these parathyroid hormone deficiency syndromes, including some rare ones will be presented in the following sections (Figure 20.1).

TABLE 20.1. Historic synopsis of significant achievements along the way to parathyroid allotransplantation.

Year	Author/s	Achievement
1884	Schiff	reversal of postthyroidectomy tetany by thyroid replantation
1924/25	Hanson	treatment of hypoparathyroidism with injections of parathyroid extract
1960	Swan et al	first attempt at parathyroid tissue immunoisolation by coating the tissue with Millipore membranes
1967	Fisher et al	increase of parathyroid graft function by choosing the anterior eye chamber as transplant site
1972	Tochilin	homologous *en bloc* transplantation of the thyroid-parathyroid complex
1973	Groth et al	first allotransplantation of the parathyroid with systemic immunosuppression in secondary hypoparathyroidism
1974	Raaf et al	first attempt to reduce parathyroid tissue immunogenicity by tissue culture passage
1977	Starling et al	first use of antidonor serum for allotransplantation of the parathyroids
1979	Ross et al	30-month survival of allotransplanted parathyroid glands with systemic immunosuppression
1979	Wells et al	first allotransplantation of the parathyroid without systemic immunosuppression in primary hypoparathyroidism
1983	Sollinger et al	15-month function of a human parathyroid allotransplant following ionizing irradiation and nude mouse *in vivo* passage of the parathyroid tissue
1989	Fu and Sun	first successful immunoisolation of parathyroid cells using microencapsulation: functioning *in vivo* for 12 weeks
1990	Kiprenskii et al	first homologous *en bloc* transplantation of parathyroid, thyroid gland, and pancreas
1991	Kunori et al	first successful treatment of postoperative hypoparathyroidism by repeated allotransplantation of parathyroid tissue particles after initial immunosuppression
1992	Segerberg et al	first successful syngenous transplantation of human parathyroids without immunosuppression
1993	Yao et al	intracerebral parathyroid allotransplants survive for 12 weeks without systemic immunosuppression
1995	Anton et al	first allotransplantation of single parathyroid cells after reduction of MHC I/II antigens; transplants functioning for 3 months with mild immunosuppression
1996	Tolloczko et al	first allotransplantation of single parathyroid cells without immunosuppression; hormonally active for up to 1 year
1997	Hasse et al	first successful clinical allotransplantation of cultured, microencapsulated parathyroid tissue without immunosuppression

Primary Hypoparathyroidism (Congenital Hypoparathyroidism)

Congenital forms of persistent hypoparathyroidism comprise three distinct entities: autoimmune hypoparathyroidism, DiGeorge syndrome and Kenny-Linarelli syndrome. Only few reports exist in the world literature, not allowing for estimates of the prevalence of these entities. They are, however, quite rare compared to the prevalence of acquired parathyroid hormone deficiency syndromes.

Autoimmune Hypoparathyroidism

Autoimmune hypoparathyroidism most commonly occurs as a part of the polyglandular autoimmune syndrome (APECED, autoimmune polyendocrinopathy candidiasis ectodermal dystrophy). The two types of autoimmune polyendocrinopathy have to be differentiated. Autoimmune hypoparathyroidism only occurs in association with Type I, in which half of the cases are familial (autosomal recessive), the other half sporadic (Windeck and Reinwein 1992). With variable

frequency, Type I is associated with pernicious anemia, Hashimoto's thyroiditis, ovarian insufficiency of early onset, diabetes mellitus Type I, and/or mucocutaneous candidiasis. Other ectodermal disorders associated to Type I autoimmune polyendocrinopathy include dental or nail dystrophy, alopecia areata, vitiligo, and keratinophathy (Kruse and Kuhlencordt 1994, Windeck and Reinwein 1992). It is in the familial cases only that one may also encounter Addison's disease (Windeck 1996). Persistent hypoparathyroidism is only seen in Type I autoimmune polyendocrinopathy, where it has an incidence of 80%. Patients suffering from Type II of the polyglandular autoimmune syndrome (Schmidt syndrome) have normal parathyroid function (Kruse and Kuhlencordt 1994).

DiGeorge Syndrome

Hypoparathyroidism in DiGeorge syndrome is due to agenesis of the parathyroid and the thymus gland caused by a failure of the development of the third and fourth branchial pouch. It may be associated with severe immunodeficiency, facial dysplasia, and

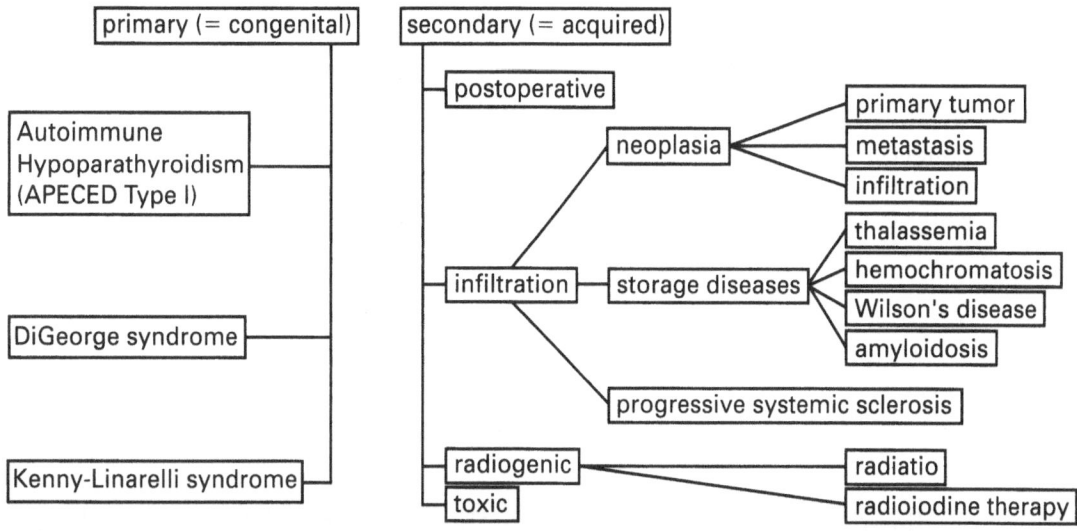

FIGURE 20.1. Etiology and classification of hypoparathyroidism.

anomalies of the heart (Kruse and Kuhlencordt 1994, Windeck 1996, Windeck and Reinwein 1992).

Kenny–Linarelli Syndrome

This autosomal recessively inherited disease is characterized by cretinism, delayed occlusion of the fontanels, myopia or hyperopy, cortical hyperostosis of cortical bones, and episodes of hypocalcemic tetanies due to hypoparathyroidism. Exogenous parathyroid hormone has been administered to patients affected by this syndrome, resulting in an increase of cAMP excretion in urine (Keck and Kuhlencordt 1992, Kruse and Kuhlencordt 1994).

Secondary Hypoparathyroidism (Acquired Hypoparathyroidism)

Injury to the parathyroid glands during operative procedures in the neck is the main reason for acquired hypoparathyroidism. The parathyroids are either inadvertently removed or their blood supply impaired by either iatrogenic hypo- or devascularization, edema, or hematoma (Ziegler 1976). The incidence of postoperative hypoparathyroidism is related to the type of surgery performed as well as to the event of prior surgery to the neck. Accordingly, it may vary considerably from 0 to 33% (Gann and Paone 1979).

Parathyroid hormone deficiency may also very occasionally be caused by neoplastic infiltration of the parathyroid glands. This has been reported for *primary tumors* of the parathyroid, local tumors of various etiologies, and metastases (Keck and Kuh-

lencordt 1992, Ziegler 1976, 1987). As with the rare disorders of the parathyroids secondary to progressive systemic sclerosis, abnormal storing of copper or iron, and amyloidosis, only very few cases have been reported or are anecdotally mentioned in textbooks (Keck and Kuhlencordt 1992, Kruse and Kuhlencordt 1994, Windeck 1996). This is also true for some isolated cases of hypoparathyroidism occurring after irradiation therapy, radioiodine therapy, and chemotherapy (Keck and Kuhlencordt 1992, Kruse and Kuhlencordt 1994).

Pathogenesis

The concentration of ionized extracellular calcium is regulated by parathyroid hormone and vitamin D (calcitriol) within tight margins of variance. Due to a reduction of parathyroid-hormone-dependent osteoclastic bone resorption, chronic parathyroid hormone deficiency invariably results in hypocalcemia. The symptoms of parathyroid hormone deficiency are caused by hypocalcemia (Ziegler 1976, Kruse and Kuhlencordt 1994). Serum calcium concentrations may drop to a minimum of 1.5 mmol/l. Below this threshold range, the gradient between the calcium concentration of the bone and that of the unsaturated extracellular fluid is such that bone-derived calcium will effectively go into solution independently of parathyroid hormone (Ziegler 1976). In this respect, the often observed hyperphosphatemia, resulting from an increase in tubulorenal reabsorption, is of minor importance.

Symptoms

The signs and symptoms of chronic hypoparathyroidism were first described in 1941 by Lachmann (Lachmann 1941). One of the main features of symptomatic hypoparathyroidism is increased neuromuscular excitability resulting from hypocalcemia (Kruse and Kuhlencordt 1994, Ziegler 1976, Wagner 1980). Hyperexcitability may cause severe muscular convulsions without loss of consciousness (Keck and Kuhlencordt 1992, Kruse and Kuhlencordt 1994). They are identifiable as either tonoclonic convulsions or isolated muscular spasms, which often occur as symmetric spasms of the extremities (Kruse and Kuhlencordt 1994, Wagner 1980). The characteristic clinical sign of carpopedal tetany induced by hypocalcemia is referred to as "obstetrician's hand" and as talipes in the case of tarsopedal tetany (Keck and Kuhlencordt 1992, Kruse and Kuhlencordt 1994, Wagner 1980, Ziegler 1976, 1987) (see also Table 20.3). Spasms of the perioral muscles cause a typical facies commonly referred to as "fish mouth appearance" (Ziegler 1987). The bronchi and the respiratory muscles may also be affected, causing laryngeal spasms and severe dyspnea (Keck and Kuhlencordt 1992, Kruse and Kuhlencordt 1994). Occasionally, hypocalcemia may also be responsible for cramping abdominal pain, hence the term visceral tetany, and for stenocardic complaints in some patients (Kruse and Kuhlencordt 1994, Windeck 1996, Ziegler 1987).

All of these spasms may occur without any appreciable precipitating cause, yet more often in situations of increased physical or psychological stress. They may last for only a few minutes or for up to several hours. Females are more often affected than males, and a periodic increment in frequency as well as severity has been reported, suggesting a relation to the menstrual cycle. Also, these spasms apparently have a tendency to occur more often during spring and autumn (Ziegler 1976).

The typical clinical appearance of a tetany is easily diagnosed; however, in only one-fourth of patients is it fully identifiable (Ziegler 1987). This may explain why less obvious cases of "latent" hypoparathyroidism are often overlooked. Muscular hyperexcitability may be discrete and only appreciated as aching, tingling, and numbness of the extremities ("tingling paresthesia") (Kruse and Kuhlencordt 1994, Wagner 1980, Ziegler 1976).

A comprehensive synopsis of all the symptoms associated with persistent hypoparathyroidism is depicted in Table 20.2.

TABLE 20.2. Symptoms of hypoparathyroidism (Ziegler 1976, Wagner 1980, Ziegler 1987, Keck and Kuhlencordt 1992, Kruse and Kuhlencordt 1994, Windeck 1996).

increased neuromuscular excitability:
—Convulsions in conscious individuals:
⇒ spasms:
  • painful, symmetric, carpopedal tetany ⇒ "obstetricians' hand"
  • carpopedal tetany = "talipes"
  • laryngeal and bronchial spasms, spasms of the respiratory musculature ⇒ dyspnea
  • spasms of perioral muscles ⇒ "fish mouth appearance" of the mouth
  • stenocardia
  • cramping abdominal pain = "visceral tetany"
⇒ Convulsions
—tingling paresthesia (facial, perioral, extremities)
psychological alterations = endocrine psychosyndrome
—depressive mood
—hallucinations
—anxiety, restlessness, confusion, agitation, irritability
—progressive mental retardation
impaired hearing, loss of hearing
pain of joints and bones
feeling of coldness of the extremities
diarrhea
coprostasis, constipation
muscular pain and weakness
exhaustion

Diagnosis

An exhaustive history and physical examination is the keystone of the diagnosis of hypoparathyroidism. Symptoms and sequelae of the disease, as described in the "Symptoms" section, are sought after; they are either reported by the patient or are found as a result of a meticulous examination.

A number of clinical tests are further employed. For instance, tapping the cheek of the patient just anterior to the ear, may cause reproducible contractions of the facial muscles, a sign commonly referred to as Chovstek's sign (Keck and Kuhlencordt 1992, Kruse and Kuhlencordt 1994, Windeck 1996, Ziegler 1987). Such contractions of the facial muscles, conducted by the third branch of the facial nerve, may also be found in as many as 15% of healthy subjects. They are, however, considered characteristic of tetanic predisposition induced by hypocalcemia, if contractions in the area of distribution of the second and third branch of the facial nerve are elicited (Ziegler 1976). Trousseau's sign is provoked by inflating a blood-pressure cuff placed around the upper arm to pressures above the mean arterial pressure for several minutes. The test is regarded as positive if carpopedal spasms are noted within 3 minutes. The test is positive in less than 1% of normocalcemic subjects (Keck and Kuhlencordt 1992, Kruse and Kuhlencordt 1994, Windeck 1996, Ziegler 1987). This sign may also be observed after up to 3 minutes of provocative hyperventilation (Kruse and Kuhlencordt 1994, Ziegler 1976). Both Chvostek's and Trousseau's sign were found in 55% to 62% of patients with postoperative hypoparathyroidism. Finally, the so-called Lust's sign is represented by reproducible extraplantar dorsiflections of the foot, in response to tapping the fibular nerve at the height of the head of the fibula (Ziegler 1976).

Laboratory tests will reveal low serum concentrations of total and ionized calcium, parathyroid hormone, and magnesium, as well as an increase in serum phosphate. Urinary excretion of calcium, phosphate, and cAMP is decreased. Normal serum levels of creatinine and urea, as well as normal levels of albumin, exclude renal insufficiency and syndromes of malassimilation as the cause of hypocalcemia, rendering the diagnosis of hypoparathyroidism very likely. Occasionally, electrocardiograms may show prolongation of the QT interval, caused by an increase of ST segments (Keck and Kuhlencordt 1992, Windeck and Reinwein 1992). Furthermore, T-wave inversion or peaking may be observed (Keck and Kuhlencordt 1992).

Radiographic examinations in patients with congenital or long-standing persistent hypoparathyroidism may demonstrate scattered skeletal condensations combined with areas of demineralized bone as well as periarticular osteophytes (Windeck and Reinwein 1992). Some 20% of patients will have evidence of calcifications of the basal ganglia or gray matter by plain radiograms or computed tomograms of the skull (Keck and Kuhlencordt 1992).

The diagnosis should be substantiated by functional tests. The diagnosis of hypoparathyroidism is very likely if administration of intact parathyroid hormone (400 IU i.v.) yields a 5- to 10-fold increase of urinary phosphate excretion. This test is called the Ellsworth-Howard test (Keck and Kuhlencordt 1992, Ziegler 1976). If latent hypoparathyroidism is suspected, an ethylenediaminetetraacetate (EDTA) test may be diagnostic. The EDTA chelate binds calcium potently. In healthy subjects, a decrease of ionized calcium following infusion of EDTA stimulates PTH liberation, resulting in a fairly rapid restoration of normal serum calcium levels. In patients with latent hypoparathyroidism, this autoregulatory loop is deficient. During the test, serum calcium levels are repetitively determined, before and after infusion of 50–70 mg sodium EDTA per kilogram body weight. In contrast to healthy individuals, a typical patient will not reach basal values of serum calcium within 12 hours after administration of the chelate (Ziegler, 1976). Another functional test can be used to differentiate pseudohypoparathyroidism from hypoparathyroidism. Healthy individuals as well as patients with primary hypoparathyroidism who were administered infusions of synthetic human 1-34 PTH (30 µg within 5 minutes) will show a fourfold increase of 1,25 dihydroxy vitamin D within 24 hours, in contrast to those with pseudohypoparathyroidism, who will have no evidence of increasing levels of vitamin D. Electromyography (EMG) is a useful study to substantiate increased neuromuscular excitability. Doublets and triplets during hyperventilation evidence tetanic predisposition (Ziegler 1976). If these signs and symptoms of hypoparathyroidism are accompanied by facial dysplasia in a child, the diagnosis of the rare DiGeorge syndrome is made. Persistent hypoparathyroidism combined with chronic mucocutaneous candidiasis leads to the diagnosis of APECED Type I (Kruse and Kuhlencordt 1994). Table 20.4 depicts a complete investigative procedure in patients with suspected hypoparathyroidism.

Therapy of Hypoparathyroidism

Symptomatic Therapy

Acutely symptomatic patients experiencing tetanic convulsions readily respond to slowly administered intravenous calcium (calcium gluconate, maximum of 2 ml of concentration per minute) (Windeck 1996). Care must be taken not to exceed a maximum dose of 15 mg calcium per kilogram body weight (Ziegler 1976). Chronic hypocalcemia due to persistent hypoparathyroidism is initially treated by continuous oral substitution of calcium. If relief of symptoms and serum calcium concentrations in the low normal range are not achieved with calcium alone, vitamin D (calcitriol) is given additionally (Windeck 1996). Until the introduction of vitamin D in the 1930s, hypoparathyroidism was a life-threatening condition. Calcitriol raises serum calcium concentrations by osteolysis and increasing

TABLE 20.3. Sequelae of hypoparathyroidism (Ziegler 1976, Wagner 1980, Ziegler 1987, Keck and Kuhlencordt 1992, Windeck and Reinwein 1992, Kruse and Kuhlencordt 1994, Windeck 1996).

psychological alterations = "endocrine psychosyndrome"
  —psychosis
  —depression
metastatic calcifications:
  —basal ganglia (symmetric, deposits of hydroxyapatite)
    ⇒ Parkinson's disease and/or
    ⇒ Fahr's disease:
      • headaches
      • speaking disability
      • dementia
      • pyramidal symptoms (e.g., epileptiform convulsions)
  —subcutaneous
  —lung
  —lens ⇒ cataract ("tetanic glaucoma")
  —inner ear ⇒ loss of hearing
  —periarticular osteophytes
ectodermal alterations:
  —trophic impairment and infections of the skin and male sex organs:
      • alopecia
      • flaky skin
      • Candidiasis
increase of bone density and calcium contents of bones
cardiomyopathy, unspecific enlargement of the heart
alterations of dentine ⇒ anomalies of the teeth*
deformations of the skeleton together with diminished growth*
osteoporosis
moonface

*mainly in children

TABLE 20.4. Diagnostic procedures in hypoparathyroidism (Ziegler 1976, Ziegler 1987, Keck and Kuhlencordt 1992, Windeck and Reinwein 1992, Kruse and Kuhlencordt 1994, Windeck, 1996).

clinical signs (refer to symptoms and sequelae of hypoparathyroidism as described in Section 2.4., Tables 20.2 and 20.3
clinical tests
  —Chvostek's sign
  —Trousseau's sign
  —provocative hyperventilation
  —Lust's sign
EKG:
  —QT increment, peaking or inversion of T-wave
laboratory parameters:
  —$[Ca^{2+}]$ i.s. ↓
  —$[iPTH]$ i.s. ↓
  —$[magnesium^{2+}]$ i.s. ↓
  —$[phosphate]$ i.s. ↑
  —$[hydroxyproline]$ i.s. ↑
  —$[Ca^{2+}]$ i.u. ↓
  —$[phosphate]$ i.u. ↓
  —$[cAMP]$ i.u. ↓
functional tests:
  —Ellsworth-Howard test
  —EDTA test
skull radiograms and computed tomography
EMG
ophthalmologic investigation: lenticular cataract formation?
psychiatric evaluation
osteodensitometry

intestinal absorption of calcium. Regular serum calcium and urinary calcium excretion tests are needed to evaluate the effectiveness of calcitriol therapy. The urinary calcium excretion must not exceed 250 mg per day (Windeck and Reinwein 1992).

The overall course of persistent hypoparathyroidism mainly depends on how early therapy was initiated, its stringent implementation, and lifelong control. Without adequate therapy and control, all of the symptoms and sequelae of the disease are inevitable. Long-term therapy of patients in whom the disease takes a rather moderate course with only minor symptoms is most challenging. A milder symptomatology is most commonly encountered in patients with postoperative hypoparathyroidism, representing by far the majority of patients with persistent hypoparathyroidism. Unspecific conditions such as bone pain, loss of energy, exhaustion, depression, and latent convulsions are often either ignored or unrecognized for many years until irre-

versible consequences arise (Ziegler 1976, 1987). It is, therefore, the permanent care of this large number of patients at risk that is of particular importance.

However, even with rigorous, long-term treatment of symptoms, sequelae of hypoparathyroidism may still occur. Chronic parathyroid hormone deficiency invariably results in a loss of activity of bone metabolism, resulting in osteoporosis and bone pain (Wagner 1980). On the other hand, calcitriol lacks the full renal calcium-retaining ability of parathyroid hormone. Patients administered vitamin D may develop hypercalciuria, although their serum calcium levels may be normal or in the low normal range. Accordingly, these patients have an increased risk of nephrolithiasis, nephrocalcinosis, and subsequent impairment of renal function (Potts 1995). Moreover, these patients have an increased risk of developing calcium deposits to the soft tissues, as well as accelerated arteriosclerosis.

Since the therapeutical range of calcium and vitamin D substitution is small, patients are at risk to experience drug overdosing at all times, which is then observed as hypercalcemia syndrome of varying degrees. Also, oral calcium and vitamin D substitution will level out the imbalances of calcium and phosphate metabolism, at best. They are, however, unable to replace the many other metabolic actions of parathyroid hormone, which remain permanently untreated. For this reason, the need to treat the underlying cause of parathyroid hormone deficiency is undisputed.

Causative Therapy

Treatment of the cause of the parathyroid hormone deficiency syndrome may involve substitution of parathyroid hormone or transgenous transplantation of parathyroid tissue.

Already in 1925, Collip described a hormone of the parathyroid glands as the main denominator of calcium homeostasis (Collip 1925). However, it was not until 1978 that the hormone was sequenced in its entirety (Keutmann et al 1978). Meanwhile, genetically engineered human recombinant 1-34 PTH has become available. First clinical trials indicate that, in principle, permanent hypoparathyroidism can be treated by recombinant 1-34 PTH (Winer et al 1996). For the time being, the short half-life and the in vivo instability of synthetic PTH do not allow for continuous, sufficient treatment. Moreover, resistance to the hormone after repetitive application has been described. Therefore, a major focus of continued research is to develop a tracer-bound hormone preparation, suitable for long-term replacement therapy.

Replacing the parathyroid organ itself is a much more promising approach if physiologic calcium homeostasis is to be achieved. Currently, parathyroid allotransplantation has only been performed under continuous systemic immunosuppression. This approach is not justified, since hypoparathyroidism is rarely a life-threatening condition, and because systemic immunosuppression harbors significant side effects. The advantages of functional replacement of the organ are undisputed. Therefore, allo- and xenotransplantation of parathyroid glands without a need for systemic immunosuppression is a matter of intense research activity. Several different ways have recently emerged from the stage of animal experimentation to that of first clinical trials. Encouraging results have been reported, particularly for those techniques that aim either to reduce the immunogenicity of the tissue to be transplanted or to immunoisolate it. At present, however, only first clinical observations are published, and none of the

methods has been established as a clinical procedure yet (Anton et al 1995, Hasse et al 1997d, Tolloczko et al 1996). Which of the methods, or combinations thereof, is best suited for long-term treatment of patients with symptomatic postoperative hypoparathyroidism awaits further clinical trials in larger series of patients, and thus remains to be seen.

Microencapsulation

Microencapsulation refers to a process by which gas, fluids, or solid active ingredients are encapsulated by a shell substance, producing microspheres of a size ranging from 1 micron to 7 millimeters, isolating the incorporated core substance from the surrounding environment (Franjione and Vasishtha 1996).

In principle, the encapsulated substance may be released in three different ways. The first is mechanical disruption of the capsule. For instance, the microcapsules applied to the back of carbonless copy paper burst and release the incorporated dye when pressure is applied to the front of the paper with a pen point. This application of the microencapsulation technology was first introduced by the American chemist Green in the 1930s (Fanger 1974). Second, the capsule may slowly erode or biodegrade. This mechanism is utilized to encapsulate protein-reactive enzymes in powder detergents, which are released under defined conditions, e.g., a certain temperature. Finally, other applications make use of the fact that the contents of a microcapsule are released slowly over time by diffusion, maintaining the integrity of the capsule. This particular aspect, together with the ability of a microcapsule to isolate the core substance from its environment, has made this technology a major focus of interest in transplant medicine.

Ennis and James in 1950 devised the first apparatus by which it was possible to obtain single droplets of a uniform size. They modified a principle of producing small amounts of aqueous and nonaqueous fluid spheres, conceived in 1947 by Lane and Levvy, and called it a "droplet sizer" (Ennis and James 1950). This apparatus was the progenitor of the microencapsulation technology from which many applications currently used in industry, basic research, and medicine have developed.

The first application of this technology in the field of medicine was proposed by Chang in 1964, who intended to develop artificial cellular systems to be used to replace blood. While investigating different carrier systems to simultaneously encapsulate hemoglobin and several different enzymes by thin nylon-like semipermeable membranes, he developed a method of microencapsulation based on interfacial polymerization (Chang 1964). However, the reaction needs to be carried out under rigorous conditions, because it results in liberation of cytotoxic byproducts. Moreover, the membranes are not entirely biocompatible, and the monomers are chemically active to the extent that necrosis of cells is induced. This prompted extensive efforts to find alternative shell substances.

Alginates

Alginates are extracted from brown algae (*Macrocystis pyrifera*), which are most commonly found on the shorelines of Scotland and Great Britain. Already in the seventeenth century it was known that the ashes of brown algae contained high amounts of soda, which was used in the production of soap and glass. In 1811, iodine was discovered in the ashes. After less expensive methods to produce soda had been established, the production of potash and iodine kept the local, algae-based industry alive. Today, alginates are used for many different purposes in industry (e.g., food, concrete, and fire-resistant synthetic materials) as well as in medicine (wound dressings, prosthetics, and orofacial surgery).

Commercially extracted alginates are a mixture of uronic acid polymers, composed of mainly mannuroic and guluronic acid. Calcium ions are used as a cross-linking agent. The polymer is a complex three-dimensional network of varying density, depending on the concentration of the two major compounds. Grant et al have referred to the ultrastructure of the cross-linked polymer as an "egg-box model" because of a similar pattern of layering (Grant et al 1973).

In 1977, Kierstan and Bucke were the first to use alginates as a shell substance in microencapsulation. For the first time, microencapsulation of viable cells was achieved. By dropwise addition of cells immersed in sodium alginate into a calcium chloride solution, stable hydrogel beads were formed (Kierstan and Bucke 1977). A drawback of this technique is the instability of the capsule. Biodegradation occurs as a consequence of dilution of calcium ions reacting with phosphates present in the body. For this reason, Lim and Sun, in 1980, proposed to coat the alginate capsules with a second membrane of poly-l-lysine (PLL-membrane) (Lim and Sun 1980). Zim-

mermann and coworkers succeeded in integrating more stable ions, e.g. barium, as cross-linking agents. Given appropriate conditions of incubation, highly stable, fully biocompatible alginates are produced (Zimmermann et al 1996).

Experimental and Clinical Transplantation of Microencapsulated Parathyroid Tissue

We adopted the well-described technique of microencapsulation with alginates already used in pancreatic islet cell transplantation and modified it extensively for transplantation of parathyroid tissue. Combining refined tissue culturing techniques and microencapsulation, iso-, allo-, and xenotransplantation of microencapsulated parathyroid tissue were reproducibly achieved in an animal model of experimental hypoparathyroidism without a need for systemic immunosuppression (Hasse et al 1994, Hasse et al 1997a).

Prior to the first clinical use, continued analysis of the alginate for the purpose of testing its biocompatibility determined mitogenic properties. Using patented techniques recently published, a purified, amitogenic alginate, which is suitable for clinical use, was isolated by the Institute of Biotechnology of Würzburg, Germany (Klöck et al 1994, Zimmermann et al 1992). The novel alginate displayed distinctly different biochemical and biophysical properties compared to the mitogenic alginate, which required several adjustments of the purification as well as the microencapsulation procedure. However, subsequent in vivo studies in syn- and allogenic animal models unequivocally showed that the same results as with the mitogenic alginate were also obtained when the novel amitogenic alginate was used (Hasse et al 1996, Hasse et al 1997b and c). These results established the rationale for a first clinical use of microencapsulated parathyroid allotransplants in patients suffering from symptomatic persistent postoperative hypoparathyroidism. We recently published the data of the first two patients treated with this method. Twelve weeks following allotransplantation of cultured and microencapsulated hyperplastic parathyroid tissue, the two patients were normocalcemic and had normal levels of parathyroid hormone without immunosuppression. Both patients reported an impressive improvement of symptoms and sequelae of hypoparathyroidism (Hasse et al 1997d).

Alternate Approaches and Perspectives

The feasibility of parathyroid allotransplantation for patients receiving systemic immunosuppression has already been outlined several times. This has repeatedly and successfully been performed during transplantation of the kidney, pancreas, heart, and liver, and, since these patients already require lifelong immunosuppression, it is an entirely justified approach (Bloom et al 1987, Groth et al 1973, Starling et al 1977, Wells et al 1979). For the time being, this approach is the only clinically established method that allows for a therapy to treat the cause of symptomatic hypoparathyroidism.

Reducing the immunogenicity of parathyroid tissue has been accomplished utilizing several different methods. However, aside from the few already-mentioned exceptions, these methods will not be of clinical relevance until long-term success is documented for clinical allografting without prolonged immunosuppression (Anton et al 1995).

It would therefore appear that the only two alternatives to allotransplantation of cultured and microencapsulated parathyroid tissue that potentially may be of relevance for the treatment of hypoparathyroid patients without posttransplant immunosuppression are substitution of parathyroid hormone and implantation of parathyroid cells pretreated by specific tissue-culturing methods aimed at reducing their antigenicity. All of these approaches are at the stage of first clinical trials.

Earlier attempts to use bovine PTH were hampered by the invariable development of antibodies resulting in hormone resistance (Melick et al 1969). Although synthetic human PTH has become available, it appears that development of hormone resistance prevails. In 1990, Stoegmann et al reported their initial experience in two patients treated with subcutaneous injections of synthetic human 1-34 PTH (400 E daily). One patient developed resistance after only 10 weeks; the other was able to maintain normal levels of serum calcium for a period of 7 months. The further course of their patients is unknown (Stoegmann et al 1990). Their observations were confirmed in another report of a larger series of patients treated for a period of 10 weeks. During this short-term trail, therapy with 1-34 synthetic PTH was successful. However, the problem of overcoming resistance during long-term treatment remains to be solved (Winer et al 1996).

In 1991, Tolloczko et al reported several cases in which allografting of cultured parathyroid cells by subfascial injection to the nondominant forearm had been performed. Their series, currently comprising 23 patients some of whom showed evidence of allograft function for up to 14 months after the transplantation without immunosuppression, has recently been updated (Tolloczko et al 1991, Tolloczko et al 1994, Tolloczko et al 1996). Unfortunately, the details of how these results were achieved are not entirely clear. At first, hyperplastic parathyroid tissue from patients with renal hyperparathyroidism was assessed for infiltration with passenger lymphocytes and expression of HLA antigens, using immunohistochemical methods. Without a detailed description of selection criteria, 10–15 particles of parathyroid tissue that seemed most appropriate were used for in vitro culture. Tissue culture was initiated by primary explant technique and maintained for 6 to 12 weeks, apparently without further attempts to select for parathyroid cells. Subsequently, a number (2×10^6 to $2-3 \times 10^7$) of viable cells, assumed to be antigen-depleted parathyroid cells, were allotransplanted (Wozniewicz et al 1996). However, during the serial subculture, an exponential decrease of PTH secretion was noted, which may be explained by either necrosis of parathyroid cells or by fibroblastic overgrowth. Cellular contamination and residual antigenicity were underscored by an increasing posttransplant donor-specific T cell reactivity, with gradual cessation of parathyroid allograft function observed in all recipients (Tolloczko et al 1994).

At present, there is no validated approach to treat the considerable number of patients with persistent hypoparathyroidism that addresses the underlying cause and is sustainable without immunosuppression. However, several recent developments have delivered promising early results. Evidently, a combination of different methods—including those that reduce the immunogenicity and immunoisolate the graft—is needed to achieve long-term in vivo graft function without posttransplant immunosuppression. To optimize these different approaches and investigate the clinical utility of their combination is a promising approach that should be further explored.

Acknowledgment. This work was supported by Deutsche Forschungsgemeinschaft (DFG: Ha 2036/03-02).

References

Anton, G; Decker, G; Stark, JH; Botha, JR; Margolius, LP. 1995. Allotransplantation of parathyroid cells. Lancet 345:124.

Bloom, AD; Economou, SG; Baker, JW; Gebel, HM. 1987. Prolonged survival of rat parathyroid allografts after preoperative treatment with cyclosporine A. Current Surgery 44:205–207.

Chang, TMS. 1964. Semipermeable microcapsules. Science, 146: 524–525.

Collip, JB. 1925. The extraction of a parathyroid hormone which will prevent or control parathyroid tetany and which regulates the level of blood calcium. J. Biol. Chem. 63:293.

Dib-Kuri, A; Revilla, A; Chavez-Peon, F. 1975. Successful rat parathyroid allografts and xenografts to the testis without immunosuppression. Transplant. Proc. 7:753–756.

Ennis, WB Jr. and James, DT. 1950. A simple apparatus for producing droplets of uniform size from small volumes of liquids. Science 112:434–436.

Fanger, GO. 1974. What good are microcapsules? Chem. Tech. 4:397–405.

Fisher, B; Fisher, ER; Feduska, N; Sakai, A. 1967. Thyroid and parathyroid implantation: an experimental re-evaluation. Surgery 62:1025.

Franjione, J and Vasishtha, N. 1996. The art and science of microencapsulation. Information from Southwest Research Institute, San Antonio, Texas, S.1.

Fu XW and Sun AM. 1989. Microencapsulated parathyroid cells as a bioartificial parathyroid. Transplantation 47:432–435.

Gann, DS and Paone, JF. 1979. Delayed hypocalcemia after thyroidectomy for Graves' disease is prevented by parathyroid autotransplantation. Ann. Surg. 190:508–513.

Gough, IR and Finnimore M. 1980. Rat parathyroid transplantation. Allograft pretreatment with organ culture and antilymphocyte serum. Transplantation 29:149–152.

Grant, GT; Morris, ER; Rees, DA; Smith, PJC; Thom, D. 1973. Biological interactions between polysaccharides and divalent cations: The egg-box model. FEBS Letters 32:195–198.

Groth, CG; Popovtzer, M; Hammond, WS; Cascardo, S; Iwatsuki S; Halgrimson, CG; Starzl, TE. 1973. Survival of a homologous parathyroid implant in an immunosuppressed patient. Lancet 1082–1085.

Hanson, AM. 1924. The hydrochloric X Sicca: A parathyroid preparation for intramuscular injection. Milit. Surgn. 54:218.

Hasse, C; Schrezenmeier, J; Stinner, B; Schark, C; Wagner, PK; Neumann, K; Rothmund, M. 1994. Successful allotransplantation of microencapsulated parathyroid in rats. World J. Surg. 18:630–634.

Hasse, C; Klöck, G; Zielke, A; Schlosser, A; Barth, P; Zimmermann, U; Rothmund, M. 1996. Transplantation of parathyroid tissue in experimental hypoparathyroidism: in vitro and in vivo function of parathyroid tissue microencapsulated with a novel amitogenic alginate. Int. J. Artificial Organs 19:1–7.

Hasse, C; Zielke, A; Klöck, G; Barth, P; Schlosser, A; Zimmermann, U; Rothmund, M. 1997a. First successful xenotransplantation of microencapsulated human parathyroid tissue in experimental hypoparathyroidism: long-term function without immunosuppression. J. Microencapsulation 14:617–626.

Hasse C; Zielke, A; Klöck, G; Schlosser, A; Zimmermann, U; Rothmund, M. 1997b. Isotransplantation of microencapsulated parathyroid tissue in rats. Exp. Clin. Endocrinol. Diab. 105:53–56.

Hasse, C; Klöck, G; Zimmermann, U; Rothmund, M. 1977c. Amitogenes Alginat—Schlüssel zum ersten klinischen Einsatz der Microenkapsulierungstechnologie. Langenbecks Arch. Chir., Chir. Forum f. exp. u. klin. Forsch. 755–759.

Hasse, C; Klöck, G; Schlosser, A; Zimmermann, U; Rothmund, M. 1997d. Parathyroid allotransplantation without immunosuppression. Lancet 350:1296–1297.

Keck, E. and Kuhlencordt, F. 1992. Krankheiten der Nebenschilddrüsen. In: Innere Medizin in Praxis und Klinik, Hrsg.: Hornbostel, H; Kaufmann, W; Siegenthaler, W. Stuttgart, New York: Thieme, S. 4.79–4.86.

Keutmann, HT; Sauer, MM; Hendy, GN; O'Riordan, JLH; Potts, JT. 1978. Complete amino acid sequence of human parathyroid hormone. Biochemistry 17:5724.

Kierstan, M. and Bucke, C. 1977. The immobilization of microbial cells, subcellular organelles and enzymes in calcium alginate gels. Biotechnol. Bioeng. 19:387–397.

Kiprenskii, IV; Priakhin, IS; Podshivalin, AV. 1990. The transplantation of a segment of pancreas and thyroid-parathyroid complex by using a microsurgical technique. Vestn. Khir. Im. I. I. Grek. 145:108–110.

Klöck, G; Frank, H; Houben, R; Zekorn, T; Horcher, A; Siebers, U; Wöhrle, M; Federlin K; Zimmermann, U. 1994. Production of purified alginates suitable for use in immunoisolated transplantation. Appl. Microbiol. Biotechnol. 40:638–643.

Kruse, H and Kuhlencordt, F. 1994. Hypoparathreoidismus. In: Innere Medizin, Hrsg: Gross, R; Schölmerich, P; Gerok, W.—Stuttgart: Schattauer, S. 883–887.

Kunori, T; Tsuchiya, T; Itoh, J; Watabe, S; Arai, M; Satomi, T; Takakura, K; Yamaguchi, H. 1991. Improvement of postoperative hypocalcemia by repeated allotransplantation of parathyroid tissue without antirejection therapy. Tohoku. J. Exp. Med. 165:33–40.

Lachman, A. 1947. Hypoparathyroidism in Denmark. Acta Med. Scand. 121:1.

Lim, F, and Sun, AM. 1980. Microencapsulated islets as bioartificial endocrine pancreas. Science 210:908.

Loeb, J. 1899–1900. On the different effects of ions upon myogenic and neurogenic rhythmical contractions and upon embryonic and muscular tissue. Amer. J. Physiol. 3:383.

MacCallum, WG and Voegtlein, C. 1909. On the relation of tetany of the parathyroid glands to calcium metabolism. J. Exp. Med. 11:118.

Melick, RA; Gill, JR; Berson, SA; Yalow, RS; Bartler, FC; Potts, JT; Aurbach, GD. 1967. Antibodies and clinical resistance to parathyroid hormone. N. Engl. J. Med. 276:144–147.

Potts, J. 1995. Hypoparathreoidismus. In: Harrisons Innere Medizin II, Hrsg.: Schmailzl, K.—New York: McGraw-Hill. S. 2151–2522.

Raaf, JH; Farr, HW; Laird Meyers, WP; Good, RA. 1974. Transplantation of fresh and cultured parathyroid glands in the rat. Am. J. Surg. 128:478–483.

Ross, AJ III; Dale, JK; Gunnells, JC; Wells, SA Jr. 1979. Parathyroid transplantation: Fate of a long-term allograft in man. Surgery 85:382–384.

Segerberg, EC; Grubb, WG; Henderson, AE. 1992. The first successful parathyroid transplant from an identical twin for the cure of permanent postoperative hypoparathyroidism. Surgery 111:357–358.

Sollinger, HW; Mack, E; Cook, K; Belzer, FO. 1983. Allotransplantation of human parathyroid tissue without immunosuppression. Transplantation 36:599–602.

Starling, JR; Fidler, R; Corry, RJ. 1977. Prolongation of survival of rat parathyroid allografts by enhancing serum and tissue culture. Surgery 81:668–675.

Stoegmann, W; Bohrn, E; Woloszczuk, W. 1990. Initial experiences with substitution treatment of hypoparathyroidism with synthetic human parathyroid hormone. Monatsschr. Kinderheilkd. 138:141–146.

Swan, H; Hallin, R; Callaghan, P. 1960. Observation on the function and morphology of the parathyroid grafts in the dog using Millipore chambers. Surg. Forum 10:87.

Tochilin, VI. 1972a. Dynamics of accumulation of radioactive iodine by transplantation and protein-bound iodine of plasma in patients following homotransplantation of the thyroid-parathyroid complex. Probl. Endokrinol. 18:32–35.

Tochilin, VI. 1972b. Homotransplantation of the thyroid-parathyroid complex in myxedema. Khirurgiia 48:20–24.

Tochilin, VI. 1972c. The technique of taking and transplanting a thyroid-parathyroid complex to patients suffering thyroid-parathyroid insufficiency. Klin. Khir. 9: 36–40.

Tolloczko, T; Sawicki, A; Woniewicz, B. 1991. Clinical results of human cultured parathyroid cell allotransplantation in the treatment of surgical hypoparathyroidism. Abstract of the 34th World Congress of Surgery, Stockholm. 304.

Tolloczko, T; Wozniewicz, B; Sawicki, A; Nawrot, I; Migaj, M; Zabitowska, T; Gorski, A. 1994. Cultured parathyroid cell transplantation without immunosuppression in the treatment of surgical hypoparathyroidism. Transplant. Proc. 26:1901–1902.

Tolloczko, T; Woniewicz, B; Sawicki, A; Gorski, A; Nawrot, I; Zawitkowska, T; Migaj, M. 1996. Clinical results of human cultured parathyroid cell allotrans-

plantation in the treatment of surgical hypoparathyroidism. Transplant. Proc. 28:3545–3546.

Wagner, PK. 1980. Konservierung und Transplantation der Nebenschilddrüsen. In: Hyperparathyreoidismus, Hrsg.: Rothmund, M.—Stuttgart, New York: Georg Thieme. S. 216–228.

Welbourn, RB. 1990. The parathyroid glands. In: The history of endocrine surgery. Welbourn, RB, ed. New York, Westport, Connecticut, London: Praeger. S. 217–221.

Wells, SA Jr; Burdick, JR; Christiansen, CL; Abe, M; Sherwood, LM; Hattler. BG; Davis, RC. 1973a. Long-term survival of dogs transplanted with parathyroid glands as autografts and as allografts in immunosuppressed hosts. Transplant. Proc. 5:769–771.

Wells, SA Jr; Burdick, JF; Ketcham, AS; Christiansen, C; Abe, M; Sherwood, L. 1973b. Transplantation of the parathyroid glands in dogs. Biochemical, histological, and radioimmunoassay proof of function. Transplantation 15:179–182.

Wells, SA Jr; Christiansen, CL. 1974. The transplanted parathyroid gland: evaluation of cryopreservation and other environmental factors which affect its function. Surgery 78:49.

Wells, SA Jr; Ross, AJ; Dale, JK; Gray, RS. 1979. Transplantation of the parathyroid glands: current status. Surg. Clin. N. Amer. 59:167.

Windeck, R. 1996. Hypoparathyreoidismus. In: Praktische Endokrinologie, Hrsg.: Allolio, B; Schulte, H.—München: Urban & Schwarzenberg. S. 289–293.

Windeck, R and Reinwein, D. 1992. Hypoparathyreoidismus. In: Klinische Endokrinologie und Diabetologie, Hrsg.: Reinwein, D; Benker, G.—Stuttgart: Schattauer. S. 163–166.

Winer, KK; Yanovski, JA; Cutler, GB. 1996. Synthetic human parathyroid hormone 1-34 vs calcitriol and calcium in the treatment of hypoparathyroidism. JAMA 276:631–636.

Wozniewicz, B; Migaj, M; Giera, B; Prokurat, A; Tolloczko, T; Sawicki, A; Nawrot, I; Gorski, A; Zabitkowska, T; Kossakowska, AE. 1996. Cell culture preparation of human parathyroid cells for allotransplantation without immunosuppression. Transplant. Proc. 28:3542–3544.

Yao, CZ; Ishazuka, J; Townsend, CM; Thompson, JC. 1993. Successful intracerebroventricular allotransplantation of parathyroid tissue in rats without immunosuppression. Transp. 55:251–253.

Ziegler, R. 1976. Hypoparathyreoidismus. In: Praktische Endokrinologie, Hrsg.: Jores, A; Nowakowski, H.—Stuttgart: Thieme. S. 122–131.

Ziegler, R. 1987. Hypoparathyreoidismus. In: Hormon- und stoffwechselbedingte Erkrankungen, Hrsg.: Ziegler, R.—Weinheim: VCH. S. 273–283.

Zimmermann, U; Klöck, G; Federlin, K; Hannig, K; Kowalski, M; Brezel. RG; Horcher, A; Entenmann, H; Siebers, U; Zekorn, T. 1992. Production of mitogen contamination free alginates with variable ratios of mannuronic to guluronic acid by free flow electrophoresis. Electrophoresis 13:269.

21
Bioartificial Livers

Jeffrey M. Macdonald, John P. Griffin, Hiroshi Kubota, Linda Griffith, Jeffrey Fair, and Lola M. Reid

Introduction

The field of research on bioartificial organs is a relatively young one. Although methods of artificial support of patients with dysfunctional livers, lungs, hearts, and kidneys have been utilized for many years, efforts to create bioartificial organs, containing biological and artificial components in the devices, have reached significant levels only within the last decade. The bioartificial organ field is a natural outcome of investigations into organ culture, cell culture, extracellular matrix, hormonally and chemically defined media, signal transduction studies, stem cell biology, and regulation of tissue-specific gene expression. Ideas and techniques from all of these fields are being woven by investigators, especially by engineers, into approaches for the creation of highly sophisticated bioartificial organs. Although efforts are ongoing to develop bioartificial organs for many tissue types, the most advanced to date are those on the liver. Therefore, this review will focus, of necessity, on bioartificial livers as representative of the field.

Bioartificial Organs Versus Artificial Organs

In contrast to bioartificial organs, artificial organs are made entirely of nonbiological materials. At present, forms of artificial organs are being used clinically in kidney, heart, and lung support. The best known of these are the kidney dialysis machines, mechanical respirators for dysfunctional lungs, and various types of mechanical pumps for heart function, all now used routinely in hospitals everywhere. Even with their success, these artificial organs are limited in their capacity to mimic the body's normal functioning and are fraught with complications for the patients. For example, kidney dialysis, one of the oldest and most successful of artificial organs, results in depletion of key factors that in turn causes numerous "side effects," including sterility and immunological deficits (Chang 1974, Hida et al 1987, Jacobs et al 1996). Similarly, the first forms of artificial liver support have been based on hemodialysis, extracorporeal liver perfusion, exchange transfusion, and hemoperfusion over adsorbents (Kasai et al 1994, Anand 1996). Although these artificial livers have been able to overcome the severe pH effects and intracranial pressure problems that lead to coma and, if not corrected, to death, the positive effects have been transient; the patients wake up but then rapidly deteriorate due to lack of other liver functions such as metabolic and endocrine functions (Hughes and Williams 1996b). The inefficacy of these early techniques for liver support has fostered an increasing interest in the potential for bioartificial organs.

The key problem in the development of all forms of bioartificial organs is how to maintain highly differentiated cells ex vivo for extended periods of time (Reid and Jefferson 1984, Wu et al 1994). Although differentiated functions persist in tissues maintained in organ culture, the tissue remains alive only for a matter of hours (Brill et al 1994, Minuth et al 1992, Reid and Jefferson 1984). Cell viability beyond 8–10 hours is achieved only in dissociated cells that are then cultured under appropriate conditions (Pollard and Walker 1997, Reid and Jefferson 1984, Wu et al 1994). However, the dissociation of tissues into cell suspensions results in rapid dedifferentiation, such that tissue-specific functions are lost entirely within a period ranging from hours to at most a week (Brill et al 1994, Jefferson et al 1984, Reid 1990). Given this technical impasse, the early efforts to correct tissue dysfunction in patients have consisted of implantation of isolated cell suspensions or of tissue aggregates. The most successful have been bone marrow transplants to correct aberrant hemopoiesis (Forman et al 1994) and skin grafts of freshly isolated cells and of cultured keratinocytes, used routinely and successfully for years by plastic surgeons for severely burned patients (Rouabhia 1997). Less successful examples have been implantation of pancreatic islets for diabetes or liver cells for all forms of liver dysfunction. Implantation of pancreatic islet cells, either as islet cells alone or as islet cells after embedding in some material to minimize transplantation rejection phenomena, have failed clinically as a result of rejection and/or rapid deterioration of the implanted cells (Lombardi et al 1992). Similarly, attempts to correct liver dysfunction have included implantation of isolated liver cells into the liver, spleen, and intraperitoneum, but none of the options showed long-term success, because of rejection and dedifferentiation of the cells injected (Sigal et al 1995b).

The tissues for which implantation approaches have failed have become the focus of efforts to create bioartificial organs, a prime example being the liver (Brunner et al 1979, Brunner and Schmidt 1981, Crepaldi et al 1997, Demetriou 1994). Indeed, there are more studies on bioartificial livers than on all other bioartificial organs combined, spanning from the 1960s to the present. The proliferation of reviews on bioartificial livers is evidence for the increasing interest in this field (Anand 1996, Bader et al 1995b, Demetriou et al 1995b, Dixit and Gitnick 1996, Fremond et al 1996, Gerlach et al 1996a, Hughes and Williams 1996a, Jauregui et al 1997, Kasai et al 1994, Nyberg et al 1992a, Reid 1997a, Sussman and Kelly 1996a, Yarmush et al 1992). Partially successful forms of bioartificial livers have been achieved by research groups around the world. Some of these bioartificial livers are in clinical trials utilizing either cell lines (e.g., Hepatix, Inc.) or normal pig liver cells (Circe Biomedical, Inc.). Thus, it is likely that the liver will be the first clinically successful bioartificial organ device. Current bioartificial livers that are in clinical trials are intended for acute liver failure (ALF), to provide temporary liver support until the liver can regenerate or until liver transplantation.

Clinical Requirements for Bioartificial Livers

In 1990, more than 27,000 deaths in the United States were caused by liver failure (Nyberg et al 1992a). The mortality rate depends on diagnostic classification of the liver disease, and the appropriate therapy will vary depending on the classification. There are at least three categories of liver disease in which a bioartificial liver device could be applied; acute liver failure, chronic liver disease, and multiorgan failure.

There are two recent diagnostic classifications for acute liver disease (Bernuau et al 1986, O'Grady et al 1993). Both are based on a correlation of prognosis with the time from onset of jaundice to encephalopathy. In general, mortality rate increases (i.e., prognosis is worse) with a longer period of time between onset of jaundice to encephalopathy (Woolf 1994). There is a 90% mortality rate with fulminant hepatic failure (FHF) associated with grade III and IV hepatic encephalopathy. FHF and subfulminant hepatic failure (SHF) are defined as <8 weeks and 8–12 weeks, respectively, between onset of jaundice to encephalopathy (Woolf 1994). There is a third classification for subacute or subchronic liver disease, late onset hepatic failure (LOHF), that has 3–6 months between onset of jaundice to encephalopathy (Gimson et al 1986). Patients have a much higher probability of developing chronic liver disease if they survive LOHF or SHF than if they survive FHF.

There are four main etiologic categories for acute liver failure: viral hepatitis, drug-induced hepatotoxicity, toxins, and miscellaneous (Woolf 1994). A small number of hepatocytes can support a patient in FHF or SHF (Watanabe and Rosenthal 1994), but enhanced functional biomass could provide the collection of synthetic and detoxification functions needed to avoid onset of cytotoxic cerebral edema, which typifies hepatic encephalopathy with grade III–IV coma. Typically, treatment of FHF and SHF focuses on patient stabilization and expectant management, until either the patient's own liver recovers or liver transplantation is performed (Watanabe and Rosenthal 1994). Therefore, a bioartificial liver that performs the myriad of liver functions would decrease the recovery period and thus the total liver damage, shorten the period of hospitalization, increase survivability, and thus decrease medical expense.

Transplantation is the therapy of choice for chronic liver desease (Gerlach et al 1996a). However, only a fraction of the patients with severe liver disease are able to receive a transplant. To qualify to be on the list for transplantation, patients must either have severe forms of acute liver failure or be at the end stage of their disease (End Stage Liver Disease or ESLD). The approximately 4500 hepatic allografts likely to become available in 1998 will provide transplants for only about half of the 9291 patients on the transplantation list (United Network of Organ Sharing Monthly Report, 1997). The number of patients who need liver transplants and yet are not on the list is much higher. There are 2000–4000 cases per year of patients with FHF and SHF, most of whom do not qualify for transplantation; a large group of ESLD patients is not able to get transplants because they cannot endure the surgical procedures, and there is an even larger number of people with chronic liver disease who are not at the end stage of the disease.

A bioartificial liver is a logical alternative for the many patients who do not or who cannot receive a liver transplant. The device can be used to bridge these patients through the acute phase of FHF or SHF until normal liver regeneration occurs or until a transplant is available. The survival rates of patients with FHF or SHF should increase dramatically. In addition, a bioartificial liver could stabilize ESLD patients before the surgery. As a larger percentage of FHF and SHF patients survive, there will be a larger number of these patients developing chronic liver disease. Also, a bioartificial liver can accommodate patients with chronic liver disease as outpatients, which will enhance their quality of life, and lower the hospital expenses. Patients with chronic liver disease have problems with portal hypertension, low prothrombin time, and bilirubin accumulation. A minor variceal bleed could lead to death. Patients with portal hypertension would be able to undergo portal-systemic decompression for relief of pain and discomfort associated with ascites. Therefore, a bioartificial liver could minimize complications caused by compromised immune and clotting systems.

Another clinical focus for bioartificial livers is the liver failure associated with multiorgan failure, a phenomenon that occurs at the end stage of many types of diseases, including sepsis, trauma, heart and kidney diseases, and cancer (Beal and Cerra 1995). In general, liver failure is the proximal cause of death in multiorgan system failure. Theoretically, a bioartificial liver could bridge liver function until the other organs regained function. There will always be a need for a bioartificial liver, given the deficiency in the number of livers available for transplantation, other delays in transplantation, and the many patients who, because of age or frailty, cannot be considered for transplant surgery.

Clinical requirements for optimizing bioartificial liver devices are that it be (1) easily maintained, (2) attached extracorporeally in a fashion similar to kidney dialysis, and (3) able to perform the many liver functions necessary for normal life. In order to achieve this, the ideal bioartificial liver will need to have a high surface-to-volume ratio, high cell density, high metabolic potential with long-term maintenance of well-differentiated liver cells, mass transport of nutrients, toxins, and proteins to and from the cells similar to that in normal liver, and a total biomass sufficient to maintain liver function for normal life, estimated to be 15–20% of the normal liver mass of the patient. With the enormous toxic load in plasma associated with chronic liver patients this biomass may need to be larger under certain circumstances.

Bioartificial Liver Devices

The most critical variable for bioartificial livers is to have a device with sufficient liver biomass to clinically sustain patients with liver failure. Maintaining 20% of normal functioning liver mass, estimated to be a biomass of 300 g of human liver, is necessary to maintain normal life for most adult patients (Hughes and Williams 1996b). Typical adult human liver cell density is 1.7×10^8 cells/g, giving a total of 2.5×10^{11} hepatocytes (Kasai et al 1994). For clinical relevance, a bioartificial liver should have cell densities similar to those of normal liver tissue and should have a large surface-area-to-volume ratio, in order to maintain an optimum exchange with patient plasma. A decrease in liver function will have the same effect as decreasing liver biomass and could compromise bioartificial liver therapy. Therefore, long-term maintenance of differentiated parenchymal cells is desired. All bioartificial livers show degrees and varying duration of biotransformation and synthetic capability.

No bioreactor design accounts for excretion of biliary products. For example, if a functional biliary tree forms, there is not a port designed in bioreactors to accommodate the biliary tree. Typically, the excretory function is performed using adsorbents or dialysis in line with bioartificial liver devices (Hughes and Williams 1996b). One problem with adsorbents, plasmapheresis, and dialysis is that they are nonspecific, and factors necessary for liver repair are also removed (Anand 1996). The issue of removal of biliary cell products remains a problem for all known bioreactor designs.

There are several common functional features that are necessary for defining success of bioartificial livers. These can be categorized as biotransformatory (detoxification, primary metabolism), synthetic (proteins, storage), and excretory (Hughes and Williams 1996b). These functions are generally determined by measuring nutrients, xenobiotics, or proteins in media streams, but NMR spectroscopy can noninvasively measure these functions in "real time" within bioreactor cultures (Gillies et al 1991, Macdonald et al 1998, Mancuso et al 1990, Mancuso et al 1995) or by in-line media stream analysis (O'Leary et al, 1987). The design of protocols to test liver functions will be difficult (Hughes and Williams 1996b), and the optimum and minimum number of tests for liver function and quality control is still being debated.

There are several other key features that one must consider when designing a bioartificial liver. Because of the high metabolic potential of hepatocytes, bioreactors used for bioartifical livers must have high mass transfer rates (Catapano and De Bartolo 1996). Typically, oxygen is a key biolimiting nutrient (Glacken et al 1983). A typical acinus oxygen gradient ranges from 70 mm Hg near the portal triad to 30 mm Hg near the central vein (Holzer and Maier 1987, McCuskey 1994). The mass transfer rates of oxygen and glucose are determined, and their concentration profiles are typically modeled across the cell mass (Catapano 1996a, Heath and Belfort 1987). These concentration profiles are inversely proportional to the hepatocyte metabolic reaction rate and proportional to the effective solute mobility through the cell mass (Catapano 1996a). Solute mobility in a bioreactor will depend on solute molecular weight,

membrane properties, and solute diffusion and convection, which are driven by concentration and pressure gradients, respectively (Catapano 1996a, Heath and Belfort 1987). The effective diffusivity of a solute increase with flow rate; however, this can damage cells as a result of shear forces (Papoutsakis and Michaels, 1993). Blood flow across the liver acinus occurs because of a pressure difference across the sinusoid (Campra and Reynolds 1988), but hepatocytes are protected in vivo from shear forces by a layer of endothelial cells. The various methods of ex vivo maintenance of cells, the issue of maturational stage of the cells, and the perfusion configurations used in bioartifical liver devices are discussed below. Table 21.1 lists the three general categories of bioartificial livers and parameters corresponding to the aforementioned requirements necessary for bioartificial liver clinical success.

Cell Biology Issues

Maturational Stage of the Cells

A critical variable likely to influence the biological properties of bioartificial livers is the maturational stage of the liver cells used to inoculate the bioreactors. To date, only mature liver cells from rodent, pig, or human tissues have been used to seed the varying forms of bioreactors. The life span of these bioartificial livers, produced with mature parenchymal cells, is approximately a week or two, and for some (see Table 21.1) only a few hours, even with optimal culture conditions or type of bioreactor.

Investigators have speculated that a better microenvironment (specific hormones, matrix substrata, basal media) is needed to increase the longevity of the bioartificial livers (Brunner and Schmidt 1981, Crepaldi et al 1997, Demetriou 1994). Certainly, these speculations are correct in that monolayer cultures of liver cells maintained under serum-free or minimal serum conditions and with defined mixes of hormones have survived for several weeks (Brill et al 1994, Block et al 1996, Reid 1990). Long-term survival of cells (months) is dependent upon the use of complex matrix substrata (e.g., Matrigel, Biomatrix) in combination with chemically and hormonally defined media (Bissell and Choun 1988, Brill et al 1994, Enat et al 1984). However, the conditions that promote long-term survival of highly differentiated cells are also conditions that inhibit cell growth (Enat

TABLE 21.1. Summary of various parameters of bioartificial livers.

Bioreactor Type	Reference	Cell type	Cell density[a] (#/ml; # cm^2)	Total biomass (total cell #;g)	Culture[b] duration (days)
SUSPENSION					
A. Kidney Dialyzers					
1. Kiil	Matsumura et al 1987	rabbit	3.3×10^7 (cells/ml)	100 g 1×10^{10} cells (1 g = 10^8 cells)	0.08
2. Cobe	Olumide et al 1977	porcine	$1-2 \times 10^6$ (cells/ml)	114–228 g $1.14-2.28 \times 10^{10}$ (1 g = 10^8 cells)	0.17
3. Gambro	Olumide et al 1977	porcine	$1-2 \times 10^6$ (cells/ml)	114–228 g $1.14-2.28 \times 10^{10}$ (1 g = 10^8 cells)	0.17
B. Hemoperfusion	Margulis et al 1989	porcine	2×10^6 (cells/ml)	4×10^7 cells (Calculated from cell density)	0.04
IMMOBILIZED					
A. Packed-Bed					
Solid Microcarriers					
1. Glass beads	Li et al 1993	S-D rat hepatocytes	5×10^7 (cells/cm^2)	1.6×10^8	15
2. Biosilon	Shynra et al 1991 & 1990	Wistar rat hepatocytes	0.75×10^5 (cells/cm^2)	$0.5-1.5 \times 10^8$	3–9
Reticulated Microcarriers					
1. Reticulated PVC	Yanagi et al 1992; Miyoshi et al 1996	Wistar rat hepatocytes	1.2×10^7 (cells/ml)	$0.3-1.1 \times 10^8$	1–7
2. Cytodex 3 collagen Type 1 coated	Demetriou et al 1986	Rat Hepatocytes	1.6×10^7 (cells/ml)	1.6×10^8	0.04
Encapsulated					
1. Alginate poly-L-lysine coated	Dixit et al 1995	Rat Hepatocytes	ND	ND	ND
2. Alginate	Fremond et al 1993	SD Rats	1×10^7 (cells/ml gel)	10^8	1
Nonwoven					
1. Wooly fiber	Naruse et al 1996a & b	porcine	*50 ml unit* 2×10^7 cells/ml *200 ml unit* 5×10^7 cells/ml	*50 mL unit* 1×10^9 *200 mL unit* 1×10^{10}	6
B. Spirally-Wound					
1.	Flendrig et al 1997	porcine	1.1×10^7 (cells/ml)	2.2×10^8	4
C. Multiplate or Flat-Bed					
1. Monolayer on glass	Uchino et al 1988	canine	2.5×10^5 (cells/ml) (lg = 10^8 cells)	80 grams (lg = 10^8 cells; from Bader et al. 1995)	14
2. Monolayer on glass with collagen sandwich	Taguchi et al 1996	Wistar rats	8.1×10^5 (cells/ml)	6.5×10^7	14
3. Rotating disk	Yanagi et al 1989	Wistar rats & N.Z. white rabbits	$1.9 \times 10^6 - 1.7 \times 10^7$ (cells/ml gel) $1.6 \times 10^6 - 1.5 \times 10^7$ (cells/ml)	1.3×10^8 -1.2×10^9	0.17
4. Bilaminar membrane with sinusoids	Bader et al 1995 & 1997	Rat & Porcine	1.74×10^5	$1.4-4 \times 10^{10}$	11
5. Flat-bed hollow fiber multiplate hybrid	Smith et al 1997	Rat & Porcine	$1.5-1.8 \times 10^6$ (cells/cm^2)	Rat $2.5-5.6 \times 10^7$ Porcine $4.3-9.7 \times 10^7$	8

Surface-Volume[c] internal (cm^{-1})	Surface-Volume[c] external (cm^{-1})	Animal		Differentiation	
		Albumin[d] pg/cell/h	Urea[d] pg/cell/h	Phase I & II[d] pg/cell/h	Vendor
SUSPENSION					
A. Kidney Dialyzers					
NA	33.3	—	—	—	Vernitron Corp.
NA	9.5	—	—	—	Cobe Labs
NA	9.5	—	—	—	Gambro, Inc.
NA (direct hemoperfusion)	NA (direct hemoperfusion)	—	—	—	None
IMMOBILIZED					
A. Packed-Bed					
Solid Microcarriers					
44	NA	0.055	0.388	—	None
286	NA	5.0 decreased by day 5	—	Bilirubin conjugation = 25	None
Reticulated Microcarriers					
NC	NC	(1 µg/h/ml) 8 with serum supplement	—	—	None
ND	ND	—	—	—	None
Encapsulated					
ND	ND	—	—	—	None
66	60	—	—	—	None
Nonwoven					
NC	NC	*200 ml unit* 0.55	*200 ml unit* 2.05	—	None
B. Spirally-Wound					
8.2	NA	—	1.5	lidocaine 34.1	Microgon
C. Multiplate or Flat-Bed					
100	NA	0.088; decreasing after 5 days (from Bader et al 1995)	0.46; decreasing after 5 days (from Bader et al 1995)	—	None
4.1	NA	—	13.3; decreasing after 5 days. (from collagen gel sandwich treatment)	—	None
0.6	12	—	NC	—	None
NC	NC	3.6	6.4–27.3	0.4-1 urapil	None
19.5	NA	—	2–3	—	None

(continued)

TABLE 21.1. (*continued*)

Bioreactor Type	Reference	Cell type	Cell density[a] (#/ml; # cm^2)	Total biomass (total cell #;g)	Culture[b] duration (days)
D. Suspension					
1. Spouted-bed	Takabatake et al 1991	SD rats	$2.5–7.5 \times 10^6$ (cells/ml gel) $1.4–4.2 \times 10^5$ (cells/ml)	$1–1.5 \times 10^5$ spheroids $1–3 \times 10^7$ cells	3.45
2. Rotating wall vessel	Khaoustov et al 1995	Primary Human	10^6 cells/ml (50 ml volume)	5×10^7	30
E. Microchannel					
1. Microchannel	Bhatia et al 1997	Lewis Rats	3.2×10^5 (cells/cm^2)	3×10^6	14
HOLLOW FIBER					
A. Conventional					
Hepatocytes in Extracapillary Space					
1. ELAD	Sussman et al 1992	Human hepatoma	NC	200 g 2×10^{10} cells (lg = 10^8 cells)	7.5–10
2. Hepatassist	Chen et al 1996; Demetriou et al 1995	porcine	2.8×10^7 (cells/ml)	5×10^9	0.25
Hepatocytes in Intracapillary Space					
1. Coallagen Type 1 entrapment	Nyberg et al 1992b; Shatford et al 1992	Human Cell line HepG2 & S.D. Rat	$3–5 \times 10^6$ (cells/ml gel)	$5 \times 10^7–10^8$	14
B. Coaxial					
1. Coaxial	Macdonald et al 1998	SD rat	1.5×10^8 (cells/ml)	6.5×10^8	0.25
2. Radially scaled-out	Goffe 1997	Porcine	5.6×10^7 (cells/ml)	10^9	35
C. Woven					
1. Woven	Gerlach et al 1994	Porcine	1.6×10^8 (Total ECS vol is ~15 ml)	2.5×10^9 cells	49 (7 weeks)

[a] Cell density calculated from the inoculation number

[b] Culture duration is the total number of days hepatocytes were maintained, and not necessarily the longest possible duration.

[c] Surface-volume ratio was calculated for (1) packed bed as: surface area of the bead = $4\pi r^2$; internal volume is the dead volume of the bioreactor, assumed to be the volume of a cube $(8r^3)$ − volume of the bead $(4/3\pi r^3)$; external volume is the volume of the bead $(4/3\pi r^3)$; packing order was assumed to be one on top of another and side-by-side. (2) For the rotating disk bioreactor, the surface area was calculated as $(2\pi \times \text{radius of the bioreactor}) \times (2.4 \text{ cm})$; the internal volume = 70 ml; the external volume = $(2\pi \times \text{radius of the bioreactor}) \times (0.2 \text{ cm}^2)$. (3) Spirally-wound bioreactor surface area is given in Flendrig et al 1997 as 15 times the volume of the polyester matrix. Matrix volume was calculated from bioreactor specifications to be 6.04 ml, and the internal

et al 1984). Thus, the investigator is constrained by having bioartificial livers in which the cell mass must be achieved at the time of inoculation. Even when the cells are introduced at subconfluence, little or no growth has been observed (Bader et al 1995b, Demetriou et al 1995b).

Recent investigations into liver stem cell biology suggest an explanation for the poor expansion potential of the mature cells in bioreactors and an alternate option for a means to achieve expansion and bioartificial livers that are stable and differentiated. Progenitor cell populations, long known to be in hemopoietic, gastrointestinal, and epidermal tissues, have been found in liver and in most other quiescent tissues (Brill et al 1995, Fiorino 1998, Reid 1997a, Sell 1990, Sigal et al 1992, Sigal et al 1994, Sigal et al 1995a). The progenitors produce daughter cells that go through a step-wise maturational process ultimately ending in terminally differentiated cells; the tissues consist, therefore, of plates of maturational lineages of cells (Sigal et al 1995a, Sigal et al, submitted). The liver's lineage is found within the acinus, the key structural and functional unit of the liver (Figure 21.1 and 21.2). It is organized like a wheel around two distinct vascular beds: six sets of portal triads, each with a portal venule, hepatic arteriole, and a bile duct, form the periphery, and the central vein forms the hub. The parenchyma, effectively the "spokes" of the wheel, consist of plates of cells lined on both sides by the fenestrated sinusoidal endothelium. Blood flows from the portal venules and hepatic arterioles at the portal triads, through sinusoids that line plates of parenchyma, to the terminal hepatic venules, the central vein. According to this mi-

Surface-Volume[c] internal (cm^{-1})	Surface-Volume[c] external (cm^{-1})	Animal		Differentiation	Vendor
		Albumin[d] pg/cell/h	Urea[d] pg/cell/h	Phase I & II[d] pg/cell/h	
D. Suspension					
3.4	60	1,8 (from Bader et al 1995)	0.018 (from Bader et al 1995)	—	None
NC	NC	—	—	—	Synthecon
E. Microchannel					
NC	NC	—	—	—	None
HOLLOW FIBER					
A. Conventional					
Hepatocytes in Extracapillary Space					
NC	NC	10	—	—	Hepatix
61	33	—	—	—	Circe Biomedical
Hepatocytes in Intracapillary Space					
NC	NC	0.6	0.156 (From Bader et al 1995)	Lidocaine 0.74	None
B. Coaxial					
61	10.2	—	—	—	Setec Inc.
61	10.3	—	—	—	Genespan Inc.
C. Woven					
NC	NC	—	—	—	None

(*continued*)

   volume is the dead volume = 11 ml. (4) Multiplate surface area is the total area for cell growth given (Taguchi et al 1996) or calculated from the plate dimension (Uchino et al 1988). Internal volume is the dead volume of the bioreactor. (5) Hollow-fiber surface area is the total fiber wall area in the intracapillary space (ICS). The internal volume is the volume of the ICS, and the external volume is the volume of the extracapillary space (ECS). Coaxial bioreactor surface area is the fiber wall area of the ICS; internal and external volume is the volume of the ICS and ECS, respectively. Scale-up coaxial surface-area-to-volume ratio was calculated from the data given.

[d] Metabolic rates were converted to pg/cell/hr when necessary, using the inoculation number or total biomass.

crocirculatory pattern, the acinus is divided into three zones: Zone 1, the periportal region; Zone 2, the central region; and Zone 3, the region surrounding the central vein or terminal hepatic venule.

Hepatocytes display marked morphologic, biochemical, and functional heterogeneity according to their zonal location (Figure 21.2). They are smaller and diploid in Zone 1, intermediate in size and a mixture of diploid and tetraploid in Zone 2, and larger and polyploid in Zone 3; there are zone-dependent variations in the morphology of mitochondria, endoplasmic reticulum, and glycogen granules. Proliferative ability of hepatocytes varies across the acinus. Gene expression across the plate can be uniform, uniform with a peak abundance in particular zones or cells, broadly zonal, or restricted to individual cells at distinct positions within one of the zones. Zonal distribution is evident for enzymes involved in glycogen degradation and gluconeogenesis (Zone 1) as opposed to glycogen synthesis and glycolysis (Zone 3) and for aspects of fatty acid, bile, and glutathione metabolism. By contrast, discrete expression is evident for connexin 26 (periportal), a-fetoprotein (periportal), major urinary protein (4–5 cells pericentrally), α-2-microglobulin (pericentral), and glutamine synthetase (single pericentral cell layer).

The potential for cell division also is variable among the parenchymal cells. Grompe and associates (Overturf et al 1997) have tested the potential of parenchymal cells to divide by injecting them into livers damaged by tyrosinemia. They found that whereas most parenchymal cells showed limited cell division, a small subpopulation of parenchymal cells has the capacity to divide more than 69 times in vivo

A   **B**

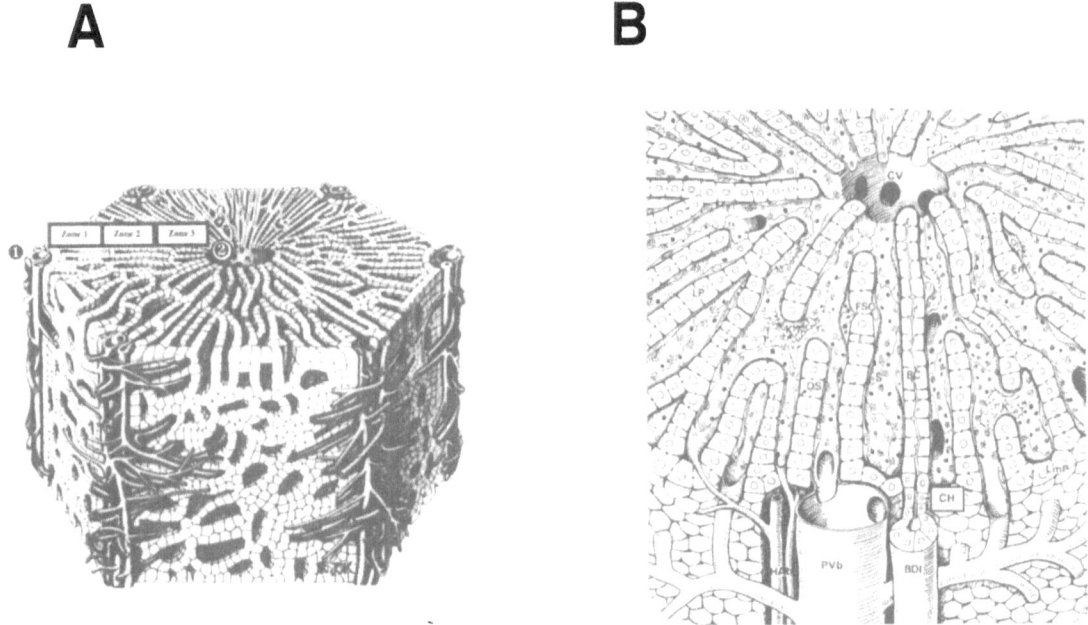

FIGURE 21.1. A three-dimensional representation of the liver lobule (A) illustrating the three zones of the liver acinus. An enlargement of the liver acinus illustrates the various anatomical features (B).

Zone 1   Zone 2   Zone 3

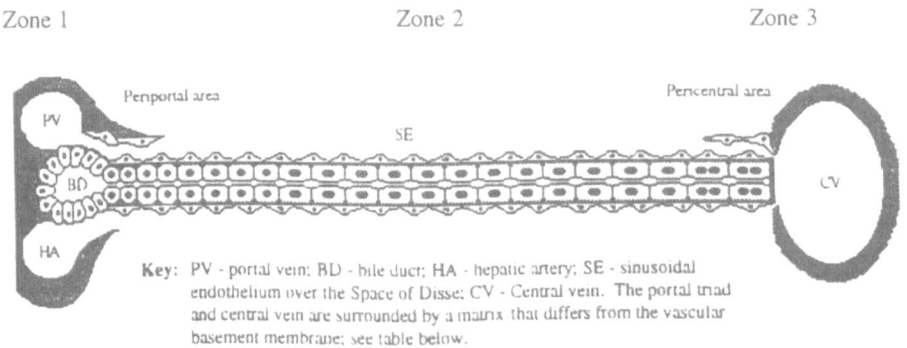

Progenitor Cell Populations	Adult Cells	Presumptive Old Cells
Diploid	Diploid + tetraploid	Tetraploid + octaploid
Embryonic matrix chemistry	<---------------Adult matrix chemistry------------------------>	
Type IV collagen, laminin	Fibrillar collagens, fibronectin	
Heparan sulfate-PGs	Gradual increase in Heparin-PGs	
Maximum growth potential	Limited growth (5–8 divisions)	Incomplete or no cell division
Early gene expression	Early adult gene expression	Late adult gene expression

FIGURE 21.2. Schematic of the liver acinus showing a liver plate extending between the portal triad and central vein. Some of the known distinctions in cell mophology, ploidy, matrix chemistry growth, and gene expression are noted.

(Grompe et al, submitted). In studies of cells in vitro, pericentral ("old") hepatocytes are either incapable of cell division or capable of nuclear division with incomplete cytokinesis resulting in polyploid cells; periportal ("young") hepatocytes are diploid and capable of perhaps 10–12 rounds of division, if given serum-free, hormonally defined medium supplemented with additives such as nicotinamide (Block et al 1996, Gebhardt 1990, Reid and Luntz 1997, Sigal et al 1995a). Complementing the in vivo investigations of Grompe and associates (Grombe et al, submitted), who identified that there are parenchymal cell subpopulations with extensive growth potential, are the findings by Reid and associates (Brill et al 1995, Brill et al, submitted, Sigal et al 1994) showing the extensive expansion potential in culture of hepatoblasts (early progenitors), isolated by multiparametric flow cytometric methods and cultured under defined culture conditions. Implicit in all of these studies is a lineage-position dependence (correlating with acinar localization) in growth potential of the parenchymal cells.

A hypothesis derived from the above studies is that early progenitor cells should be ideal for inoculation of the bioreactors to form bioartificial livers. They have the greatest expansion potential, meaning that a relatively small number are needed for seeding the bioreactors and should have the ability to mature into all the cells of the adult liver if given the appropriate culture conditions. Theoretically, the bioartificial livers created with the progenitors should have significantly greater longevity and stability than is observed with bioreactors seeded only with mature cells. Studies testing this hypothesis are ongoing.

Cellular Microenvironment

Survival, growth, and tissue-specific gene expression in eukaryotic cells have been found to be regulated by the synergistic effects of soluble signals (hormones, growth factors, nutrients, oxygen) and insoluble ones (extracellular matrix components) (Brill et al 1994, Reid 1997a). In Tables 21.2 and 21.3 are summarized known variables for long-term maintenance of parenchymal cells in a growth versus differentiation state. Some of the requirements are general for all maturational stages, and some are unique to specific stages (Reid 1997a).

The most significant developments improving ex vivo maintenance of normal cells are the use of serum-free hormonally and chemically defined media (Reid 1990) and of substrata of extracellular matrix (Brill et al 1994, Hassel 1997, Reid 1997a). The extracellular matrix is a complex of insoluble material produced by all cells and found on the lateral borders between homotypic cells (the lateral matrix), and on the basal surface of cells separating heterotypic cell types (the basal matrix). The soluble (hormones, nutrients) and insoluble (extracellular matrix) materials are interregulated and also have mutiple synergistic effects. Thus, soluble signals can induce synthesis of specific matrix components; matrix substrata can dictate output of specific soluble signals or regulate receptors; and specific matrix components can have sequences that, when released by enzymatic activity, prove to be functional as signals (Panayotou et al 1989). Extracellular matrix components turn over slowly, on the order of days to weeks, and serve to stabilize cells in specific configurations of adhesion sites, specific antigenic profiles, hormone receptors, ion channels, etc. Since the extracellular matrix is connected directly to the cytoskeleton via transmembrane molecules, changes in the conformation of one or more matrix components can directly influence cell shape. The specificity and extent of a biological response are dictated by a soluble signal, the cell type, the maturational stage of the cell, and the extracellular matrix chemistry. The chemical composition of the extracellular matrix varies with the cell type and with the maturational stage of the cells. All known progenitor populations are associated with Type IV collagen, fetal forms of laminin, fetal forms of heparan sulfate proteoglycan, and hyaluronates. Maturation of cells is associated with changes in the matrix chemistry. Three variations in the matrix chemistry changes have been identified (Reid 1997a):

1. Preservation of expression of all major classes of matrix components but with gradients in isoforms within those classes. This pattern occurs in all endodermal tissues.
2. Loss of expression of basal matrix components. Thus, the cells lose expression of collagens and basal adhesion proteins (laminin, fibronectin). The gradient in matrix chemistry occurs in the lateral matrix. The skin and nervous system are representative of this pattern.
3. Loss in expression of some collagens and shift to regulated expression of adhesion proteins encoding cell-binding domains. This pattern is evident in hemopoietic tissues.

TABLE 21.2. Requirements for ex vivo expansion of parenchymal cells.

Maturational stage	Feeder layers	Optimal substratum	Possible basal media	Proteoglycan (added to medium)	Calcium	Medium additives	Lipids	Hormones/ growth factors
Hepatoblasts	Liver-specific stroma + myeloid progenitors	Porous, flexible, and coated with Type IV collagen + laminin	RPMI 1604 or Ham's F12	Liver-derived heparan sulfate-PG or heparan sulfates	<0.5 mM	Nicotinamide putrescine	FFA on serum albumin, HDL, LDL	HGF, T3, LIF, transferrin, IGF II, bFGF
Committed Progenitors	Nonspecific stromal feeders	Porous, flexible, and coated with Type IV collagen + laminin	—	Liver-derived heparan sulfate-PG or heparan sulfates	<0.5 mM	Nicotinamide putrescine	FFA on serum albumin, HDL, LDL	HGF, T3, LIF, transferrin, IGF II, bFGF
Mature Parenchymal Cells	No feeders required	Porous, flexible, and coated with Type IV collagen + laminin	—	Liver-derived heparan sulfate-PG or heparan sulfates	<0.5 mM	Nicotinamide putrescine	Linoleic acid will suffice	Insulin, EGF, T3, growth hormone, prolactin

TABLE 21.3. Requirements for *ex vivo* differentiation of parenchymal cells.

Maturational stage	Feeder layers	Optimal substratum	Possible basal media	Proteoglycan (added to medium)	Calcium	Medium additives	Lipids	Hormones/ growth factors
Hepatoblasts	Liver-specific stroma + myeloid progenitors	Porous, flexible, and coated with Type I or III collagen + fibronectin	Serum-free RPMI 1640 or Ham's F12	Liver-derived heparan-PG or heparins	>0.5 mM	Nicotinamide putrescine	FFA on serum albumin, HDL, LDL	IGF II, transferrin, glucocorticoids, + hormones tailored to gene(s) of intere
Committed progenitors	Nonspecific stromal feeders	Porous, flexible, and coated with Type I or III collagen + fibronectin	—	Liver-derived heparan-PG or heparins	>0.5 mM	Nicotinamide putrescine	FFA on serum albumin, HDL, LDL	IGF II, transferrin, glucocorticoids, + hormones tailored to gene(s) of intere
Mature parenchymal cells	No feeders required	Porous, flexible, and coated with Type I or III collagen + fibronectin	—	Liver-derived heparan-PG or heparins	>0.5 mM	Nicotinamide putrescine	Linoleic acid will suffice	Insulin, T3, glucocorticoids + hormones tailored to gene(s) of intere

Matrix substrata are essential for cell survival and growth, and can dictate fates with respect to differentiation. The extracellular matrix components can be provided ex vivo either by soluble signals that induce the synthesis of the appropriate matrix components and/or by presenting the cells with prepared substrata of tissue extracts enriched in extracellular matrix (Matrigel, Biomatrix) or purified matrix components. The purified matrix components and the tissue extracts enriched in extracellular matrix, many of them commercially available, produce highly reproducible biological responses. In summary, for bioartificial livers, it will be essential to introduce the cells onto appropriate matrix substrata tailored for the cells of interest and for the appropriate maturational stage of the cells.

Engineering Aspects of Bioartificial Livers

The key engineering challenge in designing a bioartificial liver is maintaining adequate transport of nutrients to all cells in the bioartificial device while at the same time providing a local microenvironment conducive to the differentiated state and minimizing the total reactor volume. Hepatocytes have high metabolic rates, and consequently hepatocytes in normal liver are typically no more than a single cell diameter (a few microns) away from sinusoidal blood, rich in nutrients. This close proximity to the nutrient supply can be mimicked in vitro by culturing the cells in monolayers with perfusion above the monolayer (Bhatia et al 1994, Smith et al 1997). The differentiated function of hepatocytes, however, is typically superior in cultures in which the cells are in an aggregate or 3-D configuration. Most reactor designs under study and all reactor designs currently used in clincial trials employ some sort of 3-D culture arrangement. Three-dimensional cultures are inherently susceptible to large gradients in nutrient concentrations—particularly oxygen—whenever multilayers of cells are involved. Because so many of the bioreactors described in this review employ 3-D culture arrangements, it is useful to consider the basic physicochemical aspects of nutrient transport in 3-D cultures to provide a basis of comparison for the different reactor configurations.

The dominant mechanism of transport of low molecular weight nutrients (e.g., oxygen and glucose) within tissue or cell aggregates is diffusion. The length scale for diffusive transport (i.e., the distance over which oxygen penetrates into the tissue from the nutrient stream before it is completely consumed by the cells) depends on the volumetric concentration of metabolite in the nutrient stream, C_o, the rate at which cells consume the nutrient, Q_i, the diffusion coefficient for the metabolite in the tissue, D_t, and the system geometry. The metabolic consumption rate Q_i is generally a function of the nutrient concentration in the cell mass, C. The most common rate expression is of the Michaelis–Menten type

$$Q_i = \frac{V_{max}C}{K_m + C}$$

which reduces to zero-order kinetics for high concentrations of the nutrient (i.e., $C \gg K_m$) and to first-order kinetics for low nutrient concentrations ($C \ll K_m$). It is reasonable to consider the zero-order limit, because the condition $C \gg K_m$ is met for many important nutrients under normal physiological conditions.

An estimate of the length scale for nutrient diffusion in a 3-D cell mass can be obtained by mathematical modeling. The general equation governing the balance between steady-state diffusion and metabolic consumption is

$$D_i \nabla^2 C = Q_i \tag{1}$$

where C is the concentration of the nutrient within the cell mass. Three simple geometries amenable to analytical solutions are shown in Figure 21.3: a slab, cylinder, or sphere of cells bathed in medium containing the nutrient at concentration C_0. Expanding the gradient operator for each of these geometries, Equation 1 can be written as

slab: $\qquad D_t \dfrac{d^2 C}{dx^2} = Q_i \tag{1a}$

cylinder: $\qquad D_t \dfrac{1}{r} \dfrac{d}{dr}\left(r\dfrac{dC}{dr}\right) = Q_i \tag{1b}$

sphere: $\qquad D_t \dfrac{1}{r^2} \dfrac{d}{dr}\left(r^2\dfrac{dC}{dr}\right) = Q_i. \tag{1c}$

The manner in which the distance variables are defined in each case is shown in Figure 21.3. A stan-

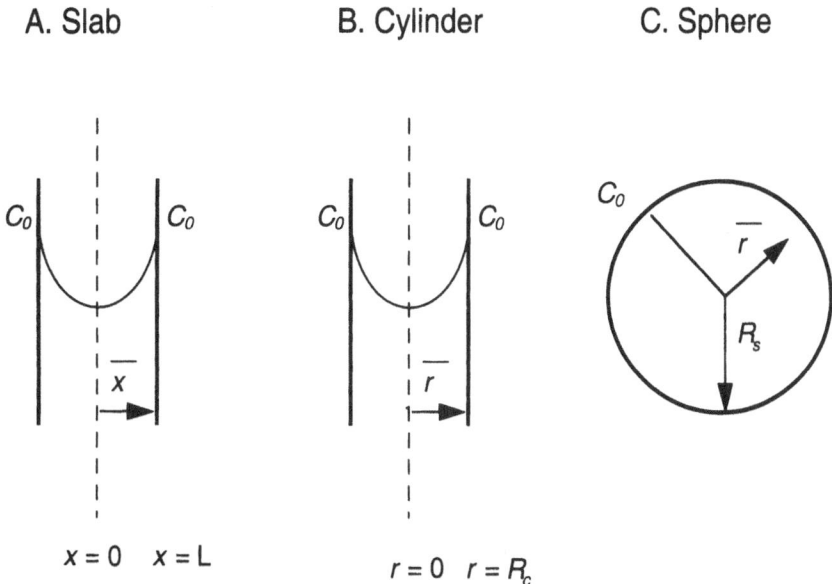

FIGURE 21.3. The three possible configurations considered by the mass transfer equations.

dard technique in the analysis of differential equations describing transport phenomena is to scale the variables so that they are dimensionless and range in value from 0–1. This allows the relative magnitude of the various terms of the equations to be evaluated easily and allows the solutions to be plotted in a set of graphs as a function of variables that are universally applicable. For all geometries, C is scaled by its maximum value, C_0; i.e., $\bar{C} = C / C_0$. Distance is scaled with the diffusion path length, so that for a slab, $\bar{x} = x / L$, for a cylinder $\bar{r} = r / R_c$, and for a sphere, $\bar{r} = r / R_s$. With these definitions, the boundary conditions for all three geometries are no flux of nutrient at the center ($d\bar{c} / d\bar{x}, d\bar{c} / d\bar{r} = 0$ at $\bar{x}, \bar{r} = 0$) and that $\bar{C} = 1$ at the surface ($\bar{x}, \bar{r} = 1$).

The use of scaling allows the solutions for all three geometries to collapse to a common form:

$$\bar{C} = 1 - \frac{\phi^2}{2} (1 - \bar{x}^2).$$

All of the system parameters are lumped together in the dimensionless parameter ϕ^2, which is often called the Thiele modulus. The Thiele modulus represents the relative rates of reaction and diffusion and is defined slightly differently for each geometry:

$$\varphi^2_{slab} = \frac{Q_i L^2}{C_0 D_t}; \quad \varphi^2_{cyl} = \frac{Q_i R_c^2}{2C_0 D_t}; \quad \varphi^2_{sph} = \frac{Q_i R_s^2}{3C_0 D_t}.$$

Figures 21.4A and 21.4B are oxygen concentration profiles in conventional and coaxial bioreactors, respectively (Custer 1988). As oxygen is consumed by the cell mass, its concentration decreases further from the media source (i.e., hollow fiber wall), and depending on the magnitude of the metabolic rate of the cell mass, there could be hypoxic regions in the bioreactor. Figures 21.4A and 21.4B demonstrate the effect of applying zero-order (i.e., linear) and first-order (i.e., nonlinear) kinetics in modeling oxygen concentration profiles, demonstrating that first-order kinetics causes a more dramatic decrease in concentration. Hypoxic regions could be avoided by increasing oxygen concentration in the media such that oxygen is not biolimiting, and zero-order kinetics can be applied. However, increased oxygen concentration results in increased toxicity due to oxidative stress. Recent bioartificial livers have integrated oxygenation into the bioreactor design, termed "integral" oxygenation. Because of the large diffusion coefficient of dissolved oxygen relative to convective flux in bioreactors. diffusion is generally dominant in determining oxygen mass transfer in bioreactors.

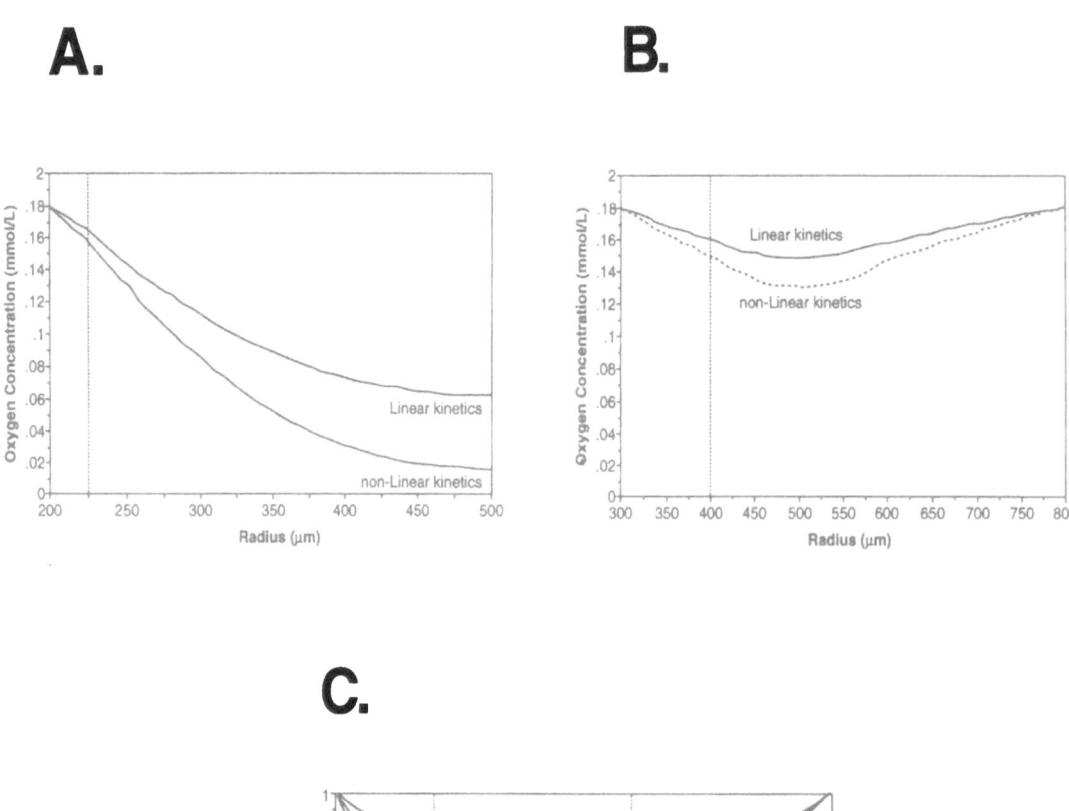

FIGURE 21.4. The oxygen concentration in conventional hollow-fiber bioreactors (A) and coaxial bioreactors (B), demonstrating the effect of zero- and first-order consumption rates. The effect of flow rate on oxygen concentration in the coaxial bioreactor (C). Reproduced by permission from Dr. Linda Custer (Custer, 1988).

One can increase the diffusivity of nutrients, especially macromolecules, by increasing convective flux. Diffusivity will depend on membrane properties and mass transport in the device (Catapano 1996). Increasing the bulk axial flow rate (q) will increase diffusivity, depending on solute mobility through the membrane and the cell mass. Solute mobility is a function of diffusion rate and convective flux, which depend on concentration and pressure gradients, respectively. The relative importance of diffusion and convection for modeling substrate profiles in bioreactors will depend on the bioreactor configuration (i.e., its fluid dynamics, cell loading and distribution, and reactor geometry) (Catapano 1996). Catapano and others, (Catapano et al 1996) found that 100 μm is the maximum distance oxygen can diffuse in hepatocyte cultures before levels become hypoxic. Diffusivity can be increased by increasing concentration gradients (i.e., diffusion rate), and pressure gradients (i.e., convective flux). For proper hepatocyte function it is important to have high mass transfer rates of nutrients to and waste products away from the cells, yet not to generate shear forces, which have been shown to decrease differentiation (Frangos et al 1993). Figure 21.4C shows the effect of flow rate on the oxygen profile for coaxial bioreactors. The dimensionless flow parameter, the Peclet number (P_e), can be expressed as

$$P_e = \frac{q}{rD_t} \qquad ([m^3/s]/[m]*[m^2/s]) \text{ axial}$$

$$P_e = \frac{Jr}{D_t} \qquad ([m/s]*[m]/[m^2/s]) \text{ radial}$$

where J is radial flux or velocity. The coaxial bioreactor is modeled only for radial flux. Figure 21.4C illustrates the effect of increasing radial flux (i.e., P_e) on the oxygen concentration profile in the coaxial bioreactor (Custer 1988). Although the conventional bioreactor is not displayed, the same effect occurs, which is that the oxygen profile is less dramatic with increasing P_e.

Consumption rate will depend on hepatocyte differentiation, cell density, and the metabolite of interest. The oxygen consumption rate of HepG2 cells ranged from 0.59 to 0.7 nmol/s^{-1}/10^6 cells (Smith et al 1996); that of cultured rat hepatocytes was 0.4 nmol/s^{-1}/10^6 cells (Rotem et al 1992); and that of isolated hepatocytes was 0.42 nmol/s^{-1}/10^6 cells (Sulieman and Stevens 1987). Depending on growth rate and other factors, the consumption rates for glutamate, pyruvate, and glucose are typically in the range of 0.03 to 0.3 nmol/s^{-1}/10^6 cells, assuming a cell density of 1.7×10^8 cells/cm^3 tissue (Cremmer et al 1981, Glacken 1987, Imamura et al 1982). The diffusion coefficients for oxygen in tissue range from $0.6-2.5 \times 10^{-5}$ cm^2/s. The diffusion coefficients for glucose and pyruvate in water at 37°C are 9.3×10^{-6} cm^2/s and 1.3×10^{-5} cm^2/s, respectively, and were correlated with tissue glucosaminoglycan concentration (Swabb et al 1974). Therefore, oxygen is generally the biolimiting nutrient in bioartificial livers not only because of its relatively high consumption rate, but especially because of its relatively low concentration in normal culture conditions, for example, its solubility at 1 atm and 37°C is 0.2 mM (solubility constant in pure water is 1.06 μM/ml/atm) while glucose is as high as 25 mN (i.e., 4.5 gr/L) under some culture conditions.

Survey of Forms of Bioartificial Livers

There are three general types of bioartificial livers used with isolated or cultured hepatocytes:

1. cells in suspension
2. cells immobilized on or in matrices other than hollow fibers (e.g., flat-bed bioreactors)
3. cells in hollow-fiber bioreactors

Although studies of perfused livers or liver slices, mostly porcine, are considered a form of bioartificial liver (Hughes and Williams 1996b), they will not be discussed. The various clinical and experimental bioreactor designs available are listed in Table 21.1 with the various descriptive features and parameters of the bioartificial liver, such as the cell type and density, total biomass, surface-to-volume ratio, duration of culture, differentiated functions, and commercial vendor. There are two surface-to-volume ratios listed in Table 21.1, the internal and external surface-to-volume ratios. The surface area is the total area that the plasma directly encounters in the bioreactor, while the internal and external volumes are those volumes contained by the plasma side or the cell side of the bioreactor, respectively. For example, in conventional hollow-fiber bioreactors, where the cells reside in the extracapillary space (ECS), the internal volume (surface-to-volume$_i$) is the area of the fiber lumen (ICS), and the external volume (surface-to-volume$_e$) is the ECS. The surface-area-to-volume ratio is a measure of mass transport in the bioreactor. As discussed previously, optimization of these parameters will fit the requirements of the ideal bioartificial liver. In most of the studies using primary hepatocyte preparations, it is assumed that there is no cell growth, since adult hepatocytes are used and/or because it is difficult to extricate the cell mass; therefore, the inoculate number is used to calculate bioartificial liver cell density and metabolic rates (Gerlach 1994, Naruse et al 1996b, Takabatake et al 1991, Yanagi et al 1992). Hepatocytes can be immobilized onto the perfusion vessel interior or onto or within a matrix and placed inside a perfusion vessel. Therefore, the bioreactor is composed of the perfusion vessel and cells and is the metabolic unit of the bioartificial liver. Bioreactors are typically modeled to determine the optimum operation conditions (Catapano 1996a, Gerlach et al

1990, Giorgio et al 1993, Glacken et al 1983, Heath and Belfort 1987, Nyberg et al 1992a). To avoid shear forces during scale-up (Glacken et al 1983), maintain differentiation, and better control the physicochemical environment, more bioartificial liver research groups are inoculating hollow-fiber reactors with hepatocytes on microcarriers, on matrix-coated microcarriers, or encapsulated in gels (Demetriou et al 1995a, Gerlach et al 1996b, Nyberg et al 1992a). Another set of researchers has tried to protect the hepatocytes from shear forces and immune response using microencapsulation techniques; typically spheroidal aggregates or encapsulates of hepatocytes are coated with a semipermeable membrane (Dixit and Gitnick 1996). However, a more recent set of researchers has addressed the oxygen demand of hepatocytes by incorporating "integral" membrane oxygenation rather than using "in-series" oxygenation (Bader et al 1995a, Flendrig et al 1997, Gerlach 1994, Goffe 1997, Smith et al 1997).

Suspension-Type Bioartificial Livers

Freshly isolated hepatocytes suspended in plate dialyzers, which are used clinically for kidney dialysis, were the first bioartificial livers used to treat human beings (Matsumura et al 1987, Margulis et al 1989), and demonstrated increased survival rates. The main problem with this bioartificial liver is that differentiated functions decrease immediately following hepatocyte isolation. For example, Olumide and et al (1977), in an animal study using anhepatic pigs (Table 21.1), demonstrated that oxygen consumption in isolated porcine hepatocytes was 62.5% of a theoretical normal liver initially, and decreased 60% after 1 hour in the suspension bioartificial liver device. In fact, hepatocytes were changed every 1 hour (Margulis et al 1989) and 4 hours (Matsumura et al 1987) during human bioartificial liver dialysis treatment because of the decrease in hepatocyte function. The fact that hepatocytes require attachment to maintain differentiation via cell-cell and cell-ECM contact has prompted many researchers to devise many specific and nonspecific immobilization matrices (Dixit and Gitnick 1996).

Bioreactors with Cells Immobilized in Specific or Nonspecific Matrices

Most of the bioartificial livers described in the literature immobilize hepatocytes (1) by mixing them with a matrix that will solidify and encapsulate them, (2) by using microcarriers or a solid support to which they attach, or (3) by using membranes that microencapsulate the cells. Other chapters in this book describe these methods in detail. We will discuss the various immobilization methods used for hepatocytes.

Bioreactors with Microencapsulation of Cells

Microencapsulation techniques used for bioartificial liver devices typically have used hepatocyte single cell suspensions or spheroids mixed with xenogenous or endogenous matrices (Fremond et al 1993, Miura et al 1986, Takabatake et al 1991, Yanagi et al 1989) and often coated with various materials that form a membrane creating a microcapsule (Chang 1986, Dixit and Gitnick 1996, Joly et al 1997, Stange and Mitzner 1996, Vacanti et al 1988). The specific materials and techniques will be discussed for the different bioartificial liver devices.

Bioreactors with Cells on Microcarriers

Microcarriers are different from the encapsulation technique, in that cells typically attach to the surface of the microcarrier. However, for microcarriers that are biodegradable or composed of collagen or some other ECM-derived material, the cells will eventually invade the carrier, and ultimately the microcarrier will disintegrate. In fact, this is the goal for the biodegradable microcarriers. Biodegradable beads have primarily been used for implantation of hepatocytes (Cima et al 1991, Vacanti et al 1988), but recently biodegradable scaffolds have been used to recreate the liver architecture (Griffith et al 1997).

Bioreactor Devices for Immobilized Cells

Once immobilized, the cells are perfused in packed-bed, multiplate or flat-bed, microchannel, suspension, spirally-wound, or hollow-fiber bioreactors. Hollow-fiber bioreactors will be discussed in detail in a later section. For the others, see the following sections.

Packed-Bed Bioreactors

Packed-bed bioreactors are perfusion devices for hepatocytes encapsulated or attached on solid support material. Typically, it is assumed in modeling mass transfer in spherical beads or microcapsules that flow is uniform throughout the reactor bed (Belfort 1987

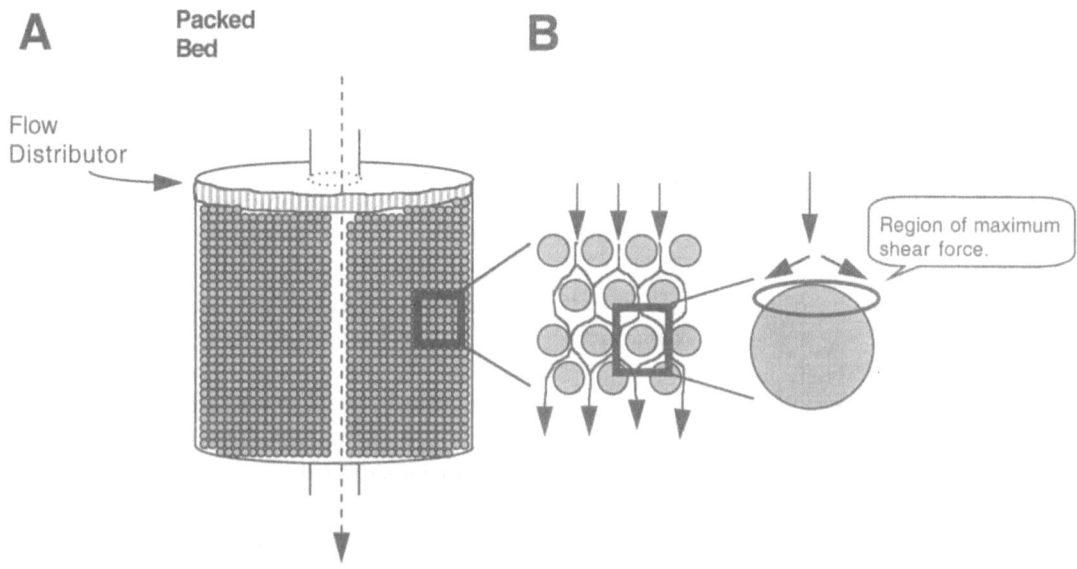

FIGURE 21.5. Schematic representation of a packed-bed bioreactor, demonstrating "channeling" caused by excessive flow rate (A). An enlargement of the beads is illustrated to show even flow around the beads and the area of maximum shear force (B).

and Heath); however, "channeling" or plug flow through the reactor bed can occur (Egan 1987), which could generate stagnant regions in the reactor bed. In fact, scale-up for glass bead cultures may require increased flow rates, causing beads to shift and damaging cells as a result of the shear forces (Glacken et al 1983). These shear forces will cause morbidity and/or loss of cellular function (Frangos et al 1993). Figure 21.5 illustrates typical flow in a packed-bed bioreactor, depicting regions of maximum shear force, and "channeling" through the packed bed caused by excessive flow rate. These problems can be avoided by using low flow rates and a flow distributor on the inlet, and by designing the perfusion vessel so that stagnant pockets are not created.

On the basis of the metabolic consumption rate and diffusivity of oxygen in hepatocyte culture, one can calculate the maximum diameter for a spheroidal encapsulate of hepatocytes. The diffusivity of a solute through the extracellular or encapsulation matrix can decrease solute diffusivity by 50–60% (Catapano 1996a, 1996b). In addition, the diffusivity in microcapsules will depend on the physicochemical properties of the solute and membrane, such as molecular weight and pore size, respectively. If the maximum diffusion distance of oxygen in hepatocyte cultures is 100 μm (Catapano 1996, Catapano and De Bartolo 1996b, Catapano et al 1996c), then the diameter of the encapsulated beads should not exceed 200 μm, or hypoxia may occur. This maximum diameter is based on the assumption that mass transfer is diffusive. However, convective flux through the sphere may become more important with microencapsulated hepatocytes since shear forces at higher flow rates may deform flexible shell walls, causing internal mixing (Heath and Belfort 1987). Microencapsulation protects cells from shear forces and immunological consequences (Joly et al 1997), and determining the optimum membrane coating has been the focus of recent research (Carturan et al 1997, Dixit and Gitnick 1996, Joly et al 1997).

Table 21.1 shows that cell density in encapsulated gels or microcapsules can reach tissue densities, and that there is a large range of surface-area-to-volume ratios, depending on the immobilization method. They can achieve the highest surface-area-to-volume ratio of all bioreactors (Dixit and Gitnick 1996, Glacken et al 1983). This range was calculated using the surface area of a sphere ($4\pi r^2$), assuming that beads pack one on top of another. There are at least five packed-bed bioartificial liver devices described in the literature that use microencapsulation or microcarrier techniques. These packed-bed bioreactors (Table 21.1) have used various hepatocyte supports such as collagen coated Cytodex (Demetriou et al 1986), glass beads (Li et al 1993), Biosilon® microcarriers (Shnyra et al 1990, Shnyra et al 1991), polyvinylformal (PVF) reticulated cubes (Miyoshi et al 1996, Yanagi et al 1992), nonwoven polyester fabric (Naruse et al 1996a, Naruse et al 1996b), and encapsulation in calcium alginate beads (Dixit 1995, Fremond et al 1993, Joly et al 1997, Mathew et al 1993) and silicate (Carturan et al 1997).

Microcarriers are typically spherical. Solid microcarriers have some of the highest mass transfer of any configuration because of their large surface-area-to-volume ratio (Table 21.1). In theory, the smaller the sphere the greater the mass transfer, mainly because the sphere packing reduces the void volume. In practice, there is a lower limit in sphere size, determined by bioreactor engineering and hepatocyte biology, and in order to maintain this relatively high mass transfer the bead must be directly hemoperfused. Hemoperfusion has problems associated with immune response that require the incorporation of design changes to reduce immunoglubulin in the perfusate (Velde et al 1997), and in the case of solid microcarrier bioartificial livers, bioreactor designs do not contain separate nutrient perfusion ports to sustain hepatocytes. Demetriou and others developed the first packed-bed bioreactor containing hepatocytes grown on collagen-coated Cytodex 3 microcarriers (Demetriou et al 1986). This bioreactor was developed to characterize these hepatocytes, which were designed for transplantation, and there is no published report of its use as an extracorporeal device. Glass beads have been used to culture primary rat hepatocytes as aggregates, and the cumulative urea and albumin synthesis increased linearly and was maintained for 15 days (Li et al 1993). Biosilon® microcarriers (160–300 μm diameter) coated with human fibronectin were inoculated with rat hepatocytes and cultured to a density of 0.75×10^5 cells/cm^2 (Shnyra et al 1990, Shnyra et al 1991). These beads were packed into an HPLC column and directly hemoperfused from blood of rats with hepatic failure induced by galactosamine and carbon tetrachloride (Shnyra et al 1991). The galactosamine and carbon tetrachloride treated rats had a 64% and 60% survival rate after 168 hr of bioartificial liver-treatment, respectively, as compared to 0% survival rate for the respective control groups. The authors demonstrate that a decrease in bilirubin may be one mechanism for the increased survival rate (Shnyra et al 1991).

Reticulated matrices have irregular architecture. Yanagi et al (1992) used cubes of PVF (2 mm $\times$ 2 mm $\times$ 2 mm) with a porosity of 100–500 μm, inoculated with 2.6–11.3×10^7 rat hepatocytes, and perfused with Williams E media. They found that with larger porosity, hepatocytes form aggregates and distribute near the surface (Yanagi et al 1992). Hepatocytes were monitored for 26 hrs, and there was a small decrease in urea synthetic rate with increased cell density. This rate was slightly less than that of monolayer culture (Yanagi et al 1992). Subsequent studies found that 60% of inoculated cells were immobilized after 1 day and 30–60% of these were still preserved by 7–9 days of culture. The decrease in albumin synthesis during culture was proposed to be due to biomass loss (Miyoshi et al 1996).

Porcine single-cell hepatocytes and spheroids were immobilized on nonwoven fabric composed of polyester and coated with collagen, which was packed into a perfusion module and perfused with media (Naruse et al 1996a, 1996b). Naruse et al found that urea and albumin synthesis was superior in spheroids as compared to single-cell hepatocytes on a per cell basis, but was less efficient in general, because of loss of cells during spheroid formation (Naruse et al 1996a). A 200 ml radial flow packed-bed bioreactor was constructed, packed with nonwoven fabric and inoculated with 10^{10} pig hepatocytes. Cultures lasted 6 days, and albumin and urea synthesis and ammonia removal were significantly better than in suspension cultures, but decreased over the 6-day period. Hepatocytes did not appear to be attached to the nonwoven material but were attached to each other and to the collagen coating the nonwoven material (Naruse et al 1996b). This bioreactor is presently being tested on animals (Naruse et al 1996b).

There have been several metabolic studies of hepatocytes encapsulated in calcium alginate (Dixit 1995, Fremond et al 1993, Mathew et al 1993, Miura et al 1990, Miura et al 1986, Takabatake et al 1991, Tompkins et al 1988). One of the problems with this technique has been that xenogeneic hepatocytes become metabolically inactive as a result of direct hemoperfusion and binding of immunoglobulins to the cell membrane of the hepatocytes (Takahashi et al 1993, Velde et al 1997), and these groups have recently investigated various microencapsulation techniques. Dixit and others used poly-l-lysine to coat a membrane on alginate beads and perfused these beads in a packed-bed bioreactor (Dixit 1995, Dixit et al 1995b). Joly et al (1997) were able to inhibit immune inactivation of the encapsulated hepatocytes by coating the encapsulate with an albumin acyl propylene glycol membrane. The hepatocytes survived for 1 week and expressed various liver-specific functions such as phase I and II metabolism, but albumin secretion was less than in conventional cultures. This was attributed to diffusional resistance caused by the alginate gel. Carturan et al (1997) have demonstrated silicate encapsulation of hepatocytes. The advantage of this over other materials is that precise porosities can be acheived, silicate provides resistance to biodegradation and mechanical stress, and mass transfer is good because capsule wall and cell surface are in close proximity.

Spirally-Wound Bioreactors

Spirally-wound bioreactors were first described by House et al (1972). Recently, Flendrig et al (1997) have modified this design and incorporated hollow fibers into the bioreactor for integral oxygenation and used nonwoven fabric as a support matrix, as did Naruse and others (Naruse et al 1996a, 1996b). The bioreactor is formed by wrapping a layer of nonwoven polyester material with a layer of evenly spaced polypropylene hollow fibers (630 μm outer diameter) and placing the roll into a polysulphon dialysis housing (1.32 inner diameter, length of 15.5 cm). Air flows through the lumen of the hollow fibers, and the medium flows in the spaces between hollow fibers; the device thus employs direct hemoperfusion of hepatocytes. The medium flows in these spaces evenly throughout the bioreactor as determined by MRI. There were 2.2×10^8 porcine hepatocytes inoculated into bioreactors and perfused with supplemented Williams E medium. Cells were allowed to attach by placing the bioreactor on a rotator and rotating the bioreactor along the longitudinal axis at 1 revolution/2 minute. Galactose elimination, urea synthesis, lidocaine and amino acid metabolism, protein synthesis, and lactate/pyruvate ratios were measured. In all cases these liver functions were significantly better than those of mono-

layer culture and were maintained for 14 days. Microscopic analysis revealed that the cell density was 2×10^7 cells/ml. Scaled-up bioreactors can be inoculated with 10^{10} cells at a density of 4×10^7 cells/ml.

Multiplate or Flat-Bed Bioreactors

There are five types of multiplate bioartificial livers; all are listed in Table 21.1. The first large-scale multiplate bioreactor was described by Weiss and Schleicher for generation of vaccine and biochemicals (Weiss and Schleicher 1968). This bioreactor design was attractive for use as a bioartificial liver because of the observation that dispersed hepatocytes cultured in monolayer remained functional for more than 2 weeks (Ichihara et al 1980). Uchino (1988) have shown that the multiplate bioreactor can sustain anhepatic dogs and rabbits (Takahashi et al 1992). Although differentiated cells are maintained for a suitable duration in monolayer culture, in comparison, collagen sandwich culture increased hepatocyte function (Dunn et al 1991). In sandwich culture, monolayers of hepatocytes are cultured between two layers of collagen to mimic the plates of the liver acinus. To enhance hepatocyte function, porcine hepatocytes were cultured between two collagen gel layers in a small prototype multiplate bioreactor (Taguchi et al 1996). Animal studies still need to be performed to compare this type of device to previous multiplate bioreactors that did not use collagen sandwich cultures (Uchino et al 1988, Takahashi et al 1992) and determine if hepatic functional enhancement correlates to enhanced animal survivability. A small version of the multiplate bioartificial liver module was used for the evaluation of perfusion (Takahashi et al 1992).

A third type of multiplate bioartificial liver is the rotating disk bioreactor (Yanagi et al 1989), which is formed by coating a mixture of hepatocytes and alginate onto steel mesh disks (60 mm diameter) in the form of an annulus 2 mm $\times$ 10 mm in dimension, thus creating a donut shape. Several of these donut-shaped disks are stacked in a 70 mm diameter cylinder and rotated. Blood enters the bioreactor through the center of the donuts and is in direct contact with the hepatocyte/alginate gel. Problems with alginate encapsulation, as noted above in packed-bed bioreactors (Miura et al 1986), are that direct hemoperfusion dissolves the gel, immune response inactivates the hepatocytes, and the loss of hepatocyte viability is too dramatic for the system to be a rational choice for a bioartificial organ (Yanagi et al 1989).

A fourth type of multiplate bioreactor is the bilaminar membrane sinusoidal bioreactor, or organotypical bioreactor (Bader et al 1995a). The aim of Bader et al is to design a bioartificial liver based on an organotypical bioreactor concept, mimicking the tissue architecture within the liver. The hepatocytes are arranged as plates, and both sinusoidal surfaces of the hepatocytes are enclosed within the matrix corresponding to the tissue anatomic landmark the space of Disse. Fenestrated endothelial cells are located on the other side of the space of Disse toward the sinusoid. For supporting monolayer hepatocytes with bipolar attachment to extracellular matrix, a sandwich cultivation method was used. Nonparenchymal cells were seeded on top of the double-gel hepatocyte culture to allow reconstruction of sinusoidal tissue analog. The hepatocytes are highly functional, and the bioreactor has been scaled to contain as many as $1.4-4 \times 10^9$ cells (Bader et al 1997).

Smith et al (1997) have developed a flat-bed hollow-fiber, multiplate hybrid bioreactor. The flat-bed hollow-fiber bioreactor (3–6 hollow fibers deep) was first described by Ku and et al (1981) and used integral membrane oxygenation to enhance mass transfer of oxygen, while medium flowed radially, directly across the cell mass that grew in the hollow-fiber bundle. Smith et al (1997) grow hepatocytes on a monolayer 350 μm away from a single layer of hollow fibers (0.28 μm diameter) in which gas flows. This is contained in a rectangular compartment (50 cm $\times$ 0.6 cm) and the medium flows directly over the monolayer culture. Therefore, mass transfer of nutrients is extremely high. These rectangular units are combined to form a system of stacked plates. Typical densities range from 0.15 to 0.18×10^6 cells/cm^2 and have been scaled up to 10^9 cells. Significant urea synthesis was measured in response to continuous ammonia loading out to 168 hr of culture.

Suspension of Immobilized Hepatocytes and Spheroids

Takabatake et al (1991) have developed a spouted-bed bioreactor for alginate-encapsulated hepatocyte spheroids. Beads were placed in funnel containing 70 ml of hormonally-defined medium, and the flow of medium was from the bottom center of the cone containing the suspended beads. The medium was evacuated from the top of the funnel. The culture maintained significantly better differentiated functions than monolayer for a duration of 3 days, and this was attributed to spheroid characteristics. Animal studies were never performed with this bioartificial liver, perhaps because of the low surface-to-volume ratio, which would require excessive quantities of plasma for hemoperfusion, or because of the inability to feasibly incorporate an animal's plasma into the spouted bed.

NASA's rotating wall vessel is a bioreactor designed for low-shear suspension cultures that mimic microgravity conditions. The bioreactor rotates so that the cells inside the container remain continuously suspended in the growth fluid rather than setting to the bottom or bumping against the container wall. Cells in this environment aggregate and have been found to differentiate (Goodwin et al 1993). By simultaneously rotating culture fluid and growing cells, the NASA bioreactor provides a low turbulence culture environment that promotes the formation of large, three-dimensional cell clusters. Because of their high level of cellular organization and specialization, samples constructed in the NASA bioreactor more closely resemble the original tumor or tissue found in the body (Ingram et al 1997). The bioreactor has a volume of 50 ml and a diameter of 9 cm. Ingram and coworkers have determined that 10^6 cells/ml is the optimum cell density, while the chamber rotates at 15 to 25 rpm. Khaoustov et al (1995) have cultured isolated human hepatocytes in serum-free medium in the microgravity bioreactors. Within 24 hrs of seeding, aggregates formed that increased to 1 cm in diameter with tissue-like architecture. Albumin was detected up to 30 days in culture when highly complex sinusoidal-like structures were observed (Khaoustov et al 1995).

Microchannel

Although this bioreactor could be considered a multiplate bioreactor, it is sufficiently different to be listed separately. At present it does not appear to be easy to scale up. The interactions of two cell types have been demonstrated in liver that contains hepatocytes and sinusoidal endothelial cells. Experimentally, primary hepatocytes have been cocultured using micropattern technology to create rows of sandwiched hepatocytes alternating with 3T3 fibroblasts (Bhatia et al 1997). This micropatterning would also produce channels for fluid flow and efficient mass transport of metabolites and waste. This coculturing system may prove useful for engineering artificial organs (Bhatia et al 1994).

Glutaraldehyde-immobilized collagen Type I micropatterns were applied with a width of 20 μm and a height of 4 nm above the aminoethylaminopropyltrimethoxysilane (AS) surface. The Type I collagen still maintained native structure as measured by immunoreactivity, and seeded rat hepatocytes localized to the collagen-derived areas with normal polygonal morphology. The hepatocytes alter their shape to remain in contact with and cover all of the collagen-derived surface. Co-cultured NIH 3T3 fibroblasts did not attach to the primary hepatocytes, but attached in a nonspecific serum-mediated modality and spread on the unmodified borosilicate glass. The attachment of both cell types was independent of cell density. Motile cells may migrate in time; however, cutting grooves in the surface, or using mitomycin C to prevent mitosis, may decrease the degree of micropattern variation (Bhatia et al 1997).

The micropatterning will precisely control the required cell-cell interactions and sinusoidal geometry. In vivo sinusoidal lengths and flow rate have been measured at 0.04 cm and 0.1 nl/s, respectively, while in the microchannel model, taking into account oxygen pressure and media viscosity, the optimal artificial sinusoidal length is 0.6 cm with a flow rate of 2 nl/s. Both values are within one order of magnitude of the in vivo values (Bhatia et al 1994).

Hollow-Fiber Bioreactors

The use of hollow fiber bioreactors as bioartificial liver devices was first demonstrated by Wolf and Munkelt (1975), who cultured hepatocytes in the extracapillary space (ECS) of a commercially available hollow-fiber bioreactor derived from the original design of Knazek et al (1972). Hollow-fiber bioreactors are essentially perfusion vessels with better controlled and more complex fluid dynamics than most other perfusion vessels. The fluid dynamics for all of the bioreactors discussed herein have been modeled under various flow configurations (Brotherton and Chau 1990, Catapano 1996a, Catapano

and De Bartolo 1996b, Cima et al 1990, Chresand et al 1988, Custer 1988, Gerlach et al 1990, Giorgio et al 1993, Nyberg et al 1992b, Robertson and Kim 1985, Tharakan and Chau 1986a, Tharakan and Chau 1986b). Hollow-fiber bioreactor designs can be categorized as two-compartment, three-compartment, or multicompartment. Unlike suspension and microcarriers, mass transfer of nutrients in hollow-fiber bioreactors is restricted by diffusion across a semipermeable membrane. Hollow fibers are arranged in the bioreactor so that convection can occur, significantly increasing mass transfer rates especially of larger molecules that have relatively small diffusion coefficients. Various convection flow configurations can operate in the various hollow fibers, and these will be discussed (Cima et al 1990, Custer 1988, Giorgio et al 1993, Tharakan and Chau 1986b). Also, the highest cell densities, similar to tissue densities, can be achieved in hollow-fiber bioreactors in comparison to other bioreactors (Glacken et al. 1983).

Analysis of Hollow-Fiber Bioreactors

Analysis of hollow-fiber bioreactors has been performed almost entirely by using mathematical models that use various assumptions regarding bioreactor fluid dynamics and metabolic rates. It is important to point out that, typically, models make assumptions to simplify the analysis, and that these assumptions are not always correct and should be critically evaluated in each bioreactor configuration (Catapano 1996, Heath and Belfort 1987). For example, some typical assumptions do not take radial or axial convection into account, use only one metabolic rate across the bioreactor, and ignore membrane resistance. These parameters can be modeled. However, there is difficulty in directly measuring these parameters empirically. Flow distribution in hollow-fiber bioreactors has been determined by use of tracers (Noda et al 1979, Park and Chang 1986) and radiopaque dyes in conjunction with x-ray computed tomography (Takasawa et al 1988), but this is difficult to perform with in situ cultures, and to the best of our knowledge, there is no report using these techniques with in situ cultures. Recently, ^{31}P and ^{13}C nuclear magnetic resonance (NMR) spectroscopy has been applied to measure growth (Gillies et al 1991) and metabolic rates (Mancuso et al 1995). Magnetic resonance imaging (MRI) has been applied to observe diffusion (Callies et al 1994, Van

Zijl et al 1991) and flow (Donoghue et al 1992, Hammer et al 1990, Zhang et al 1995) in hollow-fiber bioreactors and will be a powerful technique in analysis of bioartificial liver bioreactors and culture iteration (Constantinidis and Sambanis 1995, Donoghue et al 1992, Gillies et al 1991). Extensive NMR or MRI studies have not yet been performed with any bioartificial liver, so only modeled fluid dynamics of published bioartificial liver hollow-fiber devices will be discussed. Once metabolic rate constants and solute diffusivity (i.e., mobility of solute) are determined in a bioreactor, the oxygen and glucose concentration profiles can be generated (Figure 21.4) and bioreactor performance optimized (Glacken et al 1983, Heath and Belfort 1987).

Typically, membrane resistance is considered negligible in these models. In fact, membrane fouling occurs during cell culture, which clogs pores, and delicate fibers such as cellulose acetate will burst. This has been demonstrated experimentally by MRI (Callies et al 1994). Membrane-fouling problems can be avoided by using microcarriers and relying on the cells to generate a 3-D matrix. However, hepatocytes preferentially attach to fibers as follows: polysulfone > polyvinylidine > polypropylene = polycarbonate = polyamide > cellulose acetate > Bioflux™ (Gerlach et al 1996b, Qiang et al 1997, Yang et al 1994). The majority of published bioartificial liver devices rely on hepatocyte attachment for normal operation of the device. The various hollow-fiber bioreactor configurations will be discussed in general terms.

Conventional Hollow-Fiber Bioreactors (Two- and Three-Compartment)

The two-compartment bioreactor design is the only hollow-fiber bioreactor clinically used as a bioartificial liver. It was developed by Knazek and others in 1972 (Knazek et al 1972). It has been used to expand a number of cell types, including fibroblasts (Knazek et al 1972), breast cancer cells (Knazek et al 1977), CHO cells (Szkudlinski et al 1993), choriocarcinoma cells (Knazek et al 1974), endothelial cells (Ott 1995), hybridomas (Beck 1994), myoblasts (Kulesh et al 1994), hematopoietic stem cells (Davis et al 1995), and numerous lymphocyte cultures (Freedman et al 1994a, Freedman et al 1994b). There are essentially four flow configurations for two-compartment hollow-fiber bioreactors used for

bioartificial livers. These are depicted in Figure 21.6A–6D. For hepatocytes grown in the extracapillary space, there can be closed-shell (A), open-shell (B), and cross-flow (C) configurations. For hepatocytes grown in the lumen, or intracapillary space, flow is directed inward toward the hepatocytes and may be diffusion-driven (D) (Nyberg et al 1992b). The flow configurations in Figure 21.6A and B create two compartments, while those in 21-6C and 21-6D can create three compartments with separate functions. There can exist axial gradients, and as a general rule, axial gradients will exist if the bioreactor is longer than 25 mm (Tharakan and Chau 1986a). In response to problems of hypoxia due to axial gradients, Ku et al (1981) developed a hollow-fiber/flat-bed bioreactor. The simplest type of flat-bed bioreactor is a monolayer culture similar to the multiplate bioreactors discussed above.

Extracapillary Culture of Hepatocytes on Microcarriers

Demetriou and Rozga at UCLA, in conjunction with their industrial partner, Circe Biomedical, Inc. (formerly Grace Biomedical, Inc.), have worked for a decade to develop a clinically successful form of bioartificial liver. This bioartificial liver device, product name HepatAssist 2000®, uses a hollow-fiber biorector with macroporous fibers (0.2 μm pore size), optimized media, and cryopreserved adult porcine hepatocytes seeded into the bioreactors on Cytodex microcarriers coated with Type I collagen (Demetriou et al 1995a). Although an open-shell flow configuration resulted in superior flow (Giorgio et al 1993, Rozga et al 1993), a closed-shell flow configuration is used for this bioartificial liver (Chen et al 1996), illustrated in Figure 21.6A. The patient's blood is plasmapheresed, and the plasma flows through a charcoal cartridge and aerator, then to the bioreactor containing the adult porcine hepatocytes, and then back to the patient via the circulating plasma reservoir (Watanabe et al 1997). In ongoing clinical trials, this bioartificial liver assist device has proven quite successful for use with patients with forms of acute liver failure. However, it has not proven successful with chronic liver failure patients, both because of rejection phenomena triggered by secreted pig factors and because of the inability of the device to overcome some of the difficulties associated with chronic liver failure. Patient plasma

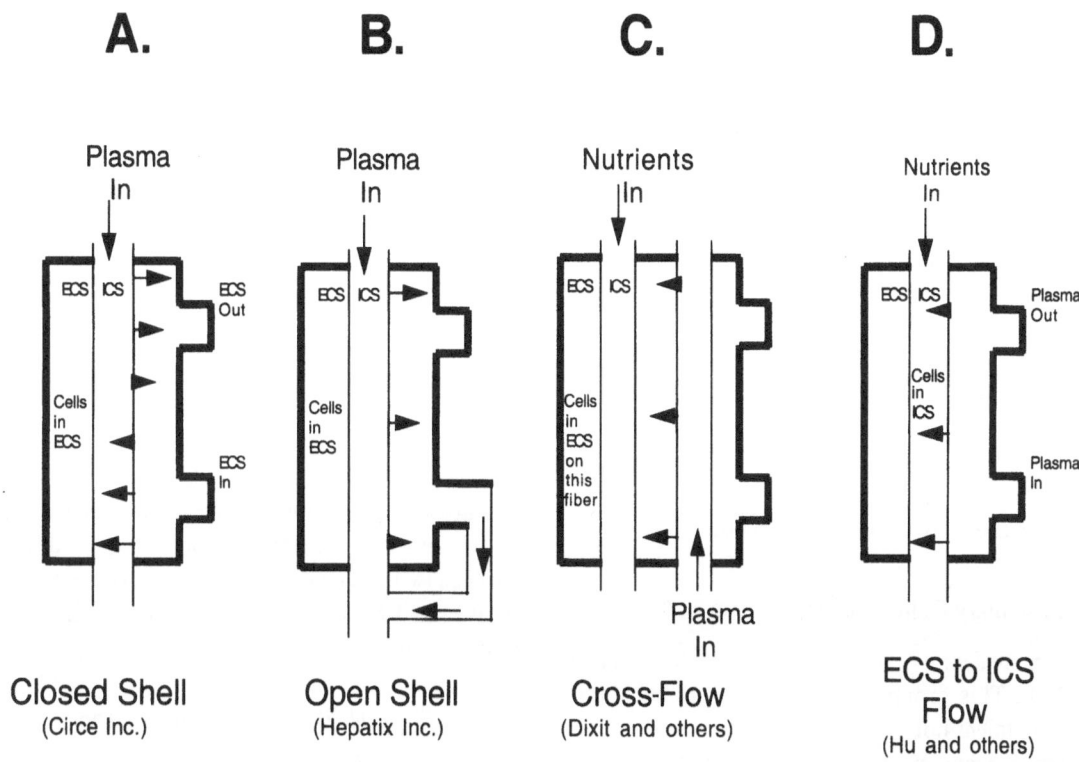

FIGURE 21.6. The various possible flow configurations for conventional hollow-fiber bioreactors. Versions of the closed shell are used by Demetriou et al 1996 (A), open shell by Sussman and Kelly 1996 (B), the cross-flow is used by Dixit 1994 (C), and ECS to ICS flow is used by Nyberg et al 1992 (D).

that was exposed to porcine hepatocytes for a single treatment of 6–7 hours showed no significant changes in xenoantibody titers. However, if patients were treated with the porcine bioartificial liver more than once, then xenoantibody titers significantly increased by a factor of 2 to 3 (Baquerizo et al 1997). Cell viability after treatment, and therefore after exposure to the patients' toxic plasma, was not different from that of pretreatment hepatocytes (Demetriou et al 1995a). The significant biochemical effects of the porcine bioartificial liver treatments included reduced serum levels of ammonia and bilirubin. Clinically, the comprehensive level of consciousness scores (CLOCS) increased; in addition, intracranial pressure was reduced, and central perfusion pressure increased. These effects lasted for the length of the treatments, and for 18 hours after the treatments (Demetriou et al 1995a). In summary, the application most likely to be approved by the FDA is a liver assist device to be used for the "bridging" of patients over the waiting period before an orthotopic

liver transplant can be performed (i.e., within a maximum of 40 hours).

Hepatic Cell Lines in the Extracapillary Space

C3A cells are an immortalized liver cell line derived from HepG2 cells. These cells express some normal liver functions, are strongly contact-inhibited, and can be maintained in bioreactors for 6–8 months. Only 2 to 5 g of C3A cells are seeded into the extracapillary space of an ultrafiltration (70 kD pore size) hollow-fiber reactor using an open-shell flow configuration, illustrated in Figure 21.6B. An in-line 0.45, μm filter is placed in the line connecting the ECS to the exit steam of the bioreactor so that macromolecules can reenter the patient's bloodstream (Sussman and Lake 1996b). The bioreactor contains 10,000 cellulose acetate fibers with a total surface area of two meters (Sussman et al 1992). The cells grow to confluence in 3–4 weeks and produce a final cell mass of 200 g per cartridge. Two cartridges have been used in series, with the equivalent of 400 g of hepatocyte function to maintain patients for up to 7 days (Sussman and Lake 1996b). Bioartificial liver function was assayed by galactose elimination capacity, and 9 of 10 patients improved with liver assist treatment (Sussman et al 1994). Current efforts are focused on redesigning the bioreactors to include macroporous fibers and media with 10% serum (Brotherton 1997) to increase the cell mass of C3A cells to opimize their efficacy in patients (Hepatix, Inc); the goal is to go into clinical trials in 1998. Another company, Amphioxus Cell Technologies Inc., has recently formed to develop a device based on the use of the C3A clone, but at the time of this writing, it is unknown whether the same bioreactor configuration as described above will be used.

Intracapillary Culture of Hepatocytes in Collagen

This bioreactor was first described by Scholz and Hu (1990). This group has made a functional three-compartment hollow-fiber bioreactor out of a two-compartment conventional bioreactor; its flow configuration is illustrated in Figure 21.6D. They circumvent mass transfer problems of hepatocyte culture in the extracapillary space by entrapping the hepatocytes in a collagen gel, inoculating this into the lumen of the hollow fibers, and shrinking the gel. This leaves an average of 40% of the lumen free so that the hepatocytes can be perfused with the medium (Shatford et al 1992). The lumen diameter is 200 μm, so mass transport by diffusion alone is sufficient. The space between the hepatocyte collagen matrix and fiber wall in the ICS is used to perfuse nutrients. Although presently no clinical studies are published, it is envisioned that the patient's plasma will be perfused in the ECS (Nyberg et al 1992a). This configuration has a greater surface-area-to-volume ratio (Table 21.1) and better diffusion distances than other conventional hollow-fiber bioreactor configurations. If one uses the outside wall of the fibers to calculate surface area, then the surface-area-to-volume ratio will always be slightly larger than that of the same bioreactor with cells grown in the ECS. Although this has proven successful in animal studies, the amount of total biomass still needs to be increased in order to reach the amount necessary to treat a human being (Table 21.1), and this would entail use of an enormous number of fibers for the bioreactor construction. In the experimental setting, long-term maintenance of differentiation is very good.

Hepatocytes in Extracapillary Space
with Cross-Flow Configuration

Knazek et al (1980) first described the idea of using different fibers with different physical characteristics in a dual-circuit configuration, with flow dynamics to mimic arterial and venous blood flow. This idea was modeled by Brotherton and Chau (1990), and Dixit applied these ideas to a bioartificial liver device. Figure 21.6C describes the flow dynamics with this design. The idea is to mimic the venous and arterial flow of the liver, using one set of fibers for nutrient delivery in a counterflow arrangement with another set of fibers that function to move waste products out of the device. Hepatocytes grow on polysulphone fibers, and waste flows into the lumen of the polysulphone fibers by diffusion because hepatocytes are proximal to these fibers (Dixit et al 1993). There have been no animal or human studies using this bioreactor.

Coaxial-Type Hollow-Fiber Bioreactors (Three-Compartment)

The coaxial design was first reported by Robertson and Kim (1985) for growth of bacteria. It permits tighter control of diffusion distance in comparison to the conventional hollow-fiber bioreactor (Cima et al 1990). The coaxial design has been radially scaled-out, using an annulus of hollow fibers encircling an inner bundle of hollow fibers (Goffe 1997).

Coaxial Bioreactor

It is possible to design a coaxial bioreactor with dimensions similar to those of the classic liver lobule. The lobule, shown in Figure 21.7A, is composed of (1) a central venule, (2) the surrounding hepatocytes, and (3) several peripheral triads consisting of the portal venule, hepatic arteriole, and bile duct. The primary physicochemical parameters (i.e., the oxygen, nutrient, hormone, and toxin gradients) of the lobule can be delineated by hemodynamically equipotential lines that radiate outward from the central venule (McCuskey 1994). For comparison, Figure 21.7B is a cross-section of a tube containing a coaxial bioreactor fiber pair, simply a smaller fiber inside a larger fiber, creating three compartments. The hepatocytes are packed in the annular space between the central fiber and the outer fiber. This coaxial design can replicate the dimensions of the lob-

A.

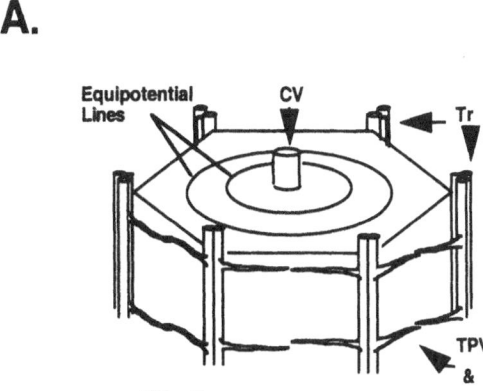

B.

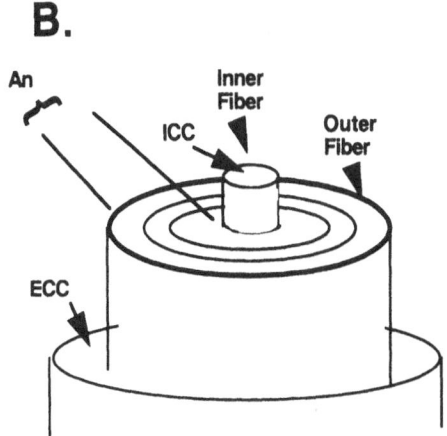

FIGURE 21.7. An illustration showing the design of the coaxial bioreactor (B) and the similarity of its dimensions to those of the liver acinus (A). Physicochemical gradients similar to those of the acinus, denoted as hemodynamically equipotential lines, can be generated in a coaxial bioreactor.

ule and generate similar hemodynamic equipotential lines (Macdonald et al 1998).

Custer (1988) compared the metabolic performance of a hollow-fiber and a coaxial bioreactor and found that the coaxial bioreactor had a fourfold greater yield of antibody per unit glucose. She attributed this to the superior perfusion in the coaxial bioreactor, which maintains greater control of the diffusion distance. Cima and others (Cima et al 1990) showed that a coaxial bioreactor can support metabolism even in the absence of radial perfusion, just by diffusion alone. We maintained isolated rat hepatocyte viability on minimal medium at human tissue densities (1.7×10^8 cells/ml) for up to 6 hrs (Macdonald et al 1998), and are presently investigating cultured hepatocytes. Clinically, one could perfuse the inner fibers with the patient's plasma and the extracapillary space with the medium.

Flow between inner and outer fibers can be counter- or co-flow configurations. The coflow configuration is depicted in Figure 21.8A. By means of the pressure gradients, convection can be directed outward from the inner fiber through the cells' mass to the outer extracapillary space or vice versa (Cima et al 1990).

Radially Scaled-Out Coaxial Design (Three-Compartment)

Goffe (1997) has radially scaled-out the coaxial bioreactor and accommodated mass transfer problems associated with the radial scale-out by using integral membrane oxygenation, effectively combining conventional and coaxial hollow fiber bioreactors.

Figure 21.8B is a diagram of the flow configuration in this bioreactor. There are two concentric rings of fibers, and the hepatocytes grow in the annulus between the fiber bundle and the extracapillary space. Air mixtures flow through the intracapillary space of the outer annulus of fibers that are composed of highly oxygen-permeable polypropylene membrane (0.2 μm pore size). The mass transfer of the air/medium interface is significantly better than that of a medium/medium interface. Additionally, the CO_2 content in the gas is used to maintain the proper medium pH (Goffe 1997). Therefore, the outer annulus of polypropylene fibers serves two purposes, oxygenation and pH regulation. The bundle of inner fibers is composed of cellulose acetate (of variable pore size) that is used for medium delivery and could be used clinically for perfusion of patient plasma. Large oxygen and nutrient gradients are generated radially between the outer and inner annuli, respectively, and these are decreased by 120 degree longitudinal rotation of the bioreactor, causing mixing in the extracapillary space. There were 10^9 porcine hepatocytes inoculated in the extracapillary space on Cytodex microcarriers, and they were maintained for 5 weeks with a viability of 90% (Table 21.1). Aside from oxygen and glucose consumption rates, no functional assays were performed (Goffe 1997).

Woven Hollow-Fiber Bioreactor (Multicompartment)

Gerlach (1994) described a bioartificial liver device woven from three types of hollow fibers comprising

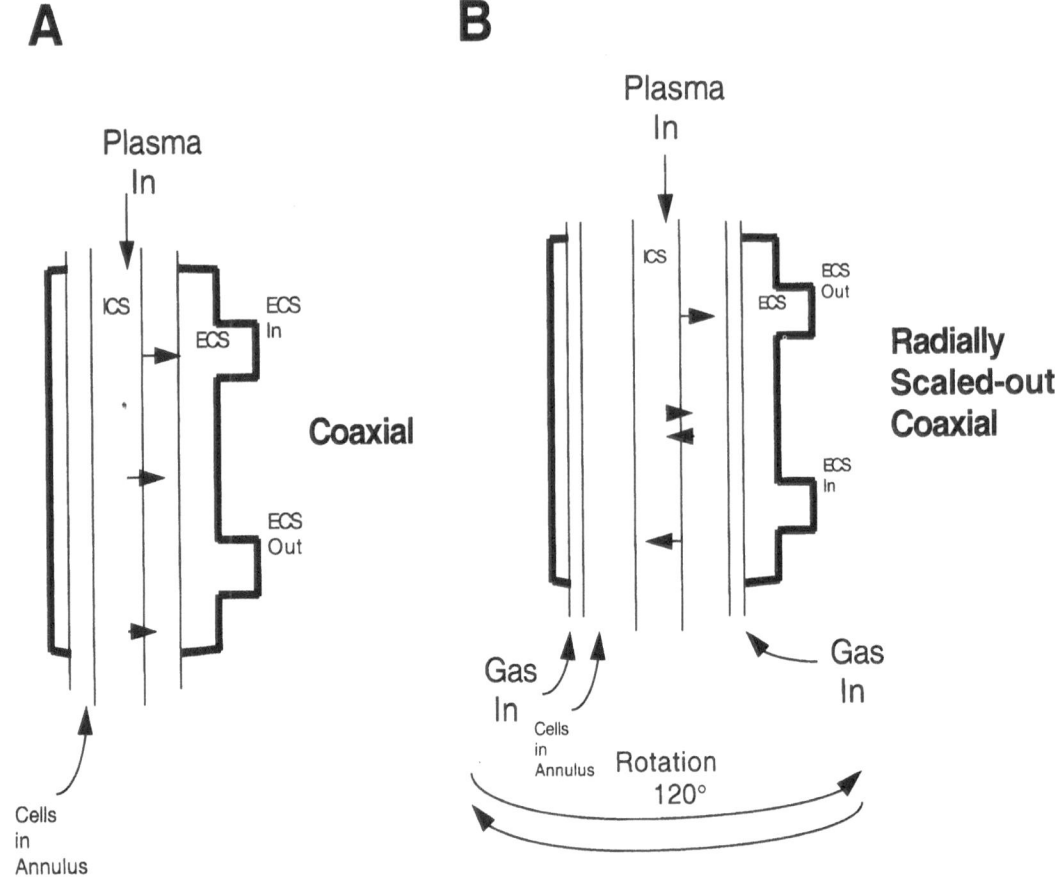

FIGURE 21.8. The configurations for coaxial (A) and radially scaled-out coaxial (B) bioreactors.

four capillary systems and used for four different functions. A model of his multiple-fiber bioreactor is illustrated in Figure 21.9. This bioartificial liver was inoculated with 10^9 cells (Table 21.1). There has been one report of this bioartificial liver device used to treat a 45-year-old patient, and sonography proved that liver size decreased with bioartificial liver treatment (Gerlach et al 1997); additional clinical trials are ongoing (Gerlach et al 1997). The four capillary systems function to provide plasma in-flow, integral oxygenation with oxygen and carbon dioxide, plasma out-flow, and co-culture (Gerlach 1994). Hydrophilic polypropylene, simply wetted with alcohol to permit water permeability, was coated with Matrigel™, and the bioreactor then was inoculated with 4×10^9 cells. Maintenance of differentiation was determined by various tests of liver metabolism (midazolam, galactose, and sorbitol elimination) and synthesis (albumin). In general these liver functions decreased by day 14. Cultures have been maintained for 7 weeks and possibly could go longer. The cell

density and total volume are not given, but the total unit volume is 790 ml and could be scaled up.

Conclusions

There have been numerous innovative bioreactor designs for bioartificial organs recently proposed. These research efforts have encompassed a broad range of scientific backgrounds, including analytical chemistry and chemical engineering, bioengineering, cell biology, and clinical studies. In general, the key needs and applications have been identified by the clinicians. The molecular and cellular biologists have contributed by identifying critical biological principles and requirements for ex vivo maintenance of cells. Finally, bioengineers have provided a bevy of new designs of bioreactors to accommodate the known cell requirements. For example, the relatively large oxygen demand of highly differentiated cells has been met by novel de-

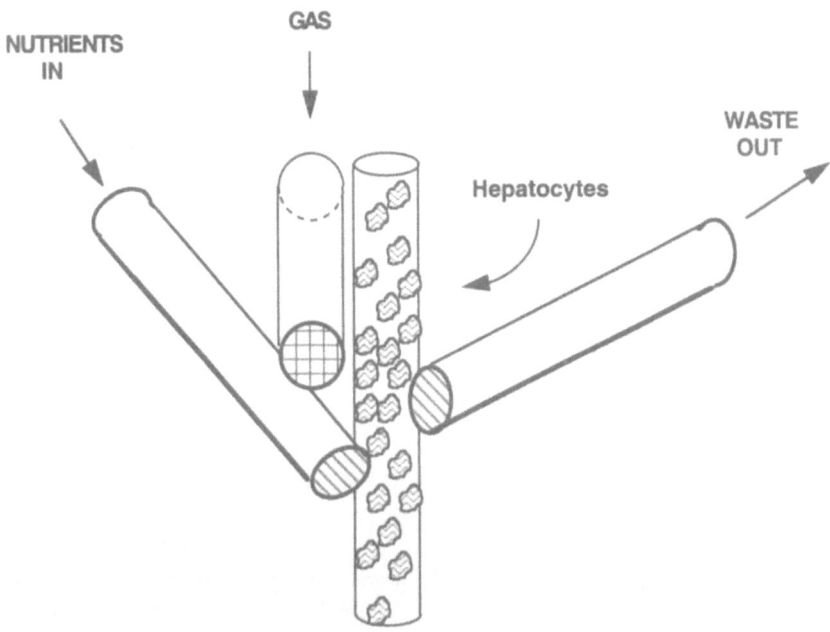

FIGURE 21.9. The smallest woven hollow-fiber unit of Gerlach's bioartificial liver design, illustrating the four different functional fibers.

signs with integral oxygenation and the use of hemoglobin. Hemoglobin permits the normal biological oxygen gradient to be high, such that diffusion can dominate, and convection—and thus shear forces—does not occur.

In the first-generation models of bioartificial organs, the ones likely to be approved for routine use with patients will include bioreactors with human cell lines and ones with pig cells, some derived from transgenic pigs to minimize immunological considerations. Those with human cell lines will offer minimal immunological concerns but also relatively minimal adult-specific functions since cell lines have mutations that preclude full differentiation of the cells. Those with pig cells will offer the entire panoply of liver functions but with exaggerated immunological problems. The likelihood of bioreactors with normal human cells is remote because of the erratic and limited supply of human tissue. Indeed, at present the only hope for human bioartificial livers, and other bioartificial organs from quiescent tissues, is likely to be the use of stem cell or committed progenitor cell populations with such great regenerative potential as to be able to overcome, or minimize, the difficulties in human tissue supply.

Even with the remaining technical and scientific difficulties, the bioartificial livers available today are sufficiently successful to enable patients to survive the waiting period for transplantation or to survive episodes of SHF and FHF. Thus, they will play an invaluable role in rescuing patients, in reducing the numbers of patients who must have transplants, and/or in reducing costs for patient care. The average cost in the United States of liver transplantation and requisite hospitalization is $250,000 (Sussman and Lake 1996b). Given these extraordinary costs, society as well as the patient will benefit from reducing the frequency of the need for liver transplantation procedures.

Future Prospects and Research Directions

Liver functions, at levels and with dynamic responses similar to those in vivo, are difficult to mimic in liver cells maintained ex vivo in current types of bioreactors. This is due, in part, to (1) remaining limitations in current engineering designs of bioreactors, (2) the need for precise extracellular matrix and hormonal conditions for the cells, and, we predict, (3) the need to seed the bioreactors with progenitor cells capable of extensive expansion and of maturing to form the needed biological infrastructure, rather than with highly differentiated cells with limited growth, precluding the establishment of the infrastructure prior to loss of viability. By addressing these themes, liver anatomic mimicry can be incorporated into bioartificial liver design.

Future research directions will include (1) investigations into bioreactor designs such as ones that allow for the elimination of bile or ones designed for progenitor cell populations, (2) studies in extracellular matrix chemistry and biology to learn how to mimic the chemically complex substratum on which the normal cells reside, (3) efforts to identify and purify growth factors and differentiation signals, (4) studies to identify and isolate stem cell/progenitor populations from diverse tissues, and (5) investigations into noninvasive methods of measurement of metabolites and microanatomy of bioartificial organs while the bioreactors are intact and functioning, such as nuclear magnetic resonance (NMR) spectroscopy and magnetic resonance imaging (MRI). Future improvements in NMR and MRI to improve their sensitivity, should greatly enhance the ability to study bioartificial organs while they are intact and functioning.

Acknowledgments. We thank Dr. Linda Custer for permission to reproduce figures from her Ph.D. dissertation and for helpful discussions. The author's studies are supported by grants from the NIH (R-01, GM44266-05 [LMR]; F32, DK09713-01 [JMM], a sponsored research grant from Renaissance Cell Technologies [LMR], and the Johns Hopkins Center for Alternatives to Animal Testing [JMM].

References

Anand AC. 1996. Bioartificial livers: The state of the art. *Trop Gastroenterol,* 17(4), 202–11.

Bader A, De Bartolo L, Haverich A. 1997. Initial evaluation of the performance of a scaled up flat membrane bioreactor (fmb) with pig liver cells. In: G Crepaldi, AA Demetriou, and M Muraca, editors, Bioartificial liver support systems. CIC Edizioni Internationali, Rome, Italy, p. 36–41.

Bader A, Knop E, Boker K, Fruhauf N, Schuttler W, Oldhafer K, Burkhard R, Pichlmayr R, Sewing KF. 1995a.

A novel bioreactor design for in vitro reconstruction of in vivo liver characteristics. *Artif Organs,* 19(4), 368–74.

Bader A, Knop E, Fruhauf N, Crome O, Boker K, Christians U, Oldhafer K, Ringe B, Pichmayr R, Sewing KF. 1995b. Reconstruction of liver tissue in vitro: geometry of characteristic flat bed, hollow fiber, and spouted bed bioreactors with reference to the in vivo liver. *Artif Organs,* 19(9), 941–950.

Baquerizo A, Mhoyan A, Shirwan H, Swensson J, Busuttil RW, Demetriou AA, Cramer DV. 1997. Xenoantibody response of patients with severe acute liver failure exposed to porcine antigens following treatment with a bioartificial liver. *Transplant Proc,* 29, 964–5.

Beal AL, and Cerra FB. 1994. Multiple organ failure syndrome in the 1990s. Systemic inflammatory response and organ dysfunction. *JAMA,* 271(3), 226–233.

Beck T. 1994. Monoclonal antibody production in the CELLMAX Artificial Capillary System. *Focus,* 17, 2–5.

Bernuau J, Rueff B, Benhamou J-P. 1986. Fulminant and subfulminant liver failure: Definitions and causes. *Sem Liver Dis,* 6, 97–106.

Bhatia SN, Toner M, Tompkins RG, Yarmush ML. 1994. Selective adhesion of hepatocytes on patterned surfaces. *Ann NY Acad Sci,* 745, 187–209.

Bhatia SN, Yarmush ML, Toner M. 1997. Controlling cell interactions by micropatterning in co-cultures: hepatocytes and 3T3 fibroblasts. *J Biomed Mater Res,* 34(2), 189–99.

Bissell DM, and Choun MO. 1988. The role of extracellular matrix in normal liver. *Scand J Gastroenterol—Suppl,* 151, 1–7.

Block GD, Locker J, Bowen WC, Petersen BE, Katyal S, Strom SC, Riley T, Howard TA, Michalopoulos GK. 1996. Population expansion, clonal growth, and specific differentiation patterns in primary cultures of hepatocytes induced by HGF/SF, EGF and TGF alpha in a chemically defined (HGM) medium. *J. Cell Biol.,* 132(6), 1133–1149.

Brill S, Holst PA, Zvibel I, Fiorino AS, Sigal SH, Somasundaran U, Reid LM. 1994. Extracellular matrix regulation of growth and gene expression in liver cell lineages and hepatomas. In: IM Arias, JL Boyer, N Fausto, WB Jakoby, DA Schachter, and DA Shafritz, editors, The liver: biology and pathobiology. Raven Press Ltd., New York, NY, p. 869–897.

Brill S, Zvibel I, Reid LM. 1995. Maturation-dependent changes in the regulation of liver-specific gene expression in embryonal versus adult primary liver. *Differentiation,* 59(2), 95–102.

Brill S, Holst P, Zvibel I, Reid LM. Analysis of rat hepatic progenitors from E14 livers (Submitted for publication).

Brotherton JD, Chau PC. 1990. Modeling analysis of an intercalated-spiral alternate-dead-ended hollow fiber bioreactor for mammalian cell culture. *Biotech Bioeng,* 35, 375–394.

Brotherton JD. 1997. Human hepatocyte cultures in various hollow fiber bioreactors in the further development of the ELAD™: An extracorporeal liver assist device. *2nd Int Sym Hollow Fiber Bioreactors Technol,* p. 22.

Brunner G, Holloway CJ, Losgen H. 1979. Large agarose beads for extracorporeal detoxification systems: Preparation and enzymatic properties of agarose-bound UDP-glucuronyltransferase. *Int J Artif Organs,* 2(3), 163–9.

Brunner G, Schmidt F. 1981. Artificial liver support, Springer-Verlag, New York.

Callies R, Jackson ME, Brindle KM. 1994. Measurements of the growth and distribution of mammalian cells in a hollow-fiber bioreactor using nuclear magnetic resonance imaging. *Biotechnol,* 12(1), 75–78.

Campra JL, Reynolds TB. 1988. The hepatic circulation. In: IM Arias, WB Jakoby, H Popper, editors, The liver: biology and pathology, Raven Press, Ltd., New York, N.Y., p. 911–947.

Carturan G, Campostrini R, Vilei MT, Dal Monte, R, Zanusso, E, Muraca, M. 1997. Encapsulation of viable animal cells by SiO$_2$: Chemical advances and prospective for hybrid bioartificial organs. In: G Crepaldi, AA Demetriou, and M Muraca, editors, Bioartificial liver support systems, CIC Edizioni Internazionali, Rome, Italy, p. 48–54.

Catapano G. 1996a. Mass transfer limitations to the performance of membrane bioartificial liver support devices. *Int J Artif Organs,* 19(1), 18–35.

Catapano G, De Bartolo L. 1996b. Importance of the kinetic characterization of liver cell metabolic reactions to the design of hybrid liver support devices. *Int J Artif Organs,* 19(11), 670–6.

Catapano G, De Bartolo L, Lombardi CP, Drioli E. 1996c. The effect of oxygen transport resistances on the viability and functions of isolated rat hepatocytes. *Int J Artif Organs,* 19(1), 61–71.

Chang TM. 1974. Artificial kidney and artificial liver. *Lancet,* 2(7894), 1451–2.

Chang TM. 1986. Experimental artificial liver support with emphasis on fulminant hepatic failure: concepts and review. *Semin Liver Dis,* 6(2), 148–158.

Chen SC, Hewitt WR, Watanabe FD, Eguchi S, Kahaku E, Middleton Y, Rozga J, Demetriou AA. 1996. Clinical experience with a porcine hepatocyte-based liver support system. *Int J Artif Organs,* 19(11), 664–9.

Chresand TJ, Gillies RJ, Dale BE. 1988. Optimum fiber spacing in a hollow fiber bioreactor. *Biotech Bioeng,* 32, 983–992.

Cima LG, Blanch HW, Wilke CR. 1990. A theoretical and experimental evaluation of a novel radial-flow hollow fiber reactor for mammalian cell culture. *Bioproc Eng,* 5, 19–30.

Cima LG, Ingber DE, Vacanti JP, Langer R. 1991. Hepatocyte culture on biodegradable polymeric substrates. *Biotech Bioeng,* 38, 145–158.

Constantinidis I, Sambanis A. 1995. Towards the development of artificial endocrine tissues: ^{31}P NMR spectroscopic studies of immunoisolated insulin-secreting AtT-20 cells. *Biotech Bioeng,* 47, 431–443.

Cremmer T, Werden K, Stevensen AFG, Lehner K, Messerschmidt O. 1981. Aging in vitro and D-glucose uptake kinetics of human diploid fibroblasts. *J Cell Physiol,* 106, 99–108.

Crepaldi G, Demetriou AA, Muraca M. 1997. Bioartificial liver support systems, CIC Edizioni Internationali, Rome, Italy, p. 132.

Custer LM. 1988. Physiological studies of hybridoma cultivation in hollow fiber bioreactors, Ph.D. Dissertation, University of California, Berkeley.

Davis TA, Kidwell WR, OKe V, Lee KP. 1995. Conditioned media from primary porcine endothelial cells alone promotes primitive human hematopoietic progenitor cell proliferation. *Exp Hematol,* 23, 847.

Demetriou AA, Whiting J, Levenson SM, Chowdhury NR, Schechner R, Michalski S, Feldman D, Chowdhury JR. 1986. New method of hepatocyte transplantation and extracorporeal liver support. *Ann-Surg,* 204(3), 259–71.

Demetriou AA. 1994. Support of the acutely failing liver, R.G. Landes Company, Austin, TX.

Demetriou AA, Rozga J, Podesta L, Lepage E, Morsiani E, Moscioni AD, Hoffman A, McGrath M, Kong L, Rosen H, et al. 1995a. Early clinical experience with a hybrid bioartificial liver. *Scand J Gastroenterol Suppl,* 208, 111–7.

Demetriou AA, Watanabe F, Rozga J. 1995b. Artificial hepatic support systems. *Prog Liver Dis,* 13, 331–48.

Dixit V, Piskin E, Denizli A, Kozluca A, Arthur M, Gitnick G. 1993. Preliminary studies on the design of an extracorporeal bioartificial liver support device with tissue engineering technology. *Hepatol,* 18(4), 329A.

Dixit V. 1995a. Transplantation of isolated hepatocytes and their role in extrahepatic life support systems. *Scand J Gastroenterol Suppl,* 208, 101–10.

Dixit V, Arthur M, Biggins S, Gitnick G. 1995b. Isolated fetal porcine hepatocytes: application in cell transplantation and bioartificial liver support. *Hepatol,* 22(4), 213A.

Dixit V, Gitnick G. 1996. Artificial liver support: state of the art. *Scand J Gastroenterol Suppl,* 220, 101–14.

Donoghue C, Brideau M, Newcomer P, Pangrle B, DiBiasio D, Walsh E, Moore S. 1992. Use of magnetic resonance imaging to analyze the performance of hollow-fiber bioreactors. *Ann NY Acad Sci,* 665, 285–300.

Dunn JC, Tompkins RG, Yarmush ML. 1991. Long-term in vitro function of adult hepatocytes in a collagen sandwich configuration. *Biotechnol Prog,* 7(3), 237–45.

Egan WM. 1987. The use of perfusion systems for nuclear magnetic resonance studies in cells. In: RK Gupta, editor, NMR spectroscopy of cells and organisms, CRC, Boca Raton, FL, 135–162.

Enat R, Jefferson DM, Ruiz-Opazo N, Gatmaitan Z, Leinwand LA, Reid LM. 1984. Hepatocyte proliferation in vitro: its dependence on the use of serum-free hormonally defined medium and substrata of extracellular matrix. Proc Natl Acad Sci USA, 81, 1411–1415.

Fiorino A, Lemischka I, Diehl A, Reid LM. 1998. Conditionally transformed rodent hepatic stem cell lines. In Vitro (in press).

Flendrig LM, la Soe JW, Jorning GG, Steenbeek A, Karlsen OT, Bovee WM, Ladiges NC, Velde AA, Chamuleau RA. (1997). In vitro evaluation of a novel bioreactor based on an integral oxygenator and a spirally wound nonwoven polyester matrix for hepatocyte culture as small aggregates. J Hepatol, 26(6), 1379–92.

Forman SJ, Blume KG, Thomas ED. 1994. Bone marrow transplantation, Blackwell Scientific, Cambridge, MA.

Frangos J. 1993. Physical forces and the mammalian cell, Academic Press, San Diego, CA, p. 400.

Freedman RS, Tomasovic B, Templin S, Atkinson EN, Kudelka A, Edwards CL, Platsoucas CD. 1994a. Large-scale expansion in interleukin-2 of tumor-infiltrating lymphocytes from patients with ovarian carcinoma for adoptive immunotherapy. J Immunol Meth 167(1–2):145–60.

Freedman RS, Edwards CL, Kavanagh JJ, Kudelka AP, Katz RL, Carrasco CH, Atkinson EN, Scott W, Tomasovic B, Templin S. 1994. Intraperitoneal adoptive immunotherapy of ovarian carcinoma with tumor-infiltrating lymphocytes and low-dose recombinant interleukin-2: a pilot trial. J Immunother, 16(3), 198–210.

Fremond B, Malandain C, Guyomard C, Chesne C, Guillouzo A, Campion JP. 1993. Correction of bilirubin conjugation in the Gunn rat using hepatocytes immobilized in alginate gel beads as an extracorporeal bioartificial liver. Cell Transplant, 2(6), 453–60.

Fremond B, Joly A, Desille M, Desjardins JF, Campion JP, Clement B. 1996. Cell-based therapy of acute liver failure: the extracorporeal bioartificial liver. Cell Biol Toxicol, 12(4–6), 325–9.

Gebhardt G, Jonitza D. 1991. Different proliferative responses of periportal and perivenous hepatocytes to EGF. Biochem Biophys Res Com, 181(3), 1201–1207.

Gebhardt G. 1992. Metabolic zonation of the liver regulation and implications of liver function. Pharmacol Ther, 53(3), 275–354.

Gerlach J, Kloppel K, Stool P, Vienken J, Muller C, Schauwecker HH. 1990. Gas supply across membranes in liver support bioreactors. Art Org, 14, 328–33.

Gerlach JC. 1994. Use of hepatocyte cultures for liver support bioreactors. Adv Exp Med Biol, 368, 165–71.

Gerlach JC, Lemmens P, Schon M, Janke J, Rossaint R, Busse B, Puhl G, Neuhaus P. 1997. Experimental evaluation of a hybrid liver support system. Transplant Proc, 29, 852.

Gerlach J, Ziemer R, Neuhaus P. 1996a. Fulminant liver failure: relevance of extracorporeal hybrid liver support systems. Int Art Organs, 19(1), 7–13.

Gerlach JC, Schnoy N, Vienken J, Smith M, Neuhaus P. 1996b. Comparison of hollow fibre membranes for hepatocyte immobilisation in bioreactors. Int J Artif Organs, 19(10), 610–6.

Gillies RJ, Scherer PG, Raghunand N, Okerlund LS, Martinez Zaguilan R, Hesterberg L, Dale BE. 1991. Iteration of hybridoma growth and productivity in hollow fiber bioreactors using 31P NMR. Magn Reson Med, 18(1), 181–92.

Gimson AES, O'Grady J, Ede RJ. 1986. Late onset hepatic failure: clinical serological and histological features. Hepatol, 6, 288–294.

Giorgio TD, Moscioni AD, Rozga J, Demetriou AA. 1993. Mass transfer in a hollow fiber device used as a bioartificial liver. ASAIO J, 39(4), 886–92.

Glacken M. 1987. Ph.D. Dissertation, MIT, Cambridge, MA.

Glacken MW, Fleischaker RJ, Sinskey AJ. 1983. Large-scale production of mammalian cells and their products: Engineering principles and barriers to scale-up. Ann NY Acad Sci, 413, 355–72.

Goffe R. 1997. High performance cell culture bioreactor and method. US Patent No. 5,622,857.

Goodwin TJ, Schroeder WF, Wolf DA, Moyer MP. 1993. Rotating-wall vessel coculture of small intestine as a prelude to tissue modeling: aspects of simulated microgravity. Proc Soc Exp Biol Med, 202, 181–192.

Griffith LG, Wu B, Cima MJ, Powers M, Chaignaud B, Vacanti JP. 1997. In vitro organogenesis of vascular liver tissue. Ann NY Acad Sci, 831.

Hammer BE, Heath CA, Mirer SD, Belfort G. 1990. Quantitative flow measurements in bioreactors by nuclear magnetic resonance imaging. Biotechnology NY, 8(4), 327–30.

Haralson MA, Hassell JR. 1995. Extracellular matrix: a practical approach. IRL Press, New York.

Heath C, Belfort G. 1987. Immobilization of suspended mammalian cells: analysis of hollow fiber and microcapsule bioreactors. Adv Biochem Eng Biotechnol, 34, 1–31.

Hida M, Takamiya T, Kitamura M, Iida T, Kitajima N, Hiraga, S, Satoh, T. 1987. Clinical reports on plasma exchange in the Kidney Center, Tokai University School of Medicine. Tokai J Exp Clin Med, 12(1), 61–6.

Holzer C, Maier P. 1987. Maintenence of periportal and pericentral oxygen tensions in primary rat hepatocyte cultures: influence on cellular DNA and protein content monitored by flow cytometry. *J Cell Physiol*, 133, 297–304.

House W, Shearer M, Maroudas NG. 1972. Method for bulk culture of animal cells on plastic film. *Exp Cell Res*, 71, 293–296.

Hughes RD, Williams R. 1996a. Assessment of bioartificial liver support in acute liver failure. *Int J Artif Organs*, 19(1), 3–6.

Hughes RD, Williams R. 1996b. Use of bioartificial and artificial liver support devices. *Semin Liver Dis*, 16(4), 435–44.

Ichihara A, Nakamura T, Tanaka K. 1980. Biochemical functions of adult rat hepatocytes in primary culture. *Ann NY Acad Sci*, 349, 77–84.

Imamura T, Crespi CL, Thilly WG, Brugengraber H. 1982. Fructose as a carbohydrate source yields stable pH and redox parameters in microcarrier cell culture. *Anal Biochem*, 124, 353–358.

Ingram M, Techy GB, Saroufeem R, Yazan O, Narayan KS, Goodwin TJ, Spauding GF. 1997. Three-dimensional growth patterns of various human tumor cell lines in simulated microgravity of a NASA bioreactor. *In Vitro Cell Dev Biol*, 33, 459–466.

Jacobs C, Kjellstrand CM, Koch KM, Winchester JF. 1996. Replacement of renal function by dialysis. fourth edition, Kluwer Academic Pub., Dordrecht, The Netherlands, p. 1536.

Jauregui HO, Mullon CJ-P, Soloman BA. 1997. Extracorporeal artificial liver support. In: RP, Lanza, WL, Chick, and R, Langer, editors, Principles of Tissue Engineering, Academic Press, San Diego, CA, p. 463–479.

Jefferson DM, Clayton D, Darnell J, Reid LM. 1984. Posttranscriptional modulation of gene expression in liver cells. *Mol Cell Biol*, 4, 1929–1934.

Joly A, Desjardins JF, Fremond B, Desille M, Campion JP, Malledant Y, Lebreton Y, Semana G, Edwards Levy F, Levy MC, Clement B. 1997. Survival, proliferation, and functions of porcine hepatocytes encapsulated in coated alginate beads: A step toward a reliable bioartificial liver. *Transplant*, 63(6), 795–803.

Kasai S, Sawa M, Mito M. 1994. Is the biological artificial liver clinically applicable? A historic review of biological artificial liver support systems. *Artif Organs*, 18(5), 348–54.

Khaoustov VI, Darlington GJ, Soriano HE, Krishnan B, Risin D, Pellis NR, Yoffe B. 1995. Establishment of 3-dimensional primary hepatocyte culture in microgravity environment. *Hepatol*, 22(4), 231A.

Knazek RA, Gullino PM, Kohler PO, Dedrick RL. 1972. Cell culture on artificial capillaries: an approach to tissue growth in vitro. *Science*, 178(56), 65–6.

Knazek RA, Kohler PO, Gullino PM. 1974. Hormone production by cells grown in vitro on artificial capillaries. *Exp Cell Res*, 84(1):251–4.

Knazek RA, Lippmann ME, Chopra HC. 1977. Formation of solid human mammary carcinoma in vitro. *J Nat Can Inst*, 58(2), 419–22.

Knazek RA, Gullino PM, Frankel DS. 1980. Method of simulation of lymphatic drainage utilizing a dual circuit, woven capillary bundle. U.S. Patent #4,206,015.

Ku K, Kuo MJ, Delente J, Wildi BS, Feder J. 1981. Development of a hollow-fiber system for large-scale culture of mammalian cells. *Biotech Bioeng*, 23, 79–95.

Kulesh DA, Anderson LH, Wilson B, Otis EJ, Elgin DM, Barker MJ, Mehm WJ, Kearney GP. 1994. Space shuttle flight (STS-45) of L8 myoblast cells results in the isolation of a nonfusing cell line variant. *J Cell Biochem*, 55(4), 530–44.

Li AP, Barker G, Beck D, Colburn S, Monsell R, Pellegrin C. 1993. Culturing of primary hepatocytes as entrapped aggregates in a packed bed bioreactor: a potential bioartificial liver. *In Vitro Cell Dev Biol*, 29a, 249–54.

Lombardi CP, Urso A, Careddu G, Ghirlanda G, Catapano G, Brisinda G, Ceriati F, Bellantone R, Doglietto GB, Crucitti F. 1992. Hybrid artificial pancreas: islet transplantation inside membrane bioreactors. *Biomat Artif Cells Immobil Biotechnol*, 20(5), 1177–92.

Macdonald JM, Grillo M, Schmidlin O, Tajiri D, James TL. 1998. NMR spectroscopy and MRI investigation of a potential bioartificial liver. *NMR Biomed*, 11, 1–12.

Mancuso A, Fernandez EJ, Blanch HW, Clark DS. 1990. A nuclear magnetic resonance technique for determining hybridoma cell concentration in hollow fiber bioreactors. *Biotechnology NY*, 8(12), 1282–5.

Mancuso A, Sharfstein ST, Tucker SN, Clark DS, Blanch HW. 1995. Examination of primary metabolic pathways in a murine hybridoma with carbon-13 nuclear magnetic resonance spectroscopy. *Biotech Bioeng*, 44, 563–585.

Margulis MS, Erukhimov EA, Andreiman LA, Viksna LM. 1989. Temporary organ substitution by hemoperfusion through suspension of active donor hepatocytes in a total complex of intensive therapy in patients with acute hepatic insufficiency. *Resuscitation*, 18(1), 85–94.

Mathew HWT, Basu S, Peterson WD, Salley SO, Klein MD. 1993. Performance of plasma perfused microencapsulated hepatocytes: prospects for extracorporeal liver support. *J Pediatr Surg* (28), 1423–1428.

Matsumura KN, Guevara GR, Huston H, Hamilton WL, Rikimaru M, Yamasaki G, Matsumura MS. 1987. Hybrid bioartificial liver in hepatic failure: preliminary clinical report. *Surgery*, 101(1), 99–103.

McCuskey RS. 1994. The hepatic microvascular system. In: IM, Arias, JL, Boyer, N, Fausto, WB, Jakoby, DA, Schachter, and DA, Shafritz, editors, The liver: biology and pathobiology, Raven Press Ltd., New York, NY, 1089–1106.

Minuth WW, Stockl G, Kloth S, Dermietzel R. 1992. Construction of an apparatus for perfusion cell cultures which enables in vitro experiments under organotypic conditions. *Eur J Cell Biol,* 57(1), 132–7.

Miura Y, Akimoto T, Kanazawa H, Yagi K. 1986. Synthesis and secretion of protein by hepatocytes entrapped within calcium alginate. *Artif Organs,* 10(6), 460–5.

Miura Y, Akimoto T, Yoshikawa N, Yagi K. 1990. Characterization of immobilized hepatocytes as liver support. *Biomater Artif Cells Artif Organs,* 18(4), 549–54.

Miyoshi H, Yanagi K, Fukuda H, Ohshima N. 1996. Long-term performance of albumin secretion of hepatocytes cultured in a packed-bed reactor utilizing porous resin. *Artif Organs,* 20(7), 803–7.

Naruse K, Sakai Y, Nagashima I, Jiang GX, Suzuki M, Muto, T. 1996a. Comparisons of porcine hepatocyte spheroids and single hepatocytes in the non-woven fabric bioartificial liver module. *Int J Artif Organs,* 19(10), 605–9.

Naruse K, Sakai Y, Nagashima I, Jiang GX, Suzuki M, Muto T. 1996b. Development of a new bioartificial liver module filled with porcine hepatocytes immobilized on non-woven fabric. *Int J Artif Organs,* 19(6), 347–52.

Noda I, Brown-West DG, Gryte CC. 1979. Effect of flow maldistribution in hollow fiber dialysis experimental studies. *J Membr Sci,* 5:209–225.

Nyberg SL, Shatford RA, Hu WS, Payne WD, Cerra FB. 1992a. Hepatocyte culture systems for artificial liver support: Implications for critical care medicine (bioartificial liver support). *Crit Care Med,* 20(8), 1157–68.

Nyberg SL, Shatford RA, Payne WD, Hu WS, Cerra FB. 1992b. Primary culture of rat hepatocytes entrapped in cylindrical collagen gels: an in vitro system with application to the bioartificial liver. Rat hepatocytes cultured in cylindrical collagen gels. *Cytotech,* 10(3), 205–15.

O'Grady J, Schalm SW, Williams R. 1993. Acute liver failure: redefining the syndromes. *Lancet,* 342, 273–275.

O'Leary DJ, Hawkes SP, Wade CG. 1987. Indirect monitoring of carbon-13 metabolism with NMR: Analysis of perfusate with a closed loop flow system. *Magn Reson Med* 5, 572–77.

Olumide F, Eliashiv A, Kralios N, Norton L, Eiseman B. 1977. Hepatic support with hepatocyte suspensions in a permeable membrane dialyzer. *Surgery,* 82(5), 599–606.

Ott MJ. 1995. Chronic flow promotes ultrastructural differential of endothelial cells. *Endothel,* 3:21–30.

Overturf K, Al-Dahalamy M, Ou CN, Finegold M, Tanguay R, Weber A, Kay M, Grompe M. 1996. Adenovirus-mediated gene therapy in a mouse model of hereditary tyrosinemia type I. *Hum Gene Ther,* 8(5), 513–521.

Overturf K, al-Dhalimy M, Ou CN, Finegold M, Grompe M. 1997. Serial transplantation reveals the stem-cell-like regenerative potential of adult mouse hepatocytes. *Am. J. Path.* 151(5), 1273–1280.

Panayotou G, End P, Aumailley M, Timpl R, Engel J. 1989. Domains of laminin with growth factor activity. *Cell,* 56(1), 93–101.

Papoutsakis ET, Michaels JD. 1993. Physical forces in mammalian cell bioreactors. In: JA Frangos, editor, Physical Forces and the Mammalian Cell, Academic Press: San Diego, CA, p.291–346.

Park JK, Chang HN. 1986. Flow distribution in the lumen side of a hollow-fiber module. *Al Chem Eng J,* 32, 1937–1947.

Pollard JW, Walker JM. 1997. Basic cell culture protocols, 2nd edition. Methods in molecular biology. Volume 75.

Qiang S, Yaoting Y, Hongyin L, Klinkmann H. 1997. Comparative evaluation of different membranes for the construction of an artificial liver support system. *Int J Artif Organs,* 20(2), 119–24.

Reid LM, Jefferson DM. 1984. Culturing hepatocytes and other differentiated cells. *Hepatol,* 4(3), 548–59.

Reid LM. 1990. Stem cell biology, hormone/matrix synergies and liver differentiation. *Cur Opin Cell Biol,* 2(1), 121–30.

Reid LM. 1997a. Stem cell/lineage biology and lineage-dependent extracellular matrix chemistry: keys to tissue engineering of quiescent tissues such as liver. In: RP Lanza, WL Chick, and R Langer, editors, Principles of Tissue Engineering, Academic Press, San Diego, CA, p. 477–509.

Reid LM, Luntz TL. 1997b. Ex vivo maintenance of differentiated mammalian cells. *Meth Mol Biol,* 75, 31–57.

Robertson CR, Kim IH. 1985. Dual aerobic hollow-fiber bioreactor for cultivation of *Streptomyces aureofaciens. Biotech Bioeng,* 27, 1012–1020.

Rojkind M, Ponce-Noyola P. 1982. The extracellular matrix of the liver. *Coll Rel Res,* 2(2), 151–75.

Rotem A, Toner M, Tompkins RG, Yarmush ML. 1992. Oxygen uptake rates in cultured rat hepatocytes. *Biotech Bioeng,* 40, 1286–1291.

Rouabhia M. 1997. Skin substitute production by tissue engineering: clinical and fundamental applications, RG Landes Company, Austin, TX.

Rozga J, Williams F, Ro MS, Neuzil DF, Giorgio TD, Backfisch G, Moscioni AD, Hakim R, Demetriou AA. 1993. Development of a bioartificial liver: properties and function of a hollow-fiber module inoculated with liver cells. *Hepatol,* 17(2), 258–65.

Scholz M, Hu W-S. 1990. A two-compartment cell entrapment bioreactor with three different holding times for cells, high and low molecular weight compounds. *Cytotech,* 4, 127–137.

Sell S. 1990. Is there a liver stem cell? *Can Res,* 50(13), 3811–3815.

Shatford RA, Nyberg SL, Meier SJ, White JG, Payne WD, Hu WS, Cerra FB. 1992. Hepatocyte function in a hollow fiber bioreactor: A potential bioartificial liver. *J Surg Res,* 53(6), 549–57.

Shnyra A, Bocharov A, Bochkova N, Spirov V. 1990. Large-scale production and cultivation of hepatocytes on Biosilon microcarriers. *Artif Organs*, 14(6), 421–8.

Shnyra A, Bocharov A, Bochkova N, Spirov V. 1991. Bioartificial liver using hepatocytes on Biosilon microcarriers: Treatment of chemically induced acute hepatic failure in rats. *Artif Organs*, 15(3), 189–97.

Sigal SH, Brill S, Fiorino AS, Reid LM. 1992. The liver as a stem cell and lineage system. *Am J Phys*, 263(2 Pt 1), G139–48.

Sigal SH, Brill S, Reid LM, Zvibel I, Gupta S, Hixson D, Faris R, Holst PA. 1994. Characterization and enrichment of fetal rat hepatoblasts by immunoadsorption "panning" and fluorescence-activated cell sorting. *Hepatol*, 19(4), 999–1006.

Sigal SH, Gupta S, Gebhard DF Jr., Holst P, Neufeld D, Reid LM. 1995a. Evidence for a terminal differentiation process in the rat liver. *Differentiation*, 59(1), 35–42.

Sigal SH, Rajvanshi P, Reid LM, Gupta S. 1995b. Demonstration of differentiation of hepatocyte progenitor cells using dipeptidyl peptidase IV deficient mutant rats. *Cell Mol Biol Res*, 41(1), 39–47.

Smith MD, Smirthwaite AD, Cairns DE, Cousins RB, Gaylor JD. 1996. Techniques for measurement of oxygen consumption rates of hepatocytes during attachment and post-attachment. *Int J Artif Organs*, 19(1), 36–44.

Smith MD, Airdrie I, Cousins RB, Ekevall E, Grant MH, Gaylor, JDS. 1997. Development and characterization of a hybrid artificial liver bioreactor with integral membrane oxygenation. In: G Crepaldi, AA Demetriou, and M Muraca, editors, Bioartificial liver support systems, CIC Edizioni Internazionali, Rome, Italy, p. 27–35.

Stange J, Mitzner S. 1996. Hepatocyte encapsulation—initial intentions and new aspects for its use in bioartificial liver support. *Int J Artif Organs*, 19(1), 45–8.

Sulieman SA, Stevens JB. 1987. The effect of oxygen tension on rat hepatocytes in short-term culture. *In Vitro Cell Dev Biol*, 23(5).

Sussman NL, Chong MG, Koussayer T, He DE, Shang TA, Whisennand HH, Kelly JH. 1992. Reversal of fulminant hepatic failure using an extracorporeal liver assist device. *Hepatology*, 16(1), 60–5.

Sussman NL, Gislason GT, Conlin CA, Kelly JH. 1994. The Hepatix extracorporeal liver assist device: initial clinical experience. *Artif Organs*, 18(5), 390–6.

Sussman NL, Kelly JH. 1996a. Artificial liver support. *Clin Invest Med*, 19(5), 393–9.

Sussman NL, Lake, JR. 1996b. Treatment of hepatic failure—1996: current concepts and progress toward liver dialysis. *Am J Kid Dis*, 27(5), 605–621.

Swabb EA, Wei J, Gullino PM. 1974. Diffusion and convection in normal and neoplastic tissue. *Cancer Res*, 34, 2814–2822.

Szkudlinski MW, Thotakura NR, Bucci I, Joshi LR, Tsai A, East-Palmer J, Shiloach J, Weintraub BD. 1993. Purification and characterization of recombinant human thyrotropin (TSH) isoforms produced by Chinese hamster ovary cells: the role of sialylation and sulfation in TSH bioactivity. *Endocrinol*, 133(4), 1490–503.

Taguchi K, Matsushita M, Takahashi M, Uchino J. 1996. Development of a bioartificial liver with sandwiched-cultured hepatocytes between two collagen gel layers. *Artif Organs*, 20(2), 178–85.

Takabatake H, Koide N, Tsuji T. 1991. Encapsulated multicellular spheroids of rat hepatocytes produce albumin and urea in a spouted bed circulating culture system. *Artif Organs*, 15(6), 474–80.

Takahashi M, Matsue H, Matsushita M, Sato K, Nishikawa M, Koike M, Noto H, Nakajima Y, Uchino J, Komai T, et al. 1992. Does a porcine hepatocyte hybrid artificial liver prolong the survival time of anhepatic rabbits? *ASAIO J*, 38(3), M468–72.

Takahashi M, Ishikura H, Takahashi C, Nakajima Y, Matsushita M, Matsue H, Sato K, Sato K, Noto H, Taguchi K, et al. 1993. Immunologic considerations in the use of cultured porcine hepatocytes as a hybrid artificial liver: Anti-porcine hepatocyte human serum. *ASAIO J*, 39(3), M242–6.

Takesawa S, Terasawa M, Sakagami M, Kobayoshi T, Hidai H, Sakai K. 1988. Nondestructive evolution by x-ray computed tomography of dialysate flow patterns in capillary dialyzers. *ASAIO Trans*, 34, 794–799.

Tharakan JP, Chau PC. 1986a. Operation and pressure distribution of immobilized cell hollow fiber bioreactors. *Biotech Bioeng*, 28, 1064–1071.

Tharakan JP, Chau PC. 1986b. A radial flow hollow fiber bioreactor for the large-scale culture of mammalian cells. *Biotech Bioeng*, 28, 329–342.

Tompkins RG, Carter EA, Carlson JD, Yarmush ML. 1988. Enzymatic functions of alginate immobilized rat hepatocytes. *Biotech Bioeng*, 31, 11–20.

Uchino J, Tsuburaya T, Kumagai F, Hase T, Hamada T, Komai T, Funatsu A, Hashimura E, Nakamura K, Kon T. 1988. A hybrid bioartificial liver composed of multiplated hepatocyte monolayers. *ASAIO Trans*, 34(4), 972–7.

United Network of Organ Sharing Monthly Report. November, 1997.

Vacanti JP, Morse MA, Saltzman MW, Domb AJ, Perez-Atayde A, Langer R. 1988. Selective cell transplantation using bioabsorbable artificial polymers as matrices. *J Ped Surg*, 23(1), 3–9.

Van Zijl P, Moonen CT, Faustino P, Pekar J, Kaplan O, Cohen JS. 1991. Complete separation of intracellular and extracellular information in NMR spectra of perfused cells by diffusion-weighted spectroscopy. *Proc Natl Acad Sci*, 88(8), 3228–32.

Velde AA, Flendrid LM, Ladiges NCJJ, Chamuleau RAFM. 1997. Possible immunological problems of bioartificial liver support. *Int J Art Org*, 20(8), 418–421.

Watanabe F, Rosenthal P. 1994. Medical therapy. In: AA Demetriou, editor, Support of the acutely failing liver, RG Landes Company, Austin, TX, p. 22–32.

Watanabe FD, Mullon CJ, Hewitt WR, Arkadopoulos N, Kahaku E, Eguchi S, Khalili T, Arnaout W, Shackleton CR, Rozga J, Solomon B, Demetriou AA. 1997. Clinical experience with a bioartificial liver in the treatment of severe liver failure. A phase I clinical trial. *Ann Surg,* 225(5), 484–91.

Weiss RE, Schleicher JB. 1968. A multisurface propagator for the mass-scale growth of cell monolayers. *Biotech Bioeng,* 10, 601–615.

Wolf CF, Munklet BE. 1975. Bilirubin conjugation by an artificial liver composed of cultured cells and synthetic capillaries. *Trans Am Soc Artif Intern Organs,* 21, 16–27.

Woolf G. 1994. Definitions and etiology. In: AA Demetriou, editor, Support of the acutely failing liver, RG Landes Company, Austin, TX, p. 5–21.

Wu JC, Merlino G, Fausto N. 1994. Establishment and characterization of differentiated, nontransformed hepatocyte cell lines derived from mice transgenic for transforming growth factor alpha. *Pro Nat Acad Sci,* 91(2), 674–8.

Yanagi K, Ookawa K, Mizuno S, Ohshima N. 1989. Performance of a new hybrid artificial liver support system using hepatocytes entrapped within a hydrogel. *ASAIO Trans,* 35(3), 570–2.

Yanagi K, Miyoshi H, Fukuda H, Ohshima N. 1992. A packed-bed reactor utilizing porous resin enables high density culture of hepatocytes. *Appl Microbiol Biotechnol,* 37(3), 316–20.

Yang MB, Vacanti JP, Ingber DE. 1994. Hollow fibers for hepatocyte encapsulation and transplantation: Studies of survival and function in rats. *Cell Transplant,* 3(5), 373–385.

Yarmush ML, Dunn JC, Tompkins RG. 1992. Assessment of artificial liver support technology. *Cell Transplant,* 1(5), 323–41.

Zhang J, Parker DL, Leypoldt JK. 1995. Flow distributions in hollow fiber hemodialyzers using magnetic resonance Fourier velocity imaging. *ASAIO J,* 41, M678–M682.

22
Renal Replacement Devices: The Development of the Bioartificial Kidney

Sudarshan Gautam and H. David Humes

Introduction

With the creation of artificial organ technology, the concept of replacement of failed natural organs has shifted focus from mere life-saving objectives to a broader goal to fulfill quality of life expectations. The kidney was the first solid organ whose function was approximated by a man-made synthetic device and also the first organ replaced by autologous transplantation on a long-term basis. Current organ substitution techniques for end stage renal disease (ESRD) involve either transplanting human kidneys or attempting substitution for functions of diseased kidneys through dialysis. Despite encouraging and improving results from organ transplantation, the preferred choice of substitution therapy, it is still hampered by a shortage of organs, chronic transplant loss, and a changed patient population (Hillebrand and Land 1996). Dialysis therapy only partially replaces the filtration component of renal function and is still an imperfect, intermittent, expensive, and time-consuming replacement of renal function. Dialysis also indiscriminately removes solutes, fails to substitute for renal hormonal and metabolic activities, and requires dietary restrictions and drug therapy to maintain the patient in a state of suboptimal health. The kidney is not simply a filter that produces urine as a waste product, but by monitoring and adjusting the concentrations of multiple compounds in the blood within very narrow limits, it regulates the internal environment to maintain near-perfect homeostasis. The realization that the complexity and amazing efficiency of the natural kidneys cannot be easily mimicked by man-made materials motivates work towards providing a biological component to the artificial kidney device. Hybridization, which means the combining of biological tissues with artificial hardware, is the most plausible solution to constructing "near-natural" kidneys. When such a neoartificial kidney is implanted, it will allow a continuous, increased level of filtration and will also replace the metabolic and endocrine functions "naturally." This approach may circumvent the shortage problem posed by renal transplantations.

Historical Review

The first attempt at blood purification seems to be documented in Ovid's *Metamorphosis,* when Medea took her unsheathed knife and cut the old king Aeson's throat, letting all his blood out, and then filled his ancient veins with a rich elixir to restore his youth. We find that from the earliest times, there has been an intuitive belief in the purification of blood, which found scientific basis in its implementation by the process of hemodialysis. The first recorded dialysis on a human being was conducted by Georg Haas of Gieszen, Germany, in 1926. For over 35 minutes, he dialyzed a 20-year-old girl in the terminal stage of uremia, using hirudin, an anticoagulant, and a continuous circulation technique. In 1938, Pim Kolff, a young doctor in the University Hospital, Groningen, constructed an artificial kidney machine suitable for human application, and is rightfully known as the Father of Artificial Organs. In 1977, K. N. Matsumura successfully treated a cancer patient with an extracorporeal artificial liver and extended a life, an event that can be considered to be the harbinger of the new generation of bioartificial hybrid organs (Matsumura 1978).

Urine Formation and Other Functions of the Kidneys

Urine is formed in the nephron, and its formation is mainly regulated by various segments of the renal tubule. It represents the sum result of three basic processes—convective glomerular ultrafiltration, highly selective tubular reabsorption, and secretion. The tubules regulate the processing of solutes independently of one another, a capability that is essential for precise control.

Glucose, amino acids, and electrolytes are completely or near completely reabsorbed from the tubules. Hydrogen ions, ammonia, and liver-processed drugs and poisons are secreted into the filtrate of the tubule. About 65% of the filtered load of sodium and water and a slightly lower percentage of filtered chloride are reabsorbed by the proximal tubule. Osmotic, active, and passive transport mechanisms all operate in the tubules. The energy for active transport of solutes against an electrochemical gradient comes from the hydrolysis of ATP by way of membrane-bound ATPases and occurs all along the renal tubule. A small change in tubular reabsorption has the potential to cause a relatively large change in urinary excretion since tubular reabsorption is quantitatively very large relative to urinary excretion for many substances. The high capacity of the highly metabolic proximal tubule epithelial cells for reabsorption results from the large number of mitochondria that support potent active transport processes, an extensive epithelial brush border with protein carrier molecules on the luminal side of the membrane, and an extensive labyrinth of intercellular and basal channels, all of which together increase the membrane surface area twentyfold for rapid transport of substances. Approximately 1000 L of blood flow through a human kidney each day, of which 180 L are processed as filtrate, and only 1.5 L are excreted as urine (Guyton and Hall 1997).

The kidney also synthesizes glutathione and free-radical-scavenging enzymes and provides gluconeogenic and ammoniagenic capabilities. Proximal tubules catabolize low molecular weight proteins, including peptide hormones, cytokines, and growth factors. Erythropoietin, renin, prostaglandins, kinins, vitamin D, and multiple cytokines, including tumor necrosis factor-α, interleukin-6, and complement factors, critical to inflammation and immunologic function, are all produced and regulated by the kidney (Humes 1995). Erythropoietin-producing cells sense decreases in red blood cell (RBC) oxygen delivery and produce erythropoietin, which binds to receptors on multipotent stem cells in the bone marrow and cues these progenitor cells to differentiate into RBCs.

Renal Tubule Stem Cells

The high metabolic demands of the renal tubular cells, the concentration of heavy metals and drugs, the transcellular transport of toxins, and the virtually absolute requirement for oxidative metabolism as an energy source by proximal tubule cells all cause susceptibility to ischemic injury. In the immediate postinjury period, regenerative mechanisms proceed to restore the structural and functional integrity of the renal tubular epithelium, to reestablish the normal cellular architecture and physiological function of the renal tubule. The proliferative repair process is responsible for the ability of the kidney to completely reverse severe organ failure. The importance of this repair process for survival has resulted in a highly evolved system dependent upon autocrine, paracrine, and endocrine pathways to deliver cytokine growth factors to the areas of injury, to activate the replicative process (Coimbra et al 1990). Studies have shown that human renal tubular cells can express cytokines, including PDGF, granulocyte-macrophage colony-stimulating factors, and interleukin-6 mRNA in vitro, and that early response genes, JK and KC, which encode for cytokines, are expressed in distal tubule cells shortly after ischemic injury. Epidermal growth factor (EGF) and its functional homolog, transforming growth factor (TGF-α), are the most potent mitogens to proximal tubule cells (Humes et al 1989). In regions of irreversible cell injury, proliferative reparative mechanisms are necessary to reline the tubular basement membrane with functional tubular epithelium. This is dependent on the presence of a resilient cell population that has retained the potential for proliferation and differentiation, the responsivity of these cells to the influence of local cytokine growth factors and extracellular matrix molecules, and the preservation of the structural integrity of the underlying basement membrane. This understanding leads to the possibility that tubule progenitor cells can be isolated from adult kidneys to expand renal cells to be used in a bioartificial organ that comprises cells and biomaterials to replace renal tubular function (Woods and Humes 1997).

Limitations of the Current Techniques for Substitution of Renal Function

Artificial kidney developers have known for years that their current artificial kidney design is suboptimal. Hemodialyzers and artificial kidney devices did not bring about the expected restoration of normality in renal failure patients. Because of the nonphysiologic manner in which these renal devices perform renal functions, patients with ESRD on these forms of therapy continue to have major medical problems, higher morbidity/mortality, and shorter life expectancy. In the United States, healthy individuals of ages 40 and 59 years have life expectancies of over 37 and 20 years respectively, while life expectancies are estimated to be only 8.8 and 4.2 years, respectively, for ESRD patients on dialysis (United States Renal Data System 1994). Current treatments should therefore not be considered as renal "replacement" therapy, but rather as renal "substitution" therapy. Renal transplantation addresses some of the issues, but immunologic barriers and organ shortages keep this approach from being ideal for a large number of ESRD patients.

Patients with acute renal failure (ARF), despite the most advanced state-of-the-art patient care, also still have an unacceptably high mortality rate of over 50%, even with dialytic or filtrative support. This suboptimal result is most likely due to the fact that current therapy does not address the crux of the pathophysiological problem, which mainly resides in the renal tubule. Careful review of the cause of death in young, healthy soldiers who develop clearly defined acute renal failure during combat injuries without concurrent preexisting chronic illnesses, provides important insights into what is missing in the present-day renal substitution devices. Analysis of the cause of mortality in these individuals demonstrates that the single most important factor is the development of infections. The contributing problems identified are (1) impairment of host defenses, (2) impaired wound healing, and (3) relative malnourishment due to high catabolic rates. It appears that the central problem resides in the injury and necrosis of renal proximal tubule cells, which causes derangement of the critical homeostatic, metabolic, synthetic, endocrine, and nutrition functions. The loss of both glutathione synthesis and production of key free-radical-scavenging enzymes diminishes host defense function. The loss of key endocrinologic function leads to alterations in growth factor metabolism and delayed wound healing. The loss of renal tubular function may aggravate uremic events and catabolic activity since renal tubule cells excrete 10–15% of the nitrogenous products of metabolism via the ammoniagenic pathway, and supply up to 50% of circulating blood glucose levels via the gluconeogenic pathway in starvation. The kidney also serves as a critically important metabolic organ, providing catabolism of low molecular weight proteins, including peptide hormones, cytokines, and growth factors, and important hormonal functions such as production and regulation of erythropoietin, vitamin D, and multiple cytokines critical to inflammation and immunologic regulation. Perhaps the loss of these important metabolic, synthetic, and endocrinologic functions is responsible for the high rates of infection, poor wound healing, and increased catabolic activity that lead to higher mortality rates in acute renal failure.

In chronic renal failure, β2 microglobulin dysmetabolism results in amyloid deposition in various body compartments and debilitating sequelae. Normally, β2 microglobulin is filtered by the glomerulus, reabsorbed by the tubular cell, and then degraded by proteases. Loss of functioning renal mass with the development of chronic renal failure reduces the rate of degradation of β2 microglobulin, and circulating levels rise to 30–50 times normal. Standard dialysis therapies fail to remove sufficient β2 microglobulin to substantially reduce elevated circulating levels. The addition of a renal tubule device may improve the clearance and metabolism of this potentially pathophysiologic build-up of amyloid. Dialysis also fails to replace the role of the kidney in calcium and phosphate metabolism. Renal osteodystrophy is universal in chronic hemodialysis patients, because conventional hemodialysis removes insufficient phosphate to compensate for dietary intake and because of lack of hydroxylation of vitamin D_3 by renal tissue. The addition of renal tubule cell metabolic activity may optimize calcium metabolism in ESRD and minimize this major problem in dialysis patients (Humes et al 1996).

Accordingly, the development of optimal supportive modalities replacing the reabsorptive, synthetic, metabolic, and endocrinologic functions of the kidney may add significant value to the current suboptimal supportive options available to treat both acute and chronic renal failure. Development of such a renal replacement device will certainly add critical renal functional components not currently substituted for by hemodialysis or hemofiltration, and thereby improve patient care in renal failure. In this context, the developing technologies of cell therapy, gene therapy, and tissue engineering may contribute to the development of a range of devices to provide alternatives to current methods of renal replacement therapy.

New Technologies

a. Cell Therapy

Modern medicine lost sight of the sixteenth century physician Paracelsus's concept of using living tissue to rebuild and revitalize ailing/aging tissue. Swiss physician Paul Niehans rediscovered the beneficial effects of live cell therapy quite by accident in 1931 when he injected fine pieces of a steer's parathyroid glands into a terminally ill woman patient with hypoparathyroidism. The woman not only had a miraculous recovery but went on to live another 30 years, thereby introducing modern cell therapy. Drugs and chemicals have to be repeatedly administered to maintain efficacy. In contrast, cell therapy with the implantation of living cells to produce a natural substance in short supply from the patient's own cells, as a result of injury and destruction, has a long-term effect due to the ability of cells, as drug delivery vectors, to use biologic processes to constantly produce a missing compound or to therapeutically treat a disease process. The current technologic hurdle to this cell therapy approach lies in the fact that xenogeneic cells/tissues will be rejected by the recipient patient. This problem can be circumvented by immunoisolating the cells within a therapeutic device.

Cell therapy is an emerging branch of tissue engineering that addresses the need for in vivo cuing

of protein production. It may be the next successful strategy to closely follow the success of recombinant genetic engineering, which produced clinically effective proteins in clinically deficient states. This approach may prove successful since many physiologic responses are due to a complex interaction of a series of cell products, rather than to a lack of one component. The blood levels of a circulating protein may be regulated by different mechanisms: at the gene level by transcriptional mechanisms, at the protein level by translational processes, or at the secretory level by cellular processes. Biocompatible/biodegradable porous polymer delivery vehicles can be used to encapsulate the cells, while simultaneously providing protection of the implanted cells from the host's immune defenses (if the cells are not autologous). The development of cell therapy modalities replacing the reabsorptive, metabolic, and endocrinologic functions of the kidney may also add significant value to the current suboptimal supportive options available to treat established acute renal failure. Similarly, the ability to isolate and grow cells with the capability of regulating and secreting the RBC growth factor, erythropoietin, may allow for an implantable drug therapy device, and thereby combat anemia associated with renal disease.

b. Gene Therapy

Medical science, with its goal to fight/prevent disease, is currently focusing its efforts towards manipulating and altering genetic material, the biologic units of heredity. Gene therapy is a revolutionary approach that works on the premise that supplying healthy copies of missing or flawed genes can help cure or prevent diseases by correcting the basic problem at the genetic level. Genes, the blueprints for the production of specific enzymes/proteins, are transferred into living cells either to deliver a gene product to a cell in which it is missing or to cause a cell to produce a foreign gene product, to promote a new function. A new gene called the transfer gene is removed, by cutting DNA at specific locations (gene splicing), and then inserted into a disabled virus vector chosen to transduce the cells that need to be altered to change the genetic makeup. These cells are allowed to multiply and then reinserted into the patient.

c. Tissue Engineering

Tissue engineering is on the cutting edge of biotechnology for the design and development of organ constructs that mimic the functions of human organs by means of a creative combination of the most advanced scientific methodologies, including stem cell culture, gene transfer, and the use of growth factors, with the scientific principles of chemical, mechanical, and biomaterials engineering. Stem, or progenitor, cells exist in the body as undifferentiated cell sources that, upon receipt of varying cuing signals from growth factors and hormones, can proliferate and differentiate into various types of cells. If stem cells of a given organ can be found and isolated, these cells could be used as a powerful tissue engineering tool in generating natural organ tissues. Until recently, renal tubule stem cells had not been found. In a series of in vitro studies, it was concluded that tubulogeneic cells exist within the adult mammalian kidney in a normally dormant state, yet manifest the ability to rapidly proliferate and differentiate in response to ischemic or toxic insult (Humes et al 1996). Tissue engineering of a biohybrid kidney comprising both the biologic renal tubule stem cells and synthetic biomaterial components probably would have substantial benefits for the patient by increasing mobility, life expectancy, and quality of life, with large savings in time and money and less risk of infection. The engineering design of a bioartificial hemofilter (BHF), by combining autologus human endothelial cells, extracellular matrix proteins, and microporous hollow fibers, is the first step toward an implantable biohybrid kidney to replace renal excretory function.

"Biological" Renal Replacement Devices

a. Design Considerations

Construction of biohybrid renal devices poses many challenges in the way of hemorheology, biomaterials, and cell culture nutrient delivery. The essential features of the natural kidney must be utilized to direct the design of a tissue-engineered biohybrid artificial kidney that replaces the critical elements of renal function. An elegant system has evolved in the design of a glomerulus to provide for efficient ul-

trafiltration with the ability to separate as much as one-third of the plasma to remove toxic wastes from the blood and yet retain important circulating components, such as albumin. However, only 5–10% of normal renal excretory function is required to sustain life. More than 98% of the glomerular ultrafiltrate is reabsorbed by the renal tubule, with nearly 65% of this amount occurring in the proximal renal tubule. This process is driven by differences of hydraulic, osmotic, and oncotic pressures within the interstitial and intracapillary spaces. The peritubular capillary hydraulic pressures are considerably lower than those in the glomerulus; they are generally below 10 mm of Hg in rats (Robertson et al 1972). The value of oncotic pressure in the blood entering the peritubular capillary network is usually about 38 mm Hg, corresponding to an efferent arteriolar protein concentration of ~9 g/100 ml (Brenner et al 1978). Thus, there are ample pressure differentials in the proximal tubule to reabsorb the majority of the filtrate.

Thus, the complete bioartificial kidney requires two main units, a filtering unit that mimics the glomerulus, and a regulatory/reabsorptive unit mimicking the renal tubule. In a hollow-fiber device, the supply of oxygen, a critical limiting factor for the viability of cells, will be obtained from diffusive transport through the permeable fiber walls from blood to the cells, and the flow of ultrafiltrate through the lumen. Flow rates, pressures, and temperature are the important parameters to monitor and adjust and keep within narrow limits. Fluid flow should be uniformly well dispersed and distributed, devoid of areas of stagnant blood flow in the extracapillary space, as identified by flow visualization studies. Computational models serve to assess the effectiveness of the hollow-fiber construct, allow for the critical study of specific parameters, and aid in the scale-up and overall design of such a device. A bioartificial proximal tubule, which reabsorbs iso-osmotically the majority of the filtrate, may be sufficient to replace required tubular function to sustain fluid electrolyte balance in a patient with ESRD, and so the other nephronal segment functions, such as the loop of Henle, may not be necessary.

The procedure of growing tubule cells on hollow fibers proves to have many inherent problems. First, it is very tedious to grow cells in absolutely sterile conditions. Second, storing such a biosynthetic device is not easy and must be used within weeks after enough cells have grown on the fibers. Wide distribution from a single production source is challenging.

b. Tissue-Engineered Bioartificial Renal Devices

b1. Bioartificial "Glomerulus"/Hemofilter

The popularity of hemofiltration has increased immensely since its first clinical application in 1976 (Baldamus and Pollok 1989). Improved treatment symptomatology, such as less fatigue and more stable posttreatment blood pressure and hemodynamic stability, remains the advantage of hemofiltration. Removal of uremic toxins, predominantly by the convective process, has several distinct advantages over hemodialysis, due to its mimicry of the glomerular process of toxin removal with increased clearance of higher molecular weight solutes/peptides (6000 to 50,000 daltons). Treatment of iron overload and/or aluminum intoxication by deferoxamine and prevention of amyloidosis in long-term ESRD by the removal of $\beta 2$ microglobulin, which is cleared with hemofiltration but not by hemodialysis, are a distinct advantage. Patients with cardiomyopathy, autonomic dysfunction, or ischemic heart disease, as well as those with excessive interdialytic weight gain who do not tolerate fluid removal during conventional hemodialysis, all benefit symptomatically from hemofiltration.

The potential for a bioartificial glomerulus has been demonstrated with the maintenance of ultrafiltration in human beings for several weeks with a single device (Golpher 1986, Kramer et al 1977). A bioartificial hemofilter may provide an adjunctive approach to provide 2–4 ml/min of filtration function during the interdialytic period, sufficient to improve solute clearance, lessen dialysis time, and optimize clearance rates of uremic toxins. However, the major limitations of a purely synthetic device include loss of filtration from concentration polarization of proteins and/or thrombotic occlusion within the hemofilter. A biologic tissue engineering approach with the use of transduced autologous, endothelial cells may overcome these problems. Since gene transfer for in vivo protein production has been clearly achieved with endothelial cells (Wilson et al 1989, Zwiebel et al 1989), gene transfer into these blood-vessel-lining cells for the production of an anticoagulant protein, such as hirudin, is clearly conceivable. Efforts to construct tissue engineered prototype bioartificial hemofilters with the confluent growth of monolayers of endothelial cells on high-flux polysulfone hollow fibers and cellular production of hirudin by gene transfer, have recently met with success. The conceptual schematization of a bioartificial glomerulus is shown in Figure 22.1. Light micrographs of the tissue engineered hollow fiber are shown in Figure 22.2, with the mesangial support layer and an overlying confluent monolayer of endothelial cells under different magnifications. This tissue engineering approach may provide a novel solution to improving the long-term hemocompatibility for in vivo continuous hemofiltrative support devices (Kadletz et al 1992, Schnider et al 1988, Shepard et al 1986). Since hirudin is a low molecular weight protein, it will be freely filtered by the bioartificial hemofilter so that systemic elevations of this anticoagulant should not occur. A scaled-up hemofilter has been constructed with seeding and growth of endothelial cells, which markedly improved the permselectivity by changing the albumin leak rate from 83% to 3%, resulting in a filtration rate of 2.2 ml/min with a filtration fraction of 45%, as shown in Table 22.1.

b2. Bioartificial Renal Tubule Assist Device (RAD)

A critical requirement for development of a bioartificial tubule is the isolation and growth in vitro of

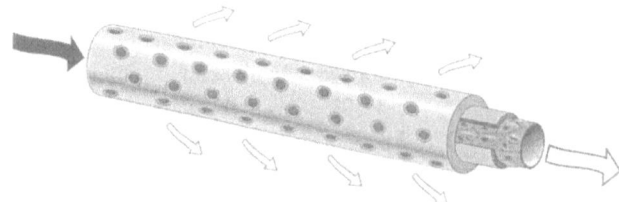

FIGURE 22.1. Conceptual schematic of a single capillary of a bioartificial hemofilter. As seen, the endothelial cells are grown as a confluent monolayer on the inner surface of a synthetic microporous hollow fiber with preadhered matrix material for attachment and growth.

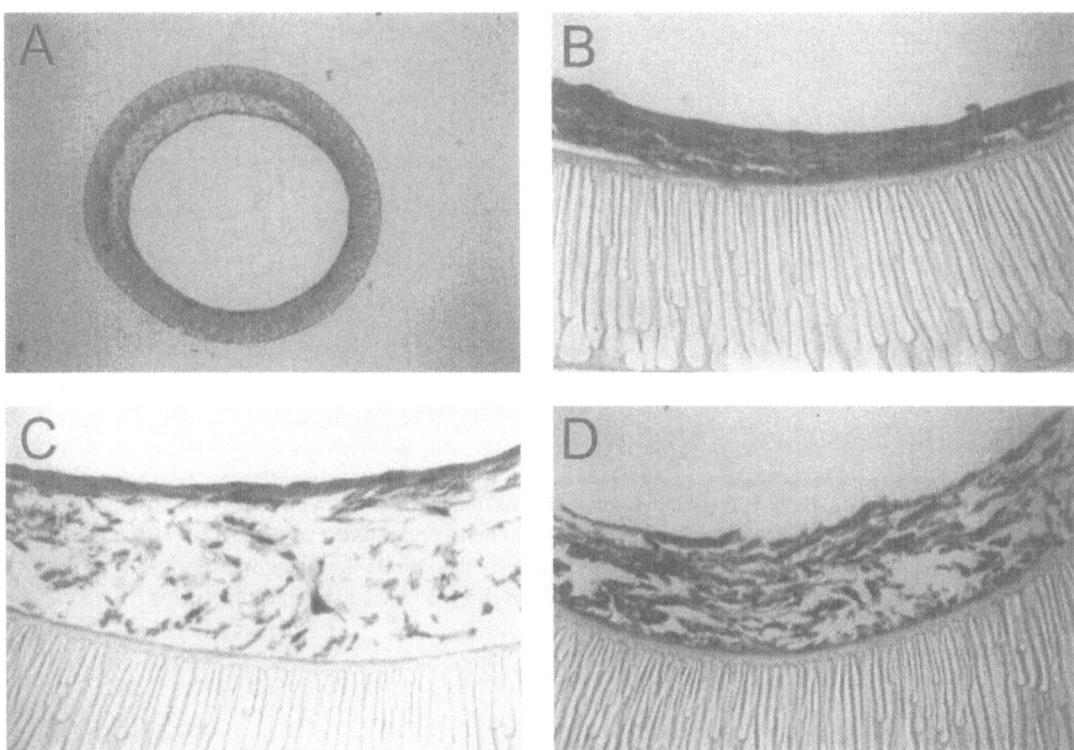

FIGURE 22.2. Light micrographs of a hollow fiber seeded with mesangial cells grown as a supporting mesenchymal layer. Panel A: Low-power cross-section of a tissue engineered hollow fiber with the mesangial support layer with an overlying confluent monolayer of endothelial cells. Upon sectioning, the upper-left area of the cell layer separated from the surface of the synthetic polymer. Panel B: Higher-power view of the lower edge of (A), demonstrating the smooth endothelial monolayer overlying the mesenchymal cell layer. Panel C: Similar to (B), except displaying the loosened area of the upper portion of (A). Panel D: High-power view of a hollow fiber constructed with only mesangial cells and without endothelial cells. As seen, compared to panels (B) and (C), a smooth covering of endothelial cells is not observed.

progenitor cells, which will replicate to form a confluent monolayer on the hollow-fiber membrane surface and also differentiate into specialized cells to provide a vectorial transport function. The technical feasibility of an implantable epithelial cell system derived from cells grown as confluent monolayers along the luminal surface of polymeric hollow fibers has been demonstrated (Mackay et al 1998).

Although successful renal tubule progenitor cell expansion has been achieved with human adult kidneys, a nonhuman animal tissue source for tubule cells has been strongly considered. Because an expensive screening process for infectious agents must

TABLE 22.1. Filtration and permselectivity of bioartifical hemofilter.

	Filtration rate (ml/min)	Filtration fraction (%)	Albumin leak rate (mg/ml)	Percentage albumin leak (%)
Non-Cell unit	3.88 ± 0.94	70.7 ± 5.7	109 ± 3.2	82.7 ± 7.6
Endothelial cell unit	2.20 ± 0.52	45.3 ± 8.5	5.0 ± 1.0	2.6 ± 0.7

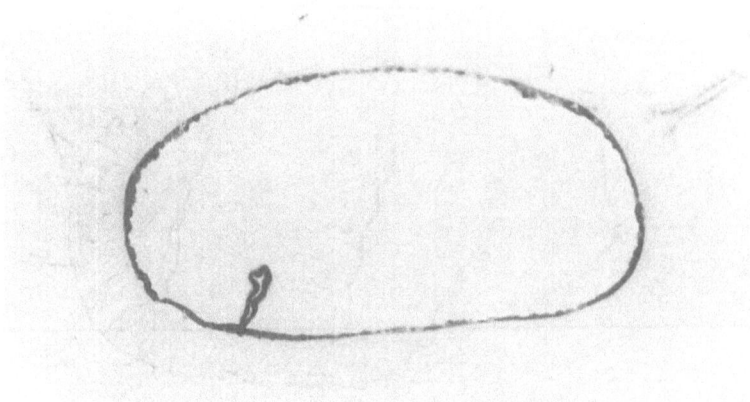

FIGURE 22.3. Light micrograph of an H & E section ($100 \times$) of hollow fiber lined with collagen-Type IV and a confluent monolayer of human renal tubule epithelial cells along the inner component of the fiber. In this fixation process, the hollow fiber is clear with the outer contour of the hollow fiber identified by the irregular line (disregard artifact in lower-left quadrant).

be accomplished to ensure the safety of a human donor source of tissue, an animal tissue source for renal tubule cells within an extracorporeal bioartificial renal tubule assist device has been chosen for development. The short-term use of this device for acute therapy in the ICU setting for the treatment of ARF allows a nonhuman tissue source as a preferred strategy. For economic and safety reasons, and also because of its anatomical and physiological similarities with human beings, the pig is currently considered the best source of organs for both human xenotransplantation and immunoisolated cell therapy devices. Pig kidneys are harvested from 4–6-week-old Yorkshire breed pigs. A full clinical profile of each donor pig for pathogens and blood and tissue pathology is accomplished to ensure the safety and noninfectivity of donor tissue. From these kidneys, renal proximal tubule segments are harvested, and renal tubule progenitor cells are expanded (Humes et al 1995). These cells were seeded into the lumen of a single hollow-fiber bioreactor. Figure 22.3 shows the light micrograph of a hollow fiber lined with a confluent monolayer of human renal tubule epithelial cells. Figure 22.4 shows the electron micrograph of the cross-section of a similar hollow fiber. The well-developed microvilli and apical tight junctions can be appreciated. The functional

confluence of the cells was demonstrated by the recovery of intraluminally perfused ^{14}C-inulin, which averaged $>98.9\%$, versus $<7.4\%$ with the control non-cell hollow fibers under identical pressure and flow conditions. The baseline absolute fluid transport rate averaged $1.4 \pm 0.4 \, \mu l/30$ min. To test the dependency of fluid flux with oncotic and osmotic pressure differences across the bioartificial tubule, albumin was added to the extracapillary space, followed by addition of ouabain, an inhibitor of Na^+-K^+ ATPase, the enzyme responsible for active transport across the renal epithelium. Addition of albumin resulted in a significant increase in volume transport to $4.5 \pm 1.0 \, \mu l/30$ min, as shown in Table 22.2. Addition of ouabain inhibited transport back to baseline levels of $2.1 \pm 0.4 \, \mu l/30$ min, as shown in Figure 22.5.

To scale up this single hollow-fiber tubule, the intraluminal surface of the hollow fibers in a hemofiltration cartridge with surface area of $0.7 \, m^2$ were coated with laminin. Renal tubule cells were then seeded at a density of 10^5 cells/ml into the intracapillary space with four cell infusions separated by 30 minutes and a $90°$ rotation of the cartridge. The seeded cartridge was connected to the bioreactor perfusion system, in which the extracapillary space was filled with culture medium, and the in-

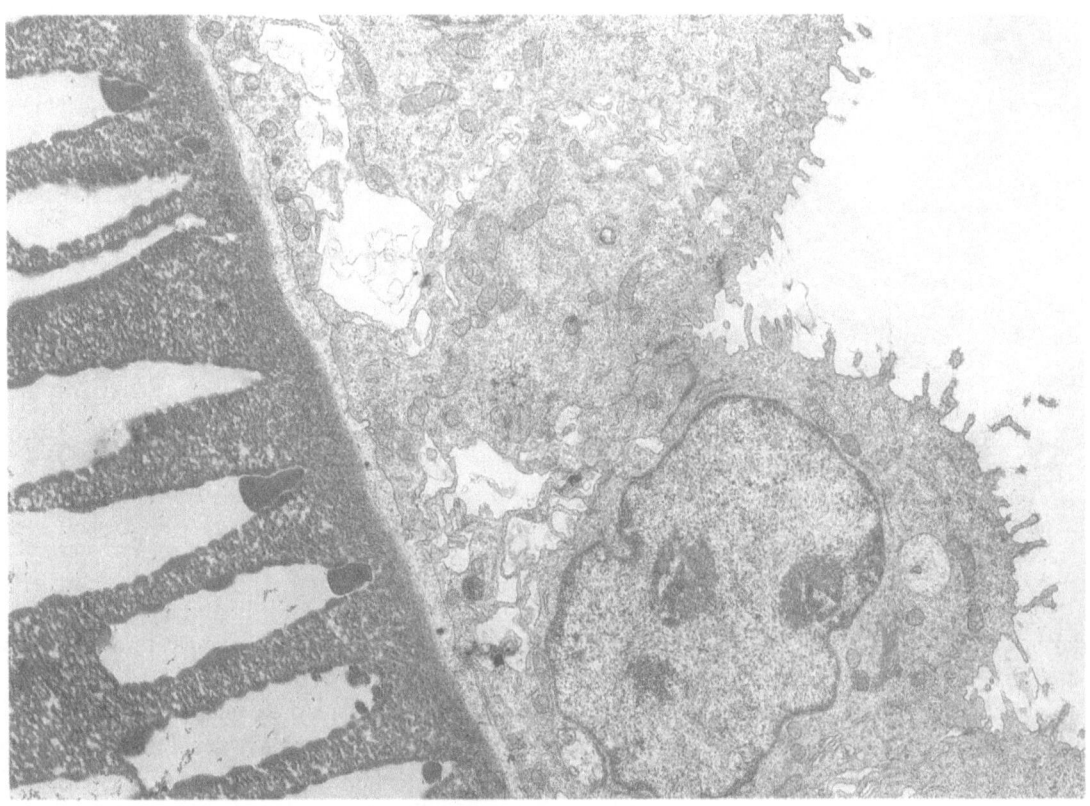

FIGURE 22.4. Electron micrograph of a single hollow fiber lined with extracellular matrix and a confluent monolayer of renal tubule cells along the inner component of the fiber. As displayed, the differentiated phenotype of renal tubule cells on the hollow fiber prelined with matrix is apparent. The well-developed microvilli and apical tight junctions can be appreciated (14,000 ×).

tracapillary and extracapillary were changed every 2–3 days to maintain adequate metabolic substrates for growth. This perfusion setup was chosen to ensure a slightly higher intraluminal hydraulic pressure within the hollow fibers, compared to extracapillary space, to simulate in vivo conditions to promote cell attachment along the internal surface of the hollow fibers. The critical parameters of oxygen delivery to cells in the device are schematized

in Figure 22.6, and the functional properties of the RAD are summarized in Table 22.3.

c. Ex Vivo Performance of the RAD

Although assessing the functionality of a RAD unit in vitro is an important component of bioreactor design, the true test of functionality and utility comes in testing the device in vivo, or ex vivo, as we shall

TABLE 22.2. Fluid reabsorption in renal tubule assist bioreactor.

Condition	Perfusion rate (ml/hr)	Reabsorption rate (ml/hr)	% Reabsorption
Baseline	25.5	0.8	3.1
Albumin (4.0 gm/dl)	25.8	5.4	20.9
Albumin + Ouabian (0.5 mM)	24.3	3.1	12.6

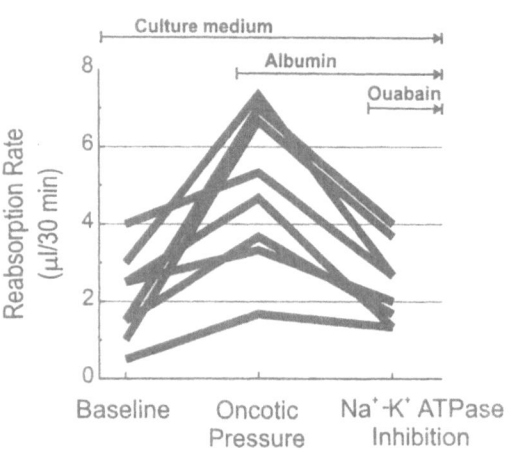

FIGURE 22.5. Single hollow-fiber bioartificial renal tubules were grown, utilizing a polysulphone hollow fiber with Madin-Darby canine kidney (MDCK) cells. After baseline measurements, bovine serum albumin (6.6 gm/dl) was added to the extracapillary space of the bioreactor. After three additional transport measurements, ouabain (0.25 mM) was added to the ECS to inhibit Na^+-K^+ ATPase activity.

describe here. Before clinical trials can be undertaken with such a device, extensive testing must be done in large animals, where the system can be evaluated under physiologic conditions and can be optimized for functionality and ease of use. The renal assist device will be used as a component in an extracorporeal circuit, in which care must be taken to ensure proper operating conditions. Conditions of operation should mimic as well as possible, physiologically or in vitro, those that the cells in the device are accustomed to. Important parameters to monitor and adjust are, for example, flow rates, pressures, and temperature.

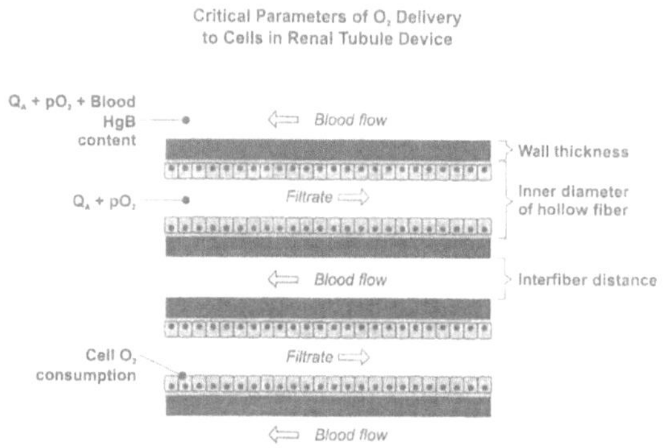

The bioartificial kidney setup consists of a filtration device (a conventional hemofilter) followed in series by the tubule unit, as shown in Figure 22.7. Specifically, blood is removed from a large animal using a peristaltic pump. The blood then enters the fibers of a hemofilter, where ultrafiltrate is formed and flows into a second pump. This pump is added to ensure a stable delivery of the flow rate of ultrafiltrate into the fibers of the tubule lumen within the RAD downstream to the hemofilter. Processed ultrafiltrate exiting the RAD is collected and can be discarded as "urine". The filtered blood exiting the hemofilter enters the RAD through the ECS port and disperses among the fibers of the device. The pressure of the blood and ultrafiltrate right before entering the RAD is monitored. Heparin is delivered continuously into

FIGURE 22.6. Schematic of the critical parameters of oxygen delivery to cells in the RAD. The blood compartment is shown in the extracapillary space. The entering oxygen content of both the hemofiltrate and the blood in the extracapillary space will have an oxygen content of mixed venous blood (pO_2). Additional input variables include the inner diameter and thickness of the hollow fibers, the interfiber distance, the flow of blood/hemofiltrate, the metabolic oxygen consumption rates of the porcine renal tubule epithelial cells, etc.

TABLE 22.3. Functional properties of the renal tubule assist device (RAD).

Transport
    Sodium—ouabain-inhibitable
    Glucose—phlorizin-inhibitable
    PAH—probenecid-inhibitable
Metabolic
    Ammoniagenesis—pH-sensitive
    Glutathione synthesis
Immunologic
    IL-10 production
Endocrinologic
    1-OH D3 activity—PTH, Pi-responsive

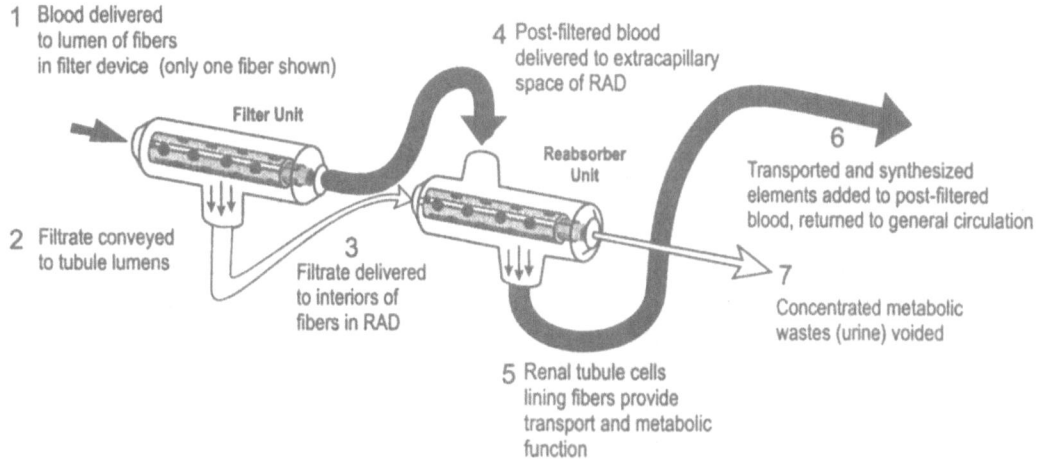

1 Blood delivered to lumen of fibers in filter device (only one fiber shown)

Filter Unit

2 Filtrate conveyed to tubule lumens

3 Filtrate delivered to interiors of fibers in RAD

4 Post-filtered blood delivered to extracapillary space of RAD

Reabsorber Unit

6 Transported and synthesized elements added to post-filtered blood, returned to general circulation

7 Concentrated metabolic wastes (urine) voided

5 Renal tubule cells lining fibers provide transport and metabolic function

FIGURE 22.7. Conceptual schematization of a tissue engineered bioartificial kidney with an endothelial-cell-lined hemofilter in series with a proximal-tubule-cell-lined renal tubule assist device (RAD).

the blood before it enters the RAD, to diminish clotting within the device. The RAD is oriented horizontally and placed into a temperature-controlled environment. The temperature of the cell compartment of the RAD must be maintained at 37°C throughout its operation to ensure optimal functionality of the cells. Maintenance of a physiologic temperature is a critical factor in the functionality of the cells.

In setting up the circuit described, initial priming of sterile tubing must be performed prior to attaching the RAD. Priming, or filling the tubing with fluid (typically from an IV bag), is a common and necessary procedure done in conventional dialysis and hemofiltration. Prior to attachment to the circuit, the cell culture medium in the RAD is thoroughly rinsed out, and the RAD itself is primed with a heparinized solution. Once the tubing is free of air, the RAD device is hooked up to the circuit, using sterile technique within a laminar flow hood. As flow is initiated, care must be taken to ensure that the pressures of the fluid entering the device remain below a design threshold.

The flow rate of the blood should be kept at a high rate so as to avoid clotting in both the hemofilter and the RAD. Excessive clotting and protein build-up can impede flow, cause increases in pressure, and lead to an added resistance or barrier to diffusion, especially critical in oxygen/nutrient delivery to the cells of the RAD. Since the ultrafiltrate is in direct contact with the cells lining the fibers of the RAD, control of its flow rate is critical. Hydraulic pressures entering the RAD, as well as transmembrane pressure gradients,

must also be tightly controlled. Functionality and cell adhesion can be adversely affected if shear forces and pressures are not controlled within allowable levels. Concentration gradients established by either active or passive transport, key epithelial cell functions, are reliant on the establishment and maintenance of tight junctions. The maximal rate of flow that a cell can withstand is cell- and environment-specific and must be determined for each system. Fluid volume losses and inputs should be monitored throughout the operations of the RAD, just as in dialysis. Adjustments should be made, and the net fluid balance can be controlled and is patient-specific. For example, fluid is lost as "urine" and is replaced via replacement fluid, as in hemofiltration therapy.

The unit is able to maintain viability because metabolic substrates and low molecular weight growth factors are delivered to the tubule cells from the ultrafiltration unit and the blood in the ECS. Furthermore, immunoprotection of the cells grown within the hollow fiber is achievable because of the impenetrability of the hollow fiber to immunologically competent cells, if the encapsulating membrane has a pore size that excludes compounds with a molecular weight greater than 150,000 daltons. Rejection of transplanted cells will, therefore, not occur. This arrangement thereby allows the filtrate to enter the internal compartments of the hollow-fiber network, lined with confluent monolayers of renal tubule cells for regulated transport function.

Given the novel nature of the described cell-based renal assist device, one can foresee some potential problems that may or may not arise with its use. Excessive transmembrane pressures or inherent defects in the cartridges could lead to the breaking of fibers, the release of xenogeneic cells, and a possible immune response. A protective mechanism can be designed in order to address this problem. The use of existing dialysis membrane biomaterials ensures minimal immunologic responses due to contact with the fibers/cartridge. Excessive flow rates or incorrect temperatures, especially of the ultrafiltrate, could lead to loss of functionality. Clotting of the cartridge could also pose a problem, and care should be taken to provide for adequate heparinization.

The use of such a device in uremic and nonuremic large animals has shown that inulin leak rates increased immediately after use, but returned to prestudy values after being maintained in culture for 2 weeks, meaning that cells remain viable and can continue growing. Preliminary data in uremic animals suggest that the RAD activates 1,25-dihydroxyvitamin D_3 and maintains plasma glutathione levels in the normal range. Table 22.3 summarizes some of the functional characteristics of the RAD. Metabolic parameters, such as ammonia production, within the RAD cartridge during large animal studies are currently being assessed. The use of scaled-up cartridges can increase metabolic production. A surface area expansion from 0.4 m^2 to 0.8 m^2 results in an increase in the number of cells from 2 $\times 10^9$ to 4 $\times 10^9$ cells. RAD cartridges have been maintained in culture in excess of 2 months after use in large animal studies. Currently, temperature and pressure conditions within the extracorporeal design are being modified to maximize the function of the RAD. These initial ex vivo studies mark significant progress towards the use of RAD cartridges to enhance current clinical treatment of renal failure.

Conclusions and Future Directions

We have described the progression of the development of a bioartificial renal tubule device (RAD), beginning from the in vitro cell isolation, through design and scale-up, to ex vivo trials in large animals. The RAD has been shown to be functional in terms of both metabolic and reabsorptive characteristics. Looking to the immediate future, further op-timization of the system and the conclusion of animal testing will lead to clinical trials in humans.

The development of a bioartificial filtration device and a bioartificial tubule processing unit would also lead to the possibility of an implantable bioartificial kidney. Figure 22.7 shows the schema for such a bioartificial kidney with the bioartificial renal tubule device in series to the bioartificial glomerulus, just as in the case of the natural kidney. The specific implant site for a bioartificial kidney will depend upon the final configuration of both the bioartificial filtration and tubule devices. As presently conceived, an autologous endothelial-cell-lined bioartificial filtration hollow-fiber cartridge can be placed into an arteriovenous circuit using the common iliac artery and vein, as in the surgical connection for a renal transplant. The filtrate is conveyed in series to a bioartificial proximal tubule device so that reabsorbate will be transported and reabsorbed into the systemic circulation. The processed filtrate exiting the tubule unit is then conveyed via tubing to the proximate ureter for drainage and urine excretion via the recipient's own urinary collecting system.

The pathophysiology of terminal renal failure and the technology for its treatment continue to be in a state of rapid progress. Artificial blood processing of uremic blood has so far been achieved by hemodialysis, hemofiltration, and hemoperfusion, the three principal mechanisms with which most current renal substitution devices operate. A fourth new principal, "bioreabsorption" which utilizes live tubule cells, is about to emerge on the clinical horizon. The development of artificial metabolic-endocrine organs achieved by harnessing the capabilities of "live" cells lays the foundation of the concept of "biosynthetic life support devices," which combines artificial materials with biologically active cells from a specific organ. A successfully functioning intracorporeal bioartificial kidney can be considered a cure rather than mere functional support. It is our hope that the kidney may well be the first organ to be available as a tissue engineered implantable device for full functional organ replacement in the human body.

References

Baldamus CA, Pollok M. 1989. Ultrafiltration and hemofiltration, practical applications in replacement of renal function by dialysis. In: Maher J, editor. 3rd ed. Boston: Kluwer Academic Publishers.

Brenner BM, Hostetter TH, Humes HD. 1978. Molecular basis of proteinuria of glomerular origin. New England Journal of Medicine 298(15):826–833.

Coimbra TM, Cieslinski DA, Humes HD. 1990. Epidermal growth factor accelerates renal repair in mercuric chloride nephrotoxicity. American Journal of Physiology 259:F438–F443.

Golper TA. 1985. Continuous arteriovenous hemofiltration in acute renal failure. American Journal of Kidney Diseases 6(6):373–386.

Guyton AC, Hall JE. 1997. Human physiology and mechanisms of disease. 6th ed. Philadelphia W.B. Saunders Company. p 212–230.

Hillebrand G, Land W. 1996. Renal transplantation: progress and prospects. Artificial Organs 20(5):403–407.

Humes HD, Cieslinski DA, Coimbra TM. 1989. Epidermal growth factor enhances renal tubule cell regeneration and repair and accelerates the recovery of renal function in postischemic acute renal failure. Journal of Clinical Investigation 84(6):1757–1761.

Humes HD, Cieslinski DA. 1992. Interaction between growth factors and retinoic acid in the induction of kidney tubulogenesis in tissue culture. Experimental Cell Research 201(1):8–15.

Humes HD. 1995. Tissue engineering of the kidney. In: Bronzino JD, editor. The biomedical engineering handbook. Boca Raton: CRC Press. p 1807–1824.

Humes HD. 1996. Application of gene and cell therapies in the tissue engineering of a bioartificial kidney (editorial). Int J Artif Organs 19(4):215–217.

Humes HD. 1996. The bioartificial renal tubule: Prospects to improve supportive care in acute renal failure. Ren Fail 18(3):405–408.

Humes HD. 1996. Tissue engineering of a bioartificial kidney: A universal donor organ. Transplant Proc 28(4):2032–2035.

Humes HD, Krauss JC, Cieslinski DA. 1996. Tubulogenesis from isolated single cells of adult mammalian kidney: Clonal analysis with a recombinant retrovirus. American Journal of Physiology 271:F42–F49.

Kramer P, Wigger W, Rieger J. 1977. Arterio-venous hemofiltration: a new and simple method for treatment of overhydrated patients resistant to diuretics. Klin Wochenschr 55:1121.

Kadletz M, Magometschnigg H, Minar, E. 1992. Implantation of in vitro endothelialized polytetrafluoroethylene grafts in human beings. J Thorac Cardiovasc Surg 104:736.

Mackay S, Funke A, Buffington D, Humes HD. 1998. Tissue engineering of bioartificial renal tubule. ASAIO: 44:179–183.

Matsumura KN. 1978. After fifteen years: artificial liver & artificial pancreas. 2d ed. Berkeley, CA: Alin Foundation Press.

Robertson CR, Deen WM, Troy JL, Brenner BM. 1972. Dynamics of glomerular ultrafiltration in the rat. 3. Hemodynamics and autoregulation. Am J Physiol 223(5):1191–1200.

Shepard AD, Eldrup-Jorgensen J, Keough EM. 1986. Endothelial cell seeding of small-caliber synthetic grafts in the baboon. Surgery 99:318.

Schnider PA, Hanson SR, Price TM. 1988. Durability of confluent endothelial cell monolayers of small-caliber vascular prostheses in vitro. Surgery 103:456.

United States Renal Data System 1994 annual data report. Prevalence and cost of ESRD therapy. American Journal of Kidney Diseases 18(5)(supp 2):21.

Wilson JM, Birinyl LK, Salomon RN. 1989. Implantation of vascular grafts lined with genetically modified endothelial cells. Science 244(4910):1344–1346.

Woods JD, Humes HD. 1997. Prospects for a bioartificial kidney. Seminars in Nephrology 17:1–6.

Zwiebel JA, Freeman SM. Kantoff PW. 1989. High-level recombinant gene expression rabbit endothelial cells transduced by retroviral vectors. Science 243(4888):220–222.

23
Encapsulation of Mammalian Embryos

Glen K. Adaniya and Richard G. Rawlins

Introduction

Encapsulation of mammalian preimplantation embryos has its origins in the work of Willadsen in the
late 1970s and in the need to develop an artificial
zona pellucida for denuded sheep, horse, and cattle
embryos used in the production of monozygotic
twins for the livestock industry. Although functional,
these early methods relied on embedding in nonbiodegradable materials, and ultimately, they saw
little widespread application. More recently the idea
of embryo encapsulation has been revisited as a possible solution to technical problems encountered
with human in vitro fertilization and embryo transfer, in hopes of improving embryonic implantation
rates. The studies reviewed here represent the current status of mammalian embryonic encapsulation
and the preliminary work done using animal models prior to attempting encapsulation for human embryos for infertility therapy.

Literature Review

The first published report of embryo encapsulation
occurred in 1979, when Willadsen was experiment-
ing with the micromanipulation of sheep, pig, and
cow embryos (Willadsen 1979). In these experiments, zonae pellucidae of fertilized embryos were
removed and the blastomeres separated. The separated blastomeres were placed into empty zonae pellucidae, and the perivitelline space filled with sheep
serum. Since the recipient zona pellucida had a hole
in it, the embryo was embedded in a small cylinder
of agar to protect it from leukocytic attack. These

small cylinders were then placed into a larger agar
plug and placed into the ligated oviduct of a ewe.
When the embryos were thought to be at the early
blastocyst stage, the cylinders were removed and the
embryos freed from the agar. The embryos were assessed for degree of cleavage and were then placed
into host animals. These studies showed that as long
as the agar coating remained intact, the embryos developed at a normal rate. The agar coating was
porous enough to allow nutrients to reach the embryos, but at the same time prevented any foreign
cells from harming the embryo.

A very different approach to "encapsulating" the
embryo was attempted by Cohan et al (1982). Since
in vitro embryonic development was found to be
slower than in vivo growth, an egg-embryo chamber was developed for the intra-abdominal culture
of monkey embryos. The cylindrical chamber was
constructed of Kel-F (trifluorochloroethylene) and
measured 7 millimeters in diameter. Two semipermeable membranes with 0.2 micrometer pore diameters were placed at the ends of the chamber to
allow passive diffusion of materials to and from the
peritoneal cavity. Eggs fertilized in vivo were removed by laparotomy, placed into the chamber, then
attached to the anterior or posterior surface of the
uterus. Another laparotomy was done after 24, 48,
or 72 hour intervals to retrieve the chamber. The
eggs were removed from the chamber and the degree of cleavage determined. This study showed that
a normal cleavage rate could be attained using the
egg-embryo chamber. The drawback in using this
nonbiodegradable chamber was that it required a
second laparotomy to remove it.

A similar method of culturing embryos in vivo
in an artificial device was tried in 1988 (Pollard and

Pineda 1988). Either whole rabbit embryos or separated blastomeres from four-cell embryos were placed into 1 cm chambers made from polymerized 2-hydroxyethyl methacrylate (pHema), which is a biocompatible and permeable hydrogel. The chambers were placed into the peritoneal cavities of mice and recovered 72 hours later. Normal cleavage rates were found for both the separated blastomeres and the intact embryos. Transfer of the intact embryos to recipient rabbits yielded a 3% pregnancy rate. The problem with this method, as with the agar plugs and the egg-embryo chamber, was that a second operation was required to remove the artificial devices.

A new microencapsulation process was developed by Lim and Sun in 1980 (Lim and Sun 1980). Islets of Langerhans, the insulin-producing cells of the pancreas, were suspended in a sodium alginate solution. Droplets of this solution were then released into a calcium chloride solution, causing them to gel. They were coated by placing them in a poly-1-lysine solution, followed by immersion in a polyethyleneimine solution. The microencapsulated islets were then cultured for a few days and injected via a hypodermic needle I.P. into diabetic rats. Normal glucose concentrations were maintained for almost 3 weeks in the experimental animals. In contrast, animals that received injections of unencapsulated islets returned to a diabetic state after only 10 days.

Further studies by Sun et al (1984) showed that the encapsulated islets were rejected as a result of biological incompatibility of the polyethyleneimine outer membrane. The polyethyleneimine coating was then replaced with an alginate membrane. Glucose concentrations of diabetic rats were kept normal for 650 days with just a single intraperitoneal injection of encapsulated islets with the new poly-1-lysine alginate capsule membrane (Sun et al 1984).

Since the above method of encapsulation resulted in a capsule that was biodegradable, allowed nutrients to reach the cells, and allowed the cell's secretions to leave, the present authors selected this method to attempt encapsulation of fertilized embryos. In 1987, Adaniya et al modified the encapsulation process using 1.1% sodium alginate for embryos. They demonstrated that the encapsulation process was not detrimental to the in vitro cleavage rate of Swiss Webster White or CB6F1 2-cell mouse embryos to the blastocyst stage (Adaniya et al 1987).

In 1990, Cosby and Dukelow encapsulated single and multiple mouse embryos in sodium alginate and found that encapsulating up to three embryos in a single capsule had no detrimental effect on the in vitro developmental rate of the embryos. They also encapsulated zona-pellucida-free embryos and found that these embryos developed at a rate comparable to unencapsulated zona-pellucida-free embryos, though at a lower rate than intact embryos (Cosby and Dukelow 1990).

In contrast to the previous study, Eaton et al showed in 1990 that encapsulation of zona-pellucida-free embryos in sodium alginate did significantly increase the number of embryos that reached the blastocyst stage, as compared to unencapsulated zona-pellucida-free embryos. They postulated that the sodium alginate provided a structural support that allowed normal cell contact during early cell division, thus increasing the percentage of embryos that developed to the blastocyst stage (Eaton et al 1990).

The first pregnancies obtained following the transfer of sodium alginate encapsulated embryos were reported by Meredith and coworkers in 1990 (Meredith et al 1990). Four- and eight-cell sheep embryos were encapsulated and transferred into the fallopian tubes of recipient sheep. Pregnancy rates were 71% for the encapsulated embryos versus 57% for the control embryos.

In 1991, Adaniya et al encapsulated eight-cell CB6F1 mouse embryos in 3% sodium alginate and transferred them to recipient mice (Adaniya et al 1991). Six of the twenty recipient mice became pregnant, with a resultant 30% pregnancy rate.

Hall and Yee (1991) performed similar experiments, transferring control and encapsulated blastocysts to recipient mice. These studies yielded a 4% implantation rate for the encapsulated embryos versus a 60% rate for the control embryos. The authors speculated that the alginate did not dissolve in time to allow the embryo to implant. Krentz and coworkers (1993) confirmed these observations when they attempted to transfer encapsulated morulae to recipient mice. No term births were seen, although later experiments showed a 17% implantation rate.

Methods

Sodium alginate is a polysaccharide derived from brown seaweeds, composed of three linked polyuronic acid blocks. The saccharide chains contain free acid groups in the sodium ion form, and upon exposure to the divalent calcium ion (or any other multivalent ion), two sodium ions are replaced by a single calcium ion bridge, resulting in ionic cross-linking of the chains. The sodium alginate also contains easily ionized carboxyl groups that react by salt formation with the polycationic polylysine to form a polyelectrolyte complex hydrogel membrane.

The original microencapsulation process described by Lim and Sun involved the suspension of the cells to be encapsulated in a 1.1% sodium alginate solution. The suspension was placed into a plastic syringe and attached to a syringe pump. The contents of the syringe were then expelled into a 1.5% calcium chloride solution, resulting in the formation of gel droplets. These droplets were placed into 1% polylysine solution, and then into a sodium citrate solution to liquefy the interior gel. This resulted in a hollow capsule containing the cells.

These microcapsules were designed to last for an extended period of time, which was not the goal for encapsulated embryos. Ideally, the capsule surrounding the embryo should dissolve by the time the embryo has reached the blastocyst stage, the stage when the embryo is able to implant in the uterine endometrium. In addition, the time the embryo is kept out of culture medium must be kept to a minimum to ensure embryo viability. Therefore, the polylysine and sodium citrate steps in the microencapsulation process were eliminated. Another consideration was the quantity of material to be encapsulated. Since embryos needed to be encapsulated one at a time, suspending the embryo in sodium alginate and expelling it from a syringe was not feasible. As a result, the embryo encapsulation technique was modified as follows:

1. Prepare a 1.1% sodium alginate solution with 0.9% saline.
2. Make a 1.5% calcium chloride solution with 0.9% saline.
3. Aspirate 0.3 μl of sodium alginate into a glass micropipette (bore size 150 μm I.D.). Aspirate one embryo, followed by 0.3 μl sodium alginate.
4. Hold the micropipette upright over a beaker containing the 1.5% calcium chloride solution, and expel the contents.
5. Let the encapsulated embryo sit for approximately 30 seconds, then transfer back to culture medium.

Effect of Encapsulation on the Development of Mouse Embryos

Ten-week-old female CB6F1 mice (housed in a 12 hour light/12 hour dark cycle) were superovulated by injection with 10 I.U. of pregnant mares serum gonadotropin I.P., followed by 10 I.U. of human chorionic gonadotropin (HCG) I.P. 48 hours later. Immediately following HCG injection, the females were mated with Swiss Webster White (SWW) male proven breeder mice by placing one female mouse per male mouse per cage. The following morning the female mice were checked for the presence of a vaginal plug to determine if mating had occurred. Forty-eight hours after HCG injection, the CB6F1 mice were sacrificed by cervical dislocation and the abdomens exposed. The oviducts were removed and placed into a tissue culture dish containing culture medium buffered with 25 mM of HEPES.

A blunted 30-gauge needle was attached to a 1.0 ml syringe, and loaded with 0.4 ml of medium. The oviducts were flushed by inserting the needle into the tubal ostium and pushing the medium through the oviducts. The flushed two-cell embryos were recovered by aspiration with a 5 μl SMI micropipetter, pooled in 1.0 ml of nonbuffered medium in a Falcon 3037 culture dish, and placed into a VWR 1810 incubator at 37° C with 5% CO_2 + air. The two-cell embryos were randomly separated into control and encapsulated groups when all the embryos were collected.

The embryos were encapsulated using the technique described previously. The encapsulated embryos were then transferred to a Falcon 3001 petri dish containing 3.0 ml of nonbuffered medium and placed into the VWR 1810 incubator at 37°C with 5% CO_2 + air. At 24, 48, and 72 hours after encapsulation, the control and encapsulated embryos were removed from the incubator and placed onto a Nikon SM 22 stereo dissecting microscope (90×). The embryos were quickly determined to be either at the two-cell stage, the four-cell stage, the eight-cell stage, the morula stage, or the blastocyst stage. The embryos were then returned to the incubator for further culture.

Effect of Encapsulation on Mouse Implantation Rates

Female CB6F1 mice (10 weeks old) were superovulated as described previously. Immediately following HCG injection, donor mice were mated with Swiss Webster White proven breeder male mice, and recipient mice were mated with vasectomized Swiss Webster White mice, one female per male per cage. Seventy-two hours later, the donor CB6F1 mice were sacrificed by cervical dislocation, and the abdomens exposed. The oviducts were then removed and placed onto a tissue culture dish containing a drop of medium buffered with 25 mM of HEPES. Eight-cell embryos were collected and encapsulated as described previously. The encapsulated embryos were transferred into culture medium and placed into an incubator until the surgery was ready to be performed.

The recipient mouse was anesthetized and the back shaved with a pair of surgical scissors. The mouse was restrained on the surgical table, and the back wiped with 70% alcohol. A single midline incision in the dorsal skin was made beginning at the level of the lowest rib and extending caudally for approximately 3 centimeters. The opening was retracted to the left side, the ovary located, and a small incision made in the abdominal wall. The uterine horn was exteriorized, and the skin flap placed over the horn prior to the transfer process.

The encapsulated embryos were aspirated into a glass micropipette using a mouth pipetter. The skin flap was uncovered, the uterine horn stretched with forceps, and a small puncture made in the cranial end of the horn with a 25-gauge needle. Once the needle penetrated the lumen of the horn, it was withdrawn and the micropipette inserted into the puncture site. The encapsulated embryos were flushed into the lumen and the micropipette removed and checked under a dissecting microscope to make sure the embryos were transferred. The uterine horn was then replaced in the abdominal cavity and the incision sutured with Dexon 4-0. The entire process was repeated with the right uterine horn, except that five unencapsulated control embryos were transferred. The skin incision was closed with wound clips, and the animal was placed into a cage to recover.

The mice were sacrificed by cervical dislocation 10 days following the embryo transfer (normal gestation for mice is between 19 and 21 days). The abdomens were exposed and the uterine horns removed and examined for the presence of implantation sites.

Effect of Encapsulation on Fetal Development

The superovulation of CB6F1 mice, and the recovery and encapsulation of eight-cell embryos were accomplished as previously described. The transfer procedure was also the same, except that encapsulated embryos were transferred to both the right and left uterine horns. The recipient mice were allowed to go to term, and the newborn mice examined for physical abnormalities. The newborn mice were housed until they reached reproductive maturity, and then allowed to mate. The pregnant females were allowed to go to term, and the newborn mice again examined for any physical abnormalities.

Results

Effect of Encapsulation on Development of Murine Embryos

The cleavage stages of the control and encapsulated embryos were compared after 72 hours of incubation in vitro. As shown in Table 23.1, 1.8% of the encapsulated embryos remained at the two-cell stage, 1.8% were four-cell embryos, 5.4% were eight-cell embryos, 19.8% were morulae, and 71.2% were blastocysts (n = 111). For the control embryos, 2.2% were two-cell embryos, 5.1% were four-cell embryos, 5.9% were eight-cell embryos, 11.8% were morulae, and 75.0% were blastocysts (n = 136). There was no significant difference in the intermediate cleavage stage reached at 72 hours ($\chi^2 = 4.65$, df = 4, p > 0.05, chi square statistics).

Effect of Encapsulation on Mouse Implantation Rates

For these experiments, control embryos were transferred to one uterine horn and encapsulated embryos to the other uterine horn of the same mouse. This allowed each mouse to act as its own control. A total of 178 control embryos were transferred to 41 CB6F1 mice, with 16 of the embryos resulting in implantation sites (see Table 23.2). For the encapsulated embryos,

TABLE 23.1. Effect of sodium alginate encapsulation on CB6F1 embryo cleavage at 72 hours.

Cleavage stage	Control embryos	Encapsulated embryos
2-cell	3 (2.2%)	2 (1.8%)
4-cell	7 (5.1%)	2 (1.8%)
8-cell	8 (5.9%)	6 (5.4%)
Morula	16 (11.8%)	22 (19.8%)
Blastocyst	102 (75.0%)	79 (71.2%)
Total	136	111

TABLE 23.2. Implantations from encapsulated and control embryos

Embryo type	Embryos transferred	# Implanted	n
Control	178	16 (9.0%)	41
Encapsulated	157	15 (9.6%)	41

157 were transferred to 41 mice and led to 15 implantations. There was no significant difference in the number of implantation sites between the control and the encapsulated groups ($\chi^2 = 0.0001$, df = 1, p > 0.05). Examining only those animals in which implantations occurred, 16 of 47 transferred control embryos (34%) and 15 of 47 (32%) transferred encapsulated embryos implanted. There was no significant difference ($\chi^2 = 0.0$, df = 1, p > 0.05) between the implantation rates for the control and encapsulated groups in animals for which implantation sites were observed.

Effect of Encapsulation on Fetal Development

To determine if the encapsulation process had a detrimental effect on fetal development, 139 encapsulated embryos were transferred to the uterine horns of recipient CB6F1 mice (n = 20) and the mice allowed to go to term. Six of the mice became pregnant (30%), with a resultant 12 live births (implantation rate of 8.6%). Examining only those animals where live births occurred, 30.8% of the encapsulated embryos resulted in offspring (12 of 39 transferred embryos). Upon gross examination, all of the offspring appeared normal, with no obvious signs of physical or anatomical abnormalities.

To determine if the encapsulation process had a detrimental effect on reproductive function, all 12 offspring were allowed to mate with mice not derived from encapsulated embryos, once reproductive maturity had been attained. All the offspring proved fertile, with a total of 75 second-generation offspring born. All 75 pups appeared to be normal, with no apparent signs of anatomical abnormalities upon gross examination (Adaniya et al 1993).

Discussion

The successful production of monozygotic twins and quadruplets in domestic animals by Willadsen through use of embryo bisection and encapsulation in agar pellets represents a major breakthrough in preimplantation embryology. Prior to Willadsen's work, only intact embryos had survived embryo transfer in these species. Although denuded chimeric embryos had been successfully transferred in mice in the 1960s, no large animal species had been produced from embryo transfer of zona-pellucida-free embryos. The use of agar to maintain the mechanical integrity of the bisected and denuded embryos by Willadsen was a novel and simple approach to providing the mechanical and immunological protection needed by the early embryo to complete implantation in utero and begin gestation. The idea of inducing monozygotic twinning was very exciting from an embryologist's point of view, but found little commercial application, despite its successful introduction in many large commercial breeding operations in the United States. For all intents and purposes, the technique was abandoned. At the time when the work by Adaniya (1987) was begun, human in vitro fertilization in the United States had become well established as a therapeutic procedure for the treatment of infertility, but the pregnancy rates obtained for patients remained relatively low, around 11% per cycle. A number of different ideas were postulated as to why the implantation rates fell far below the 50–75% per transfer seen in the cattle industry for embryo replacement in utero.

It was suggested that the cotyledonary implantation of the cow was substantially different from the invasive hemochorial implantation of the human being, and that the timing of implantation between the species was significantly different as well. Perhaps the in vitro fertilization and embryo transfer procedure in the human altered the dialog between the embryo and the uterine environment in an unknown way that impaired implantation. Perhaps the so-called "window of implantation" was closed prematurely under these artificial conditions.

The in vitro process itself was also thought to compromise the ability of the embryo to escape the zona pellucida and implant. Cleavage rates for embryos produced in vitro were significantly slower than those obtained in vivo. Also, several investigators expressed concern over unusual hardening of the zona pellucida, perhaps as a result of exposure to free oxygen radicals liberated by the demise of spermatozoa and subsequent membrane breakdown during in vitro fertilization. The latter point was of interest since it was well known that implantation rates were high for embryos of other species, such as the cow, sheep, horse, rabbit, pig, and mouse, that had the zonae pellucidae chemically or mechanically removed from the preimplantation embryo.

Finally, the mechanics of embryo transfer in the human were considered to be the problem. Collecting embryos in a microdrop of culture medium and discharging them by inserting a catheter transcervically into an unseen location in the upper third of the uterus could be at fault. The free-floating embryos might land in an inappropriate uterine location for implantation to occur (e.g., the lower third), or they might simply be expelled altogether from the uterus in the fluids, across the cervical os. If the embryos for transfer could be aggregated and confined to a single desired location within the uterus by somehow weighting or anchoring them in place, perhaps the rate of implantation could be increased.

Thus, the idea of encapsulating human embryos was born. A soft capsule would provide protection for the embryos during and after transfer, and its weight would allow the embryos to quickly settle onto the uterine wall. Its size would prevent flotation and descent to the internal os of the cervix since the capsule itself would be confined by the collapsed uterine walls in a potential anatomical space. The methods were first tested on animal models.

Prior work on embryo encapsulation had utilized either agar embedding (nondegradable) or membrane chambers, which were inappropriate for human use. However, the early work of Lim and Sun (1980) had showed that differentiated somatic cells could be successfully embedded in biodegradable sodium alginate, and this seemed to be the ideal medium to test for encapsulating preimplantation embryos.

The alginate matrix would provide for mechanical protection of the early embryo. It also would function during embryo reemplacement in utero to immediately settle the embryo onto the uterine wall. The porous matrix could protect the embryo in the uterine environment from leukocytic attack, while simultaneously permitting nutrient and waste exchange for the 3 days needed from the time of embryo transfer to implantation in the uterine wall. At that time point, the device would have to biodegrade for implantation to occur.

A serious concern in using the alginate was the need to gel the alginate solution containing embryos by adding it to a solution of calcium chloride. Calcium is a potent regulator of cytokinesis in early em-

bryonic development, and any chelation of calcium or other manipulation of calcium concentrations has the potential to completely arrest cell division and eliminate intercellular adhesion. It was critical to demonstrate that immersion in such a solution would not be detrimental to subsequent cleavage divisions or embryonic viability.

In retrospect, all of the studies reviewed here have shown that encapsulation is technically easy to perform in the laboratory and that the procedure has no negative impact on the embryos. The ability to control the rate of sodium alginate degradation has also proven important, since capsules can be designed to accommodate the temporal requirements of implantation in different animal species. Also, the biodegradable alginate eliminates the old requirement to dissect developing embryos out of agar gels. Obviously, it is impossible to return to the human uterus to dissect out the embryos at the point of implantation. Finally, the method has resulted in implantation rates at least comparable to those obtained without manipulation. All of the existing studies of sodium alginate encapsulation have shown that it is a safe procedure for the early embryo. Although there are significant differences in implantation rates, viability has been maintained, and in several studies, implantation has improved compared to control, nonencapsulated embryos.

Whether encapsulation will be used for human in vitro fertilization cases remains to be seen. The problem of excessive hardening of the zona pellucida in vitro has largely been overcome by altering the culture conditions under which preimplantation embryos are maintained in vitro. For extreme cases, implantation rates have been improved by mechanically "hatching" the embryo, using micromanipulation to create an opening in the zona pellucida with a microspray of acidified salt solution. The problem of the technical aspects of human embryo transfer remain, but no published report of embryo reemplacement in utero using encapsulation has appeared since the methods have been worked out. If there is a future for encapsulation in human assisted reproduction, it probably lies in the area of micromanipulation. New techniques for manipulation of gametes and early embryos, such as intracytoplasmic sperm injection, cytoplasmic transfusion from the oocytes of young women to those of older women, nuclear transfer, and cloning, all may require use of an artificial zona pellucida. Sodium alginate encapsulation seems the logical choice.

References

Adaniya GK, Rawlins RG, Miller IF, Zaneveld LJD. 1987. Effect of sodium alginate encapsulation on the development of preimplantation mouse embryos. J In Vitro Fertil Embryo Trans 4:343–345.

Adaniya GK, Rawlins RG, Quigg JM, Roblero L, Miller IF, Zaneveld LJD. 1991. Transfer of sodium alginate encapsulated embryos in mouse results in successful gestations and live births. Suppl of 47th Annual Meeting of the American Fertility Society.

Adaniya GK, Rawlins RG, Quigg JM, Roblero L, Miller IF, Zaneveld LJD. 1993. First pregnancies and live-births from transfer of sodium alginate encapsulated embryos in a rodent model. Fertil Steril 59(3):652–656.

Cohan BD, Harman SM, Hodgen GD. 1982. The egg-embryo chamber for intraabdominal culture. Fertil Steril 38(5):616–620.

Cosby NC, Dukelow WR. 1990. Microencapsulation of single, multiple, and zona pellucida-free mouse preimplantation embryos in sodium alginate and their development in vitro. J Reprod Fert 90:19–24.

Eaton NL, Niemeyer GP, Doody MC. 1990. The use of an alginic acid matrix to support in vitro development of isolated murine blastomeres. J In Vitro Fertil Embryo Trans 7:28–32.

Hall JL, Yee S. 1991. Implantation of zona-free mouse embryos encased in an artificial zona pellucida. Suppl of 47th Annual Meeting of the American Fertility Society.

Krentz KJ, Nebel RS, Canseco RS, McGilliard JL. 1993. In vitro and in vivo development of mouse morulae encapsulated in 2% sodium alginate or 0.1% poly-L-lysine. Theriogenology 39:655–667.

Lim F, Sun AM. 1980. Microencapsulated islets as bioartificial endocrine pancreas. Science 210:908–910.

Meredith S, Nicks DK, Kiesling DO. 1990. The effects of time-of-exposure of alginate-coated rat embryos to calcium chloride on in vitro development; and pregnancy maintenance in ewes with alginate-coated 4- or 8-cell embryos. Theriogenology 33:288.

Pollard JW, Pineda MH. 1988. Culture of rabbit embryos and isolated blastomeres in hydrogel chambers implanted in the peritoneal cavity of intermediate mouse receipients. J In Vitro Fert Embryo Trans 5(4):207–215.

Sun AM, O'Shea GM, Goosen MFA, Gharapetian H. 1984. Microencapsulation of living cells as long-term hormone and enzyme delivery systems. In: Proceedings 11th International Symposium on Controlled Release of Bioactive Materials, eds. WE Meyers, RL Dunn, eds. Controlled Release Society, Inc. pp. 55–56.

Willadsen SM. 1979. A method for culture of micromanipulated sheep embryos and its use to produce monozygotic twins. Nature (London) 277:298–300.

24
Encapsulated Plant Cells: Techniques and Applications

Wei Wen Su

Introduction

Since the early work of Brodelius (Brodelius et al 1979), immobilization techniques have been applied in many plant systems for applications ranging from secondary metabolite synthesis to synthetic seed production. Among the immobilization methods applicable to plant cells, the most common, and usually also the most effective, methods involve the entrapment or encapsulation of cells within a gel, in a solid support, or both. The scope of this chapter is thus to provide an up-to-date literature survey and to review recent advances in the techniques and applications associated with encapsulation of plant cell systems, including protoplasts, tissues, and organs, in addition to undifferentiated callus cells. Indeed, many applications other than secondary metabolite production have emerged over the years, for instance the use of encapsulation techniques for germplasm cryopreservation, synthetic seed production, and improved protoplast culture for mutant selection and transformation. In the following sections, primary applications of encapsulated plant cells will be reviewed, followed by discussions on the associated experimental protocols. Finally, future prospects of plant cell encapsulation are discussed. Since the focus of this chapter is on recent advances, for earlier work published prior to 1992, the reader is referred to the reviews by Endress (1994), Hall et al (1988), Hulst and Tramper (1989), Payne et al (1991), Scragg (1991), and Williams and Mavituna (1992), for more details.

Applications

Background

Applications of encapsulation techniques in plant cell systems have been developed in five major areas: (1) secondary metabolite and secreted protein production, (2) biotransformation, (3) protoplast cultivation, (4) cell-line storage and cryopreservation, and (5) artificial seeds/micropropagation. Other applications include pollution control (Schnabl et al 1983a), cellular and physiological studies (David et al 1995), and production of fine cell suspensions (Morris et al 1983). The benefits of encapsulation vary with the applications. For instance, cell leakage from the encapsulation matrix, which is regarded as a problem in large-scale culture for metabolite production, can be turned to an advantage in the production of single-cell suspension. However, the following features are generally considered as the benefits of cell encapsulation:

1. The encapsulation matrix can accommodate a higher cell density than is achievable in suspension cultures. Bead densities as high as 110 g dry weight/L have been reported with alginate-entrapped cells (Scragg 1991), although extremely high cell loading may not be desirable because of severe diffusion limitation.
2. Encapsulated plant cells can be reused and employed in long-term, continuous-flow processes with biomass retention. Also from a process point of view, use of encapsulated cells can alleviate cell aggregation, cell flotation, and reactor wall growth.

3. The confined growth environment can restrict fast cell division that may not be desirable, e.g. in production of most secondary metabolites. Additionally, the microenvironment surrounding the cells present in the encapsulation matrix can be manipulated.

4. The encapsulated cultures may be less susceptible to bacterial contamination than suspension cultures during cultivation, especially when antibiotics can be coencapsulated with the cells.

5. Encapsulated plant materials can be protected from chemical and physical degradation (e.g., shear, osmotic stress, oxidation, and dehydration), resulting in prolonged cell stability and shelf life.

The major limitations of encapsulation include

1. Requirement of cellular transport mechanisms or artificially altered membrane permeability for product secretion.

2. Possible alteration of cell metabolism, e.g., reduced phosphate uptake rate and lower phosphate accumulation in the vacuoles (Watts and Collin 1985).

3. Accumulation of secreted products in the encapsulation matrix prevents product release into the liquid medium.

4. The efficiency of the production process depends on the rate of release of the products rather than on the actual biosynthesis.

5. Risk of contamination during encapsulation, especially with large-scale operations.

6. It is more suitable for non-growth-associated product formation.

7. Difficulty in controlling outgrowth of plant cells from the encapsulation matrix, which may lead to breakage of the capsule. Also, cell leakage may complicate product separation. To alleviate this problem, substrate-limited continuous cultures and/or use of growth inhibitors such as 3,5-dichloro-phenoxyacetic acid (Watts and Collin 1985) can be employed.

Recent advances in the applications of encapsulated plant cells are discussed below.

Secondary Metabolite and Secreted Protein Production

Cell encapsulation has been employed as a process strategy to improve secondary metabolite and se-

creted protein production. In addition to its generic advantages, cell encapsulation has been suggested to enhance cellular differentiation through cell-cell interaction and the existence of a chemical concentration gradient in the encapsulation matrix (Payne et al 1991). In recent years, there has been a continual interest in using the encapsulation techniques to improve secondary metabolite and secreted protein production.

The simplest and most natural system of encapsulated cells is cell aggregates. The tendency of plant cells to aggregate is due to secretion of a slime layer rich in carbohydrates, proteins, and pectins (arabinogalactan, xyloglucan) (Endress 1994). Cell aggregation was found to be more serious in some cultures than others. In certain cultures, large cell aggregates were shown to produce more secondary metabolites (Hulst et al 1989a). This has been linked to plausible cellular differentiation occurring inside the cell aggregates, resulting from concentration gradients. However, not all cell aggregates display morphological differentiation. Typically, compact, nodule type aggregates are believed to possess a higher potential to undergo somatic embryogenesis or organogenesis (Bieniek et al 1995). From an engineering point of view, it is important to analyze the critical aggregate size, above which oxygen limitation may result. Ananta et al (1995) applied the diffusion-reaction analysis for heterogeneous catalysis in cell aggregates of *Solanum aviculare,* and found that a much larger critical aggregate size exists than that predicted by the heterogeneous reaction theory. They speculated that the *S. Aviculare* aggregates may possess plasmodesmata, which allow intercellular mass transfer. In addition, Ananta et al (1995) suggested that, depending on the respiratory quotient of the cells, the differential fluxes and solubilities of gases such as oxygen and carbon dioxide involved in respiration may create significant negative pressures in the gas spaces of plant tissue, thus driving further nutrient transport. The existence of gas-filled cavities in the larger *S. Aviculare* aggregates allows for pressure gradients to play a role in promoting oxygen transfer.

For entrapment in polymeric matrices, alginate gel systems remained the most common technique for plant cell encapsulation, mainly because of its mild polymerization conditions. Alginic acid is a polyuronic acid extracted from seaweeds and is composed of varying proportions of 1–4 linked β-D-

TABLE 24.1. Factors influencing the stability of alginate beads (modified from Endress 1994).

Chelating ion (Ca^{2+} being most common)
Alginate concentration
Metal ion chelators (phosphate, citrate, lactate)
Alginate composition (α-L-guluronic acid content $> 70\%$)
Molecular weight of the alginate polymers
Duration of incubation in $CaCl_2$ solution
Cell loading

mannuronic (M) and α-L-guluronic acids (G). The residues are arranged in block patterns that comprise homopolymeric regions interspersed with alternating heteropolymeric regions. The GG blocks have preferential binding sites for divalent cations, such as Ca^{2+}, and the bound ions can interact with other GG blocks to form linkages that lead to gel formation. A more detailed account of the principles of gel formation of calcium alginate can be found in Chapter 7 of this book. The major factors that affect the stability of the alginate gel are summarized in Table 24.1. Destabilization of calcium alginate is promoted by chelating agents that remove calcium ions, such as phosphate, citrate, and EDTA. Since phosphate ions are frequently used in culture media, this may pose a problem for the stability of the Ca-alginate gel beads. Moreover, the chelation of phosphates with Ca^{2+} ions makes them unusable in the metabolic reactions in the cells (Hulst and Tramper 1989b). In order to alleviate this type of gel disruption, a sufficient Ca^{2+} concentration should be used in the medium. One has to be cautious in using Ca^{2+}, however, because it is known to be an important trigger for many key metabolic processes.

A literature search has revealed that at least four studies published since 1992 used Ca-alginate for plant cell immobilization. Ramakrishna et al (1993) reported encapsulation of *Hyoscyamus muticus* for the production of solavetivone. Upon elicitation using a fungal elicitor, 53% more solavetivone was produced in immobilized culture than in suspension cultures. It was postulated that calcium (from the Ca-alginate gel) induced membrane-fluidization-facilitated oxidation of membrane fatty acids (by lipoxygenase induced via the fungal elicitation), which is believed to be involved in the transduction of signals for secondary metabolism. Suvarnalatha et al (1993) entrapped *Capsicum frutescens* in cal-

cium alginate beads for the production of capsaicinoids. The physical strength of the beads did not interfere with capsaicinoid synthesis; however, the leaching/permeation of the synthesized capsaicinoids to the medium was controlled by it. If the physical strength of the beads is high, the capsaicinoids synthesized will be retained in the beads, which in turn may retard their further production. Indeed, more capsaicinoids were found to accumulate in the beads than in the medium. Therefore, there is a compromise between gel strength and product release. Calcium alginate was also employed in the encapsulation of *Frangula alnus* (Saje et al 1995) and *Taxus cuspidata* (Seki et al 1997) for the production of anthraquinones and Taxol, respectively. While conventional alginate gel encapsulation was shown to work well in the above studies, Tanaka et al (1996) reported cell leakage in a single-layered alginate gel fiber system, and the leaked cells grew and formed cell aggregates in the broth. To overcome this problem, Tanaka et al (1996) developed a double-layered calcium alginate gel fiber system and used it for the encapsulation of *Wasabia japonica* cells that produce and secrete a large amount of chitinase into the medium. This method, originally developed for encapsulation of microbial cultures (Tanaka et al 1989), was able to prevent cell leakage from the gel matrix. The outer gel layer in the double-layered system, however, does create an additional diffusion barrier. Therefore, higher aeration or oxygen enrichment was needed to achieve a high chitinase production.

Other recently reported gel-entrapment systems for plant cells include pectin/chitosan coacervate capsules (Dörnenburg and Knorr 1996) and temperature-sensitive hydrogel (Han et al 1996). The pectin/chitosan coacervate capsules consist of a pectin core bounded by a chitosan membrane. This was modified from an earlier version consisting of a chitosan core and a Ca-alginate membrane (Daly and Knorr 1988). The high chitosan concentration (3%) used in the capsule system proposed by Daly and Knorr (1988) was found to be too toxic to plant cells (Dörnenburg and Knorr 1996). When encapsulated in the pectin/chitosan coacervate capsules, *Cruciata glabra* cells were able to maintain their viability with improved anthraquinone production for 4 weeks in a semicontinuous process employing *in situ* extraction with Miglyol (Dörnenburg and Knorr 1996). Some cell outgrowth, however, was noted with this system. In the case of temperature-sensitive hydrogel (poly(N-isopropylacrylamide)), the polymer can expand and swell when cooled below its "lower critical solution temperature" (LCST), and shrink, even collapse, when heated above the LCST. Ideally, the hydrogel-immobilized cell systems can be made to swell and shrink by manipulating the operating temperature, which in turn can enhance mass transfer of substrates and secreted products. Han et al (1996) studied immobilization of *Coleus blumei* in hydrogel beads containing 0.5% alginate (which is needed to protect plant cells from the toxicity of the chemicals used for polymer preparation), and found that cell viability could be maintained over 3 weeks. Also, the hydrogel beads with immobilized plant cells remained temperature-sensitive in aqueous media. However, no data were available to compare the performance of this system with that of other gel encapsulation systems. For more detailed accounts on the use of hydrogel or thermoresponsive polymers in enzyme and cell immobilization, the reader is referred to a recent review by Hoshino et al (1997).

Use of synthetic polymer such as polyacrylamide for plant cell entrapment has not been very successful, primarily because of the drastic conditions required for polymerization. Synthetic resins, however, do possess certain advantages over natural polymers. The physicochemical properties of encapsulation gels, such as network structure, hydrophilicity/hydrophobicity balance, and ionic property, can be optimized in the synthetic resins. Tanaka (1994) reviewed the encapsulation of plant cells (*Lavandula vera*) using a photosensitive resin prepolymer (PVA-SbQ), a derivative of poly(vinyl alcohol). *L. vera* cells entrapped in PVA-SbQ were able to produce and secrete a large amount of a blue pigment; most of the pigment, however, was retained in the gel, in contrast to the case of the calcium-alginate-entrapped cells. It was speculated that this may be due to the difference in hydrobicity/hydrophilicity balance between these gels since alginate gel has a very hydrophilic nature. When PVA-SbQ prepolymers with a high degree of saponification were used, the pigment excretion was improved.

Enhanced production of gossypol, an antifungal agent, from *Gossypium arboreum* was reported by Choi et al (1995) when the plant cells were entrapped in a cotton matrix with a spirally wound con-

figuration. Although encapsulation alone displayed less of an effect on the secondary metabolite production compared with elicitation, encapsulation apparently protected the cells, and as a result, elicitation and permeabilization were more effective in immobilized culture than in suspension cultures. In this system, the problem of the limited spontaneous adhesion of plant cells onto the cotton cloth due to its low surface tension was avoided by using the spirally wound configuration.

A fibrous network of *Luffa cylindrica* was first proposed by Iqbal and Zafar (1993) for immobilization of *Saccharum officinarum*. This simple encapsulation material was later adopted by Martinez and Park (1994) for immobilization of *Coleus blumei*. The immobilized *C. blumei* cells were successfully cultured in a spray column reactor for 52 days. Detailed characterization of this natural matrix for plant cell encapsulation, such as cell-matrix interaction, however, is still not available. Polyurethane foam, which was first used by Lindsey et al (1983) for plant cell immobilization has been applied recently in entrapping *Vitis vinifera* cells (Iborra et al 1994) and was found to enhance the accumulation of anthocyanins, although the pigment was not secreted in this system.

Biotransformation

Plant cells are capable of performing a broad range of bioconversions, including hydroxylations, methylations, and glycosylations. Encapsulation has potential advantages in protecting cells against the toxicity of the substrate and may exert positive physiological impacts on the enzymatic machinery. The encapsulation method used must allow efficient transport of the substrate into, and product out of, the capsules. However, in some reported studies, products of biotransformation were accumulated intracellularly (Pras et al 1995). In addition, it should be noted that changes in (regio)selectivity of the biotransformation may occur upon encapsulation (Vanek et al 1989). Table 24.2 summarizes some examples of recent publications dealing with biotransformation using immobilized plant cells. Calcium alginate gelling has been the most common method for encapsulation in plant cell biotransformation, although in the study by Guardiola et al (1996), grape cells entrapped in polyurethane foam particles were found to give a higher conversion of

TABLE 24.2. Some recent examples of biotransformation using encapsulated plant cells.

Plant species	Biotransformation	Reactions	Encapsulation method	Reference
Gardenia jasminoides	acetophenone → (R)-α-phenethyl alcohol (90% enantiomeric purity)	reduction & stereo-selective oxidation	alginate	Akakabe & Naoshima 1993
Papaver somniferum	L-DOPA → dopamine L-tyrosine → tyramine	decarboxylation	glutaraldehyde cross-linking	Stano et al 1995
Capsicum frutescens	ferulic acid & vanillyl-amine to capsaicin & vanillin	oxidation/oxidative deamination	alginate	Sudhakar et al 1996
Catharanthus roseus	benzoylpyridine & α-acetoxybenzyl pyridine to α-phenyl-pyridylmethanols	reduction/asymmetric hydrolysis	alginate	Takemoto et al 1996

geraniol to nerol than the Ca-alginate entrapped cells.

Protoplast Encapsulation

Encapsulation of protoplasts in gel matrices has been shown in several studies to improve cell viability and cell division efficiency. This technique also permits easier handling of cell cultures, avoids protoplast agglutination, and makes single-cell engineering routinely feasible (David et al 1995). Isa et al (1990) found that saffron protoplasts failed to divide when cultured in liquid medium, but division and plant regeneration were achieved when the protoplasts were embedded in Ca-alginate beads. Niedz (1993) also reported improved plating efficiency for Ca-alginate-embedded sweet orange protoplasts, compared with their liquid-cultured counterparts, especially at high cell loading. Schnabl et al (1983b) speculated that the alginate matrix in some ways may stimulate a cell wall, and stabilize the plasmalemma by protecting the cells from small leakage, and facilitate pressure-dependent reactions. They also suggested that the alginate matrix may dilute proteolytic enzymes released from dying cells. Besides alginate, agar and agarose have also been used for encapsulation of protoplasts. However, they have the disadvantage that individual cells or cell colonies can be recovered only at lethal high temperature (Niedz 1993). On the other hand, for alginate gel, individual cells or cell colonies can be easily recovered via nonlethal gel reversibility, by exposing the gel to a calcium sequestrant such as sodium citrate (0.1 M, pH 5.8) (Niedz 1993). In a series of studies

to characterize the effect of external matrix on wall differentiation in encapsulated flax (*Linum usitatissimum*) protoplasts (David et al 1994, 1995; Roger et al 1996), it was found that flax protoplasts immobilized in alginate behave differently than those immobilized in agarose. Cell colonies formed in agarose lacked defined outlines, since the clumps of actively dividing cells were surrounded by cells in an advanced state of lysis. In contrast, in alginate it was possible to observe the presence of compact and spherical colonies with actively dividing cells, as well as a periplasmic space with a very sinuous cell wall (David et al 1994). Because of the presence of negative charges, alginate could affect the secretion of wall polysaccharides, thus modifying the composition and architecture of the newly formed walls. Using a pectin immunoprobe, David et al (1995) showed that alginate encapsulation stimulated pectin metabolism and cell differentiation, whereas agarose did not. The pectin content of the walls of protoplast-derived cells entrapped in alginates is higher than that of agarose. David et al (1995) also speculated that since localized calcium fluxes are implicated in cell polarization, and since calcium is a universal second messenger in signal transduction, the polyelectrolytic nature of the alginate gel may modify the calcium fluxes that participate in differentiation. In a subsequent study, David et al (1996) demonstrated that with Ca-alginate encapsulation, several basic polypeptides were strongly induced and were found tightly bound to the cell wall. They then established a direct correlation of an embryo-like morphogenesis with ionically bound cell wall basic proteins. Given the benefits of Ca-alginate gel for protoplast encapsulation, one should be cautious about the potential problems associated with this method, such as chelation of phosphates with Ca^{2+} and effect of calcium on cell metabolism. The procedures for encapsulation of protoplasts are similar to those for callus cells, except that an osmoticum, such as mannitol, needs to be included in the protoplast/alginate mixture (at a concentration about 0.7 M).

Artificial Seeds/Micropropagation

Somatic embryos of several plant species including alfalfa, celery, and cauliflower have been encapsulated in alginate beads as artificial seeds. The encapsulation medium can be used to supply nutrients and necessary growth substances to the developing embryos (it should be noted, however, that certain nutrient components may chelate with calcium, as discussed earlier), and furthermore the asexual embryos are resistant to severe desiccation in certain conditions. Encapsulation is therefore a potential tool for large-scale production of clonal material for agriculture. The technique has also recently been expanded to encapsulation of shoot buds of ginger (Sharma et al 1994) and banana (Ganapathi et al 1992). This kind of capsule could be useful in exchanges of sterile material between laboratories, in germplasm conservation, or in plant propagation. Alginate is the most commonly used material for encapsulation. In most studies, alginate solution was prepared in nutrient media. Piccioni and Standardi (1995), however, have observed a superior gel formation and subsequent plantlet regeneration when deionized water, in place of a nutrient medium, was used to prepare the alginate solution. They found that the nutrient supplement to the alginate matrix reduced the viscosity of the gel and its ability to form solid capsules. To alleviate the capsule hardening problem, a higher alginate concentration (2.5–3% instead of 2%) was recommended.

Cryopreservation

Conservation of genetic resources is an important issue in plant breeding, secondary metabolite production, and germplasm preservation. A viable technology to address this need is cryopreservation. It is based on the reduction and subsequent arrest of metabolic functions of plant material by imposition of the ultra-low temperature of $-196°C$. However,

only a few biological materials in their natural state can be frozen without adverse effects on cell viability. Conventional procedures consist of preculture with sugars or polyols, applications of a cryoprotectant cocktail (DMSO, proline, sugars, polyols), slow freezing to $-40°C$ before quenching in liquid nitrogen, and rapid thawing (Withers 1987). Recently, encapsulation-dehydration techniques with sucrose alone as cryoprotectant have been employed as an alternate cryopreservation method in several studies. Cryopreservation of a *Catharanthus* cell suspension was performed after encapsulation in alginate beads (Bachiri et al 1995). Encapsulated cells were precultured in sucrose-enriched medium for several days, dried over silica gel, and directly cooled in liquid nitrogen. After rewarming in air at room temperature, alginate beads were placed on a semisolid culture medium. Following regrowth, beads transferred to liquid medium generated a new cell suspension. Cell survival and regrowth from cryopreserved encapsulated cells depended on preculture duration and residual water content after air-drying. Embryogenic aggregates of sweet potato approximately 1.5–2.0 mm in diameter were subjected to a two-step freezing protocol in liquid nitrogen following alginate encapsulation, sucrose preculture, and dehydration (Blakesley et al 1995). In the two-step freezing protocol, an initial slow cooling step prior to plunging the tissue into liquid nitrogen was found necessary to enhance tissue survival. In a subsequent paper by Blakesley et al (1996), an alternate protocol for cryopreservation of sweet potato embryogenic tissues without encapsulation was described. Although it is technically simpler, without the alginate protection, the tissue was found to be more sensitive to the evaporative dehydration. If the dehydration process is not controlled properly, the tissue can lose its embryogenic competence following freezing (Blakesley et al 1996). Paulet et al (1993) also observed that encapsulation allowed apices to withstand drastic treatments (preculture with high sucrose, desiccation) that would be harmful to naked apices.

Techniques

Background

A number of commonly used as well as recently proposed techniques for plant cell encapsulation are sum-

TABLE 24.3. Examples of commonly used and recently proposed polymeric matrices for plant cell encapsulation.

Alginate
Agar
Agarose
Carrageenan
Pectin/chitosan
Hydrogel (poly(N-isopropylacrylamide))
PVA-SbQ
Polyurethane foam
Cotton/Stainless steel woven packing
Luffa cylindrica (vegetable sponge)

marized in Table 24.3 and discussed below with descriptions of representative experimental protocols.

Gel-Entrapment Techniques

Protocol #1:

Encapsulation of plant cells in calcium alginate gel particles

1. Mix plant cells from a suspension culture (sieve the culture to remove large cell aggregates if necessary) with an equal volume of 2–8% sodium alginate solution.
2. Transfer the cell/alginate mixture to an extrusion device, such as a syringe or dropping funnel, with a means of flow control.
3. Place a beaker containing calcium chloride solution below the extrusion device and magnetically stir the solution to produce a light vortex.
4. Adjust the flow control to allow drop-wise flow through a 22-gauge needle, from a height of about 10 cm (height may affect bead shape), into a 0.1–0.2 M $CaCl_2$ solution.
5. Allow a further 20–30 min stirring before collecting the beads on a funnel, using suction.
6. Wash the beads on the funnel with an appropriate buffer or sterile water.
7. Transfer the beads to a proper culture or storage medium.

Note that by adjusting needle diameters alone, bead sizes are typically in the range of 2–5 mm. Several methods have been proposed to reduce the size of the gel beads. One way is to apply a concentric flow of sterile air while extruding the particles (Rehg et al 1986). By means of this air-jet extruder,

gel beads with ~1.5 mm diameter can be produced, as opposed to 4–5 mm without the concentric air flow treatment (Payne et al 1991). The size of the gel bead may be controlled by the air stream velocity. Another technique to produce small gel beads is the use of electrostatic droplet generation (Goosen et al 1997), which has been used for animal cell immobilization (Bugarski et al 1993). This method has yet to be tested for plant cell immobilization. The two primary advantages of electrostatic droplet generation over an air-jet extruder are the production of much smaller beads with standard needles, and ease of bead size control by varying the applied potential. It should be noted that, given the large size of plant cells, small gel beads may not provide a large enough loading capacity. For more detailed accounts of this method, the reader is referred to a recent review by Goosen et al (1997).

When a large quantity of gel beads is to be prepared for pilot or production-scale reactors or for commercial production of artificial seeds, the conventional needle technique may not be sufficient because of its low production capacity. Typically, one needle can handle a volumetric flow in the range of 100–500 cm^3 h^{-1} (Hulst and Tramper 1989b). Hulst et al (1985) have developed a technique making use of a vibration nozzle to scale up the alginate gel immobilization. The principle of this device consists of breaking up a jet of the cell-alginate mixture into uniform droplets by mechanical vibrations that are transferred to the jet. The production capacity of the vibration nozzle is two orders of magnitude larger than that of the conventional needle technique. The drop size is controlled by varying the jet diameter, vibration frequency, and jet velocity.

Protocol #2:

Encapsulation of plant tissues or organs in alginate gel beads (Piccioni and Standardi 1995)

1. Immerse each explant (embryoids, embryogenic mass, shoot tips) with a forceps in an alginate solution (2% w/v without nutrient medium components, or 2.5–3% with nutrient medium components).
2. Drop the alginate-coated explants into a complexing solution made of $CaCl_2$ (11 g l^{-1}) for 25–30 min.
3. After hardening, rinse the capsules for 10 min in distilled water to wash away calcium chloride residues.

Protocol #3:

Encapsulation of plant cells in double-layered gel fibers (Tanaka et al 1996)

The procedures are similar to those in Protocol #1, but with a different configuration for the extrusion device. Here the internal diameters of the inner and outer extrusion nozzles are 3 mm and 4 mm, respectively. Sodium alginate solution (2.5%) is used for the outer layer, while the inner layer is prepared with 2.5% Na-alginate containing 30% (v/v) plant cells. The gelling agent is a 0.1 M CaCl$_2$ solution. After the preparation, the spiral double-layered gel fibers are stabilized by maintaining them in the CaCl$_2$ solution for 30 min.

Protocol #4:

Encapsulation of plant cells in κ-carrageenan gel particles (modified from Iborra et al 1997)

1. Sieve plant cell suspension through a stainless steel mesh or a nylon cloth, using a 0.9% NaCl solution.
2. Mix the cell suspension with the carrageenan solution (15 g/L, autoclaved) to a final carrageenan concentration of 10 g/L.
3. Extrude drop-wise the mixture into a stirred 0.3 M KCl solution kept at 20°C.
4. Leave the beads for 2 hours in the aerated KCl solution.
5. Wash the beads with a fresh plant cell culture medium.

Note that compared with Ca-alginate, κ-carrageenan is less sensitive to chelating reagents in the media. However, the κ-carrageenan particles produced by the dripping technique are less spherical and uniform.

Protocol #5:

Preparation of pectin/chitosan coacervate capsules (Dörnenburg and Knorr 1996)

1. Filter enough volume of concentrated plant cell culture under sterile conditions to give ~2.7 g fresh weight.
2. Transfer the cells from the filter to 30 ml of 5% solution of low methylated pectin in a proper nutrient medium.
3. Add the mixture from Step 2 drop-wise into 80 ml of 2% chitosan solution supplemented with

0.8% calcium chloride, which is stirred continuously for 30 min.
4. Wash the capsules twice in 3% sucrose solution and dry on a filter paper.

Protocol #6:

Encapsulation in hydrogel (Han et al 1996)

1. Dissolve 14.85 g of N-isopropylacrylamide and 0.15 g of N, N′-methylenebisacrylamide in 50 ml of distilled water.
2. Mix the monomer solution with 50 ml of 1% alginate solution and bubble the mixture for 30 min with nitrogen.
3. Add 17 g (fresh weight) of plant cells into the mixture from Step 2, then add 0.25 ml N,N,N′,N′-tetraethylmethylenediamine (TEMED).
4. Extrude the cell/polymer mixture drop-wise into a 3% (w/v) CaCl$_2$/0.5% ammonium persulfate aqueous solution that is vigorously sparged with nitrogen.
5. Wash the beads with sterile medium.

Protocol #7:

Encapsulation of plant cells in PVA-SbQ gel (Tanaka 1994)

1. Mix ~0.6 g fresh weight of plant cells with 2.25 g of a 10% PVA-SbQ solution sterilized at 120°C for 10 min.
2. Pour the mixture into a plastic dish and irradiate on each side with fluorescent lamps (20,000 lux) for 30 min.
3. Cut the hardened gel into small pieces (~5 × 5 mm) and wash with 3% sucrose solution.

Other Techniques

Techniques other than gel entrapment that have been reported or employed in recent studies are summarized below.

Protocol #8:

Entrapment in a spirally wound cotton matrix (Choi et al 1995)

1. Plant cells are filtered onto commercial cotton terry cloth (~85–90 g fresh weight of cells on a 400 cm^2 cotton sheet).

2. The cotton cloth with immobilized cells is then wound spirally with a spacer (Goodloe 304 stainless steel woven packing) to provide physical strength to the immobilization unit.

3. Place the immobilized cell unit in a column reactor and initiate the medium flow.

Protocol #9:

Entrapment in polyurethane foam (Iborra et 1994, Lindsey et al 1983) and vegetable sponge (Iqbal and Safar 1993)

The foam or sponge particles are added to cell culture at the time of inoculation, and as the culture grows, the cells invade the porous network in a passive manner, become entrapped within the pores, and ultimately colonize all the available space. This initial loading stage may take 10–24 days, which becomes a drawback of the method (Hulst and Tramper 1989). Also, it should be noted that to avoid flotation of the foam or sponge pieces during cell entrapment, there should be little or no free liquid medium present after the foam pieces are saturated with the liquid medium. In this connection, trickle-bed configuration may be considered for bioreactor operation with this type of encapsulated cell system. The cells entrapped inside the porous structure may be released by means of vigorous shaking (Guardiola et al 1996). Rhodes et al (1987) have conducted a systematic study of the factors affecting plant cell immobilization in foam particles. Briefly, the major factors include the time of exposure of foam to cells, size of the foam pocket and size of cell aggregates in the suspension culture, and potential phytotoxicity associated with some commercial polyurethane foams.

Future Prospects

Encapsulation of plant cells has progressed from a process strategy to various applications in plant cell biotechnology. From the survey of recent literature it is apparent that there is a continuing interest in this field. It is important to continue to look into fundamental as well as technical issues relating to plant cell encapsulation, to fully exploit the engineering and physiological benefits of this technology. Fundamental research is exemplified by a better understanding of the interactions between external matrix and encapsulated plant cells, communications between encapsulated cells, and nutrient transport in cell aggregates. Further technical research may involve development of new polymeric materials with low toxicity and high mechanical strength, technology for large-scale capsule preparation, and new methodologies to probe the cellular metabolism of encapsulated plant cells.

References

Ananta I, Subroto MA, Doran PM. 1995. Oxygen transfer and culture characteristics of self immobilized *Solanum aviculare* aggregates. Biotechnol Bioeng 47: 541–549.

Akakabe Y, Naoshima Y. 1993. The mechanistic pathway of the biotransformation of acetophenone by immobilized cell cultures of Gardenia. Phytochem 32: 1189–1191.

Bachiri Y, Gazeau C, Hansz J, Morisset C, Dereuddre J. 1995. Successful cryopreservation of suspension cells by encapsulation-dehydration. Plant Cell Tiss Org Cult 43: 241–248.

Bieniek ME, Harrell RC, Cantliffe DJ. 1995. Enhancement of somatic embryogenesis of Ipomoea batatas in solid cultures and production of mature somatic embryos in liquid cultures for application to a bioreactor production system. Plant Cell Tiss Org Cult 41: 1–8.

Blakesley D, Al Mazrooei S, Henshaw GG. 1995. Cryopreservation of embryogenic tissue of sweet potato (*Ipomoea batatas*): use of sucrose and dehydration for cryoprotection. Plant Cell Rep 15: 259–263.

Blakesley D, Al Mazrooei S, Bhatti MH, Henshaw GG. 1996. Cryopreservation of non-encapsulated embryogenic tissue of sweet potato (*Ipomoea batatas*). Plant Cell Rep 15: 873–876.

Brodelius P, Deus B, Mosbach K, Zenk MH. 1979. Immobilized plant cells for the production and transformation of natural products. FEBS Lett 103: 93–97.

Bugarski B, Smith J, Wu J, Goosen MFA. 1993. Methods for animal cell immobilization using electrostatic droplet generation. Biotechnol Techniq 7: 677–682.

Choi HJ, Tao BY, Okos MR. 1995. Enhancement of secondary metabolite production by immobilized *Gossypium arboreum* cells. Biotechnol Prog 11: 306–311.

Daly MM, Knorr D. 1988. Chitosan-alginate complex coacervate capsules: Effects of calcium chloride, plasticizers, and polyelectrolytes on mechanical stability. Biotechnol Prog 4: 76–81.

David H, Savy C, Miannay N, Dargent R, David A. 1994. Supporting matrix influences protoplast-derived colony formation: structural analysis. Protoplasma 179: 111–120.

David H, Bade P, David A, Savy C, Demazy C, Cutsem P van. 1995. Pectins in walls of protoplast-derived cells imbedded in agarose and alginate beads. Protoplasma 186: 122–130.

Dörnenburg H, Knorr D. 1996. Semicontinuous processes for anthraquinone production with immobilized *Cruciata glabra* cell cultures in a three phase system. J Biotechnol 50: 55–62.

Endress R. 1994. Immobilization of plant cells, Ch. 7 in: Plant cell biotechnology. Berlin: Springer-Verlag. p. 256–269.

Ganapathi TR, Suprasanna P, Bapat VA, Rao PS. 1992. Propagation of banana through encapsulated shoot tips. Plant Cell Rep 11: 571–575.

Goosen MFA, Mahmud ESE, Al-Ghafri AS, Al-Hajri HA, Al-Sinani YS, Bugarski B. 1997. Immobilization of cells using electrostatic droplet generation. In: Bickerstaff GF, editor. Immobilization of enzymes and cells. Totowa: Humana Press. p. 167–174.

Guardiola J, Iborra JL, Rodenas L, Canovas M. 1996. Biotransformation from geraniol to nerol by immobilized grapevine cells (*V. vinifera*). Appl Biochem Biotechnol 56: 169–180.

Hall RD, Holden MA, Yeoman MM. 1988. Immobilization of higher plant cells. In: Bajaj YPS, editor. Medicinal and aromatic plants I. Biotechnology in agriculture and forestry 4. Berlin: Springer. p. 136–156.

Han J, Martinez BC, Ruan RR. 1996. Immobilization of *Coleus blumei* plant cells in temperature-sensitive hydrogel. Biotechnol Techniq 10: 359–362.

Hoshino K, Akakabe S, Morohashi S, Sasakura T. 1997. Immobilization of enzymes on thermo-responsive polymers. In: Bickerstaff GF, editor. Immobilization of enzymes and cells. Totowa: Humana Press. p. 101–108.

Hulst AC, Tramper J, Riet K van't, Westerbeek JMM. 1985. A new technique for the production of immobilized biocatalyst in large quantities. Biotechnol Bioeng 27: 870–876.

Hulst AC, Meyer MMT, Breteler H, Tramper J. 1989a. Effect of aggregate size in cell cultures of *Tagetes patula* on thiophene production and cell growth. Appl Microbiol Biotechnol 30: 18–25.

Hulst AC, Tramper J. 1989b. Immobilized plant cells: A literature survey. Enzyme Microb Technol 11: 546–558.

Iborra JL, Guardiola J, Montaner S, Canovas M, Manjon A. 1994. Enhanced accumulation of anthocyanins in Vitis vinifera cells immobilized in polyurethane foam. Enz Microb Technol 16: 416–419.

Iborra JL, Manjon A, Canovas M. 1997. Immobilization in carrageenans. In: Bickerstaff GF, editor. Immobilization of enzymes and cells. Totowa: Humana Press. p. 53–60.

Iqbal M, Zafar SI. 1993. Vegetable sponge: A new immobilization medium for plant cells. Biotechnol Techniq 7: 323–324.

Isa T, Ogasawara T, Kaneko H. 1990. Regeneration of saffron protoplasts immobilized in Ca-alginate beads. Jpn J Breed 40: 153–157.

Lindsey K, Yeoman MM, Black GM, Mavituna F. 1983. A novel method for the immobilization and culture of plant cells. FEBS Lett 155: 143–149.

Martinez BC, Park CH. 1994. Immobilization of *Coleus blumei* cells in a column reactor using a spray feeding system. Biotechnol Techniq 8: 301–306.

Morris P, Smart NJ, Fowler MW. 1983. A fluidized bed vessel for the culture of immobilized plant cells and its application for the continuous production of fine cell suspensions. Plant Cell Tiss Org Cult 2: 207–216.

Niedz RP. 1993. Culturing embryogenic protoplasts of 'Hamlin' sweet orange in calcium alginate beads. Plant Cell Tiss Org Cult 34: 19–25.

Paulet F, Engelmann F, Glaszmann JC. 1993. Cryopreservation of apices of in vivo plantlets of sugarcane (*Saccharum sp.* hybrids) using encapsulation/dehydration. Plant Cell Rep 12: 525–529.

Payne GF, Bringi V, Prince C, Shuler ML. 1991. Immobilized plant cells Ch. 7 in: Plant cell and tissue culture in liquid systems. Munich: Hanser. p. 177–223.

Piccioni E, Standardi A. 1995. Encapsulation of micropropagated buds of six woody species. Plant Cell Tiss Org Cult 42: 221–226.

Pras N, Woerdenbag HJ, Uden W van. 1995. Bioconversion potential of plant enzymes for the production of pharmaceuticals. Plant Cell, Tissue Organ Culture 43: 117–121.

Ramakrishna SV, Reddy GR, Curtis WR, Hymphrey AE. 1993. Stimulation of solavetivone synthesis in free and immobilized cells of *Hyoscyamus muticus* by *Rhizoctonia solani* fungal components. Biotechnol Lett 15: 307–310.

Rehg T, Dorger C, Chau PC. 1986. Trivalent cation stabilization of alginate gel for cell immobilization. Biotechnol Lett 8: 111–114.

Rhodes MJC, Smith JI, Robins RJ. 1987. Factors affecting the immobilization of plant cells on reticulated polyurethane foam particles. Appl Microbiol Biotechnol 26: 28–35.

Roger D, David A, David H. 1996. Immobilization of flax protoplasts in agarose and alginate beads. Correlation between ionically bound cell wall proteins and morphogenetic response. Plant Physiol 112: 1191–1199.

Saje L, Vunjak-Novakovic G, Grubisic D, Kovacevic N, Vukovic D, Bugarski B. 1995. Production of anthraquinones by immobilized *Frangula alnus Mill.* plant cells in a four-phase air-lift bioreactor. Appl Microbiol Biotechnol 43: 416–423.

Schnabl H, Elbert C, Youngman RJ. 1983a. Release of ethane from immobilized plant cell protoplasts in response to chemical treatment. Physiol Plant 59: 46–49.

Schnabl H, Youngman RJ, Zimmermann U. 1983b. Maintenance of plant cell membrane integrity and function by the immobilization of protoplasts in alginate matrices. Planta 158: 392–397.

Scragg A. 1991. The immobilization of plant cells. In: Stafford A, Warren G, editors. Plant cell and tissue culture. Oxford: Open University Press, p. 205–219.

Seki M, Ohzora C, Takeda M, Furusaki S. 1997. Taxol (paclitaxel) production using free and immobilized cells of *Taxus cuspidata*. Biotechnol Bioeng 53: 214–219.

Sharma TR, Singh BM, Chauhan RS. 1994. Production of disease-free encapsulated buds of *Zingiber officinale* Rosc. Plant Cell Rep 13: 300–302.

Stano J, Nemec P, Weissova K, Kovacs P, Kakoniova D, Liskova D. 1995. Decarboxylation of L-tyrosine and L-DOPA by immobilized cells of *Papaver somniferum*. Phytochem 38: 859–860.

Sudhakar Johnson T, Ravishankar GA, Venkataraman LV. 1996. Biotransformation of ferulic acid and vanillylamine to capsaicin and vanillin in immobilized cell cultures of *Capsicum frutescens*. Plant Cell Tiss Org Cult 44: 117–121.

Suvarnalatha G, Ghand N, Ravishankar GA, Venkataraman LV. 1993. Computer-aided modeling and optimization for capsaicinoid production by immobilized Capsicum frutescens cells. Enz Microb Technol 15: 710–715.

Takemoto M, Achiwa K, Stoynov N, Chen D, Kutney JP. 1996. Synthesis of optically active alpha-phenylpyridyl-methanols by immobilized cell cultures of *Catharanthus roseus*. Phytochem 42: 423–426.

Tanaka A. 1994. Immobilization of plant cells. In: Ryu DDY, Furusaki S, editors. Advances in plant biotechnology. Amsterdam: Elsevier. p. 209–219.

Tanaka H, Irle S, Ochi H. 1989. A novel immobilization method for prevention of cell leakage from the gel matrix. J Ferment Bioeng 68: 216–219.

Tanaka H, Kaneko Y, Aoyagi H, Yamamoto Y, Fukunaga Y. 1996. Efficient production of chitinase by immobilized *Wasabia japonica* cells in double-layered gel fibers. J Ferment Bioeng 81: 220–225.

Vanek T, Macek T, Stransky K, Ubik U. 1989. Plant cells immobilized in pectate gel: biotransformation of verbenol isomers by *Solanum aviculare* free and immobilized cells. Biotechnol Techniq 3: 411–414.

Watts MJ, Collin HA. 1985. Growth and nutrient uptake by immobilized tissue culture cells of celery (*Apium graveolens*). Plant Sci 42: 67.

Williams PD, Mavituna F. 1992. Immobilized plant cells. In: Fowler MW, Warren GS, volume editors; Murray Moo-Young M, editor-in-chief. Plant biotechnology: Comprehensive biotechnology, second supplement. New York: Pergamon. p. 63–78.

Withers LA. 1987. Long-term preservation of plant cells, tissues and organs. Oxford Surveys of Plant Molecular and Cell Biology 4: 221–272.

Section 2
Genetically Engineered Cells

25

Implantable Microcapsules for Gene Therapy for Hemophilia

Gonzalo Hortelano and Tracy Stockley

Introduction

Pathophysiology and Treatment of Hemophilia

Hemophilia is an X-linked recessive disorder caused by the deficiency of blood clotting factors VIII (hemophilia A) or IX (hemophilia B), which affects about 1 in 5000 live male births (Furie and Furie 1988, Hedner and Davie, 1989). Patients with severe hemophilia suffer from lifelong episodes of spontaneous bleeding. Common presentations include hematomas, bleeding into the joints, and intracranial hemorrhage, with the latter being a common cause of death. Long-term complications include chronic hemophilic arthropathy and progressive degeneration of the joints, leading to severe crippling deformity. Typically, one or more joints are affected in patients before the age of 12. Hemophilia is a debilitating disease imposing a heavy burden on both patients and families.

Plasma factors VIII or IX coagulant activity varies from 0–1% in severely affected individuals, between 1–5% in moderate hemophiliacs, and between 6–25% in mild ones (Roberts and Eberst 1993). Patients suffering from the mild deficiency require a less aggressive prophylaxis, since they have fewer bleeding episodes. Successful management of hemophilia requires that bleeding episodes be controlled by periodic factor concentrate infusions. The current generation of factor IX concentrates is prepared from pooled plasma, with the subsequent risk for viral infections, although high-purity "virus-free" factor IX concentrates are now available. However, the potential for viral infections can-

not be ruled out. Recombinant factor IX, the preferred product for reasons of viral safety, is still in human trials (Meulien 1990, White et al 1998). Recombinant factor VIII has been licensed by the FDA since December 1992 (Recombinate, Baxter), and is commercially available. The hemophiliac population has been particularly affected by HIV and hepatitis, because plasma-derived factor VIII has been used as the treatment of choice (Brownlee 1995). In fact, 70% of severe and 50% of moderate hemophiliacs with factor VIII deficiency in the U.S. are now HIV positive (AIDS Update Medical Bulletin 137, 1991). However, the cost of recombinant coagulation factors may limit its use, as is the case for other recombinant therapeutic products, such as tissue plasminogen activator and growth hormone. In fact, the current therapy for this lifelong disease comes at a high economic cost ($50 million/year in Canada). The cost to treat a severe hemophiliac is around $100,000/year. Hence, in spite of the many recent advances, the development of optimal and more cost-effective therapy for hemophilia remains to be achieved. Gene therapy could be such an alternative.

Potential for Hemophilia Gene Therapy

Hemophilia is a particularly attractive disease for potential treatment with gene therapy protocols and was identified as such even in the early days of gene therapy (Anderson 1984). Neither factor VIII nor IX is tightly regulated, and the product circulates in the bloodstream. As an additional advantage for gene therapy, there is not a tissue-specific requirement for delivery. The clinical presentation of hemophiliacs makes it apparent that delivery of 5% of the physiological level of factor VIII in plasma (5 ng/ml), or factor IX (250 ng/ml), would minimize the risk of spontaneous bleeding in patients (Fallaux et al 1995). Hence, even a small supply of the missing factor (as low as 0.1%, as noted by Kay et al 1993) can be clinically effective and sufficient to change the disease status of a severely affected individual to a moderate or mild phenotype. Furthermore, an excess of the coagulation factor in the circulation (up to 200%) does not appear to have any adverse clinical consequences. (Fang et al 1995).

Several methods of gene therapy for hemophilia have been attempted. These methods have been primarily based on in vivo gene therapy, which is the direct injection in vivo of the viral vector system. Vectors that have been used include adenovirus (Smith et al 1996), adeno-associated virus (Snyder et al 1997), and retrovirus (Nelson et al 1997). Direct injection of virus possesses some safety concerns due to the widespread distribution of the virus in vivo, the potential genetic mutations of the patient's cells, and the possibility of a modification of germ line cells, thus genetically altering future generations (Cornetta 1992). More studies must be performed to satisfy these safety concerns. As opposed to this direct injection of vectors, ex vivo gene therapy is based on the transplantation of recombinant cells (originating from the same patient in the case of autologous gene therapy) that have been engineered to secrete the therapeutic product (Yao and Kurachi 1992). Each method has its own advantages and limitations.

Immunoisolation Gene Therapy

In contrast to other gene therapy approaches, we have been developing gene therapy for hemophilia based on a novel nonautologous strategy using immunoisolation. In this approach, "universal" cell lines are engineered using molecular biology techniques to secrete a desired recombinant product. This cell line is then placed within alginate microcapsules, which act as an immunoisolation device upon implantation of the microcapsules. The selectively permeable membrane of the microcapsule allows for the release of the recombinant therapeutic product produced by the engineered cells, as well as the exchange of nutrients and waste products between the

interior and exterior of the microcapsule. However, the microcapsule membrane prevents immune mediators from entering the microcapsule and contacting the nonautologous cells, thus blocking rejection of the nonautologous cells. As a result of the immunoprotective features of microencapsulation, the same cell line can be used for all individuals requiring the particular recombinant product produced by the genetically modified cells. For use in the treatment of hemophilia, the cells for encapsulation can be engineered to secrete the desired coagulation factor needed for hemophilia disease treatment.

This approach to gene therapy has several advantages over other forms of gene therapy. Since the patient-specific genetic modifications required in ex vivo viral gene therapy protocols are not needed in this approach, the time lag before treatment is reduced, and the anticipated costs would be significantly lower. As well, the universal engineered cell line can be stored frozen for use in future patients, thus simplifying prospective treatments. Another advantage of this approach is that the safety of the cell lines used can be thoroughly evaluated since the cell line can be grown to potentially unlimited quantities for evaluation of contamination before encapsulation and implantation.

For this novel approach to gene therapy, our group has primarily used genetically modified fibroblast and myoblast cell lines encapsulated in alginate-poly-l-lysine-alginate microcapsules. Encapsulated cells maintain their viability and continue to proliferate, albeit at a reduced rate, while secreting recombinant products at undiminished levels, as shown in several instances with various products (Al-Hendy et al 1995, Awrey et al 1997, Bastedo et al 1994, Chang et al 1993). The recombinant products that can be released from the capsules can range in size up to 300 kD (Awrey et al 1997). Although the 300 kD porosity size of the microcapsule membrane suggests that IgG molecules (~160 kD) would not be excluded from the microcapsules, prevention of an immune response against the encapsulated cells is achieved. Perhaps this protection is due to other downstream molecules, such as IgM (970 kD), that cannot transverse the microcapsule membrane.

As an application of this approach to the treatment of hemophilia, we have demonstrated the delivery of coagulation factors from encapsulated recombinant cells.

Encapsulation of Fibroblasts

Although this approach is simple in concept, the choice of cells to be encapsulated is critical for the success of this gene therapy method. Encapsulated cells are not attached to a matrix, and thus the procedure is not suitable for every cell line. Several types of proliferative cells, such as hybridoma (Pelegrin et al 1998), PC12 (Campioni et al 1998) and CHO (Uludag and Sefton 1883) have been reported to survive enclosed in capsules. Historically, fibroblasts have been used since the early days of gene transfer experiments (Anson et al 1987). Their robust nature and excellent proliferation make them good candidates for encapsulation.

Indeed, fibroblasts were used in our laboratory to secrete human recombinant products from alginate microcapsules (Chang et al 1993). Ltk^- fibroblasts genetically engineered to secrete human growth hormone were encapsulated and implanted intraperitoneally into mice. Human growth hormone was detected in the bloodstream of the treated mice (Chang et al 1993). In applying the encapsulation technology to gene therapy, a primary consideration is whether recombinant fibroblasts, normally anchorage-dependent cells, can retain their growth potential and transgene expression after encapsulation.

The microcapsules are fabricated using 1.5% potassium alginate (improved Kelmar, Monsanto, MO) and have an outer diameter of between 200–500 μm. The encapsulation protocol for alginate-poly-L-lysine-alginate, described by O'Shea and Sun, (1986), was modified to optimize the expression of recombinant products from encapsulated fibroblasts (Chang et al 1994). In this study, the effect of variables such as the temperature of encapsulation, duration of the polylysine (PLL) cross-linking, duration of the alginate core liquefaction, and cell density were investigated. Particular care was given to determining, for every encapsulation condition, the number of encapsulated fibroblasts and their viability at 4 weeks after encapsulation. Particularly revealing was the drop in cell viability by day 29 when PLL treatment was extended beyond 6 minutes, and the reduced number of encapsulated cells found at day 29 when the alginate core was not sufficiently liquefied with sodium citrate. Using the optimized conditions, a viability of 80–90% was typically maintained by day 29 post-encapsulation (Chang et al 1994). In addition, the capacity of recombinant fibroblasts to continue to secrete the transgene in vitro after encapsulation, and thus their potential for gene therapy, was demonstrated.

Delivery of Factor IX from Fibroblasts

By means of the optimized encapsulation protocol, fibroblasts have been used to secrete adenosine deaminase (Hughes et al 1994) and human growth hormone (Chang et al 1993). Mouse Ltk^- fibroblasts, a transformed cell line, were transfected with plasmids carrying the cDNA of human factor IX (Liu et al 1993). Encapsulated recombinant fibroblasts secreted human factor IX in vitro at the same rate as that of unencapsulated cells. It is relevant to remember that factor IX is normally produced in the liver, and not by fibroblasts. However, approximately 70% of the factor IX secreted by the recombinant fibroblasts was biologically active, as determined by clotting assay (Liu et al 1993). Hence, biologically active factor IX (M.W. 72,000) can freely diffuse through the alginate microcapsules. Nevertheless, the level of factor IX secreted by the recombinant fibroblasts in this study was modest, approximately 10 ng/10^6 cells/day, clearly insufficient to attempt any phenotypic correction in vivo. The normal range of factor IX in human beings is around 5 μg/ml.

In a similar approach to gene therapy for hemophilia, recombinant fibroblasts have also been used to secrete human factor IX from immunoisolation implantable chambers developed at Baxter Corporation (Carr-Brendel et al 1997). Using highly effective retrovirus, Kingdon (1997) obtained recombinant fibroblasts expressing human factor IX at a rate of several μg/10^6 cells/day. After subcutaneous implantation into mice, levels of human factor IX of 200–300 ng/ml of plasma were detected. Furthermore, levels persisted for over a year (Kingdon 1997). Nonetheless, the biological activity of the secreted human factor IX found in plasma was only 30%. Hence, it is clear that fibroblasts are able to secrete large amounts of factor IX for long periods of time.

Encapsulation of Myoblasts

Despite their impressive capacity to secrete factor IX, fibroblasts share a major drawback with other proliferative cell lines. As a result of their vigorous growth within the microcapsule environment, fibroblasts eventually fill the entire capsular space. Following the lack of adequate diffusion of nutrients to the cells, cell viability ultimately plummets (Awrey et al 1996). This fact seriously restricts the use of encapsulated fibroblasts in gene therapy for the long-term delivery of required therapeutic products, such as human factor IX for hemophilia B.

Hence, alternate cell lines were actively sought. Myoblast cells are proliferative cells that have the capability to terminally differentiate into nondividing myotubes to form the muscle fibers (Blau et al 1985). Furthermore, they are easy to obtain from muscle biopsies. Muscle tissue represents around 40% of the body weight (Blau et al 1985). Hence, recombinet myoblasts may offer a feasible alternative to proliferative fibroblasts. Indeed, mouse C2C12 myoblasts do spontaneously differentiate upon encapsulation, perhaps as a result of the stressful unique conditions of the microcapsule environment (Bowie 1997). The number of encapsulated myoblast cells per capsule doubles approximately once after encapsulation, before cellular proliferation ceases. After differentiation occurs, which happens approximately 10 days after encapsulation, an increase in the length of time the capsules are cultured in vitro does not increase the number of cells per capsule. However, the viability of the encapsulated myoblasts does not decay, and can be maintained for weeks (Bowie 1997). Hence, encapsulated myoblasts are stable cells, suitable for long-term survival within alginate microcapsules.

Delivery of Factor IX from Myoblasts

Myoblasts have already been used for gene therapy of hemophilia. Indeed, both primary and C2C12 myoblasts have been shown to secrete factors VIII and IX (Roman et al 1992). Furthermore, myoblasts have the cellular machinery required to undergo the complex posttranslational modifications needed to produce a biologically active factor IX (Roman et al 1992). To assess the feasibility of using encapsulated myoblasts for the long-term secretion of factor IX, mouse C2C12 myoblasts were transfected with a plasmid containing the human factor IX cDNA. Clone 5D5 secreting 300 ng/10^6 cells/day was selected and encapsulated. After encapsulation, enclosed myoblasts were still able to secrete human factor IX at an undiminished rate (Hortelano et al 1996).

A total of 5×10^6 encapsulated myoblasts were implanted intraperitoneally into normal immunocompetent C57BL/6 mice (Hortelano et al 1996). Control mice were injected intraperitoneally with the same number of recombinant, but unencapsulated, cells. Human factor IX was detected in the plasma of all implanted mice at day 2. At the same time, control mice had no detectable levels of human factor IX (Figure 25.1), indicating the protective effect of the microcapsules. The delivery of factor IX had a peak of around 4 ng/ml (Hortelano et al 1996). This represents 0.1% of the physiological concentration present in healthy humanbeings. Although modest, a similar delivery was able to show some therapeutic effect in hemophilic dogs, and to reduce their clotting time (Kay et al 1993). However, the delivery was short-lived, and human factor IX became undetectable in the mouse plasma by day 14 (Hortelano et al 1996) (Figure 25.1). Concurrently, antibodies to human factor IX were detectable in implanted mice (Figure 25.2). Immunocompetent mice develop an immune response to a foreign antigen, such as human factor IX. The presence of antibodies was detected at stable levels until the end of the experiment (day 213, Figure 25.2). This is indirect evidence of a basal but constant level of human factor IX delivery from the microcapsules throughout the entire experiment. Implanted microcapsules were retrieved at various times postencapsulation. Interestingly, the viability of the retrieved encapsulated myoblasts was maintained, and did not drop significantly after 7 months in vivo (Figure 25.3). Thus, the encapsulated recombinant myoblasts can survive in the peritoneal cavity of implanted mice for at least 7 months (Hortelano et al 1996).

Although the protective action of the microcapsules has been established, the level of human factor IX delivered was modest. Hence, the level of factor IX expression from engineered myoblasts must

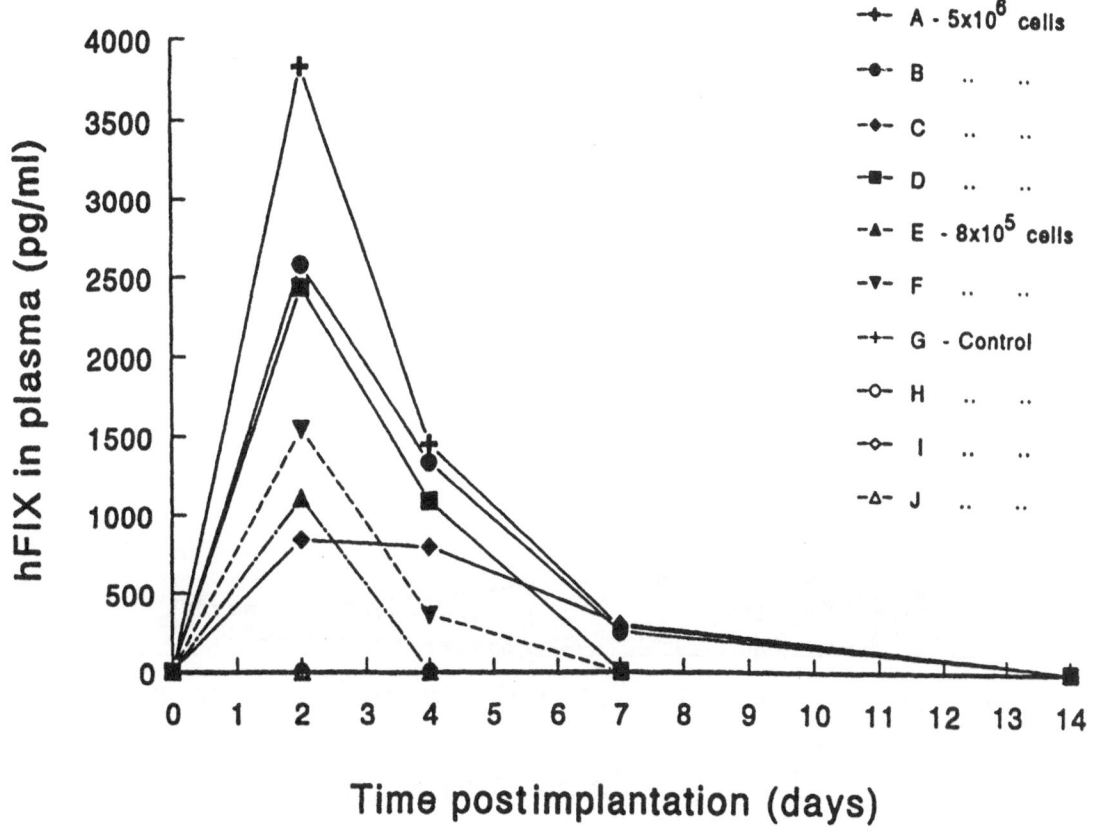

FIGURE 25.1. Human factor IX delivered in mouse plasma. Four normal C57BL/6 mice (✚,●,◆,■) were implanted IP with 5×10^6, and another two mice (▲,▼) with 8×10^5 microencapsulated recombinant myoblasts (clone 5D5). Four control mice (✛,○,✧,△) were injected with 5×10^6 unencapsulated 5D5 myoblasts. Human factor IX secretion into mouse plasma was detected by ELISA.

be increased with improved vectors, before this approach can be considered for human use.

Future Directions

Although we have demonstrated delivery of human factor IX to mice via encapsulated engineered cells, the ultimate test of the feasibility of this approach to gene therapy for hemophilia will be in the delivery of coagulation factors to large animal models of hemophilia. Large animal models are more similar in size and physiology to human, and so data obtained from large animal trials may be most indicative of the potential effectiveness of immunoisolation gene therapy in treatment of human hemophilia.

Several well-characterized canine models of hemophilia currently exist (Brinkhous et al 1973, Giles et al 1982, Pijnappels et al 1986). Both hemophilic A and hemophilic B canines are maintained by breeding programs, and the affected canines manifest a severely affected phenotype with no detectable level of coagulation factor VIII or IX, respectively (Evans et al 1989, Tinlin et al 1993). The complete lack of coagulation factors in these canine models is particularly important for their use in gene therapy treatments since current methods of treatment, including our approach, may not be able to deliver high levels of coagulation factors. Thus, the delivery of even a low level of clotting factors to the affected canines may have a significant effect on the disease state.

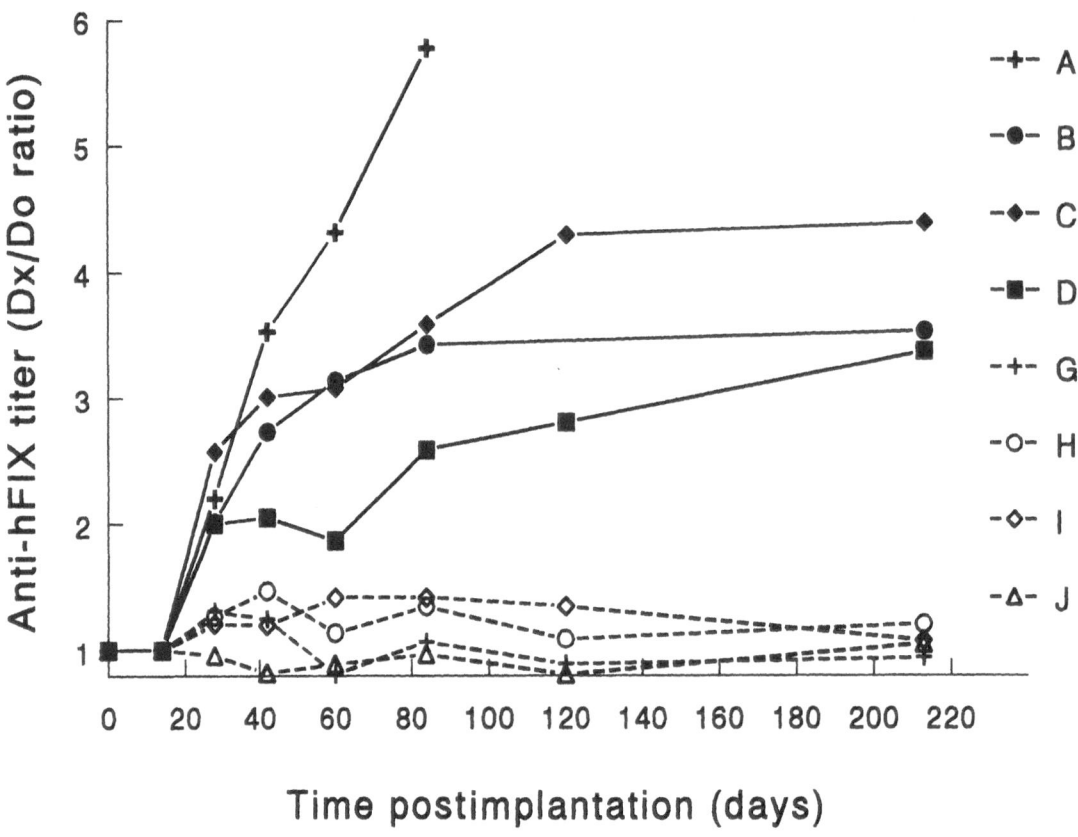

FIGURE 25.2. Development of anti-hFIX antibodies in the implanted mice. Four C57BL/6 mice (✚,●,◆,■) were implanted IP with 5×10^6 microencapsulated recombinant myoblasts, and four control mice (✚,○,◇,△) were injected with 5×10^6 unencapsulated recombinant myoblasts. Antibodies to human factor IX were detected by ELISA. The ratio of absorbance for a particular day and mouse (D_x) over that of the same mouse at day 0 (D_0) was used to indicate the antibody titer.

One area that needs to be addressed before treatment of canine hemophilia models is achieved is the stability of the microcapsules. Initial work in our lab with delivery of human growth hormone as a marker gene to normal dogs, using cells engineered to secrete human growth hormone and placed within alginate-poly-L-lysine microcapsules, demonstrated that this type of microcapsule was not stable within the dogs, and the microcapsules were lost within days, through breakage (Peirone et al 1997a). This is at odds with our results in mice, in which the microcapsules persist for months after implantation (Hortelano et al 1996). When more stable types of alginate microcapsules were implanted into canines, such as microcapsules fabricated as solid instead of hollow beads and with barium as a cross-linking agent for alginate in place of calcium, there was an

improvement in the length of delivery of human growth hormone (Peirone et al 1997b). However, even these improved capsules persisted for only two months in vivo, suggesting that delivery of products to canines will require further enhancement of microcapsule stability. To this end, we are currently testing novel capsule formulations for their suitability in recombinant product delivery, with the ultimate goal of treatment of hemophilic canines as a prelude to human clinical trials.

All the data presented here refer to the delivery of factor IX for the treatment of hemophilia B. However, the approach of immunoisolation for human gene therapy is, in theory, also applicable to the delivery of factor VIII for the treatment of hemophilia A. Nevertheless, the delivery of factor VIII presents unique challenges that have to be resolved before

Day 0   Day 28

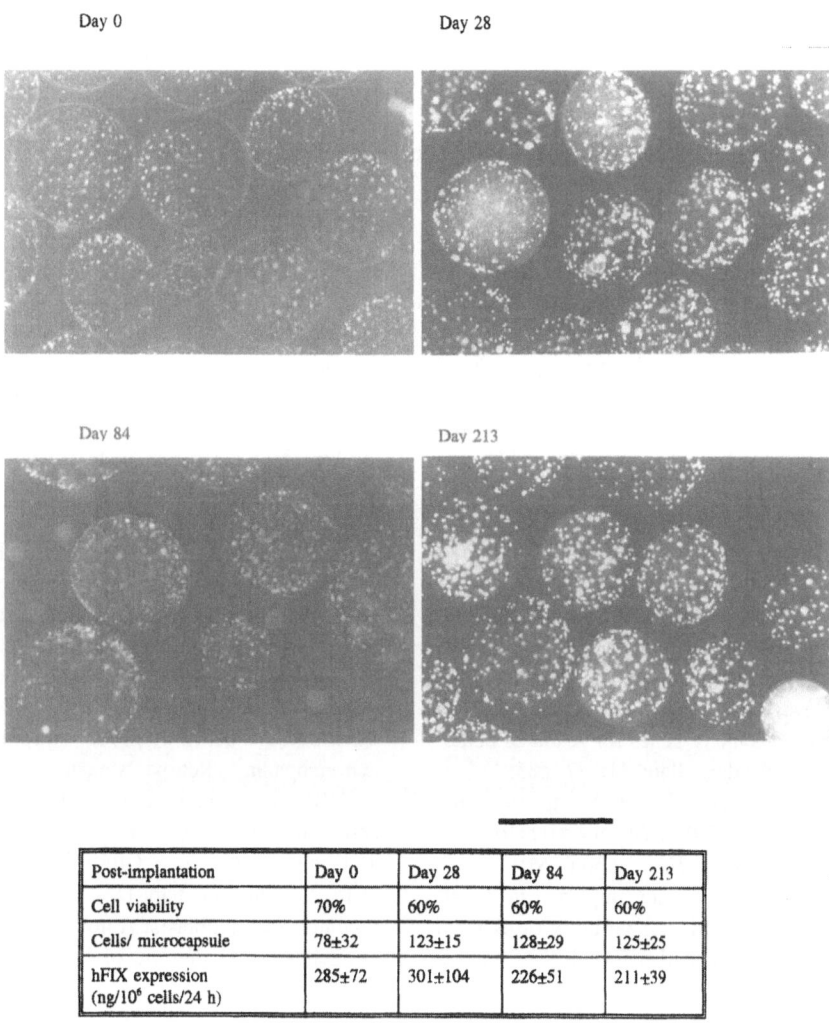

Day 84   Day 213

Post-implantation	Day 0	Day 28	Day 84	Day 213
Cell viability	70%	60%	60%	60%
Cells/ microcapsule	78±32	123±15	128±29	125±25
hFIX expression (ng/10^6 cells/24 h)	285±72	301±104	226±51	211±39

FIGURE 25.3. Retrieval of microcapsules postimplantation. Microcapsules were retrieved from the abdominal cavities of implanted mice by lavage at various time points postimplantation. The integrity and appearance of the microcapsules were recorded with dark-field photomicrography. The retrieved capsules were washed, cultured overnight, and assayed for hFIX secretion.

the approach becomes feasible. Factor VIII (Fallaux et al 1995) is an unstable, very large molecule (300 kD). Hence, its diffusion out of immunoisolation devices, as well as its effective transport to the bloodstream from the site of microcapsule implantation, still has to be proven. More recently, a new, shorter version of factor VIII (180 kD) has been described, in which the entire B domain has been deleted from the protein (Dwarki et al 1995, Hoeben et al 1992). This 180 kD protein should be more easily secreted from microcapsules. Only after these questions are

answered will the feasibility of immunoisolation devices for the delivery of factor VIII be determined.

References

Al-Hendy A, Hortelano G, Tannenbaum GS, Chang PL. 1995. Correction of the growth defect in dwarf mice: A novel approach to somatic gene therapy. Human Gene Therapy 6:165–175.

Anderson WF. 1984. Prospects for human gene therapy. Science 226:401–409.

Anson DS, Hock RA, Smith KJ, Brownlee GG, Verma IM, and Miller AD. 1987. Towards gene therapy for hemophilia B. Molecular Biology in Medicine 4(1):11–20.

Awrey DE. 1993. Alginate-poly-l-lysine-alginate microcapsules. M.Sc. Thesis. McMaster University, Hamilton, Ontario, Canada.

Awrey DE, Tse M, Hortelando G, Chang PL. 1996. Permeability of alginate microcapsules to secretory recombinant products. Biotechnology and Bioengineering 52:472–484.

Bastedo L, Sands MS, Hortelano G, Al-Hendy A, Chang PL. 1994. Partial correction of murine mucopolysaccharidosis VII with microencapsulated non-autologous recombinant fibroblasts. American Journal of Human Genetics 55(3):A210.

Blau HM, Dhawan J, Pavlath GK. 1993. Myoblasts in pattern formation and gene therapy. Trends in Genetics 9:269–274.

Bowie KM. 1997. Differentiation of recombinant myoblasts in alginate microcapsules. M.Sc. Thesis. McMaster University, Hamilton, Ontario, Canada.

Brinkhous KM, Davis PD, Graham JB, Dodds WJ. 1973. Expression and linkage of genes for X-linked hemophilia A and B in the dog. Blood 41:577–585.

Brownlee GG. 1995. Prospects for gene therapy for hemophilia A and B. British Medical Bulletin 51(1):91–105.

Campioni EG, Nobrega JN, Sefton MV 1998. HEMA/MMMA microcapsule implants in hemiparkinsonian rat brain: Biocompatibility assessment using [3H]PK11195 as a marker for gliosis. Biomaterials 19(7–9):829–37.

Carr-Brendel VE, Geller RL, Thomas TJ, Boggs DR, Yong SK, Crudele J, Martinson LA, Maryanor DA, Johnson RC, Brauker JH. 1997. Transplantation of cells in an immunoisolation device for gene therapy. Methods in Molecular Biology 63:373–387.

Chang PL. 1995. Nonautologous somatic gene therapy. In: Somatic gene therapy, PL Chang, Ed. pp 203–233. Boca Raton, Florida: CRC Press.

Chang PL, Hortelano G, Tse M, Awrey DE. 1994. Growth of recombinant fibroblasts in alginate microcapsules. Biotechnology and Bioengineering 43:925–933.

Chang PL, Shen N, Westcott AJ. 1993. Delivery of recombinant gene products with microencapsulated cells in vivo. Human Gene Therapy 4:433–440.

Cornetta K. 1992. Safety aspects of gene therapy. British Journal of Haematology 80(4):421–426.

Dwarki VJ, Belloni P, Nijjar T, et al. 1995. Gene therapy for hemophilia A: Production of therapeutic levels of human factor VIII in vivo in mice. Proceedings of the National Academy of Sciences USA 92:1023–1027.

Evans JP, Brinkhous KM, Brayer GD, Reisner HM, High KA. 1989. Canine hemophilia B resulting from a point mutation with unusual consequences. Proceedings of the National Academy of Sciences USA 86:10095–10099.

Fallaux, FJ, Hoeben RC, Briet E. 1995. State and prospects of gene therapy for the hemophilias. Thrombosis and Haemostasis 74(1):263–273.

Fang B, Eisensmith RC, Wang H, Kay MA, Cross RE, Landen CN, Gordon G, Bellinger DA, Read MS, Hu PC, Brinkhous KM, Woo SLC. 1995. Gene therapy for hemophilia B: Host immunosuppression prolongs the therapeutic effect of adenovirus-mediated factor IX expression. Human Gene Therapy 6(8):1039–44.

Furie B, Furie BC. 1988. The molecular basis of blood coagulation. Cell 53:505–518.

Giles AR, Tinlin S, Greenwood R. 1982. A canine model of hemophilic (factor VIII:C deficiency) bleeding. Blood 60:727–730.

Hedner U Davie EW. 1989. Introduction to hemostasis and the vitamin K-dependent coagulation factors. In: Scriver CR, Beaudet AL, Sly WS, et al (eds), The metabolic basis of inherited disease, vol II, ed 6. New York: McGraw-Hill, p 2107.

Hoeben RC, Einerhand MPW, Briet E, van Ormondt H, Valerio D, van der Eb AJ. 1992. Toward gene therapy in haemophilia A: Retrovirus-mediated transfer of a factor VIII gene into murine haematopoietic progenitor cells. Thrombosis and Haemostasis 67(3):341–345.

Hortelano G, Al-Hendy A, Ofosu FA, Chang PL. 1996. Delivery of human factor IX in mice by encapsulated recombinant myoblasts: A novel approach towards allogeneic gene therapy of hemophilia B. Blood 87(12):5095–5103.

Hughes M, Vassilakos A, Andrews DW, Hortelano G, Belmont JM, Chang PL. 1994. Delivery of a secretable adenosine deaminase through microcapsules: A novel approach to somatic gene therapy. Human Gene Therapy, 5(12):1445–1455.

Kay MA, Rothenberg S, Landen CN, Bellinger DA, Leland F, Toman C, M. et al. 1993. In vivo gene therapy for hemophilia B: Sustained partial correction in factor IX deficient dogs. Science 262:117–119.

Kingdon HS. 1997. Gene therapy avoidance: Implanting allogeneic cells within a device. First International Symposium on Gene Therapy for Hemophilia. September 4–6, 1997, Chapel Hill, NC, USA, p. 17.

Liu H, Ofosu FA, Chang PL. 1993. Expression of human factor IX by microencapsulated recombinant fibroblasts. Human Gene Therapy 4:291–301.

Meulien P. 1990. XIX International Congress of the World Federation of Hemophilia, Washington, DC, USA, p. 70.

Nelson DM, Metzger ME, Donahue RE, Morgan RA. 1997. In vivo retrovirus-mediated gene transfer into multiple hematopoietic lineages in rabbits without preconditioning. Human Gene Therapy 8(6):747–754.

O'Shea G, Sun A. 1986. Encapsulation of rat islets of Langerhans prolongs xenograft survival in diabetic mice. Diabetes 35:943–946.

Peirone MA, Delany K, Kwiecin J, Fletch A, Chang PL. 1997a. Delivery of recombinant gene products to canines with non-autologous microencapsulated cells. Manuscript submitted.

Peirone MA, Fitzgerald P, Chang PL. 1997b. Delivery of recombinant gene product to canines using solid core calcium-alginate gel spheres. Manuscript submitted.

Pelegrin M, Marin M, Noel D, Del Rio M, Saller R, Stange J, Mitzner S, Gunzburg WH, Piechaczyk M. 1998. Systemic long-term delivery of antibodies in immunocompetent animals using cellulose sulphate capsules containing antibody-producing cells. Gene Therapy 5(6):828–34.

Pijnappels MIM, Briet E, van der Zwet GT, Huisden R, van Tilburg NH, Eulderink F. 1986. Evaluation of the cuticle bleeding time in canine haemophilia A. Thrombosis and Haemostasis 55:70–73.

Roberts, HR, Eberst ME. 1993. Current management of hemophilia B. Hematology/Oncology Clinics of North America 7(6):1269–1279.

Roman M, Axelrod JH, Dai Y, Naviaux RK, Friedmann T, Verma IM. 1992. Circulating human or canine factor IX from retrovirally transduced primary myoblasts and established myoblast cell lines grafted into murine skeletal muscle. Somatic and Cellular Molecular Genetics 18(3):247–258.

Smith TA, White BD, Gardner JM, Kaleko M, McClelland A. 1996. Transient immunosuppression permits successful repetitive intravenous administration of an adenovirus vector. Gene Therapy 3(6):496–502.

Snyder RO, Miao CH, Patijn GA. et al. 1997. Persistent and therapeutic concentrations of human factor IX in mice after hepatic gene transfer of recombinant AAV vectors. Nature Genetics 16:270–276.

Tinlin S, Webster S, Giles AR. 1993. The development of homologous (canine/anti canine) antibodies in dogs with haemophilia. A (factor VIII deficiency): A ten-year longitudinal study. Thrombosis and Haemostasis 69(1):21–24.

Uludag H, Sefton MV. 1993. Metabolic activity and proliferation of CHO cells in hydroxyethyl methacrylate-methyl methacrylate (HEMA-MMA) microcapsules. Cell Transplantation 1993 Mar–Apr;2(2):175–82.

White G, Shapiro A, Ragni M, Garzone P, Goodfellow J, Tubridy K, Courter S. 1998. Clinical evaluation of recombinant factor IX. Seminars in Hematology 35(2 Suppl 2):33–8.

Yao S, Kurachi K. 1992. Expression of human factor IX in mice after injection of genetically modified myoblasts. Proceedings of the National Academy of Sciences USA 89:3357–3361.

26
Microencapsulation—A Novel Gene Therapy for Lysosomal Storage Diseases

Colin J.D. Ross and Patricia L. Chang

Introduction

Over 30 different lysosomal storage diseases, each associated with deficiency of a specific lysosomal enzyme, have been described in man (Gieselmann 1995). Because of the ubiquitous presence of lysosomes in almost all cell types, deficiencies of these enzymes cause multisystem anomalies, often with catastrophic consequences, and frequently result in early death. Although the incidence of individual enzyme deficiency is rare, lysosomal storage diseases collectively occur in 1 in ~10,000 births. In spite of such significant prevalence and devastating consequences, there is no cure or even definitive treatment for most of these diseases. Until recently, bone marrow transplantation has been the only experimental treatment offered with some degree of success (Krivit et al 1990, Hoogerbrugge et al 1995). However, this procedure carries high rates of morbidity (graft-versus-host disease) and mortality (10% in matched and 20–25% in unmatched donors). Furthermore, even if the risks are acceptable, most patients (70%) cannot find compatible donors (Parkman 1986). In the majority of the affected families, the only medical interventions available are palliative care until death occurs, and prevention of recurrence through prenatal diagnosis. An alternative treatment available to a subtype of one form of lysosomal storage disease is enzyme replacement. For the non-neuropathic form of Gaucher's disease, administration of glucocerebrosidase, the deficient enzyme in question, has resulted in definite clinical improvements. This therapy is now an accepted form of treatment in the U.S., but only for those who can

afford it. Under the Orphan Drug Act, with no competitive markets, the annual cost for the enzyme Ceredase alone is $380,000/70 kg (FDA recommendation, 1991). Because of the risks and the difficulty in locating donors for bone marrow transplant, and the high cost for either bone marrow transplantation (>$100,000) or enzyme replacement treatment, more cost-effective therapies for the lysosomal storage diseases need to be developed. Since the genes for many lysosomal enzymes have been cloned, somatic gene therapy for lysosomal storage diseases may offer the ultimate solution to this serious healthcare problem (Beutler 1993).

Experimental Approaches to Therapy of Lysosomal Storage Diseases

There are three experimental treatment protocols potentially available for these disorders: bone marrow transplantation, enzyme replacement, and gene therapy (Table 26.1).

Bone Marrow Transplantation

This treatment has been tried in patients with various lysosomal storage diseases (Krivit et al 1992, Parkman 1986), e.g. metachromatic leukodystrophy (Krivit et al 1990, Ladisch et al 1986), Type I Gaucher's (Rappeport and Ginns 1984) and Hurler's diseases (Krivit and Whitley 1987). Clinical improvements such as diminished visceromegaly (Hurler's)

TABLE 26.1. Therapy for lysosomal storage diseases.

Disease	Deficient enzyme	Mode of inheritance	Therapy	Recipient	Reference
MPSI-Hurler's and Scheie's syndromes	α-L-Iduronidase	AR	BMT	Human	Krivit and Whitley, 1987
			BMT	dog	Breider et al., 1989
			ER	dog	Shull et al. 1994
			Retrovirus	human	Salvetti et al, 1994
			Retrovirus	mouse	Salvetti et al, 1995
			Retrovirus	dog	Shull et al, 1996
MPS II-Hunter's syndrome	Iduronate sulfatase	X-linked recessive	BMT	human	Bergstrom et al, 1994
MPS III-Sanflilippo's syndrome type B	α-N-Acetylglucosaminidase	AR	BMT	human	Vellodi et al, 1992
MPS VI-Maroteaux-Lamy syndrome	N-Acetylgalactosamine 4-sulfatase	AR	BMT	human	Krivit et al, 1984
			BMT	cat	Gasper et al, 1984
			ER	cat	Crawley et al, 1996
			BMT	rat	Simonaro et al, 1997
MPS VII-Sly's syndrome	β-Glucuronidase	AR	BMT	mouse	Birkenmeier et al, 1991
			Herpesvirus	mouse	Wolfe et al, 1992
			Retrovirus	mouse	Wolfe et al, 1992b
			ER	mouse	Volger et al, 1993, Sands et al, 1994
			CNS-P	mouse	Snyder et al, 1995
			Adenovirus	mouse	Ohashi et al, 1997
			Retrovirus	mouse	Marachel et al, 1993, Moulier et al, 1993, Li and Davidson, 1995, Lau et al, 1995, Naffakh et al. 1996, Taylor and Wolfe, 1997
Gaucher's disease	Glucocerebrosidase	AR	BMT	human	Rappeport and Ginns, 1984
			Retrovirus	mouse	Correll et al, 1989
			ER	human	Grabowski, 1993
			Retrovirus	human	Nimgaonkar et al, 1994
α-Manosidosis	α-Mannosidase	AR	BMT	cat	Walkley et al, 1994
Metachromatic leukodystrophy	Arylsulfatase-A	AR	BMT	human	Ladisch et al, 1986, Krivit et al, 1990
Niemann-Pick	Acid sphingomyelinase	AR	BMT	human	Vellodi et al, 1987
Tay-Sachs	β-Hexosaminidase	AR	Retrovirus	mouse	Lacorazza et al, 1996

Legend: MPS: mucopolysaccharidosis, AR: autosomal recessive, BMT: bone marrow transplant. ER: enzyme replacement, CNS-P: CNS progenitor cells. (Adapted from Neufeld and Muenzer, 1995)

and relief from bone pain (Gaucher's) have been observed after allogeneic bone marrow transplantation. The disadvantages of this treatment, as noted before, are the high risks and high cost, and the difficulty in securing matched donors.

Enzyme Replacement

Supplying purified enzyme to circumvent the enzyme deficiency is the basis of enzyme replacement therapies. In spite of early failures of this approach when exogenous enzymes were quickly cleared from the circulatory system with no observable clinical benefit (Bergsma et al 1973, Desnick 1979), recent refinements of this strategy have produced remarkable clinical improvements in patients with the Type I non-neuronopathic form of Gaucher's disease (see review by Grabowski 1993). Reduction in bone pain and organomegaly, and increase in hematocrit are some of the clearly obtainable improvements. However, because of the high and, for many patients, unrealistic cost of Ceredase treatment, alternate therapeutic strategies are desperately needed (Beutler 1993), particularly for the remaining lysosomal storage diseases whose relevant enzymes have not been purified at the industrial scale.

Gene Therapy

This therapy is theoretically available for many lysosomal storage diseases, because many genes encoding lysosomal enzymes have been cloned, e.g., β-hexosaminidases in Tay-Sachs and Sandhoff's diseases (Myerowitz and Proia 1984, O'Dowd et al 1985), glucocerebrosidase in Gaucher's disease (Ginns et al 1984, Sorge et al 1985), arylsulfatase-A in metachromatic leukodystrophy (Stein et al 1989), and β-glucuronidase in mucopolysaccharidosis Type VII (Oshima et al 1987). It offers a more direct approach to therapy because the high cost of protein purification required for enzyme replacement is eliminated. The correction of the enzyme deficiency in human Gaucher's affected cells has been reported in vitro after retroviral-mediated gene transfer (Fink et al 1990), and human clinical trials are currently under way with retrovirally transduced CD34$^+$ enriched cells (Nimgaonkar et al 1994) and for mucopolysaccharidosis Type 1 with retrovirally transduced autologous fibroblast implants (Salvetti et al 1994).

Microencapsulation—An Alternate Approach to Gene Therapy

To develop a more cost-effective approach, we initiated in our laboratory a radically different form of somatic gene therapy. Instead of engineering the patient's own cells, we proposed to create standard laboratory cell lines that produce desired therapeutic products for implantation into all patients requiring the same product replacement (Chang 1995). To avoid immune rejection by the host, these nonautologous cells are enclosed in immunoisolation devices based on the technology developed for allogeneic or xenogeneic transplants (see Section 1 of Part Three in this book). The need for patient-specific genetic engineering is thus removed, and the cost of treatment should be reduced.

By using a "universal" donor cell line engineered to produce a desired therapeutic product, we have demonstrated the feasibility of this idea both in vitro and in vivo. In vitro studies showed that genetically modified fibroblasts remained viable within these immunoprotective microcapsules (Chang et al

1994a) and continued to deliver recombinant gene products such as human growth hormone, factor IX (Liu et al 1993), and lysosomal enzymes (Awrey et al 1996).

In vivo studies showed that we can successfully deliver recombinant products such as human growth hormone (Chang et al 1993) and human factor IX (Hortelano et al 1996, see also Chapter 27) into the systemic circulation of rodents for several months. The clinical efficacy of this strategy was also proven by the correction of the growth retardation of the Snell dwarf mice suffering from growth hormone deficiency (Al-Hendy et al 1995, see also Chapter 27).

The lysosomal storage diseases provide a particularly good model to test the efficacy of this approach. The diseases are devastating and in need of effective therapy. Unlike growth hormone or insulin, the required therapeutic product does not require tight regulation, and only low levels are necessary to produce therapeutic effects. In fact, many so-called pseudodeficient patients have been identified as being phenotypically normal but biochemically deficient in lysosomal enzymes. They are usually found with only ≤ 10% of the normal level of enzyme activity (Thomas 1994), suggesting that such low levels of enzymes are sufficient to limit or stop the onset of the disease (Brooks et al 1991).

Animal Models of Lysosomal Storage Diseases

Many animal models of human lysosomal storage diseases are available, providing exciting opportunities to explore different strategies in gene therapy, including that of microencapsulation. Animal models of MPS I have been reported in the cat (Haskins et al 1979), dog (Shull et al 1982, Spellacy et al 1983), and recently in the mouse (Clarke et al 1997). An animal model of MPS IIID has been described in a Nubian goat (Thompson et al 1992). Feline (Gasper et al 1984, Haskins et al 1980, Jezyk et al 1977), rat (Yoshida et al 1993), and mouse (Evers et al 1996) models of MPS VI with mutated arylsulfatase B genes have been established, with lesions that resembled those described in the human disease (McGovern et al 1985). A murine (Birkenmeier et al 1989) as well as a canine (Haskins et al 1984) model of MPS VII has also been reported. A cat model of α-mannosidosis has been well documented (Vandevelde et al 1982). An acid-sphingomyelinase-deficient mouse model generated by gene targeting closely resembles the human form of Niemann-Pick disease (Otterbach and Stoffel 1995). Canine fucosidosis models have been described in the U.K. (Littlewood et al 1983) and Australia (Hartley et al 1982; Taylor et al 1989). Transgenic murine models of Tay-Sachs disease and the related Sandhoff's disease (Liu et al 1997) have also been generated (Yamanaka et al 1994, Cohen-Tannoudji et al 1995). Among these, perhaps the murine model of the MPS VII has been most extensively studied.

Gus^{mps}/Gus^{mps} Mouse

This mouse model of the human lysosomal storage disease mucopolysaccharidosis Type VII (MPS VII) (Sly et al 1973) was discovered by Birkenmeier and his coworkers in 1989. Since then, it has been thoroughly characterized at the molecular (Sands and Birkenmeier 1993), pathological (Birkenemeier et al 1989, Vogler et al 1990), and behavioral levels (Chang et al 1994b), rendering it a superior animal model for developing strategies for treating lysosomal storage diseases. MPS VII is an autosomal recessive disorder caused by deficient β-glucuronidase activity, resulting in the failure to cleave the terminal β-glucuronic acid from various glycosaminoglycans, e.g., chondroitin sulfate, dermatan sulfate, and heparin sulfate, which accumulate in the lysosomes. The mutant mice show characteristic dysmorphic facial features, abnormal gait, skeletal and joint abnormalities, dwarfism, organomegaly, and a shortened life span (Birkenmeier et al 1989, Vogler et al 1990). The mutation is caused by a single base deletion in exon 10, resulting in a premature stop codon and complete absence of translation product, thus accounting for the total enzyme deficiency (Sands and Birkenmeier 1993).

Transgenic studies also proved that this model is a genuine counterpart of the human disease (Kyle et al 1990). Transgenic mutants, gus^{mps}/gus^{mps} mice that were genetically engineered at the embryonic stage to carry the normal human gene, produced a completely normal phenotype. Hence, therapeutic

success produced in this mutant will be directly relevant to the human disease. These mutants have neurological deficits that can be measured with behavioral tests (Chang et al 1994b).

All the experimental protocols currently available for lysosomal storage diseases have been tried on the gus^{mps}/gus^{mps} mice. As expected, bone marrow transplantation was effective in correcting the abnormal storage material in the peripheral organs. The life span of the treated mutants was prolonged from the usual 5–7 months to well over a year (Birkenmeier et al 1991), but, unfortunately, no neurological functional improvement was observed (Bastedo et al 1994). Similarly, other conventional approaches, such as enzyme replacement (Sand set al 1994, Vogler et al 1993) or gene therapy through implantation of retrovirally transfected syngeneic mutant bone marrow cells (Wolfe et al 1992a, Wolfe et al 1992b, Marechal et al 1993), skin fibroblasts (Moullier et al 1993), or myoblasts (Naffakh et al 1996) as well as direct viral infection of the CNS (Wolfe et al 1992b), have resulted in biochemical and histological improvements, resulting in diminished lysosomal storage and restoration of a low level of β-glucuronidase activity.

Methods for direct in vivo delivery of β-glucuronidase to MPS VII affected mice by several types of viral vectors have been explored (Lau et al 1995, Li and Davidson 1995, Wolfe et al 1992a). Attempts have been made at introducing β-glucuronidase directly into the CNS (see review, Sly and Vogler 1997). Herpes-infected cells expressing β-glucuronidase were found in the trigeminal ganglia and brainstems for up to 4 months (Wolfe et al 1992a), and adenoviral-mediated gene-transfer resulted in detectable β-glucuronidase activity near the implant site and scattered in the parenchyma for at least 2 weeks (Ohashi et al 1997). Immortalized neural progenitor cells expressing β-glucuronidase transplanted into the cerebral ventricles of newborn mice showed an overall average of 11% of normal β-glucuronidase activities throughout the central nervous system (CNS) in mutant mice with successful grafts (Snyder et al 1995). Implanted retroviral-corrected fibroblasts delivered up to 4.7% of normal levels of β-glucuronidase adjacent to the graft site and showed a clearance of lysosomal distention in neurons and glia near the graft site (Taylor and Wolfe 1997).

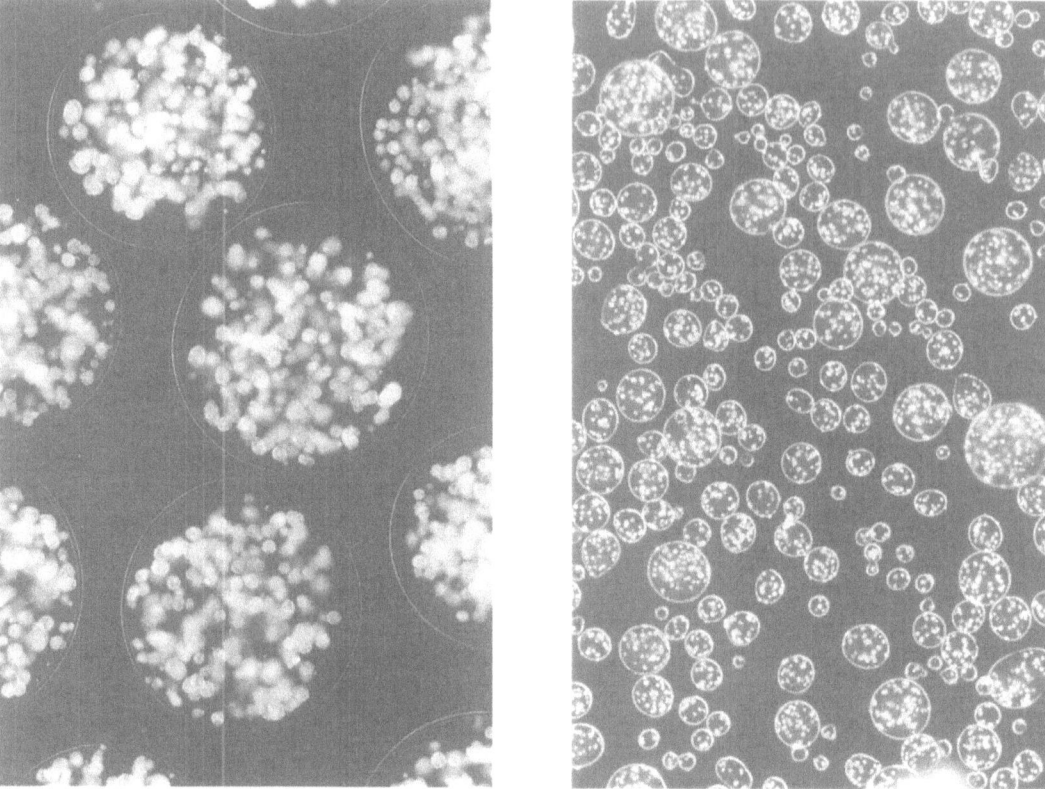

A ━━━   B ━━

FIGURE 26.1. Microcapsules enclosing 2A-50 cells. 2A-50 fibroblasts were encapsulated within alginate poly-l-lysine-alginate capsules. All cells were encapsulated as described by Chang et al (1994) and modified by Hortelano et al (1996). Large microcapsules shown 2 weeks postencapsulation (A) were used for peripheral implantation (500–1000 μm diameter, secreting 522 nmol β-glucuronidase activity/10^6 cells/h). Mice were anesthetized with isofluorane (Anaquest). Microcapsules were introduced into the peritoneal cavity with a 16-gauge catheter (4200 microcapsules or 1×10^6 cells implanted per mouse). Microcapsules for neural implantation shown 2 days postencapsulation (B) were smaller (200 μm average diameter, secreting 500 nmol β-glucuronidase activity/10^6 cells/h). Variations in concentric air flow rate delivered to the tip of the needle in the encapsulation process were manipulated to create smaller microcapsules (4 L/min for large microcapsules, 8 L/min for small microcapsules). Mice were anesthetized and 5 μl of microcapsules (~150 small microcapsules or 2.5×10^3 cells) were implanted within each of the lateral ventricles of the CNS from a 10 μl capacity glass capillary pipette (Drummond). Scale bars 500 μm.

β-Glucuronidase Delivery to Peripheral Organs

Mouse 2A50 fibroblasts expressing mouse β-glucuronidase enclosed in alginate-poly-l-lysine-alginate (APA) microcapsules were implanted into the peritoneal cavity of mice with MPS VII (Figure 26.1)

(Bastedo et al, 1994). After 24 hours, β-glucuronidase activity was detected in the plasma, reaching up to 66% of physiological levels. Plasma β-glucuronidase levels remained above background levels for at least 4 weeks (Figure 26.2). Concomitantly, urinary GAG/creatin content was decreased after 1 week and reduced to normal levels by 3 weeks postimplantation.

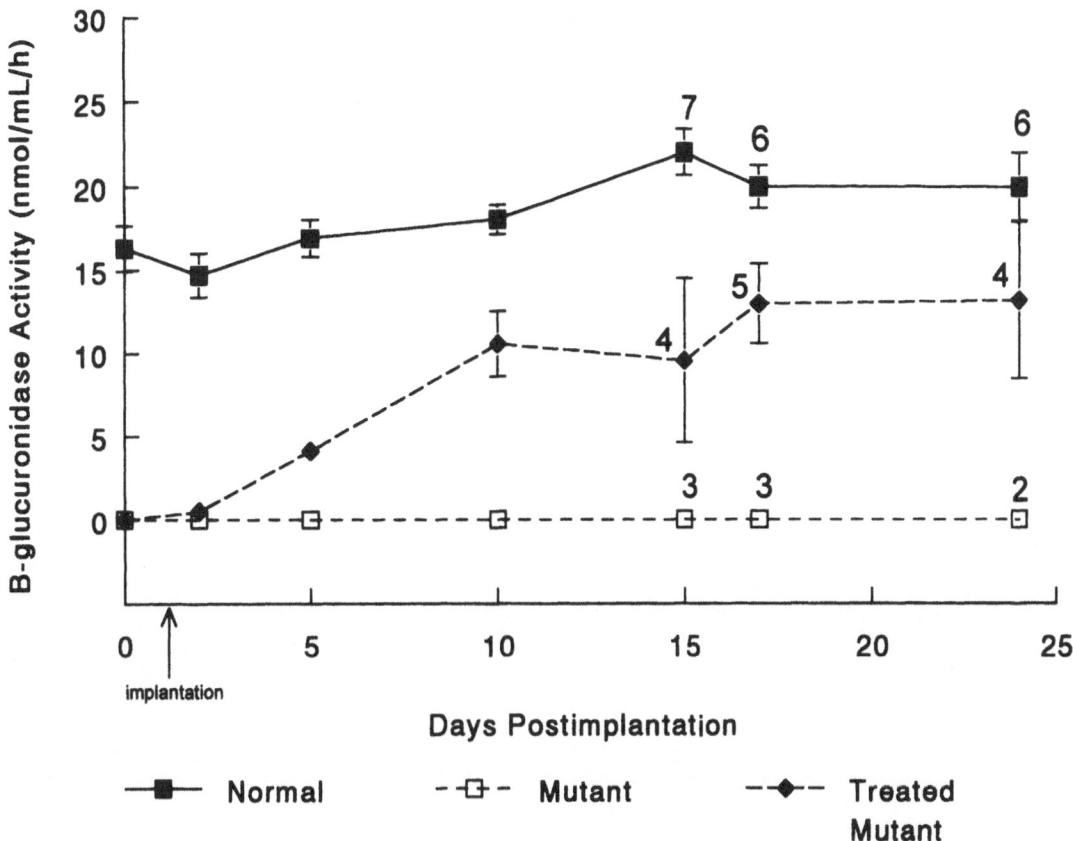

FIGURE 26.2. Delivery of β-glucuronidase to plasma. β-glucuronidase activity in the plasma of treated mutants, normal, and mutant controls. β-glucuronidase activities were assayed using the substrate 4-MU-β-glucuronide (Sigma). Day 0 represents beseline values before therapy, and Day 1 represents day of implantation. Data are the mean ± SEM or range (n < 3). Except where indicated, n = 13 normal controls, n = 9 mutant controls, and n = 9 treated mutants.

After 4 weeks, significant β-glucuronidase activity was detected in kidney (20.2% of normal), liver (43.5%), and spleen (65.8%), and at a much lower level at 8 weeks in the liver (0.3%). No significant level of β-glucuronidase was detected in the central nervous system. Tissue sections showed a dramatic reduction in lysosomal storage lesions in the spleen and pathology reversal of the liver for at least 8 weeks (Figure 26.3). However, antibodies against the recombinant β-glucuronidase were detected in implanted mutant mice at 6 weeks, concurrent with the decrease of plasma β-glucuronidase to background levels. The mounting of antibodies against a therapeutic product is another potential problem that must be resolved (Potter et al 1997) before gene or product replacement therapy can be fully effective for patients with such null mutation phenotypes.

β-Glucuronidase Delivery to the Central Nervous System

The blood-brain barrier is a protective mechanism that impedes the passage of many substances. Small nutrient molecules such as oxygen, glucose, and amino acids easily pass through the blood-brain barrier. However, larger molecules such as insulin pass through the barrier very slowly, and some molecules are almost completely excluded from the brain. The blood-brain barrier appears to restrict entry of β-glucuronidase into the CNS. Thus, attempts have been made at introducing β-glucuronidase directly into the CNS.

Immunoisolation of implanted cells confers several important advantages for a safe and economical strategy for gene therapy of diseases affecting

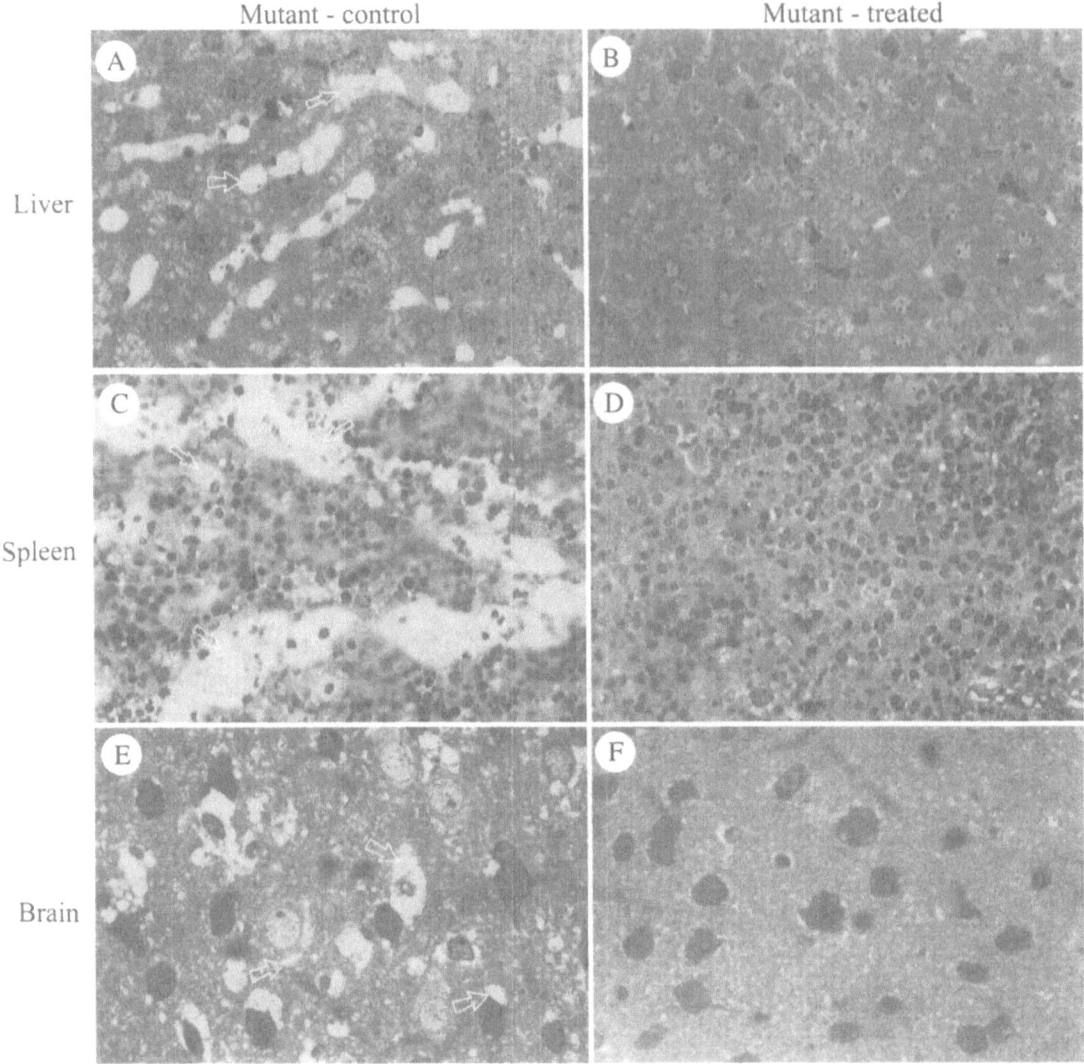

FIGURE 26.3. Histology of liver, spleen, and brain. Toluidine-blue-stained sections of mutant age-matched control liver show the white intralysosomal accumulation of GAGs (A). A reduction in liver pathology in treated mutants was evident for at least up to 8 weeks postimplantation of microcapsules within the intraperitoneal cavity (B).

Lysosomal distention in the spleen of age-matched mutant controls is apparent (C). Treated mutants show reduced lysosomal distention in the splenic sinusoidal lining cells for at least up to 8 weeks postimplantation of microcapsules within the intraperitoneal cavity (D).

Toluidine-blue-stained sections of age-matched mutant control brain show severe disease pathology of distended lysosomes (E). In CNS-treated mutant brain, 7 weeks postimplantation of microcapsules within the lateral ventricles of the CNS, there was a disappearance of disease pathology in all cells near the implantation sites (F). Original magnification 200 × (A–D), 1000 × (E, F).

the CNS. Encapsulated cell transplantation methods are not complicated by tissue availability, nor by the ethical concerns associated with the use of human fetal cell transplantation. Instead, a single "universal" recombinant cell line may be used to treat many patients who require the same recombinant gene product. Cell encapsulation can inhibit the formation of tumors (Hoffman et al 1993, Jaeger et al 1988, Schinstine et al 1995, Tresco et al 1992, Winn et al 1991), which has been seen as a result of implantation of immortalized cells within the CNS (Cunningham et al 1991, Hatton et al 1992, Jaeger 1985, Miyazono et al 1995). Furthermore, the encapsulation of cells prior to implantation in the CNS confers long-term survival of encapsulated, nonautologous cells compared to unencapsulated cells (Aebischer et al 1991b).

Intraventricular implantation of encapsulated fibroblasts secreting β-glucuronidase into the CNS was examined as a means to supply β-glucuronidase to the CNS of MPS VII affected mice. Mouse 2A50 fibroblasts secreting β-glucuronidase encapsulated within small APA microcapsules (100 μm diameter) were implanted into the left and right lateral ventricles of MPS VII mice (Figure 26.1). β-glucuronidase was detected in the CNS at both week 3 and week 8 postimplantation. Levels ranged from 3.6–7314% of normal at week 3, and 0.38–250% of normal at week 8. The highest β-glucuronidase levels were located at the implantation sites, with corresponding decreases at distances farther away from the implantation sites. There were concomitant decreases in secondarily elevated lysosomal enzyme levels that are abnormally increased in affected mice, in regions of the CNS that received β-glucuronidase. Lysosomal storage lesions were reduced in the CNS near the implantation sites (Figure 26.3) (Ross et al 1998).

To examine behavioral change as a result of β-glucuronidase delivery to the CNS, circadian rhythms of mice were examined before and after treatment. Circadian rhythms were examined by measuring mouse activity on wheels inside cages. Mice were initially placed in 12-hour cycles of light and dark for 7 days to acclimatize the mice to the cages and wheels. Mice were then placed in complete darkness for 7–10 days to determine the "pretreatment" circadian rhythms. Mice were treated and then returned to complete darkness to measure "posttreatment" circadian rhythms. Experimental mutant mice, treated with microcapsules containing β-glucuronidase-secreting fibroblasts, showed pos-

itive changes in circadian rhythms after treatment (Figure 26.4). Fragmentation was decreased, and stability of onset was more consistent. Control mutant mice treated with microcapsules that contained mouse 3521 MPS VII affected fibroblasts that do not secrete β-glucuronidase showed no change in circadian rhythms, only a slow progressive decline in overall activity over time, as observed in untreated mutant mice (Ross et al 1998).

Encapsulated cell therapy of the MPS VII mouse CNS shows the potentially high levels of enzyme delivery from a genetically modified universal cell line specifically intended to secrete high levels of a specific gene product.

Conclusion

As a "proof of principle," this approach of encapsulating nonautologous cells has been shown to be clinically effective in treating not only the somatic but also the behavioral abnormalities of lysosomal storage disease. Instead of engineering autologous patient-specific cells, as is currently practiced (Anderson 1995), a single cell line can be established as a standard reagent. It may be stored in liquid nitrogen permanently and retrieved to propagate when needed. It can be used for different patients with the same genetic disorder, thus reducing the healthcare cost and increasing the affordability of somatic gene therapy.

However, it is also particularly important to recognize that clinical application of gene therapy must be based on adequate preclinical data. With the current commercial pressure for taking gene therapy to human clinical trials, it is important to heed the recommendation from the NIH panel (December 1995) calling for more basic study on gene transfer and testing of efficacy. Hence, in spite of the current success with the mouse models, the need for testing this technology further cannot be overemphasized. It will be desirable to evaluate thoroughly the long-term risks and efficacy, particularly in larger animals models (Wilson 1996) more similar to human beings before bringing this technology to treatment of human lysosomal storage diseases.

Acknowledgment. This work was supported by the Medical Research Council of Canada and the Ontario Mental Health Foundation. We would like to thank Ms. S. Hymus for help with the manuscript.

Normal

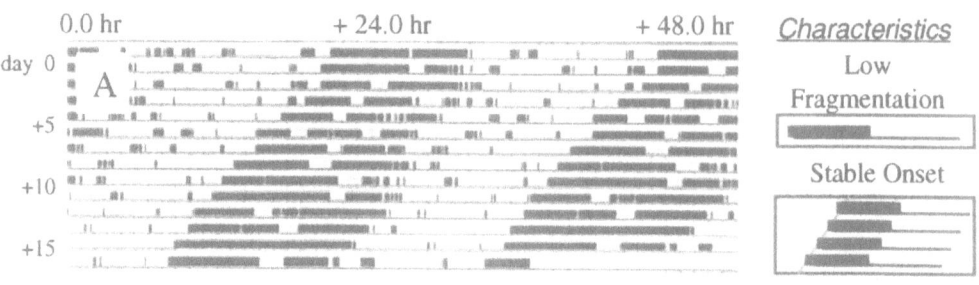

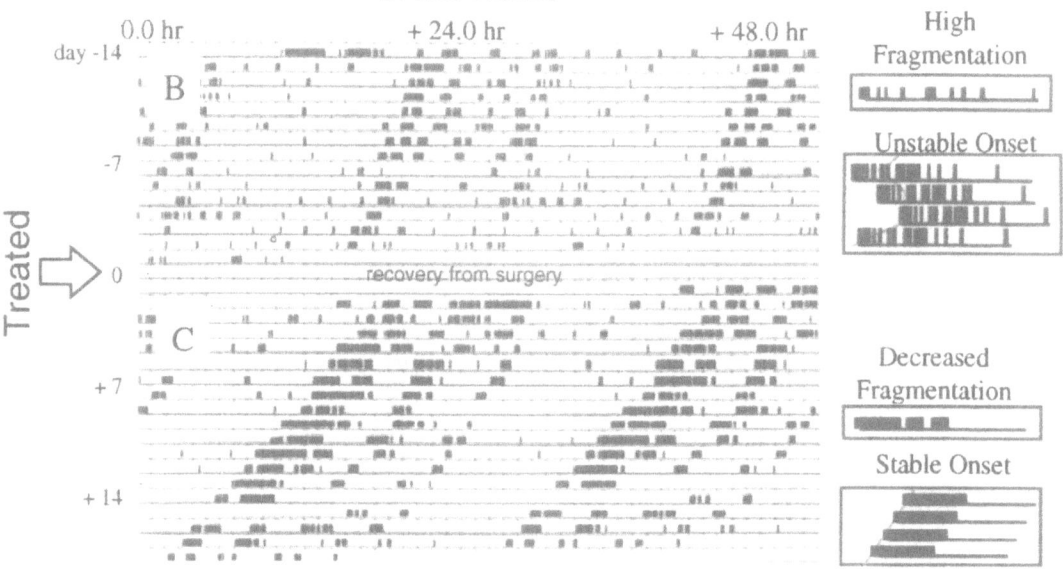

FIGURE 26.4. Circadian rhythm behavioral reversal of treated MPS VII mice. Circadian rhythms of mice were examined in (A) normal, (B) before, and (C) after treatment. The graphic analysis of circadian rhythms shows black-boxed periods of activity compared to white periods of inactivity. The data reflect a period of 48 hours (left to right) to visualize the onset and termination of daily activity. Each line down represents 1 day. Normal mice (A) have a low fragmentation of daily activity and inactivity, stable onset and termination of activity, and a stable period. Mutant mice (B) have highly fragmented activity and an unstable onset and cessation of daily activity. The circadian rhythms of treated mice (C) were compared to pretreatment patterns (B) for the same mouse to reduce the influence of individual variation. After treatment the mutant mice show an increasingly less fragmented pattern and a more stable onset and cessation of activity.

References

Aebischer P, Wahlberg L, Tresco PA, and Winn SR. 1991a. Macroencapsulation of dopamine-secreting cells by coextrusion with an organic polymer solution. Biomaterials. 12: 50–56.

Aebischer P, Tresco PA, Sagen J, and Winn SR. 1991b. Transplantation of microencapsulated bovine chromaf-fin cells reduces lesion-induced rotational asymmetry in rats. Brain Res. 560: 43–49.

Al-Hendy A, Hortelano G, Tannenbaum GS, and Chang PL. 1995. Correction of the growth defect in dwarf mice with non-autologous micro-encapsulated myoblasts—an alternate approach to somatic gene therapy. Hum. Gene Ther. 6: 165–175.

Anderson WF. 1995. Gene therapy. Sci. Am. 273: 124–128.

Awrey D, Tse M, Hortelano G, and Chang PL. 1995. Permeability of alginate microcapsules to secretory recombinant gene products. Manuscript submitted.

Bastedo L, Sands MS, Hortelano B, Al-Hendy A, and Chang PL. 1994. Partial correction of murine mucopolysaccaridosis VII with micro-encapsulated non-autologous recombinant fibroblasts. Am. J. Hum. Genet. 55:A211.

Bastedo L, Sands MS, Lambert DT, Pisa MA, Birkenmeier E, and Chang P. 1994. Behavioral consequences of bone marrow transplantation in the treatment of murine mucopolysaccharidosis type VII. J. Clin. Invest. 94: 1180–1186.

Bergsma D, Desnick R, Bernlohr RW, and Krivit W. (eds) 1973. Enzyme therapy in genetic diseases. Baltimore: Williams & Wilkins for the National Foundation-March of Dimes, BD:OAS IX(2).

Bergstrom SK, Quinn JJ, Greenstein R, and Ascensao J. 1994. Long-term follow up of a patient transplanted for Hunter's disease type IIB: a case report and literature review. Bone Marrow Trans. 14: 653–658.

Beutler E. 1993. Gaucher disease as a paradigm of current issues regarding single gene mutations of humans. Proc. Natl. Acad. Sci. USA 90: 5384–5390.

Birkenmeier EH, Davisson MT, Beamer WG, Ganschow RE, Vogler CA, Gwynn B, Lyford KA, Maltais LM, and Wawrzyniak CJ. 1989. Murine mucopolysaccharidosis type VII-characterization of a mouse with β-glucuronidase deficiency. J. Clin. Invest. 83: 1258–66.

Birkenmeier EH, Barker JE, Vogler CA, Kyle JW, Sly WS, Gwynn B, Levy B, and Pegors C. 1991. Increased life span and correction of metabolic defects in murine mucopolysaccharidosis type VII after syngeneic bone marrow transplantation. Blood. 78: 3081–3092.

Breider MA, Shull RM, and Constantopoulos G. 1989. Long-term effects of bone marrow transplantation in dogs with mucopolysaccharidosis I. Am J Pathol. 134: 677–692.

Brooks DA, McCourt PA, Gibson GJ, Ashton LJ, Shutter M, and Hopwood JJ. 1991. Analysis of N-acetylgalactosamine-4-sulfatase protein and kinetics in mucopolysaccharidosis type VI patients. Am. J. Hum. Genet. 48: 710–719.

Chang PL, Shen N, and Westcott AJ. 1993. Delivery of recombinant gene products with microencapsulated cells in vivo. Hum. Gene Ther. 4: 433–440.

Chang PL, Hortelano G, Tse M, and Awrey DE. 1994a. Growth of recombinant fibroblasts in alginate microcapsules. Biotech. Bioengin. 43: 925–933.

Chang PL, Lambert DT, and Pisa MA. 1994b. Behavioural abnormalities in a murine model of a human lysosomal storage disease. NeuroReport. 4: 507–510.

Chang PL. 1995. Non-autologous somatic gene therapy. In: Somatic gene therapy (Chang, PL, ed.) Boca Raton, Florida: CRC, Chap. 12.

Clarke L, Russell CS, Pwnall S, Warrington CL, Borowski A, Dimmick JE, Toone J, and Jirik FR. 1997. Murine mucopolysaccharidosis type I: targeted disruption of the murine α-L-iduronidase gene. Hum. Mol. Gen. 6: 503–511.

Cohen-Tannoudji M et al. 1995. Disruption of murine Hexa gene leads to enzymatic deficiency and to neuronal lysosomal storage, similar to that observed in Tay-Sachs disease. Mam. Genome. 6: 844–849.

Correll PH, Fink JK, Brady RO, Perry LK, and Karlsson S. 1989. Production of human glucocerebrosidase in mice after retoroviral gene transfer into multipotential hematopoietic progenitor cells. Proc. Nat. Acad. Sci. USA. 86: 8912–8916.

Crawley AC, Brooks DA, Muller VJ, Petersen BA, Isaac EL, Bielicki J, King BM, Boulter CD, Moore AJ, Fazzalari NL, Anson DS, Byers S, and Hopwood JJ. 1996. Enzyme replacement therapy in a feline model of Maroteaux-Lamy syndrome. J. Clin. Invest. 97: 1864–1873.

Cunningham LA, Short MP, Vielkind U, Breakefield XO, and Bohn MC. 1991. Survival and differentiation within adult mouse striatum of grafted rat pheochromocytoma cells (PC12) genetically modified to express recombinant β-NGF. Exp Neurobiol. 112: 174–182.

Desnick RJ, Dean KJ, Grabowski G, Bishop DF, and Sweeley CC. 1979. Enzyme therapy in Fabry disease-differential in vivo plasma clearance and metabolic effectiveness of plasma and splenic alpha-galactosidase A isozymes. Proc. Natl. Acad. Sci. USA 76: 5326–5330.

Evers M, Saftig P, Schmidt P, Hafner A, McLogin DB, Schmahl W, Hess B, von Figura K, and Peters, C. 1996. Proc. Nat. Acad. Sci. USA. 93: 8214–8219.

Fink JK, Correll PH, Perry LK, Brady RO, and Karlsson S. 1990. Correction of glucocerebroside deficiency after retroviral-mediated gene transfer into hematopoietic progenitor cells from patients with disease. Proc. Nat. Acad. Sci. USA. 87: 2334–2338.

Food and Drug Administration 1991. FDA Med. Bull. 21: 6–7.

Gasper PW, Thrall MA, Wenger DA, Macy DW, Ham L, Dornsife RE, McBiles K, Quackenbush SL, Kesel ML, and Gillette EL. 1984. Correction of feline arylsulphatase B deficiency (mucopolysaccharidosis VI) by bone marrow transplantation. Nature. 312: 467–469.

Gieselmann V. 1995. Lysosomal storage disease. Biochim. Biophys. Acta Mol. Basis Dis. 1270: 103–136.

Ginns EI, Choudary PV, Martin BM, Winnfield S, Stubblefield B, Mayor J, Merkle-Lehman D, Murray GJ, Bowers LA, and Barranger JA. 1984. Isolation of cDNA clones for human β-glucocerebrosidase using the λgt11 expression system. Biochem. Biophys. Res. Comm. 123: 574–80.

Grabowski GA. 1993. Gaucher disease: Enzymology, genetics, and treatment. Adv. Hum. Genet. 21: 377–441.

Hartley WJ, Canfield PJ, and Donnely TM. 1982. A suspected new canine storage disease. Acta Neuropathol. 56: 225–232.

Haskins ME, Jezyk PF, Desnick RJ, McDonaugh SK, and Patterson DF. 1979. Alpha-L-iduronidase deficiency in a cat: a model of mucopolysaccharidosis I. Pediat. Res. 13: 1294–1297.

Haskins ME, Aguirre GD, Jezyk PF, and Patterson DF. 1980. The pathology of the feline model of mucopolysaccharidosis VI. Am. J. Path. 101: 657–666.

Haskins ME, Desnick RJ, DiFerrante N, Jezyk PF, and Patterson DF. 1984. β-glucuronidase deficiency in a dog: a model of mucopolysaccharidosis type VII. Pediatr. Res 18: 980–984.

Hatton JD, Lechtman AN, and U, HS. 1992. Formation of PC12 tumors after implantation into rat brain: dependence of time course host stage. Cancer Res. 52: 1933–1937.

Hoffman D, Breakefield XO, Short MP, and Aebischer P. 1993. Transplantation of a polymer-encapsulated cell line genetically engineered to release NGF. Exp. Neurol. 122: 100–106.

Hoogerbrugge PM, Brouwer OF, Bordigoni P, Ringden O, Kapaun P, Ortega JJ, O'Meara A, Cornu G, Souillet G, Frappaz. D, Blanche S, and Fischer A. 1995. Allogeneic bone marrow transplantation for lysosomal storage diseases. Lancet 345: 1398–1402.

Hortelano G, Al-Hendy-A and Chang PL. 1996. Delivery of factor IX in mice using implantable microcapsules: towards gene therapy of hemophilia B. Blood 87: 5095–5103.

Jaeger CB. 1985. Immunocytochemical study of PC12 cells grafted into the brain of immature rats. Exp. Brain Res. 59: 615–624.

Jaeger CB, Kapoor R, and Llinas R. 1988. Cytology and organization of rat cerebellar organ cultures. Neuroscience. 26: 509–538.

Jezyk PF, Haskins ME, Patterson DF, Mellman WJ, and Greenstein M. 1977. Mucopolysaccharidosis in a cat with arylsulfatase B deficiency: a model of Maroteaux-Lamy syndrome. Science 198: 834–836.

Krivit W, Pierpont ME, Ayaz K, Tsai M, Ramsay NK Kersey JH, Weisdorf S, Sibley R, Snover D, and McGovern MM. 1984. Bone-marrow transplantation in the Maroteaux-Lamy syndrome (mucopolysaccharidosis type VI). Biochemical and clinical status 24 months after transplantation. N. Engl. J. Med. 1984 311: 1606–1611.

Krivit W, and Whitley CB. 1987. Bone marrow transplantation for genetic diseases. New Engl. J. Med. 316: 1085–7.

Krivit W, Whitley CB, Chang PN, Shapiro N, Belani KG, Snover D, Summers GC, and Blazar B. 1990. Lysosomal storage diseases treated by bone marrow transplantation: Review of 21 patients. In: Bone marrow transplantation in children (Johnson, FL and Pochedly, C, eds.) New York: Raven Press.

Krivit W, Shapiro E, Hoogerbrugge PM, and Moser HW. 1992. State of the art review: bone marrow transplantation treatment for storage diseases. Bone Marrow Transplant. 10 Suppl. 1: 87–96.

Kyle JW, Birkenmeier EH, Gwynn B, Vogler C, Hoppe PC, Hoffmann JW, and Sly WS. 1990. Correction of murine mucopolysaccharidosis VII by a human β-glucuronidase transgene. Proc. Natl. Acad. Sci. USA. 87: 3914–3918.

Lacorazza HD, Flax JD, Snyder EY, and Jendoubi M. 1996. Expression of human beta-hexosaminidase alpha-subunit gene (the gene defect of Tay-Sachs disease) in mouse brains upon engraftment of transduced progenitor cells. Nat. Med. 2: 424–429.

Ladisch S, Bayeuer E, Philipport M, and Feig S. 1986. Biochemical findings after bone marrow transplantation for metachromatic leukodystrophy: a preliminary report. Birth Defects. 22: 69–78.

Lau C, Soriano HE, Ledley FD, Finegold MJ, Wolfe JH, Birkenmeier EH, and Henning SJ. 1995. Retroviral gene transfer into the intestinal epithelium. Hum. Gene Ther. 6: 1145–1155.

Li T, and Davidson BL. 1995. Phenotype correction in retinal pigment epithelium in murine mucopolysaccharidosis VII by adenovirus-mediated gene transfer. Proc. Natl. Acad. Sci. USA 92: 7700–7704.

Liu H, Ofosu FA, and Chang PL. 1993. Expression of human factor IX by microencapsulated recombinant fibroblasts. Hum. Gene Ther. 4: 291–301.

Liu Y, Hoffmann A, Grinberg A, Westphal H, McDonald MP, Miller KM, Crawley JN, Sandhoff K, Suzuki K, and Proia RL. 1997. Mouse model of GM2 activator deficiency manifests cerebellar pathology and motor impairment. Proc. Natl. Acad. Sci. USA. 94: 8138–8143.

Littlewood JD, Herrtage ME, and Palmer AC. 1983. Neuronal storage disease in English springer spaniels. Vet. Record. 112: 86–87.

Marechal V, Naffakh N, Danos O, and Heard JM. 1993. Disappearance of lysosomal storage in spleen and liver of mucopolysaccharidosis VII mice after transplantation of genetically modified bone marrow cells. Blood. 82: 1358–65.

McGovern MM, Mandell N, Haskins M, and Desnick RJ. 1985. Animal model studies of allelism: characterization of arylsulfatase B mutations in homoallelic and heteroallelic (genetic compound) homozygotes with feline mucopolysaccharidosis VI. Genetics. 110: 733–749.

Miyazano M, Lee VM, and Trojanowski JQ. 1995. Proliferation, cell death, and neural differentiation in transplanted human embryonal carcinoma (Ntera2) cells depend on the graft site in nude and SCID mice. Lab. Invest. 73: 273–283.

Moullier P, Bohl D, Heard J-M, and Danos O. 1993. Correction of lysosomal storage in the liver and spleen of MPS VII affected mice by implantation of genetically modified skin fibroblasts. Nat. Gen. 4: 154–159.

Myerowitz R and Proia RL. 1984. cDNA clone for the a-chain of human B-hexosaminidase: deficiency of a-chain mRNA in Ashkenazi Tay-Sachs fibroblasts. Proc. Natl. Acad. Sci. 81: 5394–8.

Naffakh N, Pinset C, Montarras D, Li Z, Paulin D, Danos O, and Heard J. 1996. Long term secretion of therapeutic proteins from genetically modified skeletal muscles. Hum. Gen. Ther. 7: 11–21.

Neufeld EF and Muenzer J. 1995. The mucopolysaccharidoses. In: The metabolic and molecular bases of inherited disease. (Scriver, CR, Beaudet, AL, Sly, WS, Valle, DV, eds. New York: McGraw-Hill Inc.

Nimgaonkar, MT, Bahnson, AB, Mannion-Henderson, J, Barranger, JA, and Ball, ED. 1994. Hematopoietic stem cells as targets for gene therapy in Gaucher disease. Cold Spring Harbor Laboratory Meeting on Gene Therapy, p. 7, Sept, 1994.

O'Dowd BF, Quan F, Willard HF, Lamhonwah AM, Korneluk RG, Lowden JA, Gravel RA, and Mahuran DJ. 1985. Isolation of cDNA clones coding for the beta subunit of human beta-hexosaminidase. Proc. Natl. Acad. Sci. USA. 82: 1184–8.

Ohashi T, Watabe K, Uehara K, Sly WS, Vogler C, and Eto Y. 1997. Adenovirus-mediated gene transfer and expression of human beta-glucuronidase gene in the liver, spleen, and central nervous system in mucopolysaccharidosis type VII mice. Proc. Natl. Acad. Sci. USA. 94: 1287–1292.

Oshima AJ, Kyle W, and Miller RD, Hoffmann JW, Powell PP, Grubb JH, Sly WS, Tropak M, Guise KS, and Gravel RA. 1987. Cloning, sequencing and expression of cDNA for human β-glucuronidase. Proc. Natl. Acad. Sci. 84: 685–9.

Otterback B and Stoffel W. 1995. Acid sphingomyelinase-deficient mice mimic the neurovisceral form of human lysosomal storage disease (Niemann-Pick disease). Cell. 81: 1053–1061.

Parkman R. 1986. The application of bone marrow transplantation to the treatment of genetic diseases. Science. 232: 1373–7.

Potter MA, Hymus SA, Stockley TL, and Chang PL. 1998. Suppression of immunological response against a transgene product delivered with microencapsulated cells. Hum. Gen. Ther. 9: 1275–1282.

Rappeport JM and Ginns EI. 1984. Bone marrow transplantation in severe Gaucher's disease. New Engl. J. Med. 311: 84–8.

Ross CJD et al. 1998. Allogeneic gene therapy of the central nervous system in the murine model of mucopolysaccharidosis type VII: Treatment and behavioral recovery. Manuscript to be submitted.

Salvetti A, Moullier P, Cornet V, Brooks D, Heard JM, and Danos O. 1994. Preclinical studies for the in vivo delivery of human α-iduronidase in mucopolysaccharidosis type I. Cold Spring Harbor Meeting on Gene Therapy, September, 1994, p. 146.

Salvetti A, Moullier P, Cornet V, Brooks D, Hopwood JJ, Danos O, and Heard JM. 1995. In vivo delivery of human alpha-L-iduronidase in mice implanted with neo-organs. Hum. Gene Ther. 6: 1153–1159.

Sands MS, and Birkenmeier EH. 1993. A single-base-pair deletion in the beta-glucuronidase gene accounts for the phenotype of murine mucopolysaccharidosis type VII. Proc. Natl. Acad. Sci. U.S.A. 90: 6567–6571.

Sands MS, Vogler C, Kyle JW, Grubb JH, Levy B, Galvin N, Sly WS, and Birkenmeier EH. 1994. Enzyme replacement therapy for murine mucopolysaccharidosis type VII. J. Clin. Invest. 93: 2324–31.

Sands MS, Erway LC, Vogler C, Sly WS, and Birkenmeier EH. 1995. Syngeneic bone marrow transplantation reduces the hearing loss associated with murine mucopolysaccharidosis type VII. Blood. 86: 2033–2040.

Schinstine M, Fiore DM, Winn SR, and Emerich DF. 1995. Polymer-encapsulated schwannoma cells expressing human nerve growth factor promote the survival of cholinergic neurons after a fimbria-fornix transfection. Cell Transplant. 4: 93–102.

Shull RM, Munger RJ, Spellacy E, Hall CW, Constantopoulos G, and Neufeld EF. 1982. Canine alpha-L-iduronidase deficiency: a model of mucopolysaccharidosis I. Am. J. Path. 109: 244–248.

Shull RM, Xiaochen L, McEntree MF, Bright RM, Pepper KA, and Kohn DB. 1996. Myoblast gene therapy in canine MPS I: Abrogation by an immune response to α-L-iduronidase. Hum. Gene Ther. 7: 1596–1603.

Shull RM, Kakkis ED, McEntree MF, Kania SA, Jonas AJ, and Neufeld EF. 1994. Enzyme replacement in a canine model of Hurler syndrome. Proc. Nat. Acad. Sci. USA. 91: 12937–12941.

Simonaro CM, Haskins ME, Kunieda T, Evans SM, Visser JW, and Schuchman EH. 1997. Bone marrow transplantation in newborn rats with mucopolysaccharidosis type VI: biochemical, pathological, and clinical findings. Transplantation. 63: 1386–1393.

Sly WS, Quinton BA, McAlister WH, and Rimoin DL. 1973. β-glucuronidase deficiency: report of clinical, radiologic, and biochemical features of a new mucopolysaccharidosis. J. Pediatr. 82: 249–257.

Sly WS, and Vogler C. 1997. Gene therapy for lysosomal storage disease: a no-brainer? Transplants of fibroblasts secreting high levels of beta-glucuronidase decrease lesions in the brains of mice with Sly syndrome, a lysosomal storage disease. Nat. Med. 3: 719–720.

Snyder EY, Taylor RM, and Wolfe JH. 1995. Neural progenitor cell engraftment corrects lysosomal storage throughout the MPS VII mouse brain. Nature. 374: 367–70.

Sorge J, West C, Westwood B, and Beutler E. 1985. Molecular cloning and nucleotide sequence of human glucocerebrosidase cDNA. Proc. Natl. Acad. Sci. 82: 7289–93.

Spellacy E, Shull RM, Constantopoulos G, and Neufeld EF. 1983. A canine model of human alpha-L-iduronidase deficiency. Proc. Nat. Acad. Sci. 80: 6091–6095.

Stein C, Gieselmann V, Kreysing J, Schmidt B, Pohlman R, Waheed A, and von Figura K. 1989. Cloning and expression of human arylsulfatase A. J. Biol. Chem. 264: 1252–9.

Taylor RM, Stewart GJ, and Farrow BRH. 1989. Improvement in the neurological signs and storage lesions of fucosidoses in dogs given marrow transplants at an early age. Transplant. Proc. 21: 3818–9.

Taylor RM, and Wolfe, JH. 1994. Cross-correction of β-glucuronidase deficiency by retroviral vector-mediated gene transfer. Exp. Cell Res. 214: 606–613.

Taylor RM, and Wolfe JH. 1997. Decreased lysosomal storage in the adult MPS VII mouse brain in the vicinity of grafts of retroviral vector-corrected fibroblasts secreting high levels of beta-glucuronidase. Nat. Med. 3: 771–774.

Thomas GH. 1994. "Pseudodeficiencies" of lysosomal hydrolases. Am. J. Hum. Genet. 54: 934–940.

Thompson JN, Jones MZ, Dawson G, and Huffman PS. 1992. N-acetylglucosamine 6-sulphatase deficiency in a Nubian goat: a model of Sanfilippo syndrome type D (mucopolysaccharidosis IIID). J. Inherit. Metab. Dis. 15: 760–768.

Tresco PA, Winn SR, Tan S, Jaeger CB, Greene LA, and Aebischer P. 1992. Polymer encapsulated PC-12 cells: Long term survival and associated reduction in lesion induced rotational behavior. Cell Transplant. 1: 255–264.

Vandevelde M, Fankhauser R, Bichsel P, Wiesmann U, and Herschkowitz N. 1982. Hereditary neurovisceral mannosidosis associated with alpha-mannosidase deficiency in a family of Persian cats. Acta Neuropathol.(Berl.) 58: 64–68.

Vellodi A, Hobbs JR, O'Donnel NM, Coulter BS, and Hugh-Jones K. 1987. Treatment of Neiman-Pick disease type B by allogeneic bone marrow transplantation. Brit. Med. J. 295: 1375–1376.

Vellodi A, Young E, New M, Pot-Mees C, and Hugh-Jones K. 1992. Bone marrow transplantation for Sanfilippo disease type B.J. Inherit. Metab. Dis. 15: 911–918.

Vogler C, Birkenmeier EH, Sly WS, Levy B, Pegors C, Kyle JW, and Beamer WG. 1990. A murine model of mucopolysaccharidosis VII-gross and microscopic findings in beta-glucuronidase-deficient mice. Am. J. Pathol. 136: 207–17.

Vogler C, Sands M, Higgins A, Levy B, Grubb J, Birkenmeier EH, and Sly WS. 1993. Enzyme replacement with recombinant beta-glucuronidase in the newborn mucopolysaccharidosis type VII mouse. Pediatr. Res. 34: 837–840.

Walkley SU, Thrall MA, Dobrenis K, Huang M, March PA, Siegel DA, and Wurzelmann S. 1994. BMT corrects the enzyme defect in neurons of the central nervous system in a lysosomal storage disease. Proc. Nat. Acad. Sci. USA. 91: 2970–2974.

Wilson JM. 1996. Animal models of human disease for gene therapy. J. Clin. Invest. 97: 1138–1141.

Winn SR, Tresco PA, Zielinski B, Greene LA, Jaeger CB, and Aebischer P. (1991) Behavioral recovery following intrastriatal implantation of microencapsulated PC 12 cells. Exp. Neurobiol. 113: 322–329.

Wolfe, JH, Deshmane SL, and Fraser, NW. (1992a) Herpes virus vector gene transfer and expression of beta-glucuronidase in the central nervous system of MPS VII mice. Nat. Genet. 1: 379–84.

Wolfe JH, Sands MS, Barker JE, Gwynn B, Rowe LB, Vogler CA, and Birkenmeier EH. (1992b) Reversal of pathology in murine mucopolysaccharidosis type VII by somatic cell gene transfer. Nature. 360: 749–753.

Yamanaka S, Johnson MD, Grinberg A, Westphal H, Crawley JN, Taniike M, Suzuki K, and Proia RL. (1994) Targeted disruption of the hexa gene results in mice with biochemical and pathological features of Tay-Sachs disease. Proc. Nat. Acad. Sci. USA. 91: 9975–9979.

Yoshida M, Ikadai H, Maekawa A, Takahashi M, and Nagase S. (1993) Pathological characteristics of mucopolysaccharidosis VI in the rat. J. Comp. Pathol. 109: 141–153.

27
Growth Hormone Gene Therapy Using Encapsulated Myoblasts

Nahed Ismail, Gonzalo Hortelano, and Ayman Al-Hendy

Introduction

Increasing numbers of gene therapy protocols are being accepted for human clinical trials both for somatic and Mendelian genetic disorders. For monogenic diseases, the goal is to provide a normal level of the previously missing gene product in vivo (Morgan and Anderson 1993, Morsy et al 1993). However, for these gene products, which are subject to intricate metabolic control, treatment with somatic gene therapy protocols has not yet progressed to the clinical stage, either because our knowledge about the regulatory control of the product in question is limited, or because the technical capability to reproduce the control pathway is not available. Hence, although the gene coding for insulin was one of the first human genes cloned, treatment of diabetes with somatic gene therapy is still an unrealized goal. Secretion of insulin is tightly controlled through interaction with several metabolic pathways, and its unregulated release by implanting insulin-secreting recombinant cells could lead to severe hypoglycemia and death (Selden et al 1987).

As with to insulin, growth hormone (GH) secretion is also highly regulated. The two hypothalamic peptides, GH-releasing hormone and somatostatin, alternately stimulate and inhibit its release, respectively (Tannenbaum 1987). In addition, as with other anterior pituitary hormones, plasma GH concentration exhibits circadian rhythm in both rodents and human beings through mechanisms that are still not well understood (Macleod et al 1991, Tannenbaum and Martin 1976, Vance et al 1985).

Current Therapies for GH-Deficient Patients

The current treatment of choice for GH-deficient patients is injection with recombinant human GH (rhGH). GH is usually provided as bolus injections, as frequently as once per day. For the GH-deficient patients, GH administration continues for most of their childhood and early adolescence (Frasier 1987). Furthermore, growth hormone administration is also considered for patients without GH deficiency in conditions in which it functions as an anabolic drug. Such conditions include aging, surgical trauma, osteoporosis, renal insufficiency, thermal injuries, and athletic activities (Kaplan 1993). Because of the inconvenience of such long-term treatment, its costs, and the antigenic response to some rhGH preparations, GH gene therapy may be considered for these non-GH-deficient conditions.

An Alternative Somatic Gene Therapy for GH Deficiency

A number of gene therapy strategies have been applied to the delivery of GH in vivo. The various approaches include in vivo (Hahn et al 1996, Qiu et al 1996) as well as ex vivo (Barr and Leiden 1991, Dhawan et al 1991, Heartlein et al 1994, Hurwitz et al 1997) gene therapy. We have proposed an alternate approach described as *nonautologous somatic gene therapeutics*, in which genetically modified universal cell lines are used to deliver in vivo the therapeutic product of interest (Chang 1995). The recombinant cells are enclosed within immunoprotective alginate-poly-l-lysine-alginate microcapsules that function as a permeability barriers, preventing the rejection of the recombinant cells by the host (Christenson et al 1993). Hence, the same recombinant cell line can be used to treat different patients regardless of the patient's histocompatibility tissue type, thus eliminating the cost of patient-specific genetic manipulation. Furthermore, established cells lines can be transfected readily with chemical or physical (Chang 1994) means so that any untoward biologic consequence of viral vectors is avoided (Donahue et al 1992). In addition, such cell lines can be thoroughly characterized for safety and quality assurance before implantation, an advantage that is not available in current practices that require viral vector transduction of autologous cells (Morgan and Anderson 1993).

The feasibility of this approach was demonstrated in vitro when recombinant human factor IX (Liu et al 1993), human GH (Chang et al 1994), and human adenosine deaminase (Hughes et al 1994) were delivered from genetically modified cells enclosed within such microcapsules. Furthermore, when such allogenic cells were implanted into mice, recombinant human GH was delivered to the systemic circulation for several months (Chang et al 1993). This chapter will describe the clinical efficiency of the implantation of microencapsulated recombinant myoblasts to revert the phenotype of a GH-deficient animal model, the Snell dwarf mouse, and the importance of gene regulation in somatic gene therapy in normal mice.

The Effect of mGH-Secreting Capsules on Growth

A recombinant mouse C2C12 myoblast cell line (Myo-45) secreting 147 ng of mouse $GH/10^6$ cells/day was engineered (Al-Hendy et al 1995). Snell dwarf mice were implanted with either encapsulated mGH-secreting myoblasts (Group A) or encapsulated but untransfected C2C12 myoblasts as control (Group B). To verify the ability of the alginate microcapsules to immunoisolate the myoblasts,

we implanted into a third group (Group C) a similar number of transfected myoblasts without the protection of microcapsules (Al-Hendy et al 1995). After a latent period of 4 days, Group A started to grow faster than Group B. By the end of the 3rd week, they had gained about 1.6-fold more percentage increase in body weight (Figure 27.1A) and doubled the percentage increase in body length compared to Group B (Figure 27.1B). The third group with unprotected myoblasts was similar or even slightly retarded in growth compared to Group B (Figure 27.1A, B), suggesting that naked mGH-secreting cells were rejected after implantation and the resulting immune reaction had a negative effect on growth. After the first month postimplantation, they become indistinguishable from the control in Group B. Both Groups B and C were significantly smaller than Group A (Figure 27.2). Thus, it was clear that the allogeneic mGH-secreting myoblasts were able to deliver biologically functional GH only when protected with microcapsules. This growth enhancement was effected systematically by stimulating growth of the liver, skeletal muscles, and kidneys.

After 3–5 weeks, the growth rate of Groups A and B no longer differed from each other. When injected microcapsules retrieved at day 42 were returned to culture in vitro, encapsulated myoblasts continued to deliver mGH at a rate of 121 ng/10^6 cells/day. Therefore, the lack of further growth stimulation in Group A mice was not due to loss of the mGH transgene expression from the encapsulated cells, but was likely caused by the nonresponsiveness of the mice at this age (13–15 weeks old). Indeed, a second implantation of fresh capsules with recombinant myoblasts at day 45 failed to increase the growth rate. The lack of responsiveness of the mice to further GH stimulation at the second capsule implantation is in agreement with observations made in hGH transgenic mice (Palmiter et al 1983). These mice with hGH also stopped growing at the age of 12 weeks, despite the fact that their serum hGH was maintained at a high level throughout their life. In this experiment, the mice were followed for up to 100 days, and no more growth enhancement in the treated group was observed. However, the extra weight and length gained in Group A mice were preserved, and these animals remained significantly heavier and longer than Group B mice.

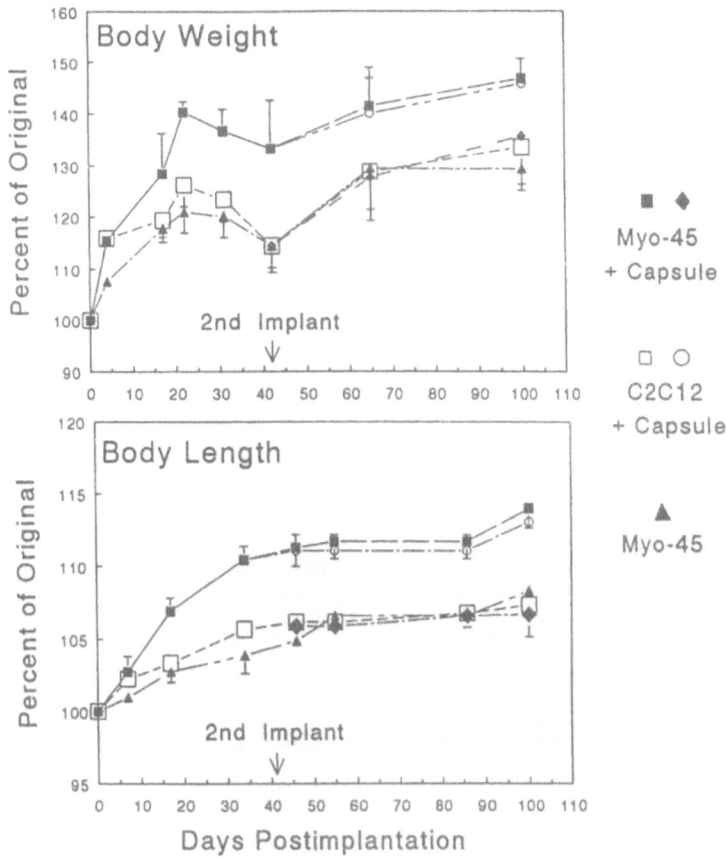

FIGURE 27.1. Encapsulated mGH-secreting myoblasts enhanced the growth of Snell dwarf mice. Dwarf mice (6–7 week old) were implanted intraperitoneally with either 3 ml of microcapsules containing 7.2×10^5 mGH-secreting Myo-45 myoblasts (Group A, ■, n = 5) or 2 ml of microcapsules containing 2.4×10^5 of untransfected C2C12 myoblasts (Group B, ❑, n = 6), or $\sim7 \times 10^5$ mGH-secreting unencapsulated myoblasts (Group C, ▲, n = 3). On day 42, mice in Groups A and B were divided into two subgroups and reimplanted with 3 ml of microcapsules containing either 4×10^5 mGH-secreting Myo-45 myoblasts (■ or ◆, n = 2) or 5.6×10^5 untransfected C2C12 myoblasts (❑ or ○), n = 2). A, The body weights are represented as means ± SD (n ≥ 3) or range (n = 2). The slight weight loss between days 22 and 42 coincided with an accidental rise in ambient temperature in the animal facilities. B, The body lengths were measured under anesthesia as the distance between the tip of the nose and the end of the tail (mean ± SD or range if n = 2). Significant differences in weight ($p < 0.025$) and length ($p < 0.0005$) between Groups A and B or C were found at day 17 postimplantation and thereafter (student's test).

Condition of the Retrieved Capsules

We also retrieved capsules at the end of the experiment after 178 days of implantation. The enclosed myoblasts continued to secrete mGH at 261 ng/10^6 cells/day. Up to 90% of the implanted capsules were retrieved in good condition with intact smooth surfaces and no indentation or inflammatory cellular adhesions. The viability of the encapsulated myoblasts was about 80% at the time of implantation. The viability then dropped to about 66% by day 42 and was maintained around that level until the end

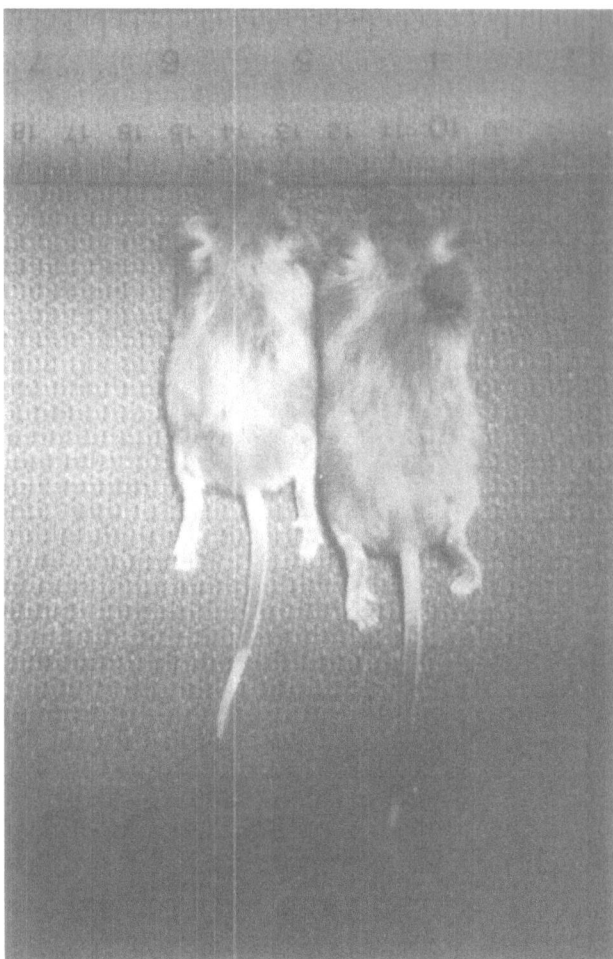

FIGURE 27.2. Representative animals implanted with microcapsules containing either the mGH-secreting myoblasts Myo-45 (right) or untransfected C2C12 (left) on day 35 postimplantation. Their weight was 10.6 and 8.9 g, while measuring 11.2 and 10.1 cm in length, respectively.

of the experiment (day 178). The retrieved recombinant myoblasts continued to secrete mGH at preimplantation levels even after 178 days in vivo.

The Capsule-Derived mGH Enhanced Skeletal Growth

Using the classical tibia test to assess the effect of the capsule-derived mGH on bone growth, we were able to show a remarkable increase in the size of the tibial epiphyseal cartilage of the treated mice versus those implanted with unmodified encapsulated myoblasts, after 35 and 42 days postimplantation. However, after 100 days, the epiphyseal cartilage in the treated group almost completely disappeared, whereas it was still of considerable width in the two

negative control groups. This suggests that the cartilage cells in the treated group, under the influence of the capsule-derived mGH, had matured, differentiated, and migrated into the bone tissues. In the negative control, however, this process was much slower, hence leaving the growth plate in a less mature state.

Plasma mGH

The accelerated growth in the treated dwarfs within the first month of implantation was paradoxically not associated with detectable increases in circulating GH level. All dwarf mice, regardless of their implantation status, had low to undetectable GH in their plasma. However, it is worth noting that although the current cell line and encapsulation protocol failed to restore in the dwarf mice a normal circulating level of GH (up to 100 ng/ml, Macleod et al 1991), a significant biologic effect (increase in body weight and length) was achieved. Indeed, some of the exogenous mGH must have reached the systemic circulation as well, as evidenced by enhancement of tibia growth. At the skeletal bone growth level, only mGH, and not systemic IGF-1, can activate the resting prechondrocytes to increase the thickness of the cartilage (Isaksson et al 1987, Nilsson et al 1986). Hence, it must be concluded that biologically active and physiologically significant levels of GH were delivered. Since GH has a half-life in the serum of only 4–20 min, the growth enhancement provided a stringent test for the continuous secretion of GH, and its access to the circulation and the end organs.

If a clinically significant therapeutic effect can be obtained in spite of subnormal, and in this case undetectable, levels of gene product delivery in vivo, the requirement for achieving a normal level of gene product replacement by gene therapy may not be as absolute as was thought before. Furthermore, this phenomenon may not be exclusive for GH deficiency. In fact, a similar experience was described when hepatectomized hemophiliac B dogs received the canine factor IX cDNA via retroviral infusion (Kay et al 1993).

Growth Retardation: Unexpected Outcome from Growth Hormone Therapy with Microencapsulated Myoblasts in Normal Mice

Using the above nonautologous approach, we addressed the question of the importance of gene regulation in somatic gene therapy. Our findings demonstrate that a constitutive supply of a low level of GH to normal mice, such as through somatic gene therapy, is ineffective in enhancing growth. On the contrary, it led to growth retardation and in some instances, death (Al-Hendy et al 1996). Within 24 hours postimplantation, normal mice implanted with encapsulated mGH-secreting myoblasts were not thriving and appeared thick. In contrast, the control mice, which received only microcapsules containing nonengineered myoblasts, appeared normal. In addition, the control mice gained more weight than

treated mice by 24 hr postimplantation (Figure 27.3A, C). However, by day 4 postimplantation, most of the treated mice stopped losing weight and gradually started to recover. By 2 weeks postimplantation, the treated mice were significantly lighter than the controls (Figure 27.3A, C). The body length of the implanted mice followed a similar pattern (Figure 27.3B, D). Infection as a cause of morbidity was excluded, because the treated and the control groups were housed together in the same environment, and mycoplasma and the six most common rodent viruses tested negative in their sera (unpublished data). The mechanisms of the observed growth retardation are unclear.

The success of our nonautologous somatic gene therapy protocol in promoting growth of the Snell dwarf mice and its unexpected growth retardation in the normal mice highlight two important issues: the effectiveness of this GH gene therapy protocol and the importance of appropriate selection of patients for such gene therapy treatment. It is clear that for those patients with defects in the pituitary or hypothalamus with genuine GH deficiency, constitutive supplement of GH through somatic gene therapy is effective. In contrast, for patients with a normal hypothalamic-pituitary axis, and for whom GH treatment is indicated, e.g., kidney transplant recipients or patients with surgical trauma, the episodic bolus injection of recombinant GH may be a more appropriate treatment than a constitutive delivery through somatic gene therapy.

Important Considerations

The therapeutic efficacy of this potentially cost-effective approach to somatic gene therapy has been demonstrated. However, several improvements need to be addressed for future clinical applications. In addition to increasing the level of gene product delivery, so as to allow the treatment of other genetic disorders with more stringent requirements for high-level product replacement, the durability of the biomaterial used for fabricating the microcapsules is an important consideration. Although with our current protocol the capsules remained intact for over 6 months, the polysaccharide membranes will eventually disintegrate after prolonged implantation (>1 year, unpublished observation). The breakdown of the capsules will lead to release of cellular debris

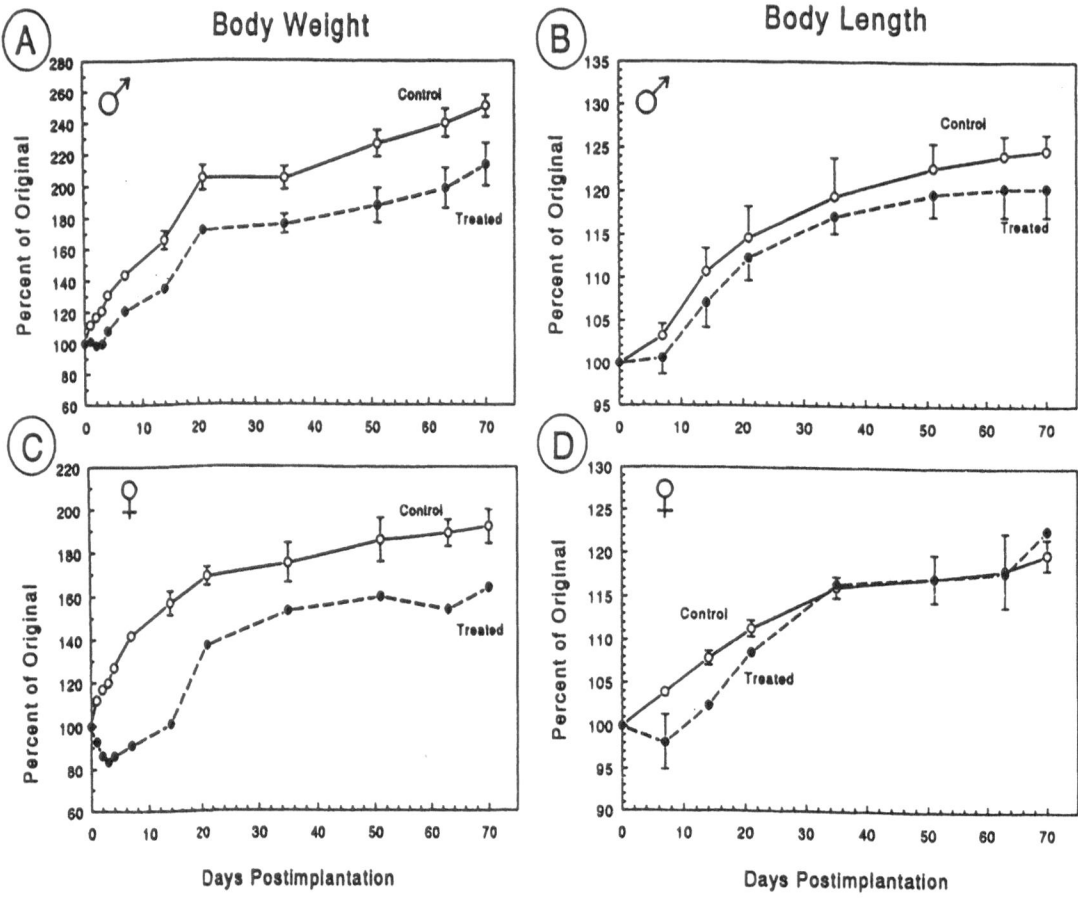

FIGURE 27.3. Effects of implanting microencapsulated mGH-secreting myoblasts on the growth of male and female mice. C57BL/6J mice 6–7 weeks old were implanted intraperitoneally with 8×10^5 microencapsulated myoblasts either from the untransfected C2C12 (control, n = 8 males per group or 3 females per group) or from the transfected Myo-61 cell line secreting 201 ng of mGH/10^6 cells/day (treated, n = 7 males per group or 3 females per group). The body weights (A and C) were measured, and expressed as percent of the original weights on day 0 of implantation. The body lengths (B and D) were measured under anesthesia and similarly expressed as the distance between the tip of the nose and the end of the tail. Each point is the mean ± SEM (n ≥ 3) or range (n = 2).

and development of antigenic responses against foreign epitopes. Alternate capsule composition may improve the durability of the microcapsules in vivo.

Another important consideration is the safety of established cell lines. It is known that over the long term, transformed cell lines (such as transfected myoblasts or Ltk-mouse fibroblasts) can lead to tumor formation (Partridge et al 1988). To overcome this *potential* problem, the use of untransformed myoblasts propagated directly from muscle tissue is particularly attractive for several reasons. First, myoblasts initiated from muscle biopsies are amenable

to large-scale propagation and thus can be used effectively as a permanent stock for transfection. Second, they can differentiate under experimentally controlled conditions into multinucleated myoblasts (Morgan et al 1994), thus further reducing the possibility of neoplasia development. Finally, muscle tumors have never been found in man (Blau et al 1993), again decreasing the potential for spontaneous transformation.

In conclusion, this mode of delivery has been shown to be clinically efficient, benign, potentially economical, highly versatile, and amenable to industrial-scale production and quality control. Furthermore, repeated implantation over time may be administered to augment the therapeutic effect. The capsules are retrievable, thus providing an additional advantage in safety and flexibility. This alternate approach to somatic gene therapy could have potential application in any disorder where systemic delivery of a therapeutic product is required.

References

Al-Hendy A, Hortelano, G, Tannenbaum, GS, and Chang, PL. 1995. Correction of the growth defect in dwarf mice: A novel approach to somatic gene therapy. Hum. Gene Ther. 6:165–175.

Al-Hendy A, Hortelano, G, Tannenbaum, GS, and Chang, PL. 1996. Growth retardation and acute IGF-1 deficiency—unexpected outcomes of growth hormone somatic gene therapy. Hum. Gene Ther. 7:61–70.

Blau, HM, Dhawan J, and Pavlath, GK. 1993. Myoblasts in pattern formation and gene therapy. Trends Genet. 9:269–274.

Barr, E, and Leiden, JM. 1991. Systemic delivery of recombinant proteins by genetically modified myoblasts. Science 254:1507–1509.

Chang, PL. 1994. Calcium phosphate-mediated DNA transfection. In: Gene therapeutics: methods and applications of direct gene transfer. JA Wolff, ed. Boston: Birkhauser, pp. 157–179.

Chang, PL. 1995. Non-autologous somatic gene therapy. In: Somatic gene therapy. PL Chang, ed. Florida: CRC Press Inc., pp. 203–223.

Chang, PL, Shen, N, and Westcott, AJ. 1993. Delivery of recombinant gene product with microencapsulated cells in vivo. Hum. Gene Ther. 4:433–440.

Chang, PL, Hortelano, G, Tse, M, and Awrey, DE. 1994. Growth of recombinant fibroblasts in alginate microcapsules. Biotech. Bioeng. 43:925–933.

Christenson, L, Dionne, KE, and Lysaght, MJ. 1993. Biomedical applications of immobilized cells. In: Fundamentals in animal cell encapsulation and immobilization, MFA Goosen, ed. Florida: CRC press Inc., pp. 43–54.

Dhawan, J, Pan, LC, Pavlath, GK, Travis, MA, Lanctot, AM, and Blau, HM. 1991. Systemic delivery of human growth hormone by injection of genetically engineered myoblasts. Science 254:1509–1512.

Donahue, RE, Kessler, SW, Bodine, D, Mcdonagh, K, Dunbar, C, Goodman, S, Agricola, B, Byrne, E, Raffled, M, Moen, R, Bacher, J, Zsebo, KM, and Neinhuis, AW. 1992. Helper virus induced T cell lymphoma in nonhuman primates after retroviral mediated gene transfer. J. Exp. Med. 176:1125–1135.

Frasier, SD. 1987. Growth hormone deficiency. In: Growth abnormalities. R Hintz and R Rosenfeld, eds. New York: Churchill Livingstone, pp. 41–140.

Hahn, TM, Copeland, KC, and Woo, SL. 1996. Phenotypic correction of dwarfism by constitutive expression of growth hormone. Endocrinology 137:4988–4993.

Heartlein, MW, Roman, VA, Jiang, JL, Sellers, JW, Zuliani, AM, Treco, DA, and Selden, RF. 1994. Long-term production and delivery of human growth hormone in vivo. Proc. Natl. Acad. Sci. USA 91(23):10967–10971.

Hughes, M, Vassilakos, A, Andrews, DW, Hortelano, G, Belmont, JW, and Chang, PL. 1994. Delivery of a secretable adenosine deaminase through microcapsules—a novel approach to somatic gene therapy. Hum. Gene Ther. 5:1445–1455.

Hurwitz, DR, Kirchgesser, M, Merrill, W, et al. 1997. Systemic delivery of human growth hormone or human factor IX in dogs by reintroduced genetically modified autologous bone marrow stromal cells. Hum. Gene Ther. 8:137–156.

Isaksson, OGP, Lindahl, A, Nilsson, A, and Isgaard, J. 1987. Mechanism of the stimulatory effect of growth hormone on longitudinal bone growth. Endoc. Rev. 8:426–438.

Kaplan, SL. 1993. The newer uses of growth hormones in adults. Adv. Intern. Med. 38:287–301.

Kay, MA, Rothenberg, S, Landen, CN, Bellinger, DA, Leland, F, Toman, C, Finegold, M, Thompson, AR, Read, MS, Brinkhous, KM, and Woo, SLC. 1993. In vivo gene therapy of hemophilia B: Sustained partial correction in factor IX-deficient dogs. Science 262:117–119.

Liu, HW, Ofosu, FA and Chang, PL. 1993. Expression of human factor IX by microencapsulated recombinant fibroblasts. Hum. Gene Ther. 4:291–301.

Macleod, JN, Pampori, NA, and Shapiro, BH. 1991. Sex differences in the ultradian pattern of plasma growth hormone concentration in mice. J. Endocrinol. 131:395–399.

Morgan, RA and Anderson, WF. 1993. Human gene therapy. Annu. Rev. Biochem. 62:191–217.

Morgan, JE, Pagel, CN, Peckham, M, Ataliotis, P, Jat, PS, Noble, MD, Farmer, K, and Partridge, TA. 1994. Myogenic cell lines derived from transgenic mice carrying a thermolabile T antigen: A model system for derivation of tissue-specific and mutation-specific cell lines. Dev. Biol. 162:486–498.

Morsy, MA, Mitani, K, Clemens, P, and Caskey, CT. 1993. Progress toward human gene therapy. J. Am. Med. Assn. 270:2238–2345.

Nilsson, A, Isgaard, J, Lindahl, A, Dahlsrom, A, Skottner, A, and Isaksson, OGP. 1986. Regulation by growth hormone of number of chondrocytes containing IGF-1 in rat growth plate. Science 233:571–574.

Palmiter, RD, Norstedt, G, Gelinas, RE, Hammer, RE, and Brinster, RL. 1983. Metallothionein-human GH fusion genes stimulate growth of mice. Science, 222:809–814.

Partridge, TA, Morgan, JE, Moore, SE, and Walsh, FS. 1988. Myogenesis in vivo from the mouse C2 muscle cell-line. J. Cell Biochem. (Suppl.). 12C:331.

Qiu, P, Ziegelhoffer, Sun, J, and Yang, NS. 1996. Gene gun delivery of mRNA in situ results in efficient transgene expression and genetic immunization. Gene Ther. 3:262–268.

Selden, RF, Skoskiewicz, MJ, Russell, PS, and Goodman, HM. 1987. Regulation of insulin-gene expression. N. Engl. J. Med. 317:1067–1076.

St. Louis, D, and Verma, IM. 1988. An alternative approach to somatic cell gene therapy. Proc. Natl. Acad. Sci. USA 85:3150–3154.

Tannenbaum, GS. 1987. Interaction of growth hormone-releasing factor and somatostatin in regulation of the thymic secretion of growth hormone. In: Growth hormone, growth factors, and acromegaly. DK Ludecke and G Tolis, eds. New York: Raven Press, pp. 37–54.

Tannenbaum, GS and Martin, JB. 1976. Evidence for endogenous ultradian rhythm governing growth hormone secretion in the rat. Endocrinology 98:562–570.

Van Wyk, JJ. 1984. The somatomedins: Biological actions and physiologic control mechanisms. In: Hormonal proteins and peptides. C. H. Li, ed. New York: Academic Press, pp. 81–89.

Vance, ML, Kaiser, DL, Evans, WS, Furlanetto, R Vale, W, Rivier, J, and Thorner, MO. 1985. Pulsatile growth hormone secretion in normal man during a continuous 24-hour infusion of human growth hormone releasing factor (I-40). J. Clin. Inves. 75:1584–1590.

28
Transplantation of Encapsulated Cells into the Central Nervous System

Jacqueline Sagen, Suzanne L. Bruhn, David H. Rein, Rebecca H. Li, and Melissa K. Carpenter

Introduction to Cell Encapsulation for Targeted Delivery to the CNS

Targeted delivery of active biomolecules to the central nervous system (CNS) has been applied as a potential therapy to treat a variety of disorders, including chronic pain and neurodegenerative diseases. This approach is advantageous for delivery of neuroactive agents that do not readily cross the blood-brain barrier. The implantable systems available for the targeted delivery of these substances to the CNS are either cell-based or non-cell-based. The non-cell-based systems include controlled release polymers and infusion via pumps and catheters. These systems involve relatively simple surgical procedures and allow high control over dose over the short term. The main disadvantages involve poor long-term release kinetics for the polymer delivery systems, high risk of infection for the catheters and pumps, and the need for highly purified and stable factors for both approaches. Cell-based systems involve the encapsulation of living secretory cells in either microencapsulation or macroencapsulation devices. The main advantage of cell-based delivery systems is the continuous production and secretion of the neuroactive agents from the encapsulated cells, thereby eliminating requirements for purified and stable therapeutic factors and the need for frequent refilling or replacement. Macroencapsulation devices have the additional advantage of high mechanical strength, ease of retrievability, and increased control over the manufacturing. This chapter will focus on the delivery of therapeutic factors to the CNS using macroencapsulation devices.

Macroencapsulation

The objective of macroencapsulation is to allow the implantation of living xenogeneic or allogeneic cells into the host species without immune rejection or the use of immunosuppression, and to allow continuous secretion of the therapeutic factors. The enabling technology involves surrounding the implant cells with a selective membrane barrier that allows free passage of small molecules such as nutrients into the device, allows release of the therapeutic factors out of the device, and hinders the passage of larger molecules and cells from the host immune system. These immunoisolatory devices can be in the form of a flat disk, hollow fiber, or intravascular tube, and all share three basic components: a permselective membrane, an internal matrix, and the living cells. These components are shown schematically in Figure 28.1 for the hollow-fiber configuration.

Recent studies have demonstrated the effectiveness of hollow-fiber immunoisolatory devices in protecting xenogeneic cells for targeted delivery into the CNS in rats (Lindner et al 1995), monkeys (Kordower et al 1995), sheep (Joseph et al 1994), and human beings (Aebischer et al 1996). Achieving functionality of these implanted devices requires selection of appropriate cells and engineering of the membrane and matrix to sustain the viability and secretion of the particular encapsulated cells in the specific host site. The following sections will review the important features of the macroencapsulation device for successful delivery to the CNS.

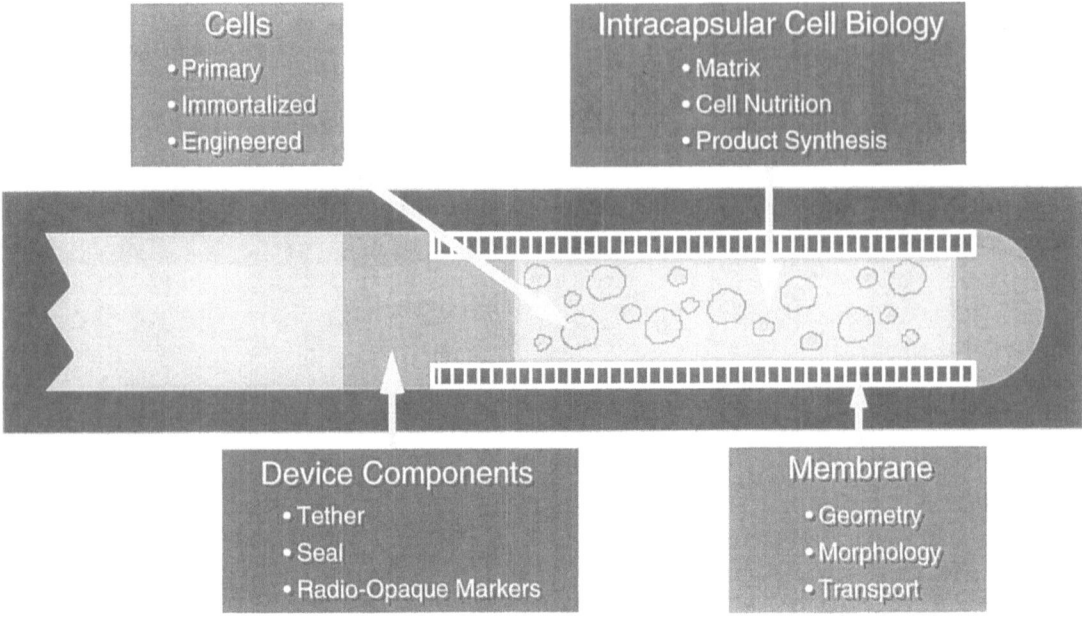

FIGURE 28.1. Components determining macrocapsule properties

Cells

Cells that are used to secrete therapeutic molecules from encapsulated devices may be primary cells, immortalized cells, or cells that have been genetically engineered. All three cell types have been successfully encapsulated. Examples of primary cells are adrenal chromaffin cells and islets of Langerhans. An example of an immortalized line is the PC12 rat pheochromocytoma cell derived from a tumor. Genetically engineered cells that have been successfully encapsulated include the C2C12 mouse myoblast line, the baby hamster kidney (BHK) line, and the Chinese hamster ovary (CHO) line. These different cell types have different requirements for survival and function in an encapsulated environment when implanted in the host CNS. For example, to genetically engineer cells, one requirement is that they be a dividing cell line. Once encapsulated, many cell lines will continue to proliferate within this environment. Thus, the proliferative nature of the cell is a primary consideration in design of encapsulated cell devices. These considerations will be explored in more detail in the sections that follow. Selection criteria for an appropriate cell type for CNS therapy include long-term survival in the CNS environment, continued secretion of the mol-

ecule(s) of interest at a therapeutic dose, and minimal risk of secretion of toxic molecules or viral transmission to the host. Examples of cells selected for several CNS indications are described in later sections.

Matrix

A key component of most encapsulated cell devices is the internal matrix. This matrix is located within the cell compartment and serves to define the microenvironment for cells once they are encapsulated. Selection of this matrix is highly dependent on the chosen cell type. In their native environment, cells inhabit a complex meshwork of extracellular proteins and polysaccharide molecules known as the extracellular matrix (ECM). The ECM serves a dynamic and complex role in regulating cell function—virtually all mammalian cells rely on the ECM for differentiated phenotypic expression.

The strategy for engineering the proper encapsulation matrix for a particular cell type generally focuses on reconstructing the optimal three-dimensional tissue environment that supports cell viability and functionality. The two major roles of the matrix are: (1) a purely mechanical role in immobilizing the cells and keeping them homogeneously distributed throughout the cell compartment and (2) a biological role by stimulating cells to secrete their own ECM, regulating cell proliferation, regulating cell secretory function, or maintaining the cells in a differentiated phenotype. A well-designed encapsulation matrix is therefore tailored to the requirements of the specific cell type.

Key considerations in the design of the encapsulation matrix have been recently reviewed (Li 1998). These include:

1. *Promotion of encapsulated cell viability and functionality.* The matrix ideally enhances cell viability but minimally must be noncytotoxic to the chosen cell type. Dose-response cytotoxicity assays are used to quickly screen for acute cell toxicity resulting from the materials used, any cross-linkers, or process conditions. One common cause of encapsulated cell death is the formation of cell aggregates. When larger cell clusters (>100 μm) form, central necrotic regions develop as those cells in the core become starved for nutrients. This may be even more critical in the CNS, especially the parenchyma, where nutrient and oxygen exchange may be poor.

2. *Cell type.* Cells normally grown in suspension culture may not need a substrate, while anchorage-dependent cells usually require a substrate on which to adhere. Certain cell types, such as fibroblasts, may secrete their own matrix. These cells may rapidly remodel their local environment, and attempts to tailor their environment by providing ECM may be futile and overridden by the remodeling process. For these cells, providing a matrix substrate that is conductive to the adhesion of cell-secreted ECM may be ideal. Other cell types that normally rely on neighboring cells in their natural environment to secrete ECM are better served by providing a more complete milieu of biological molecules within the encapsulated environment.

3. *Host-matrix and membrane-matrix compatibility.* Since the matrix does not directly contact the host, biocompatibility is not as critical a consideration as with the membrane. However, the material must be biocompatible in cases when it may degrade over time or, if leachable, small molecules from the matrix can permeate out of the membrane and elicit a host reaction. Some membrane materials may be prone to protein adsorption or "fouling." The matrix must be "compatible" with the membrane, in that matrix components may serve as additional "foulants" of the membrane pores during

any degradation processes, or during the encapsulation process if the matrix is injected into the membrane. Matrix materials must also be placed or infused into the membrane using procedures that avoid damage to the delicate membrane permselective skin layer.

4. *Implant longevity.* Naturally-derived matrix materials may be degraded by physiological enzymes secreted by the host (e.g., collagenase). Synthetic materials or cross-linking of natural materials often improves matrix longevity. Matrix longevity should be appropriately matched to implant lifetime.

5. *Diffusional transport characteristics.* Often, improved matrix longevity through increased cross-linking is obtained at the expense of matrix permeability (Li et al 1996). Ideally, the matrix should be highly permeable to small solutes and offer little diffusive resistance to nutrients and cell waste products. In some cases, the matrix may also offer secondary immunoprotection, as in the case of the hydrogel alginate, which limits diffusion of immune cells and immunoglobulins. Last, the matrix should be freely permeable in the molecular weight range of the therapeutic molecule delivered.

6. *Cell proliferation.* Cell lines may continue to proliferate postencapsulation. This may be a benefit or a drawback, depending on the rate of proliferation. A renewing population may serve to maintain the therapeutic secretion rate as cells turn over. However, if proliferation is rapid, the encapsulated cells may outstrip their nutrient supply. Specific matrices may serve to either control or encourage proliferation. For example, one advantage of alginate or agar is that they inhibit fibroblast proliferation—this is a potential benefit in primary cell encapsulation, when contaminating fibroblasts may overgrow the intracapsular space and outcompete primary cells for nutrients.

Two broad classes of matrices—synthetic and naturally-derived materials—have been applied to mimic the physiological extracellular matrix within the confines of the encapsulated microenvironment. Hydrogels are a class of polymers that swell in water without dissolving. They are gelled or cross-linked through physical forces or chemical reaction. Traditionally, matrices used in cell encapsulation devices to immobilize cells have been naturally-derived hydrogels. One of the earliest examples of the use of a hydrogel matrix material was that of

Lacy during his work with encapsulated islets (Lacy et al 1991). Islets encapsulated without an immobilizing matrix reversed the diabetic state of mice for 7–14 days, followed by resumption of their hyperglycemic state. Retrieved implants showed that individual islets had aggregated into a large necrotic mass with only a thin rim of viable cells remaining at the periphery. When an alginate hydrogel matrix was used to immobilize and separate the islets within the hollow fibers, encapsulated islets maintained normoglycemia for >60 days in this same animal model. Examples of naturally-derived hydrogels are collagen, Matrigel®, alginate, agarose, and chitosan, as detailed below.

1. *Collagen.* The three major fibrous proteins constituting physiologic ECM are collagen, elastin, and fibronectin. Collagens constitute the most substantial component of this natural ECM and therefore have been widely used to simulate natural ECM in culture and as an encapsulated cell matrix. Collagen may take several forms, including soluble collagen hydrogels, collagen sponges, and fibrillary collagen cross-linked with glutaraldehyde. Collagen hydrogel has been successfully used as a matrix for the artificial liver to entrap hepatocytes within hollow fibers (Jauregui et al 1996, Nyberg et al 1993) and in microcapsules (Dixit and Gitnick 1995). In this application, the 3-D collagen sandwich conformation was able to preserve hepatocyte function, whereas cells cultured in a monolayer rapidly lose their functionality (Dunn et al 1991). In the CNS, collagen hydrogel has also been used as a matrix for BHK cells engineered to secrete human NGF to promote survival of cholinergic neurons in a model of Alzheimer's disease (Winn et al 1994b).

2. *Matrigel®* (Collaborative Research, Bedford, MA). This is a mouse sarcoma-derived extracellular matrix complex that is predominantly laminin in composition. It has been employed as a substitute matrix for hepatocytes and other cell lines (Winn and Tresco 1994a) and shown to improve long-term viability in microcapsules (Dixit and Gitnick 1995).

3. *Alginate.* This is a negatively charged polysaccharide-based hydrogel derived from seaweed. One advantage of alginate is that it can be easily cross-linked using positively charged divalent ions (such as calcium or barium) to form polyelectrolyte hydrogels. Calcium cross-linked alginate has been used successfully as a matrix in hollow-fiber

membrane devices to immobilize bovine adrenal chromaffin cells used in the treatment of chronic pain (Aebischer et al 1994, Buchser et al 1996). In these devices, the chromaffin cells are aseptically suspended in a sterile alginate solution. As in all cell-contacting polymer solutions, the alginate solution must be brought to physiological isotonicity and pH prior to combining with cells. The cell-alginate suspension is then injected into the lumina of hollow-fiber preassembly devices. The ends of the hollow-fiber device are sealed and then the entire device immersed in a $CaCl_2$ bath for 5 minutes to cross-link the alginate and immobilize the cells in the lumen.

4. *Agarose.* Another natural polysaccharide derived from seaweed and used as a matrix is agarose (Iwata et al 1992). Compared to alginate, agarose possesses certain benefits in that its purity is more readily controlled and its cross-linking may prove more stable in vivo. Agarose also offers a long history of safe use in cell culture. Agarose hydrogels are thermally gelling—most formulations are liquid at elevated temperature and thermally gel upon cooling to room temperature. Agarose has been used as an encapsulation matrix for MIN6, a ß cell line, in macrocapsules and for various other cell lines (Hayashi et al 1996, Winn et al 1994b).

5. *Chitosan.* Chitosan is a natural polysaccharide derived from chitin. Commercially available chitosan preparations are produced by partial deacetylation of shrimp and crab exoskeleton chitin and are composed of glucosamine monomers (Chandy and Sharma 1990). Chitosan is soluble in weakly acidic solutions (below pH 6.3)—above pH 6.3 it forms a precipitate. Positively charged chitosan solution may be cross-linked with negatively charged ions to form a hydrogel. Although cross-linked chitosan has been used successfully to immobilize plant and microbial cells (Knorrs 1986, Vorlop and Klein 1987), it has not thus far been as successfully used to immobilize mammalian cells because of their greater pH-sensitivity in the acidic pH range in which chitosan remains soluble (Zielinski and Aebischer 1994). However, precipitated chitosan has been used as a successful encapsulation matrix and demonstrated to promote mammalian cell viability and functionality. For example, PC12 cells immobilized in hollow fibers with precipitated chitosan (Protan Biopolymer, Drammen, Norway) have shown improved viability compared to the same cells encapsulated without a matrix or with

cross-linked chitosan matrix (Emerich et al 1993, Zielinski and Aebischer 1994).

Although most encapsulation matrices used to date have been naturally-derived materials, synthetics offer the advantages of high purity, ready availability, and the ability to easily tailor matrix properties to fit the cell type. Recently a new class of synthetic scaffolds has been introduced as a matrix for encapsulated cell devices (Li et al 1998). The use of this synthetic foam has many potential advantages over the traditional hydrogel or hydrocolloid matrix and is based on a poly(vinyl alcohol) (PVA) sponge. Advantages of this scaffold over traditional hydrogels include: the presence of a physical surface onto which anchorage-dependent cells can lay their own extracellular matrix and adhere; proliferation control of contact-inhibited cells; improved intracapsular nutrient transport to the center of the device through the numerous porous interconnecting channels; and lastly, limited cell cluster size (restricted to the foam pore size), eliminating the tendency to form very large clusters and form necrotic centers. Control of proliferation of cell lines is highly advantageous for these devices, since rapid proliferation does not allow as predictable a dose of the therapeutic factor as does a maintenance of stable cell number. Restricted cell proliferation also improves longevity of the encapsulated devices, since buildup and shedding of antigenic debris due to cell turnover is mitigated.

PVA water-insoluble sponges are formed through reaction of aerated PVA polymer with formaldehyde vapor or glutaraldehyde solution as the cross-linker. PC12 cells have been encapsulated using either these PVA foam matrices or chitosan encapsulation matrix. Both in vitro and in vivo (rodent striatum) results showed a significantly greater number of viable cells in the foam matrix devices and up to 6-fold more efficient secretion of the neurotransmitter L-dopa (Li et al 1998). C_2C_{12} mouse myoblast cells engineered to secrete CNTF have also been encapsulated either with PVA foam matrix or in devices containing no encapsulating matrix. The devices were implanted into sheep intrathecal space to simulate the implant site for ALS treatment. The foam matrix devices had many more viable myoblast cells secreting neurotrophic factor when the devices were explanted after one month than did matrix-free or collagen matrix devices (Li, unpublished).

Membrane

A third key component of encapsulated cell devices is the membrane. The membrane provides the interface between the encapsulated cells and the host CNS. The membrane transport properties, morphology, mechanical strength, and composition are readily controlled through the formulation process, and can significantly affect the performance of the device. The membrane characteristics necessary for optimum device performance are dictated by the metabolic requirements of the encapsulated cell, the size of the therapeutic molecule to be released, the degree of required immunomodulation, and the tissue biocompatibility with the implant site (Colton 1995, Gentile et al 1995). The porosity and selectivity of the membrane must be adjusted to provide sufficient nutrient flux to maintain cell viability and flux of the therapeutic molecule, while rejecting the larger immunological species. The topography of the outer surface is captured in describing the membrane morphology, and affects the host reaction to the implant as well as the encapsulated cell viability. The membrane material and cross-sectional wall morphology impart mechanical strength to the device, and enable surgical implantation and retrievability.

A wide variety of materials can be used for macroencapsulation (Li 1998). The majority of the membranes used have been thermoplastics formed by the phase inversion process. A copolymer of polyacrylonitrile and polyvinyl chloride (PAN-PVC) has been used extensively for CNS applications (Aebischer et al 1994a, b, Joseph et al 1994, Kordower et al 1995, Lindner et al 1995), diabetes (Lacy et al 1991), and pituitary treatments (Hymer et al 1981). Poly (ether sulfone) (PES) has also been used in CNS applications and anemia (Deglon et al 1996). Polyurethane has been used for diabetes (Ward et al 1993) and pituitary treatments (Lamberton et al 1988), PTFE has been used for diabetes (Brauker et al 1992), and polypropylene has been used for liver transplantation (Takebe et al 1996). Hydrogel-based membranes have also been employed for macroencapsulation including polyvinyl alcohol for diabetes (Inoue et al 1992).

Membrane Fabrication Methods

The phase inversion process is commonly used to produce flat-sheet and hollow-fiber membranes, and has been previously described in detail (Cabasso

1980, Kesting 1985, Strathmann 1985). The process involves the precipitation of a polymer from solution by a phase transition upon exposure to a nonsolvent or a change in temperature. It is required that the solvent and nonsolvent be miscible. During the phase inversion, a two-phase system develops, consisting of a polymer-rich and polymer-lean phase. The polymer-rich phase forms the three-dimensional membrane network, and the polymer-lean phase forms the pores. The thermodynamics and kinetics of the phase transition determine the membrane structure, and can be described by a ternary phase diagram.

The phase inversion process can be applied to any polymer system that forms homogeneous solutions at specified temperatures and compositions, and separates into two phases upon changes in these conditions. A wide range in porosity and pore size distribution can be attained from ultrafiltration (5 nm to 0.1 μm pore size) to microporous (0.5 μm to 3 μm), as well as a variety of morphologies (Strathmann 1985). Spinning a hollow-fiber membrane involves the coextrusion of the polymer casting solution and nonsolvent through a coaxial nozzle. This liquid stream is then precipitated into a coagulation bath and collected onto a take-up spool. Strategies to modify membrane properties include changes in process conditions (coagulation temperature, nozzle position), changes in coagulation bath composition, and changes in the casting solution composition (% solids, additives).

Transport Characterization

A thorough understanding of the membrane transport characteristics and the specific cell requirements is necessary in optimizing the device performance. The encapsulated cells will be exposed to host immunological attack if the membrane selectivity is not sufficiently high, yet will be starved of nutrients if the flux is too low. The desired selectivity is dependent on transplanted cell type and on the implant site. Xenografts, and to a lesser extent allografts, will require restricted passage of large molecular weight immunological species, such as immunoglobulin G (MW approx. 150,000) and immunoglobulin M (MW approx. 900,000). The CNS has been described as an immunologically privileged site, so the demands for high selectivity may be relaxed compared to those for other sites, although this is controversial. Nutritional requirements will also vary with cell type, but typically those demanding high

flux will survive better in a highly permeable membrane. An exception may be with a rapidly dividing cell line where limiting the nutrient flux may prevent rapid overgrowth and eventual cell death within the device. Finally, different implant sites will induce different host reactions and thereby affect device transport. Devices implanted into the striatum may induce neovascularization and enhance nutrient flux compared to the same device in the intraventricular space.

Transport characterization of an encapsulation membrane can be determined using many different transport measurements, including hydraulic permeability (HP), solute rejection, and diffusive flux. Details of these characterization methods have been previously reported (Dionne et al 1996, Gentile et al 1995, 1996). HP and solute rejection measure the transmembrane convective flow of a bulk fluid driven by a pressure gradient. HP represents the convective water resistance and is proportional to the percentage of surface pores that are continuous through the membrane wall. This parameter represents a combination of the membrane pore size, pore size distribution, and tortuosity.

Solute rejection is defined by the ratio of concentrations of a specific molecular weight species across the pressure gradient. A range of solute sizes is typically used to generate a rejection coefficient profile representative of the pore size distribution. This profile can be characterized by the nominal molecular weight cut-off (nMWCO), defined as the solute molecular weight reduced in concentration by a log order. Solutes commonly used to generate rejection curves include globular proteins such as IgG, bovine serum albumin, ovalbumin, and polysaccharides such as dextran (Dionne et al 1996, Gentile et al 1995). The dextran rejection curve for a P(AN-VC) UF encapsulation membrane is shown in Figure 28.2. Although rejection curves do not define a unique pore size distribution, the position of the curve represents the average pore size, and the sharpness represents the pore size distribution. The objective of membrane formulation work is to shift the position of the curve and adjust the slope.

Although convective transport characterization provides information on the membrane pore structure, most immunoisolation devices operate in the diffusive mode as a result of concentration gradients. The solute diffusive flux is a combination of steric hindrance, porosity, and tortuosity. Small

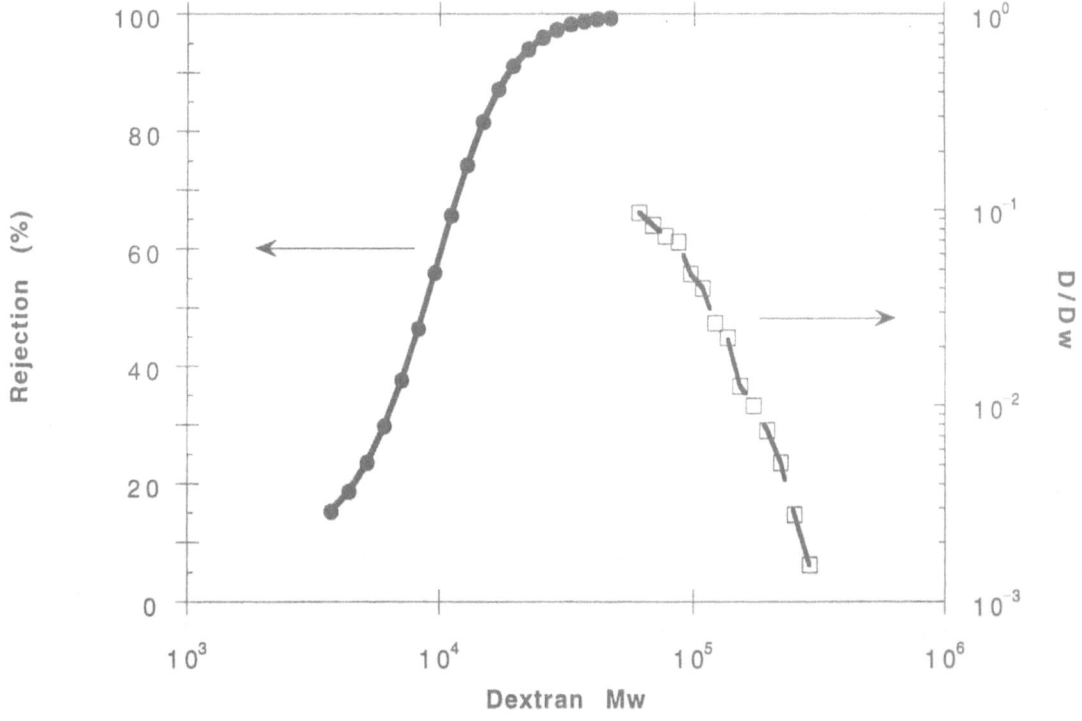

FIGURE 28.2. Rejection and diffusion curve of a typical immunoisolatory hollow fiber using polydispersed dextrans. D/Dw represents the diffusion coefficient of dextran through the membrane relative to the diffusion coefficient through water.

molecular weight species will be governed primarily by the overall porosity, while large molecular weight species will be governed by pore size. Tests have been developed to measure the diffusive flux of small molecular weight species, such as glucose, vitamin B12, and cytochrome c through hollow-fiber membranes (Dionne et al 1996, Gentile et al 1995). Similar tests were developed to measure the diffusive flux of large molecular weight species such as bovine serum albumin, IgG, and dextran species with molecular weights from 30,000 to 2,000,000 (Dionne et al 1996, Gentile et al 1995).

Figure 28.2 shows the dextran diffusion coefficients as a function of molecular weight for a P(AN-VC) encapsulation membrane. There is a steep decline in the diffusion coefficients at a molecular weight around 100,000. The objective of membrane formulation in optimizing device performance is to shift this diffusive profile, and increase the sharpness. Since there is presently a lack of clarity as to the degree of immunoisolation required to ensure cell viability and the molecular size of the species

responsible for immunological rejection (Colton 1995), this work must be done empirically.

This diffusion curve in Figure 28.2 is located at a significantly higher molecular weight than would have been estimated from the membrane nMWCO of 15,000 (the rejection curve in Figure 28.2). However, convective test measurements characterize the solute resistance of the most restrictive layer throughout the membrane wall, whereas diffusive measurements integrate the solute resistance provided by the entire wall. Depending on the overall wall morphology, convective and diffusive measurements can provide different information about the membrane structure.

When evaluating the requirements for membrane transport, interactions with the in vivo environment must be examined. Different membrane materials will induce different degrees of protein fouling to the surface of the pore structure, and this can result in a decrease of the transport flux. This decrease can become significant for medium to large molecular weight species, and has been measured to decrease the nMWCO by one log order, and to decrease the diffusion coefficient of a 70,000 MW species by 2

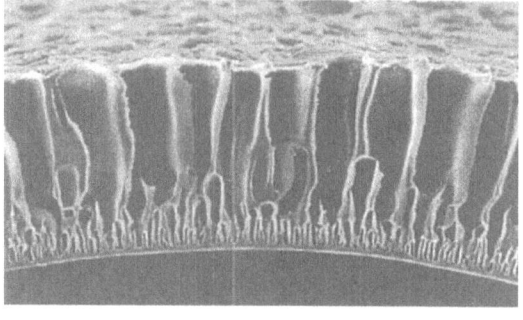
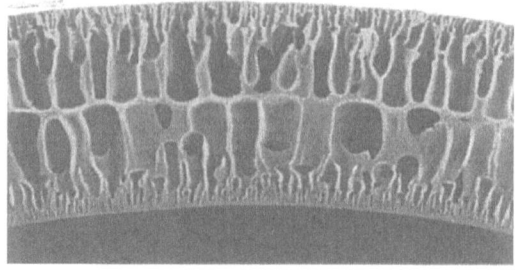
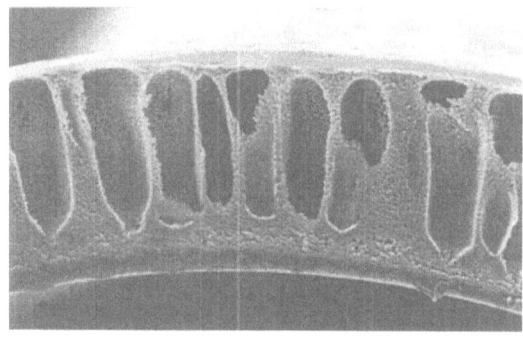
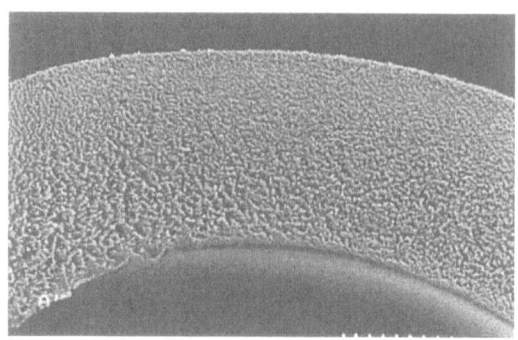

FIGURE 28.3. Scanning electron micrographs of cross-sections of typical hollow fibers used for implanted cell encapsulation. The wall thicknesses are approximately 100 μm.

log orders for P(AN-VC) membranes (Gentile et al 1995). Tissue adhesion to the outer surface of the membrane or infiltration into the wall may create additional transport resistance or nutrient consumption. Membrane surface modifications, such as PEO grafting and corona treatment, have been applied to modify tissue and protein adsorption (Kessler et al 1995, Shoichet et al 1994).

Morphological Characterization

Transport measurements should be evaluated in combination with morphological characterization to develop structure-property relations. Morphological parameters such as wall thickness, porosity, pore size distribution, and surface structures can influence the interpretation of transport results. Surface and transmembrane morphologies are routinely analyzed qualitatively using scanning electron microscopy (SEM) and less frequently using transmission electron microscopy (TEM). Examples of the range in transmembrane morphologies and outer surface porosities attained with phase inversion are shown in the SEMs of Figures 28.3 and 28.4. Re-

cently, the high resolution techniques of atomic force microscopy (AFM) and low-voltage SEM (LVSEM) have been used to image the porosity and pore size of the permselective skin of ultrafiltration membranes (Fritzsche et al 1993, Kim and Fane 1994). Although the estimation of pore size is difficult, this data can be combined with diffusive profiles or rejection curves to identify unique pore size distributions. These high-resolution morphological techniques are a valuable part of membrane characterization.

The membrane morphology also has a significant influence on cell viability when implanted into tissue sites. The outer membrane morphology (Figure 28.4) is generally either smooth with pores approx. <1 μm or rough with pores approx. >1 μm. The porous rough outer surface has been observed to induce host tissue infiltration and in some cases improve device performance, whereas the smooth outer surface elicits minimal fibrotic reaction (Lacy et al 1991). The combination of appropriate transport properties and outer morphologies has also been achieved using composite membranes (Brauker et al 1992).

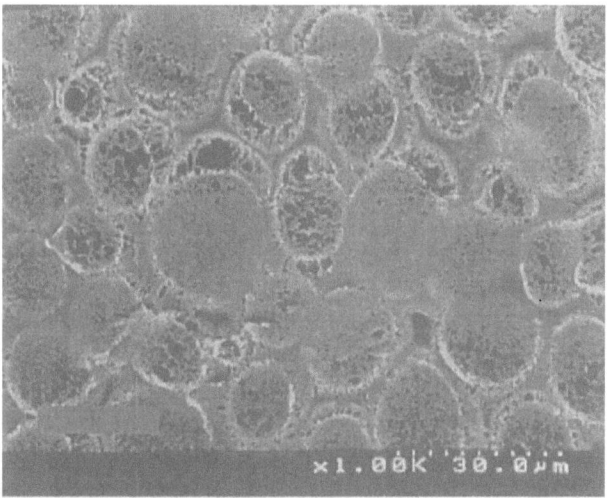

FIGURE 28.4. Scanning electron micrographs of the outer morphology of typical hollow fibers used for implanted cell encapsulation.

Implantation of Encapsulated Cells into the Spinal Intrathecal Space

The subarachnoid space of the spinal cord is a site particularly amenable to cellular implantation for delivery of therapeutic agents. Possible indications for this approach include pain, amyotrophic lateral sclerosis (ALS), spasticity, multiple sclerosis (MS), malignancies, and spinal cord injury. Since clinical tri-als have already been initiated for pain and ALS, these will be described in more detail. Factors critical to the success of encapsulated cell therapy include selection of appropriate therapeutic agents, cell sourcing, and cell delivery systems. For example, for pain management, selection (or engineering) of cells producing opioid peptides, monoamines, inhibitory neurotransmitters, excitatory amino acid antagonists, or possibly other neuroactive peptides or neurotrophic factors, might be appropriate candidates. For ALS, preclinical studies have suggested that certain neurotrophic factors (e.g., CNTF, BDNF, GDNF, NT3, NT4/5) may be useful (Hughes et al 1993b, Kato and Lindsay 1994, Magal et al 1991, Yan et al 1992, Zurn et al 1994).

Donor Cell Sources for Pain Therapy

The majority of neural transplant studies using preclinical pain models have utilized adrenal medullary tissue or isolated chromaffin cells as the source of pain-reducing neuroactive substances (Ginzburg and Seltzer 1990; Hama and Sagen 1993, 1994b; Ortega-Alvaro et al 1994, 1997; Pacheco-Cano et al 1990; Ruz-Franzi and Gonzalez-Darder 1991; Sagen et al 1986a, b, 1990, 1993; Siegan and Sagen 1997; Vaquero et al 1991; Wang and Sagen 1994; Michalewicz et al 1997; Yu et al 1997, 1998). Chromaffin cells of the adrenal medulla produce and secrete a variety of potential pain-reducing substances, the best characterized being the catecholamines and opioid peptides, which may act synergistically at host spinal adrenergic and opioid receptors to produce analgesia. Chromaffin cells also produce a "cocktail" of neurotrophic factors and neuropeptides that may provide neurotrophic support, including basic fibroblast growth factor (b-FGF), transforming growth factor β (TGF-β), interleukin-1 (IL-1), and neurotrophin (NT)-4/5 (Unsicker 1993, Unsicker et al 1996).

Other potential cell sources that have been utilized include opioid-producing cells derived from the intermediate lobe of the pituitary gland, serotonergic cells derived from the medullary raphe or pineal gland, and tumor cell lines secreting opioid peptides or catecholamines. In particular, AtT-20 cells, which were originally derived from a mouse anterior pituitary tumor, synthesize and secrete β-endorphin. These cells have been implanted in mouse and rat spinal intrathecal space (Saitoh et al 1995a; Wu et al 1993, 1994a). When implanted at lumbar levels, intrathecally administered β-adrenergic agonist isoproterenol, which is thought to increase β-endorphin release from the cells, produced antinociception, demonstrated by tail flick and hot plate tests (Wu et al 1993, 1994a). In another study, AtT-20 cells were immunologically isolated in polymer capsules and implanted in the spinal subarachnoid space at the atlanto-occipital junction (Saitoh et al 1995a). In this study, baseline responses to acute noxious stimuli (hot plate, tail pinch, neuromuscular electrical stimulation) were elevated at 2 and 4 weeks postimplantation. Another β-endorphin producing cell line, Nuero2A, a mouse neuroblastoma cell line transfected with the POMC (pro-opiomelanocortin) gene, produced similar effects (Saitoh et al 1995a, b). This was blocked by treatment with opiate antagonist naloxone. Morphological analysis of the capsules after retrieval revealed that the cells had survived and aggregated in the outer layers, while the core was necrotic, typical of capsules containing rapidly dividing cell lines.

To supply catecholamines, B16 F1C29 melanoma cells have been implanted in mouse and rat spinal subarachnoid space (Wu et al 1994b). These implants reduced substance-P-induced biting and scratching behaviors compared to control implants, and this was reversible by a2-adrenergic antagonists. In addition morphine was 10-fold more potent in mice with B16 F1C29 than in controls, as measured by the tail flick test, suggestive of synergism.

A conditionally immortalized serotonergic neuronal cell line RN46A has also been used in preclinical pain implantation studies. The cell line was derived from embryonic rat medullary raphe nucleus and transfected with brain-derived neuronotrophic factor (BDNF) to autoregulate serotonin production (Eaton and Whittemore 1996). These cells have recently been utilized for transplantation into the spinal subarachnoid space of rats with chronic constriction nerve injury (CCI), a model for neuropathic pain (Eaton et al 1997). Thermal hyperalgesia was completely reversed by serotonin-cell implants (46A-B14). Similar effects were obtained for cold allodynia and mechanical hyperalgesia. In contrast, in animals implanted with the nonserotonergic parent line (46A-V1), hyperalgesia continues for at least 6 weeks postligation.

Chromaffin Cell Implants in Preclinical Pain Studies

Published reports have utilized chromaffin cell implants in the spinal subarachnoid space in a wide variety of preclinical pain models. In our laboratory at the University of Illinois, acute analgesiometric tests have included the tail flick, paw pressure (Randall-Selitto test), and hot plate responses (Sagen et al 1986a, b, 1993; Wang and Sagen 1994). Either adrenal medullary tissue allografts from adult donors or isolated bovine chromaffin cells (either encapsulated or unencapsulated with immunosuppression) were used. Implants were placed in the spinal subarachnoid space at L1–L2 via a laminectomy and a slit in the dura. These studies have revealed that there is little change in baseline nociceptive responses using acute noxious stimuli following adrenal medullary implantation. However, following the injection of low doses of nicotine (0.1 mg/kg, s.c.), antinociception is observed using all three acute stimuli (Sagen et al 1986a, b, 1993). This most likely results from increased release from chromaffin cells following stimulation of cell surface nicotinic receptors. This antinociceptive response is most likely mediated, at least in part, by the release of catecholamines and opioid peptides from the implanted cells since it can be attenuated by pretreatment with alpha-adrenergic or opioid antagonists. In addition, increased levels of both catecholamines and opioid peptides are measured in the spinal subarachnoid space of rats with adrenal medullary implants (Sagen and Kemmler 1989, Sagen et al 1991).

Other laboratories have also reported effects of adrenal medullary implants on acute nociception, including responses to noxious thermal stimulation (Ortega-Alvero et al 1994) and noxious electric stimulation (Ruz-Franzi and Gonzalez-Darder 1991) to the tail. Recently, Michalowicz et al (1997) have reported reductions in pain responses to both A-delta and C-fiber mediated pain following implantation of isolated bovine chromaffin cells to the spinal subarachnoid space.

Our laboratory has also extensively utilized the chronic constriction injury (CCI) model as a model for chronic neuropathic pain (Bennett and Xie 1988). This model involves a unilateral injury of the sciatic nerve using loosely constrictive ligatures, and produces measurable pain behaviors reminiscent of neuropathic pain, such as thermal and mechanical hyperalgesia and tactile and cold allodynia. One week following CCI, thermal hyperalgesia is observed on the hindpaw ipsilateral to the ligation, as indicated by decreased withdrawal latencies. In animals with adrenal medullary implants at 1–2 weeks following nerve injury, thermal hyperalgesia was completely reversed and remained so throughout the rest of the study (7–9 weeks), in contrast to sustained hyperalgesia in control animals (Hama and Sagen 1993b, 1994a, b; Ibuki et al 1997;

Siegan et al 1996). Cold allodynia was also reversed by adrenal medullary transplants, and mechanical and tactile allodynia were markedly attenuated. Similar results were obtained following implants of either isolated bovine chromaffin cells in the intrathecal space of immunosuppressed rats, or encapsulated bovine chromaffin cells (Hama et al 1993a).

Using another neuropathic pain model, Ginzburg and Seltzer (1990) showed suppression of autotomy behavior following transection of sciatic and saphenous nerves in animals with adrenal medullary implants, compared to sham implanted animals. The presence of adrenal medullary transplants both delayed the onset of severe autotomy and markedly reduced the overall severity of the self-injury.

In a model for spinal-cord-injury-associated pain, findings from another group have indicated that adrenal medullary implants can reduce pain-related behaviors (Liu et al 1997). The model uses quisqualic acid injected intraspinally to produce excitotoxic spinal cord injury and results in chronic mechanical and cold allodynia combined with excessive grooming behavior. Following induction of grooming, animals were implanted intrathecally with either adrenal medullary tissue or control striated muscle tissue. Pain-related behaviors, including the progression of excessive grooming and hypersensitivity to cold and touch, were significantly reduced by adrenal medullary, but not control, transplants.

In another spinal cord injury pain model, bovine chromaffin cells, both encapsulated and unencapsulated, were reported to reduce chronic allodynia when implanted in the spinal subarachnoid space (Yu et al 1996, 1997). In this model, spinal ischemia is induced using laser irradiation aimed at spinal segment L4 following intravenous injection of erythrosin B. This results in the development of stable tactile and cold allodynia by 2–3 months following the injury. Implantation of either encapsulated or unencapsulated bovine chromaffin cells reversed mechanical allodynia and markedly reduced cold allodynia in these animals, compared with control implanted animals.

Adrenal medullary implants in the spinal subarachnoid space have also shown promise in reducing inflammatory pain, e.g., using an adjuvant arthritis model (Ruz-Franzi and Gonzalez-Darder 1991, Sagen et al 1990, Wang and Sagen 1995) and the formalin pain test (Ortega-Alvaro et al 1997, Siegan and Sagen 1997, Vaquero et al 1991).

Chromaffin Cell Implants in Clinical Pain Studies

Limited clinical trials using adrenal medullary allografts in cancer pain patients have been conducted at several centers. At the University of Illinois at Chicago, approval was obtained from the University's Institutional Review Board to enroll five patients suffering from intractable pain secondary to nonresectable cancerous lesions with prognoses of 6 months or less. Donor adrenal medullary tissue was obtained from the Regional Organ Bank of Illinois from adults ranging 20–53 years of age. Adrenal medullary tissue was dissected into 1.0–2.0 mm^3 pieces and placed in explant culture for 3–7 days to verify tissue viability by catecholamine release assays and tyrosine hydroxylase immunocytochemistry. Tissue was implanted via lumbar puncture using a 14 ga Touhy needle. A total of 1.5–2.0 ml of adrenal medullary tissue (from two adrenal glands) was implanted in each patient. The patients received intravenous fluids following the procedure and were discharged from the hospital without event the following day. Detailed case histories of the patients can be found in Winnie et al (1993). The first patient was a 61-year-old female with colon carcinoma, who developed back and lower extremity pain that increased in severity despite escalating opioid therapy. Following adrenal medullary implantation, VAS scores decreased from severe prior to implantation (8–9 on a scale of 0–10, 10 being the worst pain possible) to moderate by 30 days postimplantation and progressively decreased further following that time. By the end of the 10th week, VAS scores reached 1, and stayed at that level throughout the remainder of her life (4.5 months postimplantation). Her opioid intake also concomitantly declined over the same time period, and CSF Met-enkephalin levels were increased. The second patient was a 69-year-old male with a 4-year history of colon carcinoma. His pain was predominantly in the sacral region and was highly position-dependent, such that it was mild to moderate while he was supine in bed, but became unbearable when he attempted to sit or stand up. Following adrenal medullary implantation, the patient's pain declined rapidly and remained at low levels until his death 11 months postimplantation. In addition, the patient was able to spend increasing time out of bed and became mobile for over 12 hours daily by 2 months postimplantation. The third patient was a 49-year-old female with a 3-year history of breast carcinoma with pain in her lower back, right hip, and buttocks, which was attributed to metastatic lesions in the second and third thoracic vertebrae. After adrenal medullary implantation, the patient's pain scores were decreased to zero at 30 days postimplantation with a concomitant decrease in opioid analgesic consumption and increased catecholamine levels in the CSF. However, the patient's pain recurred, along with increased analgesic intake and progressive weakness in her lower extremities with difficulty in ambulation. This was attributed to spinal compression due to metastatic tumor invasion. Patient 4 was a 52-year-old male with a 3-year history of colon carcinoma who developed pain in the sacral area with radiation to the suprapubic region, which became unresponsive to increasing doses of opioids. Following adrenal medullary implantation, pain scores decreased to zero and remained at that level until his death at 12 months postimplantation. Opioid analgesic intake was also decreased, and CSF catecholamine and opioid peptide levels were increased. A fifth patient enrolled in the original program was a 41-year-old male with Gardner's syndrome. Following adrenal medullary implantation, the patient did not report pain relief by one month postimplantation, despite increased CSF levels of catecholamines and opioid peptides, and declined further follow-up.

A protocol similar to the one followed at the University of Illinois was conducted by Lazorthes et al (1995). Seven patients suffering from intractable pain due to cancer who had received inadequate pain control from oral morphine were included in the initial trial. Implants were performed via lumbar puncture with a 14 ga Touhy needle. Clinical follow-up, including VAS, functional activity (Karnofsky scale), and opioid analgesic intake, was done at 3 and 8 days, and then monthly postimplantation, with a mean follow-up postimplantation of 148 days. Two patients with the longest follow-up (8 and 12 months) received additional adrenal medullary implants at 4 and 2 months, respectively. A multidisciplinary pain evaluation demonstrated progressively decreased pain scores in six of the patients. Opioid analgesic intake was decreased in three of the patients, stabilized in two other patients, and increased in two patients. CSF met-enkephalin levels were found to significantly correlate with pain control. Follow-up studies in the two patients who had received a second implant and survived and remained pain free over 1 year postimplantation, revealed adrenal medullary tissue pieces distributed from T6 to S1 levels, with evidence of viable chromaffin cells by immunocytochemistry and electron microscopy (Pappas et al 1997). This study has now been extended to include 16 patients, confirming the initial promising results (personal communication).

At another center, a patient with intractable pain was implanted with adrenal medullary tissue with "striking results" (Dr. R. Drucker-Colin, Universidad Nacional Autonoma de Mexico, personal communication). This patient showed gradual reduction in somatic pain VAS scores, from most severe prior to implantation, reaching zero by approximately 1 month posttransplantation, and remaining at this level until death at 3 months. There was also a reduction in analgesic intake.

One of the limitations of allogeneic cellular implantation of primary nondividing cells for clinical application is a practical one: the short supply of allogeneic human donor tissue. The use of xenogeneic donors or cell lines is a potential approach to overcoming this limitation. Using cellular encapsulation, it is theoretically possible to immunologically isolate xenogeneic cells from host rejection (see above). Other advantages include the possibility of implant retrieval and potential use of dividing cell lines. This approach has been used for encapsulation of bovine chromaffin cells in preliminary clinical trials for pain (Aebischer et al 1994a, Buchser et al 1996, Burgess et al 1996). For these studies, adrenal glands were isolated in a sterile surgical suite from young calves, and bovine chromaffin cells were suspended in alginate matrix and loaded into PAN/PVC double-skinned membranes with molecular weight cut-off of approximately 50,000. Following loading, membranes were sealed by a methacrylate resin, and the alginate cross-linked in calcium chloride. A titanium connector, in turn attached to a silicon tether attached to the proximal end, was used for device retrieval. Isolated cells were extensively tested for viability, pathogens, and catecholamine output prior to release for implantation.

Phase I clinical trials were conducted at University of Lausanne, Switzerland to assess safety and preliminary efficacy, utilizing encapsulated xenogeneic chromaffin cells from bovine adrenal glands (Aebischer et al 1994a, Buchser et al 1996). Approximately 2 million cells were loaded into each capsule, on the basis of a linear scaling from animal results. Seven patients with severe pain (six with terminal cancer, one with unrelieved neurogenic pain secondary to thoracotomy and scoliosis) were included in the initial study. Under local anesthesia, an incision was performed at the L4–L5 level, subcutaneous tissue dissected, and the subarachnoid space punctured with a 25 ga Touhy needle. A guide wire was introduced through the Touhy needle, followed with a 4-French cannula. Following retrieval of the guide wire, the active portion of the capsule was placed in the lumbar cistern by sliding it through the cannula, and the cannula withdrawn. The external end of the silicone tether was sutured to the lumbodorsal fascia and completely covered with skin closure. Follow-up included close monitoring for adverse effects, recording of medication used, McGill and VAS pain assessments, and occasional lumbar punctures. Encapsulated cell-loaded devices were retrieved in some patients by subcutaneous dissection and withdrawal; the remainder were removed after death, following implant periods ranging from 41 to 176 days.

The implantation procedures and recovery were generally uneventful, except for occasional reports of headaches after surgery that were attributed to CSF leakage at the implant site. These resolved either spontaneously or after epidural blood patch. Four of the seven patients, who were receiving epidural morphine at the time of the implant, had decreased opioid use following implant, with either a modest improvement or no worsening in pain ratings (Figure 28.5). Three of the patients, who were not receiving oral or epidural morphine treatment at the time of the implant, showed improvement in McGill pain ratings (Figure 28.6); two showed corresponding improvements in VAS.

Examination of retrieved devices showed no fibrotic overcoats, adherence to external surfaces, or other signs of foreign body reaction. No adverse findings that could be attributed to the implantation technique or to the presence of the device in the subarachnoid space were noted with histologic examination of the spinal cords. Chromaffin cell viability

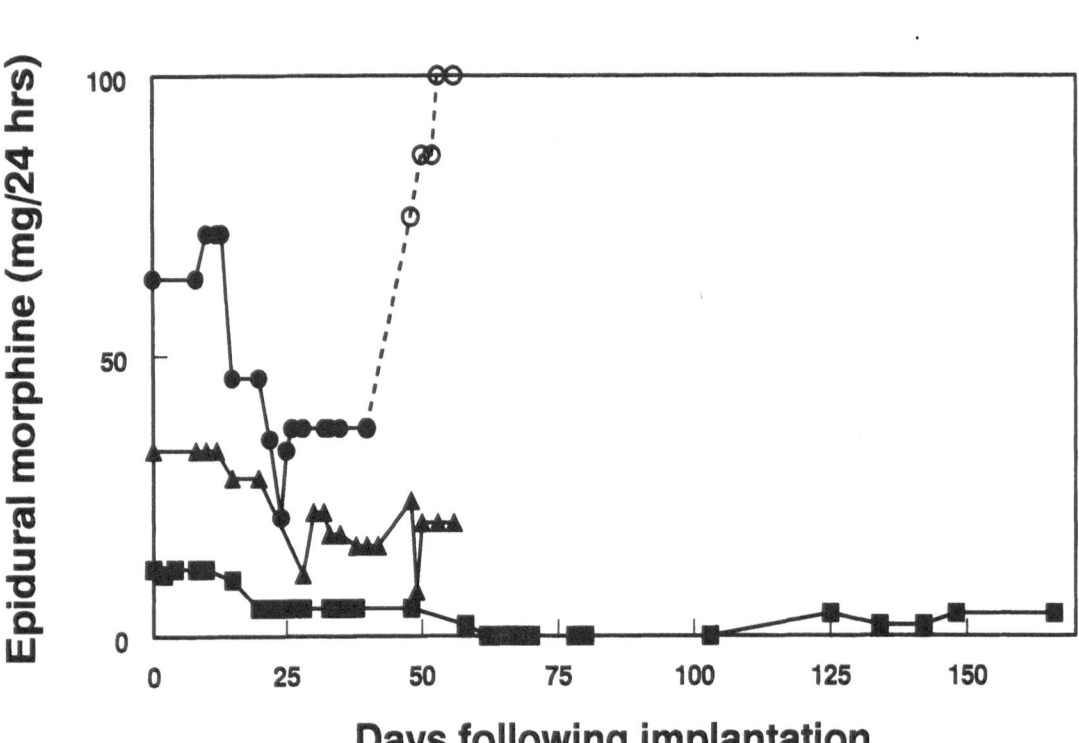

FIGURE 28.5. Decrease in opioid intake for three patients in the study. Pain ratings for these patients were stable or decreased slightly, compared with starting/baseline values. A fourth patient (Patient 1) also showed reductions in epidural morphine needs and reduced pain, but was not included because epidural bupivicaine and fentanyl therapies were initiated, and it is unclear whether improvements resulted from these agents, the implant, or both. The dotted portion of the line with open squares for Patient 2 is morphine administration after device removal; this continued to increase to an off-scale level of 135 mg/day at death on day 60. Adapted from Buchser et al., Anesthesiology 85: 1005–1012, 1996.

in the retrieved capsules was confirmed by immunostaining and catecholamine release.

Preliminary results from a similar Phase I clinical trial in the U.S. have been recently reported (Burgess et al 1996). This study included 15 patients in the initial Phase I, who received capsules containing 1×10^6 cells, and an additional four patients who received capsules containing 3×10^6 cells. Advanced cancer patients with chronic pain inadequately relieved by conventional therapies were enrolled in the study. Patients were carefully monitored for adverse experiences, vital signs, hematologic and clinical chemistry parameters, immunologic responses, and CSF biochemistry and bacteriology. Pain was rated by VAS and McGill, and medication use was recorded. Of the 15 patients implanted in the first group, capsules remained in place for an average duration of 96 days (23–220 days). The few

mild adverse experiences were similar to complications encountered with intrathecal drug administration systems, including postlumbar puncture headaches (two patients), subcutaneous fluid collections at the implant site (two patients), and subarachnoid-cutaneous fistula (one patient). Most of these resolved spontaneously or after an epidural blood patch. Minimal immunologic consequences were noted. Although the clinical picture in the patient population is complex, evidence of analgesic efficacy was suggested by reductions in VAS and MPQ scores in nine patients, and opiate reduction in eight patients of the original 15 patients (Burgess et al 1996). Recovered capsules showed chromaffin cell viability and catecholamine production. Figure 28.7 is a section of a capsule retrieved from a patient enrolled in this study 2 months postimplantation. It has been immunocytochemically stained for tyro-

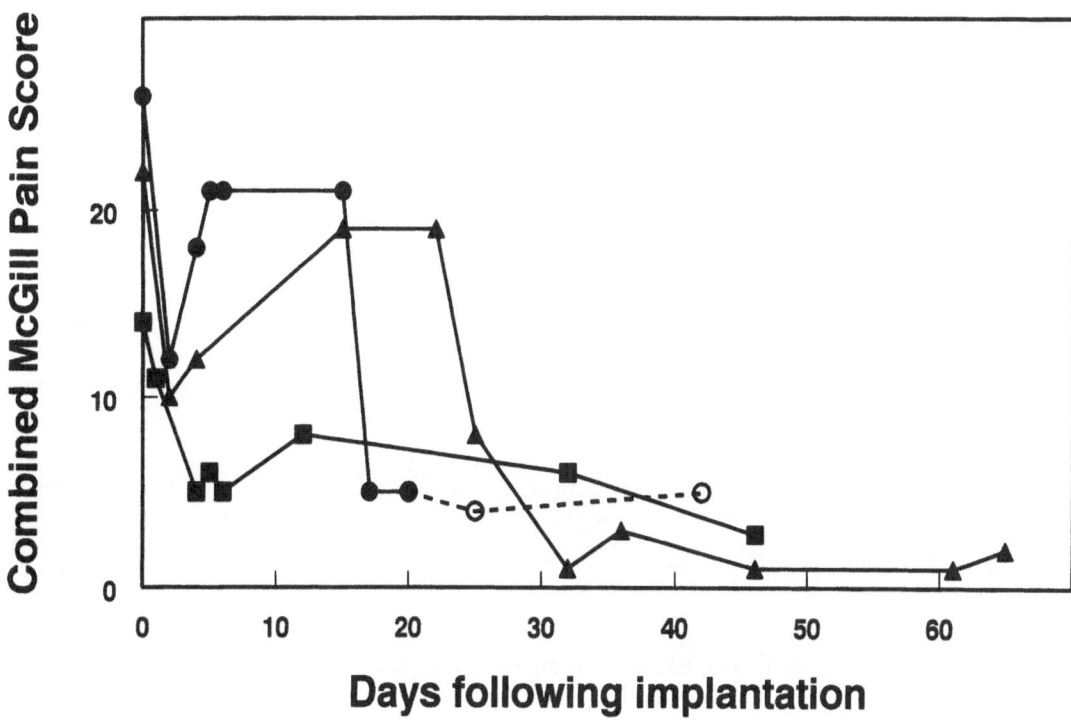

FIGURE 28.6. Decrease in combined McGill ratings of patients who did not respond to morphine (Patient 3 and Patient 7) or in whom morphine was contraindicated (Patient 4). The dotted portion of the line for Patient 3 denotes the period after which epidural morphine was initiated. Adapted from Buchser et al., Anesthesiology 85: 1005–1012, 1996.

sine hydroxylase, revealing numerous clusters of tyrosine-hydroxylase-positive chromaffin cells embedded within the alginate matrix. Phase II clinical trials in patients with cancer and neuropathic pain are currently ongoing.

Encapsulated Cell Implants in ALS

In addition to the provision of neuroactive agents via cellular implant delivery, encapsulated cell implantation offers the opportunity to provide therapeutic trophic factors that can be useful in a number of neurodegenerative diseases, traumas, and disorders. The best-studied of these using the intraspinal delivery route is ALS, which is characterized by the progressive degeneration of motor neurons in the spinal cord and at higher centers. Preclinical studies have suggested that certain neurotrophic factors, e.g., CNTF, BDNF, GDNF, NT3, NT4/5, and IGF-1, may be useful in supporting the survival of motor

neurons in ALS models (Hughes et al 1993b, Kato and Lindsay 1994, Magal et al 1991, Yan et al 1992, Zurn et al 1994). However, clinical trials using systemic administration of some of the agents, including CNTF and BDNF, in ALS patients have been discontinued because of lack of efficacy and/or because of side effects. Possible explanations include short half-lives of the active agents, route of administration requiring high doses reaching toxic levels, or low efficacy of the trophic factors themselves. A cell-based delivery system could allow for continuous slow release of these agents, possibly avoiding some of the toxicity as well as overcoming the problem of rapid degradation. In addition, using a cell-based delivery approach, it should be possible to deliver multiple trophic factors, which have been suggested to provide synergistic activity.

Baby hamster kidney (BHK) cells transfected with mouse CNTF gene were encapsulated (5×10^4 cells/capsule) and tested using preclinical models

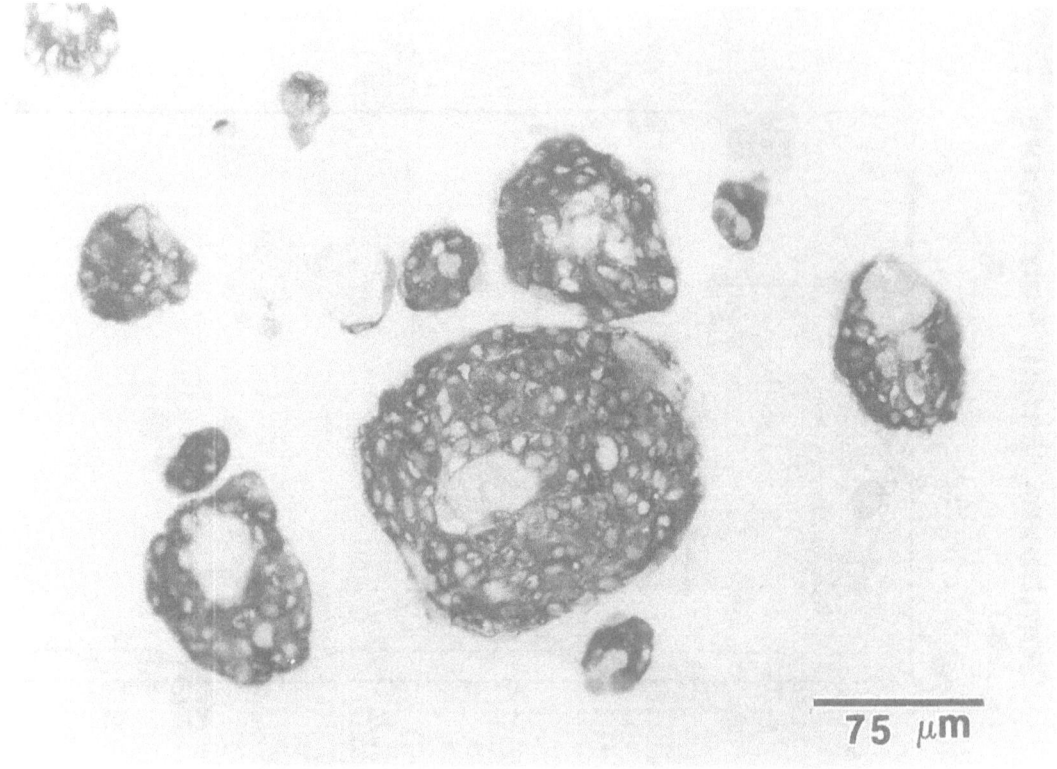

75 μm

FIGURE 28.7. Appearance of encapsulated bovine chromaffin cells following implant retrieval from the lumbar cistern 2 months postimplantation. Cells are stained with a tyrosine hydroxylase antibody, using the avidin-biotin method.

(Sagot et al 1995). The capsules were implanted s.c. in *pmn/pmn* (progressive mononeuropathy) mice. This mutant undergoes progressive hindlimb weakening, severe loss of myelinated motor fibers, significant reduction of facial motoneuron cell bodies, and death approximately 6 weeks after birth. Implantation with CNTF-producing, but not with control BHK cells, significantly prolonged the life span of these animals and improved function on motor tests (Sagot et al 1995). In addition, the number of facial motor neurons was significantly greater in the BHK-CNTF group than in controls. However, this treatment did not prevent the disease progression and ultimate death of the animals. It has been suggested that a combination of trophic factors may be more effective than single factors. For example, in wobbler mice, coadministration of CNTF and BDNF was reportedly more efficacious than CNTF alone (Mitsumoto et al 1994), and preliminary studies have shown that combinations of CNTF, BDNF, and NT-3

can produce further increased survival of *pmn/pmn* mice (Sagot et al 1995).

An open label safety trial using encapsulated BHK-hCNTF cells implanted intrathecally in ALS patients has been reported (Aebischer et al 1996). The device used in the trial consisted of approximately 100,000 cells in a collagen matrix loaded in a 5 cm hollow-fiber membrane fabricated from poly-ether-sulfone (PES, Akzo Nobel Faser AG) and connected with a silicon tether for retrieval. Capsules were monitored for sterility and CNTF production prior to implantation, and placed in the lumbar cistern as described above for the pain implants. Six patients were initially implanted for approximately 3 months, and reimplantation of a second device was offered for an additional 6 months. Progression of the disease was monitored using forced vital capacity, Tufts Quantitative Neuromuscular Evaluation, and the Norris scale. Serum and CSF CNTF levels were also determined at various times following implantation.

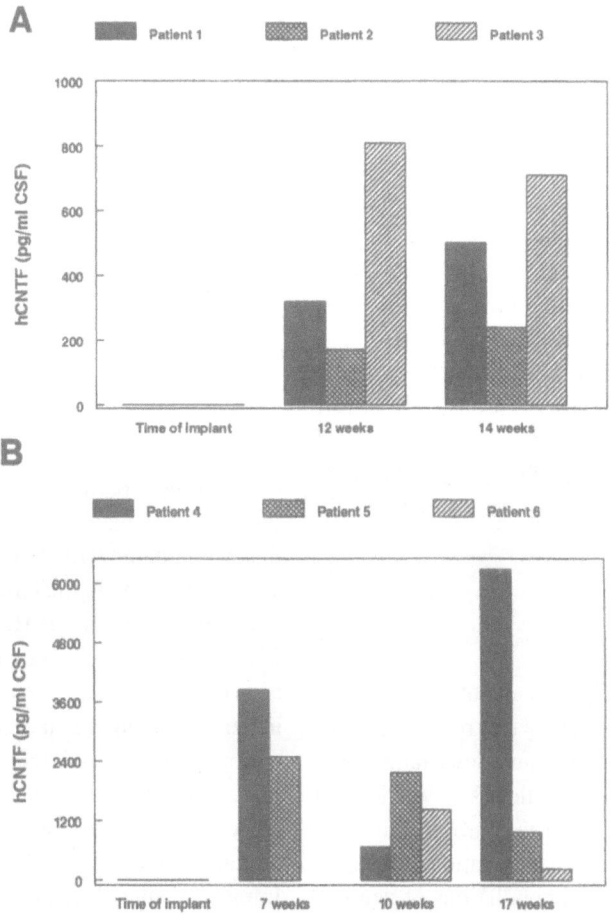

FIGURE 28.8. Human CSF CNTF levels as measured by ELISA. CSF samples were taken prior to implant, and at several intervals following implantation of encapsulated BHK-hCNTF cells into the spinal subarachnoid space of ALS patients. (A) shows results from patients 1–3, (B) shows results from patients 4–6. No CNTF was detectable in the CSF of these patients prior to implant. Adapted from Aebischer et al., Nature Med. 2: 696–699, 1996.

At the time of implantation, patients had no measurable CNTF in the CSF. All patients demonstrated measurable CSF levels of CNTF in the weeks following device implantation for as long as 17 weeks after the initial implant (Figure 28.8). There was some variability in the CSF levels of CNTF between the first and second groups of three patients, most likely due to differences in the total number of cells in the devices following different in vitro holding periods prior to implantation. No CNTF was detected in the blood at any time. Retrieved capsules secreted CNTF and contained viable cells. Adverse clinical experiences, particularly those reported in systemic delivery trials, such as cough, fever, weight loss, were not noted except for a brief episode of mild dry cough in two patients. Evaluation of Norris scale slopes before and after implantation indicated that the disease continued to progress. Nevertheless, these findings indicated that it is possible for encapsulated cellular implantation in the spinal CSF to achieve sustained delivery of neurotrophic factors.

Implantation of Encapsulated Cells in the Brain

Cellular implantation strategies for repair and restoration of CNS function has been a subject of intensive investigation for the past two decades. Using an encapsulated cell delivery approach, it may be possible to deliver therapeutically beneficial molecules, such as neurotrophic factors and neurotransmitters, to the CNS on a long-term or permanent basis. Neurodegenerative diseases, including Huntington's disease and Parkinson's disease, may be particularly amenable to this approach. However, CNS sites pose new challenges for optimization of cell survival and performance in an encapsulated environment. Preclinical findings are summarized in the following sections.

Huntington's Disease

Huntington's Disease (HD) is an autosomal dominant neurodegenerative disease that results in a progressive movement disorder and devastating psychiatric and cognitive deterioration (see Emerich and Sanberg, 1992, for review). The neuropathology associated with HD is severe degeneration of the neostriatum. Although several populations of neurons are affected, the GABAergic medium-sized spiny projection neurons are one of the major targets (Reiner et al 1988). In the later stages of the disease, the large aspiny cholinergic interneurons can also exhibit degenerative changes Roberts and DiFiglia 1989. In contrast, the local circuit NADPH-diaphorase and somatostatin neurons are relatively spared throughout the course of the disease (Beal et al 1986, 1988, 1989). The behavioral and cognitive symptoms of HD result from the selective vulnerability of these particular populations of neostriatal neurons (Albin and Greenamyre 1992, Beal 1992, Parker et al 1990, Wallace 1992). To date, there is no effective treatment for Huntington's disease.

Animal models that mimic the neural degeneration seen in HD have been used to study this disease. Intrastriatal injections of excitotoxins such as quinolinic acid (QA) or ibotinic acid (IA) mimic the pattern of selective neuronal vulnerability seen in HD (Beal et al 1986, 1988, 1989). Injection of these excitotoxins into the striatum destroys the neurons and results in motor and cognitive deficits that are among the major symptoms seen in HD (Block et al 1993, Emerich et al 1991, 1992, Sanberg et al 1981, 1989). Thus, intrastriatal injections of QA have become a useful model of HD and can serve to evaluate novel therapeutic strategies aimed at preventing, attenuating, or reversing neuroanatomical and behavioral changes associated with HD.

Growth factors that have specific actions on the vulnerable neural populations have been evaluated for their potential therapeutic effect. These factors have been delivered by direct infusion or via the use of capsules containing cells that have been genetically engineered to secrete growth factors. There are several advantages to the latter approach, particularly the ability to deliver bioactive growth factors for prolonged periods. In addition, growth factors secreted from mammalian cells show improved efficacy compared with delivery of recombinant proteins (see following material).

For the treatment of HD, there are at least two CNS sites in which implantation of capsules has been envisioned. Because the disease process results in the degeneration of the striatum, many animal experiments have been done with direct intrastriatal placement of encapsulated cells. Another possible placement site is the lateral ventricle. There are potential advantages and disadvantages at both sites. Intraventricular placement allows the capsules to be continually bathed in CSF, which may facilitate the survival of the encapsulated cells, compared to intraparenchymal sites, which may be nutrient and oxygen poor. In addition, intraventricular placement may result in a more widespread delivery of the secreted trophic factor. On the other hand, this type of delivery may result in the dilution of the growth factor to suboptimal levels at the target site. Another drawback of intraventricular placement may be delivery of the factor to inappropriate regions, increasing the potential for untoward effects. Nevertheless, preclinical findings have suggested that it is possible to achieve delivery of effective levels of trophic factors at either placement site.

One example of this is the delivery of human recombinant nerve growth factor (hNGF) into the striatum prior to a QA lesion. Within the striatum, cholinergic striatal interneurons express the low-affinity p75 NGF receptor during development and the high-affinity NGF receptor throughout life (Holtzman et al 1992, Kordower and Mufson 1993, Kordower et al 1994, Schumacher et al 1991, Yan and Johnson 1988), suggesting that the cholinergic neurons are likely to respond to NGF treatment. As predicted, infusion of NGF into the normal adult striatum resulted in hypertrophy of cholinergic striatal interneurons, indicating that NGF has a functional effect in this region (Gage et al 1989). In addition, NGF infusion spared only cholinergic neurons when infused concurrently with a lesion (Kordower et al 1994). In these experiments, neurons containing GABA, somatostatin, and NADPH-diaphorase were unaffected (Kordower et al 1994). Since the NGF-receptor-bearing cholinergic neurons are relatively spared in HD, it seems likely that NGF would not be an optimal treatment choice for this disease. However, in contrast to NGF infusion, implantation of encapsulated or unencapsulated cells genetically engineered to secrete NGF protected cholinergic, GABA, and diaphorase neurons. In addition, these cellular implants prevented the behavioral sequelae resulting from intrastriatal injections of excitotoxins including QA (Emerich et al 1994, Frim et al 1993a, b, Schumacher et al 1991). These data indicate that NGF delivery from encapsulated cells may be more effective in the treatment of HD than the infusion of recombinant proteins.

For cellular delivery of NGF, capsules containing baby hamster kidney (BHK) cells were genetically engineered to secrete hNGF and were placed unilaterally into the lateral ventricle of adult normal rats (Emerich et al 1994). Control animals received intraventricular implants of control BHK cells. Three days after implantation the animals received a unilateral QA lesion in the ipsilateral striatum. The animals were then evaluated behaviorally and histologically for the protective effects of hNGF. After 2 weeks, animals that received control capsules exhibited pronounced apomorphine-induced rotation behavior associated with the QA lesion. In contrast, animals that received hNGF-secreting capsules showed a marked reduction in apomorphine-induced rotation. In addition, 1 month after the QA lesion, histological analysis showed a dramatic loss of ChAT and NADPH-d-positive neurons in animals receiving control capsules. In contrast, in animals that received the hNGF-secreting capsules, sparing of striatal neurons was found, and this correlated with behavioral findings (Emerich et al 1994). The effective dosage of hNGF was quite low; before implantation the capsules secreted an average of 34 ng hNGF/device/24 hours. Postexplant, the capsules

were evaluated again and found to release an average of 16 ng hNGF/device/24 hours.

While the above study showed effective delivery of growth factors by encapsulated cells for 4 weeks, it is important to demonstrate that the capsules can deliver growth factors for longer periods of time in order to be useful in a therapeutic setting. In another study, normal rats were implanted in the lateral ventricle with NGF-secreting capsules for 13.5 months (Winn et al 1996). In this study, the preimplant secretion levels were 3.6 ng/device/24 hours. The postexplant values were 2.2 ng/device/24 hours, indicating that the cells in the capsules remained viable and functional even after 13.5 months in the CNS environment. In addition, there were no apparent deleterious host effects resulting from the long-term delivery of NGF. Body weights, mortality rates, motor function, and cognitive function, as measured using the Morris water maze and delayed matching to position testing, were all normal. Anatomically, cholinergic neurons in the striatum and nucleus basalis showed hypertrophy, a typical response to NGF. In addition, there was robust sprouting of cholinergic fibers within the frontal cortex and lateral septum, regions proximal to the implant.

In summary, these data indicate that the implantation of encapsulated cells in the CNS can result in the long-term delivery of bioactive hNGF. No adverse effects were identified, in contrast to the side effects often associated with the systemic delivery of growth factors. The long-term delivery of NGF via encapsulated cells did result in some behavioral and anatomical changes, supporting the potential of this approach to produce long-term therapeutic effects.

In addition to NGF, another growth factor, CNTF, has also been evaluated in this rodent model of HD. In these experiments, capsules containing BHK cells genetically modified to secrete CNTF were implanted unilaterally into the lateral ventricle (Emerich et al 1996). Subsequently, animals received unilateral QA lesions. As described above, these lesions result in apomorphine-induced rotation, which correlates to the severity of the lesion. Animals that received CNTF-secreting capsules showed significantly less rotation than animals receiving control capsules. The number of neurons on the lesioned side of the brain was compared to the number of neurons on the control/intact side of the brain. Animals that received control capsules showed marked

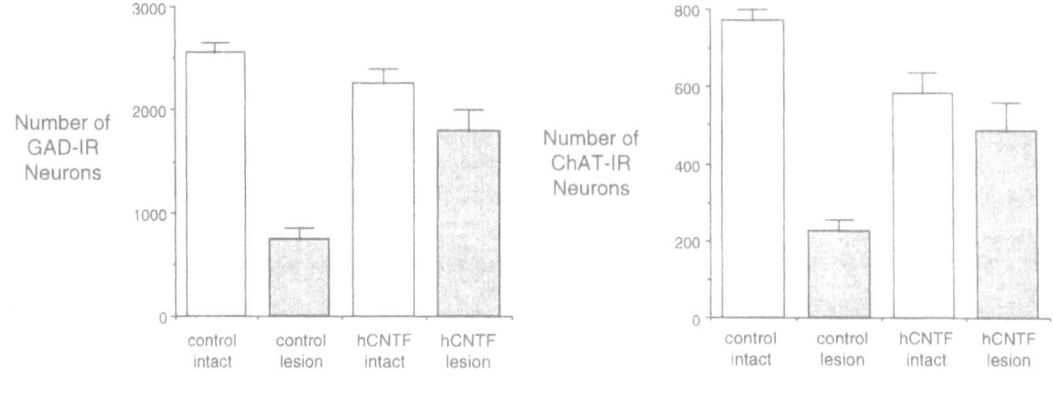

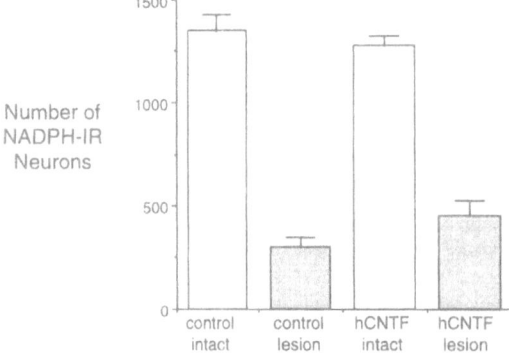

FIGURE 28.9. Quantification of ChAT-ir, GAD-ir, and NADPH-ir neurons within the striatum of quinolinic-acid-lesioned rats. Animals received either control capsules or CNTF-secreting capsules. After comparison of the implanted to the intact side of the brain, animals that received CNTF-secreting capsules showed sparing of all three types of neurons. Adapted from Emerich et al., J. Neurosci. 16: 5168–5181, 1996.

loss of cholinergic, GABAergic, and diaphorase-containing cells (Figure 28.9). In contrast, in animals that received CNTF-secreting capsules, cholinergic, GABAergic, and diaphorase-containing neurons were protected against the excitotoxic lesion (Figure 28.9).

Together, these studies demonstrated that intraventricular delivery of neurotrophic factors can prevent excitotoxin-induced degeneration in animal models of HD. A more localized delivery directly into the CNS parenchyma may also be feasible and result in improved effectiveness. In order to address this, capsules containing BHK cells genetically engineered to secrete hNGF have been implanted intrastriatally in normal adult rats (Kordower et al 1996). These animals were evaluated 1, 2, and 4 weeks after implantation. At all time points, an increase in hypertrophy of cholinergic and neuropeptide-Y-containing neurons was found, indicating that bioactive hNGF was successfully delivered into the striatum. It is particularly important to assess the ex-

tent of NGF delivery when capsules are placed into the parenchyma. In these studies, hNGF was identified by immunoreactivity 1000 μm from the capsule, indicating diffusion of at least 1000 μm through the brain parenchymal tissue (Kordower et al 1996). In addition, while the parenchymal placement might be expected to limit cell survival in the capsules as a result of hindrance of nutrient flux, hNGF secretion was found to be stable over the 1 month time course of the study. An average of 3.2 ng hNGF/device/24 hours was secreted before implantation, and an average of 2.6 ng hNGF/device/24 hours was secreted after explant.

Intrastriatal implantation has also been used in experiments that examined CNTF delivery into a primate model of HD (Emerich et al 1997). In these experiments, capsules containing BHK cells genetically engineered to secrete hCNTF or control BHK cells were unilaterally implanted into the striatum of cynomolgous monkeys. One week after implantation the monkeys received ipsilateral QA lesions.

CNTF delivered from these capsules was found to protect GABAergic, cholinergic, and diaphorase-containing neurons, which were destroyed in the control implanted lesioned animals. CNTF delivery also protected the globus pallidus and the pars reticulata of the substantia nigra, targets of these striatal projection neurons.

A clinical trial is planned to begin shortly in France in a small group of HD patients at early symptomatic stages. Patients will receive intraventricular implants of encapsulated CNTF-secreting BHK cells and will be evaluated for 6 month follow-up periods.

Parkinson's Disease

Parkinson's Disease (PD), named for James Parkinson, the physician who first described it in 1817, is a progressive degenerative motor neuron disease. The hallmarks include tremor, rigidity, akinesia, and bradykinesia, and the effects can be devastating. The physical effects appear to be the result of a disruption of the nigrostriatal system caused by the loss of dopaminergic (DA) neurons in the substantia nigra (Hughes et al 1993, Riederer and Wuketich 1976). Since striatal neurons continue to express the DA receptor, delivery of dopamine could potentially alleviate disease symptoms. Systemic administration of the dopaminergic precursor L-dopa is the current standard therapy for Parkinson's disease. However, this treatment can result in severe side effects from toxicity to nonstriatal neurons. In addition, there is a decrease in therapeutic effectiveness of this precursor with time and disease progression. Early work suggested that transplantation of embryonic mesencephalic tissue or adrenal chromaffin cells ameliorated the symptoms of PD (Dunnett et al 1981, Freed et al 1990, Perlow et al 1979, Sladek and Gash 1988, Yurek and Sladek 1990). However, the availability of fetal tissue for transplantation and ethical questions surrounding the issue make it a difficult therapeutic strategy. An alternative is the potential use of xenogeneic fetal donors, such as fetal porcine mesencephalic tissue currently being utilized in clinical PD trials (Deacon et al 1997, Edge and Dinsmore 1997). In a Phase I clinical trial, porcine ventral mesencephalic cells were transplanted into 12 Parkinson's patients maintained on cyclosporine immunosuppression. Follow-up has been up to 24

months in some patients. Examination of the striatum of one patient who died approximately 8 months after implantation (unrelated causes) revealed survival of donor-derived cells at three graft sites, some tyrosine-hydroxylase-expressing (presumably dopaminergic) neurons, and minimal MHC Class II expression or lymphocyte infiltration.

Another alternative is the use of a banked, tested cell line that could produce dopamine, which would provide a viable treatment for patients with PD, if that cell could be held in an immunoisolatory biocapsule (Emerich et al 1992, Tresco et al 1992). PC12 cells are such a cell line; originally isolated from a rat pheochromocytoma tumor (Greene and Tischler 1976, Jaeger et al 1990), these cells synthesize and release dopamine. Upon encapsulation, PC12 cells were found to release approximately 6 pg per minute per device, and remained viable in vitro for up to 6 months (Jaeger et al 1990). There was evidence of cell turnover within devices, as both cell debris and mitotic figures were observed. Some cells showed positive immunohistochemistry for tyrosine hydroxylase (TH), suggesting that cells that appeared viable histologically were also functional (Jaeger et al 1990). PC12 cells have also been placed in microcapsules that showed a similar mix of cellular morphologies, i.e., both mitotic and necrotic cells were seen among groups of healthy viable cells (Aebischer et al 1991a, Tresco et al 1992), although cells were more densely packed in microcapsules (Aebischer et al 1991a). Both microcapsules and macrocapsules showed viable, TH-positive cells upon transplantation into the striatum of nonlesioned guinea pigs (Tresco et al 1992, Winn et al 1991). In vitro, both encapsulation methods protected cells against complement-mediated lysis (Tresco et al 1992, Winn et al 1991), suggesting some immunoprotection provided by the membranes. The survival of encapsulated PC12 cells for as long as 3 months when implanted into the striatum of adult guinea pigs without immunosuppression also supports this hypothesis (Aebischer et al 1991b). In addition, the physical constraints of the encapsulation system prevented the formation of tumors that have been reported with unencapsulated PC12 cells (Tresco et al 1992). These data suggest that PC12 cells can be successfully encapsulated and remain viable for many months, a requirement for the long-term treatment necessary for PD.

There are two animal models currently in use that mimic features of Parkinson's disease in humans and are commonly used to screen potential therapies. In one model, the neurotoxin 6-OHDA is injected unilaterally into the striatum, resulting in a unilateral depletion of striatal dopamine and behavioral rotation in response to the DA-receptor agonist apomorphine (Ungerstedt et al 1974). Therapeutically active molecules result in reduced rotational responses following apomorphine administration. Efficacy has been shown in this model in studies using encapsulated PC12 cells (Tresco et al 1992, Winn et al 1991). Empty devices, implanted as controls, showed no significant effect on apomorphine-induced rotational behavior. Cells within experimental devices showed positive immunohistochemistry for TH, whereas sprouting of TH neurons was not seen in endogenous tissue (Tresco et al 1992). These data suggest that the positive behavioral effect was the product of factors secreted from encapsulated cell devices rather than an upregulation of endogenous dopaminergic neurons. PC12 cells in both microcapsules and hollow-fiber membranes showed a similar degree of functional recovery in unilaterally lesioned rats (Aebischer et al 1991a, Winn et al 1991). Encapsulated bovine chromaffin cells implanted in the striatum of lesioned rats also showed a positive behavioral effect (Aebischer et al 1991c).

In addition to acting as a minipump for restoring neurotransmitter levels, encapsulated cells may also be utilized to provide neurotrophic support for remaining host cells in progressive neurodegenerative diseases such as PD. Neurotrophic factors support the survival of many populations of neurons. Glial-cell-derived neurotrophic factor (GDNF) is a growth factor that is trophic for motor and midbrain dopaminergic neurons (Lin et al 1993), which are lost in the degenerative process of Parkinson's. In nonlesioned animals implanted with devices secreting GDNF, amphetamine-induced responses were noted (Hammang et al 1995). BHK cells genetically engineered to secrete GDNF were implanted in 6-OHDA lesioned rats (Lindner et al 1995). These encapsulated cell devices did not decrease apomorphine-induced rotations, in contrast to PC12-cell-containing devices. The GDNF capsules were not retrieved, however, and there was no measure of growth factor output from these devices. It is also possible that the severity of the lesion in this model masks some measurements of relevant clinical efficacy. A newly discovered member of the GDNF family of neurotrophic factors named neurturin (Kotzbauer et al 1996) has yet to be tested in relevant models of PD.

Another animal model of PD involves the infusion of 1-methyl-4-phenyl-1,2,3,6-tetrahydropyridine (MPTP) into nonhuman primates, which results in parkinsonian symptoms including bradykinesia, rigidity, and tremor (Burns et al 1983). However, in contrast to rodent studies, PC12 cell survival and function were much more variable in primate models. Significant improvement in motor performance was seen in four out of five unilaterally lesioned MPTP cynomolgous monkeys with encapsulated PC12 cell implants, but not bovine adrenal chromaffin cell (BAC) implants (Aebischer et al 1994b). Functional improvement was observed in monkeys implanted with BAC cells when supplemented with Sinemet (L-dopa and carbidopa), suggesting that the levels of dopamine secretion may play a role in efficacy (Aebischer et al 1994b). In another study (Kordower et al 1995), two out of three monkeys showed positive behavioral effects in a hand reaching test. The device explanted from the subject that did not show improvement in motor function had few viable PC12 cells, and had low to undetectable levels of levodopa secretion. In contrast, PC12-containing devices explanted from lesioned animals who recovered limb function had measurable levels of levodopa (in some cases higher than at implant), basal dopamine, and K+-evoked dopamine. Functional recovery was seen about 1 month after implantation and continued until the devices were retrieved at 6.5 months. Interestingly, lesioned animals did not revert to preimplant levels of motor deficit after devices were explanted. These data suggest that factors in addition to dopamine may play a role in the functional recovery seen in this model.

Implantation of Encapsulated Cells in the Eye

The neural retina is a thin piece of tissue at the back of the eye, which is derived from the neural tube early in development (Dowling 1987). As such, this piece of CNS tissue is susceptible to neurodegenerative disease processes similar to those found in other parts of the brain. Specific populations of retinal neurons are lost in diseases such as glaucoma (ganglion cells), retinitis pigmentosa (photoreceptor cells), and macular degeneration (photoreceptors). In a comparative study on the neuroprotective effects on photoreceptors in a light-damage model of degeneration (LaVail et al 1992), injection of a number of growth factors was shown to be efficacious. This work suggests that growth factor treatment, which has shown promise as a treatment in the CNS, will be applicable to the eye.

Delivery of factors to the eye faces the same challenges and requirements as the rest of the CNS. There exists within the eye a retinal barrier (the blood-retinal barrier), similar to the blood-brain barrier, which inhibits passage of molecules to the retina. Thus local delivery via a device is a strategy that is likely to be necessary in a number of disease paradigms. It will be very interesting to investigate whether a small encapsulated cell device secreting factors in the eye will prove efficacious in neuroretinal degenerations.

Conclusions

In summary, the use of encapsulated cellular implants as means of delivering therapeutic substances to the CNS has thus far yielded promising outcomes, both in preclinical and pilot clinical trials. In addition, in spite of the increased complexity of delivery from an incompletely defined biological process, thus far this approach has revealed remarkably few safety and toxicity concerns. While preliminary clinical trials utilizing this approach have been primarily targeted to the spinal CSF, this chapter has described numerous additional target sites and applications that may be amenable to therapeutic benefit using this technology. A key advantage of this approach, particularly for chronic and progressive neurological diseases, is the ability to provide therapeutic molecules on a continually renewable and long-term basis. In addition, preclinical findings have indicated that cells for encapsulation may be engineered to produce novel and perhaps multiple potent neurotrophic factors to protect and restore CNS regions damaged by disease processes. The field of encapsulated cell-based delivery to the CNS is just beginning to realize its potential. A more complete understanding of the interplay between cells, matrices, membranes, and host CNS target sites will undoubtedly lead to more effective approaches in the treatment of these debilitating CNS diseases.

References

Aebischer P, Winn SR, Tresco PA, Jaeger CB, Greene LA. 1991a. Transplantation of polymer encapsulated neurotransmitter secreting cells: Effect of the encapsulation technique. J. Biomech. Eng. 113: 178–183.

Aebischer P, Tresco PA, Winn SR, Greene LA, Jaeger CB. 1991b. Long-term cross-species brain transplantation of a polymer-encapsulated dopamine-secreting line. Exp. Neur. 111: 269–275.

Aebischer P, Tresco PA, Sagen J, Winn SR. 1991c. Transplantation of microencapsulated bovine chromaffin cells reduces lesion-induced rotational asymmetry in rats. Brain Res. 560: 43–49.

Aebischer P, Buchser E, Joseph JM, Favre J, de Tribolet N, Lysaght M, Rudnick S, Goddard, M. 1994a. Transplantation in humans of encapsulated xenogeneic cells without immunosuppression. Transplantation 58: 1275–1277.

Aebischer P, Goddard M, Signore AP, Timpson RL. 1994b. Functional recovery in Hemiparkinsonian primates transplanted with polymer-encapsulated PC12 cells. Exp. Neur. 126: 151–158.

Aebischer P, Schluep M, Déglon N, Joseph J-M, Hirt L, Heyd B, Goddard M, Hammang JP, Zurn AD, Kato AC, Regli F, Baetge EE. 1996. Intrathecal delivery of CNTF using encapsulated genetically modified xenogeneic cells in amyotrophic lateral sclerosis patients. Nature Med. 2: 696–699.

Albin RL, Greenamyre JT. 1992. Alternative excitotoxic hypothesis. Neurology 42: 733–738.

Beal MF. 1992. Does impairment of energy metabolism result in excitotoxic neuronal death in neurodegenerative illnesses? Annals of Neurology 31: 119–130.

Beal MF, Kowall NW, Ellison DW, Mazurek MF, Swartz KJ, Martin JB. 1986. Replication of neurochemical characteristics in Huntington's disease by quinolinic acid. Nature 321: 168–171.

Beal MF, Mazurek MF, Ellison DW, Swartz KJ, McGarvey U, Bird ED, Martin JB. 1988. Somatostatin and neuropeptide Y concentrations in pathologically graded cases of Huntington's disease. Annals of Neurology 23: 562–569.

Beal MF, Kowall NW, Swartz KJ, Ferranti RJ, Martin JB. 1989. Differential sparing of somatostatin-neuropeptide Y and cholinergic neurons following striatal excitotoxic lesions. Synapse 3: 119–130.

Bennett GJ, Xie Y-K. 1988. A peripheral mononeuropathy in rat that produces disorders of pain sensation like those seen in man. Pain 33: 87–107.

Block F, Kunkel M, Schwarz M. 1993. Quinolinic acid lesion of the striatum induces impairment in spatial learning and motor performance in rats. Neuroscience Letters 149: 126–128.

Brauker J, Martinson L, Hill R, Young S, Carr-Brendel V, Johnson R. 1992. Neovascularization of immunoisolation membranes: The effect of membrane architecture and encapsulated tissue. Transpl. Proc. 24: 2924.

Buchser E, Goddard M, Heyd B, Joseph JM, Favre J, Detribolet N, Lysaght M, Aebischer P. 1996. Immunoisolated xenogeneic chromaffin cell therapy for chronic pain: Initial experience. Anesthesiology 85: 1005–1012.

Burgess FW, Goddard M, Savarese D, Wilkonson H. 1996. Subarachnoid bovine adrenal chromaffin cell implants for cancer pain management. Amer. Pain. Soc. Abstr. 15: A-33.

Burns RS, Chiueh CC, Markey SP, Ebert MH, Jacobowitz DM, Kopin IJ. 1983. A primate model of parkinsonism: Selective destruction of dopaminergic neurons in the pars compacta of the substantia nigra by 1-methyl-4-phenyl-1,2,3,6-tetrahydropyridine. Proc. Natl. Acad. Sci. USA 80: 4546–4550.

Cabasso I. 1980. Hollow fiber membranes, Kirk-Othmer encyclopedia of chemical technology. Vol. 12, New York: John Wiley. 492–517.

Chandy T, Sharma C. 1990. Chitosan as a biomaterial. Biomater. Artif. Cells Artif. Organs 18: 1–24.

Colton CK. 1995. Implantable biohybrid artificial organs. Cell Transpl. 4: 415–436.

Deacon T, Schumacher J, Dinsmore J, Thomas C, Palmer P, Kott S, Edge A, Penney D, Kassissieh S, Dempsey P, Isacson O. 1997. Histological evidence of fetal pig neural survival after transplantation into a patient with Parkinson's Disease. Nature Med. 3: 350–353.

Deglon N, Henri A, Rouyer-Fessard P, Naffakh N, Beuzard Y, Aebischer P. 1996. Continuous delivery of erythropoietin in mice using encapsulated genetically engineered cell lines. Cell Transpl. 5: 52.

Dionne KE, Cain BM, Li RH, Bell WJ, Doherty EJ, Rein DH, Lysaght MJ, Gentile FT. 1996. Transport characterization of membranes for immunoisolation. Biomaterials 17: 257–266.

Dixit V, Gitnick G. 1995. Transplantation of microencapsulated hepatocytes for liver function replacement. J. Biomater. Sci. Polymer Ed. 7: 343–357.

Dowling JE. 1987. The retina: an approachable part of the brain. Cambridge: Harvard University Press.

Dunn JCY, Tompkins RG, Yarmush ML. 1991. Long-term in vitro function of adult hepatocytes in a collagen sandwich configuration. Biotechnol. Prog. 7: 234–245.

Dunnett SB, Bjorklund A, Stenevi U, Iversen SD. 1981. Behavioral recovery following transplantation of substantia nigra in rats subjected to 6-OHDA lesions of the nigrostriatal pathway. Brain Res. 215: 147–161.

Eaton M, Santiago DI, Dancausse HA, Whittemore SR. 1997. Lumbar transplants of immortalized serotonergic neurons alleviate chronic neuropathic pain. Pain 72: 59–69.

Eaton MJ, Whittemore SR. 1996. Autocrine BDNF secretion enhances the survival and serotonergic differentiation of raphe neuronal precursor cells grafted into the adult rat CNS. Exp. Neurol. 140: 105–114.

Edge ASB, Dinsmore J. 1997. Xenotransplantation in the central nervous system. Xeno. 5: 23–25.

Emerich DF, Frydel BR, Flanagan TR, Palmatier M, Winn SR, Christenson L. 1993. Transplantation of polymer encapsulated PC12 cells: Use of chitosan as an immobilization matrix. Cell Transpl. 2: 241–249.

Emerich DF, Hammang JP, Baetge EE, Winn SR. 1994. Implantation of polymer-encapsulated human nerve growth factor secreting fibroblasts attenuates the behavioral and neuropathological consequences of quinolinic acid injections into rodent striatum. Exp. Neurol. 130: 141–150.

Emerich DF, Lindner MD, Winn SR, Chen E-R, Kordower JH. 1996. Implants of encapsulated human CNTF-producing fibroblasts prevent behavioral deficits and striatal degeneration in a rodent model of Huntington's disease. Journal of Neuroscience 16: 5168–5181.

Emerich DF, Sanberg PR. 1992a. Animal models in Huntington's disease. In: Boulton AA, Baker GB, Butterworth RF, editors. Neuromethods, Vol 17, Animal models of neurological disease. Totowa, NJ: Humana.

Emerich DF, Winn SR, Christenson L, Palmatier MA, Gentile FT, Sanberg PR. 1992b. A novel approach to neural transplantation in Parkinson's disease: Use of polymer-encapsulated cell therapy. Neurosci. and Biobehav. Rev. 16: 437–447.

Emerich DF, Zubricki EM, Shipley MT, Norman AB, Sanberg PR. 1991. Female rats are more sensitive to locomotor alterations following quinolinic acid-induced striatal lesions: Effects of striatal transplants. Experimental Neurology 111: 369–378.

Emerich DF, Winn SR, Hantraye PM, Pescharski M, Chen EY, Chu Y, McDermott P, Baetye EE, Kordower JH. 1997. Positive effect of encapsulated cells producing neurotrophic Huntington's disease. Nature 386: 395–399.

Freed WJ, Poltorak M, Becker JB. 1990. Intracerebral adrenal medulla grafts: A review. Exp. Neurol. 110: 139–166.

Frim DM, Short MP, Rosenberg WS, Simpson J, Breakefield XO, Isacson O. 1993a. Local protective effects of nerve growth factor-secreting fibroblasts against excitotoxic lesions in the rat striatum. J. Neursurg. 78: 267–273.

Frim DM, Uhler TA, Short MP, Ezzedine ZD, Klagsbrun M, Breakefield XO, Isacson O. 1993c. Striatal degeneration induced by mitochondrial blockade is prevented by biologically delivered NGF. J. Neurosci. Res. 35: 452–458.

Frim DM, Yee WM, Isacson O. 1993b. NGF reduces striatal excitotoxic neuronal loss without affecting concurrent neuronal stress. NeuroReport 4: 655–658.

Fritzsche AK, Areval AR, Moore MD, O'Hara C. 1993. The surface structure and morphology of polyacrylonitrile membranes by atomic force microscopy. J. Membr. Sci. 81: 109–120.

Gage FH, Batchelor P, Chen KS, Chin D, Higgins GA, Koh S, Deputy S, Rosenberg MB, Fischer W, Björklund A. 1989. NGF receptor re-expression and NGF-mediated cholinergic neuronal hypertrophy in the damaged neostriatum. Neuron 2: 1177–1184.

Gentile FT, Doherty EJ, Rein DH, Schoichet MS, Winn SR. 1995. Polymer science for macroencapsulation of cells for central nervous system transplantation. Reactive Polymers 25: 207–227.

Gentile FT, Doherty EJ, Li RH, Rein DH, Emerich DF. 1996. Design of membrane based bioartificial organs. In: DA Butterfield, editor. Biofunctional membranes. New York: Plenum Press. 223–236.

Ginzburg R, Seltzer Z. 1990. Subarachnoid spinal cord transplantation of adrenal medulla suppresses chronic neuropathic pain behavior in rats. Brain Res. 523: 147–150.

Greene LA, Tischler AS. 1976. Establishment of a non-adrenergic clonal line of rat adrenal pheochromocytoma cells which respond to nerve growth factor. Proc. Natl. Acad. Sci. USA 73: 2424–2428.

Hama AT, Tresco PA, Aebischer P, Winn SH, Sagen J. 1993. Polymer-encapsulated bovine chromaffin cells reduce pain in rats with a painful peripheral mononeuropathy. IASP Abstr. 7: 374.

Hama AT, Sagen J. 1993. Reduced pain-related behavior by adrenal medullary transplants in rats with experimental painful peripheral neuropathy. Pain 52: 223–231.

Hama AT, Sagen J. 1994a. Alleviation of neuropathic pain symptoms by xenogeneic chromaffin cell grafts in the spinal subarachnoid space. Brain Res. 651: 345–351.

Hama AT, Sagen J. 1994b. Induction of a spinal NADPH-diaphorase by nerve injury is attenuated by adrenal medullary transplants. Brain Res. 640: 345–351.

Hammang JP, Emerich DF, Winn SR, Lee A, Lindner MD, Gentile FT, Doherty EJ, Kordower JH, Baetge EE. 1995. Delivery of neurotrophic factors to the CNS using encapsulated cells: Developing treatments for neurodegenerative diseases. Cell Trans. 4: S27–S28.

Hayashi H, Inoue K, Aung T, Tun T, Yuanjun G, Wenjing W, Shinohara S, Kaji H, Doi R, Setoyama H, Kato M, Imamura M, Maetani S, Morikawa N, Iwata H, Ikada Y, Miyazaki J. 1996. Application of a novel B cell line MIN6 to a mesh-reinforced polyvinyl alcohol hydrogel tube and three-layer agarose microcapsules: An in vitro study. Cell Transpl. 5: S65–S69.

Higgens GA, Koh S, Chen KS, Gage FH. 1989. NGF induction of NGF receptor gene expression and cholinergic neuronal hypertrophy within the basal forebrain of the adult rat. Neuron 3: 247–256.

Holtzman DM, Li Y, Parada LF, Kinsman S, Chen C-K, Valleta JS, Zhou J, Long JB, Mobley WC. 1992. p140trk mRNA marks NGF-responsive forebrain neurons: evidence that trk gene expression is induced by NGF. Neuron 9: 465–478.

Hughes AJ, Daniel SE, Blankson S, Lees AJ. 1993a. A clinicopathologic study of 100 cases of Parkinson's disease. Arch. Neurol. 50: 140–148.

Hughes RA, Sendtner M, Thoenen H. 1993b. Members of several gene families influence survival of rat motoneurons in vitro and in vivo. J. Neurosci. Res. 36: 663–671.

Hymer WC, Wilber DL, Page R, Hibbard E, Kelsey R, Hatfield J. 1981. Pituitary hollow fiber units in vivo and in vitro. Neuroendocrinology 32: 350–354.

Ibuki T, Hama AT, Wang X-T, Pappas GD, Sagen J. 1997. Loss of GABA-immunoreactivity in the spinal dorsal horn of rats with peripheral nerve injury and promotion of recovery by adrenal medullary grafts. Neuroscience 76: 845–858.

Inoue K, Fujisato T, Gu YJ, Burczak K, Sumi S, Kogire M, Tobe T, Uchida K, Nakai I, Maetani S, Ikada Y. 1992. Experimental hybrid islet transplantation: Application of polyvinyl alcohol membrane for entrapment of islets. Pancreas 7: 562–568.

Iwata H, Takagi T, Amemiya H, Shimizu H, Yamashita K, Kobayashi K, Akutsu T. 1992. Agarose for a bioartificial pancreas. J. Biomed. Mater. Res. 26: 967–977.

Jaeger CB, Greene LA, Tresco PA, Winn SR, Aebischer P. 1990. Polymer encapsulated dopaminergic cell lines as "alternative neural grafts." Prog. Brain Res. 82: 41–46.

Jauregui H, Chowdhury N, Chowdhury J. 1996. Use of mammalian liver cells for artificial liver support. Cell Transpl. 5: 353–367.

Joseph JM, Goddard MB, Mills J, Padron V, Zurn A, Zielinski B, Favre J, Gardaz JP, Mosimann F, Sagen J, Christenson L, Aebischer P. 1994. Transplantation of encapsulated bovine chromaffin cells in the sheep subarachnoid space: A preclinical study for the treatment of cancer pain. Cell Transpl. 3: 355–364.

Kato AC, Lindsay RM. 1994. Overlapping and additive effects of neurotrophins and CNTF on cultured human spinal cord neurons. Exp. Neurol. 130: 196–201.

Kessler L, Legeay G, Jesser C, Damge C, Pinget M. 1995. Influence of corona surface treatment on the properties of an artificial membrane used for Langerhans islets encapsulation: permeability and biocompatability studies. Biomaterials 16: 185–191.

Kesting RE. 1985. Synthetic polymer membranes—a structural perspective. New York: John Wiley & Sons.

Kim KJ, Fane AG. 1994. Low voltage scanning electron microscopy in membrane research. J. Membr. Sci. 88: 103–114.

Knorrs D. 1986. Chitosan gels for the entrapment of cultured plant cells. In: Muzzarelli R, Jeuniaux D, Gooday G, editors. Chitin in nature and technology. New York: Plenum Press.

Kordower JH, Mufson EJ. 1993. NGF-receptor-immunoreactivity in the developing human basal ganglia. Journal of Comparative Neurology 327: 359–374.

Kordower JH, Carles V, Bayer R, Bartus RT, Putney S, Walus L, Friden PM. 1994. Intravenous administration of an NGF conjugate prevents neuronal degeneration in a model of Huntington's disease. PNAS 19: 9077–9080.

Kordower JH, Liu Y, Winn S, Emerich D. 1995. Encapsulated PC12 cell transplants into hemiparkinsonian monkeys: a behavioral, neuroanatomical, and neurochemical analysis. Cell Transpl. 4: 155–171.

Kordower JH, Chen E-Y, Mufson EJ, Winn SR, Emerich DF. 1996. Intrastriatal implants of polymer encapsulated cells genetically modified to secrete human nerve growth factor: Trophic effects upon cholinergic and noncholinergic striatal neurons. Neuroscience 72: 63–77.

Kotzbauer PT, Lampe PA, Heuckeroth RO, Golden JP, Creedon DJ, Johnson EM Jr, Milbrandt J. 1996. Neurturin, a relative of glial-cell-line-derived neurotrophic factor. Nature 384: 467–470.

Lacy PE, Hegr OD, Gerasimidi-Vazeou A, Gentile FT, Dionne KE. 1991. Maintenance of normoglycemia in diabetic mice by subcutaneous xenografts of encapsulated islets. Science 254: 1782–1784.

Lamberton P, Lipsky M, McMillan P. 1988. Use of semipermeable polyurethane membrane hollow fiber for pituitary organ culture. In Vitro Cell Dev. Biol. 24: 500–504.

LaVail MM, Unoki K, Yasumura D, Matthes MT, Yancopoulos GD, Steinberg RH. 1992. Multiple growth factors, cytokines, and neurotrophins rescue photoreceptors from the damaging effects of constant light. Proc. Natl. Acad. Sci. USA 89: 11249–11253.

Lazorthes Y, Bès JC, Sagen J, Tafani M, Tkaczuk J, Sallerin B, Nahri I, Verdié JC, Ohayon E, Caratero C, Pappas GD. 1995. Transplantation of human chromaffin cells for control of intractable cancer pain. Acta Neurochir. 64: 97–100.

Li RH. 1998. Materials for immunoisolated cell transplantation. Advanced Drug Delivery Rev. 33:87–109.

Li R, Altreuter DH, Gentile FT. 1996. Transport characterization of hydrogel matrices for cell encapsulation. Biotech. Bioeng. 50: 365–373.

Li R, White M, Williams S, Hazlett T. 1998. PVA synthetic polymer foams as scaffolds for cell encapsulation. J. Biomaterials Sci, in press.

Lin L-FH, Doherty DH, Lile JD, Bektesh S, Collins F. 1993. GDNF: A glial cell line-derived neurotrophic factor for midbrain dopaminergic neurons. Science 260: 1130–1132.

Lindner MD, Winn SR, Baetge EE, Hammang JP, Gentile FT, Doherty E, McDermott PE, Frydel B, Ullman DM, Schallert T, Emerich DF. 1995. Implantation of encapsulated catecholamine and GDNF producing cells in rats with unilateral dopamine depletions and parkinsonian symptoms. Exp. Neurol. 132: 62–76.

Liu S, Brewer KL, Yezierski RP. 1997. Effects of adrenal medullary chromaffin cell transplants on pain related behaviors following excitotoxic spinal cord injury. Soc. Neurosci. Abstr. 23: 1537.

Magal E, Burnham P, Varon S. 1991. Effects of ciliary neurotrophic factor on rat spinal cord neuron in vitro: survival and expression of choline acetyltransferase and low-affinity nerve growth factor receptors. Dev. Brain Res. 63: 141–150.

Michalewicz P, Lu Y, Czech KA, Smalheiser N, Yeomans DC, Pappas GD. 1997. Purification of chromaffin cells allows xenotransplantation without immunosuppression. Soc. Neurosci. Abstr. 23: 1455.

Mitsumoto H, Ikeda K, Klinkosz B, Cedarbaum JM, Wong V, Lindsay RM. 1994. Arrest of motor neuron disease in wobbler mice cotreated with CNTF and BDNF. Science 265: 1107–1110.

Nyberg S, Mann H, Remmel R, Hu W, Cerra F. 1993. Pharmacokinetic analysis verifies P450 function during in vitro and in vivo application of a bioartificial liver. ASAIO J. 39: M252–M256.

Ortega-Alvaro A, Gibert-Rahola J, Chover AJ, Tejedor-Real P, Casas J, Mico JA. 1994. Effect of amitriptyline on the analgesia induced by adrenal medullary tissue transplanted in the rat spinal subarachnoid space as measured by an experimental model of acute pain. Exp. Neurol. 130: 9–14.

Ortega-Alvaro A, Gibert-Rahola J, Mellado-Fernandez ML, Chover AJ, Mico JA. 1997. The effects of different monoaminergic antidepressants on the analgesia induced by spinal cord adrenal medullary transplants in the formalin test in rats.

Pacheco-Cano MT, Garcia-Hernandez F, Hiriart M, Komisurak BR, Drucker-Colin R. 1990. Dibutyryl cAMP stimulates analgesia in rats bearing a ventricular adrenal medulla transplant. Brain Res. 531: 290–293.

Pappas GD, Czech KA, Bes JC, Lazorthes Y. 1997. Effectiveness of long-term chromaffin cell allografts for controlling intractable cancer pain: Morphological, biochemical, and clinical data from two patients. Int. Neural Transpl. Mtg. Abstr. 6: 20.

Parker WD, Boyson SJ, Luder AS, Parks JK. 1990. Evidence for a defect on the NADH:ubiquinone oxidoreductase (complex I) in Huntington's disease. Neurology 40: 1231–1234.

Perlow MJ, Freed WJ, Hoffer BJ, Seiger A, Olson L, Wyatt RJ. 1979. Brain grafts reduce motor abnormalities produced by destruction of the nigrostriatal system. Science 204: 643–647.

Reiner A, Albin DL, Anderson KD, D'Amato CJ, Penny JB, Young AB. 1988. Differential loss of striatal projection neurons in Huntington's disease. PNAS 85: 5733–5737.

Riederer P, Wuketich S. 1976. Time course of nigrostriatal degeneration in Parkinson's Disease. J. Neural Transm. 38: 277–301.

Roberts RC, DiFiglia M. 1989. Short- and long-term survival of large neurons in the excitotoxic lesioned rat caudate nucleus: A light and electron microscopic study. Synapse 3: 363–371.

Ruz-Franzi JI, Gonzalez-Darder JM. 1991. Study of the analgesic effects of the implant of adrenal medullary into the subarachnoid space in rats. Acta Neurochir. 52: 39–41.

Sagen J, Pappas GD, Perlow MJ. 1986a. Adrenal medullary tissue transplants in rat spinal cord reduce pain sensitivity. Brain Res. 384: 189–194.

Sagen J, Pappas GD, Pollard HB. 1986b. Analgesia induced by isolated bovine chromaffin cells implanted in the rat spinal cord. Proc. Natl. Acad. Sci. 83: 7522–7526.

Sagen J, Kemmler JE. 1989. Increased levels of Met-enkephalin-like immunoreactivity in the spinal cord CSF of rats with adrenal medullary transplants. Brain Res. 502: 1–10.

Sagen J, Kemmler JE, Wang H. 1991. Adrenal medullary transplants increase spinal cerebrospinal fluid catecholamine levels and reduce pain sensitivity. J. Neurochem. 56: 623–627.

Sagen J, Wang H, Pappas GD. 1990. Adrenal medullary implants in rat spinal cord reduce nociception in a chronic pain model. Pain 42: 69–79.

Sagen J, Wang H, Tresco PA, Aebischer P. 1993. Transplants of immunologically-isolated xenogeneic chromaffin cells provide a long-term source of pain reducing neuroactive substances. J. Neurosci. 13: 2415–2423.

Sagot Y, Tan SA, Baetge E, Schmalbruch H, Kato AC, Aebischer P. 1995. Polymer encapsulated cell lines genetically engineered to release ciliary neurotrophic factor can slow down progressive motor neuronopathy in the mouse. Nature Med. 2: 696–699.

Saitoh Y, Taki T, Arita N, Ohnishi T, Hayakawa T. 1995a. Analgesia induced by transplantation of encapsulated tumor cells secreting ß-endorphin. J. Neurosurg. 82: 630–634.

Saitoh Y, Taki T, Arita N, Ohnishi T, Hayakawa T. 1995b. Cell therapy with encapsulated xenogeneic tumor cells secreting b-endorphin for treatment of peripheral pain. Cell Transpl. 4 Suppl. 1: S13–S17.

Sanberg PR, Fibiger HC, Mark RR. 1981. Body weight and dietary factors in Huntington's disease patients compared with matched controls. Med. J. Aust. 1: 407–409.

Schumacher JM, Short MP, Hyman BT, Breakefield XO, Isacson O. 1991. Intracerebral implantation of nerve growth factor-producing fibroblasts protects against neurotoxic levels of excitatory amino acids. Neuroscience 45: 561–570.

Shoichet MS, Winn SR, Athavale S, Harris JM, Gentile FT. 1994. Poly(ethylene oxide) grafted thermoplastic membranes for use as cellular hybrid bioartificial organs in the central nervous system. Biotech. Bioeng. 43: 563–572.

Siegan JB, Hama AT, Sagen J. 1996. Alterations in spinal cord cGMP by peripheral nerve injury and adrenal medullary transplantation. Neurosci. Lett. 215: 49–52.

Siegan JB, Sagen J. 1997. Attenuation of formalin pain responses in the rat by adrenal medullary transplants in the spinal subarachnoid space. Pain 70: 279–285.

Sladek JR Jr, Gash DM. 1988. Nerve cell grafting in Parkinson's disease. J. Neurosurg. 68: 337–351.

Strathmann H. 1985. Production of microporous media by phase inversion processes. In: Material science of synthetic membranes. Amer. Chem. Soc. 165.

Takebe K, Shimura T, Munkhbat B, Hagihara M, Nakanishi H, Tsuji K. 1996. Xenogeneic (pig to rat) fetal liver fragment transplantation using macrocapsules for immunoisolation. Cell Transpl. 5: S31–S33.

Tresco PA, Winn SR, Aebischer P. 1992. Polymer encapsulated neurotransmitter secreting cells: potential treatment for Parkinson's disease. ASAIO J. 38: 17–23.

Ungerstedt U, Ljungberg T, Greg G. 1974. Behavioral, physiological, and neurochemical changes after 6-hydroxydopamine-induced degeneration of the nigrostriatal dopamine neurons. Adv. Neurol. 5: 421–426.

Unsicker K. 1993. The trophic cocktail made by adrenal chromaffin cells. Exp. Neurol. 123: 167–173.

Unsicker K, Krieglstein K, Bieger S, Deimling F, Huber K, Schober A, Cole T. Monaghan P, Schmid W, Schütz G, Blottner D, Flanders K, Wolf N, Kalcheim C, Schober A, Minichello L, Klein R. 1996. Molecular cues for development and maintenance of adrenal chromaffin cells and their innervation. Int. Catecholamine Sympos. Abstr. 8: 204.

Vaquero J, Arias A, Oya S, Zurita M. 1991. Chromaffin allografts into arachnoid of spinal cord reduce basal pain responses in rats. NeuroReport 2: 149–151.

Vorlop K, Klein J. 1987. Entrapment of microbial cells in chitosan. Methods in Enzym. 135: 259–268.

Wallace DC. 1992. Mitochondrial genetics: A paradigm for aging and degenerative diseases? Science 256: 628–632.

Wang H, Sagen J. 1994. Optimization of adrenal medullary allograft conditions for pain alleviation. J. Neural. Transpl. Plastic. 5: 49–64.

Wang H, Sagen J. 1995. Attenuation of pain-related hyperventilation in adjuvant arthritic rats with adrenal medullary transplants in the spinal subarachnoid space. Pain 63: 313–320.

Ward RS, White KA, Wolcott CA, Wang AY, Kuhn RW, Taylor JE, John JK. 1993. Development of a hybrid artificial pancreas with a dense polyurethane membrane. ASAIO J. 39: M261–M267.

Winn SR, Tresco PA, Zielinski B, Greene LA, Jaeger CB, Aebischer P. 1991. Behavioral recovery following intrastriatal implantation of microencapsulated PC12 cells. Exp. Neur. 113: 322–329.

Winn SR, Hammang JP, Emerich DF, Lee A, Palmiter RD, Baetge EE. 1994b. Polymer-encapsulated cells genetically modified to secrete human nerve growth factor promote the survival of axotomized septal cholinergic neurons. Proc. Natl. Acad. Sci. 91: 2324–2328.

Winn SR, Tresco PA. 1994a. Hydrogel applications for encapsulated cellular transplants. Methods in Neurosci. 21: 387–402.

Winnie AP, Pappas GD, Das Gupta TK, Wang H, Ortega JD, Sagen J. 1993. Alleviation of cancer pain by adrenal medullary transplants in the spinal subarachnoid space: A preliminary report. Anesthesiology 79: 644–653.

Wu HH, McLoon SC, Wilcox GL. 1993. Antinociception following implantation of AtT-20 and genetically modified AtT-20/hENK cells in rat spinal cord. J. Neural Transpl. Plastic. 4: 15–26.

Wu HH, Wilcox GL, McLoon SC. 1994a. Implantation of AtT-20 or genetically modified AtT-20-hENK cells in mouse spinal cord induced antinociception and opioid tolerance. J. Neurosci. 14: 4806–4814.

Wu HH, Lester BR, Sun Z, Wilcox GL. 1994b. Antinociception following implantation of mouse B16 melanoma cells in mouse and rat spinal cord. Pain 56: 203–210.

Yan Q, Johnson EM. 1988. An immunohistochemical study of nerve growth factor receptor in developing rats. Journal of Neuroscience 8: 3481–3498.

Yan Q, Elliott J, Snider WD. 1992. Brain-derived neurotrophic factor rescues spinal motor neurons from axotomy-induced cell death. Nature 360: 753–755.

Yu W, Hao J-X, Xu X-J, Saydoff J, Haegerstrand A, Wiesenfeld-Hallin Z. 1998. Long-term alleviation of allodynia-like behaviors by intrathecal implantation of bovine chromaffin cells in rats with spinal cord injury. Pain 74: 115–122, 1998.

Yu W, Hao J-X, Xu X-J, Saydoff J, Sherman S, Eriksson A, Haegerstrand A, Wiesenfeld-Hallin Z. 1997. Intrathecally implanted encapsulated bovine chromaffin cells alleviate chronic allodynia-like pain in rats with spinal cord injury. Eur. Fed. Int. Assn. Study of Pain Abstr. 2.

Yurek DM, Sladek JR, Jr. 1990. Dopamine cell replacement: Parkinson's Disease. Ann. Rev. Neurosci. 13: 415–440.

Zielinski BA, Aebischer P. 1994. Chitosan as a matrix for mammalian cell encapsulation. Biomaterials 15: 1049–1056.

Zurn AD, Baetge EE, Hammang JP, Tan SA, Aebischer P. 1994. Glial cell line-derived neurotrophic factor (GDNF), a new trophic factor for motoneurons. Neuroreport 6: 113–118.

29

Removal of Urea in Uremia and Ammonia in Liver Failure with Emphasis on the Use of Artificial Cells for Encapsulation of Genetically Engineered Cells

Thomas M.S. Chang and Satya Prakash

General

Dialysis is an effective treatment for chronic renal failure or uremia. Unfortunately, 85% of the world's uremic patients die because their countries cannot afford dialysis therapy. Even in those countries that are treating the other 15% of the uremic patients, the total healthcare cost for this has been escalating to high levels. This is partly because success in the expensive dialysis treatment can now maintain uremic patients alive. This results in continuing increases in the number of dialysis patients. In any case, most patients would prefer an oral form of therapy rather than being attached to dialysis machines or receiving peritoneal dialysis. Many groups have studied a more affordable form of oral therapy for many years. A combination of adsorbents, osmotic agents, and other oral agents can control most of the problems related to accumulation of waste metabolites and water and electrolyte imbalances. Unfortunately, these groups ended their attempts at this many years ago because of the unavailability of an effective urea removal system. The hypothesis that high levels of ammonia can lead to hepatic comas in liver failure is receiving increasing support. An effective and fast ammonia removal system would therefore be useful here also. This chapter discusses the use of artificial cells for urea and ammonia removal, with emphasis on the use of encapsulated *E. coli* DH5 cells.

Artificial Cells for Encapsulation

Research has shown the feasibility of artificial cells for biotechnology, blood substitutes, and other areas (Chang, 1964, 1972, 1995, 1997). The semipermeable membranes of artificial cells permanently retain enzymes, cells, microorganisms, or adsorbents. The enclosed material is permanently retained inside and not released. While inside, the enclosed materials act on smaller molecules that can cross the membrane rapidly. This way, artificial cells (1) containing adsorbents can remove toxins, (2) containing hemoglobin can transport oxygen, (3) containing enzymes can convert or remove metabolites and substrates, and (4) containing cells (e.g., islets and hepatocytes) and microorganisms can carry out useful functions. This review deals specifically with the use of artificial cells for the removal of urea and ammonia.

Urea Removal

Artificial Cells Containing Urease and Ammonia Adsorbent

To obtain reproducible and high urea-converting capacity, we prepared polymeric membrane artificial cells containing high concentrations of urease (Chang 1964). Ten ml of these are placed in an extracorporeal shunt perfused by blood. This lowers the systemic urea level in 10 kg dogs to 48% of its original level in 180 minutes.(Chang 1966) Artificial cells containing adsorbents remove the ammonia resulting from the urease conversion of urea. This principle of urea removal has been extended by others for routine clinical use in dialysate regeneration for dialysis patients. Our final aim is to use this principle for oral therapy. Thus, we continued with further studies showing that daily oral urease artificial cells with

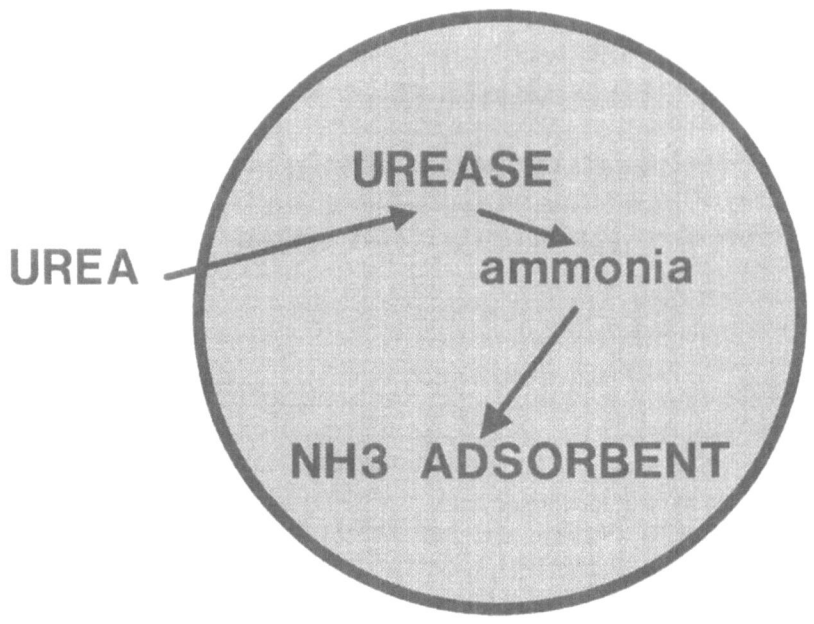

1. UREASE converts urea to ammonia
2. Adsorbent removes ammonia

FIGURE 29.1. Artificial cells containing urease and an ammonia adsorbent, zirconium phosphate. Urea is converted by urease into ammonia. Ammonia is removed by the ammonia adsorbent.

an ammonia adsorbent, zirconium phosphate (Figures 29.1 and 29.2), can lower the system urea by 40% (Chang 1976). This has been developed further by two groups with support from the NIH Chronic Uremia Program. Thus, Sparks shows that the mass transfer of urea into the gastrointestinal tract would allow an efficient oral urea removal system to lower the systemic urea level in uremia (Spark 1979). The group from the Battelle Memorial Institute has developed artificial cells containing urease and zirconium phosphate and used them in uremic dogs, followed by clinical trial in uremic patients at the Mayo Clinic (Kjellstrand 1981). Their results indicate that this approach can remove therapeutic amounts of urea. Thus, the proof of principle has been shown. Although the artificial-cell-urease component is very efficient, the 1212 grams of adsorbent needed is too large an amount for routine clinical use. This single obstacle to a practical urea removal system has re-

sulted in a temporary halt to the search for an oral therapy for uremia. We have recently rekindled this possibility because of the availability of genetically engineered *E. coli* DH5 cells with both an efficient urease system and an ammonia removal system (Prakash and Chang 1996).

Encapsulation of Genetically Engineered *E. coli* DH5 Cells

Many bacteria, fungi, and plants use small amounts of urea (Setala et al 1973). For example, normal *E. coli* uses a small amount of urea as a nitrogen source. Earlier use of bacteria to remove urea has not been successful because they do not have sufficient urea removal capacity (Setala et al 1973). More recently, with the advance of molecular biology, the urea utilization capacity of *E. coli* can be extensively enhanced by genetic engineering (Scott et al 1989). We

ARTIFICIAL CELLS (AC) IN INTESTINE

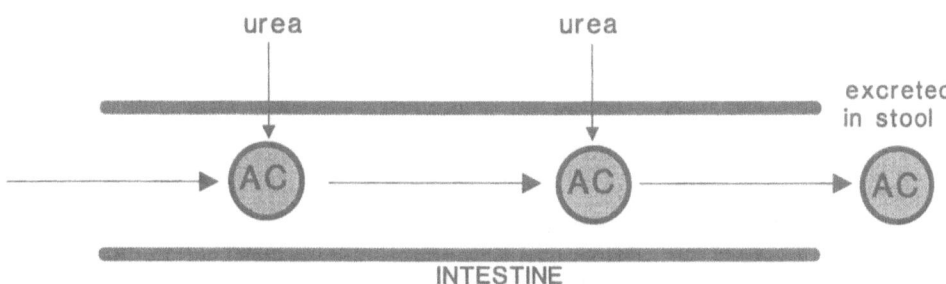

FIGURE 29.2. Artificial cells administered orally for the removal of systemic urea. Urea entering the intestine equilibrates rapidly into the artificial cells. Artificial cells containing a urea removal system can therefore remove the urea. Artificial cells with all the encapsulated material are excreted with the stool in about 24 hours. There is therefore no retention of artificial cells or the encapsulated material.

have therefore carried out studies on the microencapsulation of genetically engineered nonpathogenic *E. coli* containing the *K. aerogenes* gene (Prakash and Chang, 1993, 1995, 1996). Unlike microorganisms in the intestinal flora, this genetically engineered *E. coli* DH5 cell has a high capacity to degrade urea into ammonia, and then to use the ammonia for its nitrogen source (Figure 29.3).

When given orally inside microcapsules, *E. coli* DH5 cells remain inside the microcapsules at all times as they pass down the intestine and are then excreted with the stool (Figure 29.2). This way, no bacteria are released into the intestine. During the passage through the intestine, the encapsulated *E. coli* DH5 cells act on urea and other substrates diffusing into the microcapsules. As a result, the urea level in the treated group, but not the control group, return to normal level. This is similar to our earlier studies using artificial cells containing urease and an ammonia adsorbent. The important difference is that the amount of artificial cells containing *E. coli* DH5 cells is only 1/300th that of the urease-adsorbent system. This means that for the first time, we now have a potentially practical approach. Giving artificial cells containing *E. coli* DH5 cells orally once a day allows them to disperse throughout the entire gastrointestinal tract to have the largest contact with urea. Furthermore, the artificial cells are excreted with their content of bacteria that contain the nitrogen load from urea. This is unlike normal intestinal bacterial flora, which can effectively de-

grade urea but cannot efficiently remove the ammonia formed, even with special carbohydrate adjustments (Wrong et al 1981). The ammonia from the bacterial flora is reabsorbed into the body to be recycled.

The following sections describe more details of the in vitro studies and also in vivo studies in uremic rats.

Preparation of Encapsulated *E. coli* DH5 Cells

Alginic acid (low viscosity, lot 611994) and poly-l-lysine (MW vis 16,100, lot 11H5516) were purchased from Kelco, and Sigma Chemical Co., St. Louis, Missouri, respectively. Genetically engineered bacteria *E. coli* DH5, containing the urease gene from *K. aerogens,* was a generous gift from Professor R. P. Haussinger (Scott et al 1989). The composition of Luria Bertani (LB) medium was of 10.00 g/L bacto tryptone (DIFCO), 5.00 g/L bacto yeast extract (DIFCO), and 10.00 g/L sodium chloride (Sigma). The pH was adjusted to 7.5 by adding about 1.00 mL of 1.00 N NaOH. Media were then sterilized in Castle Labclaves for 30 min. at 250°C. Incubation was carried out in 5.00 ml LB in 16.00 ml culture tubes at 37°C in an orbital shaker shaken at 120 rpm. For the large-scale production of biomass, for microencapsulation purposes, 250 ml Erlenmeyer flasks containing 100 ml LB medium were used.

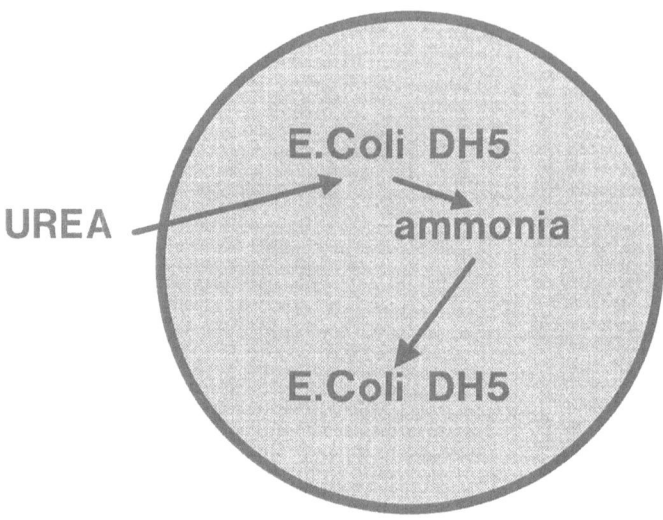

1. Urea converted to ammonia by E.Coli DH5

2. E.Coli DH5 uses ammonia for synthesis

FIGURE 29.3. Artificial cells containing *E. coli* DH5 cells. Urea is converted by urease in the *E. coli* DH5 cells into ammonia. Ammonia is then used by the *E. coli* DH5 cells for their protein synthesis.

Microcapsules containing bacteria *E. coli* were prepared, and the details are described elsewhere (Prakash and Chang 1995). Alginate beads, spherical in shape, and small in diameter, with uniform size distribution are necessary to obtain strong microcapsules with appropriate characteristics (Prakash and Chang 1995). The details of the procedure and process optimization for preparation of small-diameter, spherical, and uniformly sized alginate-poly-l-lysine-alginate (APA microcapsules) are described elsewhere (Prakash and Chang 1995). Our earlier studies show that the alginate concentration, air flow rate, and liquid flow rate are critical for obtaining microcapsules with the desired characteristics. Using the following: 2.00% (w/v) alginate, 0.0724 ml/min liquid flow rate, and 2.00 L/min air flow rate leads to superior microcapsules (Prakash and Chang 1995).

Figure 29.4 shows a group of freshly prepared APA microcapsules of 500 ± 45. The safety evaluation in terms of the mechanical strength of the microcapsules as a function of cell leakage was also studied (Prakash and Chang 1995). Results show that alginate microbeads are stable up to an agita-

tion of 190 rpm, compared to 210 rpm for APA microcapsules in terms of cell leakage (Prakash and Chang 1995).

In Vitro Plasma Urea Removal by Artificial-Cell-Entrapped Genetically Engineered *E. coli* DH5 Cells

We used bovine plasma with filter-sterilized urea added to a final urea level comparable to that of kidney failure patients. The ratio of the reactor volume to the amount of microencapsulated bacteria used was held constant. The urea depletion profiles of encapsulated bacteria were determined, and results are shown in Figure 29.5 (Prakash and Chang 1995). The log phase microencapsulated bacteria lowered $87.89 \pm 2.25\%$ of the plasma urea within 20 minutes and 99.99% of urea in 30 minutes (Figure 29.5). A comparison of the urea depletion profile of free and APA-encapsulated *E. coli* DH5 cells in modified reaction media and in plasma was carried out. The result shows that encapsulation did not decrease the rate of removal of urea (Figure 29.6). One can use encapsulated bacteria in vitro up to three cycles

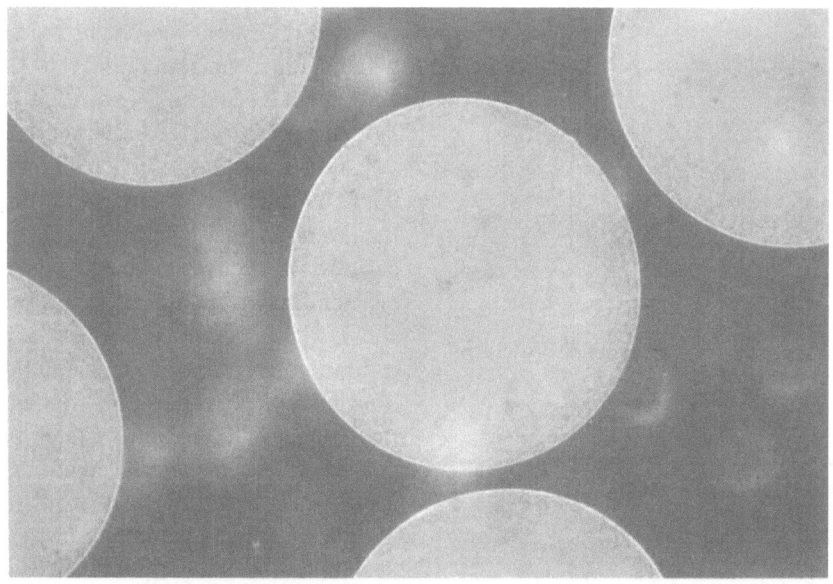

FIGURE 29.4. Photomicrograph of alginate-poly-l-lysine-alginate (APA) microcapsules containing genetically engineered *E. coli* DH5 cells after cycle 2 of in vitro plasma urea and ammonia removal studies.

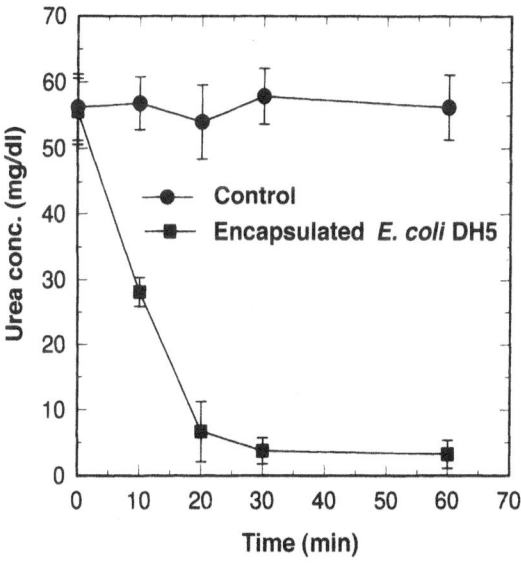

FIGURE 29.5. APA-encapsulated *E. coli* DH5 cells in in vitro plasma urea removal. (From Prakash and Chang. 1995. Biotechnology and Bioengineering 46:621–626, courtesy of John Wiley & Sons, Inc., New York.)

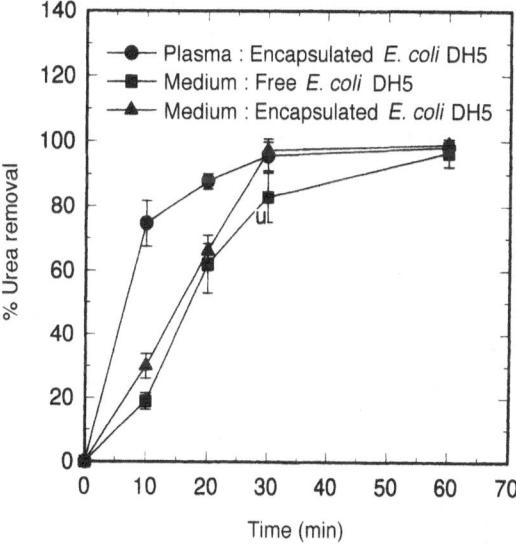

FIGURE 29.6. Comparative study of the urea depletion profile of free and APA-encapsulated *E. coli* DH5 cells in modified reaction medium and APA-encapsulated *E. coli* DH5 cells in plasma. (From Prakash and Chang, Biotechnology and Bioengineering 46, 621–626 (1995) with the courtesy of John Wiley & Sons, Inc., New York, USA.)

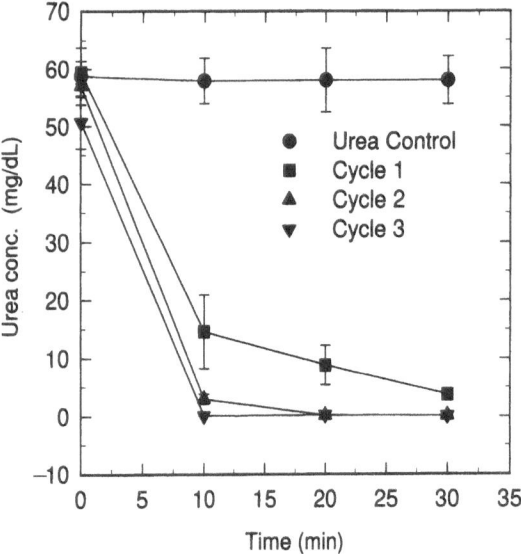

FIGURE 29.7. APA microcapsules reused, as a function of the plasma urea removal efficiency. (From Prakash and Chang Biotechnology and Bioengineering 46, 621–626 (1995) with the courtesy of John Wiley & Sons, Inc., New York, USA.)

(Figure 29.7). The APA-encapsulated bacteria plasma urea removal rate is greater in the second and third cycles than in the first. This is probably due to increases in total biomass inside microcapsules with time. There is no leakage of encapsulated bacteria in the first, second, or third cycles.

Using a single-pool model, percent of plasma urea removal efficiency by the encapsulated bacteria from a compartment containing a total of 40 L body fluid

was calculated. The urea removal capacity of encapsulated bacteria was calculated and compared with the other standard bioreactants used for urea and ammonia removal (Figure 29.8). 40.00 ± 8.60 g of APA-encapsulated bacteria can remove $87.89 \pm 2.25\%$ of the plasma urea within 20 minutes and 99.99% of urea in 30 minutes from the total 40 L fluid compartment (Figure 29.8). On the basis of this calculation encapsulated bacteria is 10 times more efficient than oxystarch. 1.00 g of oxystarch adsorbed 103.00 mg of urea at pH 7.4 at a urea concentration of 0.02 M (Prakash and Chang 1995). Thus, to remove 40 g of urea from 40 L fluid (100 mg/dl urea) 388.34 g of oxystarch is needed. Microencapsulated genetically engineered bacteria are also 30 times more efficient than microencapsulated enzyme urease-zirconium-phosphate. The encapsulated urease-zirconium-phosphate system removes 1.60 mg of urea nitrogen or 33.00 mg urea/g of microcapsules (Prakash and Chang 1995). Thus, 1212.12 g of microcapsules containing urease-zirconium-phosphate would be required to remove 40 g urea from the total body water. Overall, urea removal efficiency of microencapsulated genetically engineered bacteria is 10–30 times higher than that of existing urea removal systems (Figure 29.8).

Removal of Urea in Uremic Rats

As discussed earlier, when given orally, the *E. coli* DH5 cells remain inside the microcapsules. The microcapsules with the *E. coli* DH5 cells remain intact as they pass down the gastrointestinal tract and are

FIGURE 29.8. Comparative study. L, urea removal capacity of APA-encapsulated *E. coli* DH5 cells, oxystarch, and urease-Z.P. R, amount of bioreactant needed to lower urea concentration from 100 mg/dl to 6.86 mg/dl from the total 40 L body fluid compartment in a 70 kg adult man. (From Prakash and Chang Biotechnology and Bioengineering 46, 621–626 (1995) with the courtesy of John Wiley & Sons, Inc., New York, USA.)

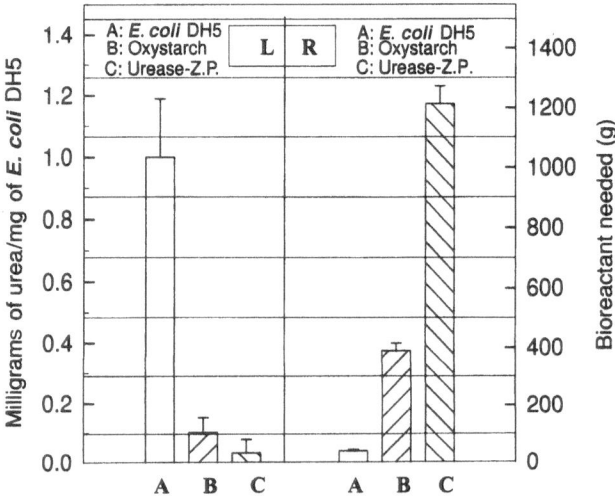

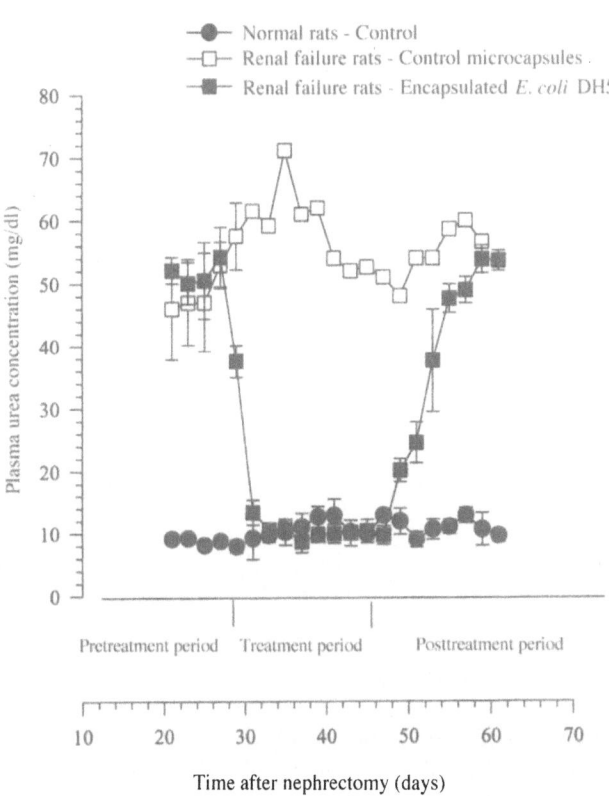

FIGURE 29.9. Plasma urea profile of uremic rats before, during, and after oral administration of APA-encapsulated *E. coli* DH5 cells. For kidney failure rats receiving empty control microcapsules no standard deviation after day 35 is reported because 50% of the rats died. (From Prakash and Chang. 1996. Nature Medicine, 2(8):893–887, with permission from the publisher.)

finally excreted intact with the stool in about 24 hours (Figure 29.2). The membranes of the intact microcapsules are selectively permeable to smaller molecules. Thus, during the passage of the intact microcapsules through the intestine, urea and ammonia can diffuse into the microcapsules. The genetically engineered *E. coli* DH5 cells in the microcapsules can therefore convert urea into ammonia, and then use ammonia for their nitrogen source (Figure 29.3). Throughout their passage through the intestinal tract, *E. coli* DH5 cells are retained inside the intact microcapsules and therefore do not enter into the body. This addresses the safety concerns of introducing genetically engineered cells into the body.

Male Wistar rats in the 300–325 g weight range were used for the surgical uremic model. These were purchased from Harlan Sprague Dawley, Inc., Indianapolis, Indiana. The surgical procedure involves two steps. The first step is to remove the right kidney, right nephrectomy. This is followed by removing most of the function of the remaining left kidney. This is done by partially ligating the left artery, vein, and ureter in such a way that the animals have elevated blood urea level but still retain some ability to adjust their own water and electrolyte balance. Throughout

the control and treatment periods, the uremic rats received water *ad lib* and a normal rat chow containing 22.5% protein (Prakash and Chang 1996).

A quantity of 11.15 ± 2.25 mg/kg body weight of log phase genetically engineered bacteria *E. coli* DH5 cells in microcapsules was administered daily to a group of 26-day-old uremic rats. These uremic rats have an initial elevated plasma urea level of 52.08 ± 2.06 mg/dl compared to the normal group with a level of 9.10 ± 0.71 mg/dl. Results in Figure 29.9 show the successful lowering of the uremic urea level from the initial 52.08 ± 2.06 to a normal level of 10.58 ± 0.85 mg/dl by day 7. Continued daily oral administration maintained normal plasma urea levels in these uremic rats for the entire test period of 21 days. With discontinuation of oral treatment, the urea level increased back to uremic level. The urea level went back to 20.11 ± 1.80 mg/dl, on the very next day, followed by 24.52 ± 3.25 mg/dl, 37.60 ± 8.21 mg/dl, 47.57 ± 2.26 mg/dl, 48.92 ± 2.09 mg/dl, 53.80 ± 2.18 mg/dl, and 53.69 ± 2.59 mg/dl on days 2, 3, 4, 5, 6, and 7, respectively, after discontinuing treatment (Figure 29.9). The fact that they were uremic is shown by the observation that 8 days after discontinuation of treatment, two animals

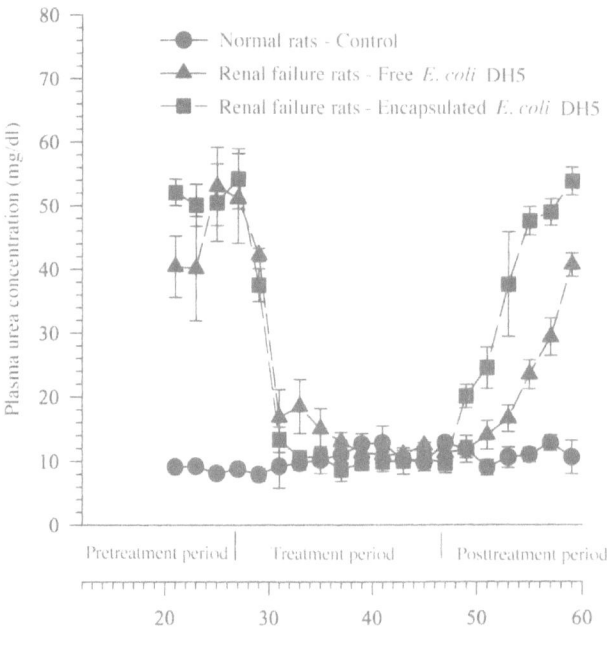

Time after nephrectomy (days).

FIGURE 29.10. Urea removal profile of free and alginate-poly-l-lysine-alginate membrane microencapsulated genetically engineered bacteria *E. coli* DH5 cells in experimental uremic rats. (From Prakash and Chang. 1996. Biomat. Artifi. Cells and Immob. Biotech., an International Journal, 24(3):201–218, courtesy of Marcel Dekker Publishers, NY.)

died. Autopsies showed a typical uremic death with symptoms of hemorrhagic gut, conjugated and collapsed lungs, and chest cavities filled with fluid. Furthermore, 50% of all uremic animals died within the first 29 days after stopping treatment. In the uremic group receiving control microcapsules containing no *E. coli* DH5 cells, there was no lowering of plasma urea level, and instead of growing, these animals lost weight and died. On the other hand, the uremic rats receiving oral encapsulated *E. coli* DH5 treatment grew at the same rate as the normal rats during the treatment period.

The Advantage of Putting Genetically Engineered Microorganisms into Artificial Cells Over Free Cells: Effects of Free Bacteria on Plasma Urea Levels

Figure 29.10 shows that free bacteria can also deplete plasma urea. During the experiment with free cells, the amount of bacteria used was the same as in the microencapsulated studies. When we compared the urea removal kinetics of free bacteria with those of encapsulated bacteria (Figure 29.10) we found the overall kinetics to be similar. However, at the outset, the rate of urea removal by free bacteria was smaller than that of encapsulated bacteria. Free bacteria removed $20.28 \pm 1.06\%$ on the first day and $68.29 \pm 4.30\%$ on the second day, compared to $36.34 \pm 4.70\%$ and $80.49 \pm 2.96\%$ by encapsulated bacteria. Also, in those uremic rats receiving free bacteria, it took a longer time for the plasma urea concentration to return to the pretreatment level upon cessation of the treatment (Figure 29.10). This shows retention of the free cells in the intestinal tract, which is not desirable. Thus, without the use of microcapsules, free microorganisms are retained in the body. This may lead to safety problems including disturbing the intestinal flora, infection, and other problems. Thus, unless they are microencapsulated inside polymeric artificial cells, using free bacteria will cause safety concerns.

Calculation of Oral Dosage Required for a 70 kg Human

Based on results obtained in this study, calculations show that oral administration of a biomass of 4.2 g/day to a 70 kg man would lower the plasma urea concentration from 100 mg/dl to 10 mg/dl. This is a very small amount to administer, compared to the amounts of about 388 g oxystarch and 1212 g of microencapsulated urease-zirconium-phosphate required to lower the plasma concentration from 100 mg/dl to 10 mg/dl in patients. The required dosages for oxystarch and microencapsulated urease-zirconium-phosphate are too large for routine daily use in patients. On the other hand, our study shows that a very small amount of microencapsulated genetically engineered *E. coli* DH5 cells can maintain a normal urea level in uremic rats.

Removal of Ammonia in Liver Failure

Artificial Cells Containing Multienzyme Systems or Hepatocytes

We have used two approaches to remove ammonia in liver failure. One is by artificial cells containing a multienzyme system (Chang and Malouf 1979). This multienzyme system can convert ammonia into the three essential amino acids that are decreased in liver failure: L-leucine, L-isoleucine, and L-valine (Chang and Malouf 1979). The conversion can be carried out using glucose as the energy source for cofactor recycling (Gu and Chang, 1988, 1990). The rate of reaction is limited by the presently available cofactor recycling enzyme systems. Another approach is to use artificial cells containing hepatocytes in the rats (Cai et al 1989, Dixit et al 1989, Wong and Chang 1986). Implantation of artificial cells containing hepatocytes is effective in increasing survival in acute liver failure. It is also effective in lowering the bilirubin levels in Gunn rats with an inborn error of metabolism in bilirubin (Bruni and Chang 1989, 1991, 1995a,b; Dixit et al 1992). The effects of hepatosimulating factor and immunoisolation were also studied (Kashani and Chang 1991a,b, Wong and Chang 1988). However, unless a very large amount of hepatocytes is used, the rate of ammonia removal is not sufficiently fast.

In Vitro Removal of Plasma Ammonia by Artificial Cells Containing Genetically Engineered Bacteria *E. coli* DH5

A study was carried out to discover whether ammonia is produced during urea utilization. The study was also carried out to see if microencapsulated genetically engineered bacteria are capable of removing high levels of plasma ammonia comparable to those found in liver failure (Prakash and Chang, 1996). Plasma ammonia was analyzed using the light-scattering Multistat III microcentrifugal analyzer. This was based on the reductive amination of 2-oxoglutarate, using glutarate dehydrogenase (GLDH), and reduced nicotinamide adenine dinucleotide (NADH). The decrease in absorbance at 340 nm due to the oxidation of NADH is proportional to the ammonia concentration. The result (Figure 29.11) shows that the bacteria are not producing ammonia during urea utilization. Furthermore, encapsulated bacteria decrease plasma ammonia concentration from 975 ± 70.15 to 81.15 ± 7.37 M in 30 minutes. It was found that ammonia removal efficiency of encapsulated bacteria in plasma is not significantly different than in the aqueous media. This efficiency of ammonia removal is better than that of currently used methods for ammonia removal.

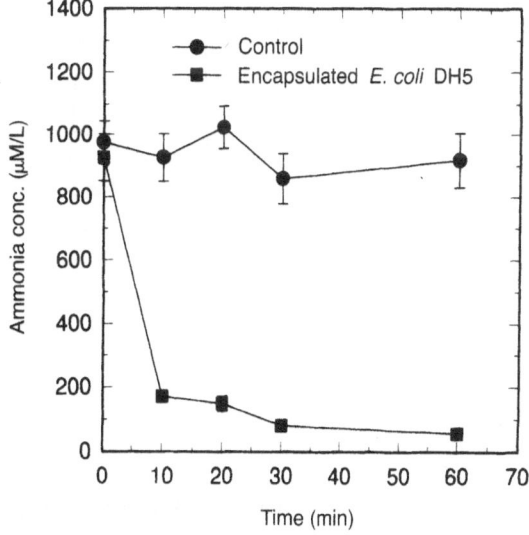

FIGURE 29.11. APA-encapsulated *E. coli* DH5 cells in in vitro plasma urea removal. (From Prakash and Chang. 1995. Biotechnology and Bioengineering 46:621–626, courtesy of John Wiley & Sons, Inc., New York.)

In Vivo Removal of Plasma Ammonia in Animal Studies Using Orally Administered Artificial Cells Containing Genetically Engineered *E. coli* DH5 Cells

Results (Figure 29.12) show that the ammonia levels in the uremic rats were always in the range of 539 ± 51 mM. Daily oral administration resulted in a decrease to 144 ± 24.70 mM, which remained constant for the entire period of treatment (Prakash and Chang, 1996). This shows that the artificial cells with genetically engineered *E. coli* DH5 cells are also effective for the removal of ammonia.

Other Factors Related to Cell Encapsulation

Potential Problems in Encapsulation of a High Concentration of Smaller Cells

In bioencapsulation of hepatocytes, there are occasionally hepatocytes incorporated into the membrane matrix, as shown in Figure 29.13 (Wong and Chang 1991a, Chang and Wong 1992). The capsular membrane over these sites appears thin and poorly formed. Cells embedded in the membrane matrix may perforate this imperfection. This is important both in implantation and in oral ingestion. In implantation, even if the membrane-embedded cells do not perforate the membrane, macrophages and lymphocytes can perforate these sites of imperfection. The activation of macrophages and lymphocytes, in both cases, could be followed by cytokine release and fibrous encapsulation of the

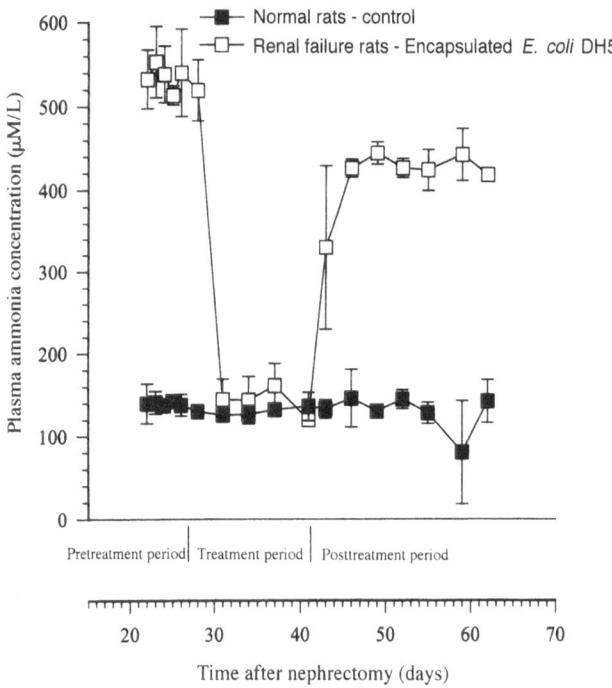

FIGURE 29.12. Plasma ammonia profile of uremic rats before, during, and after treatment by oral feeding of APA-encapsulated genetically engineered *E. coli* DH5 cells. (From Prakash and Chang. 1996. Nature Medicine 2(8):893–887, with permission from the publisher.)

PROBLEM OF STANDARD METHOD FOR ENCAPSULATION OF SMALL CELLS

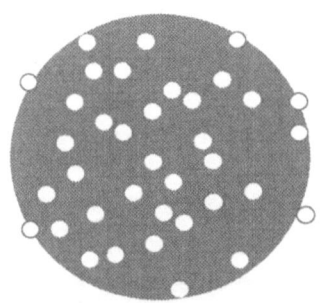 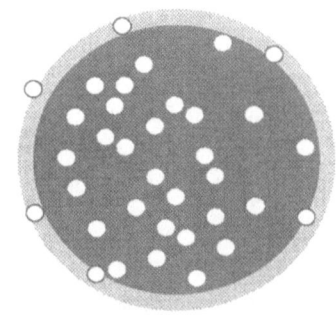 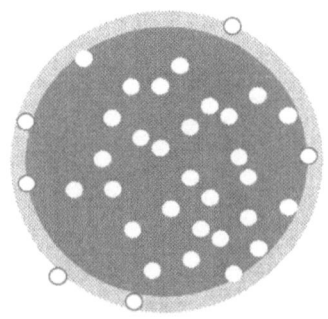

AGLINATE
GEL SPHERE   FORM MEMBRANE   DISSOLVE GEL

FIGURE 29.13. Standard method for bioencapsulation of cells. When encapsulating a large number of cells such as hepatocytes and microorganisms, a few cells may become entrapped in the membrane matrix, resulting in weakness and perforation. (From Wong and Chang. 1991. Biomat. Artif. Cells and Imm. Biotech 19:687–698, courtesy of Marcel Dekker Publishers, NY.)

implanted microcapsules. Cellular entrapment to the capsular membrane will statistically increase with increasing concentrations of cells encapsulated. Severe problems as described above are observed at hepatocyte concentrations from 10×10^6 to 20×10^6 (Wong and Chang 1991b, Chang and Wong 1992).

A Novel Approach to the Above Problem

In order to prevent the above problem, the following approach was devised (Wong and Chang 1991a, 1992). The steps, summarized in Figure 29.14, are as follows: (1) First, small calcium alginate gel microspheres containing entrapped cells were formed. (2) The small microspheres were entrapped within larger calcium alginate gel microspheres. This is an important additional step to lessen the chance that cells will protrude on the surface of the gel. (3) The alginic acid on the surface of the larger microsphere was reacted with poly-l-lysine to form a microcapsule membrane. This way, cells are less likely to be embedded in the membrane matrix. 4 The entire content of the microcapsule was then liquefied by citrate to remove calcium. This also liquefied the smaller calcium alginate gel microspheres inside the microcapsule. This way, high concentrations of smaller cells such as hepatocytes and microorganisms in the smaller gel microsphere are released to float freely in the final microcapsule. Microscopic studies show that the encapsulated cells are not embedded within the walls of the microcapsular membrane during the process of encapsulation.

Improvements in Mass Transfer, Culture Methods, and Preservation

We have devised an approach using dextran to analyze mass transfer of large molecules across these microcapsules (Corromili and Chang 1993). Dextran preparation has a wide molecular size distribution. By using chromatography, we are able to study in detail the mass transfer characteristics of different molecular weight molecules.

2 STEPS METHOD FOR
ENCAPSULATION OF SMALL CELLS

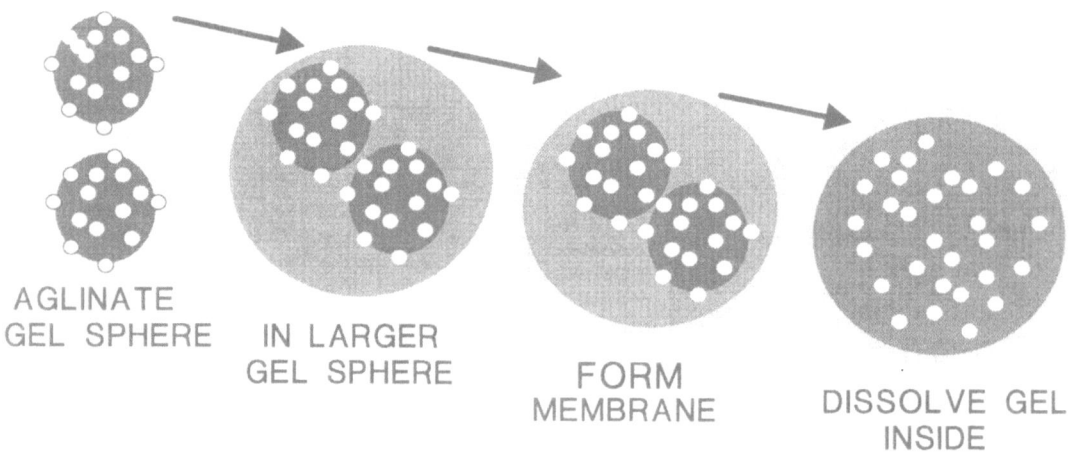

AGLINATE
GEL SPHERE   IN LARGER
                    GEL SPHERE   FORM
                                      MEMBRANE   DISSOLVE GEL
                                                         INSIDE

FIGURE 29.14. New 2-step method for bioencapsulation of a large number of cells. This helps to prevent entrapment of cells in the membrane matrix. (From Wong and Chang Biomat. Artif. Cells and Imm. Biotech, 19,687–698 (1991) with the courtesy of Marcel Dekker Publishers, NY, USA.)

Acknowledgments. T.M.S.C. gratefully acknowledges the support of the Medical Research Council of Canada and the award of Virage Center of Excellence in Biotechnology from the Quebec Ministry of Higher Education Science and Technology. The generous gift of *E. coli* DH5 from Professor R. P. Haussinger is gratefully acknowledged.

References

Bruni S, Chang TMS. 1989. Hepatocytes immobilized by microencapsulation in artificial cells: Effects on hyperbilirubinemia in Gunn rats. J. Biomat. Artif. Cells and Artif. Organs, 17: 403–412.

Bruni S, Chang TMS. 1991. Encapsulated hepatocytes for controlling hyperbilirubinemia in Gunn rats. Int. J. Artificial Organs, 14: 239–241.

Bruni S, Chang TMS. 1995b. Kientic analysis of UDP-glycoronosyl-transferase in bilirubin conjugation by encapsulated hepatocytes for transplantation into rats. Artif. Org., 19: 449–457.

Bruni S, Chang TMS. 1995a. Effect of donor strains and age of the recipient in the use of microencapsulated he-

patocytes to control hyperbilirubinemia in the Gunn rat. Int. J. Artif. Org., 18: 332–339.

Cai ZH, Shi ZQ, Sherman M, Sun AM. 1989. Development and evaluation of a system of microencapsulation of primary rat hepatocytes. Hepatology, 10: 885–860.

Chang TMS. 1964. Semipermeable microcapsules. Science, 146: 524–525.

Chang TMS. 1966. Semipermeable aqueous microcapsules ("artificial cells"): with emphasis on experiments in an extracorporeal shunt system. Trans. Am. Soc. Artif. Intern. Organs, 12: 13–19.

Chang TMS, MacIntosh FC, Mason SG. 1966. Semipermeable aqueous microcapsules: I. Preparation and properties. Can. J. Physiol. Pharmacol., 44: 115–128.

Chang TMS, MacIntosh FC, Mason SG. 1971. Encapsulated hydrophilic compositions and methods of making them. Canadian Patent, 873, 815.

Chang TMS. 1972. Artificial cells. Monograph. Springfield, Il.: Charles C. Thomas Publisher.

Chang TMS. 1976. Microcapsule artificial kidney: Including updated preparative procedures and properties. Kidney Int., 10: S218–S224.

Chang TMS, Malouf C. 1979. Effects of glucose dehydrogenase in converting urea and ammonia into amino acid using artificial cells. Artif. Organs, 31: 38–41.

Chang TMS, Wong H. 1992. A novel method for cell encapsulation in artificial cells. USA Patent No. 5,084,350.

Chang TMS. 1995. Artificial cells with emphasis on bioencapsulation in biotechnology. Biotechnology Annual Review, 1: 267–295.

Chang TMS, Prakash S. 1996. Artificial cells microencapsulated cells and genetically engineered E. coli DH 5 cells for cell therapy, gene therapy, and the removal of urea and ammonia. In: Methods in molecular biology, Vol. 63 RS Tuan (editor) New York: Humana Press, 343–358.

Chang TMS. 1997. Artificial cells. In: Encyclopedia of human biology (2nd Edition), Dulbecco, Renalo (editor-in-chief) San Diego, CA: Academic Press, Inc., pp. 457–463.

Coromili V, Chang TMS. 1993. Polydispersed dextran as a diffusing test solute to study the membrane permeability of alginate polylysine microcapsules. Biomaterials, Artificial Cells and Immobilization Biotechnology, 21: 323–335.

Dixit V, Gordon VP, Pappas SC, Fisher MM. 1989. Increased survival in galactosamine induced fulminant hepatic failure in rats following intraperitoneal transplantation of isolated encapsulated hepatocytes. In: Hybrid artificial organs, Vol 177, Baquey C, Dupuy B (editors) Paris, France: Colloque ISERM, 257–264.

Dixit V, Arthur M, Gitnick G. 1992. Repeated transplantation of microencapsulated hepatocytes for sustained correction of hyperbilirubinemia in Gunn rats. Cell Transplant. 1: 275–279.

Gu KF, Chang TMS. 1988. Conversion of ammonia or urea into L-leucine, L-valine, and L-isoleucine using artificial cell immobilizing multienzyme system and dextran-NADH+. I. Glucose dehydrogenase for cofactor recycling. ASAIO, 11: 24–28.

Gu KF, Chang TMS. 1990. Production of essential L-branched-chained amino acids, in bioreactors containing artificial cells immobilized multienzyme systems and dextran-NAD. Applied Biochemistry and Biotechnology, 26: 263–269.

Kashani S, Chang TMS. 1991b. Physical chemical characterists of hepatic stimulatory factor prepared from cell free supernatant of hepatocyte cultures. Biomaterials, Artificial Cells and Immobilization Biotechnology, 19: 565–578.

Kashani S, Chang TMS. 1991a. Effects of hepatic stimulatory factor released from free or microencapsulated hepatocytes on galactosamine induced fulminant hepatic failure animal model. Biomaterials, Artificial Cells and Immobilization Biotechnology, 19: 579–598.

Kjellstrand C, Borges H, Pru C, Gardner D, Fink D. 1981. On the clinical use of microencapsulated zirconium phosphate-urease for the treatment of chronic uremia. Trans. Am. Soc. Artif. Intern. Org., 27: 24–30.

Prakash S, Chang TMS. 1993. Microencapsulated genetically engineered E. coli cells containing genes from K. aerogens for urea and ammonia removal. Biomat. Artifi. Cells Immob. Biotechnol, 19: 687–697.

Prakash S, Chang TMS. 1995. Preparation and in-vitro analysis of genetically engineered E. coli cells, microencapsulated in artificial cells for urea and ammonia removal. Biotechnology and Bioengineering, 46: 687–697.

Prakash S, Chang TMS. 1996. Microencapsulated genetically engineered live E. coli DH 5 cells given orally to maintain normal plasma urea level in uremic rats. Nature Medicine, 2(8): 893–887.

Scott B, Mulrooney H, Stuart P, Haussinger RP. 1989. Regulation of gene expression and cellular localization of cloned K. aerogens urease. J. Gen. Microbiol, 135: 1769–1776.

Setala K, Heinonen H, Schreck-Purla I. 1973. Ingestion of lyophilized soil bacteria for alleviation of uremic symptoms. J. Int. Res. Comm. Syst., 73: 1.

Sparks RE. 1979. Review of gastrointestinal perfusion in treatment of uremia. Clin. Nephrol, 11: 81.

Wong H, Chang TMS. 1986. Bioartificial liver: Implanted artificial cells microencapsulated living hepatocytes increases survival of liver failure rats. Int. J. Artif. Organs, 9: 335–6.

Wong H, Chang TMS. 1988. The viability and regeneration of artificial cell microencapsulated rat hepatocyte xenograft transplants in mice. Biomat. Artif. Cells and Artif. Organs, 16: 731–740.

Wong H, Chang TMS. 1991b. Microencapsulation of cells within alginate poly-L-lysine microcapsules prepared with standard single step drop technique: Histologically identified membrane imperfections and the associated graft rejection. J. Biomaterials, Artificial Cells and Immobilization Biotechnology, 19: 675–686.

Wong H, Chang TMS. 1991a. A novel two step procedure for immobilizing living cells in microcapsules for improving xenograft survival. Biomaterials, Artificial Cells and Immobilizing Biotechnology, 19: 687–698.

Wrong OM, Edmonds CJ, Chadwick VS. 1981. The large intestine, Lancaster: MTP Press.

Section 3
Bioreactor Culture

30
Principles of Bioreactor Design for Encapsulated Cells

Branko Bugarski, Mattheus F. A. Goosen, and Gordana Vunjak-Novakovic

The purpose of this chapter is to give an overview of key requirements for the cultivation of microencapsulated animal cells, to discuss bioreactor design and operation, and to give representative examples of applications involving the use of encapsulated enzymes and cells. We have focused on the guidelines, rather than the details of bioreactor design and optimization. Air-lift reactors (ALR), in which a continuous liquid phase is brought into contact with gas bubbles, immobilized cells, or enzymes, and in some cases a dispersed solvent, are discussed in more detail, in order to illustrate the role of fundamental research in bioreactor design.

Introduction

The concept of artificial cells, the microcapsules containing proteins and animal cells, was first introduced by Chang (1964). The use of semipermeable biocompatible membranes resulted in the development of a new class of hydrogel microcapsules, and led to the first in vivo applications (Lim and Sun 1980). Many different types of microencapsulated animal cells have been studied and considered for application in tissue culture and transplantation (Lim 1983). The development of mass cultivation techniques of animal cells has been motivated by their exquisite ability to serve as a source of valuable products (e.g., monoclonal antibodies for use in immunotherapy for site-specific recognition and binding).

The existing high-performance bioreactors for bacterial fermentations, including those used in recombinant DNA technology, are generally not convenient for animal cell culture because of numerous differences between the two systems (Merten 1989). Metabolic rates are much lower for animal than for microbial cells. However, most animal cells are susceptible to shear, so that the required oxygen transfer rates, although relatively low, cannot be achieved in suspension cultures. In addition, most animal cells are anchorage-dependent. Ideally, large-scale animal cell systems should provide culturing conditions similar to the actual capillary-tissue environment of animal cells. If, however, the animal cells are protected from shear by a semipermeable membrane, they can be cultivated at conditions that provide more intensive mass transfer by means of increased gas throughput rates, and the overall bioreactor performance can be significantly improved. Cell and product concentrations in bioreactors operating with microcapsules can exceed those in suspension cultures by orders of magnitude (Posillico 1986).

Design of a bioreactor system is an optimization procedure by itself, in which the biological and engineering factors are balanced to obtain the required capacity and quality of the product at minimum cost. The operation of a bioreactor depends on numerous factors that interact at micro and macro levels, i.e., the level of a cell and the bioreactor as a whole (Figure 30.1). Kinetic rates of cell growth and product formation are established as a response to the local physicochemical environment. Performance of cell cultivation systems thus depends both on the characteristics of the cell line and on the bioreactor design and mode of operation. In addition, any change in the reactor size may affect the interrelations among fluid-dynamic, physicochemical, and biological parameters. Bioreactor design can take advantage of

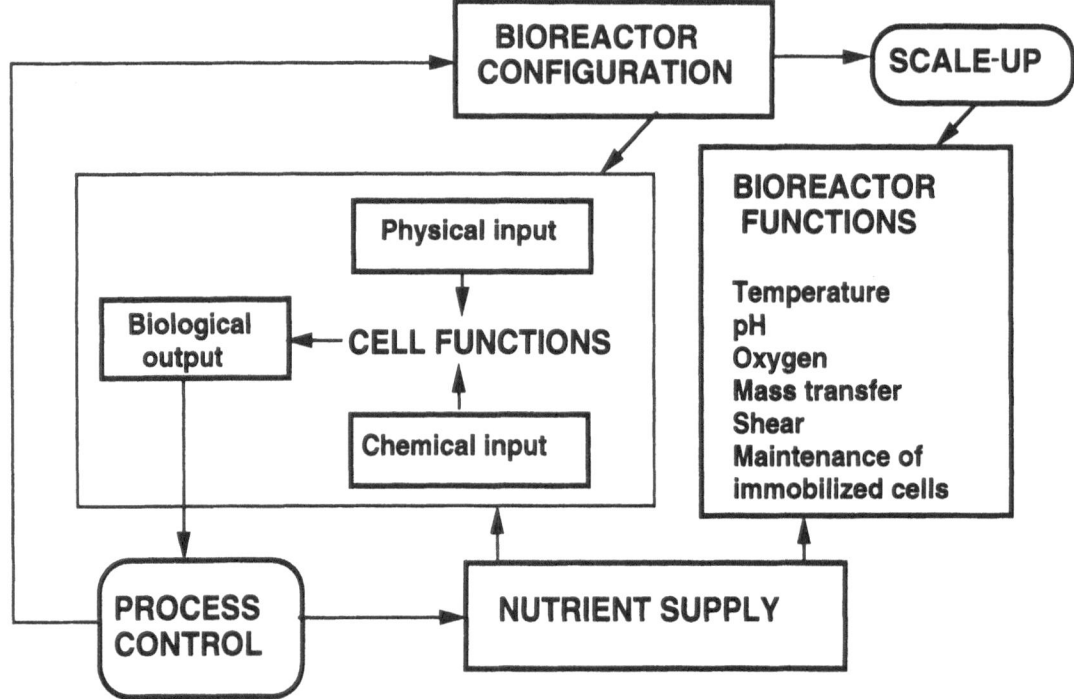

FIGURE 30.1. Overview of biological, physical, and chemical factors that may determine the design criteria and control strategy for air-lift bioreactors. From Bugarski B., G. Jovanovic, G. Vunjak-Novakovic, Bioreactor Systems Based on Microencapsulated Animal Cell Cultures. In *Fundamentals of Animal Cells Immobilization and Microencapsulation,* M.F.A. Goosen, ed. pp. 267–296. Copyright 1993 by CRC Press, Boca Raton. Reprinted with permission.

such interactions. For example, containment of the cells and cell-secreted factors within a microcapsule can be used to simulate at least some aspects of the in vivo environment and to stimulate the cells to increase synthesis rates. Moreover, the microencapsulated cell bioreactors can be designed as integrated production-recovery systems, in which the secreted product of interest is continuously removed from the system (e.g., by a solvent dispersed in the aqueous liquid phase) (Sajc et al 1995a).

Several types of microencapsulated cell bioreactors have been studied with respect to operating regimes, amounts and distributions of dispersed phases (bubbles, biocatalyst particles, solvent droplets), liquid mixing, and oxygen and nutrient transfer. Surface aeration systems were shown to support the growth of animal cells only in small culture volumes (<1 liter). The stirred tanks with mechanical agitation and air sparging have been modified in order to meet the requirements of shear-sensitive animal cells in suspension. The combination of cell microencapsulation and the use of air-lift reactors (ALR), discussed in this chapter, seems particularly suitable for the cultivation of animal cells.

Design and Operation of Air-Lift Bioreactors

Air-lift reactors (ALR) are pneumatically mixed vessels that combine the features of gas-liquid-solid fluidized beds and circulating bubble columns. Common ALR designs include: (1) the internal-loop configuration consisting of a draft tube (riser) surrounded by an annular region (downcomer) (Figure 30.2a, Vunjak-Novakovic et al 1992a,b) and (2) the external-loop configuration consisting of a riser, a downcomer, and two horizontal sections connected into a closed loop (Figure 30.2b, Bugarski et al 1991a,b). In both cases, gas and liquid are introduced at the bottom of the column to establish steady-state circulation within the reactor. In addition to the continuous liquid phase, gas bubbles, and solid particles, ALR can also contain an additional liquid phase

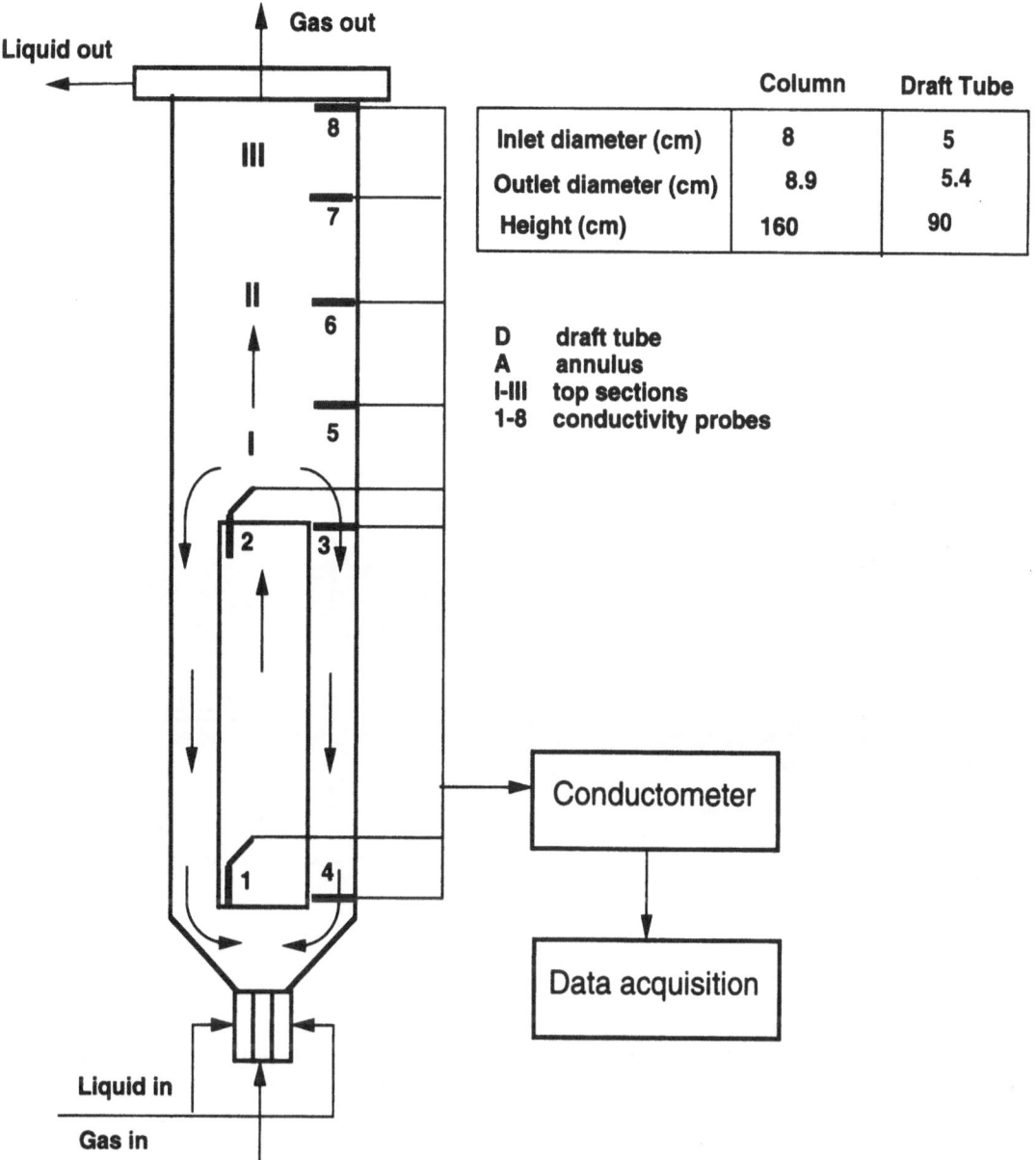

FIGURE 30.2. Schematic presentation of the air-lift reactor configurations: a, internal-loop, used in fluid-dynamic and mixing studies, (*Continued*)

dispersed to form a four-phase system (Figure 30.2c, Sajc et al 1995 a,b).

The growing application of the ALR in the chemical industry and biotechnology can be attributed to its simple design, high rates of heat and mass transfer, and efficient mixing in the liquid phase (Fan 1989). In particular, ALR is suitable for shear-sensitive biological materials (e.g., plant and mammalian cells, enzymes) immobilized in low-density gel particles or microcapsules. The cut-off pore size of a microcapsule can be adjusted to allow unrestricted diffusion of small molecules (e.g., nutrients, gases), protect cells from shear, and retain higher molecular weight products within the microcapsule, where they can reach high concentrations (Bugarski et al 1993).

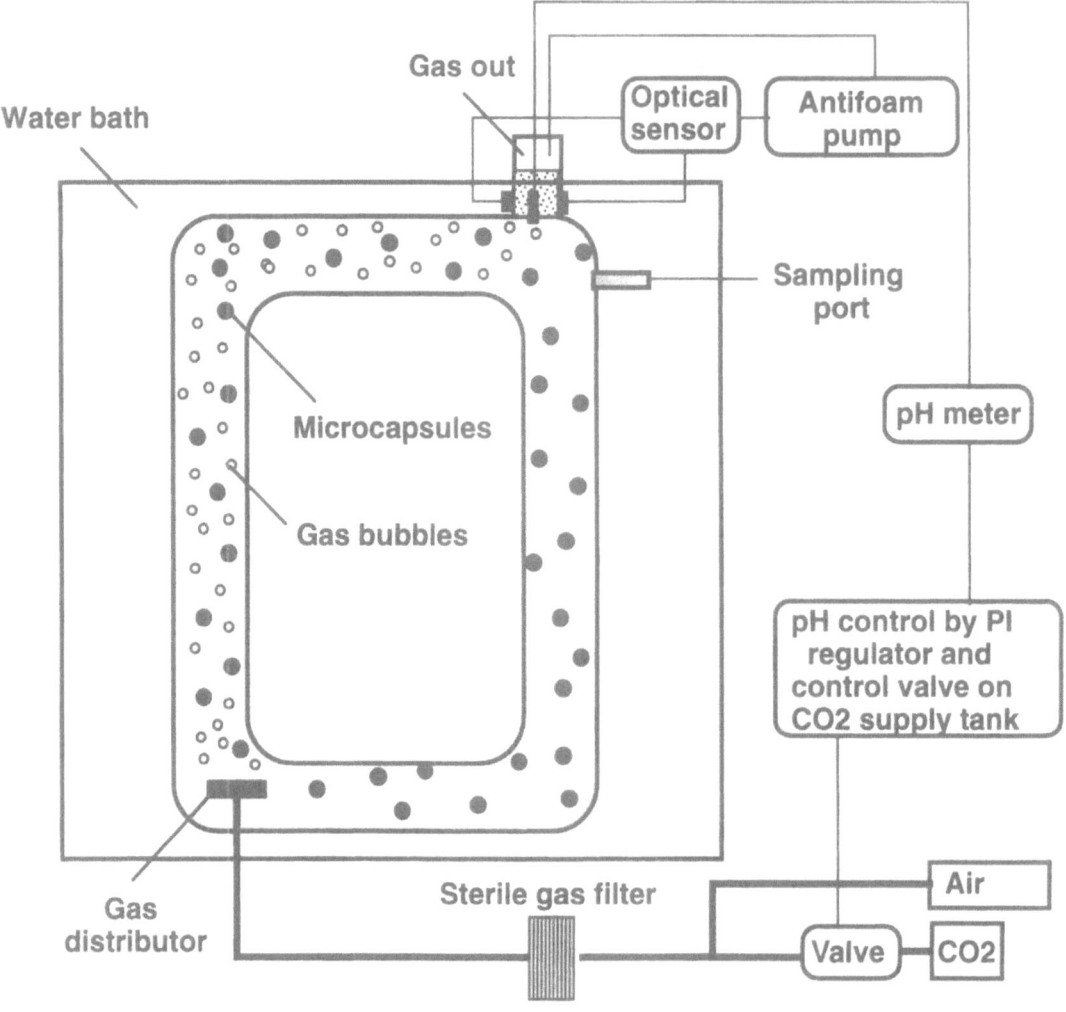

FIGURE 30.2. (*Continued*) b, external loop three-phase system, used in enzyme esterification and hybridoma cell cultures,

Here we review the research and development of ALR carried out at the University of Belgrade over the last 10 years. The reactors studied include external- and internal-loop configurations, and three- and four-phase systems. Fluid-dynamic, mixing, mass transfer, and kinetic studies resulted in mathematical models that can be used in reactor design and process scale-up. Several prototypes of the ALR equipped with process control hardware were specifically designed for application in biotechnology (e.g., lipase-catalyzed oil interesterification, production of monoclonal antibodies in hybridoma cell culture, production of anthraquinones in plant cell culture).

Three-Phase Internal-Loop Configuration

Two ALR units were used in fluid-dynamic and mixing studies: a 3-D circular column (Figure 30.2a), 7.6 L in volume, and a geometrically similar, 1 cm thick 2-D column. The 3-D column was an 8-cm-diameter, 160-cm-high cylinder with an internal draft tube (5.4/5.0 cm in diameter, 90 cm high), placed 7 cm above the top of the distributor. A convergent two-phase injector nozzle was used as a gas-liquid distributor (Vunjak-Novakovic et al 1991). Air and water were continuously introduced into the reactor, at superficial velocities of 0–1.1 cm^3/cm^2s

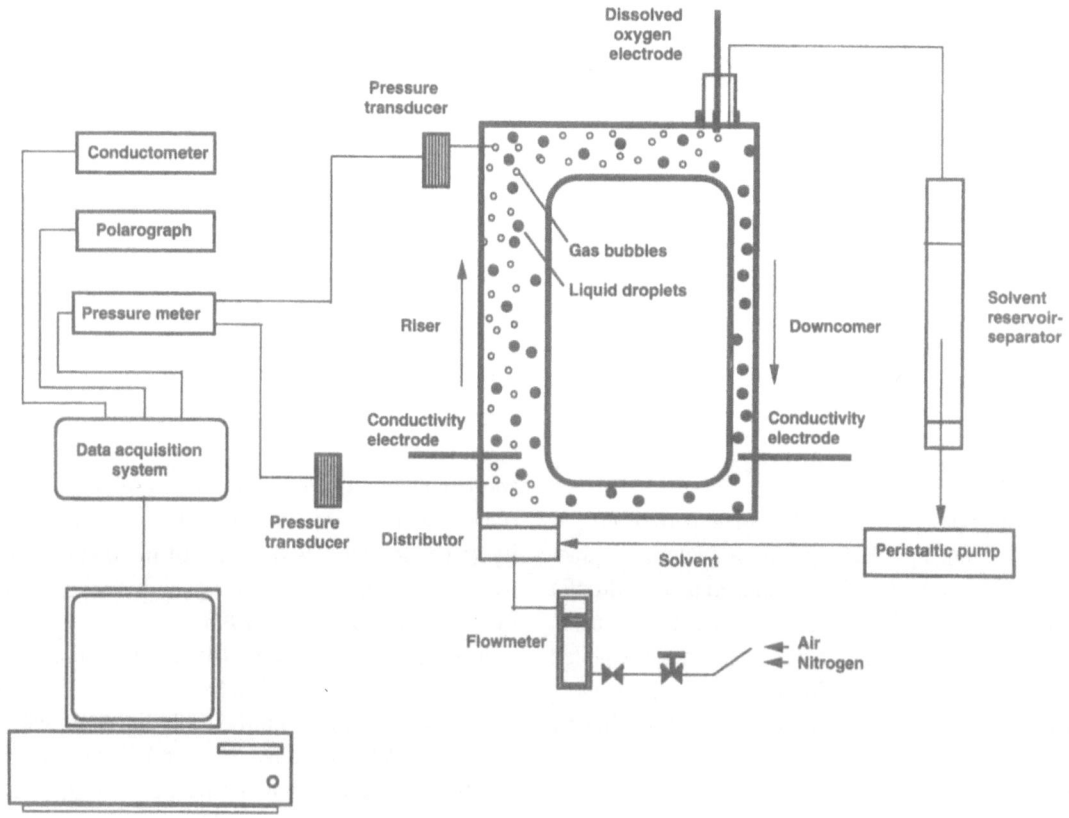

FIGURE 30.2. (*Continued*) and c, external-loop four phase system, used in plant cell cultures. From Obradovic B., A. Dudukovic, G. Vunjak-Novakovic. Local and Overall Mixing Characteristics of the Gas-Liquid-Solid Air-Lift Reactor. *Ind. Eng. Chem. Research* 33, 698–702. Copyright 1994 by the American Chemical Society. Reprinted with permission.

and 0–1 cm^3/cm^{2}s, respectively. A series of smaller size internal-loop reactors (80 ml to 3 L) was made to minimize the consumption of enzymes. For example, the reactor made for lipase-catalyzed oil esterification had an inner diameter of 3.1 cm, a draft tube diameter of 2 cm, and an overall height of 11 cm (Mojovic et al 1994).

Three-Phase External-Loop Configuration

A series of external-loop bioreactors was developed for the production of monoclonal antibodies by immobilized hybridoma cells (Figure 30.2b). The largest reactor, which was 1 L in volume, had a constant internal diameter of 40 mm, and the outer dimensions were 17 cm wide × 35 cm high. To maintain a constant temperature of 37°C in cell culture studies, the reactor was placed into a water bath. The reactor operated batchwise with respect to the liquid

phase, and with a continuous gas exchange (air/CO$_2$ mixture) for oxygen supply and pH control. Sterile gas mixture was introduced at the bottom of the riser in the form of 1–2 mm diameter bubbles, using a sintered disc sparger with 0.1–0.3 mm holes. The CO$_2$/air ratio in the gas mixture, consisting initially of 5% CO$_2$ and 95% air, was adjusted to maintain pH = 6.9–7.0 by a PID regulator coupled to the electromagnetic valve on the CO$_2$ tank. The foam control unit contained an optical foam detector, a logical circuit, and an antifoam pump.

Four-Phase External-Loop Configuration

The external-loop bioreactor developed for continuous production and extraction of extracellular metabolites by immobilized plant cells (Figure 30.2c) contained a dispersed liquid solvent (silicone oil or n-hexadecane). The reactor was 250 mL in volume, and had the overall aeration height of 27 cm. The inner diameter of the riser was 2.7 cm, while the downcomer and both horizontal sections had an inner diameter of 1.7 cm. Gas was distributed at the bottom of the riser in the form of 1–2 mm diameter bubbles, using a sintered glass plate; solvent was dispersed through five symmetrically positioned 1 mm holes within the same distributor. The gas supply consisted of two lines for air and nitrogen, needle and solenoid valves, and a flowmeter. Solvent was driven by a peristaltic pump from the top of the downcomer through a reservoir-separator to the distributor (Salc et al, 1995a, b).

Biocatalyst Particles

The solid phase consisted of alginate particles (d_s = 2.5 ± 0.1 mm, ρ_s = 1040 ± 1 kg/m^3, volumetric fraction in the reactor 0–30%), produced using a droplet generator (Bugarski et al 1994b). The cell suspension was mixed with Na-alginate and subsequently gelled for 30 min in calcium chloride. For certain applications (e.g., production of monoclonal antibodies by hybridoma cells), microcapsules were formed by reacting the alginate beads first with an aqueous solution of poly-L-lysine (MW = 22 kD) for 6 min, and then with 0.04% Na-alginate for 5 min. Alginate within the microcapsules was liquefied using 0.05 M Na-citrate for 4 min (Bugarski et al 1993).

Fluid-Dynamic Studies

Gas bubbles, solid phase, and liquid droplets were uniformly dispersed within the continuous liquid phase. A steady-state circulation of liquid phase and solid particles was maintained, with no significant recirculation of gas bubbles into the downcomer. In the three-phase internal-loop ALR, flow conditions in the riser (draft tube), downcomer (annulus), and the reactor as a whole were assessed by flow visualization (Kundakovic and Vunjak-Novakovic 1995a). Neutrally buoyant opaque alginate particles were used as liquid flow tracers in conjunction with a high-speed 16 mm camera (200 frames/s, exposure 1/800 s), frame-to-frame analyzer, data acquisition system, and image-processing software. The local velocity was calculated (average ± SD, n = 30) from series of successive positions of a tracer particle within 5 ms time intervals in a view area of 4 cm × 15 cm. Bubble velocity was calculated (average ± SD, n = 160) from successive bubble positions over 5 ms time intervals. Bubble frequency was calculated from the number of bubbles passing through a certain level within a 1.5 s time interval. Bubble diameter was assessed as an average of bubble diameters in a 4 cm × 15 cm view area. The local gas and particle holdups were measured from the respective area fractions.

In the three-phase external-loop ALR, liquid circulation rate was assessed from times required for a tracer (1% sulfuric acid, pulse signal) to travel between two fast-response pH electrodes positioned 20 cm apart, which were placed in the riser, downcomer, and the horizontal sections of the reactor. The simultaneous responses of the probes were recorded at 0.7 s intervals. Mean residence time of the tracer was calculated for each electrode from the variance of the tracer distribution. The average liquid velocity was calculated as the ratio of the distance between the electrodes and the mean residence time of the tracer (Bugarski et al 1993).

In the four-phase external-loop ALR, liquid circulation rate was calculated from the overall length of the loop and the overall circulation time, which was in turn determined by autocorrelation of the signals from the conductivity probe located right above the distributor (Sajc et al 1995b). Liquid velocities in the riser and downcomer were calculated from the respective lengths of these sections and the residence times of liquid determined by cross-correlation of signals from electrode pairs in the riser and downcomer. Gas and solvent holdups were determined by the volume expansion method, by measuring the bioreactor volume displacement by gas or solvent at steady-state conditions, after the flow of solvent or gas was switched off.

Mixing Studies

In the three-phase internal-loop ALR, mixing in the liquid phase was studied using a tracer-response technique based on fast conductivity measurements. The conductivity probes were designed as proposed by Khang and Fitzgerald (1975) and modified as suggested by Han and Kim (1990). As shown in Figure 30.2a, eight probes were used and located centrally at the inlet and outlet of the draft tube (1–2), at the inlet and outlet of the annular region (3–4) and at equal distances between the top of the draft tube and the liquid outlet (5–8). The experimental system was supported by the Burr-Brown PCI

20002M data acquisition system, and Labtech Notebook software. Probe readouts were taken simultaneously at a frequency of 10 Hz over the first 90 s, and 0.1 Hz for another 20 min, after a pulse signal of tracer (30 cm^3 of 30% NaCl) was introduced at the bottom of the column (Obradovic et al 1994).

The arrangement of conductivity probes allowed the assessment of liquid mixing in five reactor compartments: draft tube, annulus, and three top sections above the draft tube. The signal from probe 8 was used as the response of the whole reactor and to close the material balance of the tracer. Liquid residence time in each reactor compartment was assessed via cross-correlation of the respective inlet and outlet signals. The conditions of an open-open vessel (Levenspiel 1979) were provided for the signal input and output, while the reactor satisfied the closed-open boundary condition. The signals (1,000–2,000 data points per experiment) were smoothed using a 5-point moving-average filter and processed using voltage-concentration calibration curves. The liquid circulation velocity, Bodenstein number, and dispersion number, D, were evaluated by analyzing tracer-response data in time and frequency domains, using the compartmental and axial dispersion models (Obradovic et al 1997).

Mass Transfer Studies

Mass transfer rates were measured at the gas-liquid, liquid-liquid, and liquid-solid interfaces and used to determine the respective mass transfer coefficients (Bugarski et al 1993, Sajc et al 1995a, 1994b). Oxygen transfer to the microcapsule was evaluated from oxygen concentration in medium and the intracapsule cell concentration profiles. The volumetric oxygen transfer coefficient at the bubble-liquid interface in the three-phase (aqueous phase-gas-microcapsules) and four-phase (aqueous phase-solvent-gas-alginate particles) external-loop ALR was determined by the dynamic gassing-out method (Bugarski et al 1990, Sajc et al 1995b). Oxygen concentration in the aqueous phase was measured as a function of time after a step change in the gas feed from nitrogen to air. The oxygen uptake by immobilized cells was assessed in a mixed cell with nitrogen above the liquid surface, from the change in the concentration of dissolved oxygen in medium after the aeration was stopped. Mass transfer at the solvent-aqueous phase interface was assessed from the change in concen-tration of a tracer that diffused from the continuous (aqueous) into the dispersed (solvent) phase (Sajc et al 1995b).

Enzyme Reaction Studies

A lipase-catalyzed interesterification of triacylglycerols of palm oil midfraction was studied as a potential system for the production of cocoa butter equivalents for the food and cosmetic industry (Mojovic et al 1993, 1994). Lipase was immobilized by adsorption on Celite and encapsulated in lecithin reverse micelles. The reaction mixture consisted of palm oil midfraction, stearic acid dissolved in water-saturated n-hexane, and the immobilized enzyme. The reactor was sparged by nitrogen to provide mixing. Temperature was maintained at 37°C by heat exchange with the outer jacket. The top of the reactor was connected to a condenser to avoid solvent losses by evaporation.

Cell Culture Studies

In animal cell culture studies, mouse hybridoma cells were used that secrete a monoclonal antibody, IgG, specific to β-HcG. Thirty milliliters of alginate beads or alginate-PLL microcapsules containing 1.10^6 hybridoma cells/ml were loaded into the bioreactor containing 1 l of medium. Temperature, pH, and foam control had been previously set, and the gas flow rate was adjusted at a desired level. The medium was renewed once a day, to eliminate nutrient limitations and product inhibition of cell growth. Samples of medium (1 mL) and immobilized cells (0.2 mL) were withdrawn every 1–2 days over 25 days of cultivation. Cell viability was assessed by Trypan blue exclusion. Cell counts were determined by manual counts of cells released from microcapsules or alginate beads (by mechanical rupture or treatment with 0.05M Na-citrate, respectively) and from cryocut sections. An enzyme-linked immunosorbent assay (ELISA) was used to determine the concentration of IgG (Bugarski et al 1993, 1994a,b).

In plant cell cultures (Sajc et al 1994a,b, 1995a,b), callus cultures of Frangula alnus were used that produced anthraquinones. Alginate-immobilized cells were cultured batchwise over a period of 10 days. Flow rates of air and solvent were set to 0.7 vvm and 0.05 vvm, respectively, and maintained constant throughout the cultivation. Air flow rate was chosen to provide good mixing in the riser section and a sufficient supply of oxygen. Solvent flow rate was adjusted to provide complete drop coalescence at the top of the horizontal section, necessary for a continuous extraction. At these gas and solvent flow rates, the mixing time was 27 s, the mass transfer coefficient of the tracer (dantron) was 0.0125 1/s, and the rate of oxygen supply was two orders of magnitude higher than the rate of oxygen uptake by immobilized cells. The reactor operated with a continuous supply of gas and solvent, and the extracellular product was continuously removed from cell culture medium. Temperature and pH were maintained constant. A data acquisition system was used to monitor all vital process parameters.

Flow and Mixing Patterns in Air-Lift Bioreactors

Experimental data for the particle holdup reveal a relatively uniform distribution of particles within the reactor volume over a wide range of gas flow rates (Figure 30.3). At all particle loadings, the measured particle holdup in the draft tube was equal to that in the annulus. In addition, the transient from the circulating into the turbulent regime ($U_g = 0.45$ cm/s in Figure 30.3) had little effect on particle holdup (< 4% within the range of this study). Gas holdup was in turn a linear function of the gas velocity (Figure 30.3), and increased with the increase in particle loading (Kundakovic and Vunjak-Novakovic 1995b).

Liquid flow and mixing were assessed for the draft tube, annulus, and three top sections of the reactor (Figure 30.2a). At all gas flow rates, the highest liquid velocity was observed in the top Section I (Figure 30.4), which could be related to the highest degree of backmixing. With the increase in the superficial gas velocity, the region of high liquid recirculation extended to the upper sections of the reactor (Figure 30.4). The upper part of the reactor contained multiple interacting flow cells and had a flow pattern similar to that in two-phase bubbling columns (Joshi 1980). A schematic presentation of the observed flow patterns is shown in Figure 30.5 (Obradovic et al 1994).

With an increase in the superficial gas velocity, liquid recirculation rate between the draft and annular regions increased up to a certain limit (Vunjak-Novakovic et al 1992b), and a further increase in gas flow rate had no significant effect on liquid recirculation. This limiting gas velocity corresponds to the transient from the circulating into the turbulent regime, i.e. the onset of bubble recirculation into the annulus. This implies that gas velocity may be changed over a wide range without disturbing the circulating flow of liquid within the reactor. In addition, particle loading (0–30% vol.) had little effect on liquid circulation rate. Higher gas velocities resulted in extended regions of liquid recirculation in the upper sections of the reactor, as the result of a stronger effect of a "jet stream." The effect is consistent with the lower liquid mixing times observed in columns with higher upper zones (Chisti 1989, Russell et al 1994).

Fluid-Dynamic Model of the Air-Lift Bioreactor

Several approaches have been proposed to model fluid flow and mass transfer in the ALR (Chisti 1989, Fan 1989). Most of the reported models have a limited range of application, because they include empirical correlations and consider the reactor as a single compartment. So far, there has been little attempt to theoretically analyze multiphase flow in the ALR. Darton and Harrison (1975) and Chen and Fan (1990) extended the drift-flux analysis developed for two-phase flow (Wallis 1969) to three-phase systems, but did not account for the liquid-particle drift flux. Efficient mixing in the ALR was attributed to liquid recirculation (Verlaan et al 1986, Vunjak-Novakovic et al 1992b). However, single-parameter models of the overall liquid mixing gave controversial predictions (Fields and Slater 1983, Lu et al 1994a,b, Verlaan et al 1986). More recently, a fluid-dynamic model of the external-loop ALR operating with steady-state circulation of liquid and particles was proposed that accounted for momentum transfer between all three phases (Kundakovic and Vunjak-Novakovic 1995b). This model was based on the drift-flux analysis, accounting for the additional momentum transfer between the continuous liquid phase, gas bubbles, and solid particles. The model gave excellent predictions of the experimental data (Kundakovic and Vunjak-Novakovic 1995b).

Gas flow in the draft tube was considered as a bubbling flow of discrete, uniformly dispersed, noncoalescing bubbles in the liquid phase. Bubble-to-particle momentum transfer was assessed as a cu-

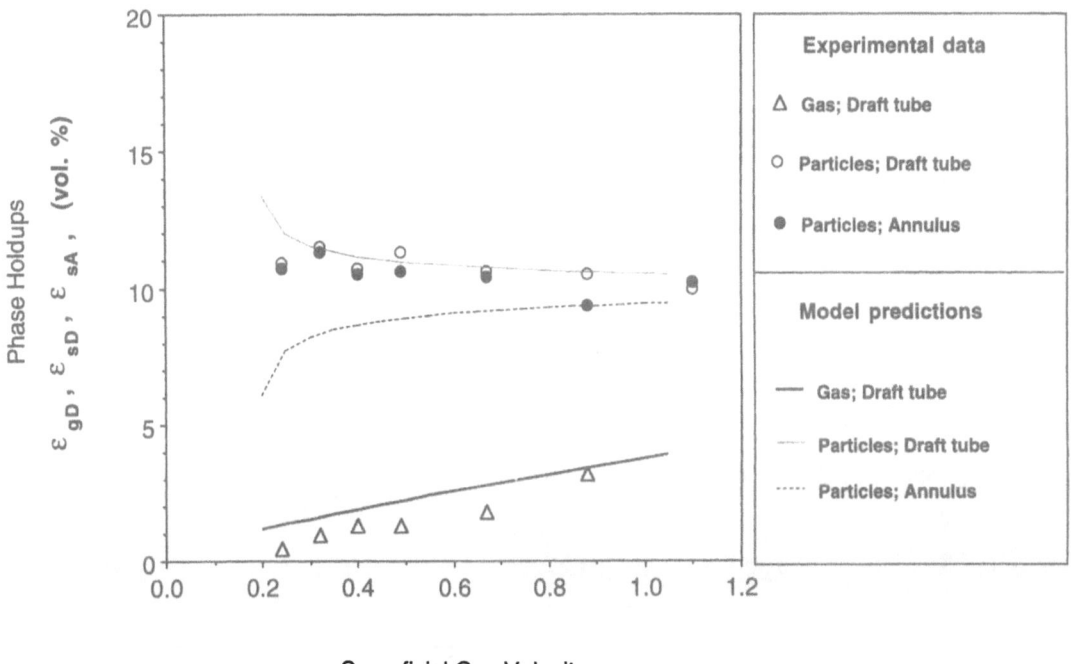

Phase Holdups
ε_{gD}, ε_{sD}, ε_{sA}, (vol. %)

Superficial Gas Velocity
Ug (cm/s)

FIGURE 30.3. Holdup of solid particles and gas in the draft tube and annulus of the internal-loop ALR; comparison of experimental data and model predictions. Kundakovic Lj., G. Vunjak-Novakovic. A Fluid Dynamic Model of the Draft Tube Gas-Liquid-Solid Fluidized Bed. *Chem. Eng. Science* 50, 3763–3775. Copyright 1995 by Elsevier Science. Reprinted with permission.

mulative effect of successive discrete changes in the relative particle-to-liquid velocity and the particle drag coefficient, and calculated in terms of the effective drag force. The momentum transfer equation for the circulating flow in the ALR,

$$\rho_l (\epsilon_{lA} - \epsilon_{lD})\, gL - \rho_s (\epsilon_{sD} - \epsilon_{sA})\, gL \tag{1}$$
$$= (-\Delta p_f)_A + (-\Delta p_f)_D + (-\Delta p) + (-\Delta p_B)$$

represents a balance of the driving force, i.e., the difference in the apparent densities of the fluid in the draft tube and the annulus, and the friction losses in the draft tube $(\Delta p_f)_D$, annulus $(-\Delta p_f)_A$, and bottom section $(-\Delta p_b)$, and at the column walls $(-\Delta p_w)$.

The friction pressure drop in the annulus, $(-\Delta p_f)_A$, was calculated as

$$(-\Delta p_f)_A = 3/4\, C_D\, L/d_s\, \rho_l\, v_{slA}^2\, \alpha_{sA} \tag{2}$$

where:

$$v_{slA} = U_t (1 - \alpha_{sA})^{ns-1} \tag{3}$$

The friction pressure drop in the draft tube, $(-\Delta p_f)_D$, was calculated as

$$(-\Delta p_f)_D = 3/4\, C_D\, L/d_s\, \rho_l\, v_{slD}^2\, \alpha_{sD} \tag{4}$$

where:

$$v_{slD} = U_t'(1 - \alpha_{sD})^{ns-1} \tag{5}$$

is the actual particle-to-liquid relative velocity in the draft tube, C_D is the corresponding drag coefficient, and Ut' is the particle settling velocity in the presence of gas bubbles.

The presence of gas bubbles resulted in a decrease in the relative liquid-particle velocity (Kundakovic and Vunjak-Novakovic 1995a):

$$U_t' = U_t - \frac{U_{br} R_b^2}{x_{max} \cdot y_{max}} \arctan\left(\frac{y_{max}}{x_{max}}\right) \tag{6}$$

where:

$$y_{max} = x_{max} = \frac{\Delta x_b - d_s}{2} \tag{7}$$

$$\Delta x_b = \frac{\Delta z\, U_b}{f_b}. \tag{8}$$

Equations (6–8) were derived from Navier-Stokes equations as proposed by Clift and Grace (1985), by assuming: (1) uniform distribution of monodisperse gas bubbles in the draft tube, (2) nonviscous flow of

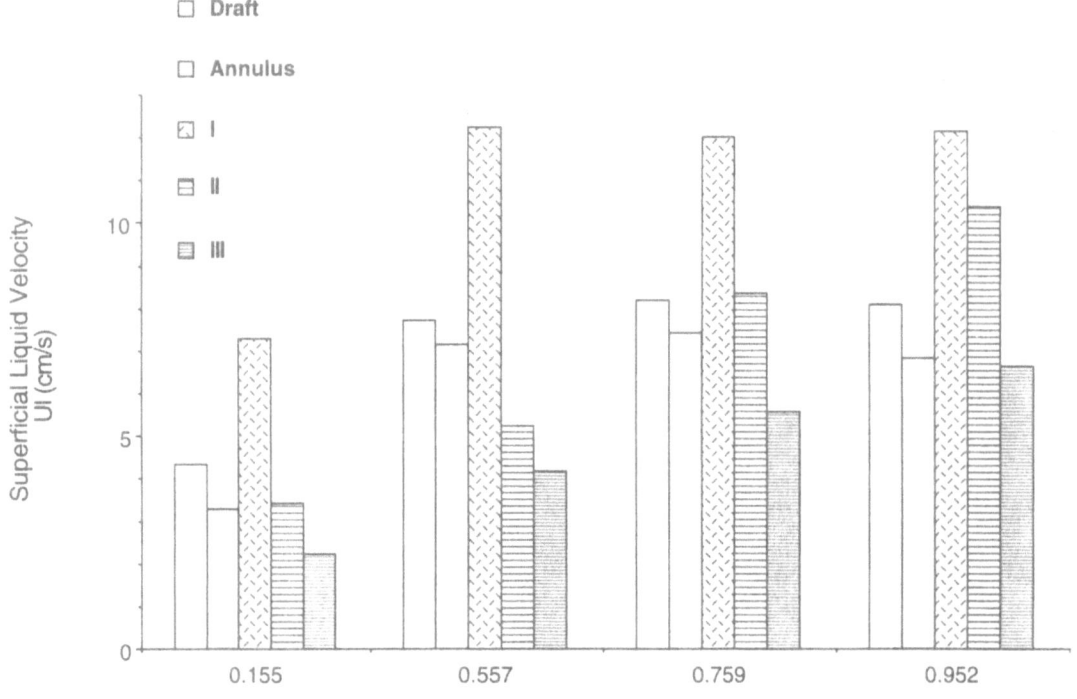

FIGURE 30.4. Local liquid flow rates in the draft, annulus, and top sections I, II, and III of the internal-loop ALR at the net liquid input of $q_1 = 1.41$ L/min ($U_1 = 0.464$ cm/s) and particle holdup $\epsilon_s = 20\%$. From Obradovic B., A. Dudukovic, G. Vunjak-Novakovic. Local and Overall Mixing Characteristics of the Gas-Liquid-Solid Air-Lift Reactor. *Ind. Eng. Chem. Research* 33, 698–702. Copyright 1994 by the American Chemical Society. Reprinted with permission.

liquid around the bubble, (3) negligible wall effects, (4) negligible horizontal component of the bubble rise velocity and (5) negligible bubble interactions. The momentum transfer equation for circulating flow between the draft and annulus was modified to take into account the pressure drops due to the wall friction ($-\Delta p_w$) and flow resistance at the bottom of the column ($-\Delta p_B$) (Equation 1). The wall friction term ($-\Delta p_w$) was calculated for two-phase flow in the heterogeneous bubbling regime, as proposed by Joshi et al (1990):

$$(-\Delta p_w) = \frac{2f\rho_l \, j_{lD}^2 \, 2L}{d} \phi_l^2 \qquad (9)$$

where:

$$\phi_l^2 = \frac{1}{\epsilon_{lD}} + \qquad (10)$$

$$\frac{2}{f j_{lD}^2} \left\{ gd \left[j_{lD} - \frac{\epsilon_{gD}}{1 - \epsilon_{gD}} - \epsilon_{gD} \, U_{br} \right]^{1/3} \right\}^2.$$

($-\Delta p_B$) was calculated as a function of the ratio of the cross-section areas of the draft and annulus (Chisti 1989):

$$(-\Delta p_B) = K_B \frac{\rho_l j_{lD}^2}{2} \left(\frac{A_D}{A_A} \right) \qquad (11)$$

where:

$$K_B = 11.4 \left(\frac{A_A}{A_B} \right)^{0.79}. \qquad (12)$$

Momentum transfer between liquid and fluidized particles in the circulating flow in draft and annular regions could be presented as

$$\frac{\alpha_{sD} \, j_{lD}}{1 - \alpha_{sD}} - U_t \, \epsilon_{sD} \, (1 - \alpha_{sD})^{n_s - 1}$$

$$= \frac{\alpha_{sA} \, j_{lA}}{1 - \alpha_{sA}} + U_l \, \alpha_{sA} \, (1 - \alpha_{sA})^{n_s - 1}. \qquad (13)$$

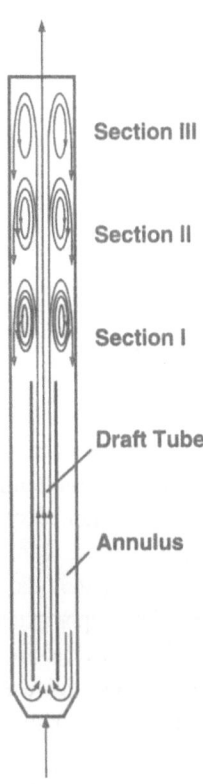

Section III

Section II

Section I

Draft Tube

Annulus

FIGURE 30.5. Schematic presentation of the observed liquid flow patterns in draft and annular regions and top sections I, II, and III of the internal-loop ALR. From Obradovic, B., A. Dudukovic, G. Vunjak-Novakovic. Local and Overall Mixing Characteristics of the Gas-Liquid-Solid Air-Lift Reactor. *Ind. Eng. Chem. Research* 33, 698–702. Copyright 1994 by the American Chemical Society. Reprinted with permission.

A corresponding equation for momentum transfer between gas bubbles and the liquid is

$$(1 - \alpha_{gD}) j_{gD} - \alpha_{gD} j_{lD} \qquad (14)$$
$$= U_{br} \epsilon_{gD} (1 - \alpha_{gD})^{n_b}.$$

A balance equation for the holdup of solids is

$$\epsilon_{sD} \frac{V_D}{V} + \epsilon_{sA} \frac{V_A}{V} = \epsilon_s. \qquad (15)$$

The liquid flow rate in the draft tube could be related to the annular liquid flow rate and the input liquid flow rate:

$$j_{lD} A_D = j_{lA} A_A + q_l. \qquad (16)$$

Equations (13)–(16) were derived from our previous studies (Vunjak-Novakovic et al 1992b, 1995a) in order to define the phase holdups as functions of the corresponding relative velocities. These equations

strictly refer to the steady-state circulation of liquid and particles between draft and annulus with gas bubbles present in the draft tube only. Equations (13)–(16) were solved in conjunction with Equations (1) and (9)–(12) to obtain liquid velocities and phase holdups in the draft and annulus. The effect of gas bubbles on particle motion was integrated into the model by predicting the relative liquid-to-particle velocity, v_{sl}, using Equations (5)–(8), and calculating the particle drag coefficient C_D as a function of v_{sl}. The model gave close predictions of the observed liquid velocities and phase holdups in the draft and annulus of the ALR operating in the circulating regime (Kundakovic and Vunjak-Novakovic 1995a). More recently, a similar approach was used in the analysis of the four-phase internal-loop ALR (Sajc et al 1995b).

Mixing in the Liquid Phase

Liquid mixing was studied by analyzing tracer-response data according to the compartmental and axial dispersion models (Obradovic et al 1994, 1997). Mixing was described by the liquid residence time (τ_r) and the Bodenstein number $Bo = v_l L/D$, which was in turn defined using the local axial liquid velocity (v_l), the characteristic mixing length (L), and the coefficient of axial dispersion (D). For each compartment, Bo and τ_r were calculated from the best fit of the experimental tracer response curve at the compartment outlet, using the model of axial dispersion for moderate to high deviation from plug flow (Levenspiel 1979). The outlet concentration profile was calculated by convoluting the experimental inlet concentration profile with the E curve. For each compartment, E curve was calculated as a function of Bo, for an open-open vessel (Levenspiel 1979), where τ_r was obtained by a cross-correlation of tracer concentration profiles at the compartment inlet and outlet. Bo was calculated for minimum standard deviation between the predicted and the experimental outlet concentration profiles (Obradovic et al 1994).

Analysis of the reactor was performed in the time and frequency domain, for the closed-open boundary conditions (Obradovic et al 1997). As expected, superficial gas velocity proved to be the key parameter that determined mixing patterns, while the volumetric fraction of particles (0–30%) had no significant effect on the axial dispersion (Chisti 1989, Joshi 1980). Liquid mixing in the reactor could be described using the axial dispersion model with Bo

numbers ranging from 0.3 to 0.9 (Figure 30.6a), which are lower than those previously reported (Field and Slater 1983, Verlaan et al 1986, Lu et al 1994a). Better liquid mixing observed in our studies is probably due to an almost optimal height of the column top section (Chisti 1989). The relatively small effect of the superficial gas velocity (U_g) on Bo is due to the simultaneous increase in the local liquid velocity (v_l) and axial dispersion coefficient (D) with the increase in U_g.

Increase in the superficial gas velocity improved mixing in top reactor sections, as seen from the decrease in Bo and increase in axial dispersion coefficient, D (Figure 30.6a,b). At a low gas velocity, liquid mixing was most efficient in the top Section I. When the gas velocity increased, the well-mixed region extended into Section II. Under all conditions, Section III had the lowest degree of mixing when compared to Sections I and II (Figure 30.6a). The reactor flow conditions were governed by liquid recirculation between the draft and annulus with plug flow of liquid in each of these regions, and a high degree of backmixing in the upper reactor zones (Sections I, II, and III).

Examples of Immobilized Cell and Enzyme Processes

Overview

A number of different processes based on immobilized cells and enzymes have been tested on a laboratory scale, and only a few of them have been scaled up to industrial scale. Some of the common characteristics of this latter group include the use of simple immobilization method and mixed populations of microorganisms, low costs of operation, large operating volumes, low substrate concentration, and easy maintenance of sterile conditions. The largest-scale operations include, in a decreasing order: (1) beer and wine fermentation with flocculating yeast in fluidized bed reactors, (2) trickle-bed acetic acid production, and (3) gel-immobilized yeast or bacteria for anaerobic ethanol production. The most notable exceptions to the above rules are animal and plant cell cultures, where the benefits of immobilization over free cell culture conditions make all the difference. The high added value of the product makes several issues of bioreactor design and operation less critical (e.g., cost of immobiliza-

tion and substrates). Representative examples of immobilized cell bioreactors are listed in Table 30.1.

Recent progress in the development of biocompatible materials has provided an array of support matrices and novel methods for immobilization of cells and enzymes. Surprisingly, there have been only a few detailed comparative studies of immobilization methods and supports (Bickerstaff 1997, Wollaert et al 1996). Therefore, no ideal support material or method of immobilization has emerged as a standard for a given area of application (e.g., animal cell culture). Selection of support material and the method of immobilization involves analysis of the specific requirements of each enzyme/cell application, the limitations and advantages of various immobilization methods, and bioreactor systems. A number of practical aspects need to be considered before embarking on experimental work to ensure that the final immobilized enzyme or cell preparation is appropriate for the planned purpose. Table 30.2 lists selected points of consideration that can assist in the choice and evaluation of the support material and immobilization method.

Reverse Micelle Enzyme System for Oil Interesterification

An external-loop ALR was used in conjunction with lipase immobilization in reverse micelles to study the production of cocoa butter equivalents by oil interesterification (Mojovic et al 1993, 1994). This system is an example of lipids that have unique structure and properties and are in high demand for use in the food and cosmetics industry. Reverse micelles allowed us to overcome several constraints of heterogeneous enzyme systems containing water-insoluble substrates. ALR provided a mild hydrodynamic environment that did not damage the immobilized enzyme, while high rates of mixing and mass transfer were maintained. Immobilized enzyme particles were fluidized in solvent suspension without suffering from regions of incomplete mixing or high shear. The equilibrium state was reached more rapidly in the ALR than in shake flasks (9 and 12 hrs, respectively), which was attributed to the differences in mass transfer and mixing between the two systems. The productivity of interesterification was 2.8-fold higher in the ALR than in the shake flask, at about 25-fold lower power input per unit volume in the ALR (Table 30.3). These results could be attributed to the combined effects of enzyme immobilization and reactor design.

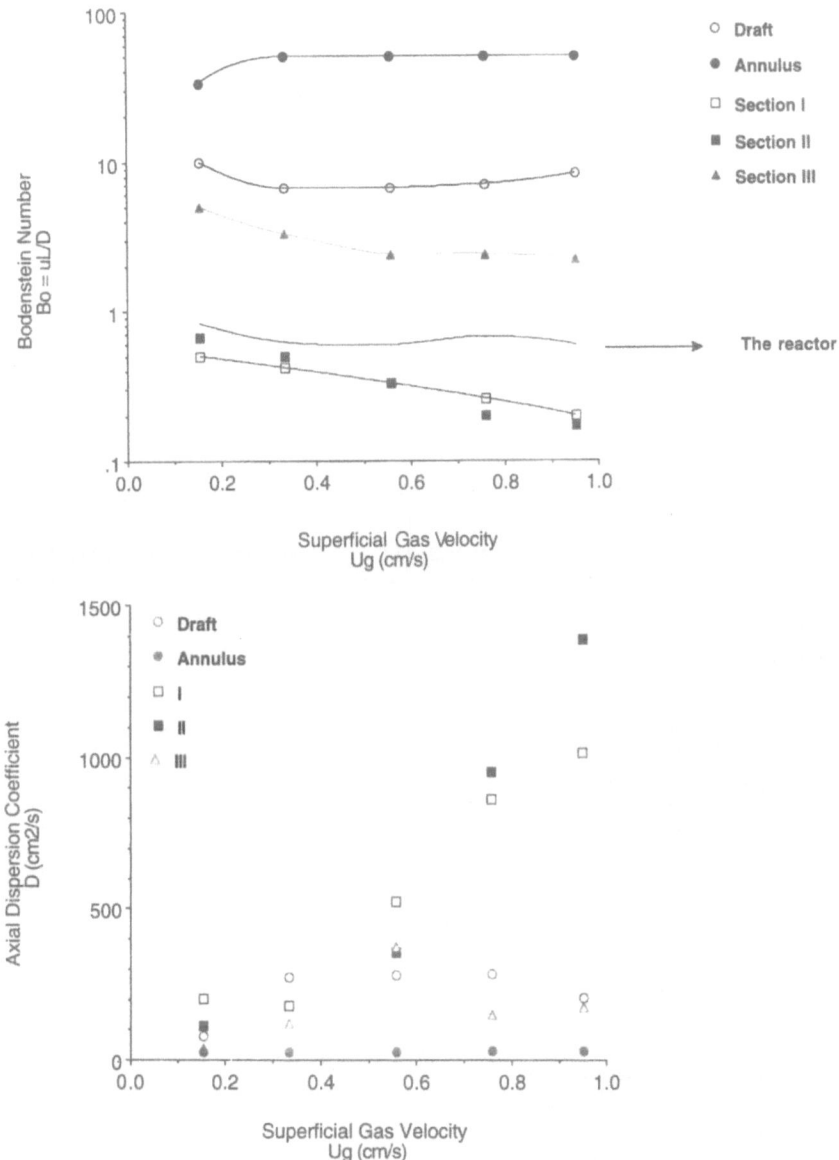

FIGURE 30.6. Effect of the superficial gas velocity on (a) *Bo* number and (b) axial dispersion coefficient, *D,* in different reactor zones of the internal-loop ALR. From Obradovic B., A. Dudukovic, G. Vunjak-Novakovic. Local and Overall Mixing Characteristics of the Gas-Liquid-Solid Air-Lift Reactor. *Ind. Eng. Chem. Research* 33, 698–702. Copyright 1994 by the American Chemical Society. Reprinted with permission.

TABLE 30.1. Examples of immobilized cell bioreactors.

Application-Product	Cell type	Support material	Bioreactor type	Reference
Hg removal	Algae	Agarose/alginate	Packed bed	Robinson and Wilkinson 1994
Citric acid	Fungi	Ca-alginate	Air-lift	Rymowicz et al 1993
Citric acid	Fungi	Cellulose sulphate/poly (dimethyldiallyl-ammonium chloride)	Fluidized bed	Forster et al 1994
1,3 propanediol	Bacteria	Polyurethane	Packed bed	Pfugmacher and Gottshalk 1994
Milk acidification	Bacteria	Chitosan	Stirred tank	Groboillot et al 1993
Milk acidification	Bacteria	Alginate-PLL	Stirred tank	Larish et al 1994
Ethanol	Fungi	κ-Carrageenan	Packed bed	Sanroman et al 1994
Cellulolytic enzymes	Biomass	Self-aggregation	Stirred tank	Lejeune and Baron 1995
Biomass	Biomass	Basalt, Glass	Air-lift	Gjaltema et al 1997
Monoclonal antibodies	Mouse Hybridoma	Alginate-PLL	Air-lift	Bugarski et al 1989 & 1993
Monoclonal antibodies	Mouse Hybridoma	Alginate-PLL	Stirred tank	Posillico 1986
Monoclonal antibodies	Hybridoma	Alginate-PEI-PLL, CMC-PLL CMC-PLL	Air-lift	Hsu and Chu 1992
Monoclonal antibodies	Hybridoma	Alginate-chitosan	Stirred tank	Overgaard et al 1991
g-Interferon	Animal cells	Glass fiber	Packed bed	Croughan et al 1995
Virus	Insect cells	Alginate-PLL	Air-lift	Goosen et al 1997
Virus	Insect cells	Alginate-PLL	Spinner flask	Goosen et al 1997

PLL poly-L-lysine
PEI polyethleneimine
CMC carboxymethylcellulose

TABLE 30.2. Selecting the support and method for cell immobilization.

Property	Points for consideration
Physical	Biomechanical properties of support material (e.g., compressive stiffness, attrition)
	Density
	Shape/form (beads/sheets/fibers)
	Available surface area
	Porosity, pore size
	Hydraulic and diffusional permeability
Chemical	Hydrophilicity (water binding by the support)
	Available functional groups for modification
	Regeneration/reuse of support
Stability	Storage conditions and shelf life
	Maintenance of cell productivity
	Maintenance of enzyme activity
	Mechanical stability
	Bio/chemical stability (e.g., pH, organic solvents, enzymes)
Biocompatibility	Maintenance of cell viability
	Toxicity of component reagents
Safety	Health and safety risks for process workers and end-product users
	Safety requirements for FDA approval for food, pharmaceutical, and medical applications
	Environmental impact
Economic	Availability and cost of support
	Required chemicals, equipment, reagents, technical skills
	Feasibility of scale-up to larger capacities
	Possibility of continuous processing
Reaction	Enzyme/cell loading
	Catalytic productivity
	Reaction kinetics, side reactions
	Diffusional limitations for mass transfer of cofactors, substrates, and products

TABLE 30.3. Comparison of the productivities of enzymatic interesterification in the lecithin reverse-micelle system in shake flasks and external-loop ALR.

	Shake flask (V = 5.6 ml)	ALR (V = 80ml)
Total amount of incorporated stearic acid (mg/120 hrs)	1437	61,210
Average productivity (mg/mL h)	2.6	7.28
Power input per unit volume (W/m³)	847.6	34.8

Hybridoma Cell Culture for the Production of Monoclonal Antibodies

The hydrodynamic performance of the external-loop ALR used in hybridoma cell studies (Figure 30.2b) was characterized with respect to gas holdup, liquid velocity, and mixing patterns of liquid and immobilized cell particles (Bugarski et al 1993). In this system, the difference in gas holdup between the riser and downcomer provided the driving force for liquid circulation. Liquid velocity was in turn the key parameter in reactor design and scale-up since it determined the volumetric mass transfer coefficients and liquid mixing, and therefore cell growth and function. A simple model, based on energy and momentum balances, was used to calculate the velocity of liquid circulation at steady-state conditions:

$$(\epsilon_R - \epsilon_D) L = K_f (V_l)^2 / 2g + \Delta p / \rho_L g \qquad (17)$$

where ϵ_R and ϵ_D are the volume fractions of liquid in the riser and downcomer, K_f is the friction coefficient, L is liquid height, g is gravitational force, ρ_L is liquid density, and Δp is the measured pressure difference between two points in the bioreactor. The mean volume fractions of liquid were obtained from the one-dimensional model of momentum transfer:

$$\epsilon = (1/2) \{1 \pm V_m / V_s \qquad (18)$$
$$- [(1 \pm V_m / V_s)^2 \pm 4 V_{gs} / V_s)^2]\}$$

where $V_m = V_{gs} + V_{Ls}$, V_{gs} is gas superficial velocity, V_s is slip velocity, and V_m is velocity of the gas-liquid mixture, and the signs (+) and (−) refer to the up- and downflow of liquid. Vs was calculated from the following equations:

$$V_s = V_{b\infty} + a + b [p(0) / p(z) - 1] \qquad (19)$$
$$p(z) = p_s(z) + a(0) p(0) \, ln \, [p(0) / p_s(0)] \qquad (20)$$

where $V_{b\infty}$ is the terminal bubble rise velocity, p_s is the hydrostatic pressure, and a and b are the parameters of bioreactor geometry. The overall gas holdup was determined from the changes in volume of the aerated and gas-free medium. The local gas holdup was calculated from differential pressures. Calculated and measured liquid velocity and gas holdup are compared in Figure 30.7.

The volumetric oxygen transfer coefficient $K_L a$ increased with an increase in the superficial gas velocity. Oxygen transfer rate in the reactor can thus be adjusted by varying the gas flow rate. In geometrically similar air-lift reactors, $K_L a$ increases with the reactor volume at a constant power input (Figure 30.8). These effects of gas flow rate and the reactor size on oxygen transfer allow the reactor scale-up at a constant $K_L a$ or constant power input (Bugarski et al 1993).

Biological outputs, such as cell density and volumetric productivity of IgG, were used to test the suitability of the external-loop ALR for monoclonal antibody production. Several mouse hybridoma cell lines were cultured batchwise for up to 2 weeks, using the 0.3 L and 1 L volume reactors. Cells were immobilized in alginate and microencapsulated in alginate-poly-l-lysine, and cultivated at a loading of 3% of the reactor volume. Microencapsulation provided retention of the cells and IgG, while allowing free diffusion of oxygen, nutrients, and smaller size metabolites (Figure 30.9). In order to maintain circulation of the immobilized cell particles under mild hydrodynamic conditions, a low gas flow rate of 0.05 vvm was used in cell cultivation studies. Even at this low gas input, the rate of oxygen transfer was adequate, and cell growth was not limited by oxygen supply (Figure 30.8). Figure 10a,b shows cell growth and production of IgG for alginate-immobilized and

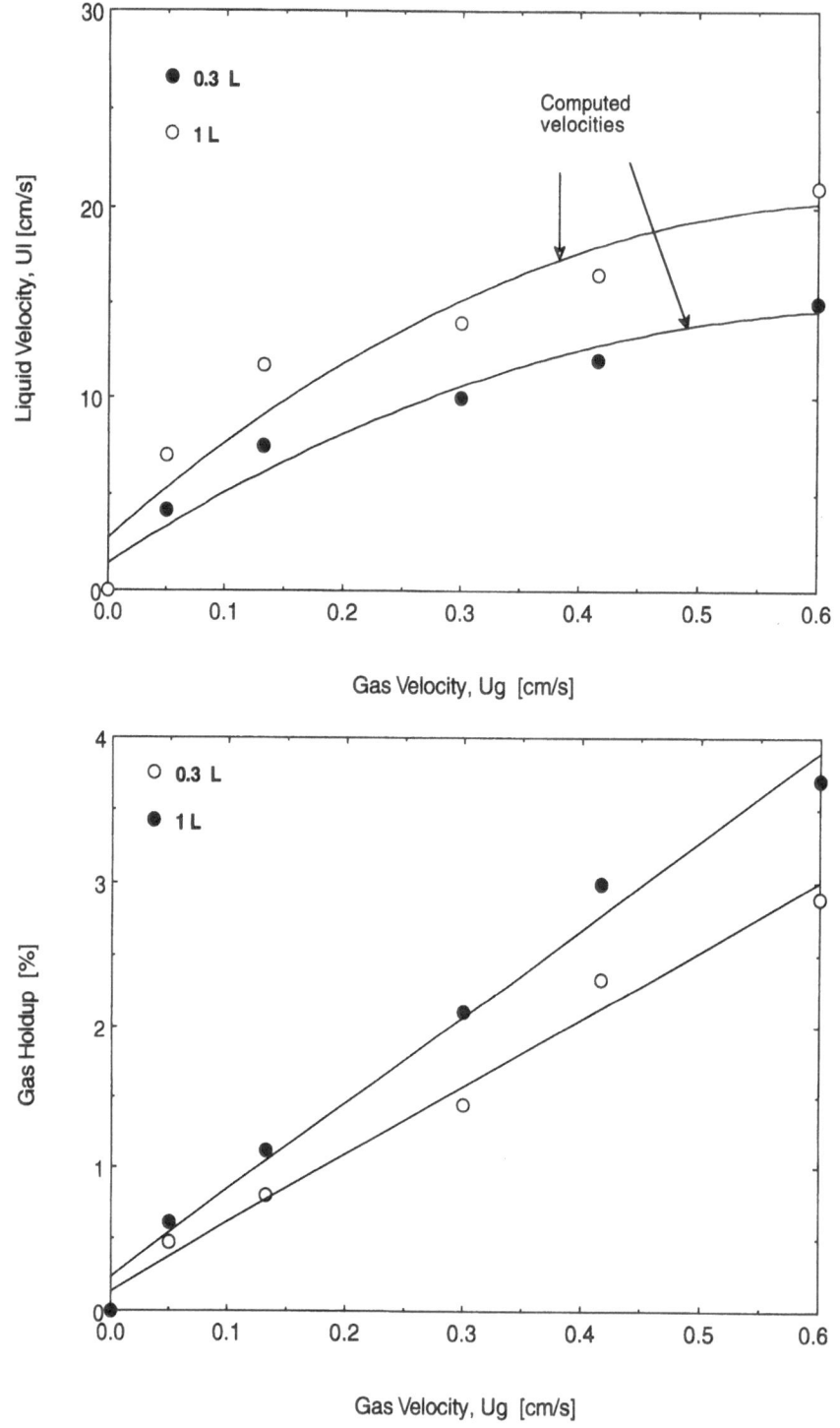

FIGURE 30.7. Effect of the superficial gas velocity on (a) liquid velocity and (b) gas holdup in the external-loop ALR. From Bugarski B., G. Jovanovic, G. Vunjak-Novakovic, Bioreactor Systems Based on Microencapsulated Animal Cell Cultures. In *Fundamentals of Animal Cells Immobilization and Microencapsulation*, M.F.A. Goosen, ed., pp. 267–296. Copyright 1993 by CRC Press, Boca Raton. Reprinted with permission.

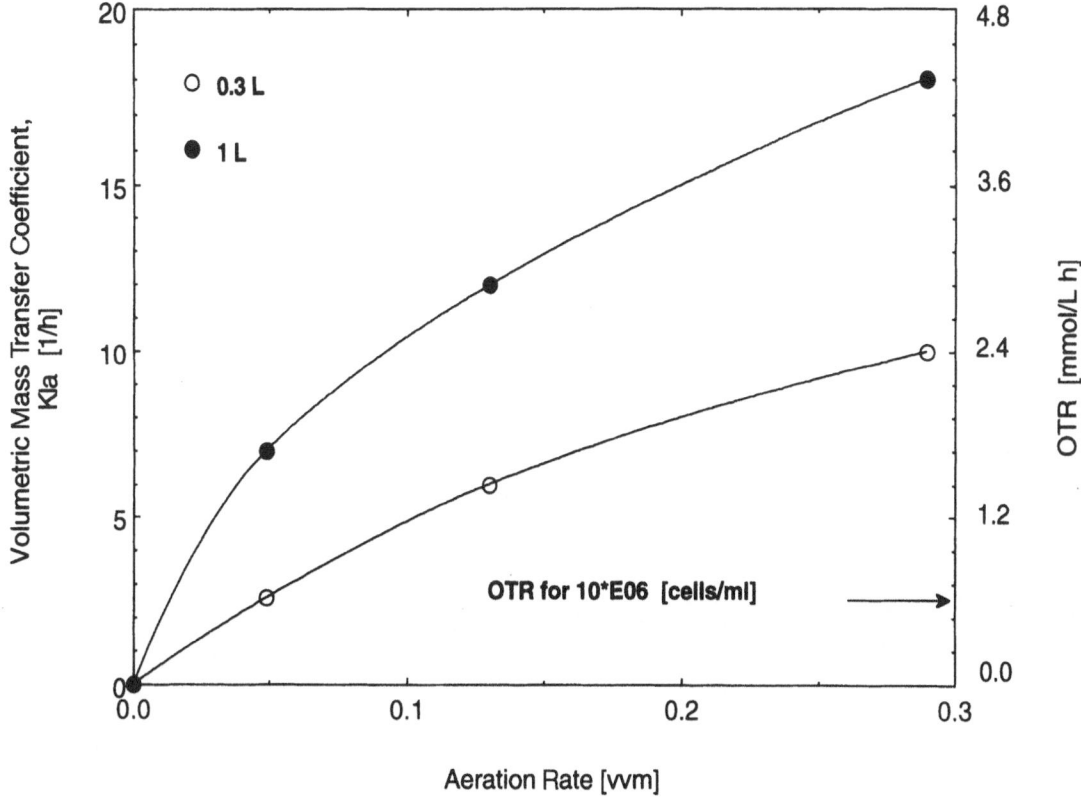

FIGURE 30.8. Effect of the aeration rate on oxygen mass transfer in the external-loop ALR. From Bugarski B., G. Jovanovic, G. Vunjak-Novakovic, Bioreactor Systems Based on Microencapsulated Animal Cell Cultures. In *Fundamentals of Animal Cells Immobilization and Microencapsulation*, M.F.A. Goosen, ed., pp. 267–296, Copyright 1993 by CRC Press, Boca Raton. Reprinted with permission.

microencapsulated hybridoma cells in the 1 L bioreactor. In the system with microcapsules, cell density reached 1.5×10^8 cells/microcapsule, and IgG concentration was 700 μg/ml of capsules. Similar kinetics of cell growth and antibody synthesis were observed in the 0.3 L bioreactor (Bugarski et al 1993).

Plant Cell Culture for the Production of Anthraquinones

The production of anthraquinones by plant cells was used to evaluate the performance of a hybrid ALR that integrated the biosynthesis and recovery processes within a single unit (Sajc et al 1994a,b, 1995a,b). The reactor was designed for simultaneous production and extraction of extracellular plant cell metabolites. *In situ* recovery of extracellular products had two major effects: enhancement of the biosynthesis due to the constant product removal, and concentration of the product in the dispersed

(solvent) phase. Aqueous phase containing nutrients was periodically replaced during cultivation, while gas and the solvent phase were continuously perfused through the reactor (Figure 30.2c).

The drop size of dispersed solvent could be controlled by superficial velocities of the solvent and gas. A decrease in the drop size to a stable and constant value of about 4 mm was observed at gas velocities that provided solvent recirculation in the reactor. For both solvents used, silicone oil and n-hexadecane, a uniform noncoalescing population of drops was obtained in the riser, while a complete coalescence of the solvent drops without emulsification was provided in the upper horizontal section. Solvent holdup had a value of 1.2% over a wide range of gas velocities and particle loadings. Gas holdup was correlated to the bubble rise velocity, which was in turn calculated from fluid-dynamic properties of the four-phase system. The difference in the gas holdup between the riser and the downcomer created a pressure difference as the driving

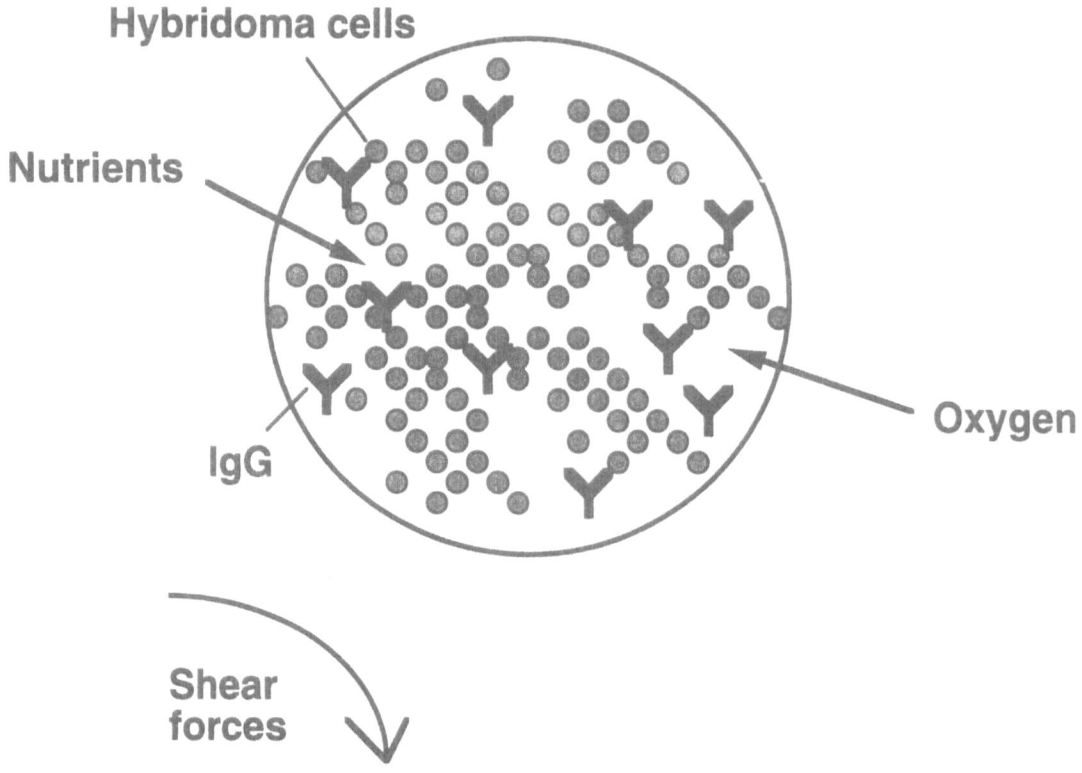

FIGURE 30.9. Microcapsules can selectively retain cells and high molecular weight products while allowing free diffusion of nutrients and gases. From Bugarski B., G. Jovanovic, G. Vunjak-Novakovic, Bioreactor Systems Based on Microencapsulated Animal Cell Cultures. In *Fundamentals of Animal Cells Immobilization and Microencapsulation*, M.F.A. Goosen, ed., pp. 267–296. Copyright 1993 by CRC Press, Boca Raton. Reprinted with permission.

force for liquid circulation. The overall liquid circulation velocity was thus essentially determined by gas holdup. Axial mixing in the liquid phase was most intensive in the riser ($Bo = 1.0–2.7$) as compared to the downcomer ($Bo = 10–50$), upper and lower horizontal sections ($Bo = 10–30$ and $Bo = 10–50$, respectively). The reactor as a whole had $Bo = 5–6$, which could be attributed to liquid circulation.

Kinetic profiles of cell growth, substrate consumption, and product formation are shown in Figure 30.11 for the external-loop ALR that operated with plant cells immobilized in alginate (35% loading, 50 g dry cell weight/L reactor volume) and silicone oil as solvent. Anthraquinone yields up to 17.5 mg/g dry cell weight were reached after 10 days of cultivation, which was 30–50% higher than in shake flasks. The time course of a batch culture of plant cells was modeled by numerically solving the balance equations for cells, substrate, and product (Sajc et al 1995). Calculated profiles gave excellent pre-

dictions of measured kinetic data (Figure 30.11). Moreover, the model provided a conceptual framework for analyzing the increase in anthraquinone production when biosynthesis is coupled to liquid-liquid extraction. The same model equations were used to evaluate the effectiveness factor and calculate the Thiele modulus for the reaction within the immobilized cells and diffusion of substrate through alginate (Sajc et al 1995b). Given the efficient mixing in the liquid phase, external mass transfer resistances were neglected. The calculated values of the effectiveness factor (~1) and Thiele moduli ($8.10^{-4}–11.10^{-4}$) corresponded to kinetically controlled systems with negligible intraparticle diffusional limitations.

Acknowledgments. We would like to acknowledge the contributions of our colleagues and students at the University of Belgrade, Biochemical Engineering Laboratory, J. Jelenkovic, G. Jovanovic, Lj.

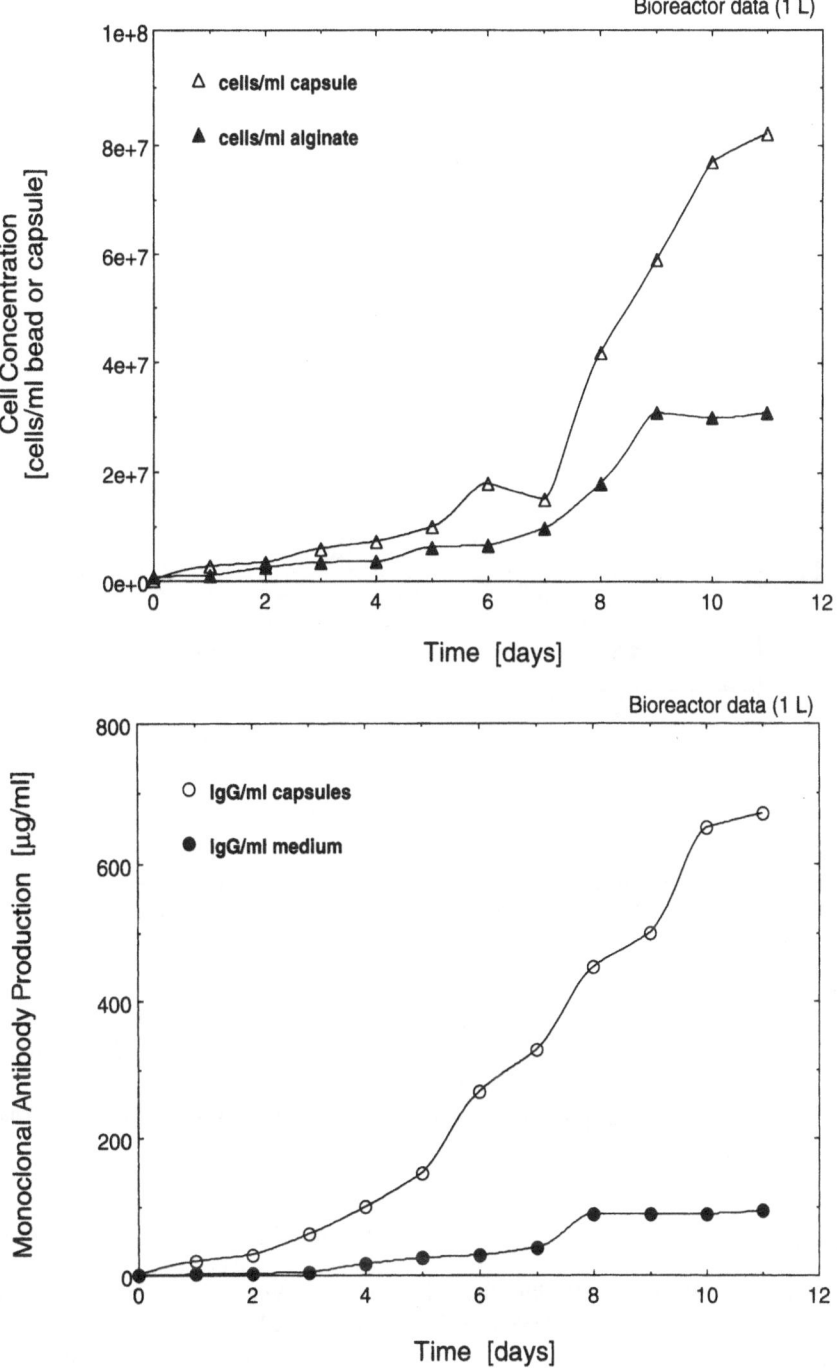

FIGURE 30.10. Time profiles of (a) hybridoma cell concentration and (b) production of mono-clonal antibodies in a 1-liter volume external-loop ALR. From Bugarski B., G. Jovanovic, G. Vunjak-Novakovic, Bioreactor Systems Based on Microencapsulated Animal Cell Cultures. In *Fundamentals of Animal Cells Immobilization and Microencapsulation,* M.F.A. Goosen, ed., pp. 267–296. Copyright 1993 by CRC Press, Boca Raton. Reprinted with permission.

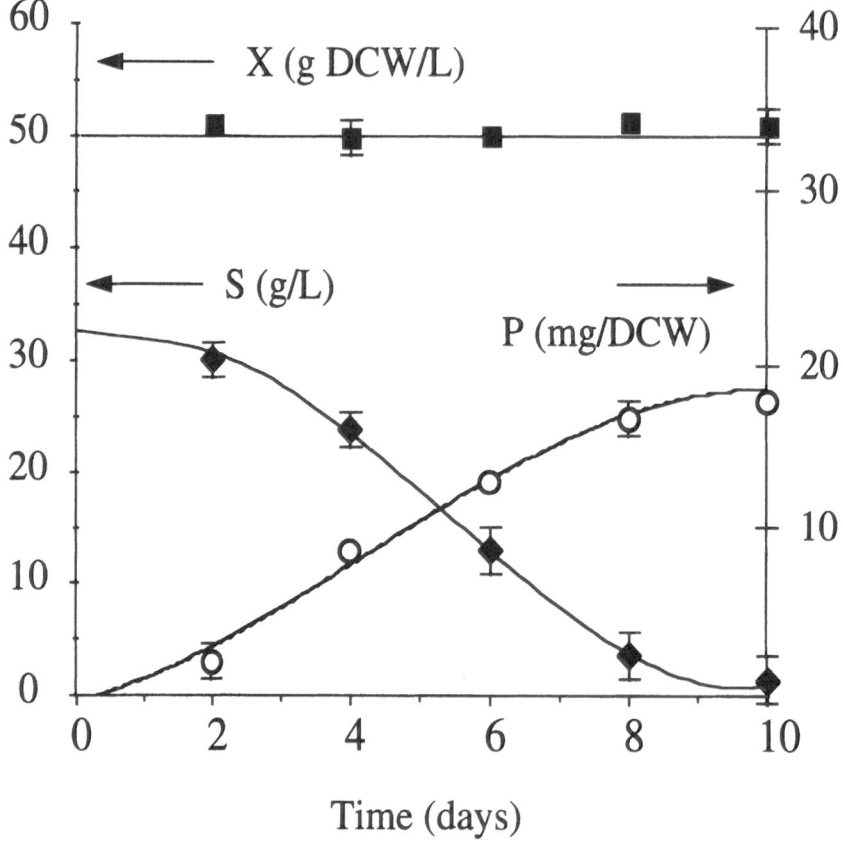

FIGURE 30.11. Model predictions (solid line) and experimentally observed profiles (ave ± SD) for cell growth, X, substrate consumption, S, and product formation, P, in the bioreactor (50 g dry cell weight per liter) and silicone oil as solvent.

Kundakovic, Lj. Mojovic, B. Obradovic, I. Pajic, and N. Vunjak, to the studies reviewed in this chapter. Funding for this work was provided by the Research Council of Serbia.

References

Baron GV, Willaert RG, de Backer L. 1996. Immobilised living cell systems: modeling and experimental methods. New York: John Wiley.

Bickerstaff GF. 1997. Immobilization of enzymes and cells. Totowa, NJ: Humana Press.

Bugarski B, King GA, Jovanovic G, Daugulis AJ, Goosen MFA. 1998. Performance of an external loop air lift bioreactor for the production of monoclonal antibodies by immobilized hybridoma cells. Appl. Microbiol. Biotechnol. 30: 264.

Bugarski B, King G, Jovanovic G, Daugulis A, Goosen MFA, Vunjak-Novakovic G. 1990. Design and operation of an air-lift reactor for production of monoclonal

antibodies by immobilized hybridoma cells. Period. Biol. 92: 121–123.

Bugarski B, Vukovic DV, Jovanovic G, Vunjak-Novakovic G, Goosen MFA. 1991a. Design and operation of the bioreactor for the production of immunochemicals. In: White MD, Reuveny S, Shafferman A, editors. Biologicals from recombinant microorganisms and animal cells, production and recovery. Philadephia: VCH. p. 69–73.

Bugarski B, Vunjak-Novakovic G, Jovanovic G, Cuperlovic K, Goosen MFA. 1991b. Operation of an air-lift bioreactor for production of immunochemicals by immobilized hybridoma cells. In: Sasake R, Ikura K, editors. Animal cell culture and production of biologicals. The Netherlands: Kluwer Academic Publishers. p. 135–141.

Bugarski B, Jovanovic G, Vunjak-Novakovic G. 1993. Bioreactor systems based on microencapsulated animal cell cultures. In: Goosen MFA, editor. Fundamentals of animal cells immobilization and microencapsulation. Boca Raton, Florida: CRC Press. p. 267–296.

Bugarski B, Amsden B, Neufeld RJ, Poncelet D, Goosen MFA 1994a. Effect of electrode geometry and charge on the production of polymer microbeads by electrostatics. Can. J. Chem. Eng. 72: 517–522.

Bugarski B, Li Q, Goosen MFA, Poncelet D, Neufeld R, Vunjak-Novakovic G. 1994b. Electrostatic droplet generation: mechanism of polymer droplet formation. AIChE J. 40: 1026–1031.

Chang TMS. 1964. Semipermeable microcapsules. Science. 146: 524

Chen Y-M, Fan L-S. 1990. Drift flux in gas-liquid-solid fluidized systems from the dynamics of bed collapse. Chem. Eng. Sci. 45: 935–945.

Chisti MY. 1989. Airlift Bioreactors. London: Elsevier.

Clift R, Grace J. 1985. Continuous bubbling and slugging. In: Davidson JF, Clift R, Harrison D, editors. Fluidization, 2nd ed. London: Academic Press. p. 73–132.

Darton RC, Harrison D. 1975. Gas and liquid hold-up in three-phase fluidization. Chem. Eng. Sci. 30: 581–586.

Fan L-S. 1989. Gas-liquid-solid fluidization engineering. Boston: Butterworth.

Fields PR, Slater NHK. 1983. Tracer dispersion in a laboratory air-lift reactor. Chem. Eng. Sci. 38: 647–653.

Gjaltema A, van Loosdrecht MCM, Heijnen JJ. 1997. Abrasion of suspended biofilm pellets in airlift reactors: effect of particle size. Biotech. Bioeng. 55: 206–215.

Goosen MFA, Mahmud ES, Al-Ghafri AS, Al-Hajri HA, Al-Sinani YS, Bugarski B. 1997. Immobilization of cells using electostatic droplet generator. In: Bickerstaff GF, editor. Immobilization of enzymes and cells. Totowa, NJ: Humana Press,—chapter 20.

Groboillot AF, Champagne CP, Darling GD, Poncelet D, Neufeld RJ. 1993. Membrane formation by interfacial cross-linking of chitosan for microencapsulation of lactococcus lactis. Biotechnol. Bioeng. 42: 1157.

Han JH, Kim SD. 1990. Radial dispersion and bubble characteristics in three-phase fluidized beds. Chem. Eng. Commun. 94: 9–26.

Hsu Y-L, Chu I-M. 1992. Poly(ethylenimine)-reinforced liquid-core capsules for the cultivation of hybridoma cells. Biotech. Bioeng. 40: 1300–1308.

Joshi JB. 1980. Axial mixing in multiphase contactors-a unified correlation. Trans. I. Chem. E. 58: 155–165.

Joshi JB, Ranade VV, Gharat SD, Lele SS. 1990. Sparged loop reactors. Can. J. Chem. Eng. 68: 705–741.

Khang SJ, Fitzgerald TJ. 1975. A new probe and circuit for measuring electrolyte conductivity. Ind. Eng. Chem. Fundam. 14: 208–213.

Kundakovic Lj, Vunjak-Novakovic G. 1995a. Mechanics of particle motion in three-phase flow. Chem. Eng. Sci. 50: 3285–3295.

Kundakovic Lj, Vunjak-Novakovic G. 1995b. A fluid dynamic model of the draft tube gas-liquid-solid fluidized bed. Chem. Eng. Sci. 50: 3763–3775.

Larish BC, Poncelet D, Champagne CP, Neufeld RJ. 1994. Microencapsulation of Lactococcus lactis subsp. cremoris. J. Microencapsulation 33: 189.

Levenspiel O. 1979. The chemical reactor omnibook. Corvallis, Oregon: Oregon State University Press.

Li Zh, Bi Zh, Zhao G, Ma W. 1988. Study on hydrodynamic behavior of a draft-tube gas-liquid-solid fluidized bed. In: Jimbo G, Beddow JK, Kwauk M, editors. Particuology '88. Beijing: Science Press. p. 280–286.

Lim F, Sun AM. 1980. Microencapsulated islets as bioartificial pancreas. Science. 210: 908.

Lim F. (1994). Microencapsulation of living cells and tissues: Theory and practice. In: F Lim (Ed), Biomedical Applications of Microencapsulation. Boca Raton: CRC Press. pp 137–154, chapter 8.

Lu W-J, Hwang S-J, Chang C-M. 1994a. Liquid mixing in internal loop airlift reactors. Ind. Eng. Chem. Res. 33: 2180–2186.

Lu W-J, Hwang S-J, Chang C-M. 1994b. Liquid mixing in two- and three-phase airlift reactors. Chem. Eng. Sci. 49: 1465–1468.

Merchuk JC. 1986. Hydrodynamics and hold-up in air-lift bioreactors. In: Cheremisinoff NP, editor. Encyclopedia of fluid mechanics. Houston, Texas: Gulf Publishers. p. 1485–1511.

Merchuk JC, Yunger R. 1990. The role of the gas-liquid separator of air-lift reactors in the mixing process. Chem. Eng. Sci. 45: 2973–2975.

Merchuk JC, Osemberg G, Siegel M, Shacham M. 1992. A method for evaluation of mass transfer coefficients in the different regions of air lift reactors. Chem. Eng. Sci. 47: 2221–2226.

Merten OW. 1989. Advanced research in animal cell technology. NATO ASI Ser. E 156: 367.

Mojovic Lj, Siler S, Kukic G, Vunjak-Novakovic G. 1993. Rhizopus arrhizus lipase catalyzed interesterification of the mid fraction of palm oil. Enzyme Microb. Technol. 15: 1–6.

Mojovic Lj, Siler-Marinkovic S, Kukic G, Bugarski B, Vunjak-Novakovic G. 1994. Rhizopus arrhizus lipase catalyzed interesterification of palm oil in a gas-lift reactor. Enzyme Microb. Technol. 16: 159–162.

Obradovic B, Dudukovic A, Vunjak-Novakovic G. 1994. Local and overall mixing characteristics of the gas-liquid-solid air-lift reactor. Ind. Eng. Chem. Res. 33: 698–702.

Obradovic B, Dudukovic A, Vunjak-Novakovic G. 1997. Response data analysis of a three phase airlift reactor. Trans. I. Chem. E. 75: 473–479.

Overgaard S, Scharer JM, Moo-Young M, Bols NC. 1991. Immobilization of hybridoma in chitosan alginate beads. Can. J. Chem. Eng. 69: 439.

Posillico EG. 1986. Microencapsulation technology for large-scale antibody production. Bio/Technology 4: 114.

Russell AB, Thomas CR, Lilly MD. 1994. The influence of vessel height and top-section size on the hydrodynamic characteristics of airlift fermentors. Biotechnol. Bioeng. 43: 69–76.

Sajc L, Bugarski B, Vukovic DV, Vunjak-Novakovic G, Vukovic J, Kovacevic N, Grubisic D. 1994a. The four-phase air-lift bioreactor with immobilized plant cells: kinetics of cell growth, substrate consumption and product formation. Biotechnology (in Russian) 9–10: 26–33.

Sajc L, Bugarski B, Vukovic DV, Vunjak-Novakovic G, Vukovic J. 1994b. The four-phase air-lift bioreactor with immobilized plant cells: synthesis and extraction of extracellular products. Biotechnology (in Russian) 11–12: 42–45.

Sajc LM, Jovanovic ZR, Vunjak-Novakovic G, Jovanovic GN, Pesic, RD, Vukovic DV. 1994c. Liquid dispersion in a magnetically stabilized fluidized bed (MSFB). Trans. I. Chem. E. 72A: 236–240.

Sajc L, Obradovic B, Vukovic DV, Bugarski B, Vunjak-Novakovic G. 1995a. Hydrodynamics and mass transfer in a four phase external loop air lift bioreactor. Biotechnol. Progress 11: 420–428.

Sajc L, Vunjak-Novakovic G, Grubisic D, Kovacevic N, Vukovic DV, Bugarski B. 1995b. Production of anthraquinones by immobilized frangula alnus mill. plant cells in a four phase air lift bioreactor. Appl. Microbiol. Biotechnol. 43: 416–423.

Sanroman A, Chamy R, Nunez MJ, Lema JM. 1994. Alcoholic fermentation of xylose by immobilized pichia stipitis in a fixed-bed pulsed bioreactor. Enzyme Microb. Technol. 16: 72.

Verlaan P, Tramper J, van't Riet K, Luyben KCAM. 1986. Hydrodynamics and axial dispersion in an air-lift loop bioreactor with two and three-phase flow. Proc. Int. Conf. Bioreactor Fluid Dynamics. 15–17.

Vunjak-Novakovic G, Jovanovic G, Vukovic DV, Vunjak N, Pajic I, Jelenkovic J. 1991. Fluid-dynamic study of the fluidized bed bioreactor with an internal draft tube. Technol. Today. 4: 216–221.

Vunjak-Novakovic G, Jovanovic G, Kundakovic Lj, Obradovic B. 1992a. Flow regimes and liquid mixing in a fluidized bed bioreactor with an internal draft tube. In: Fluidization VII. New York: Engineering Foundation. p. 433–444.

Vunjak-Novakovic G, Jovanovic G, Kundakovic Lj, Obradovic B. 1992b. Flow regimes and liquid mixing in a draft tube gas-liquid-solid fluidized bed. Chem. Eng. Sci. 47: 3451–3458.

Wallis GB. 1969. One-dimensional two-phase flow. New York: McGraw-Hill.

Warnecke H-J, Pruss J, Langemann H. 1985. On a mathematical model for loop reactors; residence time distribution, moments and equivalents. Chem. Eng. Sci. 40: 2321–2326.

Wollaert R, Baron GV, de Backer L. 1996. Immobilized living cell systems. New York: Wiley.

31
Culture of Anchorage-Dependent Cells

Delano V. Young

Introduction

There is a genuine need in the biotechnology industry for the capability of growing, cheaply and efficiently, normal, untransformed, anchorage-dependent animal cells on a large scale. This need exists for the unique products that these cells may be capable of synthesizing, and in the case of esoteric cell types, such as chondrocytes, hepatocytes, etc., for the very cells themselves, as well as for the advantages inherent in the use of normal cells in biotechnology manufacturing. Unlike transformed cells, normal cells do not secrete large amounts of proteases nor produce truncated carbohydrate side chains; they do not harbor oncogenic viruses or activated oncogenes, which pose problems of regulatory and safety issues, and, at least theoretically, they should be able to continue to produce large amounts of protein while in the quiescent, noncycling state—the ideal condition for most efficient synthetic protein production. Unfortunately, unlike transformed cells, most normal animal cells are unable to grow in suspension, but require attachment to a solid substratum or extracellular matrix (anchorage-dependence). This makes their large-scale cultivation extremely costly and technically difficult.

The major technical challenge is the maximization of the surface area for cell attachment within the culture vessel (for review, see Young 1992). A number of attempts have been made to grow anchorage-dependent cells on a large scale, starting first with roller bottles and progressing to more innovative devices such as multiple plate propagators (and other means of filling the internal volume with available surfaces, e.g., corrugated plastic sheets and tubular spiral film) and hollow-fiber cylinders (and variations on this theme, e.g., the Opticore). It soon became apparent, however, that all of these systems have one or more of these drawbacks: (1) limited potential for true scale-up, (2) difficulty in monitoring cell number and/or in sequential sampling, and (3) inability to maintain homogeneous conditions throughout the entire vessel. The microcarrier bead (Reuveny 1985), which allows for cell attachment and growth on its surface, but is small and light enough to be suspendable in low energy input impeller vessels, was invented to correct these problems and to permit animal cell growth in "suspension." Unfortunately, microcarriers still have a number of major limitations. Since the cells grow on the surface of the beads, they are targets for collisions with other beads and with the impeller blades and are subject to detrimental hydrodynamic forces such as eddy currents and fluid shear (Croughan et al 1987), which strip or lyse cells from the surface (Croughan and Wang 1989). In addition, commercially available microcarriers are prohibitively expensive when used on an industrial scale.

Porous microcarriers, which have deep cavities and an extensive internal network of channels open to the surface of the bead, represented an improvement on the first generation of microcarriers since cells could seed and grow within these internal cavities, partially protected from the damaging forces described above. Examples include the Cultispher G microcarrier, composed of cross-linked, 100% gelatin, and Verax VX100 microspheres which are spongelike lattices of bovine collagen.

Inverted Microcarriers

However, porous microcarriers still do not offer complete protection for cells since some cells still reside on the surface, and the open cavities allow exposure of the cells within to hydrodynamic forces. A superior approach would be to cultivate anchorage-dependent cells within a protective, permeable membrane or microcapsule. This "inversion" of the microcarrier should then result in total protection of the internalized cells from all collisions and destructive hydrodynamic forces.

Damon Biotech Microencapsulation Process

By the late 1970s the process of microencapsulating animal cells had been developed by Franklin Lim (Lim and Moss 1981). Soon thereafter, Damon Biotech (later Abbott Biotech) utilized the Lim microencapsulation method for large-scale production of monoclonal antibodies from hybridoma cells (Jarvis and Grdina 1983, Rupp 1985, Posillico 1986) and recombinant proteins (prourokinase, chimeric antibody) from transfected myeloma cells (Gillies et al 1989). These cells are anchorage-independent and so could grow readily in suspension (while producing their protein product) within the microcapsules' interior.

In the Damon Biotech microencapsulation process, the hybridoma (or myeloma) cells were first suspended in a saline-sodium alginate solution. The cell-alginate suspension was then forced through a multichannel jethead, which formed droplets of cells in alginate. These droplets fell into a solution of calcium ions that complexed with the alginate, resulting in gelation of these small (800 μm diameter) droplets. A semipermeable membrane was formed onto the surface of these gel spheres by the layering of poly-L-lysine (PLL) or poly-L-ornithine (PLO), which was electrostatically drawn to the negatively charged alginate. Finally, the calcium alginate gel was reliquefied by removal of the calcium ions with a chelating agent. The permeability of the polyamino acid membrane was controlled by controlling the porosity of the membrane. This property was influenced by the identity and concentration of the chosen polyamino acid, its molecular weight range, the charge density, and the duration of the actual coating process (Posillico 1986). Regulating the poros-

ity made it possible to achieve membranes permeable to small molecules (nutrients, waste products) and low molecular weight proteins (e.g., growth factors), but not to larger, multisubunit proteins (e.g., antibodies). For instance, poly-L-ornithine membranes had molecular weight cut-offs of about 90 kD, whereas poly-L-lysine membranes allowed IgG antibodies to traverse the membrane pores. In this way, a production system could be designed either to trap the protein product within the microcapsules for later harvest or to release it to the medium. Using the Lim microencapsulation process, Damon Biotech routinely produced significant quantities of monoclonal antibodies.

The major shortcoming of the Damon Biotech microencapsulation process was the inability to grow anchorage-dependent cells within the microcapsules. This was accomplished by the development of a solid substratum suitable for cell attachment—small enough to be contained within a microcapsule, but large enough to allow for cell attachment and abundant growth. One such substratum turned out to be small fragments of cross-linked gelatin ("gelatin shards"), developed at Abbott Biotech (Young et al 1989) and Karyon Technology Inc. Another suitable gelatin substratum was Gelibeads (Hazelton Research Products, Lenexa, Kansas), which were nonporous, solid, small (50 to 100 μm diameter) microbeads of gelatin. Following are detailed procedures for the preparation of gelatin shards, the attachment of cells to them, and their encapsulation.

Preparation of Gelatin Shards

One hundred grams of gelatin (bovine, Type III, Bloom no. 225, Sigma Chemical Co., St. Louis, MO or porcine, Type A, Bloom no. 225, General Foods Manufacturing, Atlantic Gelatin Div., Woburn, MA) are slowly added to 800 mL of 60–70°C preheated, deionized water, rapidly being agitated in a large industrial-grade Waring-type blender. The volume is brought to 1 L, and 20 mL of 25% glutaraldehyde is quickly added to cross-link the gelatin.

The solid gelatin is manually broken into large pieces with a large plastic spatula. The pieces are then washed three times with two volumes of deionized water and resuspended in one volume of deionized water. To break these large pieces into shards small enough to be encapsulated, the suspension is vigorously agitated in the blender for 30 sec. The gelatin shards are then washed five times by suspension in three to five volumes of deionized water and centrifugation at 3000 rpm for 5 min. Residual glutaraldehyde is removed by suspension of the washed gelatin shards in two to three volumes of a 0.1% gelatin solution and mixing in a plastic beaker with stirring overnight at room temperature. The gelatin shards are collected by centrifugation and washed three times with two to three volumes of Ca^{+2}-free and Mg^{+2}-free phosphate buffered saline (PBS). The gelatin shards are next resuspended in one volume of PBS, transferred to a stainless steel blender chamber, and autoclaved for 30 min at 120°C. The partially reannealed fragments are reblended for 15–30 sec. to break them up. These autoclaved, sterilized gelatin shards could be stored at room temperature for months.

Cell Attachment to Gelatin Shards

The gelatin shards are washed twice in complete medium (Encapsulation Medium A or EMA, Rupp and Geyer 1984, a modification of Williams Medium E, supplemented with 5% fetal bovine serum or FBS) by suspension in medium and aspi-

ration from settled gelatin shards. They are then soaked overnight in complete medium at room temperature. The next morning excess medium and any very fine shards are aspirated. An appropriate volume of gelatin shards (10 mL of shards yields 50 mL of microcapsules) is then centrifuged at 1000 rpm for 5 min. A cell suspension containing 2.5×10^6 viable cells per mL in a volume of complete medium equal to the original gelatin shard volume is added to the gelatin shards. This mixture is placed in a spinner vessel of appropriate size with the addition of just enough complete medium to permit the vessel impeller to suspend the gelatin shards. The suspension is incubated in a 37°C CO_2 incubator overnight with intermittent spinning (5 min every hour).

Microencapsulation of Cell-Loaded Gelatin Shards

The basic Damon Biotech microencapsulation procedure was modified in the following way to accommodate gelatin shards (Figure 31.1). Gelatin shards with their attached cells are centrifuged at 1000 rpm for 5 min. After aspiration of the medium,

MICROENCAPSULATION OF MAMMALIAN CELLS ON GELATIN SHARDS

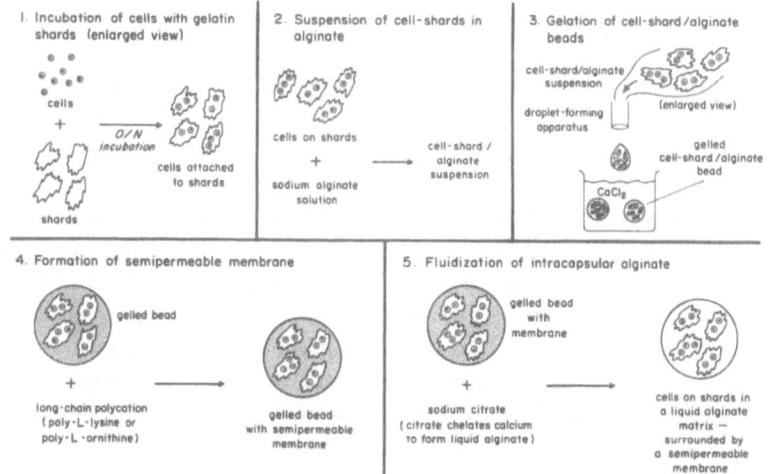

FIGURE 31.1. Microencapsulation of animal cells with gelatin shards. (Reprinted with the permission of CRC Press, Inc. from Young. 1992. Inverted microcarriers: Using microencapsulation to grow anchorage-dependent cells In: Fundamentals of animal cell encapsulation and immobilization, Goosen M, ed.)

an equal volume of 0.9% sodium chloride (saline solution) is added. The resulting suspension is then mixed with three volumes of 1.6% low-calcium, low-viscosity sodium alginate (Kelco, Clark, NJ), and the alginate-gelatin-cell suspension is then forced through a droplet-forming jethead (Artisan Industries, Waltham, MA). The rest of the microencapsulation procedure, including alginate gelation via calcium ion chelation, polyamino acid layering to form the permeable membrane, and reliquefaction of the alginate, followed the standard Damon Biotech process described above. For most of the encapsulations to be described below poly-L-lysine, which is more permeable than poly-L-ornithine, was the synthetic membrane coating. Gelatin shards prepared in this way (ranging in size from 10–50 μm), encapsulated shards with attached cells (CHO), and encapsulated Gelibeads with HaK cells are shown in the photomicrographs of Figure 31.2.

Culture Regimen for Inverted Microcarriers

When cultured on a small scale, as described in the following experiments, microcapsules with internalized cells on gelatin shards are prepared and cultivated in volumes ranging from 50 to 250 mL. The volume of EMA medium is usually four times the microcapsule volume, but this ratio can be increased (as it was for the experiment described in Figure 31.4A, which had a medium-to-microcapsule volume ratio of eight.) The microcapsules are grown in Bellco glass spinners in 37°C, CO_2 incubators with manual feeding batchwise as needed, usually once daily during which two-thirds to 100% of the medium is replaced.

To sample the microcapsules and obtain a viable cell count, one mL of settled, packed microcapsules is collected in a 15 mL conical centrifuge tube. Eight mL of 0.05% trypsin-0.53 mM EDTA (Gibco Laboratories, Grand Island, NY) are added and allowed to incubate at 37°C for 5–10 min to free the attached cells from the gelatin shards. To stop the trypsin, one mL of FBS is added. The entire suspension is transferred to a Dounce homogenizer, and the microcapsules are broken open with three strokes of the homogenizer plunger. The freed cells are counted, using Trypan Blue to distinguish between viable and dead cells. Microcapsules that do not contain gelatin shards do not require trypsinization prior to Dounce homogenization.

Growth of Anchorge-Dependent Cells in Inverted Microcarriers

Early attempts to grow anchorage-dependent, but transformed, cell lines in microcapsules without a solid substratum for attachment yielded essentially negative or ambiguous conclusions. The transformed cell lines chosen, HaK (Syrian hamster kidney cells, ATCC CCL 15) and CHO (Chinese hamster ovary cells, ATCC CCL 61), are often used in recombinant protein production because they have properties that make them attractive for large-scale manufacturing purposes. HaK cells in microcapsules initially increase in viable cell number, but eventually die (Figure 31.3A). The situation with CHO cells reveals very little growth at first, but, possibly, selection for anchorage-independent cells gives rise to a modest increase in cell number very late in the culture process (Figure 31.3B). CHO cells are known to be adaptable or selectable for growth in suspension. The encapsulation of soluble collagen with HaK cells may improve long-term survivability and provide for modest growth, but this develops very slowly (Figure 31.3A). Since collagen forms a loose hydrogel in neutral pH, it is possible that insoluble collagen hydrogels serve as a substratum for cell attachment in this case.

It is only when an insoluble gelatin substratum, either as gelatin shards or Gelibeads, is present in the microcapsules that significant, vigorous, viable cell growth for HaK and CHO cells is readily achieved. For HaK cells on Gelibeads a viable cell density of 8×10^7/mL of microcapsules (or 1.6×10^7 cells per mL of total culture volume) can be obtained. On gelatin shards, HaK cells grow to a viable cell density of 3×10^7 per mL. Furthermore, the serum level in the culture medium can be reduced from 5% to 1% with similar growth for HaK cells on gelatin shards. In the case of CHO cells in inverted microcarriers, gelatin shards are equal to Gelibeads with possible densities of viable cells reaching 7.5×10^7 per mL of microcapsules. This growth on gelatin is also sustainable for long periods of time, up to several weeks, providing that suitable feeding continues. Cell viabilities of 80% or better are often observed. Because of their small and diverse size and the exposure of their attached cells, no attempt was made to grow cells on gelatin shards without microencapsulation.

Normal, untransformed cell strains can also grow on gelatin shards in inverted microcarriers (Young et al 1989). Both human fetal lung fibroblasts (WI-38, ATCC CCL 75), which display a normal diploid karyotype, are strictly anchorage-dependent, and cannot be adapted to grow in suspension and baby hamster kidney cells (BHK-21, ATCC CCL 10), which, while not possessing the complete profile of normal cells, also are not adaptable for suspension growth, will grow in inverted microcarriers, increasing their viable cell numbers severalfold and maintaining a high percentage of viability.

Recombinant Protein Production Using Inverted Microcarriers

The ultimate aim in the development of the inverted microcarriers was the large-scale production of recombinant proteins. To demonstrate that this was possible, HaK cells were transfected with the gene for prourokinase. The enzyme, urokinase, of this proenzyme is easily assayed, and, indeed, had been successfully produced at Damon (and later Abbott) Biotech in murine myeloma cells.

The transfection/selection strategy, which is described in detail elsewhere (Young et al 1989), called for the cotransfection of HaK cells with two plasmids: one carrying both dihydrofolate reductase (dhfr) and human prourokinase, and the other carrying the neomycin-resistance (Neo^R) gene. Treatment with G418 selected for transfected clones. Clones that had taken up both plasmids were next selected for by exposure to successive rounds of increasing methotrexate (MTX) concentrations, up to 100 μM, resulting in gene amplification of both dhfr and prourokinase. Such gene-amplified clones are stable and continue to produce high levels of prourokinase even in the absence of MTX. One such clone, HDU 3.15, was chosen to test the inverted microcarrier system.

Characterization of the growth and production properties of this clone revealed a dependence on serum-containing medium (Young et al 1989). When these cells are grown without medium replenishment, they enter a true stationary phase after several days of exponential growth. Urokinase production, as measured by the appearance of the enzyme in the culture medium, shows an initial lag during early exponential cell growth; enzyme production increases dramatically in the second half of the growth curve, then ceases shortly after the cells enter stationary phase. Both cell growth and uroki-

nase production resume upon addition of fresh medium containing serum, indicating that HDU 3.15 production of urokinase depends on serum and/or fresh medium and/or cell growth.

How well these cells grow and produce urokinase when attached to gelatin shards and encapsulated in inverted microcarriers was demonstrated in the experiment depicted in Figure 31.4. In this experiment the culture was grown initially in a 1 L glass spinner vessel in a CO_2 incubator. It was batch-fed once daily. In the beginning, the 1 L spinner had 170 mL of inverted microcarriers and 400 mL of medium. The feeding schedule is described in the figure caption. Eventually, because the large cell numbers required more medium, the culture was transferred to a 3 L spinner and the total volume increased to 1.5 L.

After an initial rapid increase in cell number, the viable cell count stabilized between 1 to 2 $\times$ 10^7 cells per mL of microcarriers (Figure 31.4A). This number was maintained through the duration of the production run (65 days), although the percent viability gradually declined. Urokinase leaked out of the microcapsule and accumulated in the extracapsular medium. The assayed values in the medium, adjusted for the changes in volume due to feeding and normalized to a total volume of 700 mL, are shown in Figure 31.4B. The initial, rapid accumulation corresponded to the rapid cell growth phase; the second peak at day 20 coincided with the transfer to the larger spinner and probably resulted from the increase in medium-to-microcarrier ratio. The level of urokinase stabilized by day 25, suggesting that a steady-state level of synthesis had been achieved.

The high, sustained level of viable cell number and protein production in this experiment demonstrated the feasibility of using inverted microcarriers as a protein production system. The protection of the internalized cells by the synthetic membrane, the low cost of gelatin, and the ability of the semipermeable membrane to regulate the flow of nutrients, waste products, and protein product underscored the inherent advantages of inverted microcarriers.

Update on Related Developments

Since the cultivation of anchorage-dependent cells in inverted microcarriers was first proposed, the major use of this technology has been in the implantation of

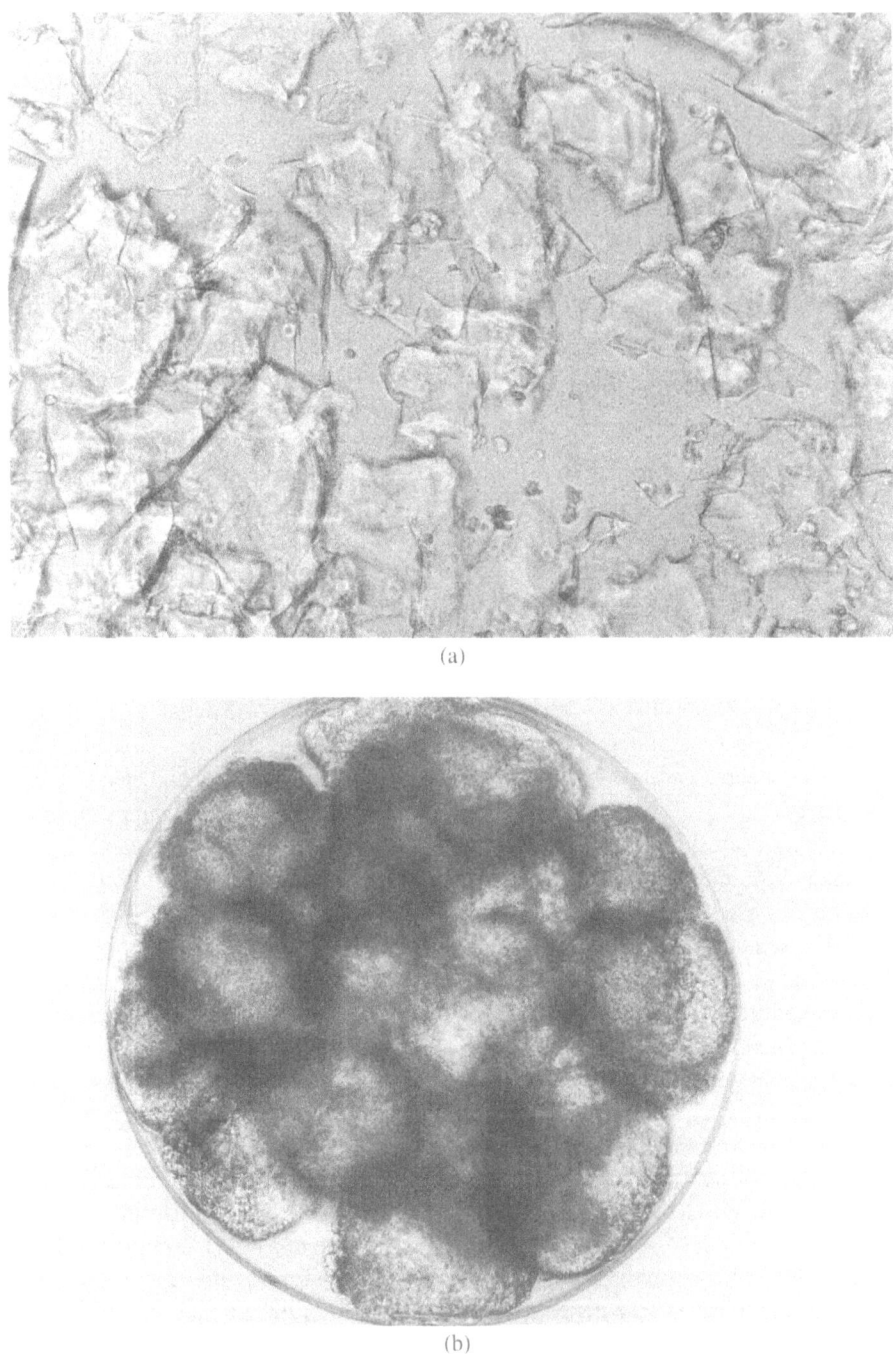

(a)

(b)

FIGURE 31.2. Photomicrographs of (a) gelatin shards, (b) Gelibeads with HaK cells in

allogeneic or xenogeneic cells into host animals. Such implanted cells would act as "artificial organs" providing needed biochemical(s) that the defective host lacked (Galletti et al 1995). Current examples of these attempts are the implantation of animal pancreatic islets for diabetes treatment (Sun et al 1993), neurons for the relief of various neurological disorders (Aebischer et al 1991), and hepatocytes as a temporary liver replacement/support system (Matthew et al 1993a). The primary objectives of such micro- and

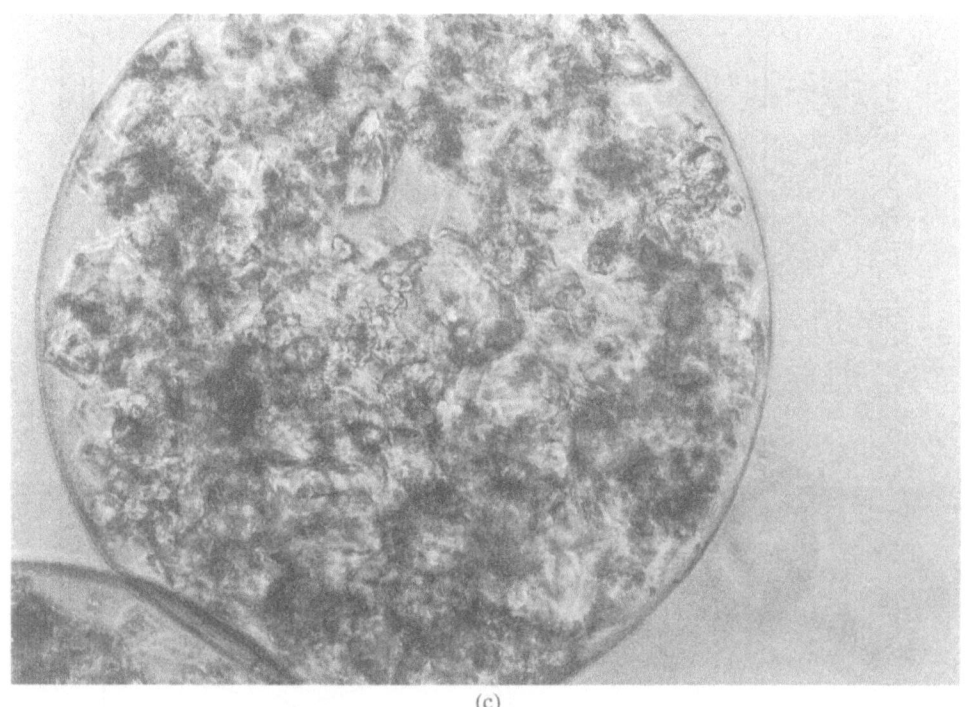

(c)

(Fig. 31.2, cont'd)
inverted microcarriers at 20 days of cultivation, and (c) gelatin shards with CHO cells attached in inverted microcarriers at day 0 of cultivation. Magnification is 100×. (Reprinted with the permission of Advanstar Communications, Inc. [formerly Aster Publishing Corp.] from Young et al. 1989. BioPharm 2 (#10):34.)

macrocapsules would be to provide, not so much for the growth of these cells (indeed, mature neurons do not divide), but for the long-term survival and maintenance of their differentiated phenotype, including secretion of the desired product, and their protection or "immunoisolation" from the host immune system.

To provide for cell survival, it is still necessary to encapsulate a suitable attachment substratum with most of these cells. The more commonly and successfully employed are alginate, Matrigel, collagen hydrogels, agarose with or without linked attachment peptides (Bellamkonda et al 1995), chitosan (e.g., precipitated chitosan, Zielinski and Aebischer 1994), and poly-L-lysine. In some cases, the need for an internal substratum may be obviated, if the inner surface of the capsule wall provides for cell attachment.

In membrane wall development, perhaps the most innovative approach has been that taken by Matthew (Matthew et al 1993b), Yoshioka (Yoshioka et al 1990), and their coworkers, in which they have produced microcapsules by the complex coacervation (miscibility interaction) of ionic polysaccharides of opposite charges. Matthew and his coworkers (Matthew et al 1993a) have observed that microcapsules formed from the mixing of carboxymethyl cellulose and chondroitin sulfate A with chitosan (and polygalacturonate as an outer coating) and enclosing Matrigel support the long-term survival and growth of liver endothelial cells and hepatocytes.

Concluding Remarks

The concept and advantages of cultivating anchorage-dependent cells in the protected, selectively-permeable environment of inverted microcarriers were demonstrated years ago. Although its use as a large-scale production system for recombinant proteins has yet to be fully realized and exploited, its inherent advantages have not gone unnoticed by the infant cell transplantation industry. Inverted microcarriers, their more recent microcapsule formulations, and the related hollow-fiber macrocapsules (Aebischer et al

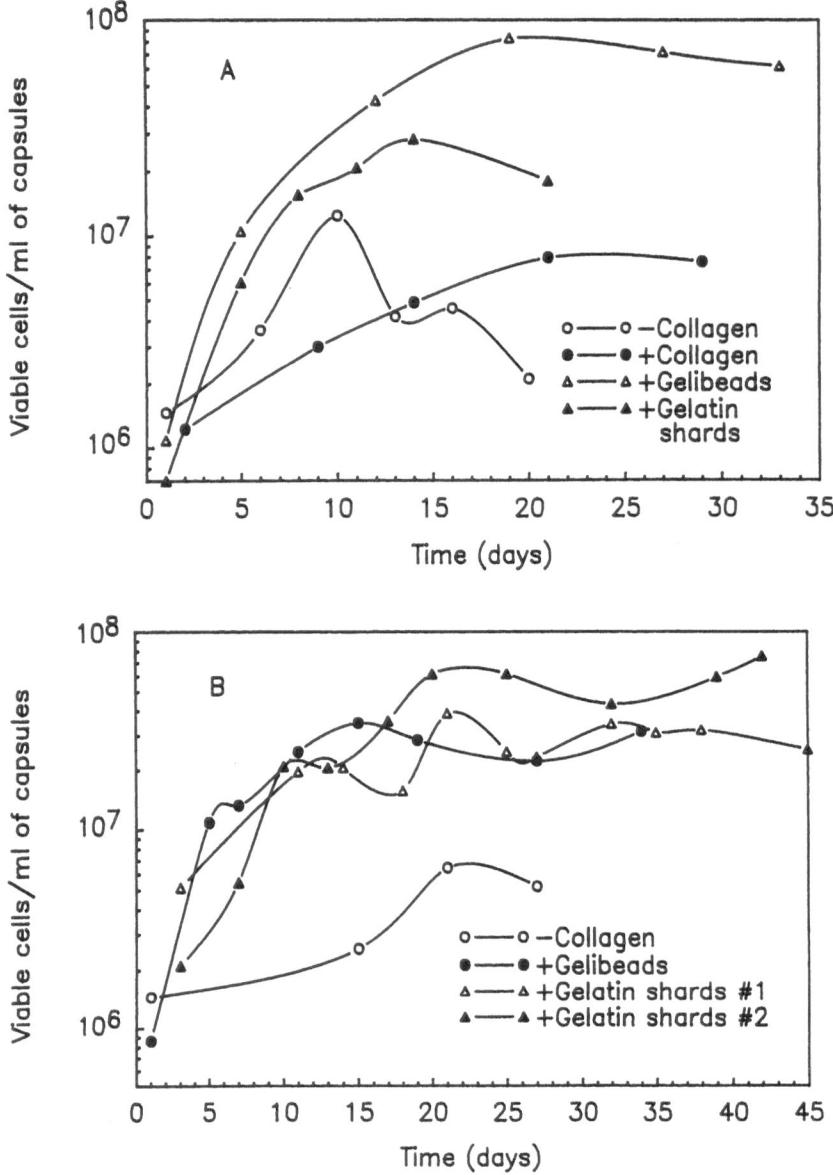

FIGURE 31.3. Growth of (A) HaK and (B) CHO cells encapsulated with or without gelatin shards, Gelibeads, or bovine collagen. To convert cell number per mL of microcapsules to cell number per mL of total culture volume, divide by 5. (Reprinted with the permission of Advanstar Communications, Inc. [formerly Aster Publishing Corp.] from Young et al. 1989. BioPharm 2 (#10):34.)

1991) can protect their enclosed cells not only from the hydrodynamic forces of the bioreactor chamber, as originally intended for microcapsules, but also from a host immune system. It may well be in this latter application that inverted microcarriers will see their greatest fruition.

Acknowledgments. The author would like to acknowledge the technical assistance of Susan Doebbels, Louise King, and Frank Deer in the conduct of this research, Steve Gillies for the plasmid constructs, and Heather Young for the preparation of this manuscript.

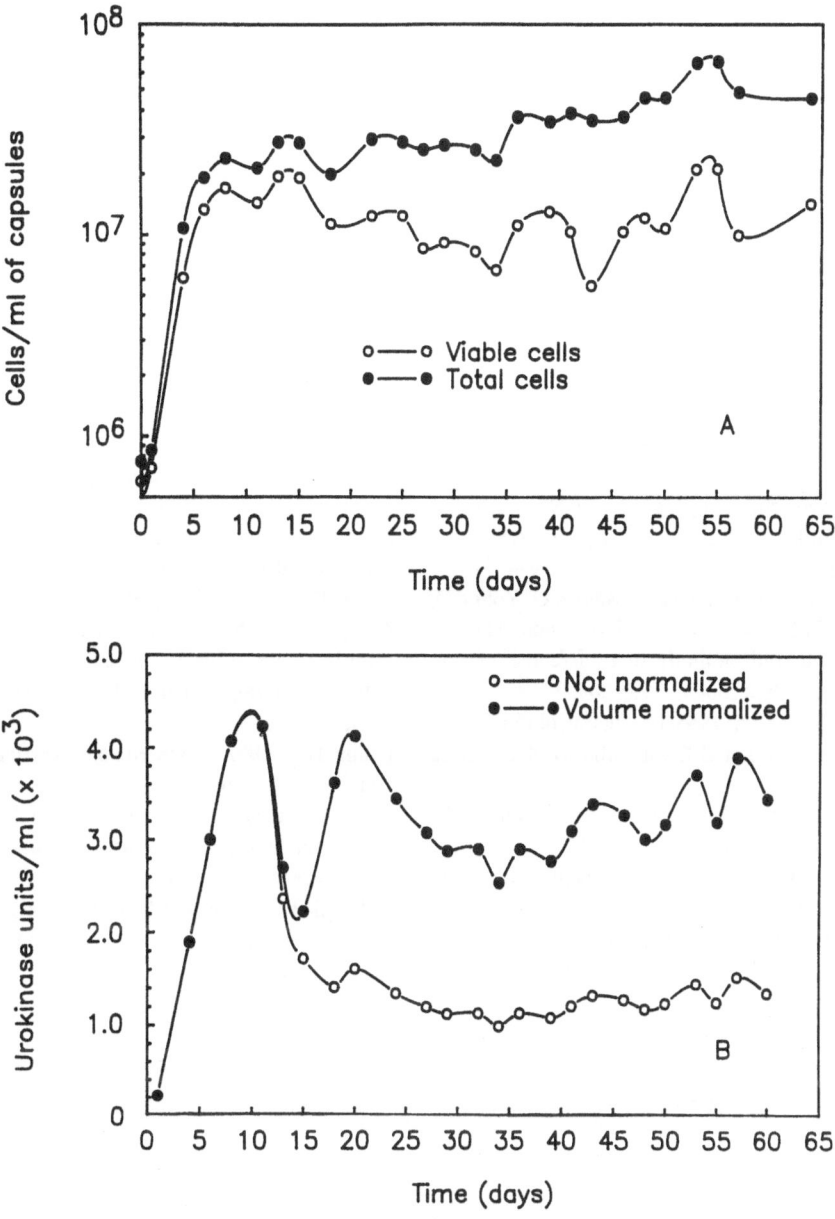

FIGURE 31.4. (A) Growth and (B) urokinase production of HDU 3.15 cells grown on gelatin
shards in inverted microcarriers. At first, 170 mL of inverted microcarriers were seeded into
400 mL of EMA + 5% FBS in a 1 L glass spinner vessel placed within a 37°C, CO_2 incubator. The total volume (medium plus microcarriers) was slowly increased, according to
the following schedule: to 700 mL (i.e., an increase of about 100 mL of medium) on day 3;
800 mL, day 12; 900 mL, day 13; 1200 mL, day 15 (with transfer to a 3 L glass spinner vessel); and, finally 1500 mL, day 16. The cells were fed once daily as follows: days 3 to 5:
200 mL; days 6 to 12: 300 mL; days 13 to 16: 600 mL; days 17 to 38: 1000 mL; and days
39 to end: 700 mL. The urokinase concentration was measured by the Urokinase Direct
Assay (Young et al 1989), which uses plasma to convert prourokinase to urokinase. The
urokinase concentrations were normalized to a volume of 700 mL, taking the above-listed
increases in volume into consideration. (Reprinted with the permission of Advanstar Communications, Inc. [formerly Aster Publishing Corp.] from Young et al. 1989. BioPharm 2
(#10):34.)

References

Aebischer P, Tresco PA, Winn SR, Greene LA, Jaeger CB. 1991. Long-term cross-species brain transplantation of a polymer-encapsulated dopamine-secreting cell line. Exptl Neurol 111:269–275.

Bellamkonda R, Ranieri JP, Aebischer P. 1995. Laminin oligopeptide derivatized agarose gels allow three-dimensional neurite extension in vitro. J Neurosci Res 41:501–509.

Croughan MS, Hamel J-F, Wang DIC. 1987. Hydrodynamic effects on animal cells grown in microcarrier cultures. Biotechnol Bioeng 29:130–141.

Croughan MS, Wang, DIC. 1989. Growth and death in overagitated microcarrier cell cultures. Biotechnol Bioeng 33:731–744.

Galletti PM, Aebischer P, Lysaght MJ. 1995. The dawn of biotechnology in artificial organs. ASAIO J 41(1):49–57.

Gillies SD, Dorai H, Wesolowski J, Majeau G, Young D, Boyd J, Gardner J, James K. 1989. Expression of human anti-tetanus toxoid antibody in transfected murine myeloma cells. Biotechnol 7:799–804.

Jarvis Jr. AP, Grdina TA. 1983. Production of biologicals from microencapsulated living cells. Biotechniques 1:24–30.

Lim F, Moss RD. 1981. Microencapsulation of living cells and tissues. J Pharm Sci 70:351–354.

Matthew HWT, Basu S, Peterson WD, Salley SO, Klein MD. 1993a. Performance of plasma-perfused, microencapsulated hepatocytes: prospects for extracorporeal liver support. J Pediatr Surg 28(11):1423–1428.

Matthew HW, Salley SO, Peterson WD, Klein MD. 1993b. Complex coacervate microcapsules for mammalian cell culture and artificial organ development. Biotechnol Prog 9:510–519.

Posillico EG. 1986. Microencapsulation technology for large-scale antibody production. Bio/Technol 4:114–117.

Reuveny S. 1985. Microcarriers in cell culture: structure and applications. Adv Cell Cult 4:213–247.

Rupp RG. 1985. Use of cellular microencapsulation in large scale production of monoclonal antibodies. In: Feder J, Tolbert W, editors. Large scale mammalian cell culture. New York: Academic Press. p 19–38.

Rupp RG, Geyer SD. 1984. Preparation of medium for large-scale hybridoma culture. J Tiss Cult Meth 8:141–146.

Sun Y-L, Ma X, Zhou D, Vacek I, Sun AM. 1993. Porcine pancreatic islets: isolation, microencapsulation, and xenotransplantation. Artif Organs 17(8):727–733.

Yoshioka T, Hirano R, Shioya T, Kako M. 1990. Encapsulation of mammalian cell with chitosan-CMC capsule. Biotechnol Bioeng 35:66–72.

Young DV, Dobbels S, King L, Deer F, Gillies SD. 1989. Inverted microcarriers: using microencapsulation to grow anchorage-dependent cells in suspension. BioPharm 2:34–46.

Young DV. 1992. Inverted microcarriers: Using microencapsulation to grow anchorage-dependent cells. In: Goosen MFA, editor. Fundamentals of animal cell encapsulation and immobilization. Boca Raton, Florida: CRC Press. p 243–265.

Zielinski BA, Aebischer P. 1994. Chitosan as a matrix for mammalian cell encapsulation. Biomaterials 15(13):1049–1056.

Index